Handbook of Experimental Pharmacology

Volume 124/I

Springer
Berlin
Heidelberg
New York
Barcelona
Budapest
Hong Kong
London
Milan
Paris
Santa Clara
Singapore
Tokyo

Drug Toxicity in Embryonic Development I

Advances in Understanding Mechanisms of Birth Defects: Morphogenesis and Processes at Risk

Contributors

R.J. Akhurst, R. Anderson, K.A. Augustine, R. Balling, S.M. Bell
J.D. Burrill, E.W. Carney, G.P. Daston, A. Gittenberger-de Groot
A.S. Goldman, M.P. Goto, M. Goulding, D.K. Hansen, C. Harris
E.S. Hunter, III, R.J. Kavlock, P.M. Kim, T.B. Knudsen, C. Lau
P.J. Linser, E.T. Liu, R. Maas, R. Markwald, P.E. Mirkes
K. Muneoka, A. Neubüser, C.J. Nicol, G.T. O'Neill, T. Parman
R. Poelmann, M. Rauchman, T.W. Sadler, H. Saueressig
C.M. Schreiner, W.J. Scott, Jr., T. Trusk, P.G. Wells, L.M. Winn
E.F. Zimmerman

Editors

R.J. Kavlock and G.P. Daston

Springer

Robert J. Kavlock, Ph.D.
Director, Reproductive Toxicology Division (MD-71)
National Health and Environmental
Effects Research Laboratory
US Environmental Protection Agency
Research Triangle Park, NC 27711
USA

George P. Daston, Ph.D.
The Procter and Gamble Company
Miami Valley Laboratories
P.O. Box 538707
Cincinnati, OH 45253
USA

With 75 Figures and 31 Tables

ISBN 3-540-61259-9 Springer-Verlag Berlin Heidelberg New York

Library of Congress Cataloging-in-Publication Data. Drug toxicity in embryonic development I: advances in understanding mechanisms of birth defects/editors, R.J. Kavlock and G.P. Daston. p. cm. – (Handbook of experimental pharmacology: vol. 124/I–II) Includes bibliographical references and index. Contents: V.I. Morphogenesis and processes at risk – v. II. Mechanistic understanding of human developmental toxicants. ISBN 3-540-61259-9 (v. I: hbk.: alk. paper). – ISBN 3-540-61261-0 (v. II: hbk.: alk. paper). 1. Drugs – Toxicology. 2. Fetus – Effect of drugs on. 3. Abnormalities, Human – Etiology. I. Kavlock. Robert J. II. Daston, George P. III. Series. QP905.H3 vol. 124/I–II [RA1238] 615′1s–dc20 [616′.043] 97–24471

Printed in Germany

Cover design: Design & Production GmbH, Heidelberg

Typesetting: Scientific Publishing Services (P) Ltd, Madras

SPIN: 10484175 27/3136/SPS – 5 4 3 2 1 0 – Printed on acid-free paper

Dedicated to
Dr. Casimer T. Grabowski, our mentor, who encouraged us to pursue careers in understanding the mechanisms by which environmental agents might cause birth defects and who gave us the foundation to be successful scientists

Preface

Having received the invitation from Springer-Verlag to produce a volume on drug-induced birth defects for the *Handbook of Experimental Pharmacology*, we asked ourselves what new approach could we offer that would capture the state of the science and bring a new synthesis of the information on this topic to the world's literature. We chose a three-pronged approach, centered around those particular drugs for which we have a relatively well established basis for understanding how they exert their unwanted effects on the human embryo. We then supplemented this information with a series of reviews of critical biological processes involved in the established normal developmental patterns, with emphasis on what happens to the embryo when the processes are perturbed by experimental means. Knowing that the search for mechanisms in teratology has often been inhibited by the lack of understanding of how normal development proceeds, we also included chapters describing the amazing new discoveries related to the molecular control of normal morphogenesis for several organ systems in the hope that the experimental toxicologists and molecular biologists will begin to better appreciate each others questions and progress.

Several times during the last two years of developing outlines, issuing invitations, reviewing chapters, and cajoling belated contributors, we have wondered whether we made the correct decision to undertake this effort. However, now that we can look back on the finished product, we are confident that our initial goal was achieved and that this volume of the *Handbook* fills a void in the literature. In particular, we feel the volume complements the excellent descriptive compendia of the development effects of chemicals currently available by its focus on assessing mechanisms and modes of action. We are very pleased to have had the pleasure to have worked with the contributing authors, and hope that they forgive us for the "can you embellish on this....can you illustrate the process....can you condense this section....have you thought about this issue..."questions that we asked of them in the review process. Already we know we have some gaps in the coverage, but believe we have accomplished a synergism in combining the three sections as we did. Hopefully the work will stand the test of time similar to Jim Wilson's

Environment and Birth Defects, which although published in 1973, is still given to new students and trainees as a starting point for how to begin assessing the potential hazards and risks of chemicals to the developing embryo.

Research Triangle Park, NC, USA
Cincinnati, OH, USA
March 1996

ROBERT J. KAVLOCK
GEORGE P. DASTON

List of Contributors

AKHURST, R.J., Department of Medical Genetics, University of Glasgow, Yorkhill, Glasgow G3 8SJ, Great Britain

ANDERSON, R., Department of Cell and Molecular Biology and the Molecular and Cellular Biology Program, Tulane University, New Orleans, LA 70118, USA

AUGUSTINE, K.A., Department of Cell Biology and Anatomy, 108 Taylor Hall, CB#7090, University of North Carolina at Chapel Hill, Chapel Hill, NC 27599–7090, USA

BALLING, R., Institut für Säugetiergenetik, GSF-Forschungszentrum für Umwelt und Gesundheit, Neuherberg, Ingolstädter Landstr. 1, 85764 Oberschleißheim, Germany

BELL, S.M., Children's Hospital Medical Center, Research Foundation, Division of Basic Science Research, 3333 Burnet Avenue, Cincinnati, OH 45229–3039, USA

BURRILL, J.D., Molecular Neurobiology Laboratory, The Salk Institute for Biological Studies, 10010 North Torrey Pines Rd., La Jolla, CA 92037, USA

CARNEY, E.W., Developmental & Reproductive Toxicology, The Toxicology Research Laboratory, 1803 Building, The Dow Chemical Company, Midland, MI 48764, USA

DASTON, G.P., The Procter and Gamble Company, Miami Valley Laboratories, P.O. Box 538707, Cincinnati, OH 45253, USA

GITTENBERGER-DE GROOT, A., Department of Anatomy and Embryology, University of Leiden, Leiden, The Netherlands

GOLDMAN, A.S., Craniofacial Research, Department of Pediatrics, University of Illinois at Chicago, College of Medicine, 840 South Wood Street, Chicago, IL 60612, USA

Goto, M.P., Craniofacial Research, Department of Pediatrics, University of Illinois at Chicago, College of Medicine, 840 South Wood Street, Chicago, IL 60612, USA

Goulding, M., Molecular Neurobiology Laboratory, The Salk Institute for Biological Studies, 10010 North Torrey Pines Rd., La Jolla, CA 92037, USA

Hansen, D.K., Division of Reproductive and Developmental Toxicology, National Center for Toxicological Research, HFT-134, Jefferson, AR 72079, USA

Harris, C., Toxicology Program, Department of Environmental and Industrial Health, University of Michigan, Ann Arbor, MI 48109–2029, USA

Hunter, E.S., III, Reproductive Toxicology Division, National Health and Environmental Effects Research Laboratory, U.S. Environmental Protection Agency, Research Triangle Park, NC 27711, USA

Kavlock, R.J., Reproductive Toxicology Division (MD-71), National Health and Environmental Effects Research Laboratory, U.S. Environmental Protection Agency, Research Triangle Park, NC 27711, USA

Kim, P.M., Faculty of Pharmacy, University of Toronto, 19 Russell Street, Toronto, Canada M5S 2S2

Knudsen, T.B., Department of Pathology, Anatomy and Cell Biology, Jefferson Medical College, 1020 Locust Street, Philadelphia, PA 19107, USA

Lau, C., Reproductive Toxicology Division, National Health and Environmental Effects Research Laboratory, U.S. Environmental Protection Agency, Research Triangle Park, NC 27711, USA

Linser, P.J., University of Florida, Whitney Laboratory, 9505 Ocean Shore Boulevard, St. Augustine, FL 32086, USA

Liu, E.T., Curriculum in Genetics, Department of Medicine, Lineberger Comprehensive Cancer Center, University of North Carolina at Chapel Hill, Chapel Hill, NC 27599, USA

Maas, R., Genetics Division, Department of Medicine, Brigham & Women's Hospital, Harvard Medical School and Howard Hughes Medical Institute, 20 Shattuck Street, Boston, MA 02115, USA

Markwald, R., Department of Cell Biology and Anatomy, Medical University of South Carolina, 171 Ashley Avenue, Charleston, SC 29425, USA

Mirkes, P.E., Birth Defects Research Laboratory, Division of Congenital Defects, Departments of Pediatrics and Biological Structure, University of Washington, Box 356320, Seattle, WA 98195, USA

Muneoka, K., Department of Cell and Molecular Biology and the Molecular and Cellular Biology Program, Tulane University, New Orleans, LA 70118, USA

Neubüser, A., Institut für Säugetiergenetik, GSF-Forschungszentrum für Umwelt und Gesundheit, Neuherberg, Ingolstädter Landstr. 1, 85764 Oberschleißheim, Germany

Nicol, C.J., Faculty of Pharmacy and Department of Pharmacology, University of Toronto, 19 Russell Street, Toronto, Canada M5S 2S2

O'Neill, G.T., BBSRC Institute for Animal Health, Kings Buildings, West Mains Road, Edinburgh EH9 3JF, UK

Parman, T., Faculty of Pharmacy, University of Toronto, 19 Russell Street, Toronto, Canada M5S 2S2

Poelmann, R., Department of Anatomy and Embryology, University of Leiden, Leiden, The Netherlands

Rauchman, M., Renal Division, Department of Medicine, Brigham & Women's Hospital, Harvard Medical School and Howard Hughes Medical Institute, 20 Shattuck Street, Boston, MA 02115, USA

Sadler, T.W., Department of Cell Biology and Anatomy, Curriculum in Genetics, University of North Carolina at Chapel Hill, 108 Taylor Hall, CB#7090, Chapel Hill, NC 27599–7090, USA

Saueressig, H., Molecular Neurobiology Laboratory, The Salk Institute for Biological Studies, 10010 North Torrey Pines Rd., La Jolla, CA 92037, USA

Schreiner, C.M., Children's Hospital Medical Center, Research Foundation, Division of Basic Science Research, 3333 Burnet Avenue, Cincinnati, OH 45229–3039, USA

Scott, W.J., Children's Hospital Medical Center, Research Foundation, Division of Basic Science Research, 3333 Burnet Avenue, Cincinnati, OH 45229–3039, USA

TRUSK, T., Department of Cell Biology and Anatomy, Medical University of South Carolina, 171 Ashley Avenue, Charleston, SC 29425, USA

WELLS, P.G., Faculty of Pharmacy and Department of Pharmacology, University of Toronto, 19 Russell Street, Toronto, Canada M5S 2S2

WINN, L.M., Faculty of Pharmacy, University of Toronto, 19 Russell Street, Toronto, Canada M5S 2S2

ZIMMERMAN, E.F., Division of Developmental Biology, Children's Hospital Research Foundation, Elland & Bethesda Avenues, Cincinnati, OH 45229, USA

Contents

CHAPTER 3

Vertebrate Limb Development

CHAPTER 4

Axial Skeleton

CHAPTER 5

Molecular Mechanisms Regulating the Early Development of the Vertebrate Nervous System

CHAPTER 6

Genetic Control of Kidney Morphogenesis

CHAPTER 7

Palate

Section II: Common Biochemical, Metabolic, and Physiological Mechanisms of Abnormal Development

CHAPTER 8

Cell Death

CHAPTER 10

Cell–Cell Interactions

CHAPTER 11

Growth Factor Disturbance

CHAPTER 12

Targeted Gene Disruptions as Models of Abnormal Development

CHAPTER 13

Nucleotide Pool Imbalance

CHAPTER 14

Interference with Embryonic Intermediary Metabolism

CHAPTER 15

Alterations in Folate Metabolism as a Possible Mechanism of Embryotoxicity

CHAPTER 16

Prostaglandin Metabolism

CHAPTER 17

Reactive Intermediates
P.G. WELLS, P.M. KIM, C.J. NICOL, T. PARMAN, and L.M. WINN

Contents of Companion Volume 124/II

Section III: Pathogenesis and Mechanisms of Drug Toxicity in Development

CHAPTER 1
Introduction

R.J. Kavlock and G.P. Daston

Between three and six of every 100 babies born are afflicted with a major congenital malformation. Congenital malformations are the leading cause of infant mortality in developed nations. The rate of congenital malformations has not changed appreciably since records began to be kept systematically in the 1930s, although the overall rate of perinatal morbidity and mortality has decreased markedly over the past 50–60 years. The decline in perinatal mortality can be attributed to factors such as the broader availability of prenatal care and the development of effective treatments for infections, neonatal respiratory distress, and other maladies that have their onset around the time of birth and can be cured by treatments administered at that time. However, malformations occur many months before birth, often before the mother is aware of her pregnancy. Once initiated, the processes leading to the development of malformations tend to be irreversible. These factors make it diffcult to decrease the malformation rate. Therefore, it is clear that prevention is the key to solving the dilemma of congenital malformations.

Prevention has many facets. First, it is necessary to identify the conditions and exposures that are teratogenic. Second, it is important to understand the nature of normal development and the mechanisms that can be perturbed that lead to abnormal development. This latter information is critical in designing and refining methods to detect human teratogens, interpreting the results of those studies, and in designing safer therapeutic agents or effective maternal treatment strategies. This book has sections that cover each of these points: Section III presents known or suspected human teratogens. Section II discusses most of the known mechanisms of teratogenesis. Section I describes the exciting new findings that are coming to light about the molecular control of organogenesis, information that should contribute greatly to our understanding of how birth defects are initiated.

The focus on prevention has had many successes: the development of vaccines for rubella has essentially eradicated this disease as a cause of birth defects; dietary supplementation has effectively prevented birth defects due to clinically severe nutritional deficiencies in developed nations; and it appears that periconceptional folic acid supplementation will significantly decrease the rate of neural tube defects. Epidemiologists and clinicians have identified a number of drugs as human teratogens, and steps have been taken to prevent or limit exposure of women of childbearing potential to these agents. Screening for developmental toxicity in animal studies also appears to be effective in preventing exposure of women to teratogens. There are only 20–30 recognized human teratogens, whereas hundreds of developmental toxicants have been identified in laboratory animals. This discrepancy strongly suggests that preclinical screening is one of the best tools of prevention. Still, the fact that we have had such successes but still not markedly decreased the historical malformation rate hints at the complexity and multifactorial causation of birth defects. While preclinical screening and epidemiological surveillance are critical factors in preventing birth defects, the web of uncertainty that envelops the root causes of malformations can only be unraveled by learning more about the fundamental controls of normal development and the mechanisms by which exogenous agents act to perturb the system.

The heightened sensitivity of the developing human fetus to exogenous chemicals has been well known for therapeutic agents since the thalidomide tragedy, and for environmental agents since the contamination of Minamata Bay with methylmercury. The list of known human developmental toxicants has expanded to include more than a score of chemicals, physical agents, and pathogens. In Section III of this work we focus on those agents, primarily drugs, for which we are now beginning to understand the critical steps in the pathogenesis of developmental anomalies. The selected agents have both academic interest as well as practical significance because many of them have produced, or are still producing, adverse effects on human development. Each contribution in this section describes the known effects or syndromes associated with the particular chemical or class, including where possible an assessment of the relative risk of an adverse birth outcome under typical exposure situations; important aspects of pharmacokinetics that may influence the induction of defects; and a survey of the experimental literature characterizing the interaction of the teratogen with the embryo at biochemical, molecular, and cellular levels of organization. Of course, the extent of the available information varies considerably from chemical to chemical, and some human developmental toxicants (e.g., cocaine, lithium, metals, chlorinated biphenyls) have been excluded because the mechanistic basis for their actions is not yet sufficiently understood. However, some aspects of the toxicity of these agents are discussed in the chapters dealing with mechanisms of abnormal development.

While thalidomide was one of the earliest documented chemical teratogens in humans, experimental research to find the mechanism of action has extended from the 1960s to the present and there has been a great deal of frustration in identifying the critical chain of events significant to teratogenesis. The inability to describe the mechanism of action of this teratogen has probably had a considerable negative impact on the field of developmental toxicology in general. The chapter by Neubert and Neubert in this work not only summarizes the rebirth of thalidomide as a potentially useful therapeutic agent but also describes exciting new hints about the action of this compound on development. Perhaps the best studied class of human developmental toxicants has been the retinoids, where the hope exists that receptor agonists can be synthesized that maintain the desirable pharmacological properties while avoiding teratogenicity. Unfortunately this has not been the case so far, but the molecular understanding of the ligand-receptor interactions, elucidated in part through genetic engineering of cells and organisms, has led to significant new insight into the regulation of normal development by endogenous retinoids, and the root causes of abnormal development by retinoid excess or deficiency. Receptor interactions have also been the key to understanding the teratogenic effects of dioxin, a ubiquitous environmental contaminant whose action is mediated by the Ah receptor, as described by Abbott. Furthermore, a variety of drugs and environmental contaminants act through the stimulation or inhibition of steroid hormone function, as described by Kelce and Gray. This topic has grown in importance with the advent of the hypothesis that reported declines in the reproductive health of men, including declining sperm count and increased incidence of cryptorchidism, hypospadias, and testicular cancer, may be attributable to hormonally active pesticides and other anthropogenic chemicals in the environment. Women may also be at risk from such agents, as we know that the potent therapeutic estrogen diethystilbestrol has adverse effects on reproductive system development. Other chapters describe how an increased understanding of maternal and developmental physiology have been important in understanding the developmental effects of cardiovascular drugs and angiotensin-converting enzyme (ACE) inhibitors. Other chapters in this section describe the effects of important human teratogens for which the mechanisms of action are only beginning to be characterized, such as ethanol (and the related alcohol, methanol), and anticonvulsants. The chapters on anesthetic gases and antiviral agents point to the particular difficulties in assessing the safety of clinically irreplaceable treatments.

Collectively these examples reiterate many of the basic principles as stated by James G. Wilson in his classic text *Environment and Birth Defects* (Wilson 1973). Thus, there are examples of how susceptibility varies with (1) genotype, such as for dioxin, ethanol and anticonvulsants, and (2) developmental stage, such as for ACE inhibitors and steroid hormone agonists. There are examples of how teratogens act in specific ways on developing receptors, cells, and tissues to initiate sequences of abnormal development, such as is described for retinoids, anticoagulants, and thalidomide. There are examples of the diverse manifestations of abnormal development and, of course, how the rate and severity of abnormalities increases with dose. A number of valuable lessons from these studies of human developmental toxicants have led to changes in methods for screening new chemicals. This includes extending the period of dosing the pregnant animal to include the critical stages of reproductive system development and enhanced observations of reproductive structures in offspring, and increased emphasis on evaluations of function of a number of organ systems. The study of mechanisms of abnormal development also sheds light on the appropriateness of animal models as surrogates for humans. These improvements in the conduct and interpretation of preclinical studies will take on increasing importance as women of childbearing potential are included more routinely in all but the earliest phases of clinical trials for new active compounds, and as the pace of these trials is accelerated. Better understanding of mechanisms of action should also improve risk assessment for environmental pollutants where the uncertainty is great in extrapolating from the high doses used in animal studies to the low doses to which people may be environmentally exposed and the cost of compliance is high.

Progress in characterizing the many mechanisms of teratogenesis has been deliberate. Just as Wilson catalogued the principles of teratology nearly 25 years ago, he also presented the principal mechanisms that had been postulated at that time by which abnormal development may arise. The pathways he identified included mutation, chromosomal aberration, mitotic interference, altered nucleic acid integrity and function, lack of metabolic precursors, altered intermediary metabolism, enzyme inhibition, osmolar imbalance, and changed membrane characteristics. While these remain intuitively important concepts, they are largely theoretical constructs lacking experimental evidence that tie the key pathogenic steps together. The ability to extract information from the embryo has been hampered by the small size, inaccessibility, and continuous change that characterize development. However, the introduction of improved analytical methods and cellular and molecular techniques has opened the door to much more sophisticated assessments of the molecular, biochemical, and morphological status of the embryo, and basic research on teratogenic mechanisms has accelerated accordingly.

In Section II the reader will find presentations of normal cellular processes in the embryo that often serve as targets for the action of toxicants, together with a review of the state of knowledge on how those processes react when perturbed. Included among these chapters are updates of long-established theories of abnormal development, including altered intracellular pH, disturbances in intermediary metabolism, altered prostaglandin metabolism, nucleotide pool imbalance, hypoxia, altered maternal physiology, and the bioactivation of chemicals to toxic intermediates. In his analysis of nucleotide pool imbalance Lau provides insight into how data from several levels of biological organization can be integrated into a quantitative, mechanistic framework for analyzing the critical embryotoxic events produced by 5-fluorouracil.

Also included in Section II are chapters that add new possibilities to the list of potential mechanisms. For example, Mirkes describes the cellular response to stress from a variety of sources. These responses, discovered in part using prokaryotic organisms, and the downstream events leading to cellular injury and death appear to be significant to normal and abnormal development of the highest eukaryotes, including humans. Knudsen picks up on this discussion by presenting the molecular events leading to necrosis and apoptosis, both of which are important in teratogenesis. Akhurst and O'Neill contribute a review of advances in understanding the function of growth factors in the TGF-β, FGF, PDGF, TGF-α, and IGF families, and Linser covers events related to the cell–cell interaction process. Woven throughout these chapters is the use of recombinant DNA techniques, which have become essential components of laboratories studying developmental biology. An excellent example is the chapter by Sadler et al. on targeted gene disruptions.

A perusal of these chapters will provide the reader with an overview of the difficulties facing the experimental teratologist. This is evident by the overlap of chemicals and perturbations across chapters. In some instances this overlap is a natural consequence of the pathogenic process, a continuum that we have artificially separated into phases that are experimentally accessible. An example is the repeated presentation of data on methotrexate, which is described in various chapters as altering folate metabolism, changing nucleotide pool sizes, and eliciting cell death. In other instances, however, multiple pathogenic pathways may be involved in teratogenesis. For example, the anti-convulsant phenytoin has effects on prostaglandin metabolism, folic acid status, and redox state, and also generates reactive intermediates and disturbs maternal cardiovascular function in some animal models. In other cases upstream events may converge into a common pathogenic mechanism. Thus, hypoxia is mentioned as a critical component in the discussions of altered intracellular pH, adverse effects of cardioactive drugs, altered intracellular redox state, altered intermediary metabolism, and maternal physiological disturbance. Throughout these chapters on pathways of pathogenesis we are constantly confronted with difficulties in describing the entire sequence of events in the mechanistic cascade.

The study of abnormal development is greatly enhanced by a better fundamental characterization of normal morphogenesis. For this reason we have included a section describing recent advances in the basic developmental biology of several organ systems with known sensitivity to teratogens. Section I presents information on the development of six organs/structures: heart, central nervous system, axial skeleton, limb, palate and kidney. The molecular control of development of these systems has been intensively studied. The recent era has been a period of intensive activity in developmental biology. Much of the progress is attributable to two factors: the broad availability of molecular biology techniques, including transgenic models, and the remarkable evolutionary conservation of developmental control genes. An increasing number of genes that have been first discovered in simpler systems such as *Caenorhabditis elegans* or *Drosophila* have been found to have structural homologs with comparable function in vertebrates. For example, Neubüser and Balling describe genes that control axial skeletal positional information that also control segmentation in *Drosophila*, and Goulding et al. describe genes that establish the dorsal-ventral axis in the mouse nervous system and have a similar function in *Drosophila*. It is not only an experimental convenience that this phylogenetic homology exists; it indicates a conservation of developmental processes that can be used to extrapolate directly to humans the results of studies in a variety of animal species (Epstein 1995).

Research on the molecular control of development has led to the identification of the causes of heritable malformations due to single gene mutation, such as Waardenburg's syndrome type I, a Pax-3 mutation (Baldwin et al. 1992; Tassabehji et al. 1992), and the Denys-Drash syndrome, a WTI mutation (Pelletier et al. 1991). It has been possible either to create animal models for these disorders through gene knock-out or to recognize that existing spontaneous animal mutants are directly homologous to human conditions. The existence of such models will greatly facilitate our understanding of the contribution of single genes or combinations of genes to normal and abnormal development. Perhaps more important is the identification of target genes that confer susceptibility to environmental agents, as it is clear that most malformations are due to complicated interactions between the environment and the genotype of both mother and embryo. For example, oral clefting appears to be more likely in the offspring of mothers who smoke and who have a variant in the transforming growth factor-α gene (Hwang et al. 1995). In this book Wells presents evidence that the embryotoxicity of phenytoin is mediated through an epoxide intermediate, and Finnell and Nau describe a correlation between low epoxide hydrolase activity in human amniocytes and susceptibility to phenytoin teratogenicity.

New discoveries about mechanisms of teratogenesis and the fundamental control of development will continue to lead to improvements in the methods by which teratogenic hazards are detected and their risks extrapolated to the human population. For example, research on the effects of steroids or agents that interfere with steroid metabolism (e.g., 5-α-reductase inhibitors) has shown that the critical period of susceptibility of the developing reproductive system extends beyond what was considered to be the limit of major organogenesis in rodent screening tests. This has led to a modification in the design of these tests. Experience with ACE inhibitors has shown that these agents act after the embryonic period, an observation that has led to the application of functional measurements of teratogenicity for this class of compounds.

Although it has long been thought that most malformations are the result of interactions between the environment and endogenous susceptibility factors, the identity of those factors is only now being discovered. Screening for gene variants that increase the risk of adverse outcome from an environmental exposure, such as TGF-α variants and maternal smoking, should greatly improve the efficiency and resolving power of epidemiology studies. The ability to measure subclinical folate deficiencies contributed to the recognition that folate supplementation can reduce the incidence of neural tube defects. It is possible that other subclinical nutritional deficiencies will be discovered to increase the risk of malformations. From a clinical perspective, it may be possible to identify individuals who are at risk for phenytoin teratogenicity by characterizing epoxide hydrolase genes. As more susceptibility factors are identified, it may be possible to individually customize therapies for pregnant women to minimize the chances of an adverse outcome.

Advances in basic developmental biology and the study of teratogenic mechanisms should ultimately improve our ability to extrapolate animal results to predict responses in the human population (KAVLOCK and SETZER 1996). Of particular relevance will be the elucidation of effects on the highly conserved molecular elements that control fundamental processes and the identification of the critical molecular and cellular events in the pathogenesis of malformations.

We expect that this book will be of use to everyone who is interested in the problem of congenital malformation: from clinicians and counselors who treat pregnant women and their children to scientists who evaluate drugs for teratogenic potential and investigate mechanisms of action. A great deal of effort has been put into this work by the leading minds in developmental biology and toxicology, and we hope that the compilation of this effort will stimulate the reader's thinking and thereby bring about still more creative research that will address the problem of abnormal development. Only through better understanding will we be able to continue to find ways to identify the causes and prevent the occurrence of birth defects.

References

Baldwin CT, Hoth CF, Amos JA, DaSilva ED, Milunsky A (1992) An exonic mutation in the *HuP2* paired domain gene causes Waardenburg's syndrome. Nature 355: 637–638

Epstein CJ (1995) The new dysmorphology: application of insights from basic developmental biology to the understanding of human birth defects. Proc Natl Acad Sci USA 92: 8566–8573

Hwang SJ, Beaty TH, Panny SR, Street NA, Joseph JM, Gordon S, McIntosh I, Francomano CA (1995) Association of transforming growth factor-alpha (TGF-alpha) Taq 1 polymorphism and oral clefts: indication of gene-environment interaction in a population-based sample of infants with birth defects. Am J Epidemiol 141: 629–636

Kavlock RJ, Setzer RW (1996) The road to embryologically based dose-response models. Environmental Health Perspectives 104 (Suppl 1): 107–121

Pelletier J, Bruening W, Kashtan CE, Maurer SM, Manivel JC, Striegel JE, Houghton DC, Junien C, Habib R, Fouser L, Fine RN, Silverman BL, Habe DA, Hausman D (1991) Germline mutations in the Wilm's tumor suppressor gene are associated with abnormal urogenital development in Denys-Drash syndrome. Cell 67: 437–447

Tassabehji M, Read AP, Newton VE, Harris R, Balling R, Gruss P, Strachan T (1992) Waardenburg's syndrome patients have mutations in the human homologue of the *Pax-3* paired box gene. Nature 355: 635–636

Wilson JG (1973) Environment and birth defects. Academic, New York

Section I
Recent Advances in Understanding Normal Development at the Biochemical and Molecular Level

CHAPTER 2

Cardiac Morphogenesis: Formation and Septation of the Primary Heart Tube

R. MARKWALD, T. TRUSK, A. GITTENBERGER-DE GROOT, and R. POELMANN

A. Introduction

Overt morphological and biochemical differentiation of the heart occurs abruptly. The formation of a single, beating, tubular heart is one of the earliest events in vertebrate embryogenesis. Acquisition of function precedes complete morphogenesis. Accordingly, it is not surprising that malformations of the heart occur frequently, approaching 1% of all births, and remain the single largest cause of infant mortality from congenital defects, exceeding cystic fibrosis, hemophilia, or childhood cancer (CLARK 1987). To be born with a congenital heart defect largely means to have (a) an abnormal communication through which blood shunts inappropriately between the two sides of the heart, (b) defective communication between the chambers on each side of the heart, or (c) faulty "plumbing" exhibited by reductions in size of the two great outlet arteries or misalignment of these arteries with ventricular chambers. In each of these situations, the developmental basis of the defect has been the cause of the failure to solve the morphogenetic riddle of heart development, i.e., how a tubular heart with a single, hollow channel is established that can be subsequently subdivided into four asymmetrical chambers while literally never missing a beat.

The purpose of this chapter is to review the molecular and cellular mechanisms by which the solutions of this riddle are being pursued. It must be underscored that cardiac development is the collective outcome of multiple fundamental processes that cannot be understood by studying a single given instant. They are dynamic, sequential, progressive, uninterrupted, and irreversible. It is the normal spatiotemporal integration of the different heart components that ultimately defines heart shape and function.

B. Establishing Heart-Forming Primordia

I. Commitment to the Heart Lineage

The molecular "decision" to restrict multipotential embryonic cells into a cardiogenic pathway has already begun by the onset of gastrulation and formation of the primitive streak (GARCIA-MARTINEZ and SCHOENWOLF 1993; ANTIN et al. 1994). In the chick, the mesodermal cells which are "specified" or intended to differentiate ultimately into myocardial or endocardial cells are localized to the rostral half of the primitive streak and possess axial identity along the anterior–posterior axis (INAGAKI et al. 1993). They are located immediately behind Hensen's node, a structure homologous to a region of the amphibian blastopore known as the Spemann organizer (JACOBSON and SATER 1988). In amphibians, contact with this region of the blastopore and endoderm is necessary and sufficient for directing gastrulating mesodermal cells into a cardiogenic pathway (SATER and JACOBSON 1990; NASCONE and MERCOLA 1995). In the chick, at stage 4, Hensen's node is a source of retinoic acid (CHEN et al. 1992), a morphogen known to influence pattern formation and differentiation, also in the heart (CHEN and SOLURSH 1993; OSMOND et al. 1991; DERSCH and ZILE 1993). The presence of retinoic acid receptors in mesodermal cells specified to a heart lineage is consistent with a role for this morphogen in directing or specifying mesoderm to a heart lineage (SMITH 1994; SUCOV et al. 1954). Since receptors for retinoic acid function as transcription factors (GUDAS 1992), it is possible that the downstream gene targets of these receptors constitute some part of the unresolved mechanism for specifying the heart lineage.

II. Formation of the Heart-Forming Fields

Once specified, presumptive heart-forming cells in the chick migrate from the streak anteriorly on either side of Hensen's node to form a pair of heart-forming fields by stage 4 (RAWLES 1943; ROSENQUIST and DEHAAN 1966). The streak regresses (literally recedes posteriorly) following the emigration of cells specified to a heart lineage, leaving behind the notochord (POELMANN 1981). At stage 4, the notochord is the principal midline structure, flanked by the two heart-forming fields. DANOS and YOST (1995) have reported that an interaction between notochord and heart-forming cells determines the dorsal–ventral and right–left axis of the future heart tube. In particular, two genes expressed by notochord, *nodal* and *hedgehog*, appear related to imprinting right–left symmetry (LEVIN et al. 1995).

The migratory pathways of the heart-forming cells from the streak can be tracked by trails of a fibrillar protein termed JB3, related to fibrillin (GALLAGHER et al. 1993; WUNSCH et al. 1994). However, the continued but restricted expression of JB3 antigen by heart-forming cells has made it possible to morphologically visualize and characterize them from among the much larger population of lateral plate mesodermal cells. Initially, at stage 4, the heart fields are a loose aggregate of mesodermal cells that progressively condense into a true epithelium (PENG et al. 1990). Epithelialization of the heart fields correlates with the polarized expression of a calcium-dependent adhesion molecule N -cadherin and an associated cytoplasmic binding partner, β-catenin (LINASK 1992). The latter has homology to *armadillo*, a segmentation/polarity gene in *Drosophila*, which also interacts with the *wingless* (Wnt-1) gene (PEIFER et al. 1992; SIEGFRIED and PERRIMON 1994; WU et al. 1995), suggesting that specification and positioning of the heart fields, including their anatomical axes (Danos and Yost 1995), may be regulated by the coordinated expression of pattern-forming genes with cell surface adhesion molecules.

Coelom formation also accompanies epithelialization and results in the splitting of the lateral plate mesoderm, including the heart fields, into somatic and splanchnic mesoderm (Linask 1992). The heart-forming fields track with the splanchnic mesoderm and, accordingly, associate with the endodermal germ layer. The potential for inductive interactions between endoderm and precardiac mesoderm have long been recognized (Orts-Llorca 1963; Lemanski et al. 1979), but the morphoregulatory roles of endoderm in heart field formation are only now becoming clear as molecular approaches are used with more defined culture systems (Antin et al. 1994; Sugi and Lough 1994).

After coelom formation (stage 6+ in the chick, day 7.5 in the mouse), commitment of the heart fields is irreversible, i.e., it has become fixed or determined (Montgomery et al. 1994), meaning it cannot be changed if grafted or transplanted elsewhere in the embryo. Thus, as recently defined by Slack (1991), commitment to a heart lineage as with other lineage pathways appears to be a two-step process: specification and determination. For the heart, the former appears to occur during gastrulation, and the latter within the heart fields. Despite enormous effort, the genes (including any retinoic acid receptor targets) which regulate either step have not yet been identified. What is clear is that the elegant cascade of transcriptional factors shown to regulate skeletal muscle differentiation (Lyons and Buckingham 1992) cannot be broadly applied to cardiac differentiation (Kern et al. 1995). Logical candidates presently under investigation include zygotic genes such as *twist* or *snail* (Kessler and Melton 1994) or segmentation/polarity genes such as those in the *Wnt* gene family, which in *Drosophila* mediate mesoderm formation, including contractile blood vessels (hearts) (Nusse and Varmus 1992; Wu et al. 1995). Other candidates are genes encoding growth factors known to be secreted by endoderm such as those of the transforming growth factor (TGF)-β gene superfamily, which induce amphibian ectodermal cells to enter a mesodermal lineage (Kessler and Melton 1994).

III. Segregation of Lineage Within the Heart Fields

Once specified and determined, the question to be answered is how the epithelium of the heart fields (i.e., the precardiac splanchnic mesoderm) gives origin to both myocardial and endocardial lineages. Is there a common (bipotential) precursor, or is the epithelium a collective of cells already committed to a single (unipotential) lineage? Microinjection of retroviral lineage tracers is the approach that will likely resolve this question. In zebrafish, results indicate the existence of a common precursor (Lee et al. 1994), whereas in chick embryos results are still inconclusive, as Mikawa et al. (1992) were able to only transfect a low percentage of the precardiac mesodermal cells, all of which became myocardial cells. However, other data indicate that stage-5 chick precardiac cells cultured in the presence of serum coexpress endothelial and myocardial markers (LINASK and LASH 1993). A common precursor is also indicated by studies using a clonal cell line, QCE-6. QCE-6 cells are a differentiation-inducible, stable cell line (EISENBERG and BADER 1995; MARKWALD et al. 1996) derived from a single, immortalized heart-forming cell isolated from a stage-4 quail heart field. They form an epithelial monolayer in culture which expresses markers characteristic of primitive, splanchnic mesodermal epithelium (e.g., vimentin and keratin). However, upon addition of retinoic acid (consistent with the role of Hensen's node) and specific growth factors (normally secreted by endoderm), QCE-6 cells form both endocardial and myocardial cells. Collectively, these data would seem to favor the existence of a common stem cell within the heart-forming mesodermal epithelium. The availability of a bipotential cardiac stem cell increases the possibility of determining the mechanism of how each lineage is established and the epigenetic influence upon those mechanisms.

As with bipotential QCE-6 cells, segregation of the two heart cell lineages occurs by an epithelium to mesenchyme transformation that begins in chick embryos in the heart fields at stage 5, prior to their fusion (SUGI and MARKWALD 1996). Those cells which transform to mesenchyme enter the basement membrane that separates the endoderm from the precardiac mesoderm of the heart field (MANASEK 1976). They express the JB3 antigen and cytotactin, a general marker for mesenchyme (not mesoderm). A definitive endothelial marker in quail embryos, QH-1, is not expressed within the epithelium of the heart fields, but only in the mesenchyme positioned beneath it. Persisting (nontransforming) epithelial cells enter a myocardial lineage, as indicated by their eventual expression of contractile proteins. To determine whether transformation to mesenchyme was a prerequisite for expression of the QH-1 marker, we isolated the heart fields from stage-4/5 embryos and cultured them on collagen gel lattices in the presence or absence of endoderm (SUGI and MARKWALD 1996). Although myocardial cell differentiation was not affected, endothelial markers, as in vivo, were not expressed except in free mesenchymal cells and only if cultured with endoderm or in medium containing serum. Thus endoderm or factors present in serum promote a transformation of precardiac mesodermal epithelium into mesenchymal cells, an event which correlates with their expression of an endothelial marker and separation from premyocardial cells. Although the inductive signals for transformation present in serum or secreted by endoderm are not known, our findings with QCE-6 cells would implicate TGF-β2 and -β3 and basic fibroblast growth factor (MARKWALD et al. 1996; EISENBERG and BADER 1995), all of which are similar to growth factors secreted by anterior endoderm (KOKAN-MOORE et al. 1991; SUGI and LOUGH 1995)

IV. Molecular Regulation of the Cardiomyogenic Lineage

Whether the hypothetical genes that are assumed to direct undifferentiated mesodermal cells into a heart lineage following gastrulation are distinct from those which regulate expression of the myocardial or endocardial phenotype is unclear. Significant progress has been made in identifying genes that enhance (but do not initiate) expression of cardiac-specific contractile proteins. For example (see Fig. 1), serum response factor (SRF) and a reciprocal inhibitor YY1 enhance/suppress actin gene expression (McQuinn and Schwartz 1995), whereas muscle-enhancing factor (MEF)-2 (Edmondson et al. 1994) and cardiocyte maturation factor (CMF-1; Litvin et al. 1993) enhance myosin expression. Although MEF-2 and CMF-1 share a basic helix–loop–helix motif with those of the skeletal muscle-regulatory gene family, e.g., myoD and myogenin (Lyons and Buckingham 1992; Lyons 1994), the cardiac myogenic regulatory proteins are not homologous to those of the skeletal myogenic family, nor do they function equivalently (Kern et al. 1995). Transfecting myoD into precardiac mesoderm does not activate myogenic programs as it does when transfected into noncardiogenic mesoderm (Miner et al. 1992;

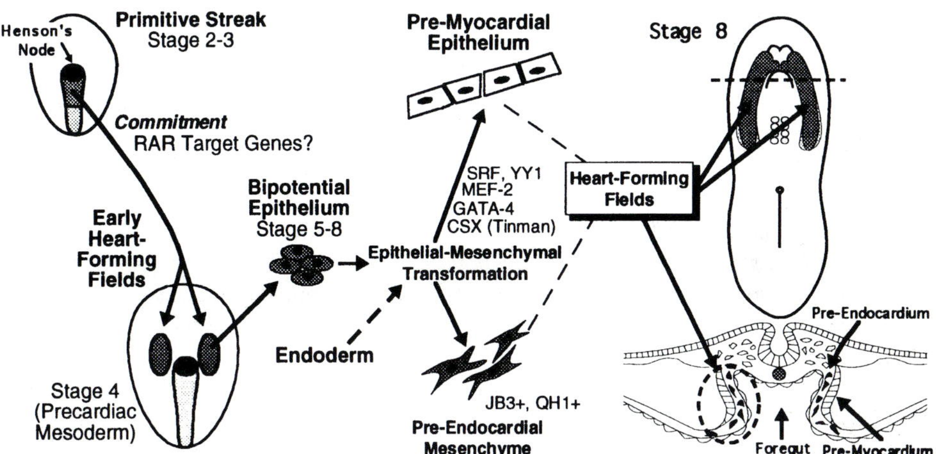

Fig. 1. Heart field formation. Stages refer to chick development. Heart fields are established from mesodermal cells of the rostral primitive streak. It is hypothesized that retinoic acid or other factors secreted from Hensen's node promote commitment (specification) to a heart lineage by controlling the expression of retinoic acid receptor (*RAR*) target genes. With coelom formation (stage 6), the heart fields become epithelialized with the potential to form endocardial and myocardial cells. Endoderm promotes a transformation to mesenchymal cells that express antigens JB3 and QH1, indicative of an endocardial lineage. More QH-1-positive cells form in the right heart field. Nontransforming cells remain epithelial and express transcriptional factors (e.g., CSX, GATA-4, see text) associated with myocardial differentiation. The heart fields progressively elongate and fuse anteriorly to form a cardiogenic crescent. *SRF*, serum response factor.

McQUINN and SCHWARTZ 1995). Thus, while CMF and MEF are expressed in the heart fields, there is no evidence they are the much sought after "myoH" (or "cardioD") which, when expressed in the heart fields, would turn on the cardiomyogenic gene programs.

Based on its expression in the heart fields, endoderm, and early heart tube, other candidates for cardiomyogenic regulation might be the vertebrate homologues of the *Tinman* gene (LINTS et al. 1993). In *Drosophila*, this muscle segmental homeobox (MSH)-type homeobox gene is expressed downstream of *twist* and *snail*. In *Tinman* mutants, *Drosophila* "hearts" (actually contractile blood vessels) do not form (BODMER 1993). The mouse homologue to *Tinman* was independently identified as *NKx-2.5* by LINTS et al. (1993) and *Csx* (cardiac-specific homeobox) by KOMURO and IZUMO (1993). While targeted mutagenesis of the transcription factor *NKx-2.5* does not prevent formation of a contractile heart tube, lethality ensued prior to septation. Interestingly, segmental identity was affected, e.g., demarcation between atrial and ventricular segments was blurred, and looping of the heart tube did not occur (LYONS et al. 1995). The fact that a beating tubular heart formed in the absence of *Csx* should not diminish enthusiasm for this gene as a leading candidate for regulating cardiomyogenesis. It is but one member of the NK gene family which might have redundant functions. Results of the *Csx* knockout are consistent with the hypothesis that the genes which direct mesoderm into a heart lineage are upstream of those regulating the cardiac myogenic program. Indeed, recent findings suggest that *NKx-2.5* and serum response factor interact to control actin gene expression and differentiation of cardiac tissues (McQUINN and SCHWARTZ 1995). Similarly, another transcription factor expressed in the heart fields, GATA-4, has been shown to initiate transcription of the gene for the α-myosin heavy chain (MOLKENTIN et al. 1994). The promoter region of GATA-4 has a retinoid response element (ARCECI et al. 1993), which provides a clue as to how retinoic acid might modify differentiation of precardiac mesoderm into atrial versus ventricular muscle phenotypes (YUTZEY et al. 1994).

V. Regulation of the Endocardial Lineage

While progress continues for the myocardial lineage, no candidate genes have been found that might function to direct heart field mesoderm into an endocardial pathway. As noted above, endocardial precursor cells are released from the precardiac mesodermal epithelium as mesenchymal cells. Eventually, they interact to establish the inner epithelial lining of the primary heart tube. By most criteria, they are in an endothelial lineage. To date, no expression of a developmentally regulated transcription factor has been reported in the mesenchymal precursors of endocardial cells. Only after they establish the endocardial epithelium is such expression observed, e.g., GATA-4/-5/-6 (KELLEY et al. 1993; LAVERRIERE et al. 1994), and Msx-1 (CHAN-THOMAS et al. 1993). However, as noted above, formation of pre-endocardial mesenchyme and their expression of the QH-1 endothelial marker were promoted by endoderm (SUGI and MARKWALD 1996). One interpretation of this finding is that endoderm induces target cell responses in competent precardiac mesoderm that lead into an endocardial lineage, suggesting the presence of endoderm response genes. We are pursuing such hypothetical genes using endoderm-induced or non-induced QCE-6 cells and subtractive hybridization protocols.

The identification of such genes could have profound implications for understanding endocardial vasculogenesis and angiogenesis. These genes may also provide insight into the mechanism by which the right–left axis is established. Handedness or asymmetry has been shown to be established by stage 6 in chick embryos, as evidenced by the dominance of the right heart field upon the direction the future heart tube bends or loops (HOYLE et al. 1992). We were recently able to provide morphological confirmation of a potential right–left asymmetry. Early in stage-7 quail embryos, the first QH-1-positive endothelial precursor cells observed in the embryo were in the heart fields, with more present in the right field than in the left. For this reason, we have hypothesized that the candidate endoderm response genes which regulate the endocardial differentiation pathway will be asymmetrically expressed in the heart fields and may be among the downstream targets for the genes regulating laterality.

VI. Fate of the Heart Fields

Each heart field migrates ventrally down the sides of the developing (open) foregut to fuse at their anterior ends by stage 8 (DRAKE and JACOBSON 1988). A resulting inverted U-shaped crescent is formed on the anteroposterior axis that histologically is composed of pre-endocardial mesenchyme and pre-myocardial epithelium (Fig. 1). Migration of the fields is an active process dependent upon fibronectin that is secreted into the basement membrane separating heart field from endoderm (LINASK and LASH 1988). At the point of fusion, a prominent three-dimensional structure, the primary heart tube, begins to emerge (Fig. 2). It extends cranially (anteriorly) as the posterior ends of each field continue to be drawn into the fusion zone of the cardiac crescent. A section through the latter (Fig. 2) reveals that the tube is assembled by the integration of pre-endocardial cells to form a single lumen or channel lined by endocardium. Endocardial vasculogenesis may begin within the unfused regions of the cardiac crescent but, unless fusion of the fields is prevented, only a single endocardial tube is formed (DE RUITER et al. 1992). It develops beneath the ventral midpharyngeal (foregut) endoderm in a region in which the endodermal epithelium is remarkably hypertrophied (Fig. 2), suggesting a potential interaction related to endocardial vasculogenesis. One potential mechanism by which this region of the endoderm could promote endocardial lumen formation is by the localized secretion of vascular endothelial cell growth factor (VEGF). VEGF has been shown to be a powerful signal for luminization in the embryo (DRAKE and LITTLE 1995; RISAU 1995). Consistent

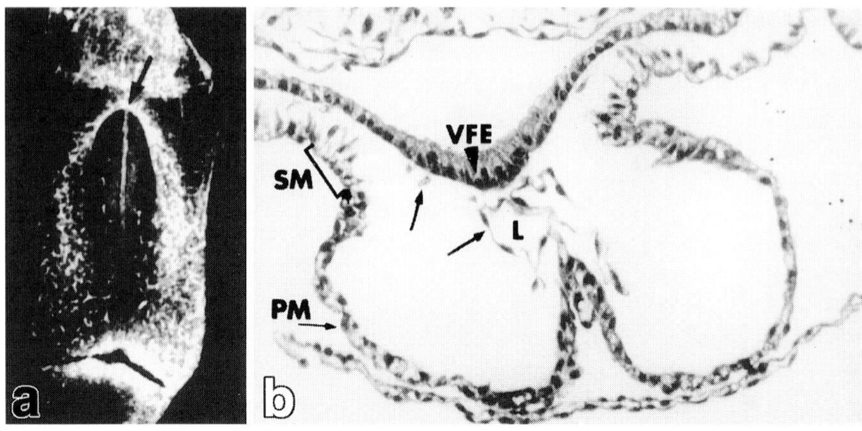

Fig. 2. a Cardiogenic crescent formed by the fusion of the heart fields (*arrow*) is visualized by QH-1 staining. ×20. **b** A routine histological cross-section through a stage 8+ embryo at the level shown by the arrow in a. *Arrows* denote pre-endocardial mesenchymal cells, some of which have assembled to form a lumen (*L*). Endocardial vasculogenesis is initiated beneath an hypertrophied region of the ventral foregut endoderm (*VFE*). The premyocardial mesoderm (*PM*) of each field fuses ventrally to encase the developing endocardial tube. Progressive fusion will bring the bracketed, noncardiogenic splanchnic mesoderm (*SM*) of each field into apposition to form a dorsal mesentery. ×80

with this hypothesis, the receptor for VEGF is expressed by endocardial endothelium (RISAU 1995).

Once established, the endocardial tube appears to act as a nidus about which the definitive myocardium is organized (VIRAGH et al. 1989; COFFIN and POOLE 1991). As shown in Fig. 3, the endocardial tube becomes encased by the premyocardial epithelium of each fusing field. Contact and union between the two premyocardial epithelia occurs dorsally and ventrally, resulting respectively in the formation of the dorsal and ventral mesocardium. The latter is a transitory structure of unknown function, whereas the extracardiac mesenchyme of the dorsal mesentery will give origin to the proepicardial organ, a mesothelial structure located between the hepatic primordia and sinus venosus. The proepicardial organ is the source or conduit for the extracardiac mesenchyme that will migrate onto the surface of the heart tube to form the epicardium (visceral pericardium) (VRANCKEN PEETERS et al. 1995). In turn, the epicardial mesenchyme contributes progenitors (endothelial and smooth muscle) to the coronary vasculature (POELMANN et al. 1993; VIRAGH et al. 1994; MIKAWA and GOURDIE 1995). Thus the fusion of the heart fields not only establishes the definitive heart tube, but also the source of its own vasculogenic mesenchyme.

Expression of contractile proteins occurs approximately at the time of field fusion (HAN et al. 1992), with sarcomeric organization and heart contractions evident immediately after formation of the early heart tube (To-

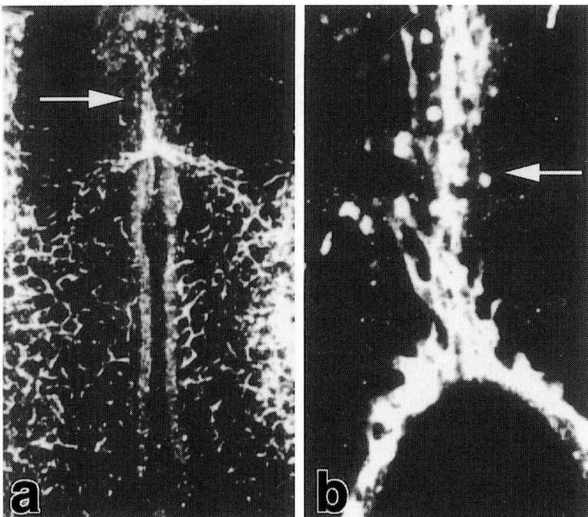

Fig. 3a,b. Confocal microscopic images of QH-1-stained, whole-mount embryos at stage 9 to show the initial formation and anterior extension of the endocardial tube (*arrows*) from the fused heart fields. Note the presence of QH-1-positive cells located at the periphery of the developing tubular heart. It is suggested that these might represent endocardial precursor cells that have formed outside the heart fields and subsequently migrated into the developing tubular heart (WUNSCH et al. 1994). **a** ×20. **b** ×150

KUYASU and MAHER 1987). The temporal and spatial program of cardiac muscle gene expression has recently been reviewed by LYONS (1994). It begins with alpha (vascular) smooth muscle actin (Fig. 4) followed by cardiac actin (MCQUINN and SCHWARTZ 1995) and a well-defined sequence of myosin isoforms (SWEENY et al. 1987). With formation of the definitive heart tube, the contractile proteins become organized into sarcomeres and myofibrils. We suggested several years ago that myocardial cell surfaces played a key organizing role in myofibrillogenesis (MARKWALD 1973), a role subsequently supported by in vitro studies on living cells by SANGER et al. (1986). Some clues as to a mechanism by which the myocardial surface might promote sarcomeric organization have come from studies on adhesion molecules, N cadherin (SOLER and KNUDSEN 1994), and a specific variant of neural cell adhesion molecule (NCAM). An NCAM variant possessing a muscle-specific domain has been localized to the cell surface in a pattern resembling zebra stripes (BYEON et al. 1995). If the exon encoding this domain is spliced out, e.g., in endocardial cells, this zebra-like pattern is abolished, suggesting to us the hypothesis that the muscle-specific domain of myocardial NCAM might function at the cell surface as a sarcomeric organizing center.

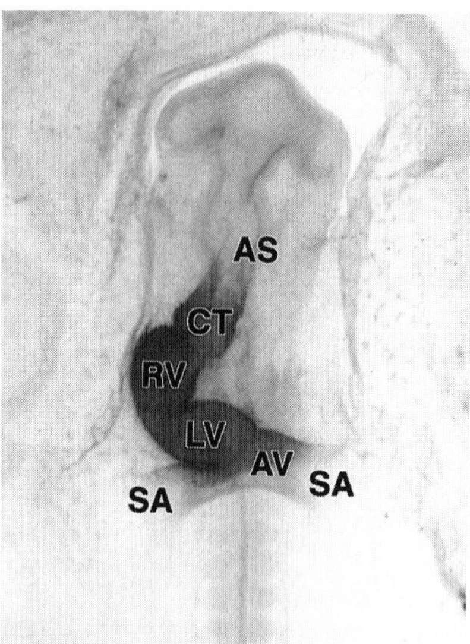

Fig. 4. Stage-11 embryo stained with anti-alpha smooth muscle antibodies. Between stage 9 (see Fig. 3) and stage 12, new segments are added to the heart tube. Symmetry is "broken" by looping to the right. Based on sequence of formation, segments of the heart tube are the right ventricle (*RV*), left ventricle (*LV*), and atrioventricular canal (*AV*). The conotruncal segment (*CT*) extends anteriorly from the right ventricle. The sinuatrial (*SA*) segments have yet to be incorporated into the tube. Note that the antigen is not expressed in the aortic sac (*AS*), consitent with its origin from the branchial arches, not the heart fields. ×100

C. Morphogenesis of the Primary Heart Tube

I. Elongation and Segmentation of the Tubular Heart

It is important to understand that the fusion process which provides the primordium for the primary tubular heart is a dynamic process beginning at stage 9 and continuing to stage 12 in the chick (DE LA CRUZ et al. 1989). As fusion proceeds, the tube extends cranially and primitive segments become evident externally within the tube along its anterior–posterior axis. Thus the primitive heart segments do not develop all at once, but rather sequentially over time. Looping or bending of the tube occurs as the segments are added (Fig. 4, 5). In chronological order the primitive segments are as follows:

1. The trabeculated region of the ventricular outlet, which becomes the future right ventricle
2. The trabeculated region of the ventricular inlet, which becomes the future left ventricle
3. The atrioventricular junction
4. The inflow tract, consisting of the future sinus venosus and atrium

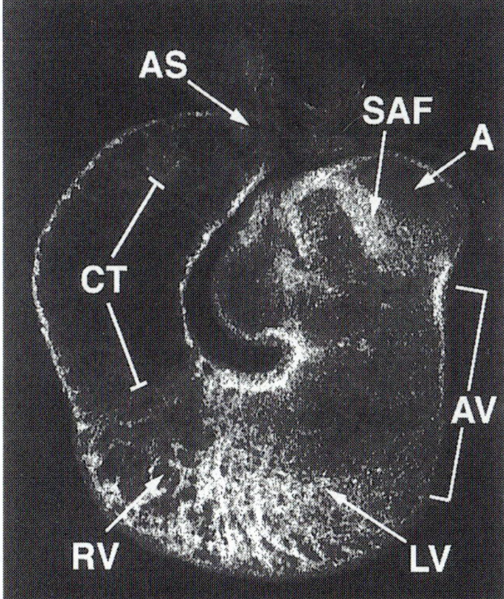

Fig. 5. Confocal imaging of a stage-18 chick heart stained as a whole mount to reveal cardiac actin expression. Segment formation and looping are complete, as indicated by the direct contact between venous and arterial poles. In the atrium (*A*), the sinuatrial folds (*SAF*) are prominent structures, whereas in the ventricles trabeculae develop. Those at the site of the primary ring, i.e., the junction between right (*RV*) and left ventricle (*LV*), are larger and more condensed. Conversely, no muscularized, trabecular invaginations are evident in the atrioventricular (*AV*) canal and conotruncus (*CT*); rather, unstained cushion mesenchyme develops. Actin staining of the myocardial epithelium stops abruptly at the junction of the conotruncus with the aortic sac (*AS*), the latter being the site in which the aortic pulmonary septum develops. ×60

The outflow tract or conotruncus develops as a cranial extension from the first segment (De la Cruz et al. 1977, 1989). The aortic sac (arterial pole), which connects the conotruncus to the pharyngeal arteries, is not derived from the heart fields and, accordingly, does not have a myocardial lining (Bartelings et al. 1989; Noden et al. 1995).

Three points are pertinent to the sequential formation of the cardiac segments by the fusion of the heart fields: (1) each has a different birthday and birthplace (position) within each heart field, (2) looping or bending of the heart tube accompanies the addition of segments, and (3) the segments have different developmental fates. However, none is directly equivalent anatomically to a specific chamber of the adult heart, i.e., chambers arise by the interaction and remodeling of primitive segments.

II. Morphology of the Primitive Segments

As shown in Figs. 5 and 6, the wall of each primitive segment is lined by two concentric epithelia – the endocardium and myocardium – which initially blend seamlessly with those of adjacent segments. Extracellular matrix, historically termed cardiac jelly, separates the two epithelia of each segment. In

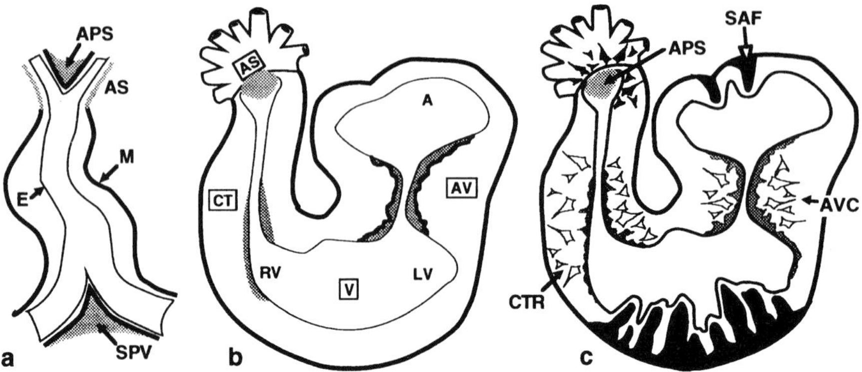

Fig. 6a–c. Three stages in the development of the tubular heart. **a** Stage 11, looping and segment formation occur. Each segment is lined by two concentric epithelia, the endocardium (*E*) and myocardium (*M*). Two wedges of extracardiac mesenchyme are produced at each pole: the aorticopulmonary septum (*APS*) and the spina vestibuli (*SPV*). **b** Stage 16. The endothelium, the atrioventricular (*AV*) canal, and conotruncus (*CT*) become activated in preparation for mesenchyme formation. **c** Stage 18, equivalent to Fig. 5. Trabeculae and sinuatrial folds (*SAF*) develop in the atrium and ventricle, mesenchymal cushions and ridges in the atrioventricular canal (*AVC*) and conotruncus, respectively. Note some cells of the APS invade the conotruncal ridges (*CTR*)

the atrioventricular and conotruncal segments, the cardiac jelly is greatly expanded. Some 50 proteins have been resolved in the cardiac jelly of ethylenediaminetetra-acetic acid (EDTA) extracts of atrioventricular segment by two-dimensional gel electrophoresis (RUNYAN and MARKWALD 1983). Many of these are characteristic of basement membranes, and most are secreted by the myocardium, suggesting that cardiac jelly represents the myocardial basement membrane (KITTEN et al. 1987). The components of cardiac jelly in the tubular heart have recently been reviewed (RONGISH and LITTLE 1995) and include collagens, fibronectin, hyaluronate, fibrillin, fibulin, proteoglycans, growth factors, and many uncharacterized glycoproteins.

Although the histological components of each segment appear homogeneous during formation of the primitive tube, molecular imaging of gene or protein expression has revealed molecular diversity in both myocardium and endocardium (Table 1). For example, in the atrioventricular (AV) and conotruncal segments, the myocardium expresses two genes that encode bone morphogenetic proteins (BMP) (JONES et al. 1991). The latter are homologous to *Drosophila* decapentaplegic, a gene family that regulates segmentation, polarity, and pattern formation. *Msx-2*, like *Csx-1*, is a homeobox gene belonging to the *Drosophila Msh-2* group. Proteins termed EDTA-soluble (ES) antigens, while initially expressed in all segments, become restricted to the AV canal and conotrunucus (MJAATVEDT et al. 1991). A gene recently identified by an insertional mutation, *heart defect* (*hd*) (M. Zhang and R. Markwald, unpublished observations), is intensely expressed in the myocardium of the

Table 1. Segmental gene expression in the primary tubular heart (chick and mice)

Myocardium	Endocardium
Atrioventricular and conotruncus	
Decapentaplegic-related[a] (*Bmp-2, -4*) (JONES et al. 1991)	JB3 (WUNSCH et al. 1994)
MSX-2 (CHAN-THOMAS et al. 1993)	MSX-1 (CHAN-THOMAS et al. 1993)
ES proteins (MJAATVEDT et al. 1991)	TGF-β1-β3 (NAKAJIMA et al. 1994)
Heart defect (*hd*)[a] (H.-Y. Zhang and R.R. Markwald, unpublished)	GATA-4[a] (KELLEY et al. 1993)
CSX-1[a] (KOMURO and IZUMI 1993)	QH-1 (COFFIN and POOLE 1991)
GATA-4/-5/-6 (LAVERRIERE et al. 1994)	
Atrial MHC (WESSELS et al. 1992)	VEGF-R[a] (RISAU 1995)
α-Smooth muscle actin (MCQUINN and SCHWARTZ 1995)	
Ventricles and atrium	
α-Smooth muscle actin	GATA-4[a]
GATA-4/-5[a]	
CSX-1[a]	QH-1
Ventricular MHC[b] (WESSELS et al. 1992)	VEGF-R[a]
Atrial MHC[b] (YUTZEY et al. 1994)	
Ventricular MLC-2v (V only) (O'BRIEN et al. 1993)	
Atrial MLC-2a (A only) (KUBALAK et al. 1994)	

Chick, Hamburger-Hamilton stages 12–19; mouse, embryonic days 9.0–12.0.
BMP, bone morphogenetic protein; MHC, myosin heavy chain; MLC, myosin light chain; V, ventricle; A, atrium; VEGF-R, vascular endothelial growth factor receptor.
[a]Data for mice only.
[b]Expressed in both atrium and ventricle early, then one becomes dominant per segment.

conotruncus, with weaker expression in the AV canal but none in the atrium or ventricles. Conversely, the myocardium of all segments expresses *Csx-1* and GATA-4 (LAVERRIERE et al. 1994). Similarly, all segments initially express both atrial and ventricular forms of myosin heavy chains, but, with time, their expression becomes restricted to the atrium or ventricle. Other myosins, e.g., ventricular myosin light chain-2c (MLC-2c) or atrial myosin light chain-2a (MLC-2a), are entirely segmentally specific (WESSELS et al. 1991; EVANS and O'BRIEN 1993; KUBALAK et al. 1994; YUTZEY et al. 1994). Endocardium similarly shows diversity when viewed by immunostaining or in situ hybridization. JB3 antigen is expressed only in some cells of the AV canal or conotruncus (WUNSCH et al. 1994). Expression of TGF-β1 and -β3 and *Msx-1* (CHAN-THOMAS et al. 1993) is spatially and temporally restricted, whereas the receptor for VEGF (VEGF-R), GATA-4, and QH-1 antigens are common to all endocardial cells (RISAU 1995).

Thus, in trying to understand the mechanisms of heart development, it is important to recognize the concept of segmentation. It begins within the heart fields and extends to the primary tube. A reasonable working hypothesis is that the developmental fate of the segments will vary because each is under different gene regulation. A corollary to this hypothesis might be that gene regulation is established at the time as the segments arise within the heart fields. Because alternating segments (ventricles and atrium versus AV canal and conotruncus) share some common candidate regulatory genes, it would be anticipated that their developmental phenotypes may be similar.

III. Developmental Fate of the Primitive Segments

1. Atrium

The atrium initially develops as part of the inflow tract segment which includes the sinus venosus. The latter is a short-lived segment in mammals whose developmental origins and fate are perhaps the least understood and most complex, particularly in mammals. The sinus venosus receives most of the major veins returning blood to the heart once circulation begins, but the sinus soon becomes "absorbed" asymmetrically into the posterior wall of the atrium. As a result, all vessels of the venous pole enter the right side of the atrium except the pulmonary veins, whose origin is controversial (GITTENBERGER-DE GROOT et al. 1995; DE RUITER et al. 1995; TASAKA et al. 1995; WESSELS 1996).

Basically, the formation of multiple invaginations of myocardial origin largely define the atrial phenotype. These invaginations, besides forming future pectinate muscles, include the sinoatrial folds or "venous valves" and the primary atrial septum (WESSELS 1996). As shown in Fig. 7, the primary atrial septum (septum primum) develops from two parts, a muscular infolding from the posterior aspect of the atrial wall and extracardiac mesenchyme termed the spina vestibuli (ASAMI and KOIZUMI 1995) The latter is a wedge of mesenchyme at the venous pole that extends between the lung buds and the two

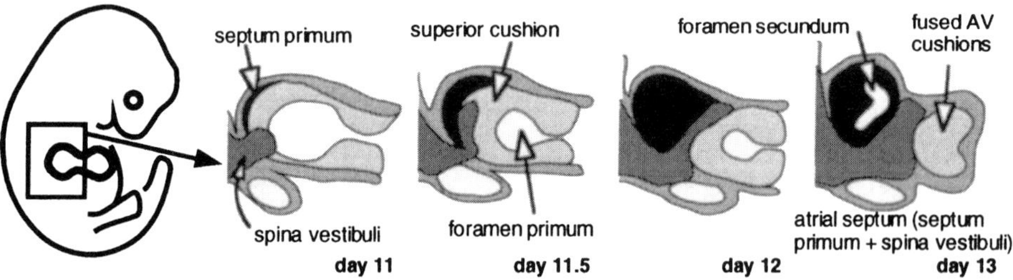

Fig. 7. Formation of the spinal vestibuli and its relationship to the development of the primary atrial septum in the mouse illustrated in a sagittal plane. The extracardiac mesenchyme of the spina vestibuli enters the posterior atrial wall and migrates to contact the fusing atrioventricular (*AV*) cushions. The atrial myocardium gives rise to an infolding that trails behind the spina vestibuli. This produces the two portions, muscular and mesenchymal, of the primary atrial septum. Contact between the spinal vestibuli portion of the septum and the fused AV cushions leads to closure of the communication (foramen primum), separating the two sides of the atrium. The communication is restored by the foramen secundum

horns of the sinus venosus. As the sinus venosus is absorbed, the spina vestibuli literally penetrates the posterior atrial wall. The ingrowth of the spina vestibuli is accompanied by an infolding of the atrial myocardium to form the primary atrial septum (Fig. 7). Asami and Koizumi (1995) have suggested that the fate of the spina vestibuli is causally linked to the septation of the atrium. Support for this hypothesis is the trisomic 16 (Down's) mouse, in which the spina vestibuli is hypoplastic and the atrial septum is defective (H. Tasaka and R. Markwald, unpublished observations). Thus, as illustrated in Fig. 6, there is an interesting analogy between the spina vestibuli and the aorticopulmonary septum (APS). Both are wedges of extracardiac mesenchyme that form at the opposite poles of the heart and play critical roles in septation. The spina vestibuli is probably derived from mesenchyme of the dorsal mesocardium, whereas the APS is derived from neural crest (Kirby 1993; Noden et al. 1995). Each contributes to septation and ultimately links up with intracardiac mesenchyme formed in the AV canal and conotruncus.

2. Ventricles

As with the atrium, the formation of myocardial invaginations into the cardiac jelly space is characteristic of the ventricular phenotype. Each ventricular invagination or trabeculum appears to be a cone-shaped, clonal expansion of a myocardial stem cell into the lumen (Mikawa et al. 1992). The trabeculae that form at the junction between the two ventricular segments are more elongated and densely packed (Fig. 5). This junction between the ventricular segments has been termed the primary ring (Wenink 1987). Consistent with the concept of segmentation, the myocardial cells of the primary ring express tissue-specific antigens, such as GIN2 (Wessels et al. 1991, 1992). The pattern of staining with anti-GIN2 antibodies and other markers is consistent with this ring being the site at which trabeculae fuse to form the muscular interventricular septum (Lamers et al. 1992). Thus, while their morphogenetic roles may vary, the formation of muscular trabeculae or internal projections of quiescent or relatively slowly growing myocardium would seem to be a developmental fate common to atrial and ventricular segments (Thompson et al. 1989).

3. Atrioventricular Canal and Conotruncus

Both the AV canal and the conotruncus, like the primary ring and sinoatrial ring, are in one sense junctions or boundaries between or flanking the more prominent atrial and ventricular segments. The endocardium of both the AV canal and conotruncus transforms into mesenchymal cells which colonize the cardiac jelly space to form "rings" of mesenchyme which circumscribe the lumen. Focalized expansions of these rings form apposing luminal swellings or ridges (Markwald et al. 1977) (Figs. 6, 8, 9). Fusion of the AV cushions forms the AV septum, which divides the original AV lumen into right and left sides (future tricuspid and mitral orifices). In the conotruncus, fusion of

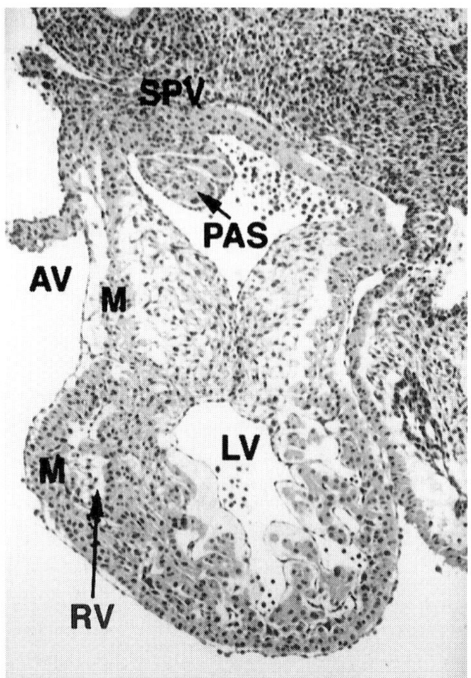

Fig. 8. Section similar to that shown by the *box* in Fig. 7. *PAS*, primary atrial septum; *SPV*, spina vestibuli. Note that most of the ventricular inlet (future left ventricle, *LV*) is derived from the posterior (inflow) half of the looped heart tube; the right ventricle (*RV*) is part of the outlet limb (see LAMERS et al. 1995). In addition, note that the myocardium (*M*) does not form invaginations within the atrioventricular (*AV*) canal

cushion ridges forms the conotruncal septum (or future "ventricular muscular outlet septum"), which provides a separate outlet for each ventricle (infundibulum for the right ventricle, aortic vestibule for the left ventricle) (VUILLEMIN and PEXIEDER 1989). Proximally, the conotruncal septum aligns with the muscular interventricular septum. Distally (downstream), the conotruncal septum directly contacts the aorticopulmonary (AP) septum, which develops within the aortic sac to separate this saddle-shaped structure into the roots of the two great arteries (BARTELINGS and GITTENBERGER-DE GROOT 1989). The AP septum is partly of neural crest origin (KIRBY 1993), whereas the conotruncal septum is of endocardial mesenchyme origin but may receive mesenchyme of extracardiac origin (NODEN et al. 1995). Together they form a spiraled septum which functions to direct ventricular blood into the aorta and right ventricular blood into the pulmonary artery. Misalignment of the conotruncal septum with either the AP septum or muscular ventricular septum results in major congenital heart defects.

In Fig. 10 we present the conceptual framework for the cellular processes and related molecular mechanisms for the segmental transformation of en-

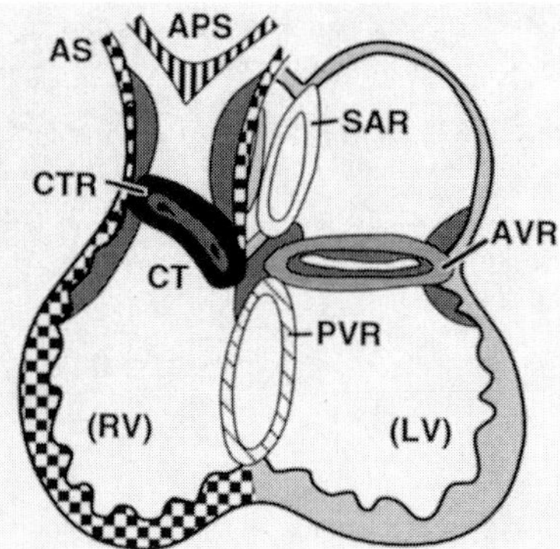

Fig. 9. Septation of the primary heart tube into four chambers. Preseptal structures are envisioned as "*rings*" which converge as a result of looping at the site of the atrioventricular ring (*AVR*). Note that the AVR, sinuatrial ring (*SAR*), and primary ventricular ring (*PVR*) form in the original inflow or cephalic limb of the looped tubular heart. Conversely, the conotruncal (*CT*) ring (*CTR*) figuratively represents the conotruncal cushion ridges, which extend most of the length of the cranial limb of the looped tubular heart. This latter ring is the most troublesome in our understanding of septation of the heart, as it is a transitional structure that becomes muscularized and its prominence in the early tubular heart is "lost" in the mature heart owing to differential growth of surrounding ventricular structures. *RV*, right ventricle; *LV*, left ventricle; *AS*, atrial septum. See text for details

docardial endothelium into cushion mesenchyme. In particular, the development of a three-dimensional culture system (BERNANKE and MARKWALD 1982) that retains the temporal and spatial specificities of the AV and conotruncal segments (RUNYAN and MARKWALD 1983) have enabled us to determine the following:

1. The AV and conotruncal myocardium, acting as a stimulator, secretes a "signal" into its expanded basement membrane (cardiac jelly) that initiates the process by which endothelial cells become mesenchymal cells. During this transformation in phenotype, competent target cells lose their cell–cell associations with neighboring nontransforming endothelial cells, undergo hypertrophy, develop elongated migratory appendages (filopodia) on their basal surfaces, actively translocate into the myocardial basement membrane as cushion mesenchyme, and finally differentiate into a valvuloseptal fibroblast (KRUG et al. 1987; MJAATVEDT et al. 1987).
2. The myocardial inductive signal is part of a multicomponent complex called an adheron (MJAATVEDT and MARKWALD 1989). Adherons are visualized as the cardiac jelly of the two mesenchyme-forming segments (MJAATVEDT et al. 1987; SINNING and MARKWALD 1992). The proteins of adherons include, but are not limited to, fibronectin and ES proteins with a molecular mass of 28, 46, 93, and 130 kDa (MJAATVEDT et al. 1991). The 130-kDa protein, ES130, has been cloned and sequenced and found to be a novel protein without homology to any known growth factor or candidate regulatory

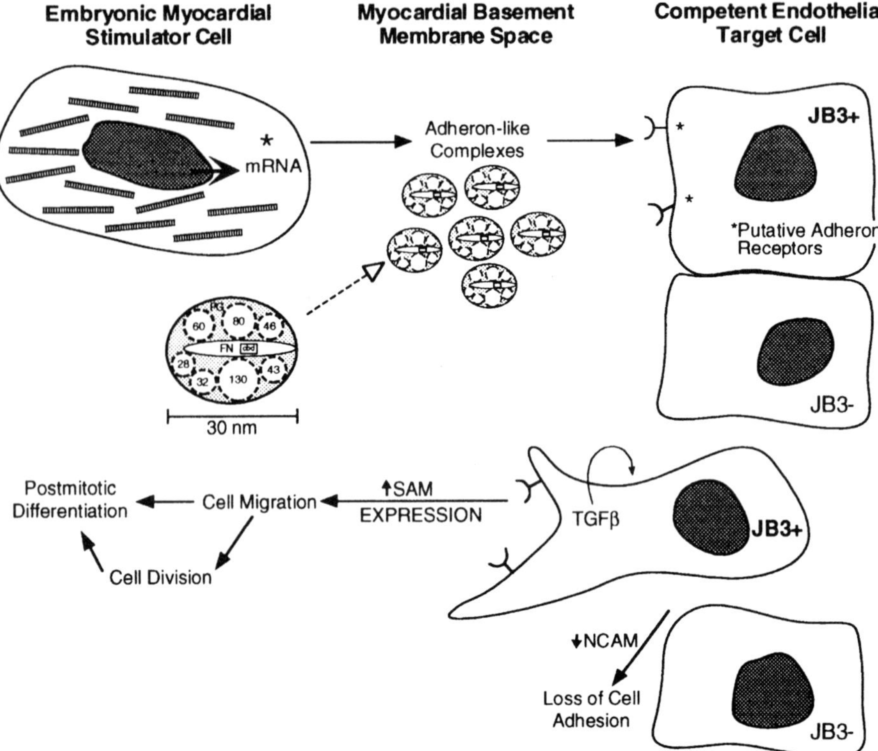

Fig. 10. Segmental basis of cushion tissue formation. Stimulator myocardial cells located in atrioventricular (AV) canal or conotruncus (CT) secrete adheron complexes that induce competent (JB3+) endocardial endothelial cells to transform to mesenchyme. Loss of adhesion and subsequent migration correlate with changes in the expression of specific adhesion molecules and transforming growth factor (*TGF*)-β. *SAM*, substrate-associated molecule; *NCAM*, neural cell adhesion molecule; *FN*, fibronectin; *PG*, proteoglycan; *cbd*, cell-binding domain. See text for details

gene (REZAEE et al. 1993). Antisense and antibody experiments indicate that ES130 is required for endothelial transformation to mesenchyme. Expression of ES130 is developmentally regulated; the protein is initially expressed in the fused heart fields and the AV and conotruncal myocardium of the primary heart tube until the onset of endothelial transformation (MARKWALD et al. 1995a). Thereafter, ES130 is iterated by transforming endothelial cells and their mesenchymal progeny, which suggests an autocrine role to sustain transformation and migration (MARKWALD et al. 1995a).

3. Competent target cells are those that express the JB3 antigen and, accordingly, arise from precursors within the heart-forming fields (WUNSCH et al. 1994). Nontransforming endocardial endothelial cells do not express JB3. The JB3-negative endocardial cells possibly originate from noncardiogenic splanchnic mesoderm (MANASEK 1976; NODEN 1991; COFFIN

and POOLE 1991; NODEN et al. 1995) and subsequently migrate into the fused heart fields or developing tubular hearts. Thus the endocardium is established from two potential subsets of pre-endocardial cells (JB3-positive and -negative). A putative receptor for adheron proteins is envisioned as the hypothetical mechanism by which JB3-positive endothelial cells develop competency to respond to a transforming signal. Evidence for such a receptor is an intracellular, phosphokinase C-dependent calcium flux triggered by myocardial activation (RUNYAN et al. 1990).

4. The loss of cell adhesion and formation of migratory appendages correlate with the diminished expression of NCAM (MJAATVEDT and MARKWALD 1989; CROSSIN and HOFFMAN 1991) and an upregulation of a serine protease, urokinase (MCGUIRE and ORKIN 1992), and substrate-associated molecules (SAM), e.g., cytotactin (CROSSIN and HOFFMAN 1991) and proteoglycans (FUNDERBURG and MARKWALD 1986). Based on their patterns of expression in target endothelial cells, some candidates for regulating changes in endothelial cell–cell or cell–matrix interactions during transformation are *Msx-1*, ES130, and TGF-β3 (CHAN-THOMAS et al. 1993; POTTS et al. 1991; MARKWALD et al. 1995a; NAKAJIMA et al. 1994). To date, evidence only exists for an autocrine-type response to TGF-β3 (NAKAJIMA et al. 1994).
5. The differentiation of mesenchyme into a valvuloseptal fibroblast includes loss of cell division, expression of alpha smooth muscle actin (V. MIRONOV and R. Markwald, unpublished observations), and secretion of a transitional matrix containing type I collagen (SINNING et al. 1988) and two microfibrillar proteins, fibulin (SPENCE et al. 1992; ZHANG et al. 1995) and JB3 antigen/fibrillin (WUNSCH et al. 1994). Transcriptional factors expressed during the early period of cushion cell differentiation include *Mox-1* (CANDIA et al. 1992) and the inhibitory to differentiation gene (*Id*) (EVANS and O'BRIEN 1993). The latter is an unusual HLH (helix-loop-helix) protein in that it lacks a DNA-binding region. It is thought that *Id* might exert its effect by interacting with other HLH or *Hox* genes to form heterodimers that could modify their regulatory activities. Interestingly, *Id* expression overlaps closely with a novel HLH protein, GbHLH1.4. However, high levels of GbHLH1.4 expression are maintained, while the expression of *Id* drops off markedly as the cushions differentiate into valves (HELMS et al. 1994). How cushion mesenchyme of the AV canal and conotruncus differentiates into valve leaflets remains one of the least-studied aspects of heart development.

D. Septation and Remodeling of the Tubular Heart into Four Chambers

The solution to the riddle of how one becomes four is complex and multifaceted. It requires (a) correct looping (D loop) of the tubular heart, (b) formation of extracardiac mesenchyme at both arterial and venous poles (the AP septum and spina vestibuli, respectively), (c) formation of muscular septa in the ventricular and atrial segments, and (d) the formation of intracardiac mesenchyme in the AV and conal segments.

I. Looping

Alignment of the internal septa depends upon proper bending of the heart tube (ICARDO and SANCHEZ DE VEGA 1991; BURN 1991). Abnormal looping can have profound adverse consequence upon septation, underscoring the importance of the first morphological asymmetry observed in the embryo. Thus it is important to recognize that looping and the progressive addition of segments occur concurrently, perhaps pointing to a causal interrelationship. As segments are added, the elongating heart tube bulges outward, to the right. At the same time, it rotates to the right on its own axis. Looping is completed when the final segments are added and the arterial and venous poles have come into direct contact. Failure of the looping process to completely bring the two poles of the heart together, as occurs if chick embryos are exposed to retinoic acid after looping has begun, results in septal misalignments even if the direction of looping is normal (BOUMAN et al. 1995). Thus looping involves at least two distinct mechanisms: handedness or direction and the mechanics by which the actual deformation in the tube is engendered. Based upon their effect on direction of looping, three candidate laterality genes have been identified: the *iv* gene (BRUECKNER et al. 1989), the *inv* gene (YOKOYAMA et al. 1993), and the gene encoding the protein core of a heparan sulfate proteoglycan (YOST 1992). In addition, the unknown downstream targets of retinoic acid receptors are likely candidates, as retinoic acid can engender L loops or A(anterior) loops (OSMOND et al. 1991; CHEN and SOLURSH 1993) as well as modify segmental identities (YUTZEY et al. 1994; KUBALAK et al. 1994; DYSON et al. 1995). As noted above, the deletion of the *NK-2.5* (*Csx*) gene also resulted in abnormal looping and loss of segmental identity, further suggesting a link between segment formation and breaking symmetry.

Regarding the mechanism by which the heart tube mechanically bends, MANASEK (1976) largely established that the motor is intrinsic to the myocardium, but was unable to determine how it worked. ITASAKI et al. (1991) and SHIRISHI et al. (1995) have proposed that looping results from specific interactions between myocardial cell surface adhesion receptors (particularly cadherins and integrins), their extracellular ligands, and actin filaments. A key question remaining is how the regulators of the motor mechanisms and the genes regulating laterality interact to move the emerging heart tube in a specific direction. Complicating this question are extracardiac factors (e.g., formation of the cephalic flexure) that modify cardiac looping (MANNER et al. 1995).

II. Integration of Septal Primordia into Adult Partitions

How the simple tubular heart with its single lumen is subdivided into one having four chambers is illustrated, for the sake of communication, admittedly, in oversimplified terms in Figs. 9 and 11 (see also GITTENBERGER-DE GROOT et al. 1995). Remodeling requires the integration of five preseptal primordia, shown figuratively as ring-like structures. Conceptually, as originally suggested by WENINK (1987), each primordium represents an original boundary interface (e.g., primary ring or sinoatrial ring) or a junctional segment, like the AV canal or conotruncus (Fig. 9). Looping brings about the convergence of the preseptal structures at the level of the AV canal, where they interact to establish four chambers. Firstly, the superior (anterior) and inferior

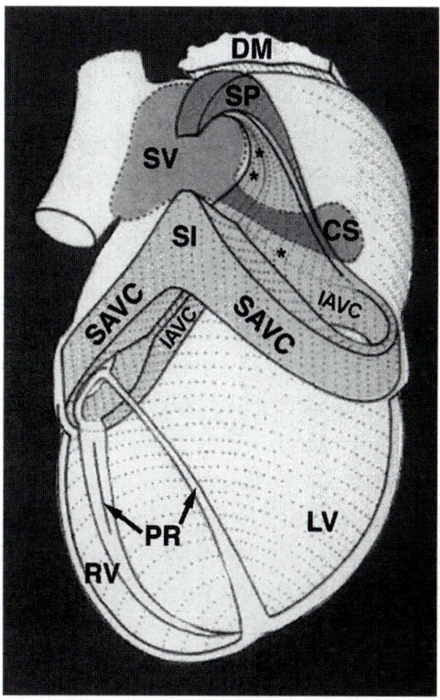

Fig. 11. More literal representation of septation which attempts, in part, to provide a three-dimensional orientation to Fig. 8. The coronary sinus (*CS*), a derivative of the sinus venosus, opens into the right side of the atrium. The dorsal mesentery (*DM*) is seen as the progenitor of the spina vestibuli (*SV*). Together with the muscular portion of septum primum (*SP*), the mesenchyme of the spina vestibuli forms the primary atrial septum in the original plane of the sinuatrial ring. Note that the spina vestibuli directly contacts two posterior extensions (*asterisks*) of the inferior atrioventricular (*AV*) cushion that lies within the original plane of the sinuatrial ring. The fused cushions are labeled SI (septum intermedium), an historical designation. Note that the primary ring (*PR*), the future site of the muscular interventricular septum, is aligned with the superior AV cushion (*SAVC*), not the septum intermedium or the inferior AV cushion (IAVC). Accordingly, after formation of the muscular septum, most of the original lumen of the AV canal will be connected to the left ventricle (*LV*). A new inlet (the tricuspid orifice) to the right ventricle (*RV*) must be remodeled from the SAVC, lateral right ventricular myocardial wall, and the muscular septum

(posterior) cushions of the AV ring fuse to form the AV septum. The primary ring is "filled in" by the formation of trabeculae, becoming the muscular interventricular septum. The muscular portion of the primary atrial septum arises from the atrial wall, apparently independent of the sinoatrial ring; its mesenchymal portion, as noted above, is of extracardiac origin (Fig. 11). However, until the development of markers which specifically recognize the sinoatrial ring or the sinus venosus segment, it may be impossible to determine the role of the sinoatrial junction in atrial septation. The cylindrical conotruncal "ring" is defined by the elongated conal ridges (shown unfused in Fig. 9). This ring is the only one to form entirely in the outlet (cephalic) limb of the looped heart tube. As looping proceeds, the fused conotruncal ridges of this ring are brought into direct contact with the AV cushion ring and the primary ventricular ring. The alignment of these septal rings is critical for closing off direct vascular communication between the ventricles and for directing blood from each ventricle into a separate outlet leading to either the aortic or pulmonary trunk (VUILLEMIN and PEXIEDER 1989).

Thus it is primarily the fused AV cushions which serve as the "glue" which knits together the various preseptal, ring-like structures. However, to complete formation of a four-chambered heart, the mesenchymal tissues of the AV canal and conotruncus and their surrounding myocardial tissues must be remodeled extensively to form the definitive chamber inlets or outlets (mitral, tricuspid, aortic, and pulmonic) and their associated valvular leaflets. For example, as revealed by G1N2 staining (see Fig. 11), most of the narrowed, primitive inlet into the right ventricle must be extensively remodeled to make a tricuspid orifice and to insure a separate outlet for the left ventricle (LAMERS et al. 1992, 1995) The molecular mechanisms for how this or other postseptation events occur, such as myocardialization of the conotruncal septum (BARTELINGS et al. 1986; LAMERS et al. 1995; THOMPSON and FITZHARIS 1985; THOMPSON et al. 1985), is to our knowledge completely unknown. Hence, the processes required to "fine-tune" the septated heart into the fully formed heart are as complex and important as any aspect of heart development (ANDERSON 1989).

Basically, failure of these rings (literally the tissues that form at these sites) to develop normally or to properly interact is the basis of congenital malformations. For example, double-outlet right ventricle results when the AP and conotruncal septa do not align properly with the primary ventricular septum (GITTENBERGER-DE GROOT et al. 1995). Failure of either the fused AV or conotruncal septum to contact with the ventricular muscular septum will result in a ventricular septal defect. Failure of the primary atrial septum ("sinuatrial ring") to anchor onto the fused AV cushions engenders atrial septal defects. Because valve tissue differentiates from cushion mesenchyme (WUNSCH et al. 1994), deficits in valve leaflet structure or organization (e.g., mitral or tricuspid atresia, aortic or pulmonic stenosis, bicuspid outlet valves) may result from alterations in the normal segmental pattern of endocardial transformation to mesenchyme (HILTGEN et al. 1996). In a masterful summation, Pexieder (1995) has presented a conceptual review and working paradigm of how perturbations (genetic or environmental) in the development of outflow septal structures predispose to specific cardiac malformations, which reemphasizes the fact that understanding normal development at molecular, cellular, or organ levels remains the key to understanding abnormal development.

E. Conclusion

While most current studies emphasize molecular genetics, we must not forget that the anatomical organization of the heart is the outcome of multiple genetic events operating at different regulatory levels. Precardiac mesoderm (Lough et al. 1990) and embryonic stem cells (Robbins et al. 1992) growing in culture will differentiate into myocardial cells, but they do not form a heart tube. As we have endeavored to show in this review, the potential to learn more about cardiogenesis is as dependent upon appreciating and integrating new morphological data as on identifying new genes (Anderson 1989; Wenink et al. 1994). For example, future morphological studies on the spina vestibuli or dorsal mesentery may greatly assist in understanding the function of a hypothetical "new" gene that is *not* expressed in the heart fields or tube, but yet whose mutagenesis results in atrial septal defects or abnormal pulmonary venous return. Transgenic animal models obviously hold enormous potential for advancing the field through a combination of molecular tools and three-dimensional imaging systems.

References

Anderson RH (1989) The present-day place of correlations between embryology and anatomy in the understanding of congenitally malformed hearts. In: Aranega A, Pexieder T (eds) Correlations between experimental cardiac embryology and teratology and congenital cardiac defects. University of Granada Press, Granada, pp 265–295

Antin PB, Taylor RG, Yatskievych T (1994) Precardiac mesoderm is specified during gastrulation in quail. Devel Dynam 200: 144–154

Arceci R, King AM, Simon MC, Orkin SH, Wilson DB (1993) GATA-4: a retinoic acid-inducible GATA binding transcription factor expressed in endodermally derived tissues and heart. Mol Cell Biol 13: 2235–2246

Asami I, Koizumi K (1995) Development of the atrial septal complex in the human heart: contribution of the spina vestibuli. In: Clark EB, Markwald RR, Takao A (eds) Developmental mechanisms of heart disease. Futura, Armonk, pp 255–260

Bartelings MM, Gittenberger-de Groot AC (1989) The outflow tract of the heart – embryologic and morphologic correlations. Int J Cardiol 22: 289–300

Bartelings MM, Wenink ACG, Gittenberger-de Groot AC, Oppenheimer-Dekker A (1986) Contribution of the aortopulmonary septum to the muscular outlet septum in the human heart. Acta Morphol Neerl-Scand 24: 181–192

Bernanke DH, Markwald RR (1982) Migratory behavior of cardiac cushion tissue cells in a collagen lattice culture system. Dev Biol 91: 235–245

Bodmer R (1993) The gene *tinman* is required for specification of the heart and visceral muscles in Drosophila. Development 118: 719–729

Bouman HGA, Broekhuizen MLA, Baasten MJ, Gittenberger-de Groot AC, Wenink ACG (1995) A spectrum of looping disturbances in stage 34 chicken hearts after retinoic acid treatment. Anat Rec 243: 101–108

Brueckner M, D'Eustachio P, Horwich AL (1989) Linkage mapping of a mouse gene, iv, that controls left-right asymmetry of the heart and viscera. Proc Natl Acad Sci USA 86: 5035–5038

Burn J (1991) Disturbance of morphological laterality in humans. CIBA Found Symp 162: 282–296

Byeon MK, Sugi Y, Markwald RR, Hoffman S (1995) NCAM polypeptides in heart development: association with Z discs of isoforms that contain the muscle-specific domain. J Cell Biol 128: 209–221

Candia AF, Hu J, Crosby J, Lalley PA, Noden D, Nadeau JH, Wright CVE (1992) *Mox-1* and *Mox-2* define a novel homeobox gene subfamily and are differentially expressed during early mesodermal patterning in mouse embryos. Development 116: 1123–1136

Chan-Thomas P, Thompson RP, Robert B, Yacoub MH, Barton PJ (1993) Expression of homeobox genes *Msx-1* (*Hox-7*) and *Msx-2* (*Hox-8*) during cardiac development in the chick. Devel Dynam 197: 203–216

Chen Y, Solursh M (1993) Comparison of Hensen's node and retinoic acid in secondary axis induction in the early chick embryo. Devel Dynam 195: 142–151

Chen Y, Huang L, Russo AF, Solursh M (1992) Retinoic acid is enriched in Hensen's node and is developmentally regulated in the early chicken embryo. Proc Natl Acad Sci USA 89: 10056–10059

Clark EB (1987) Mechanisms in the pathogenesis of congenital cardiac malformations. In: Pierpont ME, Moller JH (eds) Genetics of cardiovascular disease. Nijhoff, Boston, pp 3–11

Coffin JD, Poole TJ (1991) Endothelial cell origin and migration in embryonic heart and cranial blood vessel development. Anat Rec 231: 383–395

Crossin KL, Hoffman S (1991) Expression of adhesion molecules during the formation and differentiation of the avian endocardial cushion tissue. Dev Biol 145: 277–286

Danos MA, Yost HJ (1995) Linkage of cardiac left-right symmetry and dorsal-anterior development in *Xenopus*. Development 121: 1467–1474

De la Cruz MV, Sanchez-Gomez C, Arteaga MM, Arguello C (1977) Experimental study of the development of the truncus and the conus in the chick embryo. J Anat 123: 661–686

De la Cruz MV, Sanchez-Gomez C, Palomino MA (1989) The primitive cardiac regions in the straight tube heart (stage 9^{-}) and their anatomical expression in the mature heart: an experimental study in the chick heart. J Anat 165: 121–131

Dersch H, Zile MH (1993) Induction of normal cardiovascular development in the vitamin A-deprived quail embryo by natural retinoids. Dev Biol 160: 424–433

De Ruiter MC, Poelmann RE, Vanderplass-deVries I, Mentink MMT, Gittenberger-de Groot AC (1992) The development of the myocardium and endodcardium in mouse embryos: fusion of two heart tubes? Anat Embryol 185: 461–473

De Ruiter MC, Gittenberger-de Groot AC, Wenink ACG, Poelmann RE, Mentink MMT (1995) In normal development pulmonary veins are connected to the sinus venosus in the left atrium. Anat Rec 243: 84–92

Drake CJ, Jacobson AG (1988) A survey by scanning electron microscopy of the extracellular matrix and endothelial components of the primordial chick heart. Anat Rec 222: 339–400

Drake CJ, Little CD (1995) Exogenous VEGF induces malformed and hyperfused vessels. Proc Natl Acad Sci USA 92: 7657–7661

Dyson E, Sucov HM, Kubulak SW, Schmid-Schonbein GW, DeLano FA, Evans RM, Ross J Jr, Chien KR (1995) Atrial-like phentotype is associated with embryonic ventricular failure in retinoid X receptor α-/- mice. Proc Natl Acad Sci USA 92: 7386–7390

Edmondson DG, Lyons GE, Martin JF, Olson EN (1994) *Mef2* gene expression marks the cardiac and skeletal muscle lineages during mouse embryogenesis. Development 120: 1251–1263

Eisenberg C, Bader D (1995) QCE-6: a clonal cell line with cardiomyogenic and endothelial cell potentials. Dev Biol 167: 469–481

Evans SM, O'Brien TX (1993) Expression of the helix-loop-helix factor *ID* during mouse embryonic development. Dev Biol 159: 485–499

Funderburg FM, Markwald RR (1986) Conditioning of native substrates by chondroitin sulfate proteoglycans during cardiac mesenchymal cell migration. J Cell Biol 103: 2475–2487

Gallagher BC, Sakai LY, Little CD (1993) Fibrillin delineates the primary axis of the early avian embryo. Devel Dynam 196: 70–78

Garcia-Martinez V, Schoenwolf GC (1993) Primitive steak origin of the cardiovascular system in avian embryos. Dev Biol 159: 706–719

Gittenberger-de Groot AC, Bartelings MM, Poelmann RE (1995) Cardiac morphogenesis. In: Clark EB, Markwald RR, Takao A (eds) Developmental mechanisms of heart disease. Futura, Armonk, pp 157–168

Gudas LJ (1992) Retinoids, retinoid-responsive genes, cell differentiation, and cancer. Cell Growth Differ 3: 655–662

Han Y, Dennis JE, Cohen-Gould L, Bader DM, Fischman DA (1992) Expression of sarcomeric myosin in the presumptive myocardium of chicken embryos occurs within six hours of myocyte commitment. Dev Dynam 193: 257–265

Helms JA, Kuratani S, Maxwell GD (1994) Cloning and analysis of a new developmentally regulated member of the basic helix-loop-helix family. Mech Dev 48: 93–108

Hiltgen G, Litke LL, Markwald RR (1995) Morphogenetic alterations during endocardial cushion development in trisomy 16 (Down's) mouse. Ped Cardiol 17: 21–30

Hoyle C, Brown NA, Wolpert L (1992) Development of left/right handedness in the chick heart. Development 115: 1071–1078

Icardo JM, Sanchez de Vega MJ (1991) Spectrum of heart malformations in mice with situs solitus, situs inversus, and associated visceral heterotaxy. Circulation 84: 2547–2558

Inagaki T, Garcia-Martinez V, Schoenwolf GC (1993) Regulative ability of the prospective cardiogenic and vasculogenic areas of the primitive streak during avian gastrulation. Dev Dynam 197: 57–68

Itasaki N, Nakamura H, Sumida H, Yasuda M (1991) Actin bundles on the right side in the caudal part of the heart tube play a role in dextro-looping in the embryonic chick heart. Anat Embryol 183: 29–39

Jacobson AG, Sater AK (1988) Features of embryonic induction. Development 104: 341–359

Jones CM, Lyons KM, Hogan BLM (1991) Involvement of *bone morphogenetic protein-4* (BMP-4) and *Vgr-1* in morphogenesis and neurogenesis in the mouse. Development 111: 531–542

Kelley C, Blumberg H, Zon L, Evans T (1993) GATA-4 is a novel transcription factor expressed in endocardium of the developing heart. Development 118: 818–824

Kern MJ, Argao EA, Potter SS (1995) Homeobox genes and heart development. Trends Cardiovasc Med 5: 47–54

Kessler DS, Melton DA (1994) Vertebrate embryonic induction: mesodermal and neural patterning. Science 266: 596–604

Kirby ML (1993) Cellular and molecular contributions of the cardiac neural crest to cardiovascular development. Trends Cardiovasc Med 3: 18–23

Kitten GT, Markwald RR, Bolender DL (1987) Distribution of basement membrane antigens in cryopreserved early embryonic hearts. Anat Rec 217: 379–390

Kokan-Moore NP, Bolender DL, Lough JL (1991) Secretion of inhibin $beta_A$ by endoderm cultured from early embryonic chicken. Dev Biol 146: 242–245

Komuro I, Izumo S (1993) Csx: a murine homeobox-containing gene specifically expressed in the developing heart. Proc Natl Acad Sci USA 90: 8145–8149

Krug EL, Mjaatvedt CH, Markwald RR (1987) Extracellular matrix from embryonic myocardium elicits an early morphogenetic event in cardiac endothelial differentiation. Dev Biol 120: 348–355

Kubalak SW, Miller-Hance WC, O'Brien TX, Dyson E, Chien KR (1994) Chamber specification of atrial myosin light chain-2 expression precedes septation during murine cardiogenesis. J Biol Chem 269: 16961–16970

Lamers WH, Wessels A, Verbeek FJ, Moorman AFM, Viragh S, Wenink ACG, Gittenberger-de Groot AC, Anderson RH (1992) New findings concerning ventricular septation in the human heart. Circulation 86: 1194–1205

Lamers WH, Viragh S, Wessels A, Moorman AFM, Anderson RH (1995) Formation of the tricuspid valve in the human heart. Circ Res 91: 111–121

Laverriere AC, MacNeill C, Mueller C, Poelmann RE, Burch JB, Evans T (1994) GATA-4/5/6, a subfamily of three transcription factors transcribed in developing heart and gut. J Biol Chem 269: 2377–23184

Lee RKK, Stanier DYR, Weinstein BM, Fishman MC (1994) Cardiovascular development in the zebrafish II. Endocardial progenitors are sequestered within the heart field. Development 120: 3361–3366

Lemanski LF, Paulson DH, Hill CS (1979) Normal anterior endoderm corrects the heart defect in cardiac mutant salamanders (Ambystoma mexicanum). Science 204: 860–862

Levin M, Johnson RL, Stern CD, Kuehn M, Tabin C (1995) A molecular pathway determining left-right asymmetry in chick embryogenesis. Cell 82: 803–814

Linask KK (1992) N-cadherin localization in early heart development and polar expression of Na^+, K^+ ATPase and integrin during pericardial coelom formation and epithelialization of the differentiating myocardium. Dev Biol 151: 213–224

Linask KK, Lash JW (1988) A role for fibronectin in the migration of avian precardiac cells. I. Dose-dependent effects of fibronectin antibody. Dev Biol 114: 87–101

Linask KK, Lash JW (1993) Early heart development: dynamics of the endocardial cell sorting suggests a common origin with cardiocytes. Dev Dynam 195: 62–69

Lints TJ, Parson LM, Hartley, Lyons I, Harvey RP (1993) *NKX-2.5*: a novel murine homeobox gene expressed in early heart progenitor cells and their myogenic descendants. Development 119: 419–431

Litvin J, Montgomery MO, Goldhamer DJ, Emerson CP, Bader DM (1993) Identification of DNA-binding protein(s) in the developing heart. Dev Biol 156: 409–417

Lough JW, Bolender DL, Markwald RR (1990) A culture model for cardiac morphogenesis. Ann NY Acad 588: 421–424

Lyons GE (1994) In situ analysis of the cardiac muscle gene program during embryogenesis. Trends Cardiovasc Med 4: 70–77

Lyons G, Buckingham ME (1992) Developmental regulation of myogenesis in the mouse. Semin Dev Biol 3: 243–253

Lyons I, Parsons LM, Hartley L, Li R, Andrews JE, Robb L, Harvey RP (1995) Myogenic and morphogenetic defects in the heart tubes of murine embryos lacking the homeobox gene *Nkx2–5*. Genes Dev 9: 1654–1666

Manasek FJ (1976) Heart development: interactions involved in cardiac morphogenesis. In: Poste C, Nicolson G (eds) The cell surface in animal embryogenesis and development. Elsevier, New York, pp 545–598

Markwald RR (1973) Distribution and relationship of precursor Z material to organizing myofibrillar bundles in embryonic rat and hamster ventricular myocytes. J Mol Cell Cardiol 5: 341–350

Markwald RR, Fitzharris TP, Manasek FJ (1977) Structural development of endocardial cushions. Am J Anat 148: 85–120

Markwald RR, Rezaee M, Nakajima Y, Wunsch A, Isokawa K, Litke L, Krug EL (1995a) Molecular basis for the segmental pattern of cardiac cushion mesenchyme formation: role of ES/130 in the embryonic chick heart. In: Clark EB, Markwald

RR, Takao A (eds) Developmental mechanisms of heart disease. Futura, Armonk, pp 185–194

Markwald RR, Eisenberg L, Eisenberg C, Trusk T, Sugi Y (1996) Epithelial mesenchymal transformations in early avian heart development. Acta Anat (in press)

McGuire PG, Orkin RW (1992) Urokinase activity in the developing avian heart: a spatial and temporal analysis. Dev Dynam 193: 24–33

McQuinn TC, Schwartz RJ (1995) Vascular smooth muscle-specific gene expression. In: Mecham R, Schwartz S (eds) The vascular smooth muscle cell. Academic, New York (in press)

Mikawa T, Borisov A, Brown AMC, Fischman DA (1992) Clonal analysis of cardiac morphogenesis in the chicken embryo using a replication-defective retrovirus. I. Formation of the ventricular myocardium. Dev Dynam 195: 133–141

Mikawa T, Gourdie RG (1995) Coronary smooth muscle cells originate from pericardial mesoderm: a lineage distinct from neural crest-derived aortic smooth muscle. Dev Dynamics (in press)

Miner JH, Miller JB, Wold BJ (1992) Skeletal muscle phenotypes initiated by ectopic MyoD in transgenic mouse heart. Development 114: 853–860

Mjaatvedt CH, Markwald RR (1989) Induction of epithelial-mesenchymal transition by an in vivo adheron-like complex. Dev Biol 136: 118–128

Mjaatvedt CH, Lepera RC, Markwald RR (1987) Myocardial specificity for initiating endothelial-mesenchymal cell transition in embryonic chick heart correlates with a particulate distribution of fibronectin. Dev Biol 119: 59–67

Mjaatvedt CH, Krug EL, Markwald RR (1991) An antiserum (ESI) against a particulate form of extracellular matrix blocks the transformation of cardiac endothelium into mesenchyme in culture. Dev Biol 145: 219–230

Molkentin JD, Kalvakolanu DV, Markham BE (1994) Transcription factor GATA-4 regulates cardiac muscle-specific expression of the alpha-myosin heavy chain gene. Mol Cell Biol 14: 4947–4957

Montgomery MO, Litvin J, Gonzalez-Sanchez A, Bader D (1994) Staging of commitment and differentiation of avian cardiac myocytes. Dev Biol 164: 63–71

Nakajima Y, Krug EL, Markwald RR (1994) Myocardial regulation of transforming growth factor-beta expression by outflow tract endothelium in the early embryonic chick heart. Dev Biol 165: 615–626

Nascone N, Mercola M (1995) An inductive role for the endoderm in *Xenopus* cardiogenesis. Development 121: 515–523

Noden DM (1991) Origins and patterning of avian outflow tract endocardium. Development 111: 867–876

Noden DM, Poelmann RE, Gittenberger-de Groot (1995) Cell origins and tissue boundaries during outflow tract development. Trends Cardiovasc Med 5: 69–75

Nusse R, Varmus HE (1992) *Wnt* genes. Cell 69: 1073–1087

O'Brien TX, Lee KJ, Chien KR (1993) Positional specification of ventricular myosin light chain 2 expression in the primitive murine heart tube. Proc Natl Acad Sci USA 90: 5157–5161

Orts-Llorca F (1963) Influence of the endoderm on heart differentiation during the early stages of development of the chicken embryo. W Roux Arch Entwickl 154: 533–551

Osmond MK, Butler AJ, Voon FCT, Bellairs R (1991) The effects of retinoic acid on heart formation in the early chick embryo. Development 113: 1405–1417

Peifer M, McCrea PD, Green KJ, Wieschaus E, Gumbiner BM (1992) The vertebrate adhesive junction proteins beta catenin and plakoglobin and the *Drosophila* segment polarity gene *armadillo* form a multigene family with similar properties. J Cell Biol 118: 681–691

Peng I, Dennis JE, Rodriguez-Boulan E, Fischman DA (1990) Polarized release of enveloped viruses in the embryonic chicken heart: demonstration of epithelial polarity in the presumptive myocardium. Dev Biol 141: 164–172

Pexieder T (1995) Conotruncus and its septation at the advent of the molecular biology era. In: Clark EB, Markwald RR, Takao A (eds) Developmental basis of heart disease. Futura, Amronk, pp 227–247

Poelmann RE (1981) The head-process and the formation of the definitive endoderm in the mouse embryo. Anat Embryol 162: 41–46

Poelmann RE, Gittenberger-de Groot AC, Mentink MMT, Bökenkamp R, Hogers B (1993) Development of the cardiac coronary vascular endothelium, studied with antiendothelial antibodies, in chicken-quail chimeras. Circ Res 73: 559–568

Potts JD, Dagle JM, Walder JA, Weeks DL, Runyan RB (1991) Epithelial-mesenchymal transformation of embryonic cardiac endothelial cells is inhibited by a modified antisense oligodeoxynucleotide to transforming growth factor β3. Proc Natl Acad Sci USA 88: 1516–1520a

Rawles ME (1943) The heart-forming areas of the early chick blastoderm. Physiol Zool 16: 22–42

Rezaee M, Isokawa K, Krug EL, Markwald RR (1993) Identification of a 130kDa protein potentially involved in cardiac morphogenesis. J Biol Chem 268: 14404–14411

Risau W (1995) Differentiation of endothelium. FASEB J 9: 926–933

Robbins J, Doetschman T, Jones WK, Sanchez A (1992) Embryonic stem cells as a model for cardiogenesis. Trends Cardiovasc Med 2: 44–50

Rongish BJ, Little CD (1995) Extracellular matrix in heart development. Experentia 51: 873–882

Rosenquist GC, DeHaan RL (1966) Migration of precardiac cells in the chick embryo: a radiographic study. Carnegie Inst Washington Publ 625. Contrib Embryol 38: 111–121

Runyan RB, Markwald RR (1983) Invasion of mesenchyme into three-dimensional gels: a regional and temporal analysis of interaction in embryonic heart tissue. Dev Biol 95: 108–114

Runyan RB, Potts JD, Sharma RV, Loeber CP, Chiang JJ, Bhalla RC (1990) Signal transduction of a tissue interaction during embryonic heart development. Cell Regul 1: 301–313

Sanger JM, Mittal B, Pochapin MB, Sanger JW (1986) Myofibrillogenesis in living cells microinjected with fluorescently labeled alpha-actinin. J Cell Biol 102: 2053-2066

Sater AK, Jacobson AG (1990) The role of the dorsal lip in the induction of heart mesoderm in *Xenopus laevis*. Development 108: 461–470

Shirishi I. Takamatsu T, Onouchi Z, Fujita S (1995) Three dimensional observation of F-actin and expression of N-cadherin and fibronectin during cardiac looping of the chick embryo using CLSM. In: Clark EB, Markwald RR, Takao A (eds) Developmental basis of heart disease. Futura, Amronk, pp 471–776

Siegfried E, Perrimon N (1994) *Drosophila wingless*: a paradigm for the function and mechanism of *Wnt* signaling, Bioessays 16: 395–404

Sinning AR, Markwald RR (1992) Multiple glycoproteins localize to a particulate form of extracellular matrix in regions of the embryonic heart where endothelial cells transform into mesenchyme. Anat Rec 232: 285–292

Sinning AR, Lepera RC, Markwald RR (1988) Initial expression of type I procollagen in chick heart mesenchyme is dependent upon myocardial induction. Dev Biol 130: 167–174

Slack JMW (1991) From egg to embryo: regional specification in early development, 2nd edn. Cambridge University Press, Cambridge

Smith SM (1994) Retinoic acid receptor isoform beta2 is an early marker for alimentary tract and central nervous system positional specification in the chicken. Dev Dynam 200: 14–25

Soler AP, Knudsen KA (1994) N-cadherin involvement in cardiac myocyte interaction and myofibrillogenesis. Dev Biol 162: 9–17

Spence SG, Argraves WS, Walters L, Hungerford JE, Little CD (1992) Fibulin is localized at sites of epithelial-mesenchymal transitions in the early embryo. Dev Biol 151: 73–484

Sucov HM, Dyson E, Gumeringer CL, Price J, Chien KR, Evans RM (1994) RXRα mutant mice establish a genetic basis for vitamin A signaling in heart morphogenesis. Genes Dev 8: 1007: 1018

Sugi Y, Lough JL (1994) Anterior endoderm is a specific effector of terminal cardiac myocyte differentiation of cells from the embryonic heart forming region. Dev Dynam 200: 155–162

Sugi Y, Lough JL (1995) Activin-A and FGF-2 mimic the inductive effects of anterior endoderm on terminal cardiac myogenesis in vitro. Dev Biol 168: 567–574

Sugi Y, Markwald RR (1996) Formation and early morphogenesis of endocardial precursor cells and the role of endoderm. Dev Biol (in press)

Sweeney L, Zak R, Manasek FJ (1987) Transitions in cardiac isomyosin expression during differentiation of the embryonic chick heart. Circ Res 61: 287–295

Tasaka H, Krug EL, Markwald RR (1995) Origin of the orifice of the pulmonary vein in the mouse. In: Clark EB, Markwald RR, Takao A (eds) Developmental mechanisms of heart disease. Futura, Armonk, pp 347–350

Thompson RP, Fitzharris TP (1985) Division of cardiac outflow. In: Ferrans V, Rosenquist G, Weinstein C (eds) Cardiac morphogenesis, vol V. Elsevier, New York, pp 169–180

Thompson RP, Sumida H, Abercrombie V, Satow Y, Fitzharris TP, Okamoto N (1985) Morphogenesis of human cardiac outflow. Anat Rec 213: 578–586

Thompson RP, Lindroth JR, Wong YM (1989) Regional differences in DNA-synthetic activity in the preseptation myocardium of the chick. In: Clark E, Takao A (eds) Developmental cardiology: morphogenesis and function. Futura, New York, pp 219–234

Tokuyasu TK, Maher PA (1987) Immunocytochemical studies of cardiac myofibrillogenesis in early chick embryos. II. Generation of alpha-actinin dots with titin spots at the time of the first myofibril formation. J Cell Biol 105: 2795–2802

Viragh S, Szabo E, Challice CE (1989) Formation of the primitive myo- and endocardial tubes in the chicken embryo. J Mol Cell Cardiol 21: 123–137

Viragh S, Gittenberger-de Groot AC, Poelmann RE, Kalman F (1994) Early development of quail heart epicardium and associated vascular and glandular structures. Anat Embryol 188: 381–393

Vrancken Peeters M-P, Mentink MM, Poelmann RE, Gittenberger-de Groot AC (1995) Cytokeratins as a marker for epicardial development. Anat Embryol 191: 503–508

Vuillemin M, Pexieder T (1989) Normal stages of cardiac organogenesis in the mouse. II. Development of the internal relief of the heart. Am J Anat 184: 114–128

Wenink ACG (1987) Embryology of the heart. In: Anderson RH, Macartney FJ, Shinebourne EA, Tynan M (eds) Paediatric cardiology. Churchill Livingstone, Edinburgh, pp 83–107

Wenink ACG, Wisse BJ, Groenendijk PM (1994) Development of the inlet portion of the right ventricle in the embryonic rat heart: the basis for tricuspid valve development. Anat Rec 239: 216–223

Wessels A (1996) The development of the atrial chambers in the human heart. In: Anderson R (ed) Syndromes of isomerism in the human. Futura, New York (Perspectives in pediatric cardiology) (in press)

Wessels A, Vermeulen JLM, Viragh S, Lamers WH, Moorman AFM (1991) Spatial distribution of "tissue-specific" antigens in the developing human heart and skeletal muscle. II. An immunohistochemical analysis of myosin heavy chain isoform expression patterns in the embryonic heart. Anat Rec 229: 355–368

Wessels A, Vermeulen JLM, Verbeek FJ, Viragh S, Kalman F, Lamers WH, Moorman AFM (1992) Spatial distribution of "tissue-specific" antigens in the developing human heart and skeletal muscle. III. An immunohistochemical analysis of the distribution of the neural tissue antigen GIN2 in the embryonic heart; implications for the development of the atrioventricular conduction system. Anat Rec 231: 097–111

Wu X, Golden K, Bodmer R (1995) Heart development in Drosophila requires the segment polarity gene *wingless*. Dev Biol 169: 619–628

Wunsch A, Markwald RR, Little CD (1994) Cardiac endothelial heterogeneity defines valvular development as demonstrated by the diverse expression of JB3 antigen, a fibrillin-like protein of the endocardial cushion tissue. Dev Biol 165: 585–601

Yokoyama T, Copeland NG, Jenkins NA, Mongomery CA, Elder FFB, Overbeek PA (1993) Reversal of left-right symmetry: a situs inversis mutation. Science 260: 679–682

Yost HJ (1992) Regulation of vertebrate left-right asymmetries by extracellular matrix. Nature 357: 158–161

Yutzey KE, Rhee JT, Bader DM (1994) Expression of the atrial-specific myosin heavy chain AMHC1 and the establishment of anteroposterior polarity in the developing chicken heart. Development 120: 871–883

Zhang H-Y, Chu M-L, Te-Cheng P, Sasaki T, Timple R, Ekblom R (1995) Extracellular matrix protein fibulin-2 is expressed in the embryonic endocardial cushion tissue and is a prominent component of valves in adult heart. Dev Biol 167: 18–26

CHAPTER 3

Vertebrate Limb Development

K. Muneoka and R. Anderson

A. Introduction

The developing vertebrate limb is one of the most carefully studied model systems for examining how complex patterns of differentiated tissues form in secondary embryonic fields. The mature vertebrate limb contains a large number of distinct cell types that develop side by side in a highly coordinated and orchestrated manner, and it is the molecular underpinnings of this patterning mechanism that lie at the heart of our understanding of how genetic alterations and toxicological insults result in developmentally defective organs. The limb is an excellent experimental model system because it is a nonessential organ that can be easily manipulated. Because limb-specific defects do not generally result in embryolethality, a large number of mutations in man, mouse, and chicken that alter limb morphogenesis have been characterized, and in some cases studied in some detail. For similar reasons, the limb is also a highly visible target for toxicology studies. In addition, the developing limb bud has been used extensively as a source of cells to study issues concerning cellular differentiation, the control of cell proliferation, and the production of morphogenetic signals; thus the behavior of limb bud cells in culture is reasonably well characterized.

Our understanding of how limb morphogenesis is controlled stems primarily from studies on chick embryos due to the accessibility of the developing limb bud for embryonic manipulation. In recent years, however, the development of transgenic technology, as well as the large number of limb mutants available in the mouse, has provided important new insights into limb patterning mechanisms. Fortunately, this dichotomy in experimental approach between mouse (genetics) and chick (embryology) is gradually breaking down. In the chick, the development of retroviral constructs to facilitate gene transfer allows for the ectopic expression of specific genes (Morgan et al. 1992; Riley et al. 1993), and with the development of culture techniques for avian eggs (Perry 1988; Ono et al. 1994) the future creation of transgenic birds looks promising. In the mouse, the development of whole embryo culture techniques (see Cockroft 1990) and surgical techniques for directly manipulating the mouse embryo in situ (see Muneoka et al. 1990; Papaioannou 1990) have opened the door for embryological studies.

Despite clear morphological differences between the limbs of birds and mammals, there is a growing body of evidence indicating that limb outgrowth among all tetrapod vertebrates is regulated by similar underlying mechanisms. For example, cross-species tissue grafting studies demonstrate that posterior (zone of polarizing activity, ZPA) mouse limb bud tissues can induce the formation of chick-derived supernumerary digits in a manner analogous to homospecific (chick to chick) ZPA tissue grafting studies (TICKLE et al. 1976; WANEK and BRYANT 1991; ANDERSON et al. 1993, 1994). Similarly, position-specific gene expression by mouse mesenchymal cells has been shown to be inducible by chick-derived signals in cross-species grafting studies (DAVIDSON et al. 1991; IZPISUA-BELMONTE et al. 1992a; REGINELLI et al. 1996). Thus, while there are obvious differences in morphology between the wing of a chicken and the leg of a mouse, there is considerable support for the hypothesis that the production and response to intercellular signals used for limb patterning are evolutionarily conserved.

One of the more amazing aspects of limb development is the ability of limb bud cells to alter their developmental program to form a duplication of an existing limb structure or to regenerate missing limb structures. This process is called pattern regulation, and the study of the mechanisms driving this process forms the basis of many, if not all, embryological studies of the limb. The initial limb bud is established with the information necessary to form a single limb, and relatively simple tissue manipulations can cause the limb bud cell population to form additional copies of the limb (called supernumerary limbs). This ability of limb bud cells to undergo pattern regulation has intrigued developmental biologists for many years and has been the topic of much research. An important assumption that is made in these pattern regulation studies is that the molecular underpinnings of the patterning response to tissue manipulation will reflect the molecular basis of normal limb formation. At present, there is no reason to question this assumption (see STOCUM 1991 for an alternative view); however, it is important to recognize that its validity is the key to interpreting how the study of pattern regulatory events is advancing our understanding of limb development. With respect to developmental toxicology studies, it is obvious that toxicological insults that result in dysmorphogenesis must interface with the cellular and molecular biology of limb development and pattern regulation (see, for example, RUTLEDGE et al. 1994; SHUEY et al. 1994). Understanding developmental toxicology is, however, complicated by the pattern regulatory response, since uncovering the action of the toxicological insult must take into account the nature of the response. For example, because limb outgrowth and formation of the skeletal pattern are dependent upon interactions between the bud ectoderm and the mesoderm, toxicological insults or genetic mutations that specifically compromise the morphogenetic activity of the ectoderm result in severe skeletal abnormalities of the limb, without adversely influencing the differentiative capacity of the epidermis itself. Thus the tissue type sensitive to the insult (epidermis) would appear unrelated to the tissue type in which the dysmorphogenesis occurs (skeletal). This example serves as an important reminder that understanding the influence of toxicological agents on development goes hand in hand with an understanding of the complex interplay that occurs during the course of normal limb formation.

B. Developmental Anatomy of the Limb

Most experimental studies that utilize limb morphology as an indicator of morphogenesis rely on the pattern of skeletal elements that is initially laid down as the bud elongates (see Fig. 1). This skeletal pattern is established by the condensation of chondrogenic cells, which serves as a template for the final ossified skeleton. The early skeletal pattern is useful for morphological studies, since it is relatively easy to assess because of clear anatomical differences between the various skeletal components. The limb pattern is segmented into three regions that are considered homologous among tetrapod vertebrates (Fig. 1): the stylopodium (humerus of the arm or wing, femur of the leg), the zeugopodium (radius/ulna of the arm or wing, tibia/fibula of the leg), and the autopodium (carpal/metacarpals/phalanges of the hand or wing, tarsal/metatarsals/phalanges of the foot). The long axis of the limb is called the proximal–distal axis; proximal is at the base of the limb, and distal is at the tip of the digits. The axis defined by the organization of the digits is called the

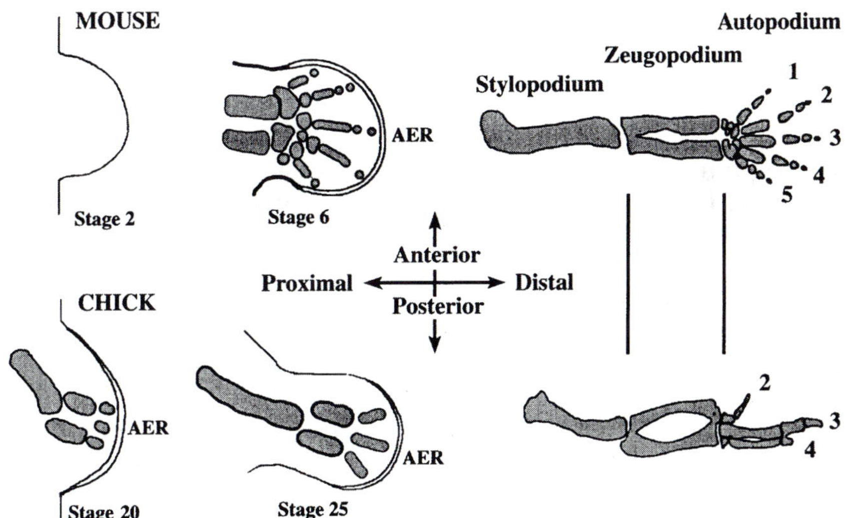

Fig. 1. Developmental anatomy of the mouse and chick limb (see text for details). Mouse limb bud stages are from WANEK et al. (1989b), and chick embryo stages are from HAMBURGER and HAMILTON (1951). No fate map is available for the early-stage mouse limb bud. The mouse stage-6 fate map was redrawn from MUNEOKA et al. (1989), and the chick fate maps were redrawn from BOWEN et al. (1989). *AER*, apical ectodermal ridge

anterior–posterior axis; anterior is the thumb or big toe (digit 1), and posterior is the little finger or toe (digit 5). The number of digits is highly variable among tetrapod vertebrates, e.g., the legs of the mouse possess five digits, whereas the chick wing is composed of only three digits (numbered 2, 3, and 4; see Fig. 1), The dorsal–ventral axis of the limb defines the top and bottom of the hand or foot with the palm or sole ventral.

The limb initially forms as an outgrowth of the flank that results from the cessation of growth by non-limb-forming flank cells (Searls and Janner 1971). The limb bud that forms is an accumulation of mesenchymal cells enclosed within an ectodermal jacket. In the chick, the earliest stages of limb bud formation are characterized by the presence of an ectodermal specialization called the apical ectodermal ridge (AER) at the apex of the bud (see Fig. 1). The AER is known to be induced by an interaction with the underlying mesenchymal cells of the limb-forming region (see Carrington and Fallon 1986). As it turns out, the AER serves a vital role in directing outgrowth of the limb bud (see below); however, the initial formation of the bud is an AER-independent event. In the mouse, the early limb bud forms without an AER; the AER forms between stages 2 and 3 (Wanek et al. 1989b). Studies on the limbless mutation in the chick show that the AER fails to form, yet early bud formation is not affected by the absence of the AER (Carrington and Fallon 1988). The limb field in the chick embryo prior to the formation of the limb bud has been mapped and the limb-forming potential of the limb field is in place as early as stage 12, about 1 day prior to the initial appearance of the limb bud (Stephens et al. 1993). Prior to stage 12, there is evidence from explantation studies that the limb field is influenced by axial tissues, such as Hensen's node, the developing neural tube, somites, mesonephros, and the lateral plate (Stephens et al. 1991). Thus the limb field is initiated very early in development and progressively develops into a self-differentiating domain prior to the first appearance of the bud.

The AER is a developmentally transient ectodermal specialization at the distal tip of the limb bud. Whereas the AER forms at distinctly different stages in the chick and the mouse, the AER disappears at comparable limb stages in both species (see Wanek et al. 1989b). The proliferating population of mesenchymal cells subjacent to and under the influence of the AER is called the "progress zone" (See Fig. 2A), which is proposed to be the site at which pattern specification occurs during limb outgrowth (Summerbell et al. 1973). It should be noted that the progress zone is not defined by any specific anatomical characteristics, but is recognized as a developmentally important region of the limb bud. Evidence linking the function of the AER to limb outgrowth and pattern specification includes AER removal studies, which result in distally truncated limbs (Fig. 2B), the degree of which corresponds to the embryonic stage at the time of removal (Saunders 1948; Summerbell 1974a; Rowe and Fallon 1982). In addition, supernumerary limb structures are induced by ectopic grafts of the AER (Saunders and Errick 1976) and also in the chick *eudiplopodia* mutant, in which an abnormal secondary AER

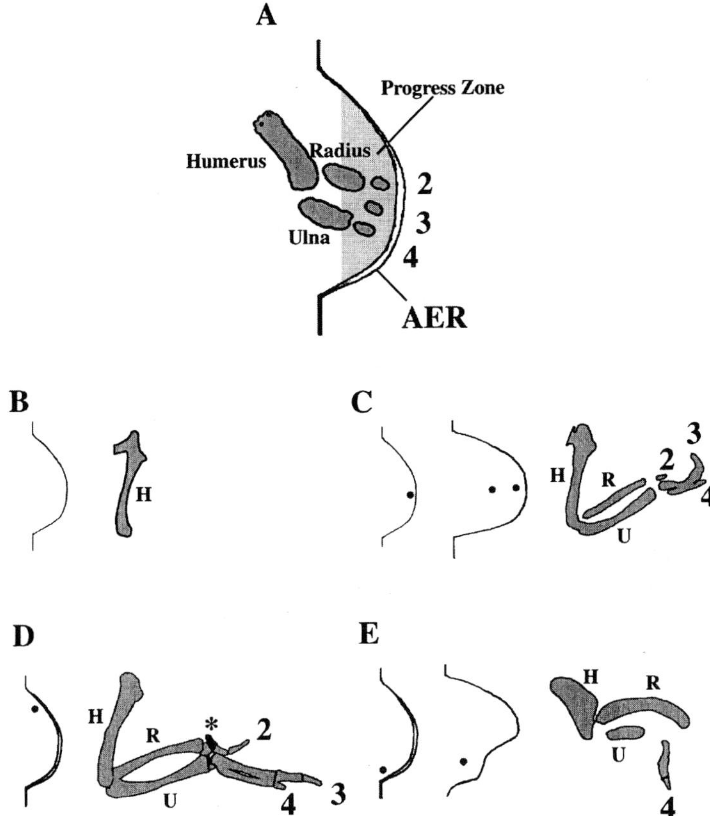

Fig. 2. A The fate map of an early-stage chick wing bud is shown with the "progress zone" of unspecified cells indicated in *dark shading*. At this stage, only the most proximal skeletal elements are specified. **B** Following apical ectodermal ridge (AER) removal at an early stage, only the most proximal skeletal element differentiates (redrawn from NISWANDER et al. 1993). **C** Following AER removal, the sequential application of fibroblast growth factor (FGF)-2-loaded beads to the distal tip of the limb bud replaces the outgrowth maintenance function of the AER (redrawn from FALLON et al. 1994). **D** Implantation of an FGF-2-loaded bead into the anterior limb bud (with the AER intact) results in a relatively minor alteration of the limb pattern. The formation of a supernumerary skeletal structure (*asterisk*) associated with the base of digit 2 is shown (redrawn from RILEY et al. 1993). **E** Implantation of an FGF-2-loaded bead into the posterior limb bud (with the AER intact) induces a dramatic change in limb bud morphology, resulting in the loss of digits and an overall reduction in the length of all skeletal elements. (Redrawn from LI et al. 1996)

develops (GOETINCK 1964). Thus, the AER is required for cell proliferation and pattern specification, and pattern specification itself occurs in a proximal to distal sequence. What emerges is the view that, as the limb forms, cells must first acquire information for the specification of the most proximal region of the limb and, as development proceeds, the progeny of these cells acquire information for the specification of more distal regions of the limb. The process of pattern specification ends when the most distal region of the limb (e.g., digit tips) is specified. The necessity of the AER for distal outgrowth in the mouse limb bud has not been extensively studied, although the AER has been shown to be important for distal elongation of the limb bud in explant cultures (NISWANDER and MARTIN 1993a), and the characterization of the *legless* mutation shows that the legless phenotype is associated with the failure to form an AER during early bud formation (SINGH et al. 1991).

A comparative chart of limb development in the mouse and the chick is presented in Fig. 1. Chick stages are based on the wing bud using whole embryo stages according to HAMBURGER and HAMILTON (1951), and mouse stages are limb-specific stages according to WANEK et al. (1989b). The sequence of vertebrate limb formation defines temporal and spatial gradients of cell proliferation and cellular differentiation. Since the developing limb responds to experimental manipulations in a stage- and position-dependent manner, understanding these developmental gradients provides important clues concerning morphogenetic processes. Once initiated, limb bud outgrowth occurs primarily in a distal direction, although circumferential growth of the limb bud also occurs. Distal outgrowth of the limb bud establishes proximal–distal differences in cell proliferation, with distal cells maintaining a uniformly high mitotic index, while some proximal cells fall out of the cell cycle and terminally differentiate. A more accurate depiction of cell proliferation in the early bud is that growth is enhanced peripherally and distally forming a cone of proliferating cells surrounding cells located centrally and proximally that are undergoing terminal differentiation (SUMMERBELL and WOLPERT 1972). As development continues, distal cells remain undifferentiated and maintain a high growth rate that drives limb outgrowth, while differentiation of the limb tissues continues in a proximal to distal sequence. In addition to a proximal to distal developmental gradient, an anterior to posterior developmental sequence exists that is more subtle but is present nonetheless. The chick wing possesses three digits, and the developmental sequence of their differentiation is from posterior to anterior; digit 4 differentiates first and digit 2 differentiates last. The mouse limb possesses five digits and displays a similar sequence of development, although in this case it is not precisely a posterior to anterior sequence. The order of digit differentiation during mouse limb development begins with digit 4 and progresses sequentially with digit 3, digit 2, digit 5, and finally digit 1 (WANEK et al. 1989b). Thus the developmental sequence of homologous digits between mouse and chicks is identical; however, the absence of digits 1 and 5 in the chick wing results in a developmental sequence that is precisely posterior to anterior.

Fate maps of the chick wing bud are available for the earliest of limb stages (see BOWEN et al. 1989), and comparable fate maps of the mouse limb bud have been developed for later stages (MUNEOKA et al. 1989). Fate maps are generated by marking a group of cells of the limb bud with a vital marker (such as carbon particles) and tracing the fate of these cells to later stages in development. Marked regions of the bud expand exclusively in a distal direction. Similarly, when labeled collagen is injected into the early limb bud it becomes reorganized to form fibers that extend along the proximal–distal limb axis (Stopak et al. 1985). Fate mapping studies also show that there is an anterior–posterior asymmetry in limb development. This is particularly clear in the chick, in which the posterior half of the limb bud contributes to the majority of the limb proper; the anterior half limb bud contributes to the shoulder and parts of the stylopod (see Fig. 1). In the mouse, the posterior limb bud contributes to three of the five digits (digits 3, 4, and 5), while the anterior bud forms the anterior two digits (digits 1 and 2). The significance of this anterior–posterior asymmetry in the fate of limb bud cells is at present unclear; however, it suggests that growth along the anterior–posterior axis of the limb bud is spatially regulated. Evidence that limb bud cells from different anterior–posterior regions of the limb bud display distinct growth characteristics in vitro has been obtained for both the chick and the mouse (Aono and Ide 1988; Shi and Muneoka 1992; Paulsen et al. 1994).

C. Pattern Regulation

The process by which embryonic cells alter their developmental course to regenerate missing structures or to form supernumerary structures is called pattern regulation, and its study lies at the heart of understanding mechanisms of pattern formation. Studies on the formation of the vertebrate limb have been instrumental in molding current theories about how pattern develops in the embryo, as well as how pattern regenerates in the adult. The fact that limb cells respond to perturbation by altering their developmental path in a defined and reproducible manner demonstrates that cells rely on intercellular interactions for normal development. Understanding the types of interactions and the molecules involved in such interactions is crucial to uncovering pattern formation mechanisms and also to understanding how extrinsic influences, such as environmental or pharmacological teratogens, can cause dramatic and devastating changes in the way that limbs develop. During limb development, two types of cell–cell interactions have been extensively studied: (1) the AER's influence on inducing mesenchymal outgrowth, and (2) the ability to reorganize the anterior–posterior pattern of the limb by grafts of ZPA signaling tissue. Recent studies have identified AER-derived signals important for limb outgrowth as members of the fibroblast growth factor (FGF) family, and a ZPA-derived signal as the product of a vertebrate homologue of the *Drosophila* hedgehog segment polarity gene, *sonic hedgehog* (*Shh*). In addition, there is clear evidence that signals from the AER and the dorsal ectoderm influence the signaling ability of the ZPA; thus it is possible to begin to formulate a signaling cascade that regulates the morphogenesis of the vertebrate limb. Defining this cascade is necessary in order to understand morphogenesis and dysmorphogenesis of the limb.

I. The Apical Ectodermal Ridge and Fibroblast Growth Factors

Recent studies in which FGF-2 or FGF-4 is loaded onto carrier beads and pinned onto chick limb buds (stage 20) in which the AER has first been surgically removed have demonstrated that FGF-2 or FGF-4 can replace AER function and maintain distal outgrowth (NISWANDER et al. 1993; FALLON et al. 1994; Fig. 2C). While limb morphogenesis in these experiments is not perfect, the limbs that form appear to contain a complete set of skeletal elements (see Fig. 2C). In addition, local application of FGF-2 induces a regenerative response following amputation of the nonregenerating chick limb bud, thus demonstrating that FGF-2 can stimulate limb bud cells to alter their developmental course to form more distal limb structures (TAYLOR et al. 1994). Finally, ectopic FGF-2 application (COHN et al. 1995) or ectopic expression of *Fgf-4* (MIMA et al. 1995) can induce the formation of supernumerary limb structures. The interpretation of results from the utilization of carrier beads to apply FGF to the limb bud should be considered with some caution, since these beads are loaded at a very high concentration of growth factor (generally 1.0 mg/ml). Since it is unlikely that endogenous FGF is present in the limb bud at such a high concentration, it is possible that limb cells are displaying a toxicological response to FGF. With this caveat in mind, there has been considerable excitement over the possibility that FGF-2 and/or FGF-4 represent the long sought after morphogenetic signals produced by the AER. This conclusion is also supported by the demonstration that FGF is endogenously produced in the limb bud (see below).

Three members of the FGF family have been shown to be expressed by early limb bud cells (see Table 1): FGF-2 (SAVAGE et al. 1993; DONO and ZELLER 1994; FALLON et al. 1994), FGF-4 (NISWANDER and MARTIN 1992; SUZUKI et al 1992), and, recently, FGF-8 (HEIKINHEIMO et al. 1994; OHUCHI et al. 1994; CROSSLEY and MARTIN 1995). All members of the FGF family are characterized by high-affinity binding to heparan sulfate and are thought to be present in association with the extracellular matrix. Thus far, only FGF-2 is known to be present in the limb bud. FGF-2 is present in particularly high concentrations in the early limb bud (MUNAIM et al. 1988; SEED et al. 1988), and it is immunolocalized to cells of the periphery of the limb bud associated with the ectoderm, where it is found to be extracellular, cytosolic, and nuclear (SAVAGE et al. 1993; DONO and ZELLER 1994). Three isoforms of FGF-2 have been identified in the chick embryo during limb bud stages (DONO and ZELLER 1994), although the relationship between these isoforms and the function of FGF-2 during limb development is unknown. The other FGF family members expressed during limb development have been characterized based on in situ hybridization studies. *Fgf-4* transcripts are found associated with cells of the AER in both the chick and the mouse limb bud (NISWANDER and MARTIN 1992; SUZUKI et al. 1992; LAUFER et al. 1994; NISWANDER et al. 1994). In the mouse, *Fgf-8* transcripts are found associated with the ventral ectoderm in the early bud and become restricted to the AER once it forms (OHUCHI et al. 1994;

Table 1. Selected genes expressed during limb development

Gene	Comments	References
HoxD 11 12 13	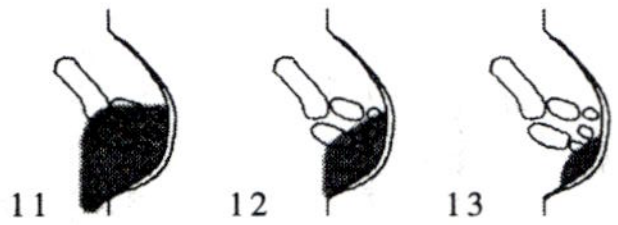 5′*HoxD* genes display nested expression in posterior/distal cells The 5′*HoxD* genes are induced following grafts of ZPA signaling tissue or RA *HoxD11* knockout results in minor defects of the autopod and distal zeugopod *HoxD11* overexpression induces homeotic digit transformations *HoxD13* knockout results in relatively minor autopodial defects	Dolle et al. 1989a, 1993; Izpisua-Belmonte et al. 1991; Nohno et al. 1991; Morgan et al. 1992; Davis and Capecchi 1994
HoxA 11 13	5′*HoxA* genes display nested expression in distal cells *HoxA13* expression is distal and induced by the AER *HoxA11* expression is restricted to the zeugopod at later stages *HoxA11* knockout results in minor autopodial defects	Yokouchi et al. 1991, 1993; Small and Potter 1993
HoxB8 *RARB2*	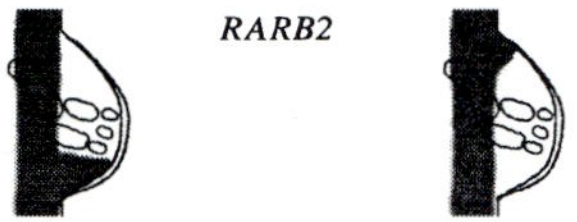 *HoxB8* is expressed in the flank and posterior limb bud *RAR*β*2* is expressed in the flank and anterior limb bud Ectopic expression of *HoxB8* by the *RAR*β*2* promoter induces an anterior *Shh* expression domain and supernumerary digits	Mendelsohn et al. 1991; Rossant et al. 1991; Reynolds et al. 1991; Balkan et al. 1992; Charite et al. 1994
Msx 1 2	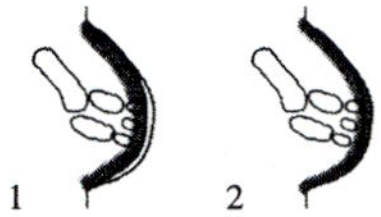 *Msx-1* expression is mesodermal; *Msx-2* expression is meso- and ectodermal *Msx-1* knockout results in normal limbs *Msx-2* mutation is associated with minor digit defects Both are induced by AER, expressed after AER regression, and in the nailbed Both are expressed during digit tip regeneration in the mouse	Hill et al. 1989; Robert et al. 1989, 1991; Davidson et al. 1991; Yokouchi et al. 1991; Coelho et al. 1991, 1992, 1993; Ros et al. 1992; Jabs et al. 1993; Satokata and Maas 1994; Reginelli et al. 1995

Table 1 (*Contd.*)

Gene	Comments	References
Fgf 2 4 8	FGF-2 is immunolocalized to the AER and distal mesenchyme *Fgf*-4 and *Fgf-8* transcripts are restricted to the AER FGF-2, FGF-4, and FGF-8 replace the AER in stimulating outgrowth FGF-2 and FGF-4 maintain *Shh* expression after AER removal	Niswander and Martin 1992; Suzuki et al. 1992; Niswander et al. 1993, 1994; Savage et al. 1993; Fallon et al. 1994; Crossley and Martin 1995; Mahmood et al. 1995
Shh	*Shh* expression is coincident with ZPA signaling tissue Ectopic *Shh* expression induces supernumerary digits SHH induces supernumerary digits *Shh* expression is dependent on the AER and FGF	Echelard et al. 1993; Riddle et al. 1993; Roelinck et al. 1994; Laufer et al. 1994; Niswander et al. 1994; Lopez-Martinez et al. 1995
Wnt-7a	*Wnt-7A* expression is restricted to the dorsal ectoderm *Wnt-7A* knockout results in a partial double ventral phenotype *Wnt-7A* knockout causes posterior digit loss and a reduced *Shh* domain Dorsal ectoderm removal causes posterior digit loss and a reduced *Shh* domain	Dealy et al. 1993; Parr et al. 1993; Parr and McMahon 1995; Yang and Niswander 1995

ZPA, Zone of polarizing activity; RA, retinoic acid; AER, apical ectodermal ridge; RAR, RA receptor; *Shh*, *Sonic hedgehog;* FGF, fibroblast growth factor.

Crossley and Martin 1995). Recent experiments in which mouse limb buds denuded of their AER are cultured in the presence of FGF-8 indicate that FGF-8 can stimulate in vitro growth of the limb bud, thus suggesting that, like FGF-2 and FGF-4, FGF-8 plays a role in AER function (Mahmood et al. 1995). There are four known FGF receptors: FGFR1 (formerly flg), FGFR2 (formerly bek or Cek), FGFR3 (formerly Cek-2), and FGFR4 (see Johnson and Williams 1993). *FgfR1* and *FgfR2* are expressed in the early mouse limb bud (Orr-Urtreger et al. 1991; Peters et al. 1992).

Ectopic application of FGF-2 to different positions of the limb bud demonstrates that, in the presence of the AER, FGF-2 can have either a stimulatory or an inhibitory influence on limb outgrowth. Ectopic application of FGF-2 to the anterior limb bud of the chick results in either normal development (Hayamizu et al. 1991) or the induction of a supernumerary skeletal structure associated with the anterior digit (Riley et al. 1993; Fig. 2D). Conversely, Li et al. (1996) have demonstrated that applying beads containing FGF-2 to the posterior chick limb bud does not stimulate the formation of supernumerary skeletal elements, but results in the loss of digits and an overall reduction in the length of skeletal elements (Fig. 2E). One possible explanation for this position-dependent effect of FGF-2 is related to the fact that most of the contribution to the final limb pattern is derived from cells in the posterior limb bud (see Sect. B). To investigate this possibility, Li et al. (1996) carried out fate mapping studies using the lipophilic dye 1,1' -dioctadecyl-3, 3, 3′, 3′-tetramethyl-indocarbocyanine perchlorate (DiI) and found that FGF-2 application to the posterior limb bud misdirects cells, causing them to move in a proximal rather than a distal direction. Thus FGF-2 appears to compete with an endogenous AER-derived signal that directs the normal distalward movement of cells during limb outgrowth.

One of the activities of the AER and FGF on limb bud cells is the stimulation of cell proliferation. Following AER removal, the rate of cell proliferation in the limb bud is reduced, as is the rate at which limb outgrowth proceeds (see Summerbell 1977). Both FGF-2 and FGF-4 stimulate cell proliferation in organ cultures of mouse limb buds lacking an AER, whereas another growth factor expressed by AER cells, bone morphogenetic protein-2a, has a growth-inhibitory effect (Niswander and Martin 1993a). Studies on the growth dynamics of cultured chick and mouse limb bud cells, however, suggest that the mitogenic action of FGF-2 is indirect. Cells from different regions of the early limb bud display position-specific growth characteristics in vitro; posterior cells fail to proliferate under conditions that stimulate proliferation of other limb bud cells (Aono and Ide 1988; Shi and Muneoka 1992). In the chick limb bud, FGF-2 has been shown to be mitogenic for anterior and central limb bud cells, but fails to stimulate posterior cell growth (Aono and Ide 1988). Because the posterior region of the chick limb bud undergoes extensive proliferation during normal limb development, the observation that FGF-2 stimulates limb outgrowth but fails to stimulate posterior cell proliferation suggests that the activity of FGF-2 in the limb bud is indirectly linked to the control of cell proliferation. Since FGF-2 is also involved in regulating cell movements during outgrowth, one possibility is that the mitogenic action of FGF-2 is somehow linked to its role in stimulating cell movements (LI et al. 1996).

II. The Zone of Polarizing Activity and the Supernumerary Response

The observation that grafts of posterior limb bud tissue into the anterior margin of the chick limb bud induce supernumerary digits which form a mirror-image duplication of the normal skeletal digit pattern (SAUNDERS and GASSELING 1968) has proved to be the most telling experimental avenue for elucidating the molecular mechanisms governing pattern regulation. Because of its signaling ability, this region of the limb bud is known as the ZPA (MACCABE et al. 1973). Cell contribution studies indicate that ZPA signaling cells themselves make a minor contribution to the supernumerary response (HONIG 1983a; JAVOIS and ITEN 1986); the ZPA signal causes anterior limb bud cells to alter their normal developmental fate to form a duplicated set of posterior skeletal elements. Thus it has been proposed that the ZPA produces a morphogenetic signal that specifies the anterior–posterior limb pattern; placement of the ZPA signaling region into the anterior region results in the specification of anterior cells to form more posterior limb structures (WOLPERT 1969; TICKLE et al. 1975). While it is clear that the ZPA produces a signal that is involved in altering pattern specification of the limb, it remains uncertain whether the ZPA signal actually instructs anterior cells to alter their developmental fate versus permitting anterior cells to carry out an endogenous developmental program that is terminated during normal development (ANDERSON et al. 1994; see below). In the chick wing bud, assessment of specification along the anterior–posterior axis is based on the quality of the digits that form; the digit sequence of the chick wing is 2-3-4, and each digit has a distinct morphology (Fig. 3). Grafting of chick ZPA signaling cells into the anterior wing bud results in the formation of supernumerary digits in mirror symmetry with the normal digit pattern; the digit sequence is **4-3-2**/2-3-4 (bold numbers represent the supernumerary digits; Fig. 3B). An attenuated response is one in which the posterior-most digits are not induced and fewer supernumerary digits form, i.e., the digit sequence **3-2**/2-3-4, **3**/2-3-4, or **2**/2-3-4 (Fig. 3C,D). The observation that the morphogenetic response to the ZPA signal can be attenuated has been interpreted as evidence that the ZPA signal acts in a concentration-dependent manner (SMITH et al. 1978; TICKLE 1981).

The ability of distal or anterior limb bud tissues to respond to the ZPA signal has been used as an assay to develop maps of ZPA signaling regions of the chick and the mouse limb bud (Fig. 4). The experimental strategy for spatially and temporally mapping ZPA signaling regions is to use a constant-stage host limb bud (Stage 19–21) and a constant graft site (anterior limb bud) to measure the ability of tissues taken from different regions of different-stage limb buds to induce a supernumerary response. Based on this assay, ZPA

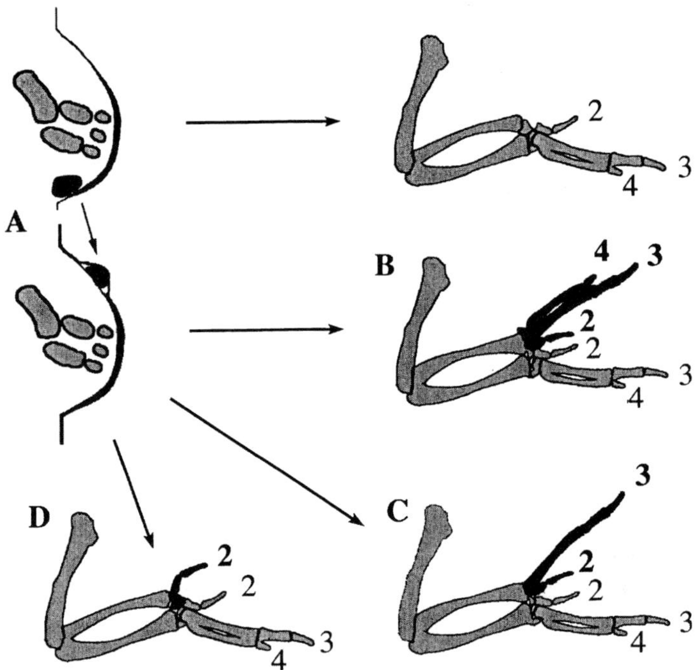

Fig. 3. A Zone of polarizing activity (ZPA) signaling is established empirically based on the ability of cells or tissue to induce a supernumerary response when grafted into the anterior of a host chick wing bud. The normal three-digit wing pattern results if grafts lack ZPA signaling. ZPA signaling grafts induce a varied response, some of which are depicted here. Note that the ZPA signaling response results in the formation of host-derived digits, regardless of the origin of the ZPA signaling cells. **B** The most extreme response is the formation of a full set of three supernumerary digits (*bold numbers*) arranged in perfect mirror-image symmetry with the host digits. Attenuated ZPA signaling responses include **C** the induction of two supernumerary digits or **D** a single supernumerary digit, which represents a partial duplication of the host digit pattern

signaling has been quantified using two methods. First, a strength of activity index has been developed that measures ZPA signaling based on the proposal that higher levels of ZPA signaling result in the induction of more posterior supernumerary digits (Honig et al. 1981). Thus the induction of a digit 4 is given a maximum score, the induction of a digit 3 is given an intermediate score, and the induction of a digit 2 is given a minimum score. Over the years, various laboratories have modified the scoring index and ZPA signaling strength is therefore not easily comparable from study to study (e.g., see Honig et al. 1981; Wanek and Bryant 1991; Vogel and Tickle 1993). An alternative approach that is not based on any assumptions concerning the nature of the ZPA signal is to use the frequency of the supernumerary response to quantify ZPA signaling (see Anderson et al. 1993, 1994), and in this review we have adopted this approach to quantify ZPA signaling. Since there

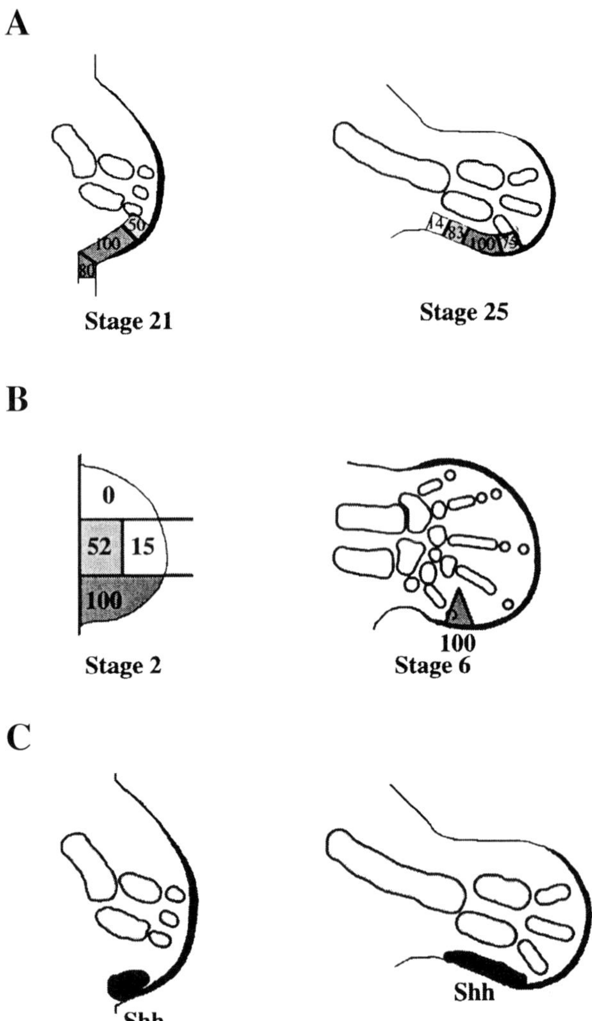

Fig. 4. A Zone of polarizing activity (ZPA) signaling is quantified based on the frequency of a supernumerary response in the ZPA signaling assay shown in Fig. 3. In the chick, both the early-stage and late-stage bud display ZPA signaling in posterior tissue. ZPA signaling is highest in tissue that is along the posterior boundary just proximal to the tip of the bud and corresponds well with the expression domain of *Sonic hedgehog* (*Shh*) in the chick limb bud. ZPA signaling is reduced in tissue that is either distal or proximal to this region of high signaling (redrawn from data from HONIG and SUMMERBELL 1985). **B** In the early-stage mouse limb bud, regions of the entire limb bud have been assayed for ZPA signaling. **C** The extreme posterior limb bud is associated with *Shh* expression and displays the highest level of ZPA signaling. ZPA signaling is also found not associated with *Shh* expression, in central–proximal tissue, and to a lesser extent in central–distal tissue. Only the anterior limb bud of the mouse lacks endogenous ZPA signaling ability. In the late-stage limb bud, ZPA signaling is present in posterior tissue that is proximal to the tip of the bud. (Redrawn from WANEK and BRYANT 1991)

are apparent species-specific differences between ZPA signaling maps for the chick and mouse, we will review them separately.

The spatial and temporal distribution of ZPA signaling tissue in the chick wing bud has been mapped to the posterior limb field and part of the flank prior to bud formation and to the extreme posterior edge of the bud after bud formation (see HONIG and SUMMERBELL 1985; Fig. 4A). Peak levels of ZPA signaling can be mapped to the posterior edge of stage -19 to -25 limb buds with activity declining as the limb bud matures. ZPA signaling maps have also been generated based on grafts to the distal tip of the limb bud (MACCABE et al. 1973); differences between the ZPA signaling maps using these two different assay sites have been attributed to self-differentiation of the grafted tissue (HONIG and SUMMERBELL 1985). In general, ZPA signaling maps in the developing chick wing bud define a small region of the posterior bud that possesses high signaling and show that adjacent tissues along the periphery of the bud display a reduced level of signaling.

ZPA signaling by mouse limb bud tissue has been demonstrated in vivo (WANEK et al. 1989a) and has been characterized using heterospecific (mouse into chick) tissue grafts into the anterior chick wing bud (Fig. 4B); thus the assay is identical to that used to map ZPA signaling in the chick limb bud. Since the supernumerary digits that are induced are of a chick wing skeletal morphology, any confusion concerning self-differentiation of the graft is avoided. A stagewise analysis of posterior mouse limb bud tissue indicates that ZPA signaling is high from stage 3 to stage 6 and declines as the bud matures (WANEK and BRYANT 1991). The results from these temporal studies largely parallel those found for the chick. A detailed analysis of ZPA signaling from different regions of the stage-2 limb bud of the mouse, however, differs dramatically from the ZPA signaling map of a comparable stage in the chick (R. Anderson, unpublished observation). ZPA signaling in the mouse limb bud is highest in the posterior region of the limb bud; however, signaling is also present in central–proximal tissue and, to some extent, in central–distal tissues (Fig. 4B). Only tissue of the anterior mouse limb bud displays no ZPA signaling; thus the ability of mouse limb bud cells to produce a ZPA signal is graded across the early-stage mouse limb bud. This is in striking contrast to the chick limb bud studies, in which the data point to the existence of a unique morphogenetic organizing center (the ZPA) at the posterior boundary of the bud. The reason for this difference may be related to differences in the developmental anatomy of chick and mouse limb buds: the difference between a three-digit limb versus a five-digit limb or the presence (chick) or absence (mouse) of an AER in the early limb bud. Another possibility is that subtle differences in the ZPA signaling assay might contribute to some differences in the signaling maps; if ZPA signaling is widespread across the limb bud, using an assay with poor sensitivity would result in a restricted region of only the highest level of signaling, and enhancing the sensitivity of the assay would result in an expansion of the signaling region. On the other hand, it is interesting to note that ZPA maps generated for the chick wing bud have focused primarily on the peripheral boundary of the bud and have largely ignored central–proximal regions of the bud. Thus one possibility is that the posterior restriction of the ZPA signal in the chick limb bud may exist simply because other limb bud tissues were not assayed. The important conclusion from mapping the ZPA signal in the mouse is that cells outside the posterior region of the limb bud produce a morphogenetic signal important for organizing the digit pattern of the limb.

ZPA signaling has also been found in a number of other tissues of the chick embryo, including Hensen's node (HORNBRUCH and WOLPERT 1986), the floorplate of the neural tube (WAGNER et al. 1990), and the genital tubercle (DOLLE et al. 1991), indicating that a similar signaling mechanism regulates pattern formation during the development of other organs. In these studies, the quality of the supernumerary response is attenuated relative to ZPA signaling grafts, suggesting that the level of the ZPA-like signal in other embryonic regions differs from that in the limb bud or that the signal itself is similar, but not identical.

III. *Sonic Hedgehog* and Zone of Polarizing Activity Signaling

Studies designed to elucidate biochemical characteristics of the ZPA signal have provided evidence that the signaling molecule is a secreted protein (HONIG et al. 1981). There is now evidence that the ZPA signal is the product of the vertebrate homologue of the *Drosophila hedgehog* (*hh*) segment polarity gene, called *Sonic hedgehog* (*Shh*), a signaling protein important for patterning during *Drosophila* development (see INGHAM and MARTINEZ-ARIAS 1992). *Shh* is expressed by cells in the posterior limb bud of the mouse and chick that corresponds to ZPA signaling regions (Fig. 4C, Table 1), and its predicted amino acid sequence suggests that it is a secreted protein (ECHELARD et al. 1993; KRAUSS et al. 1993; RIDDLE et al. 1993; CHANG et al. 1994; ROELINK et al. 1994). It is also noteworthy that *Shh* expression during embryogenesis corresponds with some, but not all, tissues that are known to possess ZPA-like signaling; *Shh* is expressed in Hensen's node and in the floor plate of the neural tube, but it is not expressed in the genital tubercle. In addition to the correlative evidence that *Shh* expression is associated with ZPA signaling in the chick, there is direct evidence that ectopic expression of *Shh* can convert non-ZPA signaling cells into ZPA signaling cells (RIDDLE et al. 1993; CHANG et al. 1994; LAUFER et al. 1994; NISWANDER et al. 1994). These findings provide convincing evidence that *Shh* expression is involved in ZPA signaling; thus elucidation of the activity of the SHH protein and how its production by ZPA signaling cells is controlled is crucial to our understanding of how the vertebrate limb forms.

The SHH protein undergoes an autoproteolytic cleavage, resulting in the production of a 19-kDa amino-terminal fragment and a 27-kDa carboxy-terminal fragment (CHANG et al. 1994; LEE et al. 1994; BUMCROT et al. 1995; PORTER et al. 1995). To date, only the amino-terminal fragment has been shown to be biologically active (Fan et al. 1995; Lopez-Martinez et al. 1995; Roelink et al. 1995), whereas the carboxy-terminal fragment is required for proper autoproteolysis (Lee et al. 1994). In the limb bud, the signaling amino-terminal fragment displays matrix-binding characteristics and remains associated with limb bud cells that express *Shh* transcripts (Marti et al. 1995). An important conclusion that can be drawn is that the ZPA signal, or a component of the ZPA signal, is locally transmitted. Because the ZPA signal has long been hypothesized to be a diffusible signal (Tickle et al. 1975), we can either expect to find that other components of the signal are diffusible or that SHH is part of a cascade of local signals that are sequentially transmitted so as to appear to behave as a diffusible molecule (see Bryant and Muneoka 1986; Riddle et al. 1993).

IV. Regulation of Zone of Polarizing Activity Signaling/*Sonic Hedgehog* Expression

A relationship between the activities of the AER and the ZPA signal have been demonstrated in tissue grafting studies; grafts of ZPA signaling tissues into the anterior of the chick wing bud are only effective in inducing a supernumerary response when they are placed in association with the AER (Summerbell 1974b). A more direct relationship between the AER and ZPA signaling has emerged from studies in which the stability of ZPA signaling in cultured posterior limb bud cells has been analyzed. Such studies demonstrate that ZPA signaling by posterior cells declines rapidly and precociously when compared to the stagewise decline of ZPA signaling in vivo; thus the ZPA signal is lost from explant cultures, monolayer cultures, or micromass cultures of ZPA tissue from either chick, quail, or mouse (Honig 1983b; Carlson 1984; Hayamizu and Bryant 1992; Anderson et al. 1993). Modifying conditions for tissue dissociation and culture of mouse limb bud cells demonstrated that a ZPA signaling maintenance activity is associated with the limb bud ectoderm and the extracellular matrix of the posterior mesenchyme and that FGF-2 could function to maintain ZPA signaling under nonpermissive culture conditions (Anderson et al. 1993). Additional studies have demonstrated that, in the chick, ZPA signaling in vivo is dependent on AER and that application of FGF-2 or FGF-4 following AER removal can maintain ZPA signaling (Vogel and Tickle 1993; Li et al. 1995). Similarly, *Shh* expression by posterior limb bud tissue of the chick and mouse has been shown to be dependent on the AER, and application of FGF-2 or FGF-4 following AER removal maintains *Shh* expression (Laufer et al. 1994; Niswander et al. 1994; Li et al. 1996; Reginelli et al. 1996). Thus FGF appears to play an important role in regulating patterning of both the anterior–posterior axis and the proximal–distal axis.

Recent studies show that patterning of the dorsal–ventral axis and the anterior–posterior axis is regulated in part by the production of WNT-7a by the dorsal ectoderm. The WNT family of proteins consists of secreted signaling factors that are associated with the cell surface and extracellular matrix and are implicated in the regulation of oncogenesis, the determination of embryonic cell fate, and pattern formation (NUSSE and VARMUS 1992). During embryonic development, members of the *Wnt* gene family are expressed in both a tissue-specific and position-specific manner; for example, *Wnt-7a* is specifically expressed by cells of the dorsal ectoderm during limb bud stages (DEALY et al. 1993; PARR et al. 1993; see Table 1). The ectoderm-specific expression of *Wnt-7a*, along with tissue recombination studies that have linked the ectoderm to the control of dorsal–ventral pattern formation (MACCABE et al. 1974; STARK and SEARLS 1974; PAUTOU 1977; GEDUSPAN and MACCABE 1987, 1989), suggest a role for WNT-7a in limb patterning. Eliminating WNT-7a signaling by knocking out the *Wnt-7a* gene in transgenic mice results in the formation of limbs that are partially symmetric in the dorsal–ventral axis; ventral structures replace dorsal structures, resulting in ventrally symmetrical limbs (PARR and MCMAHON 1995). In addition to influencing dorsal–ventral patterning, patterning of the anterior–posterior axis is modified in that *Wnt-7a*$^{-/-}$ limbs lack posterior structures, and there is a decrease in the expression of a number of posterior limb bud genes, including *Shh* (PARR and MCMAHON 1995). In the chick, surgical removal of the dorsal ectoderm also results in limbs lacking posterior structures (YANG and NISWANDER 1995), although killing the dorsal ectoderm using ultraviolet (UV) irradiation rarely results in limbs with skeletal defects (MARTIN and LEWIS 1986). Following surgical removal of the dorsal ectoderm, YANG and NISWANDER (1995) have found that ZPA signaling and *Shh* expression by posterior tissue declines and that this decline can be reversed by implantation of cells that express WNT-7a. Two important conclusions emerge from these studies. First, WNT-7a produced by the dorsal limb bud ectoderm is an important patterning signal for both the dorsal–ventral and the anterior–posterior limb axes. Second, the finding that the primary regulator of the anterior–posterior axis (ZPA signal/*Shh* expression) is influenced by signals important for pattern formation of the proximal–distal (AER/FGF) and dorsal–ventral (dorsal ectoderm/WNT-7a) limb axes leads to the conclusion that limb outgrowth involves cooperative interactions between signaling mechanisms in all axes of the limb.

V. Retinoic Acid and the Zone of Polarising Activity Signal

For the past decade, retinoic acid (RA) has been proposed to be the ZPA signaling morphogen; parallels between the activity of exogenously applied RA and the effect of ZPA grafts are extensive. First, local application of RA to the anterior limb bud results in the induction of mirror-image digit duplications that mimic the response to ZPA tissue grafts (TICKLE et al. 1982; SUMMERBELL 1983). Second, the RA response is dose dependent in a fashion that parallels the attenuation of the ZPA signal achieved by titering the number of ZPA signaling cells present in the graft (TICKLE 1981; TICKLE et al. 1985). Third, both exogenous RA and a graft of ZPA signaling tissue induce the expression of the *5′HoxD* genes in cells of the anterior limb bud (IZPISUA-BELMONTE et al. 1991). Fourth, in experiments in which the signaling source is removed at varying time points after grafting, both RA and ZPA tissue grafts stimulate a similar sequential specification of supernumerary digits with the initial specification of a supernumerary digit 2, followed by a supernumerary digit 3 and, finally, a supernumerary digit 4 (see TICKLE and EICHELE 1994). Fifth, an analysis of endogenous retinoids within the chick limb bud indicates that RA is present at a slightly higher concentration in the posterior bud when compared to the anterior bud (THALLER and EICHELE 1987; SCOTT et al. 1994). These similarities between the action of exogenous RA and ZPA signaling tissue has led to the proposal that RA is the endogenous morphogen produced by ZPA signaling cells (see TICKLE and EICHELE 1994).

There is evidence that the activities of the ZPA following grafting into the anterior limb bud are, in fact, distinct from that of RA. First, WANEK et al. (1991) used a functional assay for ZPA signaling to demonstrate that RA induces anterior (nonsignaling) cells to begin producing the ZPA signal. Since grafts of ZPA signaling tissue are not associated with a similar conversion of anterior cells into ZPA signaling cells (SMITH 1979), the conclusion drawn from this study was that RA was acting in a manner that was distinct from the ZPA signal and, therefore, was not the ZPA signal. Consistent with the results from this study, REGINELLI et al. (1996) has found that grafting ZPA signaling tissue does not result in the induction of *Shh* expression in anterior cells, whereas the administration of RA does induce *Shh* expression in anterior cells (RIDDLE et al. 1993; NISWANDER et al. 1994). Second, in response to RA bead implantation into the anterior chick wing bud, transcription of the RA-inducible gene RA receptor β, *RARβ*, is upregulated in association with the RA bead; however, following ZPA signaling tissue grafts, no such upregulation of *RARβ* is observed (NOJI et al. 1991). Thus the fact that grafts of ZPA signaling tissue do not result in a RA-induced response indicates that the ZPA signal is not mediated by endogenous RA. Together, these studies provide clear evidence that RA is not a morphogen produced by posterior limb bud cells. What, then, is RA doing to limb bud cells? Current evidence favors the hypothesis that RA causes a conversion of nonsignaling anterior limb bud cells to become ZPA signaling cells. Thus, following RA bead implantation and prior to supernumerary digit formation, two events occur: (1) the induction of a secondary ZPA signaling region in the anterior limb bud, and (2) ZPA signaling by anterior cells to alter the developmental fate of nonconverted anterior cells (WANEK et al. 1991; see BRYANT and GARDINER 1992).

The possibility that exogenous RA can induce an in vivo conversion of anterior cells into ZPA signaling cells raises questions concerning the ability of anterior cells to develop ZPA signaling ability. In the mouse limb bud, at least three positional states can be defined based on ZPA signaling maps (see Fig. 4B). First, the extreme posterior limb bud possesses a high level of ZPA signaling ability that is associated with the expression of *Shh*. Second, cells in the central–proximal region of the limb bud produce a ZPA signal that is not associated with the expression of *Shh*. Third, cells in the extreme anterior limb bud do not produce an assayable ZPA signal. Two different experimental approaches provide evidence that anterior cells in both chick and mouse buds possess the ability to take on the characteristics of more posterior limb regions. First, experiments in which anterior cells adjacent to an RA bead are assayed for ZPA signaling ability show that at 12 h no ZPA signaling is detectable, at 15 h a low level of ZPA signaling is observed, and at 24 h a high level of ZPA signaling is found (Wanek et al. 1991). Since the induction of *Shh* expression by RA does not occur until 24 h after grafting (Riddle et al. 1993; Niswander et al. 1994), the progressive development of ZPA signaling by anterior cells appears to follow the three signaling states identified by mapping ZPA signaling in the mouse limb bud: ZPA signaling without *Shh* expression followed by ZPA signaling with *Shh* expression. Second, when anterior limb bud cells from the mouse are cultured in the absence of exogenous RA, and also in serum that has been treated to remove heparin-binding proteins, they become converted into ZPA signaling cells within 24 h (Anderson et al. 1994). This ZPA signaling is not associated with *Shh* expression, but is associated with the induction of an ectopic expression domain of *Hoxd11* (discussed in Sect. D below) in anterior limb bud cells (A.D. Reginelli and R. Anderson, unpublished). This type of ZPA signaling corresponds to that of the central–proximal region of the mouse limb bud, suggesting that RA is not required for this type of conversion. We also find that FGF-2 acts to inhibit this in vitro anterior to ZPA signaling conversion; however, once this conversion has occurred, FGF-2 maintains these converted cells in a signaling state. The activity of FGF-2 on anterior limb bud cells suggests that FGF-2 maintains positional states throughout the limb bud; in the posterior region, FGF-2 maintains ZPA signaling/*Shh* expression, and in the anterior region FGF-2 maintains an anterior non-ZPA signaling state (Anderson et al. 1993, 1994). These studies point to a view of limb development in which the ZPA signaling states of mesenchymal cells are highly dynamic and are regulated by growth factors produced by the ectoderm.

D. Homeobox-Containing Genes and Positional Information

A growing number of homeobox-containing genes are known to be expressed in the early limb bud in a position-specific manner. These include members of the *Hox* gene clusters (*HoxA*, *HoxB*, *HoxC*, *HoxD*), as well as orphan genes such as the *Msx* genes (*Msx-1* and *Msx-2*) and *Evx-1*. The expression patterns of many of these genes have been linked to the activities of the AER and /or the ZPA and are thus implicated in the process of pattern specification. A summary of this subset of position-specific limb genes is provided in Table 1.

A number of genes are expressed at the distal tip of the limb bud and are known to be dependent on the AER; AER removal results in the downregulation of expression and/or the AER stimulates the upregulation of gene expression. These genes include both *Msx* genes (*Msx-1* and *Msx-2*), *Evx-1*,

HoxA13, HoxD11, and *HoxD13* (Coelho et al. 1991, 1993; Davidson et al. 1991; Robert et al. 1991; Ros et al. 1992; Niswander and Martin 1993b; Yokouchi et al. 1993; Hayamizu et al. 1994). Of these AER-related genes, there is evidence that the expression patterns of *Msx-1*, *Evx-1*, and *HoxD13* are maintained in the absence of the AER by FGF-2 or FGF-4 (Niswander and Martin 1993b; Watanabe and Ide 1993; Wang and Sassoon 1995; G. Taylor and A.D. Reginelli, unpublished). While the expression of these apical genes can be shown to be dependent on the AER during early limb bud stages, in some cases expression is maintained during later limb stages after the AER has disappeared. For example, both *Msx-1* and *Msx-2* remain expressed in cells at the apex of the limb during the formation of the distal digit tip and continue to be expressed by cells of the nailbed (Reginelli et al. 1995b). These observations lead to the important conclusion that the regulation of these apical genes changes with developmental stage.

The 5′ genes in the *HoxD* gene clusters that are expressed in the developing limb bud have been most extensively studied with respect to the anterior–posterior patterning of the limb because these genes are known to be responsive to the ZPA signal. The 5′ genes of the *HoxD* cluster are expressed in defined and overlapping domains along the anterior–proximal to posterior–distal limb axes during both mouse and chick development (See Duboule 1992). The *HoxD* genes in this cluster that are expressed in the limb include *HoxD9, HoxD10, HoxD11, HoxD12,* and *HoxD13*. During development, activation of these genes is sequential and colinear with the position of each gene in the cluster, and the spatial pattern of expression of *HoxD* genes in the limb bud nested with the 3′ genes displaying the largest region of expression and each gene located more 5′ displaying a more restricted pattern. This pattern of expression spans both the anterior–posterior and proximal–distal limb axes, with the 5′ genes expressed in progressively more posterior–distal regions. Thus the anterior–proximal limb bud is defined by the expression of only *HoxD9* and the posterior–distal limb bud is defined by the expression of all five members of this gene cluster. ZPA grafting studies show that the expression pattern of the 5′ *HoxD* genes is altered in association with the formation of supernumerary digits; grafts of ZPA tissue into an anterior site result in the mirror-image expression patterns of the *HoxD* genes that precede the formation of the mirror-image pattern of supernumerary digits (Izpisua-Belmonte et al. 1991; Nohno et al. 1991). The induction of the 5′ *HoxD* genes in response to the ZPA signal does not occur in the absence of the AER (Izpisua-Belmonte et al. 1992b) although it is not clear at this time whether this is because ZPA signaling is dependent on the AER or whether anterior cells require both the AER signal and the ZPA signal to induce the *HoxD* genes.

It has been proposed that expression of *Hox* genes within the limb bud establishes a system of positional information important for pattern specification (Yokouchi et al. 1991; Tabin 1992). In this view, positional values in the limb bud would be defined by the combinatorial expression of *Hox* genes and would specify specific morphological characteristics of the limb. The existence of distinct positional information within the limb is known primarily from results in which the final anatomy of the limb has been experimentally altered. The general concept that *Hox* genes are implicated in positional information is supported by the observation that changes in *Hox* gene expression are predictable and precede the formation of supernumerary limb structures resulting from grafts of ZPA signaling tissue or the AER. Further support for this view comes from experiments in which the ectopic expression of two *Hox* genes, *HoxD11* and *HoxB8*, results in dramatic changes in the limb pattern. *HoxD11* is normally expressed in the posterior half of the limb bud, and *HoxB8* is transiently expressed in the posterior limb bud at early stages. Misexpression of *HoxD11* across the entire chick limb bud results in a transformation of hindlimb digit 1 into a digit 2 (digit sequence of **2**-2-3-4), suggesting that the combinatorial expression domains of the *HoxD* genes specify digit identity (Morgan et al. 1992). Ectopic expression of *HoxB8* under the control of the *RARβ2* promoter in transgenic mouse embryos results in the misexpression of *HoxB8* in the anterior of the early limb bud, the formation of an anterior *Shh* expression domain, and the production of supernumerary digits (Charite et al. 1994). This latter study suggests that the involvement of the *HoxB8* gene in digit specification is indirect; *HoxB8* induces a ZPA signaling region that leads eventually to digit specification. These ectopic expression studies suggest two ways in which *Hox* genes may be involved in digit specification. The first is a direct influence in which the *HoxD* genes act in a cell autonomous manner to specify digit identity. The second is an indirect influence in which the *HoxB8* gene acts to alter cell signaling so as to influence the types of cell–cell interactions that would lead eventually to the specification of digits.

Studies in which specific *Hox* genes are knocked out by gene targeting suggest that *Hox* genes do not directly influence pattern specification, but rather guide cell–cell interactions that are important for the eventual morphogenesis of limb structures. Mutant transgenic mice with a disruption of *HoxA11*, *HoxD11*, or *HoxD13* have been created, and in each case there is no evidence for homeosis of limb skeletal elements. Instead, mutant limbs display defects in the morphogenesis of skeletal elements that are restricted to regions of the limb that fate map to the respective *Hox* expression domains in the developing limb bud (Dolle et al. 1993; Small and Potter 1993; Davis and Capecchi 1994). The types of skeletal defects include malformations of the size, shape, and articulation of specific elements rather than the loss or transformation of specific skeletal elements. Of these three mutants, disruption of *HoxD13* results in limbs which display the most severe phenotype; abnormalities are variable and include the formation of a posterior ectopic digit, loss or reduced phalangeal elements, reduced metatarsal/metacarpals, and joint fusion (Dolle et al. 1993). The most significant conclusion that arises from an analysis of these mutant mice is that individual *Hox* genes are not directly linked to the specification of specific skeletal components. Thus the model that has emerged is that *Hox* genes modulate regional growth during limb formation, the result of which is a modification of the skeletal pattern (see SONG et al. 1993; DAVIS and CAPECCHI 1994; DUBOULE 1994). The evidence that *Hox* genes are involved in establishing positional information and co-ordinating the pattern of cell proliferation within the limb bud supports an intercalation model for limb patterning in which *Hox* genes define positional information and the interaction between limb cells with different positional information regulates cell proliferation. Such an intercalation model for vertebrate limb formation has been proposed within the context of the polar coordinate model (see BRYANT and MUNEOKA 1986).

E. Digit Morphogenesis

Later stages of limb development are characterized by the formation and separation of the digits. At this stage of development, the AER has regressed and the specification of the entire skeletal pattern, including digits, is thought to be largely complete. As in the early limb bud, the timing of pattern specification is deduced from AER removal studies in the chick, and since AER function is linked to the activities of FGF family members, the timing of pattern specification hinges on the availability and stability of FGF in the limb bud. Our understanding of the localization of FGF proteins during later stages of limb formation, however, is very limited (see JOSEPH-SILVERSTEIN et al. 1989; SAVAGE et al. 1993; DONO and ZELLER 1994) and represents an important area for future studies. As limb development proceeds, the autopodial plate becomes dorsally/ventrally flattened and fans out with chondrogenic rays that represent the forming digits. In the mouse, most of the proximal regions of each digital ray fate-map to the metatarsal/metacarpal of the mature digit, with the phalangeal elements forming from only the distal regions of each ray (MUNEOKA et al. 1989). In the mouse and chick, individual digit rudiments can develop in isolation, indicating that each digit rudiment is an autonomously developing morphogenetic field. The primary events of digit morphogenesis are the distal elongation of the digit blastema that results from apical growth in combination with the programmed death of cells within the interdigital regions that separate the digits. In the mouse, following digit separation there is a secondary epidermal fusion event, so that at birth all of the digits are fused with one another (MACONNACHIE 1979). Five days after birth, digit formation is completed with the reseparation of the digits.

In the mouse, digit morphogenesis includes distal growth of the digit rudiment, the proximal to distal differentiation of phalangeal skeletal elements, and the formation of terminal digit structures (the nail on the dorsal surface and the fat pad that is distal and ventral at the digit apex). Within a few days after birth, hair develops along the dorsal surface of each digit. Digit morphogenesis is associated with the patterned expression of a number of homeobox-containing genes, although only the expression patterns of the two

Msx genes, *Msx-1* and *Msx-2* have been characterized in any detail (Reginelli et al. 1995b). *Msx-1* is expressed apically in the limb bud and continues to be expressed at the apex of each developing digit rudiment. Expression is observed in the distal mesenchyme during digit elongation and becomes restricted to the mesodermally derived cells of the loose connective tissue of the nailbed in the differentiated digit. The mesenchymal component of *Msx-2* expression during digit formation partially overlaps with the *Msx-1* expression domain, but is more distally restricted. *Msx-2* expression is associated with distal mesenchymal cells and also the cells of the distal ectoderm at the digit apex. As digit formation is completed, the expression of *Msx-2* persists and becomes localized to the ectodermally derived cells of the nailbed matrix and also to the mesodermally derived cells of the loose connective tissue of the nailbed. Thus the expression domains of both *Msx-1* and *Msx-2* are apically restricted during digit morphogenesis and become localized specifically to the nailbed in the mature digit.

Digit morphogenesis also involves a response to the programmed death of cells that occurs in the interdigital regions. The interdigital regions of cell death are apoptotic (Garcia-Martinez et al. 1993; Naruse et al. 1994; Tone et al 1994; Zakeri et al. 1994) and characterized by an invasion of monocyte-derived macrophages that clear away cellular debris (Hopkinson-Woolley et al. 1994). The decision for interdigital cells to proceed along an apoptotic course appears to occur relatively late in development, since distinct differences between digit-forming regions and interdigital regions with respect to the cell death repressor gene *Bcl-2* are not apparent until autopodial rays have formed (Novack and Korsmeyer 1994). In experiments in the chick, ectopic digit formation by interdigital cells can be induced by a variety of manipulations including simple injury, indicating that interdigital cells possess the capacity to form digit tissues (see Hinchliffe and Horder 1993; Ganan et al. 1993). Similarly, interdigital cells of the developing mammalian limb bud have the potential to differentiate into skeletal and loose connective tissues when allowed to develop in an ectopic site (Lee et al. 1993). Thus the cells of the interdigital region have the potential to participate in digit formation, although many, if not all, of these cells die as the individual digits separate. The cellular signals that modulate cell death within the developing limb bud are unknown, although RA and FGF-2 have been suggested as important regulatory components (see Dolle et al. 1989b; MacCabe et al. 1991). With respect to digit morphogenesis, it is unclear how the lateral regions of each digit develop and what role, if any, interdigital cells play in this process. Are some cells of the interdigital region active participants in morphogenesis or do they all die, resulting in a type of developmental wound that undergoes healing by digit-forming cells? Since a number of developmentally important genes that are implicated in the control of morphogenesis are also expressed by interdigital cells, e.g., *Msx-1* and *Msx-2* (see Reginelli et al. 1995), and *RARβ2* (see Smith et al. 1995), it is possible that some interdigital cells avoid death and participate in digit formation.

As a final note, an interesting and poorly studied characteristic of mammalian digits is their ability to regenerate following distal amputation. While digit regeneration is known to occur in humans and primates (see ILLINGSWORTH 1974; SINGER et al. 1987), digit regeneration in the mouse is the best characterized example (BORGENS 1982; REVARDEL and CHEBOUKI 1987; REGINELLI et al. 1995; ZHAO and NEUFELD 1995). Mammalian digit regeneration is restricted to amputation of tissues distal to the nailbed, and ablation studies suggest that the presence of nailbed tissue at the amputation wound is critical for successful regeneration (ZHAO and NEUFELD 1995). The discovery that the cells of the nailbed express *Msx-1* and *Msx-2,* coupled with the finding that both genes are expressed by digit cells during regeneration, implicates these two apically expressed homeobox-containing genes in the regenerative process (REGINELLI et al. 1995). When we consider that higher vertebrates, such as mammals, are largely regeneration defective, it is amazing that cells present in the tips of your digits can respond to traumatic injury by healing without the formation of scar tissue and by restoring a functional and perfect replacement of the digit tip. The observation that mammalian digits can regenerate is much more than a curiosity; it provides the clearest demonstration that mature cells have the capability to respond to injury by reiterating developmental events important for the coordination of growth and pattern respecification.

F. Conclusions

The developing limb is one of the best-studied vertebrate organ systems and, although our understanding of how a limb forms is far from complete, we can begin to put together a rudimentary network of molecular interactions important for limb formation (Fig. 5). This molecular network exists against a backdrop in which limb bud cells are dynamically labile, capable of changing their positional character in a proximal to distal and anterior to posterior sequence. This change in positional character can be modulated by RA, or other active retinoids, and by FGF. In the model shown in Fig. 5, the control of cell proliferation is proposed to be key to the regulation of limb outgrowth, and cell proliferation is regulated by positional interactions that occur between cells that express different combinations of *Hox* genes. The three known signaling regions of the limb bud are the AER, the dorsal ectoderm, and the ZPA signaling posterior bud, and in each case a signal important for limb morphogenesis has been identified. The AER is the site of production of three members of the FGF family of growth factors, FGF-2, FGF-4, and FGF-8, and these FGFs are implicated in the regulation of proximal–distal outgrowth and anterior–posterior patterning. The dorsal ectoderm is the site of WNT-7a production, and WNT-7a appears to control patterning in the dorsal–ventral and anterior–posterior limb axes. The expression of *Shh* is associated with ZPA signaling cells at the extreme posterior border of the limb bud and, since ZPA signaling can occur without *Shh* expression, we anticipate the existence of

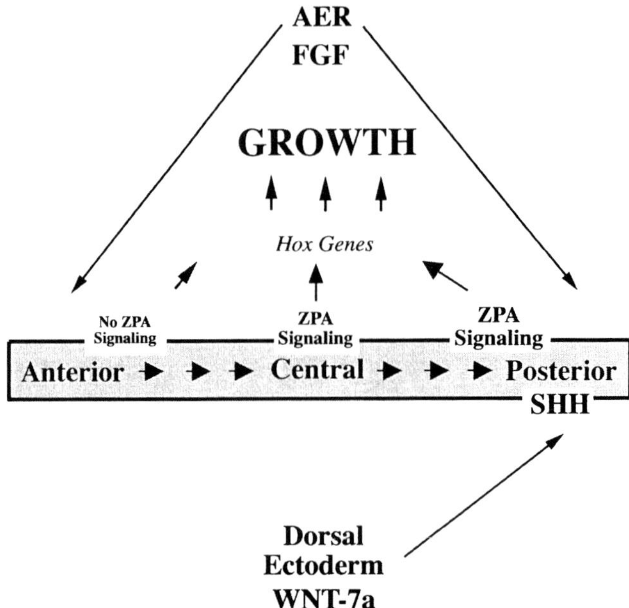

Fig. 5. A network of molecular interactions is emerging from patterning studies on the developing vertebrate limb. Limb outgrowth is regulated by morphogenetic signals produced by the apical ectodermal ridge (AER), the dorsal ectoderm, and the zone of polarizing activity (ZPA). Three members of the fibroblast growth factor (FGF) family, FGF-2, FGF-4, and FGF-8, have been implicated in the initiation and maintenance of limb bud outgrowth by the AER, although it is unclear at present whether each of these factors has a distinct role. The dorsal ectoderm is the site of production of WNT-7a, a member of the WNT family of secreted signals which determines some of the characteristics of dorsal limb pattern. The sonic hedgehog protein SHH, is produced and secreted by posterior limb bud cells that map to regions of highest ZPA signaling. A role for SHH in limb pattern formation is deduced from experiments showing that the ectopic expression of SHH in the anterior limb bud causes the formation of a supernumerary set of digits. Both FGF and WNT-7a are known to regulate the expression of *Shh* by posterior limb bud cells, thus pattern formation of all three limb axes appear to be interdependent. These three signaling molecules are proposed to regulate the spatial pattern of *Hox* gene expression, which defines regions of distinct positional values within the limb bud. During limb outgrowth, cell proliferation is thought to be stimulated at the bud apex by intercalary interactions between cells with distinct positional values. In this manner, the expression of *Hox* genes is indirectly linked to the regulation of cell proliferation during limb development

additional components of the ZPA signal. The action of FGF is proposed to be crucial to limb outgrowth, since FGF stabilizes the positional character of limb bud cells; they are indirectly implicated in the control of *Hox* gene expression and in the regulation of cell proliferation. Once the basic pattern of the limb is specified, morphogenesis of the limb continues until digit formation is complete. Digit formation is highlighted by the apical expression of the *Msx* genes that continue to be expressed in nailbed tissue and appear to play a role in the promotion of regeneration following amputation of the digit tip. In conclusion, studies on the developing vertebrate limb have identified a number of key components of a highly complex signaling network that modulate how cells utilize spatial information during morphogenesis. As we begin to unravel the details of this molecular network, we will begin to better understand how intrinsic and extrinsic agents interface with this network to alter normal limb morphogenesis.

Acknowledgements. We thank our colleagues Angela Reginelli, Shaoguang Li, and Gail Taylor for thoughtful discussions and for allowing us to present data prior to publication. This work was supported by HD 23921 and by a Grant-in-Aid from the American Heart Association, Louisiana Affiliate, Inc.

References

Anderson R, Landry M, Muneoka K (1993) Maintenance of ZPA signaling in cultured mouse limb bud cells. Development 117: 1421–1433

Anderson R, Landry M, Reginelli A, Taylor G, Achkar C, Gudas L, Muneoka K (1994) Conversion of anterior limb bud cells to ZPA signaling cells in vitro and in vivo. Dev Biol 164: 241–257

Aono H, Ide H (1988) A gradient of responsiveness to the growth promoting activity of ZPA (zone of polarizing activity) in the chick limb bud. Dev Biol 128: 136–141

Balkan W, Colbert M, Bock C, Linney E (1992) Transgenic indicator mice for studying activated retinoic acid receptors during development. Proc Natl Acad Sci USA 89: 3347–3351

Borgens RB (1982) Mice regrow the tips of their foretoes. Science 217: 747–750

Bowen J, Hinchliffe JR, Horder TJ, Reeve AMR (1989) The fate map of the chick forelimb-bud and its bearing on hypothesized developmental control mechanisms. Anat Embryol 179: 269–283

Bryant SV, Gardiner DM (1992) Retinoic acid, local cell-cell interactions, and pattern formation in vertebrate limbs. Dev Biol 152: 1–25

Bryant SV, Muneoka K (1986) Views on limb development and regeneration. Trends Genet 2(6): 153–159

Bumcrot DA, Takada R, McMahon AP (1995) Proteolytic processing yields two secreted forms of sonic hedgehog. Mol Cell Biol 15: 2294–2303

Carlson BM (1984) The preservation of the ability of cultured quail wing bud mesoderm to elicit a position-related differentiative response. Dev Biol 101: 106–115

Carrington JL, Fallon JF (1986) Experimental manipulation leading to induction of dorsal ectodermal ridges on normal limb buds results in a phenocopy of the eudiplopodia chick mutant. Dev Biol 116: 130–137

Carrington JL, Fallon JF (1988) Initial limb budding is independent of apical ectodermal ridge activity; evidence from a limbless mutant. Development 104: 361–367

Chang DT, Lopez A, von Kessler DP, Chang C, Simandl RK, Zhao R, Selden MF, Fallon JF, Beachy PA (1994) Products, genetic linkage, and limb patterning activity of a murine *hedgehog* gene. Development 120: 3339–3353

Charite J, de Graaff W, Shen S, Deschamps J (1994) Ectopic expression of *Hoxb-8* causes duplication of the ZPA in the forelimb and homeotic transformation of axial structures. Cell 78: 589–601

Cockroft D (1990) Dissection and culture of post-implantation embryos. In: Copp A, Cockroft D (eds) Post-implantation mammalian embryo. IRL Press, Oxford, pp 15–40

Coelho CND, Krabbenhoft KM, Upholt WB, Fallon JF, Kosher RA (1991) Altered expression of the chicken homeobox-containing genes GHox-7 and GHox-8 in the limb buds of *limbless* mutant chick embryos. Development 113: 1487–1493

Coelho CND, Sumoy L, Kosher RA, Upholt WB (1992) GHox-7: a chicken homeobox-containing gene expressed in a fashion consistent with a role in patterning events during embryonic chick limb development. Differentiation 49: 85–92

Coelho CND, Upholt WB, Kosher RA (1993) Ectoderm from various regions of the developing chick limb bud differentially regulates the expression of the chicken homeobox-containing genes Ghox-7 and Ghox-8 by limb mesenchymal cells. Dev Biol 156: 303–306

Cohn MJ, Izpisua-Belmonte J-C, Abud H, Heath JK, Tickle C (1995) Fibroblast growth factors induce additional limb development from the flank of chick embryos. Cell 80: 739–746

Crossley PH, Martin GR (1995) The mouse *Fgf8* gene encodes a family of polypeptides and is expressed in regions that direct outgrowth and patterning in the developing embryo. Development 121: 439–451

Davidson DR, Crawley A, Hill RE, Tickle C (1991) Position-dependent expression of two related homeobox genes in developing vertebrate limbs. Nature 352: 429–431

Davis AP, Capecchi MR (1994) Axial homeosis and appendicular skeleton defects in mice with a targeted disruption of *hoxd-11*. Development 120: 2187–2198

Dealy CN, Roth A, Ferrari D, Brown AMC, Kosher RA (1993) *Wnt-5a* and *Wnt-7a* are expressed in the developing chick limb bud in a manner suggesting roles in pattern formation along the proximodistal and dorsoventral axes. Mech Dev 43: 175–186

Dolle P, Izpisua-Belmonte J-C, Brown JM, Tickle C, Duboule D (1991) *Hox-4* genes and the morphogenesis of vertebrate genitalia. Genes Dev 5: 1767–1776

Dolle P, Izpisua-Belmonte J-C, Falkenstein H, Renucci A, Duboule D (1989a) Coordinate expression of the murine Hox-5 complex homeobox-containing genes during limb pattern formation. Nature 342: 767–772

Dolle P, Ruberte E, Kastner P, Petkovich M, Stoner CM, Gudas LJ, Chambone P (1989b) Differential expression of genes encoding alpha, beta, and gamma retinoic acid receptors and CRABP in the developing limbs of the mouse. Nature 342: 702–705

Dolle P, Dierich A, LeMeur M, Schimmang T, Schuhbaur B, Chambon P, Duboule D (1993) Disruption of the *HoxD-13* gene induces localized heterochrony leading to mice with neotenic limbs. Cell 75: 431–441

Dono R, Zeller R (1994) Cell-type-specific nuclear translocation of fibroblast growth factor-2 isoforms during chicken kidney and limb morphogenesis. Dev Biol 163: 316–330

Duboule D (1992) The vertebrate limb: a model system to study the Hox/HOM gene network during development and evolution. Bioessays 14(6): 375–384

Duboule D (1994) How to make a limb? Science 266: 575–576

Echelard Y, Epstein DJ, St-Jacques B, Shen L, Mohler J, McMahon JA, McMahon AP (1993) *Sonic hedgehog*, a member of a family of putative signaling molecules, is implicated in the regulation of CNS polarity. Cell 75: 1417–1430

Fallon JF, Lopez A, Ros MA, Savage MP, Olwin BB, Simandl BK (1994) FGF-2 apical ectodermal ridge growth signal for chick limb development. Science 264: 104–107

Fan M-M, Porter JA, Chiang C, Chang DT, Beachy PA, Tessler-Lavigne M (1995) Long-range sclerotome induction by *sonic hedgehog*: direct role of the amino-terminal cleavage product and modulation by the cyclic AMP signaling pathway. Cell 81: 457–465

Ganan Y, Macias D, Garcia-Martinez V, Hurle JM (1993) In vivo experimental induction of interdigital tissue chondrogenesis in the avian limb bud results in the formation of extradigits. Effects of local microinjection of staurosporine, zinc chloride and growth factors. In: Goetinck P, Fallon J, Stocum D, Kelley R (eds) 4th international conference on limb development and regeneration. Wiley, New York, pp 127–139

Garcia-Matinez V, Macias D, Ganan Y, Garcia-Lobo JM, Francia MV, Fernandez-Teran MA, Hurle JM (1993 Internucleosomal DNA fragmentation and programmed cell death (apoptosis) in the interdigital tissue of the embryonic chick leg bud. J Cell Sci 106: 201–208

Geduspan JS, MacCabe JA (1987) The ectodermal control of mesodermal patterns of differentiation in the developing chick wing. Dev Biol 124: 398–408

Geduspan JS, MacCabe JA (1989) Transfer of dorsoventral information from mesoderm to ectoderm at the onset of limb development. Anat Rec 224: 79–87

Goetinck PF (1964) Studies on limb morphogenesis. II. Experiments with the polydactylous mutant *eudiplopodia*. Dev Biol 10: 71–91

Hamburger V, Hamilton HL (1951) A series of normal stages in the development of the chick embryo. J Morphol 88: 49–92

Hayamizu TF, Bryant SV (1992) Retinoic acid respecifies limb bud cells in vitro. J Exp Zool 263: 423–429

Hayamizu TF, Sessions SK, Wanek N, Bryant SV (1991) Effects of localized application of TGFβ1 on developing chick limbs. Dev Biol 145: 164–173

Hayamizu TF, Wanek N, Taylor G, Trevino C, Shi C, Anderson R, Gardiner DM, Muneoka K, Bryant SV (1994) Regeneration of *HoxD* expression domains during pattern regulation in chick wing buds. Dev Biol 161: 504–512

Heikinheimo M, Lawshe A, Shackleford GM, Wilson DB, MacArthus CA (1994) *Fgf-8* expression in the postgastrulation mouse suggests roles in the development of the face, limbs and central nervous system. Methods Dev 48: 129–138

Hill RE, Jones PF, Rees AR, Sime CM, Justice MJ, Copeland NG, Jenkins NA, Graham E, Davidson DR (1989) A family of mouse homeobox-containing genes: molecular structure, chromosomal location, and developmental expression of *Hox 7.1*. Genes Dev 3: 26–37

Hinchliffe JR, Horder TJ (1993) Lessons from extra digits. In: Goetinck P, Fallon J, Stocum D, Kelley R (eds) 4th international conference on limb development and regeneration. Wiley, New York, pp 113–126

Honig LS (1983a) Does anterior (non-polarizing region) tissue signal in the developing chick limb? Dev Biol 97: 424–432

Honig LS (1983b) Polarizing activity of the avian limb examined on a cellular basis. In: Fallon JF, Caplan AI (eds) Limb development and regeneration, part A. Liss, New York, pp 99–108

Honig LS, Summerbell D (1985) Maps of strength of positional signalling activity in the developing chick limb bud. J Embryol Exp Morphol 87: 163–174

Honig LS, Smith JC, Hornbruch A, Wolpert L (1981) Effects of biochemical inhibitors on positional signalling in the chick limb bud. J Embryol Exp Morphol 62: 203–216

Hopkinson-Woolley J, Hughes D, Gordon S, Martin P (1994) Macrophage recruitment during limb development and wound healing in the embryonic and foetal mouse. J Cell Sci 107: 1159–1167

Hornbruch A, Wolpert L (1986) Positional signaling by Hensen's node when grafted to the chick limb bud. J Embryol Exp Morphol 94: 257–265

Illingsworth CM (1974) Trapped fingers and amputated fingertips in children. J Pediatr Surg 9: 853–858

Ingham PW, Martinez-Arias A (1992) Boundaries and fields in early embryos. Cell 68: 221–235

Izpisua-Belmonte J-C, Tickle C, Dolle P, Wolpert L, Duboule D (1991) Expression of the homeobox *Hox-4* genes and the specification of position in chick wing development. Nature 350: 585–589

Izpisua-Belmonte J-C, Brown JM, Crawley A, Duboule D, Tickle C (1992a) *Hox-4* gene expression in mouse/chicken heterospecific grafts of signalling regions to limb buds reveals similarities in patterning mechanisms. Development 115: 553–560

Izpisua-Belmonte J-C, Brown JM, Duboule D, Tickle C (1992b) Expression of *Hox-4* genes in the chick wing links pattern formation to the epithelial-mesenchymal interactions that mediate growth. EMBO J 11: 1451–1457

Jabs EW, Müller U, Li X, Ma L, Luo W, Haworth IS, Klisak I, Sparkes R, Warman ML, Mulliken JB, Snead ML, Maxson R (1993) A mutation in the homeodomain of the human MSX2 gene in a family affected with autosomal dominant craniosynostosis. Cell 75: 443–450

Javois LC, Iten LC (1986) The handedness and origin of supernumerary limb structures following 180° rotation of the chick wing bud on its stump. J Embryol Exp Morphol 91: 135–152

Johnson DE, Williams LT (1993) Structural and functional diversity in the FGF receptor multigene family. Adv Cancer Res 60: 1–41

Joseph-Silverstein J, Consigli SA, Lyser KM, Ver Pault C (1989) Basic fibroblast growth factor in the chick embryo: immunolocalization to striated muscle cells and their precursors. J Cell Biol 108: 2459–2466

Krauss S, Concordet J-P, Ingham PW (1993) A functionally conserved homolog of the *Drosophila* segment polarity gene *hh* is expressed in tissues with polarizing acivity in zebrafish embryos. Cell 75: 1431–1444

Laufer E, Nelson CE, Johnson RL, Morgan BA, Tabin C (1994) *Sonic hedgehog* and Fgf-4 act through a signaling cascade and feedback loop to integrate growth and patterning of the developing limb bud. Cell 79: 993–1003

Lee JJ, Ekker SC, von Kessler DP, Porter JA, Sun BI, Beachy PA (1994) Autoproteolysis in *hedgehog* protein biogenesis. Science 266: 1528–1537

Lee KKH, Chan WY, Sze LY (1993) Histogenetic potential of rat hind-limb interdigital tissues prior to and during the onset of programmed cell death. Anat Rec 236: 568–572

Li S, Anderson R, Reginelli AD, Muneoka K (1996) FGF-2 influences cell movements and gene expression during limb development. J Exp Zool 274: 234–247

Lopez-Martinez A, Chang DT, Chiang C, Porter JA, Ros MA, Simandl BK, Beachy PA, Fallon JF (1995) Limb-patterning activity and restricted posterior localization of the amino-terminal product of *Sonic hedgehog* cleavage. Curr Biol 5: 791–796

MacCabe AB, Gasseling MT, Saunders JW Jr (1973) Spatiotemporal distribution of mechanisms that control outgrowth and anteroposterior polarization of the limb bud in the chick embryo. Mech Ageing Dev 2: 1–12

MacCabe JA, Blaylock RL Jr, Latimer JL, Pharris LJ (1991) Fibroblast growth factor and culture in monolayer rescue mesoderm cells destined to die in the developing avian wing. J Exp Zool 257: 208–213

MacCabe JA, Errick J, Saunders JW Jr (1974) Ectodermal control of the dorsoventral axis in the leg bud of the chick embryo. Dev Biol 39: 69–82

Maconnachie E (1979) A study of digit fusion in the mouse embryo. J Embryol Exp Morphol 49: 259–276

Mahmood R, Bresnick J, Hornbruch A, Mahony C, Morton N, Colquhoun K, Martin P, Lumsden A, Dickson C, Mason I (1995) A role for FGF-8 in the initiation and maintenance of vertebrate limb bud outgrowth. Curr Biol 5: 797–806

Marti E, Takada R, Bumcrot DA, Sasaki H, McMahon AP (1995) Distribution of *Sonic hedgehog* peptides in the developing chick and mouse embryo. Development 121: 2537–2547

Martin P, Lewis J (1986) Normal development of the skeleton in chick limb buds devoid of dorsal ectoderm. Dev Biol 118: 233–246

Mendelsohn C, Ruberte E, LeMeur M, Morriss-Kay G, Chambon P (1991) Developmental analysis of the retinoic acid-inducible RAR-b2 promoter in transgenic animals. Development 113: 723–734

Mima T, Ohuchi H, Noji S, Mikawa T (1995) FGF can induce outgrowth of somatic mesoderm both inside and outside of limb-forming regions. Dev Biol 167: 617–620

Morgan BA, Izpisua-Belmonte J-C, Duboule D, Tabin CJ (1992) Targeted misexpression of *Hox 4.6* in the avian limb bud causes apparent homeotic transformations. Nature 358: 236–239

Munaim SI, Klagsbrun M, Toole BP (1988) Developmental changes in fibroblast growth factor in the chicken embryo limb bud. Proc Natl Acad Sci USA 85: 8091–8093

Muneoka K, Wanek N, Bryant SV (1989) Mammalian limb bud development: in situ fate maps of early hindlimb buds. J Exp Zool 249: 50–54

Muneoka K, Wanek N, Trevino C, Bryant SV (1990) *Exo utero* surgery. In: Copp A, Cockroft D (eds) Post-implantation mammalian embryo. IRL Press, Oxford, pp 41–60

Naruse I, Keino H, Kawarada Y (1994) Antibody against single-stranded DNA detects both programmed cell death and drug-induced apoptosis. Histochemistry 101: 73–78

Niswander L, Martin GR (1992) Fgf-4 expression during gastrulation, myogenesis, limb and tooth development in the mouse. Development 114: 755–768

Niswander L, Martin GR (1993a) FGF-4 and BMP-2 have opposite effects on limb growth. Nature 361: 68–71

Niswander L, Martin GR (1993b) FGF-4 regulates expression of *Evx-1* in the developing mouse limb. Development 119: 287–294

Niswander L, Tickle C, Vogel A, Booth I, Martin GR (1993) FGF-4 replaces the apical ectodermal ridge and directs outgrowth and patterning of the limb. Cell 75: 579–587

Niswander L, Jeffrey S, Martin GR, Tickle C (1994) A positive feedback loop coordinates growth and patterning in the vertebrate limb. Nature 371: 609–612

Nohno T, Noji S, Koyama E, Myokai F, Kuroiwa A, Saito T, Taniguchi S (1991) Involvement of the *Chox-4* chicken homeobox genes in determination of anteroposterior axial polarity during limb development. Cell 64: 1197–1205

Noji S, Nohno T, Koyama E, Muto K, Ohyama K, Aoki Y, Tamura K, Ohsugi K, Ide H, Taniguchi S, Saito T (1991) Retinoic acid induces polarizing activity but is unlikely to be a morphogen in the chick limb bud. Nature 350: 83–86

Novack DV, Korsmeyer SJ (1994) Bcl-2 protein expression during murine development. Am J Pathol 145: 61–73

Nusse R, Varmus H (1992) Wnt genes. Cell 69: 1073–1087

Ohuchi H, Yoshioka H, Tanaka A, Kawakami Y, Nohno T, Noji S (1994) Involvement of androgen-induced growth factor (FGF-8) gene in mouse embryogenesis and morphogenesis. Biochem Biophys Res Commun 204: 882–888

Ono T, Murakami T, Mochii M, Agata K, Kino K, Otsuka M, Ohta M, Mizutani M, Yoshida M, Eguchi G (1994) A complete culture system for avian transgenesis, supporting quail embryos from the single-cell stage to hatching. Dev Biol 161: 126–130

Orr-Urtreger A, Givol D, Yayon A, Yarden Y, Lonai P (1991) Developmental expression of two murine fibroblast growth factor receptors, *flg* and *bek*. Development 113: 1419–1434

Papaioannou VE (1990) In utero manipulations. In: Copp A, Cockroft D (eds) Post-implantation mammalian embryo. IRL Press, Oxford, pp 61–80

Parr BA, McMahon AP (1995) Dorsalizing signal *Wnt-7a* required for normal polarity of D-V and A-P axes of mouse limb. Nature 374: 350–353

Parr BA, Shea MJ, Vassileva G, McMahon AP (1993) Mouse *Wnt* genes exhibit discrete domains of expression in the early embryonic CNS and limb buds. Development 119: 247–261

Paulsen DF, Chen W-D, Pang L, Johnson B, Okello D (1994) Stage- and region-dependent chondrogenesis and growth of chick wing-bud mesenchyme in serum-containing and defined tissue culture media. Dev Dyn 200: 39–52

Pautou P-P (1977) Dorso-ventral axis determination of chick limb bud development. In: Ede DA, Hinchliffe JR, Balls M (eds) Vertebrate limb and somite morphogenesis. Cambridge University Press, Cambridge, pp 257–266

Perry MM (1988) A complete culture system for the chick embryo. Nature 331: 70–72

Peters KG, Werner S, Chen G, Williams LT (1992) Two FGF receptor genes are differentially expressed in epithelial and mesenchymal tissues during limb formation and organogenesis in the mouse. Development 114: 233–243

Porter JA, Ekker SC, Young KE, von Kessler DP, Lee JJ, Moses K, Beachy PA (1995) The product of *hedgehog* autoproteolytic cleavage active in local and long-range signaling. Nature 374: 363–366

Reginelli AD, Wang Y, Sassoon D, Muneoka K (1995) Digit tip regeneration correlates with regions of msx1 (formerly Hox7.1) expression in fetal and newborn mice. Development 121: 1065–1076

Reginelli AD, Anderson R, Li S, Van Way SM, Muneoka K (1996) Regulation of sonic hedgehog expression in ZPA signaling tissue (submitted for publication)

Revardel J-L, Chebouki F (1987) Etude de la reponse à l'amputation des phalanges chez la souris: role morphogenetique des epitheliums, stimulation de la chondrogenese. Can J Zool 65: 3166–3176

Reynolds K, Mezey E, Zimmer A (1991) Activity of the b-retinoic acid receptor promoter in transgenic mice. Mech Dev 36: 15–29

Riddle RD, Johnson RL, Laufer E, Tabin C (1993) *Sonic hedgehog* mediates the polarizing activity of the ZPA. Cell 75: 1401–1416

Riley BB, Savage MP, Simandl BK, Olwin BB, Fallon JF (1993) Retroviral expression of FGF-2 (bFGF) affects patterning in the chick limb bud. Development 118: 95–104

Robert B, Sassoon D, Jacq B, Gehring W, Buckingham M (1989) Hox-7, a mouse homeobox gene with a novel pattern of expression during embryogenesis. EMBO J 8(1): 91–100

Robert B, Lyons G, Simandal B-K, Kuroiwa A, Buckingham M (1991) The apical ectodermal ridge regulates Hox-7 and Hox-8 gene expression in developing limb buds. Genes Dev 5: 2363–2374

Roelink H, Augsburger A, Heemskerk J, Korzh V, Norlin S, Ruizi Altaba A, Tanabe Y, Placzek M, Edlund T, Jessel TM, Dodd J (1994) Floor plate and motor neuron induction by vhh-1, a vertebrate homologue of *hedgehog* expressed by the notochord. Cell 76: 761–775

Roelink H, Porter JA, Chiang C, Tanabe Y, Chang DT, Beachy PA, Jessell TM (1995) Floor plate and motor neuron induction by different concentrations of the amino-terminal cleavage product of *sonic hedgehog* autoproteolysis. Cell 81: 445–455

Ros MA, Lyons G, Kosher RA, Upholt WB, Coelho CND, Fallon JF (1992) Apical ridge dependent and independent mesodermal domains of GHox-7 and GHox-8 expression in chick limb buds. Development 116: 811–818

Rossant J, Zirngibl R, Cado D, Shago M, Giguere V (1991) Expression of a retinoic acid response element-hsplacZ transgene defines specific domains of transcriptional activity during mouse embryogenesis. Gene Dev 5: 1333–1344

Rowe DA, Fallon JF (1982) The proximodistal determination of skeletal parts in the developing chick leg. J Embryol Exp Morphol 68: 1–7

Rutledge JC, Shourbaji AG, Hughes LA, Polifka JE, Cruz YP, Bishop JB, Generoso WM (1994) Limb and lower-body duplication induced by retinoic acid in mice. Proc Natl Acad Sci USA 91: 5436–5440

Satokata I, Maas R (1994) *Msxl* deficient mice exhibit cleft palate and abnormalities of craniofacial and tooth development. Nature Genet 6: 348–357

Saunders JW (1948) The proximo-distal sequence of origin the parts of the chick wing and the role of the ectoderm. J Exp Zool 108: 363–403

Saunders JW Jr, Errick J (1976) Inductive activity and enduring cellular constitution of a supernumerary apical ectodermal ridge grafted to the limb bud of the chick embryo. Dev Biol 50: 16–25

Saunders JW Jr, Gasseling MT (1968) Ectodermal mesenchymal interactions in the origin of limb symmetry. In: Fleischmajer R, Billingham R (eds) Epithelial-mesenchymal interactions. Williams and Wilkins, Baltimore, pp 78–97

Savage MP, Hart CE, Riley BB, Sasse J, Olwin BB, Fallon JF (1993) Distribution of FGF-2 suggests it has a role in chick limb bud growth. Dev Dyn 198: 159–170

Scott WJ Jr, Walter R, Tzimas G, Sass JO, Nau H, Collins MD (1994) Endogenous status of retinoids and their cytosolic binding proteins in limb buds of chick vs mouse embryos. Dev Biol 165: 397–409

Searls RL, Janner M (1971) The initiation of limb bud outgrowth in the embryonic chick. Dev Biol 24: 198–213

Seed J, Olwin BB, Hauschka SD (1988) Fibroblast growth factor levels in the whole embryo and limb bud during chick development. Dev Biol 128: 50–57

Shi C, Muneoka K (1992) Position-specific growth of mouse limb bud cells in vitro. Dev Biol 151: 9–17

Shuey DL, Lau C, Logsdon TR, Zucker RM, Elstein KH, Narotsky MG, Setzer RW, Kavlock RJ, Rogers JM (1994) Biologically based dose-response modeling in developmental toxicology: biochemical and cellular sequelae of 5-fluorouracil exposure in the developing rat. Toxicol Appl Pharmacol 126: 129–144

Singer M, Weckesser EC, Geraudie J, Maier CE, Singer J (1987) Open finger tip healing and replacement after distal amputation in Rhesus monkey with comparison to limb regeneration in lower vertebrates. Anat Embryol 177: 29–36

Singh G, Supp DM, Schreiner C, McNeish J, Merker HJ, Copeland NG, Jenkins NA, Potter SS, Scott W (1991) *legless* insertional mutation: morphological, molecular, and genetic characterization. Genes Dev 5: 2245–2255

Small KM, Potter SS (1993) Homeotic transformations and limb defects in *Hoxa-11* mutant mice. Genes Dev 7: 2318–2328

Smith JC (1979) Evidence for a positional memory in the development of the chick wing bud. J Embryol Exp Morphol 52: 105–113

Smith JC, Tickle C, Wolpert L (1978) Attenuation of positional signaling in the chick limb by high doses of γ-radiation. Nature 272: 612–613

Smith SM, Kirstein IJ, Wang Z-S, Fallon JF, Kelley J, Bradshaw-Rouse J (1995) Differential expression of retinoic acid receptor-b isoforms during chick limb ontogeny. Dev Dyn 202: 54–66

Song K, Wang Y, Sassoon D (1993) Expression of *Hox-7.1* in myoblasts inhibits terminal differentiation and induces cell transformation. Nature 360: 477–481

Stark RJ, Searls RL (1974) The establishment of the cartilage pattern in the embryonic chick wing, and evidence for a role of the dorsal and ventral ectoderm in normal wing development. Dev Biol 38: 51–63

Stephens TD, Spall R, Baker WC, Hiatt SR, Pugmire DE, Shaker MR, Willis HJ, Winger KP (1991) Axial and paraxial influences on limb morphogenesis. J Morphol 208: 367–379

Stephens TD, Baker WC, Cotterell JW, Edwards DR, Pugmire DS, Roberts SG, Shaker MR, Willis HJ, Winger KP (1993) Evaluation of the chick wing territory as an equipotential self-differentiating system. Dev Dyn 197: 157–168

Stocum DL (1991) Limb regeneration: a call to arms (and legs). Cell 67: 5–8

Stopak D, Wessells NK, Harris AK (1985) Morphogenetic rearrangement of injected collagen in developing chicken limb buds. Proc Natl Acad Sci (USA) 82: 2804–2808

Summerbell D (1974a) A quantitative analysis of the effect of excision of the AER from the chick limb bud. J Embryol Exp Morphol 32: 651–660

Summerbell D (1974b) Interaction between the proximo-distal and antero-posterior coordinates of positional value during the specification of positional information in the early development of the chick limb bud. J Embryol Exp Morphol 32: 227–237

Summerbell D (1977) Reduction of the rate of outgrowth, cell density, and cell division following removal of the apical ectodermal ridge of the chick limb-bud. J Embryol Exp Morphol 40: 1–21

Summerbell D (1983) The effects of local application of retinoic acid to the anterior margin of the developing chick limb. J Embryol Exp Morphol 78: 269–289
Summerbell D, Wolpert L (1972) Cell density and cell division in the early morphogenesis of the chick wing. Nature 239: 24–26
Summerbell D, Lewis JH, Wolpert L (1973) Positional information in chick limb morphogenesis. Nature 244: 492–496
Suzuki HR, Sakamoto H, Yoshida T, Sugimura T, Terada M, Solursh M (1992) Localization of *Hst1* transcripts to the apical ectodermal ridge in the mouse embryo. Dev Biol 150: 219–222
Tabin CJ (1992) Why we have (only) five fingers per hand : *Hox* genes and the evolution of paired limbs. Development 116: 289–296
Taylor G, Anderson R, Reginelli AD, Muneoka K (1994) FGF-2 induces regeneration of the chick limb bud. Dev Biol 163: 282–284
Thaller C, Eichele G (1987) Identification and spatial distribution of retinoids in the developing limb bud. Nature 327: 625–628
Tickle C (1981) The number of polarizing region cells required to specify additional digits in the developing chick wing. Nature 289: 295–298
Tickle C, Eichele G (1994) Vertebrate limb development. Annu Rev Cell Biol 10: 121–152
Tickle C, Shellswell G, Crawley A, Wolpert L (1976) Positional signalling by mouse limb polarizing region in the chick wing bud. Nature 259: 396–397
Tickle C, Summerbell D, Wolpert L (1975) Positional signalling and specification of digits in chick limb morphogenesis. Nature 254: 199–202
Tickle C, Albert B, Wolpert L, Lee J (1982) Local application of retinoic acid to the limb bond mimics the action of the polarizing region. Nature 296: 564–566
Tickle C, Lee J, Eichele G (1985) A quantitative analysis of the effect of all-trans-retinoic acid on the pattern of limb development. Dev Biol 109: 82–95
Tone S, Tanaka S, Minatogawa Y, Kido R (1994) DNA fragmentation during programmed cell death in the chick limb buds. Exp Cell Res 215: 234–236
Vogel A, Tickle C (1993) FGF-4 maintains polarizing activity of posterior limb bud cells in vivo and in vitro. Development 119: 199–206
Wagner M, Thaller C, Jessell T, Eichele G (1990) Polarizing activity and retinoid synthesis in the floor plate of the neural tube. Nature 345: 819–822
Wanek N, Bryant SV (1991) Temporal pattern of posterior positional identity in the mouse limb buds. Dev Biol 147: 480–484
Wanek N, Muneoka K, Bryant SV (1989a) Evidence for regulation following amputation and tissue grafting in the developing mouse limb. J Exp Zool 249: 55–61
Wanek N, Muneoka K, Burton R, Holler-Dinsmore G, Bryant SV (1989b) A staging system for mouse limb development. J Exp Zool 249: 41–49
Wanek N, Gardiner DM, Muneoka K, Bryant SV (1991) Conversion by retinoic acid of anterior cells into ZPA cells in the chick limb bud. Nature 350: 81–83
Wang Y, Sassoon D (1995) Ectoderm-mesenchyme and mesenchyme–mesenchyme interactions regulate MSX-1 expression and cellular differentiation in the murine limb bud. Dev Biol 168: 374–382
Watanabe A, Ide H (1993) Basic FGF maintains some characteristics of the progress zone of chick limb bud in cell culture. Dev Biol 159: 223–231
Wolpert L (1969) Positional information and the spatial pattern of cellular differentiation. J Theor Biol 25: 1–47
Yang Y, Niswander L (1995) Interaction between the signaling molecules WNT7a and SHH during vertebrate limb development: dorsal signals regulate anteroposterior patterning. Cell 80: 939–947
Yokouchi Y, Sasaki H, Kuroiwa A (1991) Homeobox gene expression correlates with the bifurcation process of limb cartilage development. Nature 353: 443–445
Yokouchi Y, Yamamoto M, Toyota T, Sasaki H, Kuroiwa A (1993) Regulatory interaction of positional signaling on coordinate expression of homeobox genes in

developing limb buds. In: Fallon JF, Goetinck PF, Kelley RO, Stocum DL (eds) Limb development and regeneration. Wiley-Liss, New York, pp 71–78
Zakeri Z, Quaglino D, Ahuja HS (1994) Apoptotic cell death in the mouse limb and its suppression in the hammertoe mutant. Dev Biol 165: 294–297
Zhao W, Neufeld DA (1995) Bone regrowth in young mice stimulated by nail organ. J Exp Zool 271: 155–159

CHAPTER 4
Axial Skeleton

A. NEUBÜSER and R. BALLING

A. Introduction

The development of the axial skeleton is a complex process that requires the coordinate regulation of many molecular and cellular events. Although we are still far from understanding it in detail, the availability of molecular markers and the use of in vitro culture techniques has recently opened up a new avenue towards a molecular analysis and has already added much to our understanding of the process.

Many of the molecular markers which are now being used to study vertebrate development were originally identified in the fruitfly *Drosophila melanogaster*. In extensive mutagenesis screens, a large number of *Drosophila* mutants with embryonic defects were isolated. The affected genes were subsequently identified and used in conjunction with the mutants as a basis for a genetic and molecular analysis of *Drosophila* embryogenesis (NÜSSLEIN-VOLHARD and WIESCHAUS 1980; ST. JOHNSTON and NÜSSLEIN-VOLHARD 1992). Surprisingly, the principal mechanisms underlying the formation of the body plan of both *Drosophila* and vertebrates turned out to be more similar than expected, despite the apparent differences between insects and vertebrates. In both organisms, an initially unsegmented body axis subsequently becomes segmented, leading to the formation of body segments (in the case of *Drosophila*) and for example, the vertebral column, the most obviously segmented structure in vertebrates. With the advances in the vertebrate field that have been achieved, it now becomes increasingly obvious that gene networks controlling segmentation as well as other fundamental processes are highly conserved between insects and vertebrates, although the segmentally organized tissues are not always homologous (reviewed by NOLL 1993).

In order to isolate new, vertebrate-specific genes and to produce a large number of mutants as a tool to study the function of the affected genes, just recently the first saturation mutagenesis screen in a vertebrate, the zebrafish (*Brachydanio rerio*), was initiated (MULLINS and NÜSSLEIN-VOLHARD 1993). In the mouse, systematic breeding for more than 100 years has produced a great variety of skeletal mutants, which either arose spontaneously or were induced during radiation and mutagenesis experiments (LYONS and SEARLE 1989). Isolation and maintenance of these mutants was greatly facilitated by the fact that skeletal abnormalities are easily recognized and often do not interfere with the viability and the reproductive performance. A phenotypic and embryological analysis of various mouse skeletal mutants allowed them to be grouped into different classes reflecting the different steps of skeletal development affected (GRÜNEBERG 1963). Positional cloning techniques now enable us to isolate the genes underlying observed defects and were recently used successfully to isolate the genes affected in, for example, the skeletal mutants *Brachyury* (*T*) (HERMANN et al. 1990), *short ear* (*se*) (KINGSLEY et al. 1992) and *brachypodism* (*bp*) (STORM et al. 1994).

A complementary approach is frequently used to analyze the function of genes which are known to be expressed during formation of the axial skeleton and are expected to be required. For these genes, loss or gain of function mutants can be created with the help of homologous recombination in embryonic stem cells and the subsequent production of mice completely derived from these cells (JOYNER 1993). A phenotypical and molecular analysis of the defects in the created mutants can then be used as a basis to elucidate the function of the affected gene. In addition, as the phenotypes of many mouse mutants resemble the malformations observed in a variety of human skeletal dysplasias, mouse mutants can often serve as a model system to understand the molecular steps involved in human dysplasias as well as to test potential therapies.

However, not only genetic but also environmental factors can influence skeletal development by interfering with one of the many steps involved and can sometimes give rise to malformations that resemble the ones caused by a genetic defect. Such similarities suggest that in both cases the same step of skeletal development is affected and can help to identify the molecular basis of the observed malformations. An example is the action of retinoic acid, which, by altering the expression pattern of *Hox* genes, leads to homeotic transformations of vertebrae similar to the ones caused by mutations in *Hox* genes (KESSEL 1992).

In this chapter we will give an overview of the development of the axial skeleton, starting with an outline of the morphological events. We will then review the current knowledge on the molecular basis of skeletal development, including a discussion of some mouse skeletal mutants and human skeletal dysplasias. We will finish with a short section on how environmental factors might interact with the complex genetic network controlling skeletal development.

B. Morphogenesis

The development of any skeletal element starts with a cascade of events in which the cells which are going to initiate skeletal formation are specified and positioned at the right place. This step is usually not apparent on a morphological level. Subsequently, mesenchymal cells proliferate and give rise to mesenchymal condensations, which already roughly outline the later shape of the skeletal structures to be formed. Condensed mesenchymal cells then differentiate into cartilage cells, which undergo hypertrophy and are finally replaced by immigrating osteoblasts secreting the organic bone matrix, which calcifies to form the mature bone. In the following paragraphs, these steps are described in detail for the formation of the vertebral column.

I. Formation of the Primitive Streak and the Notochord

The development of the vertebrate axial skeleton is intimately linked to the formation of the primary body axis. At about day 6.5 of mouse development (primitive streak stage), epithelial cells from the primitive ectoderm begin to delaminate and migrate through the primitive streak to form mesoderm (reviewed in HOGAN et al. 1994). With the establishment of the primitive streak, the future anteroposterior axis of the embryo is fixed. With exception of craniofacial structures (COULY et al. 1992; COULY and LE DOUARIN 1985, 1987), the skeleton is entirely derived from mesoderm (CHRIST and ORDAHL 1995, and references therein); the major decisions that determine the fate of mesodermal cells are made during or shortly after gastrulation. Immediately after gastrulation, two essential structures for the development of the axial skeleton can be recognized: the notochord and the paraxial mesoderm. The notochord arises concomitantly with the regression of the primitive streak. The importance of the notochord during embryogenesis cannot be overestimated; it not only marks the anteroposterior axis of the embryo, but also acts as an organizer for adjacent embryonic tissues (PLACZEK et al. 1990; VAN STRAATEN et al. 1988; WATTERSON et al. 1954; YAMADA et al. 1991). In most vertebrates, the notochord is present only in early stages of embryogenesis. In primitive chordates, e.g., *Amphioxus* or the primitive vertebrates of the cyclostomata class, the notochord persists throughout life and no vertebral bodies develop. In vertebrates, the notochord is essential for dorsoventral patterning of both the neural tube and the paraxial mesoderm (PLACZEK et al. 1990; VAN STRAATEN et al. 1988; YAMADA et al. 1991). In the neural tube it induces the floor plate in a contact-dependent manner (PLACZEK et al. 1990, 1991, 1993; VAN STRAATEN et al. 1988), whereas induction of motor neurons requires the presence of a notochord and/or a floor plate but is contact-independent (TANABE et al. 1995; YAMADA et al. 1993). The notochord also plays a key role during sclerotome development (BRAND-SABERI et al. 1993; KOSEKI et al. 1993; POURQUIÉ et al. 1993), from which the bones, cartilage, and connective tissue of the axial skeleton are derived (reviewed in CHRIST and ORDAHL 1995) and which will be discussed below.

II. Segmentation of the Paraxial Mesoderm into Somites

After the end of gastrulation, the mesoderm is organized in three main blocks each with a different fate (Fig. 1). The paraxial mesoderm consists of two unsegmented strands of mesoderm flanking the notochord, which are laterally connected to the intermediate mesoderm. Laterally, the intermediate mesoderm itself is connected to the lateral plate mesoderm, consisting of somatopleura and splanchnopleura layers (Fig. 1B). The axial skeleton derives entirely from the paraxial mesoderm (CHRIST and ORDAHL 1995, and references therein; CHRIST and WILTING 1992). Starting around day 8 in the mouse, somitogenesis proceeds craniocaudally by budding off at the cranial end of the unsegmented paraxial mesoderm (Fig. 1A). The molecular mechanisms un-

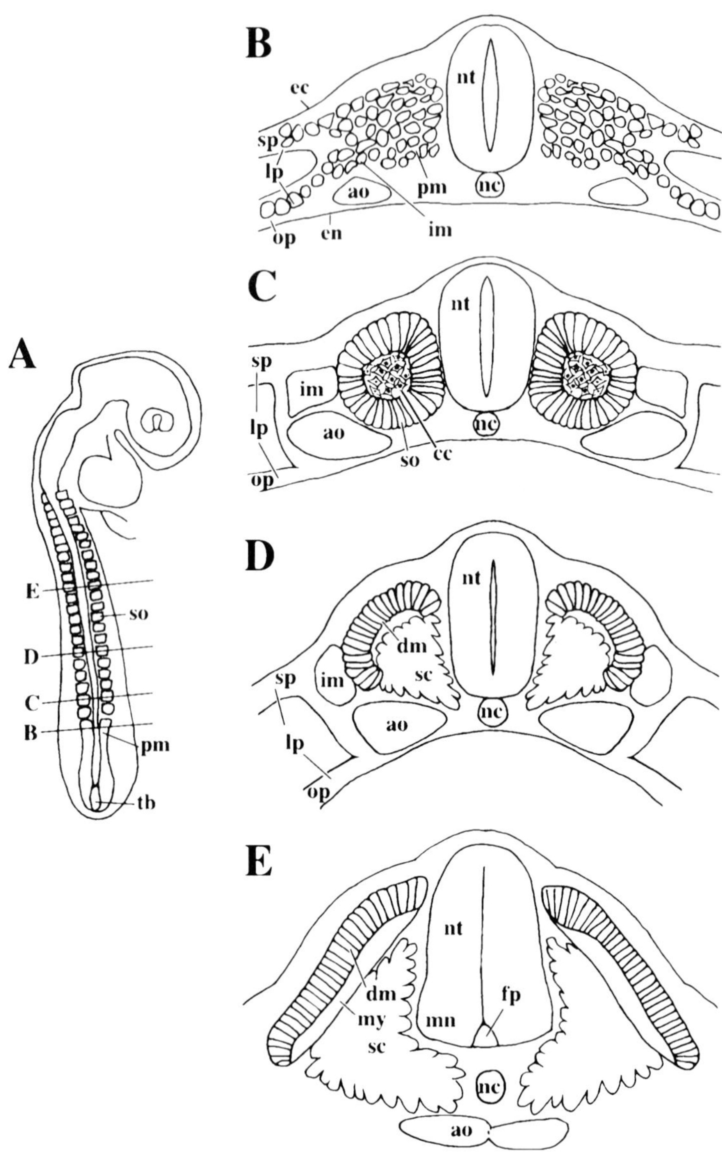

derlying the segmentation of the paraxial mesoderm into somites are not yet understood. Segmentation of the paraxial mesoderm seems to be independent of the notochord and also occurs in an autonomous fashion during in vitro culture of paraxial mesoderm (BELLAIRS 1963, 1979; CHRIST et al. 1972, 1973; MENKES and SANDOR 1977; PACKARD 1978).

Shortly after their formation, somites can be described as epithelial balls of columnar cells enveloping a central cavity, the somitocoele, which is filled with mesenchymal somitocoele cells (Figs. 1C, 2A). The outer side of the somite is surrounded by a basement membrane, and the somitic cells are polarized such that their apical sides face the somitocoele (SOLURSH et al. 1979).

III. Differentiation of Somites into Dermomyotome and Sclerotome

Soon after their formation, somites start to differentiate. Although the fate map of somitic cells has not been completely established, the principal fate of different parts of the somite has been determined using the chick quail chimera system (CHRIST et al. 1977; CHRIST and ORDAHL 1995; CHRIST and WILTING 1992; HUANG et al. 1994; ORDAHL and LE DOUARIN 1992). The dorsal half of the somite gives rise to the dermomyotome, whereas from the ventral half the sclerotome develops (Fig. 1D,E). The dermomyotome, is the origin of the dermis of the back and of all skeletal muscles. Epaxial muscles develop from

---

Fig. 1A–E. Differentiation of the paraxial mesoderm. **A** 26-somite stage chick embryo. The four most anterior somites de-epithelialize and fuse shortly after their formation to form the anlage of the basioccipital bone and are not shown. The *lines* indicate the level of the transverse sections schematically shown in Fig. 1B–E. As formation and segmentation of the paraxial mesoderm (*pm*) into somites (*so*) and their subsequent differentiation proceeds from cranial to caudal, there is a cranial to caudal gradient of development maturity within one embryo. Whereas somites at the anterior end of the embryo have already started differentiation, the most posterior, newly formed somites are still epithelial. Even further posterior, the paraxial mesoderm is still unsegmented. At the very caudal end of the embryo, in the tail bud (*tb*), gastrulation is still continuing. **B** Transverse section through the anterior end of the unsegmented paraxial mesoderm. The paraxial mesoderm flanks the neural tube (*nt*) and notochord (*nc*) and is laterally connected to the intermediate mesoderm (*im*). The intermediate mesoderm itself is laterally connected to the splanchnopleuric (*sp*) and somatopleuric (*op*) layers of the lateral plate mesoderm (*lp*). Ventrally, the paraxial mesoderm is flanked by the aorta (*ao*) and endoderm (*en*), and dorsally by the ectoderm (*ec*). **C** Transverse section through an epithelial somite. Shortly after their formation, somites are epithelial balls of cells with a central cavity which is filled with mesenchymal somitocoele cells (*cc*). **D** Transverse section through a somite which has just started differentiation. At the medial side of the somites, an epithelial–mesenchymal transition leads to the formation of the sclerotome, whereas the dorsal sides of the somites retain their epithelial organization and differentiate into the dermomyotomes (*dm*). **E** Transverse section through a slightly later stage of somite differentiation. Medially from the epithelial dermomyotome, a myotome (*my*) layer has formed. Sclerotome cells are migrating towards the notochord to form the perichordal tube. Within the neural tube, a floor plate (*fp*) and motorneurons (*mn*) have been induced by the notochord (Courtesy of Ralf Spörle, GSF-Neuherberg)

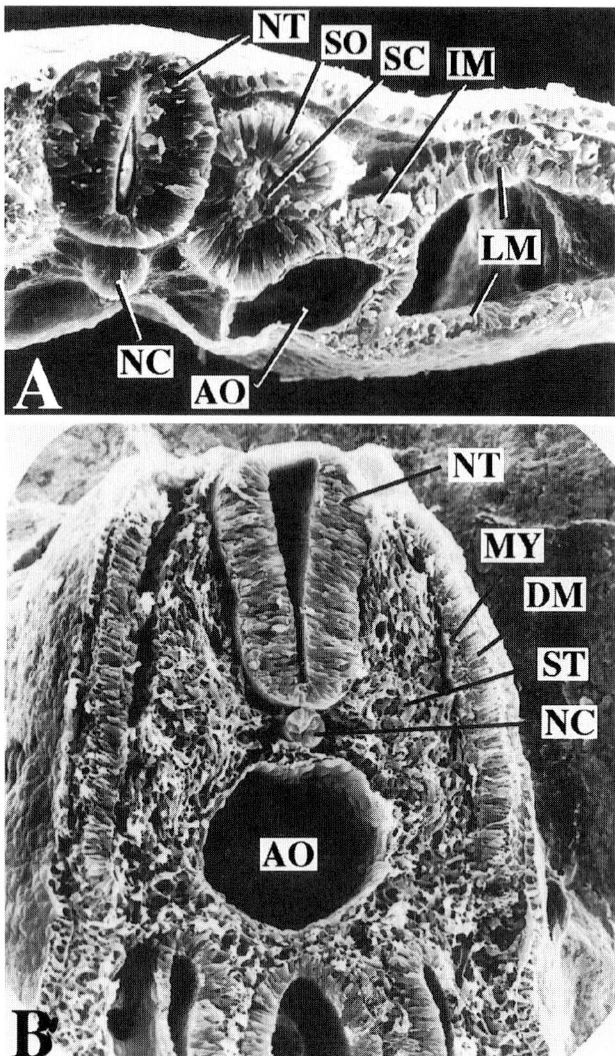

Fig. 2A,B. Scanning electron micrographs of transverse fractures of **A** day-2 and **B** day-3 chicken embryos. *AO*, aorta; *DM*, dermomyotome; *IM*, intermediate mesoderm; *LM*, lateral plate mesoderm; *MY*, myotome; *NC*, notochord; *NT*, neural tube; *SC*, somitocoele cells; *SO*, somites. (Courtesy of Prof. Bodo Christ, Institute of Anatomy, University of Freiburg)

dorsomedial "quadrants" of the somites, whereas dorsolateral quadrants give rise to hypaxial muscles. The fate of sclerotome cells depends on their relative position within the sclerotome (VERBOUT 1985). Cells located ventromedially contribute to intervertebral disks and vertebral bodies. Lateral sclerotome cells participate in the formation of the ribs, the pedicles, and lamina as well as of connective tissue surrounding the dorsal root ganglia. Just before sclerotome formation, the ventral epithelial surface of the somites bulges towards the notochord. An epithelial–mesenchymal transition then leads to the formation of sclerotome cells (JACOB et al. 1975). The somitocoele cells become part of the posterior sclerotome and contribute to the intervertebral disks and the ribs. The mechanisms underlying the breaking up of the basement membrane and the transformation of ventral epithelial cells into mesenchymal sclerotomal cells are not yet known.

IV. From Sclerotomes to Mesenchymal Prevertebrae

Following the de-epithelialization, medially located sclerotome cells migrate medioventrally towards the notochord (Figs. 1E, 2B). The morphologically apparent metameric organization of these cells along the craniocaudal axis is lost, and the perinotochordal tube is formed, consisting of loosely arranged mesenchymal cells with no apparent segmentation. Shortly thereafter, centers of condensations arise within the perinotochordal tube. These will develop into intervertebral disks, whereas from the loose intermediate zones the vertebral bodies form. In contrast to the medial region, lateral sclerotome cells maintain their segmental organization. Cell density within the lateral sclerotome is higher in the caudal half, from which the pedicles, laminae, and ribs develop, and lower in the cranial half, which form the perineural connective tissues (also called "neurotome" by BLECHSCHMIDT 1961) surrounding the spinal nerves and ventral part of the dorsal root ganglia (VERBOUT 1985). The precise relationship between the metameric organization of the somites and the secondary metameric pattern of the perinotochordal tube is the subject of a long-standing argument. The concept of "resegmentation," first proposed by REMAK (1855), suggests that cells from one somite contribute to two vertebrae. Experimental evidence from chick quail transplantation studies and experimental manipulations of somite halves seem to support this concept, as far as the vertebral body, the ribs, and the laminae are concerned (BAGNALL et al. 1988, 1989). The pedicles and intervertebral discs seem to be derived from only one somite (CHRIST and WILTING 1992, and references therein). Neither the molecular mechanisms that lead to sclerotome cell migration nor the basis of cell condensation is known. However, notochord ablation and transplantation experiments in the chick and a molecular analysis of mouse notochord mutants have demonstrated that the notochord is essential for the specification of ventral sclerotome derivatives (BRAND-SABERI et al. 1993; DIETRICH et al. 1993; GOULDING et al. 1994; KOSEKI et al. 1993; POURQUIÉ et al. 1993). Notochord ablations lead to the loss of ventral sclerotome derivatives, whereas transplantation of a notochord to an ectopic position next to the unsegmented paraxial mesoderm induces ectopic sclerotome formation and inhibits dermomyotome development in the vicinity. After long-term incubation, ectopic cartilage is formed.

V. Chondrification of Prevertebrae

After the major components of the vertebral column have been laid down as mesenchymal condensations, also called blastema, around day 13 of mouse embryogenesis, chondrification of the vertebral column begins, proceeding in a craniocaudal direction. Chondrification starts from chondrification centers in the middle of the vertebral bodies, within the costal processes, and from two eccentric positions within the pedicles (PATTEN 1953; THÖNDURY 1958). At that time, the mesenchymal condensations forming the primordia for the lamina of the neural arches have not yet reached the dorsal midline; as a result the neural arches are still wide open and only become closed after day 17 (ROUGH 1990). With the chondrification of the vertebral bodies, the notochord is deformed such that its diameter decreases at the site of a forming vertebral body, whereas it enlarges at the site of a developing intervertebral disk (CHRIST and WILTING 1992). There, the notochordal cells participate in the formation of the nucleus pulposus of the intervertebral disk, whereas at the periphery of the intervertebral disk the annulus fibrosus develops. Remnants of the notochord can still be found at birth, but usually disappear during adulthood (RAUBER and KOPSCH 1987).

VI. Ossification of Prevertebrae

Ossification of the vertebral column starts around day 14.5 of mouse embryogenesis (ROUGH 1990). In each vertebra, three ossification centers appear, one in the center of the vertebral body and one in each pedicle (PATTEN 1953). In addition, one ossification center also appears in each costal process. In different regions of the vertebral column, the order of appearance of the ossification centers differs, and ossification starts at slightly different timepoints. Cervical neural arches ossify first, followed by the thoracic and upper lumbar vertebral bodies, the thoracic and lumbar neural arches, and finally the cervical and the lower lumbar vertebral bodies (ROUGH 1990). Within the vertebral bodies, ossification follows the typical endochondral pattern, beginning with hypertrophy of the chondrocytes, mineralization of the cartilage matrix, and invasion of blood vessels, followed by cartilage erosion, invasion of osteoblasts, and replacement of the cartilage by bone (BLOOM and FAWCETT 1968). Within the neural arches, a layer of perichondral bone is produced before cartilage erosion and endochondral ossification starts. At birth, the three independently ossified parts of the vertebrae are still separated by cartilage. Cartilage is also present on both cranial and caudal surfaces of the vertebral bodies (RAUBER and KOPSCH 1987).

C. Genes Involved in Axial Skeleton Formation

For the development of the axial skeleton as well as for any other developmental process, a great variety of different gene products are required. Diffusible signaling molecules and peptide growth factors as well as membrane-bound ligands and the corresponding receptors mediate intercellular interactions. Components of the signal transduction machinery, such as protein kinases, then have to transduce signals arriving at the cell surface to the nucleus, finally leading to changes in gene expression, which itself is controlled by transcription factors. Components of the extracellular matrix can bind peptide growth factors and modulate their action (reviewed in ADAMS and WATT 1993; RUOSLAHTI and AMAGUCHI 1991). In addition, extracellular matrix molecules can themselves serve as signals, triggering specific cellular responses in cells attached to the matrix (JULIANO and HASKILL 1993). The extracellular matrix is also important as a substrate for cell migration. Direct adhesive cell–cell contact necessary for the integrity of tissues and the modulation of this adhesive interaction in order to allow cell migration requires the differential expression of cell adhesion molecules such as cadherins. Finally, the acquirement of the terminal differentiated stage requires the expression of the molecules typical for this differentiated stage, such as specific proteoglycans and collagens in the case of cartilage and bone.

In this section we will describe what is known about the molecular basis of the different steps of vertebral column development outlined in the previous section.

I. Gastrulation and Formation of Paraxial Mesoderm

The vertebral column is entirely derived from mesoderm, and its formation is intimately linked to the establishment of the primary body axis. Therefore, genes involved in the formation and patterning of mesoderm during gastrulation frequently also affect the specification and development of early precursors of the axial skeleton, e.g., the paraxial mesoderm. Defects in these genes, however, often lead to gross embryonic alterations and early embryonic lethality, making an analysis of their specific function for vertebral column development very difficult, especially during later steps. Progress in understanding the genetic control of gastrulation in the mouse has been recently reviewed (FAUST and MAGNUSON 1993; STERN 1992), and in the present review only genes specifically affecting paraxial mesoderm development will be discussed.

1. Growth Factors, Signaling Molecules, and Their Receptors

Among the genes involved in formation and patterning of mesoderm are peptide growth factors of the WNT family (MCMAHON 1992; PARR and MCMAHON 1994, and references therein), the fibroblast growth factor (FGF) family (MASON 1994, and references therein), and the transforming growth factor (TGF)-β superfamily (KINGSLEY 1994a, and references therein) and their receptors (ATTISANO et al. 1994; GIVOL and YAYON 1992).

Inactivation of the *Wnt-3a* gene impairs development of dorsal, somitic mesoderm and leads to a disrupted notochord, a lack of caudal somites, and a failure of tail bud formation, resulting in a truncation of the body axis at the level of the forelimbs (AUGUSTINE et al. 1993; TAKADA et al. 1994). Other members of the *Wnt* gene family, which in the mouse currently comprises 12 members, are also expressed during gastrulation and later on during somite development (CHRISTIANSEN et al. 1995; MCMAHON 1992). Their function, however, has yet to be determined.

Fgf genes are expressed in complex, partially overlapping patterns during embryogenesis. The function of several of the nine identified mouse *Fgf* genes has been studied by gene inactivation. Inactivation of *Fgf-4* affects the proliferation of the inner cell mass and leads to embryonic death shortly after implantation, thereby preventing the analysis of possible later functions (FELDMAN et al. 1995). Inactivation of *Fgf-5* leads to mice with an increased hair length (angora mice), and so far no mesoderm defects have been reported for these mutants, in spite of *Fgf-5* being expressed during gastrulation (HÉBERT et al. 1994). Loss of *Fgf-3* function causes malformations of caudal vertebrae and a dorsal curling of the tail, which was tracked back to a defective organization of the tail at day 11.5 (MANSOUR et al. 1993). Inactivation of the FGF receptor-1 gene (*Fgfr-1*), one of four known murine FGF receptors, results in an expansion of axial mesoderm at the expense of paraxial mesoderm, resulting in a lack of somitogenesis and embryonic death before day 9.5 (DENG et al. 1994; YAMAGUCHI et al. 1994).

Several members of the TGF-β superfamily have been implicated in being required during gastrulation. Results from ectopic expression of activin subunits as well as of their receptors and truncated receptors in the frog *Xenopus laevis* have suggested that activins are essential for mesoderm induction (WOODLAND 1993). Recent results from targeted disruption of activin subunit genes in mice, however, demonstrate that zygotic expression of activins is not required for mesoderm development in the mouse, leaving open the possibility that maternal activin-β_A may cross the placenta and may partially resolve the zygotic deficiency (MATZUK et al. 1995a,b). The TGF-β-related gene *nodal* has recently been shown to be affected in the retroviral insertion mouse mutation 413-d (ZHOU et al. 1993). Homozygous mutants are unable to form axial structures, including dorsal mesoderm and notochord, during gastrulation, in line with an expression of *nodal* in cells at the periphery of the node.

2. Transcription Factors

The transcription factor *Brachyury* (*T*) (HERRMANN and KISPERT 1994, and references therein) is expressed in the primitive ectoderm, the adjacent mesoderm, and later on in the notochord, consistent with the malformations seen in embryos homozygous for mutant *T* alleles: axial elongation is blocked and structures posterior to the forelimb buds do not form, while the anterior portion of the notochord initially forms but does not persist as a coherent structure (CHESLEY 1935; GLUECKSOHN-SCHOENHEIMER 1944; HERRMANN 1991; SPIEGELMAN 1976; WILKINSON et al. 1990). *T* heterozygotes show loss of the tail and skeletal abnormalities, suggesting the requirement of an increasing amount of *T* gene product for the development of the notochord and formation of mesoderm along the craniocaudal axis (SCOTT et al. 1993). Several homeobox-containing transcription factors, among them *goosecoid* (BLUM et al. 1992) and *Mox-1* (CANDIA et al. 1992) are also expressed during gastrulation and are thought to be involved in mesoderm specification. As loss of function mutants for these genes are not yet available, their function remains to be elucidated.

II. Segmentation of the Paraxial Mesoderm into Somites

The mechanism underlying the sequential mesenchymal–epithelial transition of blocks of cells at the cranial end of the paraxial mesoderm, which results in the formation of somites as the first obviously segmented structures of the body axis, is currently still poorly understood. Several models for somite formation have been proposed and are discussed by KEYNES and STERN (1988), who themselves proposed a model in which segmentation is suggested to be coupled to the cell cycle, in order to explain multiple periodic vertebral abnormalities caused by a single heat shock applied to chicken embryos (PRIMMETT et al. 1988, 1989). Recently, CONLON et al. (1995) suggested a new model for somitogenesis, which critically depends on the function of the transmembrane receptor *Notch-1* and which is discussed below.

1. Intercellular Signaling

Notch-1 is a member of the Notch protein family, members of which have been identified in several species (GREENWALD and RUBIN 1992, and references therein; LARDELLI et al. 1994; LARDELLI and LENDAHL 1993; REAUME et al. 1992). These proteins are large transmembrane receptors with tandem epidermal growth factor (EGF)-type repeats and Notch family-specific repeats at the extracellular side and tandem ankyrin repeats in the intracellular domain. Members of the Notch family have been shown to be essential for the correct implementation of cell fate decisions in a large number of developmental processes in both *Drosophila* and *Caenorhabditis elegans* by interacting with ligands of the Delta family displayed at the cell surface of adjacent cells (GREENWALD and RUBIN 1992, and references therein). The isolation of a mammalian Delta homologue Dll1 has just been reported (BETTENHAUSEN et al. 1995).

The expression of high levels of *Notch-1* in a somite-sized domain at the anterior end of the presomitic mesoderm and downregulation of *Notch-1* after somite formation led Reaume et al. (1992) to suggest a role of *Notch-1* in the promotion of cell association preceding somite formation. Recent results from targeted disruption of *Notch-1* support this idea (Conlon et al. 1995; Swiatek et al. 1994). In *Notch-1*-deficient embryos, somitogenesis is delayed and uncoordinated on contralateral sides, whereas specification of somitic subpopulations is apparently not disturbed (Conlon et al. 1995). Conlon et al. (1995) suggest that, in analogy to the function of *Notch* in *Drosophila* and *C. elegans*, within the presomitic mesoderm a periodic arrangement of *Notch-1*- and Notch-ligand (Delta-like protein)- expressing cells might be established through an amplification of initial differences in ligand and receptor activity. Upregulation of *Notch-1* at the anterior end of the presomitic mesoderm should then promote the physical association and epithelialization of these cells through interaction of the membrane-bound *Notch-1* receptor and its ligands followed by a separation of this group of cells from the ones posterior to them. *Notch-1* expression at the anterior end of the presomitic mesoderm is thought to be controlled by extrinsic factors. In the mouse, two other *Notch* genes have been identified (Lardelli et al. 1994; Lardelli and Lendahl 1993). One of them, *Notch-2*, is also expressed in the anterior presomitic mesoderm (Swiatek et al. 1994). One might therefore speculate that in a double mutant for *Notch-1* and *Notch-2*, coordination of somite formation might be even more severely disrupted. The existence of periodic prepattern within the presomitic mesoderm was already suggested by Meier and colleagues (Jacobson and Meier 1986; Meier 1979, 1984), who reported the metameric arrangement of cells in so-called somitomeres visible by scanning electron microscopy.

The expression of *Fgfr-1*, which was already mentioned in the context of mesoderm patterning, is strongly upregulated at the anterior end of the presomitic mesoderm, similar to *Notch-1*, in a block of cells roughly corresponding to one somite in size (Yamaguchi et al. 1992). After epithelialization, however, *Fgfr-1* is not downregulated, but becomes restricted to the cranial half of the newly formed epithelial somites. Upon somite differentiation, the expression pattern of *Fgfr-1* becomes more complex (Orr-Urtreger et al. 1991; Peters et al. 1992; Yamaguchi et al. 1992). As in *Fgfr-1*-deficient mice even early mesoderm patterning is defective and paraxial mesoderm never forms (Deng et al. 1994; Yamaguchi et al. 1994), the specific role of signaling through the *Fgfr-1* for somite formation has not been analyzed so far. Expression of *Fgfr-1* in *Notch-1*-deficient mice is unaffected, suggesting that *Fgfr-1* is apparently not downstream of the *Notch-1* signaling pathway (Conlon et al. 1995). It is unclear which of the FGF family members *Fgfr-1* interacts with during somitogenesis.

The receptor protein tyrosine kinase *sek*, a member of the *eph* receptor kinase family originally identified as an orphan receptor, has been implicated in the segmental patterning of both the hindbrain and the presomitic mesoderm (Becker et al. 1994, and references therein; Nieto et al. 1992). It is expressed in early mesoderm, is then downregulated, and is subsequently sharply upregulated at the anterior end of the presomitic mesoderm. There, *sek* transcripts are present in two stripes: the cranial narrower stripe corresponding to the cranial part of the next somite to form and the broader posterior stripe comprising the cells of the next but one somite to condense. After epithelialization of the somite, *sek* is again sharply downregulated (NIETO et al. 1992). Recently, the membrane-anchored protein Elf-1 was identified as a potential ligand for the sek receptor kinase, and preliminary data suggests that it is expressed in presomitic mesoderm and somites (CHENG and FLANAGAN 1994).

2. Cell Adhesion

Somite formation is accompanied with an increase in cell adhesion between the cells at the cranial end of the presomitic mesoderm in comparison to cells located further caudally (BELLAIRS et al. 1978; CHENEY and LASH 1984). It has been shown that the expression of both the calcium-dependent adhesion molecule N-cadherin and the neural cell adhesion molecule (NCAM) is upregulated just before somite formation. Whereas expression of N-cadherin is subsequently downregulated in the ventromedial part of the somite just before the epithelial–mesenchymal transition, giving rise to the sclerotome, NCAM expression continues throughout the somite (DUBAND et al. 1987). Monoclonal antibodies against N-cadherin, but not against NCAM have been shown to cause disaggregation of chick somites in vitro (DUBAND et al. 1987). In mice with a targeted disruption of the NCAM gene, so far no vertebral column defects have been detected (CREMER et al. 1994).

3. Extracellular Matrix Components

LASH et al. (1984, 1987) have implicated the extracellular matrix component fibronectin in the epithelialization of somites. These authors demonstrated that the addition of fibronectin or a peptide corresponding to the cellular recognition site of fibronectin can stimulate aggregation of disaggregated paraxial mesoderm cells in vitro (LASH et al. 1984, 1987). Targeted disruption of the fibronectin gene leads to mesoderm defects during gastrulation and embryonic lethality after day 9.5 (GEORG et al. 1993). In embryos deficient of fibronectin, a notochord and somites never form. Because of the early and rather global mesoderm defect in these embryos this phenotype, however, is not informative in regard to the specific role of fibronectin for somite epithelialization.

4. Transcription Factors

The search for new cell type-specifically expressed members of the basic helix–loop–helix (bHLH) transcription factor family which could dimerize with E12, one of the ubiquitous bHLH proteins, recently led to the identification of two new family members, *paraxis* and *scleraxis*, with interesting expression patterns during somite development (Burgess et al. 1995, and references therein; Cserjesi et al. 1995). Before the onset of somitogenesis, *paraxis* is weakly expressed in part of the primitive mesoderm and later it is highly expressed in the cranial part of the presomitic mesoderm, equivalent to approximately two somites in size. Expression of *paraxis* continues throughout the uncompartimentalized epithelial somite. After compartimentalization, *paraxis* is shut down in the myotome soon after its formation, and expression in the other somite derivatives gradually declines (Burgess et al. 1995). Expression of *scleraxis*, on the other hand, does not starts until after compartimentalization of the somite within the lateral part of the sclerotome (Cserjesi et al. 1995) and will be discussed in the next section.

III. Patterning of Somites

1. Craniocaudal Patterning: Cranial and Caudal Somite Halves

Even in the early epithelial somites, differences between the cranial and caudal somite halves exist (Norris et al. 1989), which are thought to be responsible for generating the segmental arrangement of the peripheral nervous system (Keynes and Stern 1988; Stern et al. 1991, and references therein). Rotation experiments of the presomitic paraxial mesoderm in the chick indicated that this craniocaudal polarity of the somites is already determined before their segmentation (Keynes and Stern 1984). The restricted expression of *Fgfr-1* only in the cranial half of the newly formed epithelial somite has already been mentioned (Yamaguchi et al. 1992). Interestingly, in the chick embryo *Fgf-3* is expressed in a complementary pattern only in the caudal half (Mahmood et al. 1995). After sclerotome formation, motor axons and neural crest cells migrate exclusively through the cranial half of the sclerotome (Bronner-Fraser 1986; Loring and Erickson 1987; Rickmann et al. 1985; Teillet et al. 1987). Stern and Keynes (1987) were able to show that cranial and caudal sclerotome cells are unable to mix and thereby maintain their distinct segmental positions. Caudal sclerotome cells bind peanut agglutinin and secrete chondroitin-6-sulfate, both of which are thought to act as barriers to axon invasion and express T cadherin. In the cranial sclerotome half, butyryl cholinesterase and the extracellular matrix glycoprotein tenascin are expressed (Keynes and Stern 1988, and references therein). The relevance of the localized tenascin expression for guidance of neural crest cells and motor neurons, however, is unclear, as mice homozygous for a targeted disruption of the tenascin gene develop normally without displaying any malformations (Saga et al. 1992). How the early differences between cranial and caudal somite halves are established is not known.

2. Dorsoventral Patterning

Lineage tracing studies and somite rotation and transplantation experiments have demonstrated that, in the presomitic mesoderm and newly segmented somites, a dorsoventral axis, in contrast to the craniocaudal axis, is not yet determined. Early somitic cells are not committed to a dermamyotomal or a sclerotomal lineage. Dorsoventral patterning of the somites has been shown to be controlled by interactions between the somites and adjacent structures such as the notochord, the neural tube, and the surface ectoderm (reviewed by CHRIST and ORDAHL 1995).

a) Signaling Molecule Sonic Hedgehog

Whereas the nature of the signals which are thought to emanate from the ectoderm and the neural tube to control dermomyotome and myotome development is unclear (see also Sect. C.IV), sonic hedgehog (SHH), secreted from the notochord and floor plate of the neural tube, has recently been shown to mediate sclerotome induction (FAN et al. 1995; FAN and TESSIER-LAVIGNE 1994; JOHNSON et al. 1994). *shh* is the vertebrate homologue of the *Drosophila* segment polarity gene *hedgehog* (INGHAM 1994, and references therein). Initially, it is secreted only from the notochord. It is also involved in patterning of the neural tube, where it is responsible for the induction of the floor plate, which subsequently also expresses *shh*. It is furthermore involved in motor neuron induction (ECHELARD et al. 1993; ROELINK et al. 1994, 1995). SHH mediates both the inductions of motor neurons and of sclerotome via a long-range effect, whereas the floor plate induction requires contact between the neural tube and the notochord (FAN et al. 1995; JOHNSON and TABIN 1995; ROELINK et al. 1995). Similar to the *Drosophila* hedgehog protein, the vertebrate SHH is autocatalytically cleaved to yield two fragments (FAN et al. 1995; PORTER et al. 1995; ROELINK et al. 1995). Only the aminoterminal (SHH-N) but not the carboxyterminal fragment has inducing activity and is thought to be responsible for both the short- and long-range inductive properties of the notochord, which seem to require different threshold concentrations of SHH-N. The threshold concentrations required for long-range inductions have been shown to be a factor of 3 lower than the concentration required for short-range induction. As the majority of the autocatalytically cleaved SHH-N is apparently retained at the cell surface by an as yet unknown mechanism, the threshold concentration for the short-range induction of the floor plate might only be reached if the inducing tissue and the induced tissue are in direct contact with each other.

b) Interpretation of the Inductive Sonic Hedgehog Signal: Transcription Factors of the Pax Gene Family

The inductive SHH signal leads to changes in the expression pattern of *Pax* genes within the somite, which are thought to be causally related to its dorsal ventral patterning (FAN et al. 1995; WALLIN et al. 1994). *Pax* genes are transcription factors characterized by a 384-bp paired box sequence which was originally identified in the *Drosophila* segmentation and segment polarity genes *paired* and *gooseberry* and which encodes the DNA-binding domain of the proteins (GRUSS and WALTHER 1992, and references therein). In humans and mice, nine *Pax* genes have been identified so far, four of which, *Pax-1* (DEUTSCH et al. 1988; WALLIN et al. 1994), *Pax-3* (GOULDING et al. 1994; WILLIAMS and ORDAHL 1994), *Pax-7* (JOSTES et al. 1991), and *Pax-9* (NEUBÜSER et al. 1995; WALLIN et al. 1993), are known to be expressed during somite development. *Pax-3* is already expressed throughout the unsegmented presomitic mesoderm. As a result of interaction with the notochord, it is downregulated in the ventral half of the somite soon after epithelialization and continues to be expressed in the dermomyotome (GOULDING et al. 1994; WILLIAMS and ORDAHL 1994). Concomitantly, *Pax-1* is induced in the ventral region of the epithelial somite and the mesenchymal core and continues to be expressed in the sclerotome after its formation by de-epithelialization (Fig. 3A; BRAND-SABERI et al. 1993; EBENSBERGER et al. 1995). Upon further differentiation of the sclerotome, *Pax-1* expression concentrates in the caudal half of the sclerotome and eventually marks cells that give rise to the prospective intervertebral disk (Fig. 3B) and the perichondrium surrounding the vertebral bodies in the midaxial region. *Pax-1* is also expressed in early progenitors of the developing pediculi and lamina of the neural arch, the transverse processes, and the proximal part of the ribs. A recurrent theme concerning *Pax-1* is its expression in mesenchymal cells and downregulation upon terminal differentiation (DEUTSCH et al. 1988).

Deficiency in *Pax-1* function leads to severe malformations of the vertebral column, as is dramatically documented by the phenotype of the three different alleles of *undulated* (*un*), in all of which the *Pax-1* gene is affected

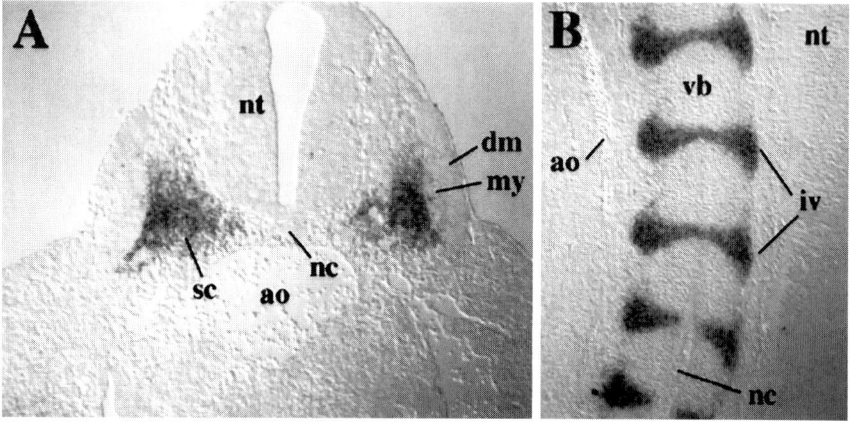

Fig. 3A,B. Expression of *Pax-1* in the developing sclerotome (nonradioactive in situ hybridizations). **A** Transverse section through a day-11.5 mouse embryo, showing *Pax-1* expression in the sclerotome (*sc*). **B** Saggital section through the developing vertebral column of a day-13.5 mouse embryo, showing *Pax-1* expression in intervertebral discs (*iv*). *ao* aorta; *dm* dermomyotome; *iv*, intervertebral disc; *my*, myotome; *nc*, notochord; *nt*, neural tube; *sc*, sclerotome, *vb*, vertebral body

(DIETRICH and GRUSS 1995; WALLIN et al. 1994). In the classical *un* mouse, a point mutation within the paired box replaces a highly conserved Gly with a Ser, strongly reducing the DNA-binding affinity. In *undulated extensive* (un^{ex}), the last of five exons of the *Pax-1* gene, and in *Undulated short tail* (Un^{S}) the complete *Pax-1* locus are deleted. *un* and un^{ex} are recessive mutations, whereas Un^{S} displays a semidominant phenotype. This suggests that *Pax-1* may be haploid insufficient and leads to perinatal lethality in the homozygous state. *Undulated* mice have severe abnormalities both in the intervertebral disks as well as the vertebral bodies, whereas dorsal sclerotome derivatives, such as the laminae of the neural arches, are less affected. In the Un^{S} allele, no intervertebral disks can be found in homozygous embryos. The remaining tissue in the perinotochordal region is still segmentally organized, suggesting that *Pax-1* might not be involved in segmentation or sclerotome induction per se but might be important for the specification of ventral sclerotome derivatives.

Deficiency in *Pax-3* causes the *Splotch* phenotype in the mouse (EPSTEIN et al. 1993; GOULDING et al. 1993; VOGAN et al. 1993) and Waardenburg's syndrome in humans. In *Splotch*, mainly the neural tube, neural crest derivatives, and the limb muscles are affected, whereas the axial skeleton is normal. However, for the interpretation of the phenotypes of both *undulated* and *Splotch* with regard to possible functions of *Pax-1* and *Pax-3*, it has to be kept in mind that for both genes a closely related *Pax* gene, *Pax-9* in the case of *Pax-1* (NEUBÜSER et al. 1995) and *Pax-7* in the case of *Pax-3* (JOSTES et al. 1991), is expressed in an overlapping fashion within the somites and its derivatives and might partially substitute for the loss of *Pax-1* or *Pax-3*. For an understanding of the combined action of *Pax-1* and *Pax-9* for sclerotome development and of *Pax-3* and *Pax-7* for dermomyotome development, the creation of double mutant mice has to be awaited.

c) Connection Between the Inductive Sonic Hedgehog Signal and the Nuclear Response Mediated by Pax Genes: Signal Transduction Via Cyclic Adenosine Monophosphate-Dependent Protein Kinases

In vitro culture and ectopic expression experiments have demonstrated that SHH is involved in both repression of *Pax-3* and induction of *Pax-1* in the dorsal and ventral half of the somite respectively. In *Drosophila*, the cyclic adenosine monophosphate (cAMP)-dependent protein kinase (PKA) and hedgehog act antagonistically in the signal transduction pathway that regulates expression of *decapentaplegic* (*dpp*), a TGF-β-related molecule, in the appendages and eye discs (LEPAGE et al. 1995; PAN and RUBIN 1995; STRUTT et al. 1995). Loss of function of PKA resembles the phenotype of ectopic *hedgehog* expression, suggesting that hedgehog induces *dpp* by inhibition of the catalytic subunit of PKA. Recent results suggest that an interaction of PKA with the hedgehog signaling pathway also play a role in the development of the mouse sclerotome. Modulation of PKA activity by drugs dramatically influenced the induction of *Pax-1* by SHH-N in presomitic explants. Increasing the activity of PKA, i.e., by forskolin, inhibited the induction of *Pax-1* and released the repression of *Pax-3* (FAN et al. 1995). These results suggest that the ventralizing effect of the notochord and the dorsalizing signals from the neural tube or surface ectoderm act antagonistically and influence the fate of somite cells by perhaps modulating the level of PKA activity. Future work will have to dissect the signaling pathways mediated by cAMP-dependent kinases during sclerotome development.

IV. Further Development of the Somite Compartments

In *Drosophila*, the paired box-containing transcription factors *paired* and *gooseberry* regulate the expression of the secreted factor *wingless* (*wg*) (KILCHHERR et al. 1986; LI and NOLL 1993), which itself regulates the expression of the homeobox-containing transcription factor *engrailed* (*en*) in neighboring cells (reviewed in MCMAHON 1992; NOLL 1993). The interaction of *wg* and *en* is involved in numerous developmental processes during *Drosophila* embryogenesis and is frequently part of the genetic hierarchy which leads to the establishment of boundaries (reviewed in INGHAM 1988; PEIFER and BEJSOVEC 1992). The mammalian homologues of *wg* are the members of the *Wnt* family (CHRISTIANSEN et al. 1995; MCMAHON 1992; PARR and MCMAHON 1994) already mentioned because of their role during gastrulation. The homologues of *en* are the two genes *engrailed-1* (*En-1*) and *engrailed-2* (*En-2*) (HILL et al. 1987; JOYNER and MARTIN 1987). Several members of the *Wnt* family are expressed during the development of the vertebral column. Their precise expression patterns, however, are poorly characterized. Of the two *engrailed* genes only *En-1* is known to be expressed in somite derivatives. *En-1* is first detected in the dermomyotome after its formation. Later, at day 12.5 of mouse embryogenesis, *En-1* transcripts can be detected medially in the loosely organized tissue of the primordia of the vertebral bodies and in the developing neural arches as well as laterally, in the developing costal processes (DAVIDSON et al. 1988; DAVIS et al. 1991). In *En-1*-deficient mice, however, so far no gross vertebral column defects except for a truncation of the 13th rib have been detected (WURST et al. 1994). Further analysis has to establish whether *En-1*-deficient mice display more subtle, so far undetected defects, or whether *En-1* is completely dispensable for vertebral column development, still leaving open the possible requirement of other as yet unknown *engrailed* genes. Interestingly, *Wnt-11*, a new member of the *Wnt* family was recently reported to be expressed at the junction of dermomyotome and myotome, immediately adjacent to the region where *En-1* is expressed (CHRISTIANSEN et al. 1995). It is furthermore important to know that several *Wnt* genes are expressed in dorsoventrally restricted patterns within the neural tube and are thought to be involved in its dorsoventral patterning (reviewed in MCMAHON 1992; NOLL 1993). Members of the *Wnt* gene family therefore also qualify as candidates for the neural tube signals implicated in influencing somite development. The cross-breeding of different *Wnt* gene knockout mice will help to reveal possible redundancies and help to elucidate the function of *Wnt* genes during somite development. A family of cell type-specific bHLH proteins including MyoD, myogenin, Myf-5, and MRF4/herculin/Myf-6 (MyoD family; muscle regulatory factors, MRF) has recently been shown to be expressed during early steps of skeletal myogenesis and to play an important role in the regulation of myogenesis in vitro, as demonstrated by their ability to activate muscle gene expression when expressed ectopically in nonmuscle cells (reviewed by WEINTRAUB 1993; WEINTRAUB et al. 1991). The MRF are thought to act in concert with additional ubiquitous and muscle-specific transcription regulators to activate transcription of many muscle-specific genes, leading to muscle cell determination and differentiation. The first MRF transcripts to be detected in the mouse are transcripts of *Myf-5* in the dorsomedial quadrant of the epithelial somite. Subsequently, after differentiation of the somite in dermomyotome and sclerotome, expression of *Myf-5* is restricted to the dermomyotome (OTT et al. 1991). As myotome cells emerge from the cranial lip of the dermomyotome, they activate multiple MRF in a characteristic pattern (reviewed by BUCKINGHAM 1992). Several of the MRF genes have been inactivated by homologous recombination (BRAUN et al. 1992; HASTY et al. 1993; NABESHIMA et al. 1993; RUDNICKI et al. 1992, 1993; BRAUN and ARNOLD 1995; ZHANG et al. 1995). Targeted disruption of the *Myf-5* and *MRF-4* genes resulted in unexpected phenotypes. *Myf-5*-deficient mice lack the distal part of the ribs, whereas *MRF-4*-deficient mice display multiple rib anomalities, including extensive bifurcations, fusions, and supernumerary processes. Surprisingly, in both mutants muscle development is not disrupted (BRAUN et al. 1992). Creation of mice deficient for both *MyoD* and *Myf-5* revealed that either *Myf-5* or *MyoD* is required for myoblast formation (RUDNICKI et al. 1993). The unexpected rib phenotypes of *Myf-5- and MRF-4*-deficient mice suggest that their expression is either required in sclerotome precursors of the rib rudiments, in concordance with a recent report of Myf-5 protein in a small subset of sclerotomal cells (T.H. SMITH et al. 1994), or the mutations may act in early myotomal cells and disturb a crucial interaction between these cells and cells of the sclerotome compartment. In order to decide between these two models, further analysis of the *Myf-5*- and *MRF-4*-deficient phenotypes on a molecular basis is required. In this context, it is interesting to note that several members of the TGF-β and FGF families are expressed in the myotome which could possibly influence sclerotome development (FLORINI et al. 1991; HAUB and GOLDFARB 1991; LYONS et al. 1990; NISWANDER and MARTIN 1992).

With the recent cloning of *paraxis* and *scleraxis*, two bHLH genes with restricted expression patterns in somite derivatives outside the dermomyotome and myotome were identified (BURGESS et al. 1995; CSERJESI et al. 1995). The expression of *paraxis* was already discussed in connection with segmentation. *Scleraxis* expression starts after compartmentalization of the somite within the sclerotome. Subsequently, scleraxis is expressed at high levels within mesenchymal precursors of the axial and appendicular skeleton and in cranial mesenchyme in advance of chondrogenesis and is downregulated when ossification is initiated, suggesting that it might regulate gene expression required for determination and differentiation of progenitor cells for cartilage and connective tissue. Whether *scleraxis* plays a similar role for the regulation of chondrogenic cell fate as the MRF for myogenic fates, e.g., whether expression of *scleraxis* is sufficient to induce chondrogenesis in vitro, has yet to be determined.

V. Regionalization Along the Craniocaudal Axis and *Hox* Genes

Although somites at different axial levels resemble each other morphologically and in the expression pattern of many genes, there are early regional differences among them, which manifest themselves later on in the formation of the specific patterns of the different parts of the vertebral column. Transplantation experiments in the chick, for instance, suggested that the rib-forming potential is restricted to the thoracic level (Jacob et al. 1975; Kieny et al. 1972). Transplantation of presomitic mesoderm from the thoracic level to the cervical level resulted in the formation of cervical ribs, indicating that the presomitic mesoderm was already regionally determined before somites were actually formed (Chevallier 1975; Kieny et al. 1972).

The transcriptional regulators encoded by the genes of the *Hox* clusters are believed to be responsible for the establishment of positional information along the craniocaudal body axis as well as along the proximodistal and craniocaudal axis of the limbs (reviewed by Krumlauf 1993, 1994; McGinnis and Krumlauf 1992; Morgan and Tabin 1993). The vertebrate *Hox* genes are organized in four clusters and share a 180-bp homeobox sequence with the homeotic genes from the *Antennapedia* and *Bithorax* complexes in *Drosophila melanogaster*, which were identified because of their remarkable ability to change the identity of larval segments when ectopically expressed or disrupted. In vertebrates, cells begin to express a particular *Hox* gene upon leaving the primitive streak, resulting in nonidentical, overlapping expression domains of different *Hox* genes along the craniocaudal axis, including the sclerotomes. Combinations of different *Hox* genes, the "*Hox* code," were suggested to specify the identity of vertebral segments (Kessel and Gruss 1991). Ectopic expression or targeted disruption of individual *Hox* genes was shown to result in homeotic transformations in which the identity of individual vertebrae was changed in such a way that they took on the identity of vertebrae anterior or posterior to their position (Charité et al. 1994; Krumlauf 1994, and references therein; Suemori et al. 1995). For example, eight instead of seven cervical vertebrae are formed in transgenic mice in which *Hoxa-7* is expressed ectopically under the control of a β-actin promoter, caused by a posterior transformation of part of the basioccipital bone into an additional cervical vertebrae (Balling et al. 1989; Kessel et al. 1990). Targeted inactivation of *Hoxd-13* resulted in the transformation of the fourth sacral vertebrae into a bone resembling the third and in limb abnormalities (Dollé et al. 1993).

Administration of retinoic acid (RA) severely affects vertebrate morphogenesis. Among other defects, both anterior and posterior homeotic transformations of vertebrae are observed, depending on the timepoint of administration (reviewed by KESSEL 1992, and references therein). Further analysis revealed, that RA induces alterations of *Hox* gene expression (KESSEL 1992; KESSEL and GRUSS 1991; KRUMLAUF 1994, and references therein). It is thought that the effects of RA are mediated through two families of receptors, the RA receptors (RAR) and the retinoid X receptors (RXR), which act as inducible transcriptional regulatory proteins and belong to the superfamily of nuclear receptors (reviewed in CHAMBON 1994). These receptors activate transcription by binding to RA response elements of target genes predominantly in the form of RAR/RXR heterodimers. The RAR and RXR families are each encoded by three genes designated RAR-α, -β, and -γ and RXR-α, -β, and -γ, respectively. Each gene generates multiple receptor isoforms with different N-terminal regions of the protein (α_1 and α_2, β_1–β_4, and γ_1 and γ_2) through differential usage of alternative promoters and differential splicing. Recently, gene targeting has been used to generate mice deficient of all isoforms of RAR-α, RAR-β and RAR-γ as well as mice deficient for the individual RAR-α_1, -β_2, and -γ_2 isoforms (LI et al. 1993; LOHNES et al. 1993; LUFKIN et al. 1993; MENDELSOHN et al. 1994b; LUO et al. 1995). Interestingly, mutants deficient in individual isoforms appeared to be completely healthy (LI et al. 1993; MENDELSOHN et al. 1994b). Mutants deficient for all isoforms of RAR-β were viable and fertile; however, they showed fusion of the ninth and tenth cranial ganglia with low penetrance (LUO et al. 1995). Mutants deficient for all RAR-α and RAR-γ isoforms displayed early postnatal lethality and some abnormalities such as testis degeneration in RAR-α mutant mice (LUFKIN et al. 1993) and homeotic transformations of cervical vertebrae in RAR-γ mutants (LOHNES et al. 1993). When compound mutants of different RAR isoforms are generated by cross-breeding of different RAR mutant mice, double RAR mutants are not viable and display multiple defects, among them the typical malformations associated with the vitamin A deficiency syndrome (LOHNES et al. 1994; MENDELSOHN et al. 1994a). These experiments clearly establish that RAR participate in the transduction of the RA signal in vivo, but also suggest extensive functional redundancy between members of the RAR family. Unfortunately, the expression of *Hox* genes has so far only been analyzed in RAR-β mutants (which show no homeotic transformations) and were found to be unchanged (LUO et al. 1995). Therefore, further analysis of RAR mutants, particularly of those displaying homeotic transformations, has to be awaited to determine the extent to which RA is involved in the establishment and maintenance of *Hox* gene expression in vivo.

VI. Local Control of Bone Shape During Embryogenesis and Growth

1. Bone Morphogenetic Proteins

Whereas the *Hox* genes seem to specify the overall skeletal pattern during early stages of embryogenesis, in a second step secreted paracrine and autocrine factors apparently control the local shape of individual skeletal elements by regulating mesenchymal condensation and coordinating growth (reviewed in Erlebacher et al. 1995). The bone morphogenetic proteins (BMP), secreted proteins belonging to the TGF-β superfamily (except for *BMP-1*), have been implicated in this task (Reddi 1992, 1994, and references therein). The BMP were originally identified as the factors responsible for the remarkable ability of lyophilized, decalcified extracellular bone matrix to induce ectopic cartilage and bone at nonskeletal sites upon transplantation. To date, eight mammalian *Bmp* genes have been identified, which belong to three different subfamilies: (1) *Bmp-2* and *Bmp-4* are homologous to the *Drosophila* gene *decapentaplegic* (*dpp*), which is involved in multiple signaling processes during *Drosophila* embryogenesis; (2) *Bmp-3*; and (3) *Bmp-5* to *Bmp-8*, which are homologous to *Drosophila 60A*. *Bmp-1* is not a member of TGF-β superfamily, but encodes a protein with protease activity and EGF-like repeats (Kingsley 1994a; Reddi 1994, and references therein). The temporal and spatial expression patterns of BMP suggest that they are involved in multiple aspects of development, including endochondral bone formation, which apparently requires the coordinate expression of different members of the BMP family (Kingsley 1994b).

The importance of BMP for the regulation of mesenchymal condensation has recently been highlighted with the identification of *Bmp-5* as the gene affected in the mouse skeletal mutant *short ear* (*se*) (Kingsley et al. 1992). In *se* mice, the size, shape, and number of many small bone and cartilage elements as well as some soft tissues are affected. Among the observed defects are the reduction of the size of the external ear, the altered size and shape of the sternum, the loss of one pair of ribs and of two ventral processes unique to the sixth cervical vertebrae, and a reduced ability to repair rib fractures (King et al. 1994; Kingsley et al. 1992). Interestingly, the long bones of the appendicular skeleton are not affected in *se*, whereas exactly these bones are reduced in size in the *brachypodism* (*bp*) mouse mutant. The abnormalities observed in *bp* mice were recently attributed to mutations in the growth and differentiation factor-5 (*Gdf-5*) gene (Storm et al. 1994). The GDF are also members of the TGF-β family and are closely related to the BMP (Kingsley 1994a). In both *se* and *bp*, the abnormalities were traced back to altered size or shape already apparent in the mesenchymal condensation for the affected skeletal elements. Consistent with a role in local control of mesenchymal condensation, *Bmp-5* is expressed in mesenchymal tissues already prior to the formation of obvious condensations, continues to be expressed in the condensed mesenchyme, and becomes restricted to the perichondrium upon chondrification. *Bmp-5* transcripts, however, are not restricted to the skeletal elements affected in *se* but are, for example, present in all vertebral bodies, neural arches, and ribs, suggesting the existence of a compensatory mechanism for the loss of BMP-5 in many skeletal elements, perhaps relying on the function of different members of the BMP family (King et al. 1994). The expression of *Gdf-5* during skeletogenesis seems to be restricted to the structures affected in *bp*, the appendicular long bones, where it is expressed in precartilaginous condensation and in the perichondrium (Storm et al. 1994).

2. Fibroblast Growth Factors and Their Receptors

In bones that develop via endochondral ossification, ossification must be coordinated with both the longitudinal growth at the epiphyseal growth plate and the radial growth of the periosteum. Incomplete coordination of these processes can result in achondroplasia, the most frequent cause of human dwarfism. In achondroplasia patients, overgrowth of the periosteum relative to the epiphyses results in wide, short bones (Erlebacher et al. 1995). Recently, mutations in the transmembrane region of the FGF receptor-3 (*FGFR-3*) were identified as the cause for this dominantly inherited disease (Shiang et al. 1994). In achondroplasia patients, only bones which develop via endochondral ossification are affected, consistent with *FGFR-3* being exclusively expressed in resting chondrocytes and not in the periosteum. In addition, three other dominantly inherited human skeletal disorders were recently correlated with mutations in FGF receptors. Different mutations in *FGFR-2* are responsible for the malformations observed in patients with Crouzon syndrome and Jackson-Weiss syndrome (Jabs et al. 1994), whereas mutations in *FGFR-1* were identified as the cause of Pfeiffer's syndrome (Muenke et al. 1994). In all three syndromes, premature closure of the sutures of the skull results in craniosynostosis, which in the latter two syndromes is accompanied by limb defects, whereas the axial skeleton is largely unaffected. The dominant pattern of inheritance of these syndromes, and the discrepancy between the observed malformations in patients with Pfeiffer's syndrome and the lack of any malformations in mice heterozygous for a targeted disruption of the *Fgfr-1* gene (Deng et al. 1994; Yamaguchi et al. 1994), suggest either that the mutations in the FGF receptor genes in these human skeletal dysplasias are gain of function mutations, or alternatively that the mutated receptors can function in a dominant negative fashion. Taken together, these data suggest that specific FGF's might regulate growth rates of different parts of the skeleton as well as different zones of individual bones by autocrine and paracrine signaling. The local action of FGF (and other factors) might then be integrated to overall skeletal growth by systemic factors such as estrogens (Federman 1994; E.P. Smith et al. 1994) and growth hormone, which predominately acts through the induction of insulin-like growth factor IGF-1 (Baker et al. 1993).

3. Parathyroid Hormone-Related Peptide

Recent results from targeted inactivation of the parathyroid hormone-related peptide (*PTHrP*) gene suggest that PTHrP is another factor involved in the regulation of skeletal growth (Karaplis et al. 1994). In PTHrP-deficient mice, proliferation of chondrocytes in the epiphyseal plates is reduced and differentiation of chondrocytes occurs prematurely, resulting in a greatly increased rate of endochondral ossification and an almost completely ossified skeleton at the time of birth. In contrast to BMP and FGF, PTHrP seems not to be involved in the initial steps of mesenchymal condensation and differentiation into chondrocytes, which is in line with the expression of its receptor

(PTHrPR) only after chondrification in proliferating chondrocytes (KARPERIEN et al. 1994). Upon chondrocyte hypertrophy, *PTHrPR* expression is shut down, suggesting that PTHrP and its receptor might be involved in the regulation of the rate of chondrocyte maturation. The phenotype observed in the PTHrP-deficient mice is reminiscent of the symptoms observed in several human osteochondrodysplasia syndromes. Whether the *PTHrP* or its receptor genes are affected in these syndromes has yet to be investigated.

VII. Collagens and the Extracellular Matrix

Collagens are the major proteins in bone and cartilage matrices. They form a large family of at least 25 members, some of which are expressed at specific places and stages during endochondral ossification (reviewed by VAN DER REST 1991). Collagen type II is the main collagen of cartilage, while collagens IX and XI are two minor cartilage collagens. Collagen X is specifically expressed in hypertrophic cartilage (REICHENBERGER et al. 1991), and type I collagen is the most abundant protein in bone. Procollagen monomers associate to form trimers with triple-helical structure, which after secretion aggregate to form larger structures such as fibrillas or networks, depending on the collagen type. Minor mutations in procollagens frequently lead to a dominant negative effect, as aberrant molecules which are still capable of multimerizing can inhibit the function of unaffected monomers derived from the wild-type allele, whereas complete disruption of the genes usually gives rise to a mild phenotype in the heterozygous state (ERLEBACHER et al. 1995). Numerous mutations in collagens have been detected, and these are frequently associated with skeletal malformations. Mutations in collagen type I lead to osteogenesis imperfecta (reviewed by BYERS and STEINER 1992; WILLING et al. 1993), whereas in different chondrodysplasias collagen type II is affected (reviewed by VIKKULA et al. 1994). Recently, mutations affecting mouse and human collagen XI have been identified as the molecular basis underlying a recessive chondrodysplasia phenotype of a mouse mutant strain (*cho* mice) and both recessive and autosomal dominant human osteochondrodysplasias (LI et al. 1995; VIKKULA et al. 1995). In patients with spondylometaphyseal dysplasia and with metaphyseal dyplasias, in which the hypertrophic cartilage of the growth plates is compressed and bone formation is reduced, mutations in collagen type X have been detected. Expression of a mutated type X collagen in transgenic mice leads to a similar phenotype (reviewed by JACENKO et al. 1994).

The gene affected in diastrophic dysplasia, an autosomal recessive osteochondrodysplasia characterized by dwarfism, spinal deformation, and specific joint abnormalities, was recently identified via positional cloning and was demonstrated to encode a novel sulfate transporter (HÄSTBACKA et al. 1994). Disruption of this gene leads to undersulfation of proteoglycans such as chondroitin sulfate. The overall phenotype of diastrophic dysplasia shares some similarity with the malformations caused by mutations in the FGF receptor genes mentioned above, suggesting some sort of interaction between the FGF signaling pathway and the function of this transporter, which has yet to be verified.

VIII. Regulation of Bone Maintenance and Bone Remodeling: Osteopetrosis and Osteoporosis

For the development of the final bone structure and for its maintenance, the coordinated deposition of new bone matrix by osteoblasts and the resorption of old bone matrix by osteoclasts, which belong to the hematopoietic cell lineage, is necessary. If bone resorption is impaired, the bone marrow space normally present in the long bones is completely filled by dense, homogenous, trabeculation, whereas the density of cortical bone is reduced. These are the characteristic symptoms of osteopetrosis. Several mouse models for osteopetrosis have been identified, in which defects in very different genes affect osteoclast differentiation. Among them are the *osteopetrotic* mouse (*op*), in which a defect in the gene for the colony-stimulating factor-1 (*Csf-1*) results in a nonautonomous early block of osteoclast differentiation (Wiktor-Jedrczejczak et al. 1990; Yoshida et al. 1990), whereas in the *microphthalmia* mouse (*mi*) a mutation in a novel HLH zipper protein impairs osteoclast differentiation due to an intrinsic defect in these cells (Hodgkinson et al. 1993). Mutations in c-*fos* and c-*src* are other examples of defects in osteoclast precursors which result in osteopetrosis (Grigoriadis et al. 1994; Lowe et al. 1993; Soriano et al. 1991).

Whereas in osteopetrosis excess of bone is formed, in osteoporosis the bone mass is reduced as a result of an imbalance between bone resorption and bone formation. The equilibrium between these two processes is to a large extent controlled via endocrine factors such as estrogen and corticosteroids. In addition, a common allele of the vitamin D receptor was recently associated with reduced bone mass resulting in a higher risk of osteoporosis, especially for postmenopause women (Morrison et al. 1994). The vitamin D receptor, which belongs to the nuclear steroid hormone receptor family, and its ligand, vitamin D, are known to be key regulators of bone and calcium homeostasis, consistent with a potential involvement of vitamin D receptor alleles in the development of osteoporosis.

IX. Influence of Teratogens on Axial Skeleton Development

As discussed in the preceding sections, numerous genetic defects may lead to skeletal malformations. However, malformations of the axial skeleton are also common observations in chemical teratogenesis experiments. The chemicals which may cause vertebral column and rib defects include widely used compounds such as sodium salicylate, methyl salicylate, methanol, and caffeine (Warkany and Takacs 1959; Kavlock et al. 1985; Rogers et al. 1993; Connelly and Rogers 1994) The mechanistic basis for their action is usually not known. In order to explain the frequent occurrence of extra-lumbar ribs after chemical exposure of pregnant females, it was suggested that this phenotype may be the result of maternal intoxication rather than a direct effect on the embryo. In support of this hypothesis, Kavlock et al. (1985) demonstrated that seven out of ten different chemicals tested in their study led to the development of supernumerary lumbar ribs and that the occurrence of this rib phenotype showed a strong correlation with an alteration in maternal health status.

Administration of methanol to pregnant mice before day 9 of embryonic development produces skeletal alterations, particularly of cervical vertebrae, that are reminiscent of the changes in positional identity caused by mutations in *Hox* genes or by the administration of RA to pregnant mice (Rogers et al. 1993; Connelly and Rogers 1994). Thus a chemical that must be primarily regarded as a nonspecific toxicant can produce very specific defects. Further experiments are needed to establish whether the control of *Hox* genes or upstream factors is exquisitely sensitive not only to the action of chemicals such as RA which specifically interact with the *Hox* pathway, but also to nonspecific toxic insults by substances such as methanol.

References

Adams JC, Watt FM (1993) Regulation of development and differentiation by the extracellular matrix. Development 117: 1183–1198

Attisano L, Wrana JL, López-Casillas F, Massagué J (1994) TGF-β receptors and actions. Biochim Biophys Acta 1222: 71–80

Augustine K, Liu ET, Sadler TW (1993) Antisense attenuation of Wnt1 and Wnt3a expression in whole embryo culture reveals roles for these genes in craniofacial, spinal cord, and cardiac morphogenesis. Dev Genet 14: 500–520

Bagnall KM, Higgins S, Sanders EJ (1988) The contribution made by a single somite to the vertebral column: experimental evidence in support for resegmentation using the chick-quail chimera model. Development 103: 69–85

Bagnall KM, Higgins S, Sanders EJ (1989) The contribution made by a single somite to tissues within a body segment and assesment of their interaction with similar cells from adjacent segments. Development 107: 931–943

Baker J, Liu JP, Robertson EJ, Efstratiadis A (1993) Role of insulin like growth factors in embryonic and postnatal growth. Cell 75: 73–82

Balling R, Mutter G, Gruss P, Kessel M (1989) Craniofacial abnormalities induced by ectopic expression of the homeobox gene *Hox-1.1* in transgenic mice. Cell 58: 337–347

Becker N, Seitanidou T, Murphy P, Matteí MG, Topilko P, Nieto MA, Wilkinson DG, Charnay P, Gilardi-Hebenstreit P (1994) Several receptor tyrosine kinase genes of the *Eph* family are segmentally expressed in the developing hindbrain. Mech Dev 47: 3–17

Bellairs R (1963) The development of the somites in the chick embryo. J Embryol Exp Morphol 51: 697–714

Bellairs R (1979) The mechanism of somite segmentation in the chick embryo. J Embryol Exp Morphol 51: 227–243

Bellairs R, Curtis ASG, Sanders EJ (1978) Cell adhesiveness and embryonic differentiation. J Embryol Exp Morph 46: 245–252

Bettenhausen B, Hrabe de Angelis M, Simon D, Guénet J-L, Gossler G (1995) Transient and restricted expression during mouse embryogenesis of *Dll1*, a murine gene closely related to *Drosophila Delta*. Development 121: 2407–2418

Blechschmidt E (1961) Die vorgeburtlichen Entwicklungsstadien des Menschen. Karger, Basel

Bloom W, Fawcett DW (1968) A textbook of histology. Saunders, Philadelphia, pp 223–262

Blum M, Gaunt SJ, Cho KWY, Steinbeisser H, Blumberg B, Bittner D, De Robertis EM (1992) Gastrulation in the mouse: the role of the homeobox gene *goosecoid*. Cell 69: 1–20

Brand-Saberi B, Ebensberger C, Wilting J, Balling R, Christ B (1993) The ventralizing effect of the notochord. Anat Embryol 188: 239–245

Braun T, Arnold HH (1995) Inactivation of *Myf-6* and *Myf-5* genes in mice leads to alterations in skeletal muscle development. EMBO J 14: 1176–1186

Braun T, Rudnicki MA, Arnold H-H, Jaenisch R (1992) Targeted inactivation of the muscle regulatory gene *Myf-5* results in abnormal rib development and perinatal death. Cell 71: 369–382

Bronner-Fraser M (1986) Analysis of the early stages of trunk neural crest cell migration in avian embryos using monoclonal antibody HNK-1. Dev Biol 115: 44–55

Buckingham M (1992) Making muscle in mammals. Trends Genet 8: 114–149

Burgess R, Cserjesi P, Ligon KL, Olson EN (1995) *Paraxis*: a basic helix-loop-helix protein expressed in paraxial mesoderm and developing somites. Dev Biol 168: 296–306

Byers PH, Steiner RD (1992) Osteogenesis imperfecta. Annu Rev Med 42: 17–24

Candia AF, Hu J, Crosby J, Lalley PA, Noden D, Nadeau JH, Wright CVE (1992) *Mox-1* and *Mox-2* define a novel homeobox gene subfamily and are differentially expressed during mesodermal patterning in mouse embryos. Development 116: 1123–1136

Chambon P (1994) The retinoid signaling pathway: molecular and genetic analyses. Sem Cell Biol 5: 115–125

Charité J, de Graaff W, Shen S, Deschamps J (1994) Ectopic expression of *Hoxb-8* causes duplication of the ZPA in the forelimb and homeotic transformations of axial structures. Cell 78: 589–601

Cheney CM, Lash JW (1984) An increase in cell-cell adhesion in the chick segmental plate results in a meristic pattern. J Embryol Exp Morph 79: 1–10

Cheng H-J, Flanagan JG (1994) Identification and cloning of Elf-1, a developmentally expressed ligand for the *Mek4* and *Sek* receptor tyrosine kinases. Cell 79: 157–168

Chesley P (1935) Development of the short-tailed mutant in the house mouse. J Exp Zool 71: 429–459

Chevallier A (1975) Rôle du mésoderme somitique dans le developpement de la cage thoracique de l'embryon d'oiseau. I. Origine du segment sternal et méchanismes de la différentation des côtes. J Embryol Exp Morph 33: 291–311

Christ B, Oradhl CP (1995) Early stages of chick somite development. Anat Embryol 191: 381–396

Christ B, Wilting J (1992) From somites to vertebral column. Ann Anat 174: 23–32

Christ B, Jacob HJ, Jacob M (1972) Experimentelle Untersuchungen zur Somitenentwicklung beim Hühnerembryo. Z Anat Entwicklungsgesch 138: 82–97

Christ B, Jacob M, Jacob HJ (1973) Weitere Befunde zur Differenzierung des achsennahen Mesoderms junger Hühnerembryonen. Anat Anz Ergänz-H zum Bd 113: 175–182

Christ B, Jacob HJ, Jacob M (1977) Experimental analysis of the origin of the wing musculature in avian embryos. Anat Embryol 150: 171–186

Christiansen JH, Dennis CL, Wicking CA, Monkley SJ, Wilkinson DG, Wainwright B (1995) Murine *Wnt-11* and *Wnt-12* have temporally and spatially restricted expression patterns during embryonic development. Mech Dev 51: 341–350

Conlon RA, Reaume AG, Rossant J (1995) *Notch1* is required for the coordinate segmentation of somites. Development 121: 1533–1545

Connelly LE, Rogers JM (1994) Methanol causes posterization of cervical vertebrae in the mouse fetus. Teratology 49: 393

Couly GF, Le Douarin MN (1985) Mapping of the early neural primordium in quail chick chimeras. I. Developmental relationships between placodes, facial ectoderm, and prosencephalon. Dev Biol 110: 422–439

Couly GF, Le Douarin MN (1987) Mapping of the early neural primordium in quail chick chimeras. II. The prosencephalic neural plate and neural folds: implications for the genesis of cephalic human congenital abnormalities. Dev Biol 120: 198–214

Couly GF, Coltey PM, Le Douarin MN (1992) The developmental fate of the cephalic mesoderm in quail-chick chimeras. Development 114: 1–15

Cremer H, Lange R, Chritoph A, Plomann M, Vopper G, Roes J, Brown R, Baldwin S, Kraemer P, Scheff S, Barthels D, Rajewsky K, Wille W (1994) Inactivation of the N-CAM gene in mice results in size reduction of the olfactory bulb and deficits in spatial learning. Nature 367: 455–459

Cserjesi P, Brown D, Ligon KL, Lyons GE, Copeland NG, Gilbert DJ (1995) *Scleraxis*: a basic helix-loop-helix protein that prefigures skeletal formation during mouse embryogenesis. Development 121: 1099–1110

Davidson D, Graham E, Sime C, Hill R (1988) A gene with sequence similarity to *Drosophila engrailed* is expressed during the development of the neural tube and vertebrae in the mouse. Development 104: 305–316

Davis CA, Holmyard DP, Millen KJ, Joyner AL (1991) Examining pattern formation in mouse, chicken and frog embryos with an *En*-specific antiserum. Development 111: 287–298

Deng C-X, Wynshaw-Boris A, Shen MM, Daugherty C, Ornitz DM, Leder P (1994) Murine FGFR-1 is required for early postimplantation growth and axial organization. Genes Dev 8: 3045–3057

Deutsch U, Dressler GR, Gruss P (1988) *Pax-1*, a member of a paired box homologous murine gene family, is expressed in segmented structures during development. Cell 53: 159–172

Dietrich S, Gruss P (1995) *undulated* phenotypes suggest a role of Pax-1 for the development of vertebral and extravertebral structures. Dev Biol 167: 529–548

Dietrich S, Schubert FR, Gruss P (1993) Altered Pax gene expression in mouse notochord mutants: the notochord is required to initiate and maintain ventral identity in the somite. Mech Dev 44: 189–207

Dollé P, Dierich A, LeMeur M, Schimmang T, Schuhbaur B, Chambon P, Duboule P (1993) Disruption of the *Hoxd-13* gene induces localized heterochrony leading to mice with neotenic limbs. Cell 75: 431–441

Duband J-L, Dufour S, Hatta K, Takeichi M, Edelman GM, Thierry JP (1987) Adhesion molecules during somitogenesis in the avian embryo. J Cell Biol 104: 1361–1374

Ebensberger C, Wilting J, Brand-Saberi B, Mizutani Y, Christ B, Balling R, Koseki H (1995) *Pax-1*, a regulator of sclerotome development is induced by notochord and floor plate signals in avian embryos. Anat Embryol 191: 297–310

Echelard Y, Epstein DJ, St-Jacques B, Shen L, Mohler J, McMahon JA, McMahon AP (1993) Sonic hedgehog, a member of a family of putative signaling molecules, is implicated in the regulation of CNS polarity. Cell 75: 1417–1430

Epstein DJ, Vogan KJ, Trasler DG, Gros P (1993) A mutation within intron 3 of the *Pax-3* gene produces aberrantly spliced mRNA transcripts in the *Splotch* (Sp) mouse mutant. Proc Natl Acad Sci USA 90: 532–536

Erlebacher A, Filvaroff EH, Gitelman SE, Derynck R (1995) Toward a molecular understanding of skeletal development. Cell 80: 371–378

Fan CM, Tessier-Lavigne M (1994) Patterning of mammalian somites by surface ectoderm and notochord: evidence for sclerotome induction by a hedgehog homologue. Cell 79: 1175–1186

Fan CM, Porter JA, Chiang CC, DT, Beachy PA, Tessier-Lavigne M (1995) Long range sclerotome induction by sonic hedgehog: direct role of the aminoterminal cleavage product and modulation by the cyclic AMP signaling pathway. Cell 81: 457–465

Faust C, Magnuson T (1993) Genetic control of gastrulation in the mouse. Curr Opin Genet Dev 3: 491–498

Federman DD (1994) Life without estrogen. New Engl J Med 331: 1088–1089

Feldman B, Poueymirou W, Papioannou VE, DeChiara TM, Goldfarb M (1995) Requirement of FGF-4 for postimplantation mouse development. Science 267: 246–249

Florini JR, Ewton DZ, Magri KA (1991) Hormones, growth factors and myogenic differentiation. Ann Rev Physiol 53: 201–216

Georg EL, Georges-Labouesse EN, Patel-King RS, Rayburn H, Hynes RO (1993) Defects in mesoderm, neural tube and vascular development in mouse embryos lacking fibronectin. Development 119: 1079–1091

Givol D, Yayon A (1992) Complexity of FGF receptors: genetic basis for structural diversity and functional specificity. FASEB J 6: 3362–3369

Gluecksohn-Schoenheimer S (1944) The development of normal and homozygous Brachy (T/T) mouse embryos in the extraembryonic coelom of the chick. Proc Natl Acad Sci USA 30: 134–140

Goulding M, Sterrer S, Fleming J, Balling R, Nadeau J, Moore KJ, Brown SDN, Steel KP, Gruss P (1993) Analysis of the *Pax-3* gene in the mouse mutant *Splotch*. Genomics 17: 355–363

Goulding M, Lumsden A, Paquette AJ (1994) Regulation of *Pax-3* expression in the dermomyotome and its role in muscle development. Development 120: 957–971

Greenwald I, Rubin GM (1992) Making a difference: the role of cell-cell interactions in establishing separate identities for equivalent cells. Cell 68: 271–281

Grigoriadis AE, Wang Z-Q, Cecchini MG, Hofstetter W, Felix R, Fleisch HA, Wagner EF (1994) c-Fos: a key regulator of osteoclast-macrophage lineage determination and bone remodeling. Science 266: 443–448

Grüneberg H (1963) The pathology of development. Blackwell, Oxford

Gruss P, Walther C (1992) Pax in development. Cell 69: 719–722

Hästbacka J, de la Chapelle A, Mahtani MM, Clines G, Reeve-Daly MP, Daly M, Hamilton BA, Kusumi K, Trivedi B, Weaver A, Coloma A, Lovett M, Buckler A, Kaitila I, Lander ES (1994) The diastrophic dysplasia gene encodes a novel sulfate transporter: positional cloning by fine structure linkage disequilibrium mapping. Cell 78: 1073–1087

Hasty P, Bradley A, Morris JH, Edmondson DG, Venuti JM, Olso EN, Klein WH (1993) Muscle deficiency and neonatal death in the mice with a targeted mutation in the myogenin gene. Nature 364: 501–506

Haub O, Goldfarb M (1991) Expression of the fibroblast growth factor 5 gene in the mouse embryo. Development 112: 397–406

Hébert JM, Rosenquist T, Götz J, Martin GR (1994) FGF5 as a regulator of the hair growth cycle: evidence from targeted and spontaneous mutations. Cell 78: 1017–1025

Hermann BG (1991) Expression pattern of the Brachyury gene in whole-mount T^{Wis}/T^{Wis} mutant embryos. Development 113: 913–917

Hermann BG, Kispert A (1994) The *T* gene in embryogenesis. Trends Genet 10: 280–286

Herrmann BG, Labeit S, Poustka A, King TR, Lehrach H (1990) Cloning of the T gene required for mesoderm formation in the mouse. Nature 343: 617–622

Hill RE, Hall AE, Sime CM, Hastie ND (1987) A mouse homeobox-containing gene maps near a developmental mutation. Cytogenet Cell Genet 44: 171–174

Hodgkinson CA, Moore KJ, Nakayama A, Steingrimsson E, Copeland NG, Jenkins NA, Arnheiter H (1993) Mutations at the mouse microphthalmia locus are asso-

ciated with defects in a gene encoding a novel basic-helix-loop-helix-zipper protein. Cell 74: 395–404
Hogan B, Beddington R, Costantini F, Lacy E (1994) Manipulating the mouse embryo. A laboratory manual. Cold Spring Harbour Laboratory Press, Cold Spring Harbour
Huang R, Zhi Q, Wilting J, Christ B (1994) The fate of somitocoel cells in avian embryos. Anat Embryol 190: 243–250
Ingham PW (1988) The molecular genetics of embryonic pattern formation in *Drosophila*. Nature 335: 25–34
Ingham PW (1994) Hedgehog points the way. Curr Biol 4: 347–350
Jabs EW, Li X, Scott AF, Meyers G, Chen W, Eccles M, Mao J-i, Charnas LR, Jackson CE, Jaye M (1994) Jackson-Weiss and Crouzon syndromes are allelic with mutations in fibroblast factor receptor 2. Nature Gent 8: 275–279
Jacenko O, Olsen BR, Warman ML (1994) Of mice and men: heritable skeletal disorders. Am J Hum Genet 54: 163–168
Jacob M, Christ B, Jacob HJ (1975) The early differentiation of the perinotochordal connective tissue. A scanning and transmisson electron microscopic study on chick embryos. Experientia 31: 1083–1086
Jacobson A, Meier S (1986) In: Bellairs R, Ede DA, Lash JW (eds) Somites in developing embryos. Plenum, New York, pp 1–16
Johnson RL, Tabin C (1995) The long and short of *hedgehog* signaling. Cell 81: 313–316
Johnson RL, Laufer E, Riddle RD, Tabin C (1994) Ectopic expression of *sonic hedgehog* alters dorsal-ventral patterning of somites. Cell 78: 1165–1173
Jostes B, Walther C, Gruss P (1991) The murine paired box gene *Pax7*, is expressed specifically during the development of the nervous and muscular system. Mech Dev 33: 27–38
Joyner AL (1993) Gene targeting; a practical approach. Oxford University Press, Oxford
Joyner AL, Martin GR (1987) *En-1* and *En-2*, two mouse genes with sequence homology to the *Drosophila engrailed* gene: expression during embryogenesis. Genes Dev 1: 29–38
Juliano RL, Haskill S (1993) Signal transduction from the extracellular matrix. J Cell Biol 120: 577–585
Karaplis AC, Luz A, Glowacki J, Bronson RT, Tybulewicz VLJ, Kronenberg HM, Mulligan RC (1994) Lethal skeletal dysplasia from targeted disruption of parathyroid hormone-related peptide gene. Genes Dev 8: 277–289
Karperien M, van Dijk TB, Hoeijmakers T, Cremers F, Abou-Samra A-B, Boonstra J, de Laat SW, Defite LHK (1994) Expression pattern of the parathyroid hormone/parathyroid hormone related peptide receptor mRNA in mouse postimplantation embryos indicates involvement in multiple developmental processes. Mech Dev 47: 29–42
Kavlock RJ, Chernoff N, Rogers EH (1985) The effect of acute maternal toxicity on fetal development in the mouse. Teratogen Carcinogen Mutagen 5: 3–13
Kessel M (1992) Respecification of vertebral identities by retinoic acid. Development 115: 487–501
Kessel M, Gruss P (1991) Homeotic transformations of murine prevertebrae and concomitant alteration of Hox codes induced by retinoic acid. Cell 67: 89–104
Kessel M, Balling R, Gruss P (1990) Variations of cervical vertebrae after expression of a Hox-1.1 transgene in mice. Cell 61: 301–308
Keynes RJ, Stern CD (1984) Segmentation in the vertebrate nervous system. Nature 310: 786–789
Keynes RJ, Stern CD (1988) Mechanisms of vertebrate segmentation. Development 103: 413–429
Kieny M, Mauger A, Sengel P (1972) Early regionalization of the somitic mesoderm as

studied by the development of the axial skeleton of the chick embryo. Dev Biol 28: 142–161

Kilchherr F, Baumgartner S, Bopp D, Frei E, Noll M (1986) Isolation of the *paired* gene of *Drosophila* and its spatial expression during early embryogenesis. Nature 321: 493–499

King JA, Marker PC, Seung KJ, Kingsley DM (1994) *BMP5* and the molecular, skeletal and soft-tissue alterations in *short ear* mice. Dev Biol 166: 112–122

Kingsley DM (1994a) The TGF-β superfamily: new members, new receptors, and new genetic tests of function in different organisms. Genes Dev 8: 133–146

Kingsley DM (1994b) What do BMPs do in mammals?: clues from the mouse short-ear mutation. Trends Genet 10: 16–21

Kingsley DM, Bland AE, Grubber JM, Marker PC, Russell LB, Copeland NG, Jenkins NA (1992) The mouse short ear skeletal morphogenesis locus is associated with defects in a bone morphogenetic member of the TGFβ superfamily. Cell 71: 399–410

Koseki H, Wallin J, Wilting J, Mitzutani Y, Ebensberger C, Christ B, Balling R (1993) *Pax-1* as a mediator of notochordal signals in the dorsoventral specification of vertebrae. Development 119: 649–660

Krumlauf R (1993) Mouse *Hox* genetic functions. Curr Opin Genet Dev 3: 621–625

Krumlauf R (1994) *Hox* genes in vertebrate development. Cell 78: 191–201

Lardelli M, Lendahl U (1993) *MotchA* and *MotchB*-two mouse *Notch* homologues coexpressed in a wide variety of tissues. Exp Cell Res 204: 264–272

Lardelli M, Dahlstrand J, Lendahl U (1994) The novel *Notch* homologue mouse *Notch 1* lacks specific epidermal growth factor-repeats and is expressed in proliferating neuroepithelium. Mech Dev 46: 123–136

Lash JW, Seitz AW, Cheney CM, Ostrovsky D (1984) On the role of fibronectin during the compaction stage of somitogenesis in the chick embryo. J Exp Zool 232: 197–206

Lash JW, Linask KK, Yamada KM (1987) Synthetic peptides that mimic the adhesive recognition signal of fibronectin: differential effects on cell-cell and cell-substratum adhesion in embryonic chick cells. Dev Biol 123: 411–420

Lepage T, Cohen S, Diaz-Benjumea FJ, Parkhurst S (1995) Signal transduction by cAMP-dependent protein kinase A in *Drosophila* limb patterning. Nature 373: 711–714

Li E, Sucov HM, Lee K-H, Evans RM, Jaenisch R (1993) Normal development of mice carrying a targeted disruption of the α1 retinoic acid receptor gene. Proc Natl Acad Sci USA 90: 1590–1594

Li X, Noll M (1993) Role of the gooseberry gene in *Drosophila* embryos: maintenance of wingless expression by a wingless-gooseberry autoregulatory loop. EMBO J 12: 4499–4510

Li Y, Lacerda DA, Warman ML, Beier DR, Yoshioka H, Ninomiya Y, Oxford JT, Morris NP, Andrikopoulos K, Ramirez F, Wardell BB, Lifferth GD, Teuscher C, Woodward SR, Taylor BA, Seegmiller RE, Olsen BR (1995) A fibrillar collagen gene, *Col11a1* is essential for skeletal morphogenesis. Cell 80: 423–430

Lohnes D, Kastner P, Dietrich A, Mark M, LeMeur M, Chambon P (1993) Function of retinoic acid receptor γ in the mouse. Cell 73: 643–658

Lohnes D, Mark M, Mendelsohn C, Dollé P, Dietrich A, Gorry P, Gansmuller A, Chambon P (1994) Function of the retinoic acid receptors (RARs) during development. I. Craniofacial and skeletal abnormalities in RAR double mutants. Development 120: 2723–2748

Loring JF, Erickson CA (1987) Neural crest cell migratory pathways in the trunk of the chick embryo. Dev Biol 121: 220–236

Lowe C, Yoneda T, Boyce FF, Chen H, Mundy GR, Soriano P (1993) Osteopetrosis in Src-deficient mice is due to an autonomous defect of osteoclasts. Proc Natl Acad Sci USA 90: 4485–4489

Lufkin T, Lohnes D, Mark M, Dierich A, Gorry P, Gaub M-P, LeMeur M, Chambon P (1993) High postnatal lethality and testis degeneration in retinoic acid receptor α mutant mice. Proc Natl Acad Sci USA 90: 7225–7229

Luo J, Pasceri P, Conlo RA, Rossant J, Giguère V (1995) Mice lacking all isoforms of retinoic acid receptor β develop normally and are susceptible to teratogenic effects of retinoic acid. Mech Dev 53: 61–71

Lyons KM, Pelton RW, Hogan BLM (1990) Organogenesis and pattern formation in the mouse: RNA distribution patterns suggest a role for bone morphogenetic protein-2A (BMP-2A). Development 109: 833–844

Lyons MF, Searle AG (1989) Genetic variants and strains of the laboratory mouse. Oxford University Press, Oxford

Mahmood R, Kiefer P, Guthrie S, Dickson C, Mason I (1995) Multiple role for FGF-3 during cranial neural development in the chicken. Development 121: 1399–1410

Mansour SL, Goddard JM, Capecchi MR (1993) Mice homozygous for a targeted disruption of the proto-oncogene *int-2* have developmental defects in the tail and inner ear. Development 117: 13–28

Mason IJ (1994) The ins and outs of fibroblast growth factors. Cell 78: 547–552

Matzuk MM, Kumar TR, Bradley A (1995a) Different phenotypes for mice deficient in either activins or activin receptor type II. Nature 374: 356–360

Matzuk MM, Kumar TR, Vassalli A, Bickenbach JR, Roop DR, Jaenisch R, Bradley A (1995b) Functional analysis of activins during development. Nature 374: 354–356

McGinnis W, Krumlauf R (1992) Homeobox genes and axial patterning. Cell 68: 283–302

McMahon AP (1992) The *Wnt* family of developmental regulators. Trends Genet 8: 236–242

Mendelsohn C, Lohnes D, Decimo D, Lufkin T, LeMeur M, Chambon P, Mark M (1994a) Function of retinoic acid receptors during development. II. Multiple abnormalities at various stages of organogenesis in RAR double mutants. Development 120: 2749–2771

Mendelsohn C, Mark M, Dollé P, Dierich A, Gaub MP, Krust A, Lampron C, Chambon P (1994b) Retinoic acid receptor β2 (RARβ2) null mutant mice appear normal. Dev Biol 166: 246–258

Meier S (1979) Development of the chick embryo mesoblast. Formation of the embryonic axis and establishment of the metameric pattern. Dev Biol 73: 25–45

Meier S (1984) Somite formation and its relationship to metameric patterning of the mesoderm. Cell Differ 4: 235–243

Menkes B, Sandor S (1977) Somitogenesis: regulation potencies, sequence determination and primordial interactions. In: Ede DA, Hinchliffe JR, Bails M (eds) Vertebrate limb and somite morphogenesis. Cambridge University Press, Cambridge, pp 405–420

Morgan BA, Tabin CJ (1993) The role of homeobox genes in limb development. Curr Opin Genet Dev 3: 668–674

Morrison NA, Cheng Qi J, Tokita A, Kelly PJ, Crofts L, Nguyen TV, Sambrook PN, Eisman JA (1994) Prediction of bone density from vitamin D receptor alleles. Nature 367: 284–287

Muenke M, Schell U, Hehr A, Robin NH, Losken WH, Schinzel A, Pulleyn LJ, Rutland PR, Malcom S, Winter RM (1994) A common mutation in the fibroblast growth factor receptor 1 gene in Pfeiffer syndrome. Nature Genet 8: 269–274

Mullins MC, Nüsslein-Volhard C (1993) Mutational approaches to studying embryonic pattern formation in the zebrafish. Curr Opin Genet Dev 3: 648–654

Nabeshima Y, Hanaoka K, Hayasaka M, Esumi E, Li S, Nonaka I, Nabeshima Y (1993) Myogenin gene disruption results in perinatal lethality because of a severe muscle defect. Nature 364: 532–535

Neubüser A, Koseki H, Balling R (1995) Characterization and developmental expression of *Pax9*, a paired box containing gene related to *Pax1*. Dev Biol 170: 701–716

Nieto MA, Gilardi-Hebenstreit P, Charnay P, Wilkinson DG (1992) A receptor protein tyrosine kinase implicated in the segmental patterning of the hind brain and mesoderm. Development 116: 1137–1150

Niswander L, Martin GR (1992) *Fgf-4* expression during gastrulation, myogenesis, limb and tooth development in the mouse. Development 114: 755–768

Noll M (1993) Evolution and role of Pax genes. Curr Opin Genet Dev 3: 595–605

Norris WE, Stern CD, Keynes RJ (1989) Molecular differences between the rostral and caudal halves of the sclerotome in the chick embryo. Development 105: 541–548

Nüsslein-Volhard C, Wieschaus E (1980) Mutations affecting segment number and polarity in Drosophila. Nature 287: 795–801

Ordahl CP, Le Douarin NM (1992) Two myogenic lineages within the developing somite. Development 114: 339–353

Orr-Urtreger A, Givol D, Yayon A, Yarden Y, Lonai P (1991) Developmental expression of two murine fibroblast growth factor receptors, *flg* and *bek*. Development 113: 1419–1434

Ott M-O, Bober G, Lyons G, Arnold H-H, Buckingham M (1991) Early expression of the myogenic regulatory gene, *myf-5*, in precursor cells of skeletal muscle in the mouse embryo. Development 111: 1097–1107

Packard DS (1978) Chick somite determination: the role of factors in young somites and the segmental plate. J Exp Zool 203: 295–306

Pan D, Rubin GM (1995) cAMP dependent protein kinase and hedgehog act antagonistically in regulating *decapentaplegic* transcription in *Drosophila* imaginal discs. Cell 80: 543–552

Parr BA, McMahon AP (1994) *Wnt* genes and vertebrate development. Curr Opin Genet Dev 4: 523–528

Patten BM (1953) Human embryology, 2nd edn. Blackiston, New York

Peifer M, Bejsovec A (1992) Knowing your neighbors: cell interactions determine intrasegmental patterning in *Drosophila*. Trends Genet 8: 243–248

Peters KG, Werner S, Chen G, Williams LT (1992) Two FGF receptor genes are differentially expressed in epithelial and mesenchymal tissues during limb formation and organogenesis in the mouse. Development 114: 233–243

Placzek M, Tessier-Lavigne M, Jessell T, Dodd J (1990) Mesodermal control of neural cell identity: floor plate induction by the notochord. Science 250: 985–988

Placzek M, Yamada T, Tessier-lavigne M, Jessell TM (1991) Control of dorso-ventral pattern in the vertebrate neural development: induction and polarizing properties of the floor plate. Development 113 [Suppl 2]: 105–122

Placzek M, Jessell TM, Dodd J (1933) Induction of floor plate differentiation by contact dependent, homeogenetic signals. Development 117: 205–218

Porter JA, von Kessler DP, Ekker SC, Young KE, Lee JJ, Moses K, Beachy PA (1995) The product of *hedgehog* autoproteolytic cleavage active in local and long-range signaling. Nature 374: 363–366

Pourquié O, Coltey M, Teillet MA, Ordahl C, Le Douarin NM (1993) Control of dorsoventral patterning of somitic derivatives by notochord and floor plate. Proc Natl Acad Sci USA 90: 5342–5246

Primmett DRM, Stern CD, Keynes RJ (1988) Heat shock causes repeated segmental anomalies in the chick embryo. Development 104: 331–339

Primmett DRM, Norris WE, Carlson GJ, Keynes RJ, Stern CD (1989) Periodic segmental anomalies induced by heat shock in the chick embryo are associated with the cell cycle. Development 105: 119–130

Rauber AA, Kopsch F (1987) Anatomie des Menschen. Lehrbuch und Atlas. Thieme, Stuttgart

Reaume AG, Conlon RA, Zirngibl R, Yamaguchi TP, Rossant J (1992) Expression analysis of a *Notch* homologue in the mouse embryo. Dev Biol 154: 377–387

Reddi AH (1992) Regulation of cartilage and bone differentiation by bone morphogenetic proteins. Curr Opin Cell Biol 4: 850–855

Reddi AH (1994) Bone and cartilage differentiation. Curr Opin Genet Dev 4: 737–744

Reichenberger E, Aigner T, von der Mark K, Stöß H, Bertling B (1991) In situ hybridization studies on the expression of type X collagen in fetal human cartilage. Dev Biol 148: 562–572

Remak R (1855) Untersuchungen über die Entwicklung der Wirbeltiere. Reimer, Berlin

Rickmann M, Fawcett JW, Keynes RJ (1985) The migration of neural crest cells and the growth of motor axons through the rostral half of the chick somite. J Embryol Exp Morph 90: 437–455

Roelink H, Augsburger A, Heemskerk J, Korzh V, Norlin S, Ruizi Altaba A, Tanabe Y, Placzek M, Edlund T, Jessell TM, Dodd J (1994) Floor plate and motor neuron induction by *vhh-1*, a vertebrate homologue of hedgehog expressed by the notochord. Cell 76: 761–775

Roelink H, Porter JA, Chiang C, Tanabe Y, Chang DT, Beachy PA, Jessell TM (1995) Floor plate and motor neuron induction by different concentrations of the amino-terminal cleavage product of sonic hedgehog autoproteolysis. Cell 81: 445–455

Rogers JM, Mole ML, Chernoff N, Barbee BD, Turner CI, Logsdon TR, Kavloc RJ (1993) The developmental toxicity of inhaled methanol in the CD-1 mouse, with quantitative dose response modeling for estimation of benchmark doses. Teratology 47: 175–188

Rough R (1990) The mouse. Its reproduction and development. Oxford University Press, Oxford

Rudnicki MA, Braun T, Hinuma S, Jaenisch R (1992) Inactivation of *MyoD* in mice leads to up-regulation of myogenic HLH gene *Myf-5* and results in apparently normal muscle development. Cell 71: 383–390

Rudnicki MA, Schnegelsberg PNJ, Stead RH, Braun T, Arnold H-H, Jaenisch R (1993) *MyoD* or *Myf-5* is required for the formation of skeletal muscle. Cell 75: 1351–1360

Ruoslahti E, Amaguchi Y (1991) Proteoglycans as modulators of growth factor activities. Cell 64: 867–869

Saga Y, Yagi T, Ikawa Y, Sakakura T, Aizawa S (1992) Mice develop normally without tenascin. Genes Dev 6: 1821–1831

Scott D, Kispert A, Herrmann BG (1993) Rescue of the tail defect of *Brachyury* mice. Genes Dev 7: 197–203

Shiang R, Thompson LM, Zhu Y-Z, Church DM, Fielder TJ, Bocian M, Winokur ST, Wasmuth JJ (1994) Mutations in the transmembrane domain of FGFR3 cause the most common genetic form of dwarfism, achondroplasia. Cell 78: 335–342

Smith EP, Boyd J, Frank GR, Takahashi H, Cohen RM, Specker B, Williams TC, Lubahn DB, Korach KS (1994) Estrogen resistance caused by a mutation in the estrogen-receptor gene in man. N Engl J Med 331: 1056–1061

Smith TH, Kachinsky AM, Boone Miller J (1994) Somite subdomains, muscle cell origins, and the four muscle regulatory factor proteins. J Cell Biol 127: 95–105

Solursh M, Drake C, Meier S (1979) The role of extracellular matrix in the formation of the sclerotome. J Embryol Exp Morphol 54: 75–98

Soriano P, Montgomery C, Geske R, Bradley A (1991) Targeted disruption of the c-src protooncogene leads to osteopetrosis in mice. Cell 64: 693–702

Spiegelman M (1976) Electron microscopy of cell associations in *T*-locus mutants. Ciba Found Symp 40: 199–220

St Johnston D, Nüsslein-Volhard C (1992) The origin of pattern and polarity in the *Drosophila* embryo. Cell 68: 201–219

Stern CD (1992) Vertebrate gastrulation. Curr Opin Genet Dev 2: 556–561

Stern CD, Keynes RJ (1987) Interactions between somite cells: the formation and maintenance of segment boundaries in the chick embryo. Development 99: 261–272

Stern CD, Jaques KF, Lim T-M, Fraser SE, Keynes RJ (1991) Segmental lineage restrictions in the chick embryo spinal cord depend on the adjacent somite. Development 113: 239–244

Storm EE, Huynh TV, Copeland NG, Jenkins NA, Kingsley DM, Lee SJ (1994) Limb alterations in brachypodism mice due to mutations in a new member of the TGFβ-superfamily. Nature 368: 639–643

Strutt DI, Wiersdorff V, Mlodzik M (1995) Regulation of furrow progression in the *Drosophila* eye by cAMP dependent protein kinase A. Nature 373: 705–709

Suemori H, Takahashi N, Noguchi S (1995) *Hoxc-9* mutant mice show anterior transformation of the vertebrae and malformation of the sternum and ribs. Mech Dev 51: 265–273

Swiatek PJ, Lindsell CE, Frnaco del Amo F, Weinmaster G, Gridley T (1994) *Notch1* is essential for postimplantation development in mice. Genes Dev 8: 707–719

Takada S, Stark KL, Shea MJ, Vassileav G, McMahon JA, McMahon AP (1994) *Wnt-3a* regulates somite and tail bud formation in the mouse embryo. Genes Dev 8: 174–189

Tanabe Y, Roelink H, Jessell TM (1995) Induction of motor neurons by Sonic hedgehog is independent of floor plate differentiation. Curr Biol 5: 651–658

Teillet M-A, Kalcheim C, Le Douarin NM (1987) Formation of the dorsal root ganglion in the avian embryo: segmental origin and migratory behavior of neural crest progenitor cells. Dev Biol 120: 329–347

Thöndury G (1958) Entwicklungsgeschichte und Fehlbildungen der Wirbelsäule. Hippokrates, Stuttgart

van der Rest M (1991) Collagen family of proteins. FASEB J 5: 2814–2823

van Straaten HWM, Hekking JWM, Wiertz-Hoessels EJLM, Thors F, Drukker J (1988) Effect of the notochord on the differentiation of a floor plate area in the neural tube of the chick embryo. Anat Embryol 177: 317–324

Verbout AJ (1985) The development of the vertebral column. Adv Anat Embryol Cell Biol 89: 1–122

Vikkula M, Metasaranta M, Ala-Kokko L (1994) Type II collagen mutations in rare and common cartilage diseases. Ann Med 26: 107–114

Vikkula M, Mariman ECM, Lui VCH, Zhidkova NI, Tiller GE, Goldring MB, van Beersum SEC, de Wal Malfijt MC, van den Hoogen FHJ, Ropers H-H, Mayne R, Cheah KSE, Olson BR, Warman ML, Brunner HG (1995) Autosomal dominant and recessive osteochondrodysplasias associated with the *COL11A2* locus. Cell 80: 431–437

Vogan KJ, Epstein DJ, Trasler DG, Gros P (1993) The *Splotch-delayed* (Sp^d) mouse mutant carries a point mutation within the paired box of the *Pax-3* gene. Genomics 17: 364–369

Wallin J, Koseki H, Imai K, Moriwaki K, Miyshita N, Balling R (1993) A new Pax gene, *Pax-9*, maps to mouse chromosome 12. Mammal Genome 4: 354–358

Wallin J, Koseki H, Wilting J, Ebensberger C, Christ B, Balling R (1994) The role of *Pax-1* in the development of the axial skeleton. Development 120: 1109–1121

Warkany J, Takacs E (1959) Experimental production of chemical malformations in rats by salicylate poisoning. Am J Pathol: 315–331

Watterson RL, Fowler I, Fowler BJ (1954) The role of the neural tube and notochord in development of the axial skeleton of the chick. Am J Anat 95: 337–400

Weintraub H (1993) The MyoD family and myogenesis: redundancy, networks, and thresholds. Cell 75: 1241–1244

Weintraub H, Davis R, Tapscott S, Thayer M, Grause M, Benezra R, Blackwell T, Turner D, Rupp R, Hollenberg S, Zhuang Y, Lassar A (1991) The MyoD gene family: nodal point during specification of the muscle cell lineage. Science 251: 761–766

Wiktor-Jedrczejczak W, Bartocci A, Ferrante AWJ, Ahmed-Ansari A, Sell KW, Pollard JW, Stanley ER (1990) Total absence of colony stimulating factor 1 in the macrophage deficient osteopetrotic (*op/op*) mouse. Proc Natl Acad Sci USA 87: 4828–2832

Wilkinson DG, Bhatt S, Hermann BG (1990) Expression pattern of the mouse *T* gene and its role in mesoderm formation. Nature 343: 657–659

Williams BA, Ordahl CP (1994) *Pax-3* expression in segmental mesoderm marks early stages in myogenic cell specification. Development 120: 785–796

Willing MC, Pruchno CJ, Byers PH (1993) Molecular heterogeneity in osteogenesis imperfecta type I. Am J Med Genet 45: 223–227

Woodland HR (1993) Identifying the three signals. Curr Biol 3: 27–29

Wurst W, Auerbach AB, Joyner AL (1994) Multiple developmental defects in *engrailed-1* mutant mice: an early mid-hindbrain deletion and patterning defects in forelimbs and sternum. Development 120: 2065–2075

Yamada T, Paczek M, Tanaka H, Dodd J, Jessell TM (1991) Control of cell pattern in the developing nervous system: polarizing activity of the floor plate and notochord. Cell 64: 635–647

Yamada T, Pfaff SL, Edlund T, Jessell TM (1993) Control of cell pattern in the neural tube: motor neuron induction by diffusible factors from notochord and floor plate. Cell 73: 673–686

Yamaguchi TP, Conlon RA, Rossant J (1992) Expression of the fibroblast growth factor receptor *FGFR-1/flg* during gastrulation and segmentation in the mouse embryo. Dev Biol 152: 75–88

Yamaguchi TP, Harplan K, Henkeeyer M, Rossant J (1994) *fgfr-1* is required for embryonic growth and mesodermal patterning during mouse gastrulation. Genes Dev 8: 3032–3044

Yoshida H, Hayashi S, Kunisada T, Ogawa M, Nishikawa S, Okamura H, Sudo T, Shultz LD, Nishikawa S (1990) The murine mutation osteopetrosis is in the coding region of the macrophage colony stimulating factor gene. Nature 345: 442–444

Zhang W, Behringer RR, Olson EN (1995) Inactivation of the myogenic bHLH gene MRF4 results in up-regulation of myogenin and rib anomalities. Genes Dev 9: 1388–1399

Zhou X, Sasaki H, Lowe L, Hogan BLM, Kuehn MR (1993) Nodal is a novel TGF-β-like gene expressed in the mouse node during gastrulation. Nature 361: 543–547

CHAPTER 5

Molecular Mechanisms Regulating the Early Development of the Vertebrate Nervous System

J.D. Burrill, H. Saueressig, and M. Goulding

The vertebrate nervous system arises from a pseudostratified epithelium, the neural plate, that is formed from the dorsal ectoderm of the embryo. Development of the nervous system is marked by a number of morphological events that involve first the induction of "neural tissue", and then the patterning and refinement of cell fate within this neural tissue. The end result of these morphogenetic events is the generation of topologically defined classes of neurons each with distinct patterns of interconnections and neurotransmitter phenotypes, together with various glial cell types and other accessory cells such as the ependymal cells that line the meninges.

In recent years a number of striking advances have begun to unravel the molecular mechanisms that control the development of the nervous system. This has been due to the advent of molecular biology and the identification of developmental regulatory genes, many of which are related to genes that have demonstrated roles in invertebrate neurogenesis in organisms such as the fruit fly *Drosophila melanogaster* and the nematode *Caenorhabditis elegans*. This review will outline some of the recent advances in our understanding of the molecular mechanisms underlying vertebrate neurogenesis.

A. Early Development of the Nervous System

The nervous system in vertebrates is derived from ectodermal cells positioned dorsally in the developing embryo that have the ability to differentiate as either epidermal cells or neural plate cells. Neural development begins with neural induction, whereby cells in the presumptive neural plate acquire the ability to undergo differentiation into neurons and are no longer able to develop as surface ectoderm. Experimental evidence suggests that neural induction may represent the default pathway. In *Xenopus* embryos, for example, dissociated dorsal ectodermal cells, undergo neuronal differentiation in the absence of further signaling (Godsave and Slack 1989; Hemmati-Brivanlou et al. 1994). Recent studies indicate that many of the molecules capable of acting as "neural inducers," such as *noggin* and *follistatin*, act by blocking signaling by members of the TGF-β family, in particular BMP-4, which is thought to repress neural induction (Hemmati-Brivanlou et al. 1994; Hawley et al. 1995). In early embryos BMP-4 expression is excluded from the presumptive neural plate. For example, in *Xenopus*, BMP-4 is expressed in the marginal zone and is absent from the dorsal region of the embryo (Fainsod et al. 1994). In gastrula stage chick embryos, BMP-4 is expressed in the epiblast in presumptive ectoderm and mesoderm derivatives but is excluded from the neural plate (A. Bang and M. Goulding, unpublished results).

Studies in *Drosophila* have identified a gene, *shortened gastrulation* (sog), that is required for the correct dorsoventral patterning of the blastoderm (Francois et al. 1994). *Sog* acts by antagonizing the activity of the dorsally expressed morphogen *decapentaplegic (dpp)*, the *Drosophila* homologue of BMP-4 (Holley et al. 1995). In *Xenopus* embryos, a protein sharing sequence similarity with *sog*, *chordin*, has been shown to negatively regulate the activity of BMP-4 in frogs (Sasai et al. 1994). *Chordin* and *sog* are functionally equivalent, since *chordin* can partially rescue *sog* mutations in *Drosophila*. Interestingly, *chordin* is expressed in cells in the dorsal marginal zone that are known to be the source of the neuralizing signal in *Xenopus* embryos. In the chick, *chordin* is also expressed in Henson's node (the equivalent of the dorsal blastopore lip), consistent with *chordin* acting as neuralizing signal in vertebrates. Transcripts of *noggin* are also present in this node, or "organizer," and as such *noggin* may be a component of the "neuralizing signal" that emanates from the organizer.

B. Neurogenesis: The Role of the Helix–Loop–Helix Proneural Genes

In *Drosophila* a family of basic helix-loop-helix (bHLH) transcription factors encoded by the achaete-scute complex (ASC) and the *atonal* gene are necessary for the development of neurons. The genes are known as proneural genes and regulate decisions regarding neural versus epidermal cell fate. A number of genes encoding bHLH proteins with homology to *Drosophila* ASC or *atonal* have now been described in vertebrates, including *Mash-1* in rat and mouse; *Xash-1*, *Xash-3* and *neuroD* in *Xenopus*; *Cash-1* in chick; and *Math-1* and *Math-2* in mouse. These proteins probably function as heterodimers with the ubiquitously expressed E-type bHLH factors, which are related to *daughterless*. Overexpression studies in *Xenopus,* as well as loss-of-function studies in mice, indicate that these factors promote determination and differentiation of neural fate. In *Drosophila* the gene *atonal* is required for the formation of chordotonal organs in the peripheral nervous system (Jarman et al. 1993). *lin-32,* a *C. elegans* gene with sequence similarity to *atonal*, is required for the specification of neuroblast fate in peripheral sense organs (Zhao and Emmons 1995), indicating that the generation of neural precursors may be regulated by conserved mechanisms throughout the animal kingdom.

Xash-3 is one of the earliest known genes expressed after neural induction in *Xenopus* and is thus a good candidate for playing an actual "proneural" role, mediating neural versus epidermal cell fate decisions. *Xash-3* expression is detected transiently in proliferating neural precursors, primarily restricted to longitudinal domains along the mediolateral axis of the presumptive neural plate. Overexpression of *Xash-3* in developing *Xenopus* embryos leads to an expansion (but not overproliferation) of the neural plate at the expense of surrounding epidermal tissue. Dorsal ectoderm explants, or "animal caps", from blastula stage *Xenopus* embryos will normally develop as ectoderm unless exposed to a neural inducer. Animal caps isolated from *Xash-3* injected embryos transiently express markers of neurogenesis such as NCAM, but this expression is not stable unless the neural inducing factor noggin is co-injected. The authors postulate that *Xash-3* is unable to convert ectoderm to neural tissue and stably induce neurogenesis in animal caps because it may be sensitive to specific inhibitors in the ectoderm. Thus, neural induction would relieve these inhibitors, allowing *Xash-3* to act in determination of neural cell fate.

Mash-1 is transiently expressed in a number of unrelated subsets of proliferating neural precursors in the spinal cord, autonomic ganglia, nasal epithelium, and forebrain (Lo et al. 1991). Although it is unclear whether *Mash-1* plays a proneural function in the selection of neural versus non-neural cell-fate, *Mash-1* is required for the development of olfactory receptor neuronal progenitors, as well as neuronal progenitors in the sympathetic and enteric ganglia, because these cell types fail to develop in *Mash-1* -/- mutant embryos (Guillemot et al. 1993). The chick homologue of *Mash-1*, *Cash-1*, is also expressed in sympathetic ganglia, where its expression is independent from signals from the notochord that are known to regulate the catecholaminergic phenotype of sympathetic neurons (Groves et al. 1995). This suggests that both *Mash-1* and *Cash-1* may act by promoting "pan-neuronal" differentiation rather than specifying particular neuronal cell types. It is likely that other proneural-type bHLH genes similar to *Mash-1*/*Cash-1* are likely to be found in the developing nervous system, since both of these genes are expressed only in a subset of neuronal precursor cells. Interestingly, a family of murine bHLH genes that encode for atonal-like proteins have recently been identified, and these are expressed in cells that are distinct from *Mash-1* expressing cells (Shimizu et al. 1995; Akazawa et al. 1995).

Using a two-hybrid screen, Lee et al. (1995) recently identified mouse *neuroD*, a novel bHLH family member that also shares some similarity to *Drosophila atonal* bHLH transcription factor. In contrast to *Xash-3* and *Mash-1*, both *Xenopus* and mouse *neuroD* are expressed in post-mitotic neurons. *NeuroD* is detected transiently in a variety of regions, including the spinal chord, nasal epithelium, retina, and cranial ganglia during active neuronal differentiation, but ceases to be expressed in differentiated neurons. Overexpression of *neuroD* in *Xenopus* embryos promotes premature as well as ectopic neural differentiation over widespread regions. Interestingly, ectopic expression of *neuroD* in *Xenopus* animal caps results in conversion of ectoderm into neurons. Thus, unlike *Xash-3*, *neuroD* can induce *Xenopus* ectoderm to form neural tissue. *NeuroD's* ability to induce neurogenesis could stem from a relative insensitivity (compared with *Xash-3*) to postulated inhibitors present in the ectoderm. Since *neuroD* is expressed postmitotically it is more likely to promote neural differentiation than to act as a "proneural" determination factor. As such, neural differentiation in vertebrates may be achieved by the combined action of the early-acting AC-S proteins and proteins such as neuroD that can maintain the "differentiated neuronal" state.

C. Establishing Identities Along the Anterior–Posterior Axis

Forty years ago Nieuwkoop proposed the activation-transformation model for the patterning of the embryonic axis (Nieuwkoop and Nigtevecht 1954). In this model, cells are first activated to become neural and are later transformed to become specific types of neurons. Recent studies provide molecular evidence for this model. Only those genes that are characteristic for anterior cell types in the nervous system are expressed in "unpatterned" neural tissue. Consequently, "anterior" appears to represent the ground state of the neural plate. In *Xenopus*, when dorsal ectoderm (animal caps) is treated with *noggin* or *follistatin* to induce neural tissue, the induced neural tissue expresses the homeobox gene *Otx2*, which is normally restricted to the early midbrain and forebrain (Lamb et al. 1993; Pannese et al. 1995), but not posterior genes such as *Krox20* or *Xhlbox6*. In order to generate cell types that are characteristic for the posterior neuroaxis, further "transformation" signals that modify the ground state are required. Two molecules able to mimic the transformation signal in *Xenopus* embryos are FGF and retinoic acid. The growth factor, basic FGF, can induce forebrain tissue from *Xenopus* embryos to express markers that are characteristic for the hindbrain and spinal cord. In these assays bFGF appears to mimic a signal that is normally derived from the posterior mesoderm that acts to posteriorize the overlying neural plate (Cox and Hemmati-Brivanlou 1995). Retinoic acid (RA) is a second molecule that is able to posteriorize the embryo, including cells in the nervous system. Signaling by RA is mediated by a family of nuclear steroid receptors known as the retinoic acid receptors (RARs) and the retinoid X receptors, RXR α, β and γ (reviewed in Kastner et al. 1995; Mangelsdorf and Evans 1995). A number of posterior transformations are seen in the nervous system of embryos treated with RA. Treatment of *Xenopus* embryos results in the loss of forebrain structures with an associated increase in the hindbrain. In the mouse, RA is able to induce posterior transformations in structures derived from both the mesoderm and neuroectoderm; however, these transformations are not as dramatic as those observed in *Xenopus* (see Kessel 1993 and references therein). The homeobox gene *Otx2* is a very early marker of cells that are fated to give rise to anterior tissue. In both chick and mouse embryos, *Otx2* is expressed at neural plate stage before any morphological changes that mark the regionalization of the neural plate. The early expression of *Otx2* in the anterior regions of the embryo suggests that *Otx2* may be part of the early regulatory cascade that controls the anterior-posterior (A-P) regionalization of the neural plate. Interestingly, treatment of gastrula stage embryos with RA causes the downregulation of *Otx2* in the neuroectoderm consistent with the posteriorization of these tissues by RA (Pannese et al. 1995).

While the downstream targets that mediate the effects of RA are largely unknown, the vertebrate Hox genes appear to be targets of RA action. Retinoic acid response elements (RAREs) have been identified in a number of Hox genes, including *Hoxa-1* and *Hoxb-1*. Further evidence that the Hox genes mediate some of the effects of RA on A-P specification of cells come from studies showing a gradient of RA responsiveness of genes in the Hox gene cluster. This gradient of sensitivity, such that genes expressed in the posterior embryo are more sensitive to RA, parallels the posteriorizing effects of RA on the embryo during gastrulation and neuralation. Indeed there is evidence that Henson's node may be source of endogenous RA in the embryos and that a gradient of RA emanating from the node may contribute to both the A-P patterning and the regulation of Hox genes along the embryonic axis (Sundin and Eichele 1992; Kessel and Gruss 1991). Additional evidence that the Hox genes are targets of RA action comes from studies showing that treatment of embryos with RA can induce ectopic expression of Hox genes. This has been observed in both the neuroectoderm and underlying mesoderm (Conlon and Rossant 1992; Kessel and Gruss 1991; Kessel 1993; Marshall et al. 1992). The net result of these induced changes in Hox expression is that tissues derived from a particular region of the embryo develop abnormally, taking on the morphological characteristics of structures that arise from more posterior regions of the embryo.

An example of this is the effect RA treatment has on the patterning of the hindbrain in the mouse (Kessel 1993). When mid-gastrula mouse embryos were exposed to RA, the expression domains of *Hoxa-1*(*Hox 1.6*) and *Hoxb-2* (*Hox 2.9*) were shifted anteriorly. These changes in Hox gene expression were closely associated with morphological changes in which cells from rhombomere 3 assumed a more posterior identity. The most striking example was the alteration of the projection patterns of the trigeminal motor neurons. Motor neurons in rhombomere 3 (r3) normally project anteriorly to their exit point, which is located in r2, but after RA treatment they instead project abnormally to r4, where the exit point for the facial motor nucleus is located. The cells that make up the facial motor nucleus are located in rhombomeres 4 and 5. RA treatment of zebrafish embryos also causes cells in the hindbrain and midbrain to be respecified (Hill et al. 1995). The ectopic appearance of Mauthner neurons in r2 of RA-treated embryos, in addition to those that are normally present in r4, is similar to the posterior transformation of r3 observed in RA-treated mouse embryos.

D. Segmentation and Patterning of the Hindbrain

During embryonic development the vertebrate hindbrain exhibits a transiently segmented morphology. The eight transverse segments that are apparent as bulges of the neuroepithelium are known as rhombomeres and form lineage-restricted compartments within the hindbrain. The peripheral nervous system is also segmented in the trunk region of the embryo; however, the mechanism that generates this segmentation differs between the hindbrain and the spinal cord. In the hindbrain, segmentation of the neural tube occurs by an intrinsic mechanism, whereas in the spinal cord segmentation is imposed by the adjacent somitic mesoderm.

Neurons in the hindbrain exhibit a segmental pattern of organization. The formation of rhombomeric boundaries within the hindbrain is an important mechanism controlling the segmental pattern of the cranial nerves and the organization of the branchial arches. Evidence for hindbrain segmentation comes from the organization of the cranial ganglia. The trigeminal motor neurons and neural crest derivatives of the first branchial arch are derived solely from rhombomeres 2 and 3, while the facial motor nucleus and second arch neural crest cells arise from rhombomeres 4 and 5. An additional indication of this segmentation comes from the analysis of cell division in the hindbrain neuroepithelium of the chick. Cells at the rhombomeric boundaries or closely adjoining the boundaries have significantly lower mitotic indices than cells in the middle of each rhombomere, suggesting these cells have specialized functions that require their early differentiation. Indeed many of the early axons in the hindbrain are located at the boundaries between each rhombomere, indicating that cells at the boundaries may provide an environment for axon outgrowth. In addition, reticulospinal neurons are initially generated in rhombomeres 2, 4, and 6, then subsequently in the odd-numbered rhombomeres (LUMSDEN 1990).

Transplantation experiments in the chick have addressed the cellular nature of these rhombomeres. Transplantation of an odd rhombomere next to an even-numbered rhombomere is sufficient to generate a rhombomere boundary, whereas placing two odd-numbered or two even-numbered rhombomeres together results in a single, large, fused rhombomere (GUTHRIE and LUMSDEN 1991). Specialized interactions between the cells that will populate adjoining rhombomeres appear to be the driving force that generates the rhombomere boundary. Furthermore, these experiments demonstrate that specialized boundary cells are not generated de novo at fixed positions in the hindbrain. Rather, the generation of boundaries (and specialized boundary cells) results from interactions between neuroepithelial cells that are prespecified with respect to their axial identity. The nature of these interactions at the cellular level is not known. Cell mixing experiments suggest that neuroepithelial cells derived from different rhombomeric levels in the hindbrain show preferential adhesion to cells derived from the same hindbrain level (A. Lumsden, unpublished results). This suggests that the differential expression of cell surface molecules in the hindbrain may contribute to the segregation and development of rhombomere boundaries. A number of Eph-family receptor tyrosine protein kinases are expressed during the segmentation of the hindbrain, including *Sek1*, which is expressed in rhombomeres 3 and 5. Experiments injecting mRNAs that encode for a truncated form of the Sek1 protein, lacking the kinase domain, indicate Sek1 is required for the correct expression of *Krox20* in r3 and r5 and for the proper segmentation of the hindbrain (Xu et al. 1995). Interestingly, Krox20 function is also required for correct segmentation of the hindbrain (Schneider-Maunoury et al. 1993; see below).

A number of genes have been isolated exhibiting expression domains that coincide with these rhombomere boundaries. For example, many of the Hox genes have anterior borders of expression that map to rhombomere boundaries, indicating that the Hox genes (1) may have some role in specifying the A-P identity and/or the phenotype of cells in rhombomeres and (2) may control the expression of downstream genes that regulate the formation of rhombomeres. The *Hoxb-1* gene provides a striking example of restricted Hox gene expression in the hindbrain. Initially *Hoxb-1* is expressed throughout the caudal hindbrain with an anterior expression border marking the r3/r4 boundary. Subsequently *Hoxb-1* expression becomes restricted to rhombomere 4. By transposing presumptive rhombomeres 2 and 4, Guthrie et al. (1992) have shown that *Hoxb-1* is expressed autonomously in rhombomere 4 and that the identity of r4 is established prior to the appearance of rhombomere boundaries in the hindbrain. The establishment of A-P identity within the hindbrain occurs early and precedes the dorsoventral specification of cells (Simon et al. 1995). The role of various Hox genes in controlling cell identity in the hindbrain has also been addressed in the mouse by gene knockout experiments. These experiments provide compelling evidence that the Hox genes are essential for establishing A-P cell identity within the hindbrain. *Hoxa-1*, a paralogue of the *hoxb-1* gene, is normally expressed in the spinal cord and hindbrain with an anterior limit of expression at the r3/r4 boundary. In mice lacking *Hoxa-1* function a number of defects in the hindbrain are observed. Particularly noteworthy is the loss of the facial motor nucleus, which normally arises in r4 and r5. In addition there are defects in the neural crest that arise from these two rhombomeres. While the primary segmentation of the hindbrain does not appear to be affected in *Hoxa-1* null mice, the identity of cells derived from r4 and r5 appears to be incorrectly specified (Lufkin et al. 1991).

While the Hox genes play important roles in specifying cell fates within rhombomeres, there is no evidence that they are responsible for generating rhombomeres. Studies in the mouse indicate that two genes encoding transcription factors, *Krox-20* and *kreisler*, are required for boundary formation and the correct segmentation of the hindbrain. The zinc-finger transcription factor *Krox-20* is expressed in rhombomeres 3 and 5 prior to segmentation and the formation of rhombomeres. By generating a null allele of *Krox-20* by homologous recombination, Schneider-Maunoury et al. (1993) have shown that Krox-20 is necessary for the formation of rhombomeres 3 and 5. In mutant embryos the region encompassed by rhombomeres 2-6 is smaller and shows no evidence of segmentation. This is consistent with earlier transplantation studies showing that alternating odd- and even-numbered rhombomeres are required for rhombomere boundaries to form (GUTHRIE and LUMSDEN 1991). The patterning of the cranial ganglia in *Krox-20* null mice is also consistent with the loss of r3 and r5, with reductions in the size of both the trigeminal and the abducens nucleus.

A second mutation, *kreisler,* also shows defects in the segmentation of the hindbrain (MCKAY et al. 1994). In *Kreisler* mice, the hindbrain posterior to the r3/r4 boundary is unsegmented. This defect appears to be largely due to the loss of rhombomeres 5 and 6. Cells that normally give rise to r5 and r6 contribute to an enlarged r4. Recently the *kreisler* gene has been isolated and found to encode a leucine zipper-containing bZIP transcription factor (CORDES and BARSCH 1994). Two alleles of *kreisler* have been identified. The original *kreisler* allele has a breakpoint inversion close to the *kreisler* gene, and in a second chemically induced *kreisler* mutation, a conserved asparagine residue in the basic domain is altered to a serine. *kreisler* functions in the hindbrain to specify the identity of cells in r5 and r6. Interestingly, the *kreisler* protein is related to a protein identified in *Drosophila* that interacts with the homeodomain protein deformed to specify segmental identity. This raises the possibility that *kreisler* acts in combination with a Hox or Hox proteins to impart an A-P identity to cells in the r5/r6region of the hindbrain. Furthermore, the hindbrain phenotype of *kreisler* demonstrates that the process of hindbrain segmentation depends on cells in the hindbrain acquiring their correct A-P identity.

E. Generation of the Midbrain–Hindbrain Junction: The Role of *Wnt1*, *En1* and *Pax2*

One of the earliest morphological landmarks in the developing brain is the generation of the junction between the hindbrain and midbrain. This junction is known as the rhombencephalic isthmus and coincides with the anterior limit of the hindbrain and of the underlying notochordal process. The midline cells that underly the forebrain and midbrain are derived from the prechordal plate mesoderm. A number of transcription factors have been strongly implicated in the generation of the rhombencephalic isthmus. Two genes, *En1* and *Wnt1*, are expressed in a band of cells located in the prospective caudal midbrain at early neural plate stages in a variety of vertebrate organisms (WILKINSON et al. 1987; DAVIS et al. 1991; PATEL et al. 1989). The function of both genes in the development of this region has been examined by generating null alleles in mice. The development of the midbrain-hindbrain junction is severely disturbed. Both midbrain structures and the cerebellum, which is derived from the anterior hindbrain, are deleted in mice that lack *Wnt1* gene function

(McMahon and Bradley 1990; Thomas and Capecchi 1990). No *En1* expression is present in the caudal midbrain of *Wnt1* null mice, suggesting that *Wnt1* is required for the induction or continued expression of *En1* in these cells, or for their survival and proliferation (McMahon et al. 1992). A phenotype similar to that observed in mice lacking *Wnt1* function is found with *En1* null mice (Wurst et al. 1994). In E9.5 homozygous mutant embryos, the colliculus and cerebellum are missing. In addition, cranial nerves III (oculomotor) and IV(trochlear) are missing, consistent with the loss of cells from the midbrain-hindbrain region.

A second *engrailed*-like gene, *En2*, is also expressed in the midbrain of vertebrates. Mice lacking the *En2* gene have a limited hindbrain-midbrain phenotype with defects in the foliation of the cerebellum. *En2* is expressed after *En1*, and loss of *En2* does not cause a concomitant loss of *Wnt1* and *En1* expression in the midbrain. Thus it appears that the early expression of *En1* is required for generating the midbrain-hindbrain junction. This is consistent with recent experiments showing that *En1* can be replaced with *En2*. The midbrain develops normally in mice expressing the *En2* gene under the control of *En1* regulatory sequences, such that *En2* rather than *En1* is now expressed first in the midbrain (Hanks et al. 1995).

In zebrafish the early morphological regionalization of the brain is preceded by the expression of two paired domain transcription factors, *Pax6* (*Pax*[*zf-a*]) and *Pax2* (*pax*[*zf-b*]). *Pax6* is expressed in the presumptive diencephalon, while *Pax2* exhibits a stripe of expression that coincides with where the midbrain-hindbrain junction will form (Krauss et al. 1992). The expression of *Pax2* in the midbrain precedes the expression of *Wnt1* and *En1*, suggesting that expression of *Pax2* is an early step in the formation of this region. Experiments in zebrafish are consistent with this hypothesis. Zebrafish embryos injected with antibodies specific for *Pax2* exhibit disruptions of the rhombencephalic isthmus. In addition, *Wnt1* expression is lost from the midbrain-hindbrain, suggesting either that these cells have been lost or fail to proliferate or that *Pax2* regulates the expression of *Wnt1* in the caudal midbrain (Krauss et al. 1992). However, in *Pax2* null mice the midbrain-hindbrain appears to be relatively normal (Torres et al. 1995). One possible reason for this discrepancy is that a second gene, *Pax5,* that is closely related to *Pax2* is expressed in the midbrain from E8.5 onwards. *Pax5* may be able to rescue the loss of *Pax2* function in the midbrain. *Pax5* null mice show some defects in the anterior cerebellum and a loss of cells in the inferior colliculus (Urbanek et al. 1994). The limited defects observed in Pax5 null mice may reflect the late onset of *Pax5* expression in the midbrain-hindbrain compared to *Wnt1* and *En1* expression.

F. Dorsoventral Patterning: The Opposing Roles of *Sonic Hedgehog* and BMPs

The position of a cell along the dorsoventral axis of the neural tube is also a major determinant of cell fate. In the midbrain, hindbrain and spinal cord, motor neurons differentiate from the basal plate neuroepithelium, while the first commissural neurons appear in the dorsal alar plate. Neural crest cells migrate early from the dorsal neural tube to give rise to sympathetic and sensory neurons. The latter, in turn, project back to the dorsal horns. In addition a number of discrete types of neurons appear to arise at distinct dorsoventral locations in the spinal cord. e.g. interneurons that express the transcription factors, *En1* and *Evx1* (DAVIS et al. 1991; BASTIAN and GRUSS 1990) The first studies of dorsoventral patterning in the embryonic nervous system date from the middle of this century, when Holtfreter and Watterson described the effects of notochord manipulations on the development of the spinal cord (HOLTFRETER 1934; WATTERSON 1965). A systematic analysis of the role that the notochord plays in patterning the spinal chord in the chick embryo was undertaken first by VAN STRAATEN and HEKKING (1991) and then, with more precise markers, by YAMADA et al. (1991). Grafting a notochord in the vicinity of the closed neural tube at embryonic stages 12–15 results in an enlargened spinal cord on the side of the graft. Histological analysis of the spinal cord adjacent to the graft revealed an increase in the number of early-differentiating ventral neurons that stained positive for acetylcholinesterase activity. When notochords were grafted into stage 9–11 chick embryos, an ectopic floor plate was induced together with extra acetylcholinestaerse-positive cells. Taken together, these results suggest that the notochord is able to induce either an ectopic or enlarged floor plate and the increased proliferation/differentiation of cells in the spinal cord adjacent to the ectopic notochord. Deletion of the notochord has the opposite effect: the spinal cord is smaller and lacks both the floor plate and early-differentiating acetylcholinesterase-positive cells (VAN STRAATEN and HEKKING 1991; YAMADA et al. 1991).

Analysis of the changes in cell patterning in the spinal cord induced by the notochord led to the hypothesis that signals from the notochord are required for the induction of ventral cell types (YAMADA et al. 1991, 1993). In the absence of a notochord, motor neurons and floor plate cells fail to differentiate, while an ectopic notochord is able to induce both cell types in dorsal regions of the spinal cord. Recently a vertebrate gene, *sonic hedgehog*, has been isolated that encodes a protein highly homologous to the *Drosophila* hedgehog protein. *Sonic hedgehog (shh)* is expressed in Henson's node and then in notochord cells that are derived from the node. In addition to its expression in the notochord, *shh* is also induced in the overlying neural plate, in floor plate cells. The expression of *shh* in the floor plate is particularly interesting, since the floor plate is capable of inducing an ectopic floor plate and motor neurons both in vitro and in vivo in a similar fashion to the notochord (YAMADA et al. 1991; ROELINK et al. 1994). Indeed recent experiments using purified *shh* protein indicate that *shh* alone is able to mimic the effects of the notochord on isolated neural plate tissue (ROELINK et al. 1994; MARTI et al. 1995). In addition to its expression in the notochord and floor plate, *shh* is expressed in the basal regions of the diencephalon and midbrain (ECHELARD et al. 1993; ROELINK et al. 1994). The role of *shh* in anterior regions of the neural tube is reminiscent of its role in the hindbrain and spinal cord. The differentiation of a number of cell types in the ventral midbrain appears to be *shh* dependent. Dopaminergic neurons in the ventral midbrain are induced by *shh*, and this effect can be antagonized by increasing protein kinase A activity, which is known to block the action of *hedgehog* in *Drosophila* (HYNES et al. 1995). Recent analysis of mice lacking the *shh* gene indicate that *shh* is required for the differentiation of these dopaminergic cells in the midbrain and for the development of motor neurons and floor plate cells in the hindbrain and spinal cord (C. Chiang and H. Westphal, unpublished results).

Signals from the ventral midline regulate the expression patterns of a number of genes that are expressed in the early spinal cord. Three Pax genes, *Pax3*, *Pax6* and *Pax7*, are expressed in dorsoventrally restricted regions of the early spinal cord and hindbrain (GOULDING et al. 1993, 1994). The expression domains of all three genes in the spinal cord are restricted to progenitor cells prior to the time when they begin to differentiate. In a series of experiments, GOULDING et al. (1993) have shown that signals from the notochord and floor plate alter the expression of these three genes in spinal cord progenitors in a manner that is completely concordant with the latter changes in neuronal cell differentiation that occur. Implantation of a notochord adjacent to the dorsal spinal cord represses Pax3 and Pax6 expression while promoting the generation of motor neurons in cells that no longer express both genes. This suggests that the spatial regulation of these three Pax genes (and other genes) in the hindbrain and spinal cord may underly the later changes in cell fate specification that occur in response to ventralizing signals from the midline. Normally *Pax3* and *Pax7* are only expressed in the dorsal alar plate; however, when the notochord is removed cells throughout the spinal cord begin to express *Pax3* and *Pax7*. In addition to expressing *Pax3* and *Pax7*, these cells express other dorsal markers, including *Msx1* and AC4 (YAMADA et al. 1993; LIEM et al. 1995). This suggests that the default differentiation pathway for cells in the spinal cord is "dorsal" and that signals from the ventral midline are required for the generation of ventral cell types.

The LIM homeodomain protein *Isl1* is expressed in motor neuron precursors in the ventral spinal cord (THOR et al. 1991). *Isl1* expression in the ventral spinal cord is dependent on notochord signaling and the notochord, notochord-conditioned media or purified *shh* are capable of inducing *Isl1* in isolated dorsal neural plate tissue (ERICSON et al. 1992). While *shh* appears to act directly to induce floor plate cells and motor neurons, the role of *shh* in generating other ventral cell types in unknown. Both motor neurons and *En1*$^{+}$ interneurons are lost from mouse embryos that lack the *Isl1*. The observation that *Isl1* expression is restricted to motor neuron precursors and that these cells die early in the *Isl1* homozygous mutant embryos suggests that the differentiation of $En1^{+}$ interneurons may require a signal from early motor neurons (Pfaff et al. 1996).

Signals from the dorsal midline are also likely to be important for establishing dorsoventral polarity in the neural tube. A number of studies have indicated that interactions between the surface ectoderm and adjacent neural plate are sufficient and necessary for the induction of neural crest cells (Moury and Jacobson 1990; Selleck and Bronner-Fraser 1995). A recent study has provided evidence that two members of the TGFβ family, BMP-4 and BMP-7, are expressed in the surface ectoderm adjacent to the dorsal neural plate/tube and are able to induce dorsal cell types in the caudal neural plate (Liem et al. 1995). Consequently, the acquisition of cell fates within the neural tube may require the ventralizing activity of *shh* and possibly *chordin* together with the dorsally expressed morphogens BMP-4 and BMP-7. It is unclear if BMP-4 and BMP-7 alone are sufficient to induce the complete dorsalization of the neural plate. Signals from the underlying lateral plate and somitic mesoderm may also act in combination with signals from the ectoderm to induce the dorsal differentiation of cells in the neural tube.

References

Akazawa C, Ishibashi M, Shimizu C, Nakanishi S, Kageyama R (1995) A mammalian helix-loop-helix factor structurally related to the product of *Drosophila* proneural gene *atonal* is a positive transcriptional regulator expressed in the developing nervous system. J Biol Chem 270: 8730–8738

Bastian H, Gruss P (1990) A murine even-skipped homologue, Evx 1, is expressed during early embryogenesis and neurogenesis in a biphasic manner. EMBO J 9: 1839–1852

Conlon RA, Rossant J (1992) Exogenous retinoic acid rapidly induces anterior ectopic expression of murine *Hox-2* genes in vivo. Development 116: 357–368

Cordes SP, Barsh GS (1994) The mouse segmentation gene *kr* encodes a novel basic domain-leucine zipper transcription factor. Cell 79: 1025–1034

Cox WG, Hemmati-Brivanlou A (1995) Caudalization of neural fate by tissue recombination and bFGF. Development 121: 4349–4358

Davis CA, Holmyard DP, Millen KJ, Joyner AL (1991) Examining pattern formation in mouse, chicken and frog embryos with an En-specific antiserum. Development 111: 287–298

Echelard Y, Epstein DJ, St-Jacques B, Shen L, Mohler J, McMahon JA, McMahon AP (1993) Sonic hedgehog, a member of a family of putative signaling molecules, is implicated in the regulation of CNS polarity. Cell 75: 1417–1430

Ericson I, Thor S, Edlund T, Jessell TM, Yamada T (1992) Early stages of motor neuron differentiation revealed by expression of homeobox gene *Islet-1*. Science 256: 1555–1560

Fainsod A, Steinbeisser H, DeRobertis EM (1994) On the function of BMP-4 in patterning the marginal zone of the *Xenopus* embryo. Embo J 13: 5015–5025

Francois V, Solloway M, O'Neill JW, Emery J, Bier E (1994) Dorsal-ventral patterning of the *Drosophila* embryo depends on a putative negative growth factor encoded by the short gastrulation gene. Genes Dev 8: 2602–2616

Godsave SF, Slack JM (1989) Clonal analysis of mesoderm induction in *Xenopus laevis*. Dev Biol 134: 486–490

Goulding MD, Lumsden A, Gruss P (1993) Signals from the notochord and floor plate regulate the region-specific expression of two Pax genes in the developing spinal cord. Development 117: 1001–1016

Goulding M, Lumsden A, Paquette AJ (1994) Regulation of *Pax-3* expression in the dermomyotome and its role in muscle development. Development 120: 957–971

Groves AK, George KM, Tissier-Seta JP, Engel JD, Brunet JF, Anderson DJ (1995) Differential regulation of transcription factor gene expression and phenotypic markers in developing sympathetic neurons. Development 121: 887–901

Guillemot F, Lo LC, Johnson JE, Auerbach A, Anderson DJ, Joyner AL (1993) Mammalian achaetescute homolog 1 is required for the early development of olfactory and autonomic neurons. Cell 75: 463–476

Guthrie S, Lumsden A (1991) Formation and regeneration of rhombomere boundaries in the developing chick hindbrain. Development 112: 221–229

Guthrie S, Muchamore I, Kuroiwa A, Marshall H, Krumlauf R, Lumsden A (1992) Neuroectodermal autonomy of Hox-2.9 expression revealed by rhombomere transpositions. Nature 356: 157–159

Hanks M, Wurst W, Anson-Cartwright L, Auerbach AB, Joyner AL (1995) Rescue of the En-1 mutant phenotype by replacement of En-1 with En-2. Science 269: 679–682

Hawley SH, Wunnenberg-Stapleton K, Hashimoto C, Laurent MN, Watabe T, Blumberg BW, Cho KW (1995) Disruption of BMP signals in embryonic *Xenopus* ectoderm leads to direct neural induction. Genes Dev 9: 2923–2935

Hemmati-Brivanlou A, Kelly OG, Melton DA (1994) Follistatin, an antagonist of activin, is expressed in the Spemann organizer and displays direct neuralizing activity. Cell 77: 283–295

Hill J, Clarke JD, Vargesson N, Jowett T, Holder N (1995) Exogenous retinoic acid causes specific alterations in the development of the midbrain and hindbrain of the zebrafish embryo including positional respecification of the Mauthner neuron. Mech Dev 50: 3–16

Holley SA, Jackson PD, Sasai Y, Lu B, DeRobertis EM, Hoffmann FM, Ferguson EL (1995) A conserved system for dorsal-ventral patterning in insects and vertebrates involving *sog* and *chordin*. 376: 249–253

Holtfreter J (1934) Arch Exp Zellforsch 15: 281–301

Hynes M, Poulsen K, Tessier-Lavigne M, Rosenthal A (1995) Control of neuronal diversity by the floor plate: contact-mediated induction of midbrain dopaminergic neurons. Cell 80: 95–101

Jarman AP, Grau Y, Jan LY, Jan YN (1993) *atonal* is a proneural gene that directs chordotonal organ formation in the *Drosophila* peripheral nervous system. Cell 73: 1307–1321

Joyner AL, Herrup K, Auerbach BA, Davis CA, Rossant J (1991) Subtle cerebellar phenotype in mice homozygous for a targeted deletion of the En-2 homeobox. Science 251: 1239-1243

Kastner P, Mark M, Chambon P (1995) Nonsteroid nuclear receptors: what are genetic studies telling us about their role in real life? Cell 83: 859–869

Kessel M (1993) Reversal of axonal pathways from rhombomere 3 correlates with extra Hox expression domains. 10: 379–393

Kessel M, Gruss P (1991) Homeotic transformations of murine vertebrae and concomitant alteration of Hox codes induced by retinoic acid. Cell 67: 89–104

Krauss S, Maden M, Holder N, Wilson SW (1992) Zebrafish *pax* [*b*] is involved in the formation of the midbrain-hindbrain boundary. Nature 360: 87–89

Lamb TM, Knecht AK, Smith WC, Stachel SE, Economides AN, Stahl N, Yancopolous GD, Harland RM (1993) Neural induction by the secreted polypeptide noggin. Science 262: 713–718

Lee JE, Hollenberg SM, Snider L, Turner DL, Lipnick N, Weintraub H (1995) Conversion of *Xenopus* ectoderm into neurons by NeuroD, a basic helix-loop-helix protein. Science 268: 836–844

Liem Jr. KF, Tremml G, Roelink H, Jessell TM (1995) Dorsal differentiation of neural plate cells induced by BMP-mediated signals from epidermal ectoderm Cell 82: 969–979

Lo LC, Johnson JE, Wuenschell CW, Saito T, Anderson DJ (1991) Mammalian achaete-scute homolog 1 is transiently expressed by spatially restricted subsets of early neuroepithelial and neural crest cells. Genes Dev 5: 1524–1537

Lufkin T, Dierich A, LeMeur M, Mark M, Chambon P (1991) Disruption of the *Hox-1.6* homeobox gene results in defects in a region corresponding to its rostral domain of expression. Cell 66: 1105–1119

Lumsden A (1990) The cellular basis of segmentation in the developing hindbrain. Trends Neurosci 13: 329–335

Mangelsdorf DJ, Evans RM (1995) The RXR heterodimers and orphan receptors. Cell 83: 841–50

Marshall H, Nonchev S, Sham MH, Muchamore I, Lumsden A, Krumlauf R (1992) Retinoic acid alters hindbrain Hox code and induces transformation of rhombomeres 2/3 into a 4/5 identity. Nature 360: 737–741

Marti E, Bumcrot DA, Takada R, McMahon AP (1995) Requirement of 19K form of *sonic hedgehog* for induction of distinct ventral cell types in CNS explants. Nature 375: 322–325

McKay IJ, Muchamore I, Krumlauf R, Maden M, Lumsden A, Lewis J (1994) The *kreisler* mouse: a hindbrain segmentation mutant that lacks two rhombomeres. Development 120: 2199–2211

McMahon AP, Bradley A (1990) The *Wnt-1* (*int-1*) proto-oncogene is required for development of a large region of the mouse brain. Cell 62: 1073–1085

McMahon AP, Joyner AL, Bradley A, McMahon JA (1992) The midbrain-hindbrain phenotype of Wnt-1/Wnt-1-mice results from stepwise deletion of *engrailed*-expressing cells by 9.5 days postcoitum. Cell 69: 581–595

Moury JD, Jacobson AG (1990) The origins of neural crest cells in the axolotl. Dev Biol 141: 243–253

Nieuwkoop PD, Nigtevecht GV (1954) Neural activation and transformation in explants of competent ectoderm under the influence of fragments of anterior notochord in urodeles. J Embryol Exp Morphol 2: 175–193

Pannese M, Polo C, Andreazzoli M, Vignali R, Kablar B, Barsacchi G, Boncinelli E (1995) The *Xenopus* homologue of *Otx2* is a maternal homeobox gene that demarcates and specifies anterior body regions. Development 121: 707–720

Papalopulu N, Clarke JD, Bradley L, Wilkinson D, Krumlauf R, Holder N (1991) Retinoic acid causes abnormal development and segmental patterning of the anterior hindbrain in *Xenopus* embryos. Development 113: 1145–1158

Patel NH, Martin-Blanco E, Coleman KG, Poole SJ, Ellis MC, Kornberg TB, Goodman CS (1989) Expression of engrailed protein in arthopods, annelids and chordates. Cell 58: 955–968

Pfaff SL, Mendelsohn M, Stewart CL, Edlund T, Jessell TM (1996) Requirement for LIM homeobox gene *Isl1* in motor neuron generation reveals a motor neuron-dependent step in interneuron differentiation. Cell 84: 309–320

Roelink H, Augsburger A, Heemskerk J, Korzh V, Norlin S, Ruiz-i-Altaba A, Tanabe Y, Placzek M, Edlund T, Jessell TM et al. (1994) Floor plate and motor neuron induction by *vhh-1*, a vertebrate homolog of *hedgehog* expressed by the notochord. Cell 76: 761–775

Sasai Y, Lu B, Steinbeisser H, Geissert D, Gont LK, De-Robertis EM (1994) *Xenopus chordin*: a novel dorsalizing factor activated by organizer-specific homeobox genes. Cell 79: 779–790

Selleck MAJ, Bronner-Fraser M (1995) Origins of the avian neural crest: the role of neural plate-epidermal interactions. Development 121: 525–538

Shimizu C, Akazawa C, Nakanishi S, Kageyama R (1995) MATH-2, a mammalian helix-loop-helix factor structurally related to the product of *Drosophila* proneural gene *atonal*, is specifically expressed in the nervous system. Eur J Biochem 229: 239–248

Schneider-Maunoury S, Topilko P, Seitandou T, Levi G, Cohen-Tannoudji M, Pournin S, Babinet C, Charnay P (1993) Disruption of Krox-20 results in alteration of rhombomeres 3 and 5 in the developing hindbrain. Cell 75: 1199–1214

Simon H, Hornbruch A, Lumsden A (1995) Independent assignment of antero-posterior and dorsoventral positional values in the developing chick hindbrain. Current Biol 5: 205–214

Sundin O, Eichele G (1992) An early marker of axial pattern in the chick embryo and its respecification by retinoic acid. Development 114: 841–852

Thomas KR, Capecchi MR (1990) Targeted disruption of the murine *int-1* proto-oncogene resulting in severe abnormalities in midbrain and cerebellar development. Nature 346: 847–850

Thor S, Ericson J, Brannstrom T, Edlund T (1991) The homeodomain LIM protein Isl-1is expressed in subsets of neurons and endocrine cells in the adult rat. Neuron 7: 881–889

Torres M, Gomez-Pardo E, Dressler GR, Gruss P (1995) Pax-2 controls multiple steps of urogenital development. Development 121: 4057–4065

Urbanek P, Wang ZQ, Fetka I, Wagner EF, Busslinger M (1994) Complete block of early B cell differentiation and altered patterning of the posterior midbrain in mice lacking Pax5/BSAP. Cell 79: 901–912

van Straaten HW, Hekking JW (1991) Development of floor plate, neurons and axonal outgrowth pattern in the early spinal cord of the notochord-deficient chick embryo. Anat Embryol Berl 184: 55–63

Watterson RL (1965) Structure and mitotic behaviour of the early neural tube. In: DeHaan RL, Ursprung H (eds) Organogenesis. Holt, Rinehart and Winston, New York, pp 129–159

Wilkinson DG, Bailes JA, McMahon AP (1987) Expression of the proto-oncogene *int-1* is restricted to specific neural cells in the developing mouse embryo. Cell 50: 79–88

Wurst W, Auerbach AB, Joyner AL (1994) Multiple developmental defects in *Engrailed*-1 mutant mice: an early mid-hindbrain deletion and patterning defects in forelimbs and sternum. Development 120: 2065–2075

Xu Q, Alldus G, Holder N, Wilkinson DG (1995) Expression of truncated Sek-1 receptor tyrosine kinase disrupts the segmental restriction of gene expression in the *Xenopus* and zebrafish hindbrain. Development 121: 4005–4016

Yamada T, Placzek M, Tanaka H, Dodd J, Jessell TM (1991) Control of cell pattern in the developing nervous system: polarizing activity of the floor plate and notochord. Cell 64: 635–647

Yamada T, Pfaff SL, Edlund T, Jessell TM (1993) Control of cell pattern in the neural tube: motor neuron induction by diffusible factors from notochord and floor plate. Cell 73: 673–686

Zhao C, Emmons SW (1995) A transcription factor controlling development of peripheral sense organs in *C. elegans*. Nature 373: 74–78

CHAPTER 6

Genetic Control of Kidney Morphogenesis

R. MAAS and M. RAUCHMAN

A. Kidney Development as a Paradigm for Organogenesis

The vertebrate excretory unit, the nephron, is functionally conserved and found in all vertebrates, even dating back to the fossil record (TORREY 1965). As might well be expected from this degree of functional conservation across species, the genes thus far implicated in nephrogenesis are also conserved at the molecular level, with sequence conservation extending even to genes expressed in the primitive excretory organ of the fruit fly, *Drosophila melanogaster*. The purpose of this chapter is to review recent advances in the molecular genetics of kidney development and to correlate this information with a rich body of information on nephrogenesis derived from experimental embryology and developmental biology. A definitive monograph on the cell biology of kidney organogenesis has been published (SAXÉN 1987), and two recent minireviews and a "kidney gene-expression database" are also available (CLAPP and ABRAHAMSON 1993; BARD et al. 1994, 1996).

A prototypical nephron is illustrated in Fig. 1 and consists of the nephric duct (ND), the tubule (T), and the glomerulus (G) and its capsule (Ca), the latter including the glomerular basement membrane. The embryonic development of the nephron involves a large number of different steps, including cell lineage commitment, inductive tissue interactions, cell migration, and angiogenesis. Largely due to the pioneering efforts of classical experimental embryologists and developmental biologists, considerable information is available about some of these developmental events, and a powerful organ culture system exists for studying the epithelial–mesenchymal transition that plays a pivotal role during nephrogenesis. Studies using this organ culture system and related techniques have provided significant insight into the requirements for metanephric induction. The more recent advent of molecular genetic techniques, allowing the selective mutation or insertion of specific genes into the mouse germline, has provided further insight into the identities of genes required for normal nephrogenesis. The convergence of these two experimental approaches makes the embryonic kidney a powerful system for studying the general role that inductive interactions play in vertebrate organogenesis.

A central theme in vertebrate organogenesis is the inductive interaction between two adjacent cell layers, typically an epithelium and a mesenchyme,

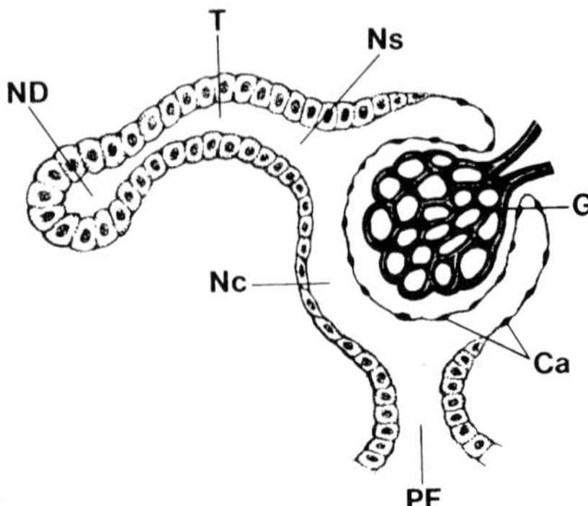

Fig. 1. Basic components of a prototypical nephron. *Ca*, glomerular capsule; *G*, glomerulus; *Nc*, nephrocoele; *ND*, nephric duct; *Ns*, nephrostoma; *PF*, peritoneal funnel; *T*, tubule. After TORREY 1954 and SAXÉN 1987. (Reprinted with permission of Cambridge University Press)

resulting in the morphologic transformation of one or both tissues. The vertebrate kidney depends critically upon inductive interactions for its initial formation. A pivotal step in early kidney morphogenesis is the contact between the ureteric bud, an epithelial outgrowth of the Wolffian or mesonephric duct, and the metanephric mesenchyme or blastema (see Fig. 3). Successful formation of the kidney depends upon the proper execution of this developmental step. This contact triggers a conversion of mesenchyme to epithelium, followed by connection of the distal pole of epithelium to the ureteric bud, forming the collecting duct system. The proximal pole of the epithelium develops into the glomerular crevice. The intervening epithelium becomes the tubular epithelium, which undergoes further differentiation into proximal and distal tubules specialized for ion transport. This process is subsequently reiterated some 10^6 times in the genesis of the mature kidney, whose formation can thus be understood in terms of the formation of the individual nephron. In the mouse, where embryonic development normally spans approximately 19 days, metanephric induction commences at embryonic day (E)11.5, and at E12.5 in the rat. The equivalent developmental event occurs at 5 gestational weeks in the human embryo.

B. Experimental Approaches for Studying Kidney Morphogenesis

I. Transfilter Recombination System

Classical experiments, performed by BOYDEN in chick and GRUENWALD in amphibian embryos in the 1920s, 1930s, and 1940s provided unequivocal evidence for an inductive interaction between the epithelial nephric duct and the nephric mesenchyme in the formation of the meso- and metanephros (Boyden 1927; reviewed in Gruenwald 1952). In 1927, Boyden demonstrated that obstruction of the chick Wolffian duct blocked nephric development. Subsequently, Gruenwald showed that extirpation of the nephric duct by cautery resulted in a failure of induction and that blocking the tip of the nephric duct with extraneous tissue prevented metanephric induction. Lastly, Gruenwald found that juxtaposition of neural tissue with metanephric mesenchyme resulted in successful, albeit limited metanephric induction.

In 1953, Grobstein developed a technique for separating the metanephric mesenchyme and the ureteral epithelium and then recombining them (Fig. 2; Grobstein 1953a,b, 1955, 1956). In Grobstein's original experiments, the mouse metanephric mesenchyme was microdissected at E11.5, along with the ureteric bud. Although the ureteric bud has invaded the mesenchyme at this stage, it is still rudimentary and has bifurcated only once, and it can be separated from the mesenchyme by treatment with 3% trypsin or by treatment using 23-gauge needles in the presence of Versene buffer (phosphate-buffered solution, PBS, containing 0.02% ethylenediaminetetraacetate, EDTA). Induction of tubules in the mesenchyme was found to occur only when the mesenchyme was apposed with certain specific inducing tissues, notably the ureteric bud or embryonic dorsal spinal cord, although embryonic submandibular salivary epithelium, and jaw, bone, and tail mesenchyme were also effective (Saxén 1987; Grobstein 1956). Thus, the specificity of the induction resides in the juxtaposition of certain specific combinations of epithelium and mesenchyme. The fact that heterologous tissues can induce the metanephric mesenchyme to form renal tubules has led Saxén (1987) to classify this par-

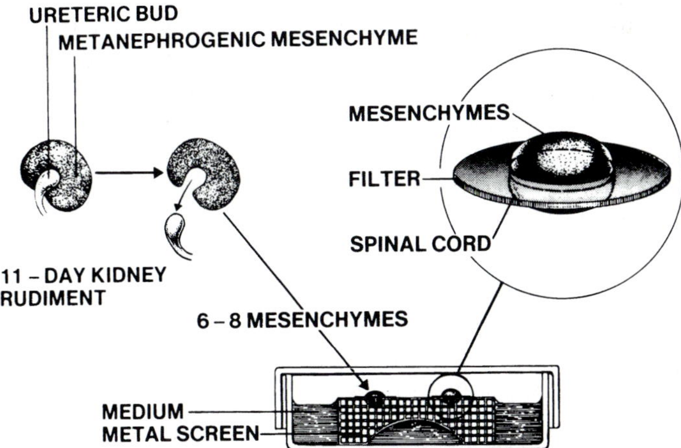

Fig. 2. Basic features of the transfilter recombination system used for studying kidney morphogenesis in vitro. (Reprinted with permission of Cambridge University Press from Saxén 1987)

ticular induction as "permissive" instead of "directive". We now know that the particular spectrum of inducing tissues that are effective in inducing metanephric mesenchyme can be rationalized in terms of the pleiotropic expression patterns frequently observed for candidate signaling molecules, such as growth factors or their receptors, in different embryonic tissues. Presumably, the tissues which are active for metanephric induction are those which express a certain signaling molecule or a certain specific combination of signaling molecules.

A major technical innovation in the study of nephric induction was the introduction of an interposed porous filter between the inducing and responding tissues, to which both tissues could be fixed with agar or, in the original experiments, with a clot (Fig. 2). The filter provides not only a stable and experimentally consistent interface between the two tissues, but also a system in which such parameters as pore size and length of contact time between the tissues can be varied. In the years since its introduction, the transfilter recombination technique, illustrated in Fig. 2, has been further refined and utilized with remarkable success by SAXÉN and colleagues (reviewed in SAXÉN 1987). Currently, polycarbonate Nuclepore filters (Costar, Cambridge, MA) are used, which offer the advantage of highly consistent pore sizes and densities, with available pore sizes ranging from 0.1–1.0 μm. The filter is placed on a triangular wire screen (Industrial Wire Products, Los Angeles, CA) in a 60-mm organ culture dish (Becton Dickinson, NJ) and the center well filled with culture medium until the surface of the filter just wets from the underside; the rudiment on the surface of the filter is bathed by capillarity. Typically Dulbecco's Modified Eagle's Medium (D-MEM) containing 10% fetal calf serum (FCS) is used, followed by incubation at 37 °C in 5% CO_2, although defined media have been developed (EKBLOM et al. 1991). The course of nephrogenesis can now be followed over a period of several days, using standard histological, immunochemical, and metabolic techniques. Although it exhibits a progressive delay relative to in vivo nephrogenesis, the in vitro system results in tubule formation by day 3 in culture and in the formation of nonvascularized glomeruli by day 7 (BERNSTEIN et al. 1981). This experimental system has proven exceptionally important to our understanding of how the vertebrate kidney develops. In recent years, it has been further complemented by a second set of equally important techniques, those of contemporary mouse molecular genetics.

II. Genetic Analysis of Nephrogenesis Using Knockout, Transgenic, and Spontaneous Mouse Mutations

The ability to mutate known genes by transgenic and knockout techniques in the mouse embryo has greatly facilitated our understanding of the role that certain genes play in embryonic development. It is first useful to distinguish between these two basic techniques. In classical transgenic technology (HOGAN et al. 1986), a gene is introduced into the mouse germline by direct pronuclear injection, either under the control of its own promoter or of a heterologous promoter. The resulting transgene integrates at random, typically as a concatenate consisting of a large number of copies. In addition, there may be some chromosomal damage, not infrequently a deletion, occurring with the integration event. The normal consequence of preparing a transgenic mouse line which overexpresses, or ectopically expresses, a particular gene product is a dominant, gain of function phenotype. In the rare case in which endogenous genes are mutated or deleted by the transgene insertion event, a recessive, loss of function mutation, unrelated to the molecular identity of the introduced transgene, may also be introduced. However, such rare, artifactually induced recessive mutations, in contrast to the dominant effects of the transgene, will ordinarily only be apparent when the transgenic line is crossed to homozygosity. Under ordinary circumstances, the transgene can be transmitted to progeny and the phenotypic effects can be assessed in the context of the embryonic or adult mouse.

In knockout technology, a known gene is selectively targeted for disruption in embryonic stem (ES) cells by the principle of homologous recombination (Capecchi 1989, 1994). In homologous recombination, a targeting vector is prepared which consists of a genomic DNA fragment containing the gene of interest. Typically, exonic sequences of the gene are specifically mutated, most often by insertion of a selectable marker such as the neomycin resistance gene (*neo*^{R}) into an essential portion of the coding region of the gene. The introduction of a selectable marker serves a dual purpose, introducing a mutation into the gene and providing a means to select for ES cell lines which have taken up the targeting vector. A second selection is often utilized to further enrich for homologous recombination events, as opposed to illegitimate integration events. If a marker encoding viral (herpes simplex virus, HSV) thymidine kinase (HSV-*tk*), conferring gancyclovir sensitivity, is placed at the ends of the targeting vector, the double reciprocal crossover involved in homologous recombination will result in loss of these sequences when the recombination event occurs. In contrast, a random integration event will often preserve one or both *tk* genes. In the presence of the nucleoside analogue gancyclovir, the integration of the viral *tk* gene causes randomly integrated ES cells to die. This simplifies the task of identifying homologous recombinant ES lines, which must be accomplished by polymerase chain reaction (PCR) assay or by Southern blot analysis. The pluripotent, homologous recombinant ES cells identified by these screening methods are heterozygous for the desired mutation and can be clonally expanded and used to generate a chimeric mouse by blastocyst injection (Robertson 1987; Bradley 1987). Under favorable circumstances, the introduced mutation will be transmitted through the germline of the chimera and can be used to establish a mutant line of mice which carry the desired mutation. These mice can then be bred to homozygosity, and the phenotypic consequences of loss of gene function can be assessed by isolating embryos from timed pregnancies.

Several points regarding this technique deserve special attention. First, while knockout technology is enormously powerful for determining the developmental function of a known gene, a knockout may not prove informative if the gene whose function is eliminated is required for embryonic viability prior to the time at which the developmental process being studied occurs, i.e., in the case of kidney morphogenesis, prior to E11.5. This problem is being circumvented by the advent of conditional knockout systems, such as the Cre-lox recombinase system (GU et al. 1994). Second, an emerging theme in knockout experiments is that many embryonic tissues which express the gene may not exhibit a phenotype. In some cases, this is because of functional redundancy between related, coexpressed members of a multigene family or because of other compensatory mechanisms. In kidney development, a good example of this is the combined deficiency of the Src family members *fyn* and *yes*, in which the double homozygotes develop diffuse segmental glomerulosclerosis (STEIN et al. 1994). Nonetheless, with these caveats in mind, a surprising number of mouse mutants have been generated which express specific morphogenetic defects in nephrogenesis (Table 1). Taken together, these mutants offer the eventual hope of illuminating the discrete genetic pathways involved in nephrogenesis.

A last genetic resource which deserves mention are the naturally occurring, spontaneous mouse mutations which exhibit phenotypic defects in renal development. For many years these mouse mutations have been useful only for embryologic and genetic studies. However, the advent of efficient mapping strategies using recombinant inbred (RI) and interspecific backcross (*Mus spretus–Mus musculus*) techniques, the increasing saturation of the mouse genetic map, and improvements in positional cloning techniques have afforded increased opportunities to identify the genes responsible. In addition, an increasingly frequent finding is that some artificially generated knockout mice turn out, unexpectedly, to be alleles of previously known classical mutations. Table 2 lists the mouse kidney developmental mutations for which the gene is unknown, along with their chromosomal locations.

Several spontaneous mouse mutations deserve special mention. These include *Danforth's short tail* (*Sd*), *limb deformity* (*ld*), and a large number of independent recessive mutations that result in polycystic kidneys (see Tables 1, 2). The specific phenotypes of some of these mutants are further discussed later in the text, in connection with the developmental steps which seem to be most seriously affected in them.

III. Cell Culture Systems

A final system which has been employed to study kidney morphogenesis is the use of continuous and primary cultured cell lines. Cell culture systems have been useful for studying: (a) the establishment of apical–basolateral polarity and intercellular junctions, (b) tubulogenesis and branching morphogenesis, and (c) characterization of renal progenitor cells.

The process of mesenchymal to epithelial transformation that occurs during nephrogenesis involves the conversion of a round, nonpolarized mesenchymal cell to a columnar epithelial cell that segregates the apical (urinary or luminal) space from the basolateral (blood) space. The ability of mature renal tubules to absorb or secrete solutes and fluid depends on polarization of the tubular epithelia and on the intercellular junctions which make up selectively permeable paracellular pathways. Madin-Darby canine kidney (MDCK) cells, a well-characterized renal epithelial cell line, have proven to be especially valuable. The Ca^{2+} switch model has been used to study mechanisms of junctional biogenesis and establishment of cell–cell contacts in MDCK cells. When a confluent monolayer of MDCK cells is incubated in low Ca^{2+} medium ($< 5\ \mu M$), intercellular junctional complexes are disrupted and cells become rounded and lose their polarity (GONZALEZ-MARISCAL et al. 1985). When Ca^{2+} in the medium is again increased, the reestablishment of polarity and junctional complexes can be studied. A limitation of this model is that it utilizes a polarized, differentiated epithelial cell and therefore may not recapitulate the events of nephrogenesis in which a mesenchymal cell differentiates into a epithelial cell. This model may thus better represent the maintenance of polarity and cellular junctions in differentiated cells, rather than their initial formation. Furthermore, these experiments are performed on polycarbonate filters and therefore effects of extracellular matrix, the composition of which changes significantly during nephrogenesis, are not taken into account. Cell–matrix interactions have been shown to be important in the generation and maintenance of polarity. Nonetheless, some similarities with the process of mesenchymal cell differentiation in vivo have been demonstrated and undoubtedly others also exist (GUMBINER et al. 1988).

Separate from the differentiation of mesenchymal cells or ureteric bud cells into mature renal epithelial cells, the developing kidney must also organize these cells spatially into a specialized filtering unit, the glomerulus, and into tubular structures with a particular segmental orientation. The formation of the collecting duct from the ureteric bud epithelium involves multiple branching events and ultimately connection to the mesenchyme-derived distal tubule. In contrast to the collecting duct, the mesenchyme-derived tubules do not undergo extensive branching during their formation (OSATHANONDH and POTTER 1963a,b).

Both the formation of tubular structures from individual cells and branching morphogenesis can be studied in cells cultured in three-dimensional type I collagen gels. This was first described for MDCK cells, which form cystic structures in collagen gels in the presence of serum, but can be induced to form branching tubular structures in the presence of hepatocyte growth factor (HGF) or when cocultured with induced metanephric mesenchyme (MONTESANO et al. 1991a,b; SANTOS et al. 1994). Tubulogenesis and branching morphogenesis have also been demonstrated for a murine inner medullary collecting duct cell line (mIMCD-3) in three-dimensional gels (CANTLEY et al. 1994). This model has been used to study the effects of different combinations

Table 1. Genes implicated in nephric development

Gene name	Type of mutation	Phenotype	Reference
Activin type IIBr	ko	Renal agenesis	E. Li, p.c.
ADA	ko	Developmental retardation in formation of the kidney	S. Migchielsen, p.c.
α_3-integrin	ko	Dilated tubules in medulla, glomerular podocytes without foot processes	J. Kreidberg, p.c.
bcl-2	ko	Cystic kidneys and renal hypolasia	Veis et al. 1993
BF-2	ko	Reduced nephron number and collecting system	Hatini et al. 1996
BMP-7	ko	Renal agenesis	Dudley et al. 1995 Luo et al. 1995
c-*myc*	tg(o)	Polycystic kidneys	Trudel et al. 1991
COX-2	ko	Reduced nephron number	Morham et al. 1995
c-ret	ko	Renal agenesis/dysplasia	Schuchardt et al. 1994
Egfr	ko	Cystic dilation of collecting ducts	Threadgill et al. 1995
GDNF	ko	Renal agenesis	Sanchez et al. 1996 Pichel et al. 1996 Moore et al. 1996
HNF1	ko	Strong glucose wasting and polyuria	Pontoglio et al. 1996
Hoxa11/d11	ko	Renal agenesis	Davis et al. 1995
Hoxb7	tg(o)	Double ureter and pelvis	Argao et al. 1995

Hoxb8	tg(o)	Small kidneys	Charité et al. 1994
hph-2	enu	Hyperaminoaciduria	D. Symula, p.c.
ld	ko	Renal agenesis	A. Wynshaw-Boris, p.c.
laminin B2	ko	Congenital nephrotic syndrome	Noakes et al. 1995
Lim 1	ko	Renal agenesis	Shawlot and Behringer 1995
N-*myc*	ko	Reduced number of mesonephric tubules	Charron et al. 1992
Notch-2	ko	Lack of mature glomeruli	T. Gridley, p.c.
p53	tg(o)	Reduced nephron number	Godley et al. 1996
Pax-2	tg(im)	Aplasia and cystic abnormalities	Keller et al. 1994
Pax-2	ko	Renal and ureter agenesis	Torres et al. 1995
Pax-2	tg(o)	Congenital nephrotic syndrome	Dressler et al. 1993
PDGF-β	ko	Absence of mesangial cells and reduction in glomerular capillary network	Levéen et al. 1994
PDGF-βR	ko	Absence of mesangial cells and reduction in glomerular capillary network	Soriano et al. 1994
TG737	tg(im)	ARPKD	Moyer et al. 1994
Wnt-4	ko	Renal dysgenesis	Stark et al. 1994
WT1	ko	Renal agenesis; Wilms and WAGR syndrome in man	Kriedberg et al. 1993

Abbreviations: ko, knockout; tg(o), transgenic overexpression mutant; tg(im), transgene induced insertional mutation disrupting the gene indicated; enu, ethylnitrosourea induced mutant; p.c., personal communication.

Table 2. Classical mouse mutations affecting kidney development (for which the gene is unknown)[a]

Gene symbol	Gene name	Map location	Kidney defect
bd	Bradypneic	Chr 12	Dilated distal tubules
bl	Blebbed	Chr 5	Uni-, bilateral renal agenesis
bpk	Biliary and polycystic kidney disease	—	Polycystic kidneys
Br	Brachyrrhine	—	Small, pale kidneys, few glomeruli
c^{3H}	Albino locus	Chr 7	Very small kidneys
cph	Congenital progressive hydronephrosis	Chr 15	Ureteropelvic junction obstruction
cpk	Congenital polycystic kidneys	Chr 12	Polycystic kidneys
eb	Eye Blebs	Chr 10	Hypoplastic, cystic kidneys renal agenesis
Fu	Fused	Chr 17	Renal agenesis
Hyp	Hypophosphatemia	Chr X	Tubular brush border defect
jck	Juvenile cystic kidneys	Chr 11	Polycystic kidneys
jcpk	Juvenile congenital polycystic kidney	Chr 10	Polycystic kidneys (–/–); glomerulocystic disease (+/–)
lx	Luxate	Chr 5	Horseshoe kidney, hydronephrosis, hydroureter
my	Blebs	Chr 3	Renal agenesis
nm1633		Chr 8	Polycystic kidneys
ol	Oligodactyly	Chr 7	Horseshoe or cystic kidneys, renal agenesis
Os	Oligosyndactylism	Chr 8	Small kidneys, marked reduction in glomeruli
pcy	Polycystic kidney disease	Chr 9	Adult-onset polycystic kidneys
Sd	Danforth's short tail	Chr 2	Renal and ureteral agenesis
spk	Sponge kidney	—	Polycystic kidneys
SR2–3		Chr 11	Polycystic kidneys
Xpl	X-linked polydactyly	Chr X	Renal agenesis, hydronephrosis
Xid	X-linked immune deficiency	Chr X	Polycystic kidneys

Dominant mutations begin with an upper case letter, recessive mutations with a lower case letter. Mouse mutants expressing metabolic diseases resulting in kidney disease are not listed, nor are inbred strains with a high incidence of renal anomalies. Mouse mutations not cited in Lyon and Searle are from the following: *bpk*, Nauta et al. 1993; *jck*, Iakoubova et al. 1995; *jcpk*, Flaherty et al. 1995; *nm 1633*, cited in Reeders 1992; *pcy*, Takahashi 1991; *spk*, M.T. Davisson, personal communication; *SR2–3*, Boulter et al. 1992.
[a]From Lyon and Searle (1990).

of growth factors along with matrix composition on tubulogenesis and branching (see below). Since tubulogenesis and branching occur in the development of other organs, data from mammary and salivary gland, lung, and keratinocytes may complement that obtained with renal epithelial cells. As in the Ca^{2+} switch model, the use of differentiated epithelial cells to study developmental processes in vitro may be an important limitation, and therefore caution is warranted in interpreting results obtained with these models. Since this cell culture model can be manipulated to form cysts or tubules, it could also be used to study the pathogenesis of renal cystic diseases.

Cell culture systems of undifferentiated renal progenitor cells could be very useful to dissect the mechanisms which stimulate cell proliferation soon after metanephric induction. This proliferative phase is followed by exit from the cell cycle and subsequent terminal epithelial differentiation. These efforts have been limited by the inability of isolated metanephric mesenchymal cells to proliferate in culture in the absence of inducers such as ureteric bud or dorsal spinal cord, which also cause differentiation. Burrow and Wilson (1993) have reported primary culture of nephroblasts derived from human fetal kidney. Proliferation of these cells is sustained by conditioned media from a Wilms tumor cell line. Oncogenic transformation of renal progenitor cells by transfection of mesenchymal cells from uninduced mesenchyme with a temperature-sensitive SV40 large T has enabled continuous cell lines to be established (Karp et al. 1994). These cells can be induced in culture with embryonic spinal cord to express the early markers of epithelial differentiation E -cadherin and cytokeratin. Another powerful approach is the development of cell lines from targeted oncogenesis in transgenic animals. This method utilizes standard transgenic technology, as outlined above; however, regulatory sequences are chosen which reproducibly drive expression in the kidney (for a review, see Briand et al. 1995). This method is proving more successful at establishing continuous cell lines than in vitro infection of primary cultures with a transforming virus. Using this approach, cell lines can be derived from microdissected tubules or embryonic kidney rudiments such that the site of origin is known. In contrast, even the well-characterized MDCK cell line is of uncertain origin (distal tubule versus cortical collecting duct). For example, Larsson et al. (1995) isolated a WT1-expressing mesonephric cell line from mice transgenically expressing polyoma large T. The collecting duct cell line referred to above (mIMCD-3) used to study tubulogenesis and branching in vitro was derived from microdissected tubules of a transgenic mouse expressing SV40 large T (Rauchman et al. 1993).

Established epithelial cell lines, primary cultures, or cell lines derived from Wilms tumor or renal cell carcinoma often display intermediate phenotypes which can be used to examine specific steps or functions of specific gene products in the differentiation process. There are also numerous renal epithelial cell lines available which exhibit differentiated characteristics of specific nephron segments (for a review, see Gstraunthaler 1988). These cell lines can be used to study the development of specific properties of terminally differentiated epithelial cells or the expression of developmental genes, such as *Antp*-class *Hox* genes, which are expressed in a proximal tubule cell line (WOLF et al. 1991).

C. Early Morphogenetic Events

I. Pronephros, Mesonephros, and Metanephros

Although their relative importance differs among species, nephrogenesis has been divided into three phases: pronephros, mesonephros, and metanephros. The pronephros is largely vestigial in mammals and will receive only brief attention here. The mesonephros, or intermediate kidney, varies widely in size and functionality between different species. In male embryos, the mesonephros transforms into the anlage (primordium) of the epididymus, with the mesonephric duct becoming the vas deferens. In females, these structures undergo almost complete regression, becoming the paroophoron, epoophoron, and the vestigial duct of epoophoron (Gartner's duct). Although the mesonephros has received relatively little attention in mammalian development, its development and induction by the nephric duct appear similar to that of the metanephros by the ureteric bud. This suggests the hypothesis that similar molecular pathways may be utilized in the formation of both the meso- and metanephros. Indeed, many of the same genes are expressed in both tissues. By examining mesonephric development in mouse mutants in which metanephric induction is disturbed, it should be possible to gain evidence for or against this hypothesis.

In the mouse, mesonephric development commences at E8.5. Mesenchymal condensates form in response to an inductive stimulus provided by the Wolffian duct, in a process very analogous to that described in more detail later for the metanephros. The mesonephros subsequently undergoes a segmentation process not unlike that for the metanephros, including the formation of a prospective glomerulus, a tripartite proximal tubule, a distal tubule, and a collecting duct. Tubules form first in the most cranial portions of the mesonephros and may even undergo regression in some species while additional nephrons are forming more caudally (Fig. 3). In the pig, the 50 or more mesonephric glomeruli subsequently become vascularized by 11 paired arteries emanating from the abdominal aorta, each supporting up to 15 glomeruli (TIEDMANN and EGERER 1984). The functional ability of the mesonephros has been documented in amphibian and avian species by the ability to concentrate vital dyes or by alterations in the specific gravity of the embryo following mesonephric extirpation. In mammals, functional impairment of the mesonephros may not have an obvious effect during embryogenesis because of redundancy provided by maternal renal clearance.

In mouse, formation of the metanephros can be viewed as starting at E11, when the ureteric bud (U) emerges from the caudal end of the Wolffian duct (ND; Fig. 3). This epithelial structure grows dorsally toward and then into the

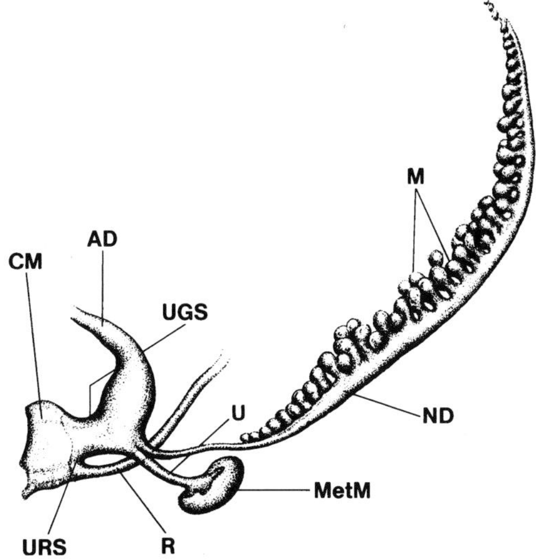

Fig. 3. Arrangement of the mesonephros (*M*), the metanephric mesenchyme (*MetM*), the Wolffian or nephric duct (*ND*), the urogenital sinus (*URS*), the rectum (*R*), the ureter (*U*), the allantoic duct (*AD*), and the cloacal membrane (*CM*). *Note*: The ureteric bud is no longer defined as a bud once ingrowth into the metanephric mesenchyme occurs. (Reprinted with permission of Cambridge University Press from SAXÉN 1987)

metanephric mesenchyme or blastema (MetM), a dense collection of mesenchyme situated at the caudal end of the nephric cord. As described further below, the kidney can thus be viewed as originating from these two epithelial and mesenchymal lineages, to which a third lineage, the developing vasculature, is subsequently added.

II. Commitment to Nephrogenic Fate

Embryologists use the terms "determination" or "commitment" to refer operationally to a tissue whose ultimate fate can no longer be influenced by transfer to a new environment. Fate-mapping experiments in amphibian embryos at the midneurula stage indicate that the pronephric region, which resides in mesoderm in the lateral body wall, is determined by this time, since transplants of this region to heterotopic sites will form nephric tubules, whereas other tissues have lost this ability. In contrast, the mesonephric region is not completely determined by this time, because the type and arrangement of tubules can still be influenced by the surrounding environment. The nephric duct also derives from the mesoderm, although how its communication with the pronephros is established is unclear. In mammals, the metanephric mesenchyme originates from the intermediate mesoderm, situated medial to the lateral plate mesoderm destined to form the somites and dermomyotome. The first known marker for the uninduced mesenchyme is the Wilms tumor gene,

WT1, which is expressed in the mesenchyme prior to metanephric induction (MUNDLOS et al. 1993). Wilms tumor is a pediatric malignancy affecting approximately one in 10 000 infants (reviewed in HASTIE 1994). It may be either hereditable or sporadic and is associated in 10%–15% of cases with loss of heterozygosity (LOH), either by cytogenetic deletions of chromosome 11p13 or by somatic mutation. The Wilms tumor gene, *WT1*, was identified in 1990 by positional cloning approaches (CALL et al. 1990) and encodes a protein containing four C-terminal Zn fingers and an N-terminal proline–glutamine-rich transregulatory domain. WT1 encodes four protein isoforms created via alternative splicing: exon 5, encoding 17 amino acids, is variably present, as are the three amino acids KTS located between Zn fingers 3 and 4. The Zn finger DNA-binding domain of WT1 recognizes the DNA sequence motif 5′-CGGGGGCG-3′ (RAUSCHER et al. 1990), similar to that recognized by another Zn finger transcription factor, Egr-1, which is also expressed in the developing kidney (SUKHATME 1990). Because loss of *WT1* function contributes to tumorigenesis, it is believed to function as a tumor suppressor gene (reviewed in HASTIE 1994; HABER et al. 1993).

Recent data indicate that the +KTS and –KTS isoforms are localized differently within the nucleus (LARSSON et al. 1995). The –KTS isoform distribution parallels that of classical transcription factors, while the +KTS isoform preferentially associates with specific components of the splicing apparatus. These data suggest that WT1 may not only function as a transcription factor, but that it may also exert a post-transcriptional role in kidney and gonadal morphogenesis. Denys-Drash syndrome, a renal and gonadal dysmorphosis, is associated with two specific classes of WT1 mutations: (1) missense and nonsense mutations which abolish DNA binding (PELLETIER et al. 1991) and (2) splicing mutations affecting the production of the +KTS isoform (BRUENING et al. 1992). The fact that these individuals primarily suffer developmental defects, and Wilms tumor only to a lesser extent, would imply such a second function for WT1.

Although both mice and rats contain a Wilms tumor gene, only the rat is susceptible to Wilms tumor (SHARMA et al. 1994). In fact, tumorigenesis in mice cannot even be induced when spontaneous mouse mutants containing chromosomal deletions of *WT1* (e.g., Sey^{Dey}) are treated with mutagens such as ethyl nitrosourea (ENU) or ionizing radiation in order to effect a loss of heterozygosity (T. Glaser and R. Maas, unpublished observations). This and other data have led to the conclusion that multiple genetic pathways can lead to Wilms tumor in humans and that these pathways differ in rodents and humans (reviewed in HASTIE 1994). Knockout of *WT1* in mice, however, does prove a central role for *WT1* in induction of the metanephric mesenchyme, since *WT1*-deficient mice form a metanephric mesenchyme, but one in which 10%–50% of cells (note: approximately 1% in wild type) undergo apoptosis (KREIDBERG et al. 1993). Consequently, *WT1* homozygous deficient embryos, which survive until E15.5, express a renal agenesis phenotype. Interestingly, mesonephric tubules form in *WT1*-deficient mice, but with a two- to threefold lower efficiency than in wild type. Theoretically, the molecular basis for the renal agenesis phenotype could reside in either the mesenchyme, the ureteric bud, or both, since the ureteric bud is absent in all *WT1* homozygotes. The fact that *WT1*-deficient metanephric mesenchyme cannot be induced by spinal cord, normally a powerful inducer of tubulogenesis in metanephric mesenchyme, indicates that *WT1* is required for the mesenchyme to respond to inducer.

The *WT1* mutation can be compared to the phenotype observed in the recessive mouse mutation *ld*, which expresses a variably penetrant renal agenesis phenotype, occasioned by defective outgrowth of the ureteric bud (MAAS et al. 1994). In this case, both the frequent occurrence of normal kidneys in multiple independent *ld* mutations, including some presumptive null alleles, and the finding that spinal cord can induce tubulogenesis in *ld*/*ld* mesenchymes suggest that the renal agenesis results from defective ureteric bud outgrowth. However, since *ld* transcripts and Ld protein are found in both the metanephros and the ureteric bud, the molecular basis for the defect in ureteric bud outgrowth could reside in either of these tissues (TRUMPP et al. 1992; MAAS et al. 1994). A role for stochastic factors in ureteric bud outgrowth is indicated by the fact that different degrees of ureteric bud outgrowth, and hence of renal agenesis, are observed even in inbred genetic backgrounds.

A few potential targets of WT1 have been identified, including the insulin-like growth factor (IGF) II, IGFI receptor (IGF IR), and platelet-derived growth factor (PDGF)-A chain genes, and also the gene encoding the early growth response gene *Egr-1*. Although definitive in vivo evidence is lacking, all four genes have been proposed to play roles during nephrogenesis (DRUMMOND et al. 1992; WANG et al. 1992; WERNER et al. 1994; SUKHATME 1992). The case is relatively persuasive for IGF II: IGF II levels are increased ten to 100-fold in Wilms tumor samples (REEVE et al. 1985), and levels of IGF II, a known mitogen, decrease as WT1 levels increase during metanephric induction (YUN et al. 1993). A model has been proposed which places WT1, Egr-1, and IGF II in a regulatory loop. Egr-1 is turned on by a mitogenic signal from a growth factor such as IGF II. IGF II contains WT-1/Egr-1 DNA-binding sites in its promoter region and in the absence of transcriptional repression by WT-1, Egr-1 can induce expression of IGF II. Thus Egr-1 and IGF II may serve to promote proliferation of undifferentiated mesenchyme through an autocrine loop (SUKHATME 1992). However, as differentiation proceeds, WT1 expression progressively increases, leading to transcriptional repression of both Egr-1 and IGF II, both of which contain WT1-binding sites in their promoters, thereby breaking the autocrine loop and inhibiting proliferation. In addition, recent data indicate that the *Pax-2* gene, also implicated in maintaining the proliferative state, is a direct target of transcriptional repression by the WT1 protein (RYAN et al. 1995). The potential role of genetic pathways involving growth factors and growth factor receptor genes and *Pax-2* in metanephric induction will be further discussed below.

III. Nephric Duct and Ureteric Bud Outgrowth

In an elegant series of experiments conducted in the axolotl salamander, pronephric duct primordia were transplanted to sites distant from the primary duct and their behavior examined by scanning electron microscopy (SEM) (POOLE and STEINBERG 1981, 1982). Under these conditions, the ectopically transplanted nephric duct is able to reestablish contact with the primary duct. This has been interpreted to imply the existence of a gradient in the mesoderm capable of providing directional signals to the ductal epithelium. This idea finds some support from the expression of *Hox* genes in the periductal mesenchyme, further discussed below in connection with the related problem of ureteric bud outgrowth.

Despite its importance as an inducer of metanephric mesenchyme, little is known about the events which guide the initial outgrowth of the ureteric bud from the nephric duct. One gene that appears to participate in this process is that encoding the c-ret receptor tyrosine kinase (RTK) (PACHNIS et al. 1993). The *c-ret* gene is expressed in the nephric duct between E8.5 and E10.5, in the ureteric bud epithelium but not the surrounding metanephric mesenchyme at E11.0–11.5, and at the tips of the renal collecting ducts at E13.5–E17.5. It is not expressed in the renal vesicles, which derive from the metanephric mesenchyme, or in the subcortical portion of the collecting ducts. Direct genetic evidence supports the view that the *c-ret* gene is involved in mediating a directive signal required for ureteric bud outgrowth. In *c-ret*-deficient knockout mice, outgrowth of the ureter is abnormal; it either fails to form, is arrested in its outgrowth, or engages in abnormal contact with the metanephric mesenchyme, resulting in renal agenesis in the former cases and in dysplastic kidneys in the latter (SCHUCHARDT et al. 1994). Given that WT1 knockout mice display defective ureteric bud outgrowth, these data suggest a model whereby WT1 expression in the undifferentiated metanephric mesenchyme regulates, either directly or indirectly, the expression of a signal capable of interacting with the c-ret RTK expressed in posterior Wolffian duct, to initiate ureteric bud outgrowth. Consistent with this, recombination experiments between wild-type ureteric bud and c-*ret*-deficient metanephric mesenchyme result in ureteric bud growth and branching, whereas the reciprocal combination fails (SCHUCHARDT et al. 1996).

Recent studies have identified GDNF, a member of the TGF-beta superfamily, as the ligand of the c-ret RTK (DURBEC et al. 1996; TRUPP et al. 1996). At E11.5 GDNF is expressed in the metanephric mesenchyme adjacent to the c-ret expressing cells at the tips of the invading ureteric bud (DURBEC et al. 1996). Furthermore, GDNF deficient knockout mice are a phenocopy of c-ret deficient mice (PICHEL et al. 1996; MOORE et al. 1996; SANCHEZ et al. 1996), and recombinant GDNF can rescue the GDNF −/− mutant ureteric bud in organ culture (PICHEL et al. 1996) but not c-ret −/− mutant ureteric bud (DURBEC et al. 1996). Together these results indicate that GDNF is a mesenchymal signal which mediates outgrowth and branching of the ureteric bud through the c-ret RTK. GDNF association with and activation of the c-ret RTK is mediated by a novel high affinity GDNF receptor (GDNFR-alpha) which forms a multimeric complex with c-ret RTK and the ligand GDNF (JING et al. 1996; TREANOR et al. 1996).

A second signaling molecule potentially relevant to ureteric bud outgrowth is the RTK encoded by the c-*ros* RTK gene. However, this RTK likely recognizes a membrane-bound ligand, thus requiring cell–cell contact. The c-*ros* gene, therefore, appears to be a candidate for mediating the inductive events associated with contact between the ureteric bud and metanephric mesenchyme, as opposed to the long-range interactions presumed to mediate bud outgrowth. This molecule is therefore considered separately later, in the developmental context of metanephric induction. Lastly, it should be noted that other genetic loci besides *WT1*, *ld*, and c-*ret* are also likely to be important in regulating ureteric bud outgrowth: the spontaneous mutant *Sd* (Table 2) shows abnormalities of ureteric bud morphology similar to those reported in the c-*ret* knockout mice, and *Sd* mice also manifest renal agenesis (GLUECKSOHN-SCHOENHEIMER 1943, 1945; GLUECKSOHN-WAELSCH and ROTA 1963). Similarly, mice homozygous at the *Fused* locus also express defects in ureteric bud outgrowth and a variably penetrant renal agenesis phenotype (THEILER and GLUECKSOHN-WAELSCH 1956).

IV. Role of *Hox* Genes

A seminal event in developmental biology occurred in 1984 with the discovery of the homeobox, an evolutionarily conserved 180-bp DNA sequence found in several *Drosophila* genes which specify segment identity (MCGINNIS et al. 1984; SCOTT and WEINER 1984; reviewed in MCGINNIS and KRUMLAUF 1992). Mutations in these genes, termed homeotic genes, affect segment identity and result in homeotic transformations in which one body part is transformed into another. The homeobox encodes a 60-amino acid DNA-binding domain, termed the homeodomain, and homeotic genes have been shown to encode transcription factors, acting in conjunction with other accessory proteins, including those of the *extradenticle* family, which includes the human *pbx* proto-oncogenes (VAN DIJK and MURRE 1994; CHEN et al. 1994; RAUSKOLB et al. 1993), and MADS box (MCM-1, AG, DEFA, and SRF; SCHWARZ-SOMMER et al. 1990) families.

In mammals, 38 *Hox* genes have been identified. These reside in four main chromosomal clusters, termed *Hoxa, Hoxb, Hoxc,* and *Hoxd*, and define 13 paralogous groups (Fig. 4). Diverse other vertebrate genes also contain homeoboxes, but in general these genes are dispersed to different chromosomal locations in the genome and, although they may be referred to as "homeobox genes," they should not be referred to as "*Hox* genes," which implies a specific evolutionary relationship to the *Drosophila* homeotic complex, or *HOM-C*. Remarkably, the relative order of mammalian *Hox* genes in each cluster parallels the order of the related genes in the *Drosophila* HOM-C,

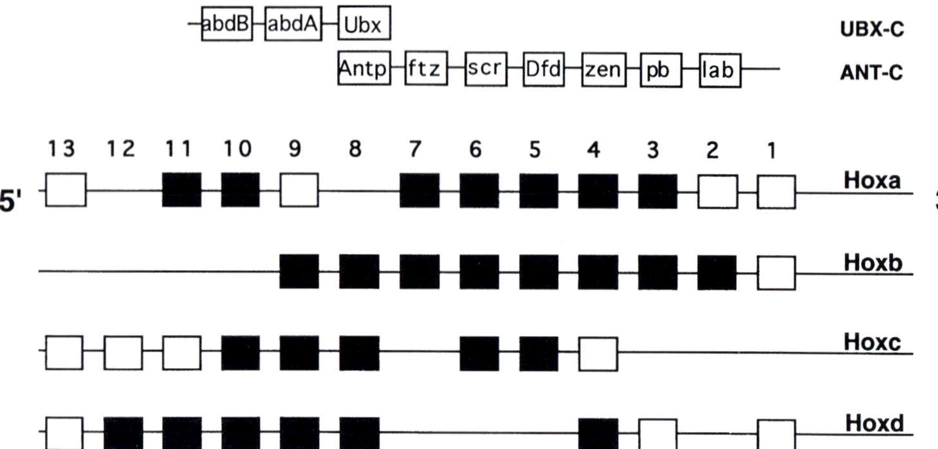

Fig. 4. Genomic organization of the *Drosophila* (*top*) and mammalian (*bottom*) *Hoxa, Hoxb, Hoxc,* and *Hoxd* clusters. In mammals, the 38 genes are organized into 13 paralogous groups in four separate chromosomal clusters. Note the conserved order of closely related *Drosophila* and mammalian *Hox* genes. Genes shown to be expressed in developing kidney are in *black*

a paralogous relationship which permits the grouping of mammalian *Hox* genes into the 13 discrete groups shown in Fig. 4. Thus, in approximately 600 Myr separating vertebrate and invertebrates, homeobox genes have been both conserved and amplified, possibly in connection with the greater complexity of the vertebrate body plan. Within each mammalian *Hox* cluster, individual *Hox* genes display several unique properties, including temporal co-linearity and nested patterns of expression, with genes at the 5′ end of each cluster exhibiting progressively more posteriorly restricted anterior limits of expression.

As indicated above, the *Hox* genes located at the 5′ end of the clusters display common properties which are a function of their specific location within the *Hox* clusters. The 12 genes belonging to paralogous groups 9–13 are all most closely related to the *Drosophila Abdominal B* (or *AbdB*) gene, from which they evolved collectively. They thus are expected to share similar properties and some overlap in function. For example, these more 5′-located *AbdB Hox* genes tend to be expressed relatively late in embryonic development, are insensitive to induction by retinoic acid, prefer a different core DNA recognition sequence than more 3′-located genes, and are expressed in a graded fashion restricted to the posterior or caudal end of the embryo (reviewed in McGINNIS and KRUMLAUF 1992). Thus, the posterior *AbdB Hox* genes tend to be expressed preferentially in the developing urogenital system (Fig. 4). Nonetheless, many *Antp*-class *Hox* genes (those in paralogous groups 1–8) are also expressed in the developing kidney (Fig. 4). However, while overexpression of *Hoxb8* results in renal hypoplasia (CHARITÉ et al. 1994), neither *Hoxa3* nor *Hoxc8* deficiencies cause a renal phenotype. It is possible that the *Antp-Hox* genes are involved in kidney differentiation per se, rather than the regulation of ureteric bud outgrowth. The expression of a number of *Hox* genes in the adult kidney, in some cases in a segment-specific manner (e.g., *Hoxb7* is restricted to the medullary collecting duct), could imply a role in maintaining the differentiated state of these epithelia (KRESS et al. 1990; DESCHAMPS et al. 1987). Ectopic expression of *Hoxb7* results in multiple malformations including double ureter and pelvis (ARGAO et al. 1995). *Hox* genes which have been shown to be expressed in the developing kidney are indicated in Fig. 4.

Relevant to the question of ureteric bud outgrowth, the *AbdB Hox* genes which are expressed in the developing mammalian excretory system are found not only in the meta- and mesonephros, but intriguingly also in the periureteral mesenchyme surrounding the ureteric bud and Wolffian duct. These same posterior *AbdB* genes also display a nested pattern of expression in periductal mesenchyme in both the male and female developing reproductive tracts, with graded anterior boundaries (DOLLÉ et al. 1991). This is strikingly reminiscent of similar nested patterns of expression observed for *Hoxd AbdB* genes in the developing limb bud and for *Hoxa* and *Hoxd AbdB* genes in the developing gut mesenchyme, suggesting that similar "*Hox* codes" may be used to instruct axial structures in general (KESSEL and GRUSS 1991; reviewed in HUNT and KRUMLAUF 1991). Although it is not clear whether these domains of expression are in place prior to mesonephric duct or ureteric bud outgrowth, these results could suggest the existence of a signaling system between the periureteral mesenchyme and ureteric bud or nephric duct epithelium, specifying positional identity or differentiation cues to the epithelium.

Recently, ES cell technology has been employed to determine the functional consequences of homozygous loss of function mutations in several *AbdB Hox* genes. Despite the strong expression of *Hoxa10* and *Hoxa11* in the midgestation metanephros, mutations in neither gene lead to renal anomalies, and homozygotes have normal life expectancies (SATOKATA et al. 1995; HSIEH et al. 1995). This result suggests functional redundancy between paralogous *Hox* genes, such that closely related gene products may be able to compensate for one another's absence. In support of this, while neither *Hoxa11* nor *Hoxd11* homozygous mutants express a kidney phenotype, mutant mice homozygous deficient for both genes express bilateral renal agenesis or, in some cases, an enlarged "spongy" kidney (DAVIS et al. 1995). Some such mice can survive several days postnatally, implying a small amount of residual renal function, since complete renal agenesis causes neonatal lethality within 24 h of birth. These results indicate the involvement of *Hox* genes in early steps of metanephric induction, and since *Hox* genes are expressed at progressively more rostral levels of periureteral mesenchyme, this may perhaps even be at the level of controlling ureteric bud outgrowth.

D. Ureteric Bud and Metanephric Induction

Once induction of the metanephric mesenchyme by the ureteric bud occurs in mouse at E11.5, a number of discrete morphologic events occur in a specific temporal sequence. Saxén (1987) has divided these events into the following stages: (a) proliferation (12–24 h), (b) loss of interstitial proteins from the extracellular matrix (also within the first 12–24 h, concomitant with the stimulation in DNA synthesis), and (c) synthesis of epithelial specific extracellular and cytoskeletal proteins, which occurs throughout the entire induction period, from initial mesenchymal precondensation to final tubular segmentation. First, we will consider the nature of the inducing signal which is transmitted from the ureteric bud to the metanephric mesenchyme. Then we will consider each of these phases in the induction process, including what is known about the molecular events involved.

I. Transmission of the Inductive Signal from Ureteric Bud to Metanephric Mesenchyme

The transfilter culture technique has provided some insight into the requirements for metanephric induction (Grobstein 1953a,b, 1955, 1956; Unsworth and Grobstein 1970; reviewed in Saxén 1987). Although metanephric induction can occur across a filter, detailed studies have revealed that cytoplasmic processes are extruded across the filter by the inducing tissue, making contact with the mesenchyme (Wartiovaara et al. 1972, 1974; Saxén and Lehtonen 1978). Filters with pore sizes small enough to exclude these processes do not support induction. In addition, antibodies to cell surface molecules such as ganglioside G_{D3} (Sariola et al. 1988) and extracellular matrix molecules such as laminin A (Klein et al. 1990) or its $\alpha6\beta1$ integrin receptor (Sorokin et al. 1990) block metanephric induction, suggesting that cell–cell contact is required. A role for cell adhesion molecules (reviewed in Saxén 1987) and gap junction proteins such as connexins has also been suggested (Sainio et al. 1992).

Soluble factors are capable of modulating metanephric induction; transforming growth factor-β (TGF-β) (Avner and Sweeney 1990) and the cytokine leukemia inhibitory factor (LIF) (Bard and Ross 1991) act as inhibitors of metanephric induction. Antibodies to IGF I and IGF II block metanephric induction, implying a role for these factors, which are present in developing metanephros, in metanephric induction (Rogers et al. 1991). Although concerns exist about the fidelity of antisense oligonucleotides, such experiments suggest a role for nerve growth factor (NGF) in kidney morphogenesis (Sariola et al. 1991) and are consistent with the expression pattern of neurotrophin receptors in the developing kidney (Durbeej et al. 1993). HGF has also been implicated as an effector of tubular morphogenesis in cell culture and organ culture, as discussed later (Montesano et al. 1991a,b). As noted above, it is possible to culture kidney rudiments in defined media supplemented only with transferrin. However, in the absence of an inducer such as dorsal spinal cord, no single factor can promote metanephric differentiation in organ culture, though increased DNA synthesis may occur. In one study, pituitary extract was able to promote metanephric differentiation in rat but not mouse (PERATONI et al. 1991), and FGF2 was subsequently shown to account in part for this inducing activity (PERATONI et al. 1995). It is probably unfair to conclude on this basis, however, that these factors are largely dispensable, since many of these growth factors may be produced and act locally within the rudiment, being sequestered by extracellular matrix and capable of signaling only over short distances. In knockout mice which lack individual endogenous growth factor genes, exogenous serum-derived growth factors may be sufficient in vivo to permit normal nephrogenesis to ensue.

Elegant studies by SAXÉN and colleagues, conducted over a number of years, have served to distinguish between the possibilities of cell surface versus diffusable induction mechanisms. The interested reader is urged to consult the original references discussed in SAXÉN's excellent monograph on this important subject (SAXÉN 1987). Evidence against the free diffusion model for transmission of the inductive signal is quite strong. First, the interposition of additional filters prolongs the induction period in a manner inconsistent with free diffusion. Second, electron microscopic analyses reveal a basement membrane discontinuity at the inductively active tip of the ureteric bud, permitting close apposition of epithelium and mesenchyme. Moreover, inductive potential in the transfilter assay correlates with the ability of the inducing tissue to penetrate the pores in the filter. These results support the view that cell–cell contacts, between inducing ureteric bud cells and responding mesenchymal cells, play an essential role in transmission of the inductive signal(s) involved in metanephric induction.

One gene which might encode a product capable of providing this signal is that encoding the c-ros RTK (TESSAROLLO et al. 1992). The expression of this gene in the developing kidney is restricted to the epithelium of the ureteric bud and later to the collecting duct. Beginning at E11.5, c-*ros* is expressed in the epithelial tip of the ureteric bud. As the ureter branches, c-*ros* expression is maintained, but only at the tips of the individual branches of the collecting system. This pattern is similar to the pattern of inductively active regions of the branching ureter. Interestingly, the ros RTK is a homologue of the protein encoded by the *Drosophila sevenless* gene, which is involved in the specification of R7 photoreceptor cell fate in the omatidia of the *Drosophila* compound eye, by the adjacent R8 photoreceptor. The R8 photoreceptor expresses a ligand, *bride of sevenless* or *boss,* a seven-transmembrane-spanning domain protein which provides the signal to the R7 precursor cell required for R7 differentiation. The fact that boss is a membrane-bound ligand thus implies that its activity is normally expressed via cell–cell contact. Although a mammalian boss homolog has not been identified, these data suggest that a protein homologous to boss might be expressed in the metanephric mesenchyme. Since other examples of seven-transmembrane-spanning proteins exist which interact with G proteins to effect signal transduction, c-ros could participate in the transmission of signals not only from the mesenchyme to the ureter via the c-ros RTK, but also from the epithelium to mesenchyme, via the as yet unidentified c-ros ligand. Such a reciprocal inductive model is consistent with the classical view, enumerated above, that the ureter provides the signal required for induction of the metanephric mesenchyme and that, reciprocally, the mesenchyme provides a signal required for the continued branching of the ureter. The idea that c-ros might be an effector of this interaction, however, must be reconciled with the fact that c-*ros* is not expressed in embryonic spinal cord, brain, or other tissues which can induce metanephric mesenchyme. Moreover, the knock out of c-*ros* has recently been accomplished and analysis of the phenotype does not demonstrate a defect in kidney development (SONNENBERG-RIETHMACHER et al. 1996). Thus, while c-*ros* might participate in metanephric induction, it cannot be necessary. Lastly, recent work indicates that a novel *wingless* gene family member, Wnt-11, is expressed at the tip of the arborizing ureteric bud and is therefore also a potential candidate for an inducing molecule (A. McMahon, personal communication).

II. Propagation of the Inductive Signal Within the Mesenchyme

Once the initial inductive interaction between ureteric bud and metanephric mesenchyme occurs, two spreading mechanisms have been proposed for the propagation of the inductive signal within the induced mesenchyme: (1) an assimilatory, homeogenetic ("like begets like") mechanism, in which induced cells in turn induce their neighbors, and (2) a migratory model, in which the induced cells migrate into the deeper layers of the mesenchyme. Two experimental results effectively rule out the homeogenetic model. First, an induced mesenchyme cannot pass on the inductive signal to another uninduced mesenchyme through an intervening filter, and second, in admixtures between karyotypically distinguishable induced and uninduced mesenchymes, followed by isolation and karyotyping of the tubular cells, all of the tubule-forming cells are derived from the initially induced mesenchyme (SAXÉN and SAKSELA 1971).

The most persuasive evidence in favor of the cell migration model derives from chick–quail chimeras, using the distinctive Feulgen-positive quail nucleolus to distinguish between the two cell populations. In explants in which quail ureteric bud was deliberately excised along with a layer of quail mesenchymal cells and then recombined with chick metanephric mesenchyme, quail tubules were identified in the chick mesenchyme, frequently at a considerable distance many cell diameters away from their original location attached to the ureter (SAXÉN 1984). This peripheral migration of the induced cells would be expected to allow new, uninduced cells to come in contact with the ureter and undergo a fresh cycle of induction. Moreover, the peripheral migration of induced mesenchymal cells occurs on a time scale consistent with the kinetics observed for spread of the inductive signal throughout the mesenchyme. Taken together, these studies lead to the conclusions that the transmission of inductive signals is mediated by short-range interactions (< 0.005 μm) between the ureteric bud and the metanephric mesenchyme and that the propagation of the inductive signal within the mesenchyme is effected by cell migration. While these principles almost certainly will not prove general to all examples of organogenesis that occur by epithelial–mesenchymal interactions, they will undoubtedly serve as important landmarks and provide a precedent for the methodical dissection of possible mechanisms in other developing organ systems.

The vertebrate signaling molecule Wnt-4 is a secreted glycoprotein homologous to the gene product encoded by the *Drosophila wingless* gene and is expressed in condensed mesenchyme on both sides of the ureter stalk at the site where the first pretubular aggregates form (STARK et al. 1994). This expression is repeated in newly forming aggregates, with expression persisting in comma bodies, before becoming restricted in S-shaped bodies to the region where epithelial fusion with the collecting duct occurs, then disappearing entirely after fusion. Interestingly, Wnt-4 is expressed in analogous fashion during development of the chick mesonephros, further supporting the view that mesonephric and metanephric induction may utilize similar molecular pathways.

Mice lacking Wnt-4 activity fail to form pretubular aggregates, exhibiting small kidneys consisting of undifferentiated mesenchyme interspersed with branches of collecting duct epithelium (STARK et al. 1994). Although initial induction of the mutant mesenchyme occurs in these mutants, as judged by the initial formation of mesenchymal condensates, by E15 the mutant kidneys are growth retarded and most of the kidney mesenchyme is undifferentiated and lacks comma or S-shaped bodies. Markers for pretubular aggregation and tubule formation, such as N-myc and Pax-8, are either not expressed or expressed only poorly. In contrast, the initial aspects of mesenchymal and ureteric bud development are unaffected in *Wnt-4* mutants, as judged from histology and preserved mesenchymal expression of WT1 and Pax-2 in *Wnt-4* mutant mesenchymes and of c-ret and c-ros in the ureteric bud. It has therefore been proposed that Wnt-4 may function as an autoinducer of the mesenchyme to epithelium transition that occurs during nephrogenesis. In *Drosophila, wingless* signaling requires the activity of *armadillo*, a β-catenin homolog likely to interact with the cell adhesion molecule E cadherin, and several Wnt gene products can modulate cell adhesion (BRADLEY et al. 1993; HINK et al. 1994; SHIMAMURA et al. 1994). It is possible that propagation of the inductive signal throughout the metanephric mesenchyme by Wnt-4 is achieved through the modulation of cell adhesion molecules. This speculation is also consistent with in vitro data indicating that Pax-8 can regulate the expression of neural cell adhesion molecule (NCAM) (HOLST et al. 1994) and that E cadherin is expressed in the kidney after expression of Wnt-4 and Pax-8. Unfortunately, the E cadherin knockout mouse is not informative, since these mutants exhibit preimplantation lethality (RIETHMACHER et al. 1995). NCAM is expressed only early after induction, but does not persist in mesenchymal cells committed to an epithelial fate, in spite of persistent Pax-8 expression (Plachov et al. 1990). However, NCAM expression persists longer in mesenchymal cells destined to become interstitial cells, implying a possible role in binary cell fate decisions in differentiating metanephric mesenchyme (Ekblom and Weller 1991).

Unlike the mesenchyme of other organs, the metanephric mesenchyme gives rise to both the stroma and the epithelia of the kidney. BF-2, a Winged Helix transcription factor, is an early marker of the stromal lineage in the developing metanephros (Hatini et al. 1996). Interestingly, BF-2 expression at E12.5 is restricted to a ring of cells that surrounds the induced metanephric mesenchyme. In BF-2 deficient knockout mice, the metanephric mesenchyme condenses normally but differentiation into tubular epithelium is affected. These mice also display impaired elongation and branching of the ureter and collecting system. These data indicate that the stroma provides a factor(s) which is critical for the propagation of the inductive signal by influencing both the metanephric mesenchyme and the ureteric bud epithelium (Hatini et al. 1996).

Although the knockouts of activins βA and βB do not express renal defects, knockout of the activin type IIB receptor (ActIIB-r) does result in a variably penetrant renal agenesis phenotype, with about 30% of homozygotes expressing unilateral renal agenesis (E. Li, personal communication). In organ culture, activin can disrupt branching morphogenesis, but only at relatively high doses (Ritvos et al. 1995). In *Drosophila*, evidence exists to support the ability of the homologous receptor to transduce a signal from *decapentaplegic* (*dpp*), a Bmp homolog and, like activin, a TGF-β superfamily member. Interestingly, among the six known Bmp (excluding the GDF subclass of Bmp), Dpp is most closely related to Bmp-2 and Bmp-4, both of which are strongly expressed, along with Bmp-7, a member of the 60A Bmp subclass (Lyons et al. 1995), in the condensing metanephric mesenchyme. Since *Bmp-4*-deficient embryos die at gastrulation (B.L.M. Hogan, personal communication), and *Bmp-2* embryos die before E11.5 (Bradley and Zhang, submitted), genetic tests of the hypotheses that Bmp-2 and Bmp-4 participate in the propagation of metanephric induction must await the generation of a conditional knockouts. *Bmp-7* knockout mice, however, die at birth with dysplastic, cystic kidneys and hydroureter. Histological analyses indicate that wild-type and mutant kidneys appear similar at E14, but that subsequently a marked depletion of cortical nephrons and an excess proliferation of centrally disposed collecting ducts commences in the mutant. Molecular studies indicate that amongst Pax-2, Pax-8, c-ret, and Wnt-4, only Wnt-4 expression is disturbed at E14, being abnormally expressed around the collecting ducts and lost from the cortical zone (Dudley et al. 1995; Luo et al. 1995). These data might support a model whereby initial inductive events occur normally in the Bmp-7 knockout, but with a subsequent block in propagation of induction and a secondary accumulation of collecting ducts due to loss of an inhibitory signal normally expressed by mesenchymal condensates.

III. Epithelial–Mesenchymal Transformation

Following a 24-h induction of the metanephric mesenchyme by inducer, one of the first events to occur, compared to uninduced mesenchymes, is a marked stimulation of [^{3}H]thymidine uptake. This stimulation of DNA synthesis is closely correlated with the induction process itself: experiments employing interrupted inductions or carried out under various transfilter recombination conditions (e.g., different pore sizes or transfilter to different tissues) show that incorporation of [^{3}H]thymidine occurs only under conditions that also result in induction. SAXÉN et al. (1983) suggest that this increase in DNA synthesis is attributable to a decreased cell cycle length, instead of an increase in the number of cells in S phase. As might be expected, the proliferation of mesenchymal cells occurring after metanephric induction coincides with the expression of "immediate early genes" or IEG, defined as genes such as *Egr-1* whose expression is induced by mitogenic stimuli in the absence of de novo protein synthesis. As mentioned earlier, proliferation of undifferentiated renal progenitor cells can be sustained in primary culture by a factor, as yet unidentified, secreted by a Wilms tumor cell line (BURROW and WILSON 1993). Epidermal growth factor (EGF), IGF I and II, PDGF, acidic and basic fibroblast growth factor (FGF), TGF-α, TGF-β, and LIF failed to stimulate proliferation of these cells in culture. In the absence of this factor and in the presence of serum, these cells do express early markers of epithelial differentiation such as E cadherin and cytokeratin. Thus this putative growth factor may be involved in sustaining proliferation of committed renal stem cells and inhibiting their differentiation by other growth factors present in serum. The model discussed earlier implicating WT1 and Egr-1 in a regulatory loop with mitogenic growth factors suggests that Egr-1 likely induces expression of this factor which sustains proliferation of mesenchymal cells in the absence of functional WT1, which would be expected to act as a transcriptional repressor of this putative mitogen.

An important series of experiments address the question of whether the different epithelial lineages found in the mature kidney derive from a multipotent stem cell or are determined as independent lineages prior to metanephric induction (HERZLINGER et al. 1992; reviewed in AL-AWQATI 1992). Using retroviral vectors expressing the *Escherichia coli lacZ* gene, AL-AWQATI and colleagues found that infection of uninduced E13 rat metanephric mesenchymes at limiting dilution, followed by analysis for β-galactosidase activity after 7 days in organ culture, yielded clones consisting of glomerular, proximal, and distal tubular epithelial cells. Additionally, individual clones were distributed amongst different nephrons. This indicates that mesenchymal progenitors can contribute to more than one nephron and implies that these original progenitors or their immediate descendants can migrate over short distances. In contrast, when the infections are carried out after a 24-h induction period using E13 dorsal spinal cord as the inducer, a significant restriction in differentiation capacity is observed, such that labeled cells are only observed in individual nephrons and are restricted to individual nephron segments, e.g., glomerulus, proximal, or distal tubule. In addition, only a small number of cells, typically six or less, were present in each clone, compared to approximately 50 in the experiments in which the infection was carried out prior to induction, suggesting that, in these experiments, the cells became marked after the rapid phase of cell proliferation occurring after induction. Interestingly, the number of cells present in labeled clones analyzed 3 days after infection cannot account for the higher number of labeled cells when observations are carried out at 7 days after infection, unless an increased rate of proliferation after induction is inferred. These observations are consistent with the studies described above in which DNA synthesis and, therefore, cell proliferation are markedly increased following induction of the metanephric mesenchyme.

As described earlier, most of the metanephric mesenchyme is thought to be induced by an active migratory process which continually brings uninduced mesenchyme into contact with the ureter. Uninduced mesenchymal cells normally undergo apoptosis, an effect which can be prevented by addition of EGF (Koseki et al. 1991; Coles et al. 1993). WT-1 has been shown to induce apoptosis in osteosarcoma cells through repression of EGF receptor (EGFR) transcription and consequent reduction in EGFR synthesis (Englert et al. 1995). Bcl-2, a protein which serves to prevent the execution of programmed cell death (for a review, see Hockenbery 1994), is expressed in induced metanephric mesenchyme around E12.5, in ureteric bud and primitive tubular structures, and is a logical candidate to regulate the apoptotic elimination of uninduced metanephric mesenchyme (LeBrun et al. 1993). This functional role is substantiated by the *bcl-2* knockout mouse. Organ culture of day E12 metanephric mesenchyme from *bcl-2*-deficient mice confirms a marked increase of apoptosis as the mechanism for renal hypoplasia observed in these mice (Veis et al. 1993; Sorenson et al. 1995). Other bcl-2 family members include both positive and negative regulators of cell death which participate in homo- and heterodimeric protein–protein interactions. These other family members are also likely to be involved in regulating apoptosis in the metanephros.

After invading the mesenchyme, the ureteric bud begins to divide dichotomously. Osathanondh and Potter (1963a,b) have examined this pattern of ramifications in great detail in the human kidney and found that both symmetric and asymmetric divisions occur. After the initial induction event, the ampulla of the collecting duct is able to induce an additional nephric vesicle; in primates and rodents, this process is repeated so that the three or four initially formed juxtamedullary or midcortical nephrons are all connected to the same collecting duct to form what has been termed a "nephron arcade" (Fig. 5). A final phase of nephron induction occurs even after the collecting duct has ceased to branch. As it continues to grow, it induces a set of terminal subcapsular nephrons, five to seven in the human and two in the rat, which join the collecting duct directly (Fig. 5).

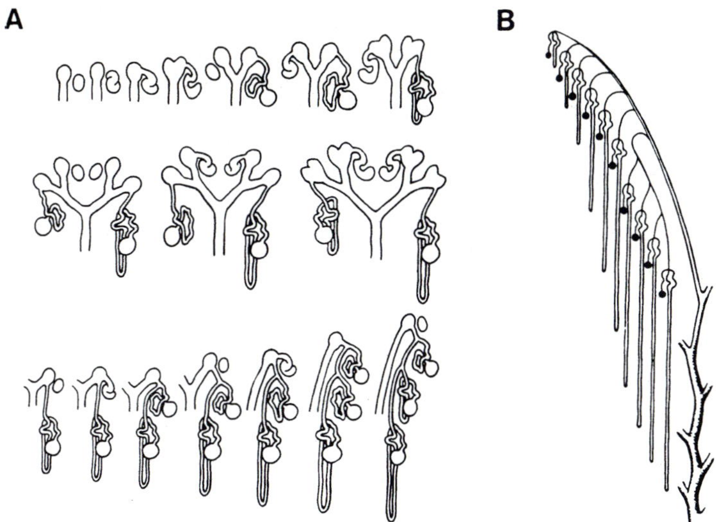

Fig. 5. A Early nephron development. Note the ability of the terminal collecting duct to induce multiple nephrons. **B** Resulting formation of nephron arcades. (Reprinted with permission of Cambridge University Press from SAXÉN 1987, based on the observations of OSATHANONDH and POTTER 1963a,b)

The formation of the epithelial component of the nephron involves a stereotyped series of morphologic events resulting in a conversion of mesenchyme to epithelium and the adoption of features characteristic of the mature nephron: a polarized epithelium, which organizes into the proximal and distal tubules. This process, which has been carefully studied, is illustrated in Fig. 6. As the ureter divides, mesenchyme condenses around the tip or ampulla of the ureter, leading to the formation of individually discernible mesenchymal condensates or renal vesicles (Fig. 6B). A striking feature of this condensation, preceding even overt morphologic changes, is the disappearance from prospective condensates of various components of the extracellular matrix, including type I and type III collagen and fibronectin, which are uniformly expressed in the uninduced mesenchyme (EKBLOM et al. 1981). Although it seems plausible that these events occasion the condensation process, this remains speculative. These condensates exhibit strong mitotic activity and assume a curved appearance, a stage termed the "comma body" stage (Fig. 6C). Subsequently, mitotic activity decreases and cellular histogenesis commences such that cells farthest from the collecting duct at the proximal pole elongate to form a crevice, which will become the future site for glomerular development and vascular ingrowth. At the distal pole, a second slit forms and joins the collecting duct. This latter stage constitutes the "S-shaped body" stage (Fig. 6D).

It is important to appreciate that this pattern of inductive spread throughout the mesenchyme results in a centripetal appearance, with centrally located nephrons being more developed than those which are peripherally

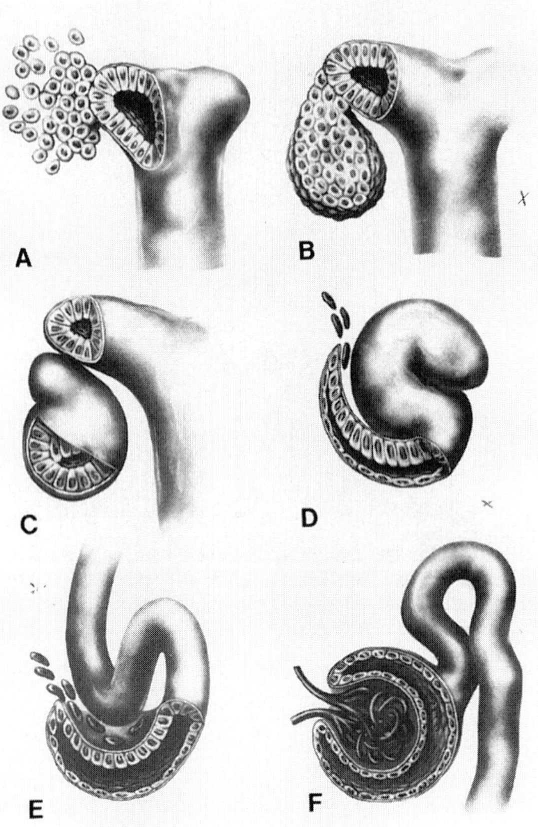

Fig. 6. Formation of the nephron, from the initial mesenchymal condensation to form the renal vesicle (**A,B**), comma body (**C**), S-shaped body (**D**), and then subsequent vascular ingrowth (**D–F**). (Reprinted with permission of Cambridge University Press from SAXÉN 1987)

situated. This means that if we follow epithelial histogenesis temporally during kidney development, we expect to observe a particular pattern of nephron development originating near the center of the organ and spreading peripherally as a function of time. This pattern recurs frequently in molecular studies using the technique of radioactive in situ hybridization to examine the expression of genes which are believed important in nephrogenesis.

IV. Role of *Pax-2*

A critical regulatory role for the mammalian *Pax-2* gene in controlling mesenchymal–epithelial transformation in kidney development is indicated by several lines of evidence. First, the expression pattern of *Pax-2* (DRESSLER et al. 1992) is highly suggestive of an important regulatory role. Second, important evidence comes from experiments blocking its function in culture (Rothenpieler and Dressler 1993) and from analysis of mouse *Pax-2* (Keller et al. 1994; Torres et al. 1995) and human *PAX2* mutants (Sanyanusin et al. 1995). Pax-2 is a member of the paired box or Pax family of transcription factors, which are unified by the presence of a 128-amino acid DNA-binding domain, the paired domain, first identified on the basis of homology with the *Drosophila paired* gene product. The mammalian *Pax* family consists of nine physically unlinked genes, two of which, *Pax-2* and *Pax-8*, are expressed in the developing kidney (reviewed in Strachan and Read 1994). Interestingly, a role for the gene product encoded by *Sd* (*Danforth's short tail*) in regulating *Pax-2* expression is suggested by the observation that *Pax-2* expression is lost in *Sd* metanephric mesenchyme soon after epithelial differentiation begins (Phelps and Dressler 1993).

In mouse at E9–10, *Pax-2* is expressed in pronephric tubules, but only after extension of the nephric cord and Wolffian duct, which also express *Pax-2* (Dressler et al. 1992). Similarly, at E11, the ureteric bud expresses *Pax-2* as it undergoes outgrowth from the Wolffian duct, while the surrounding mesenchyme does not. After the ureteric bud undergoes initial bifurcation in the mesenchyme, however, both it and the metanephric mesenchyme around the branched ureter strongly express *Pax-2*. At E12, this pattern continues, with *Pax-2* expression in condensed but not uncondensed mesenchymal cells and in the collecting duct. *Pax2* expression then shifts to the tubules which form last and which are located at the periphery of the kidney and declines in the mature, centrally located tubules; it is not detected at all in the adult kidney. Interestingly, however, *Pax-2* expression can be reinduced in adult kidney tubular epithelium after insult with mercuric chloride, suggesting a role in tubular regeneration occurring after injury (Grone et al. 1993). Although *Pax-2* is not expressed in the WT1-deficient metanephric mesenchyme, it cannot be concluded that *Pax-2* requires WT1 for its expression, since expression of *Pax-2* normally requires induction by the ureteric bud, which does not occur in this knockout. As argued elsewhere, the available data in fact suggest that, at least in the context of glomerular epithelial differentiation, WT1 is likely to act as a repressor of *Pax-2* expression (Ryan et al. 1995). The expression pattern for *Pax-2* strongly implicates it in the transmission of inductive signal(s) between renal epithelium and mesenchyme. In addition, Pax-2 may be involved in maintenance of the proliferative state of the mesenchyme, prior to terminal differentiation. Transcriptional repression of p53 by *Pax-2* resulting in an inhibition of apoptosis may, at least in part, mediate this effect (Stuart et al. 1995). Overexpression of p53 in transgenic mice resulted in high rates of apoptosis in uninduced metanephric mesenchyme (Godley et al. 1996). Surprisingly, this was due to misexpression of p53 in the ureteric bud, a site of *Pax-2* expression, with secondary effects on the adjacent mesenchyme, since the transgene is only expressed in ureteric bud epithelium and not the mesenchyme (Godley et al. 1996). Finally, the fact that *Pax-2* is expressed in both the inducing tissue and, after induction, in the induced tissue is a feature common to *Pax* gene expression in other organ systems, and suggests a reiteratively utilized genetic pathway.

Loss of *Pax-2* function in kidney organ cultures has been accomplished by antisense oligonucleotides, which specifically inhibit Pax-2 protein accumulation in kidney mesenchymal cells (ROTHENPIELER and DRESSLER 1993). These mesenchymal cells fail to aggregate and do not undergo the sequential morphologic changes associated with epithelial cell formation or form S-shaped or comma bodies, an effect reversed by removal of the antisense oligonucleotide. The ureteric bud in these cultures also exhibits reduced branching, a result suggested as secondary to the block in mesenchymal condensation, since in culture, as mentioned earlier, the induced condensing mesenchyme can reciprocally induce the ureter to branch. Markers for epithelial development, such as laminins B1, B2, and A and uvomorulin, are also greatly reduced in the antisense-treated specimens, but these are not thought to be direct targets for induction, since a 24-h delay occurs between induction and the onset of expression for these markers.

A semidominant transgenic-induced deletion mutant, *Krd* (kidney and retinal defects), encompasses the mouse *Pax-2* gene (KELLER et al. 1994). Heterozygotes manifest a significant incidence (three out of 23 mice) of unilateral or bilateral renal agenesis and an incidence of unilateral renal cysts and megaureter of about 25%. Upon histological examination at postnatal day 1 (P1), *Krd* /+ mice exhibit immature renal mesenchymal tissue, particularly at the corticomedullary junction, a decreased nephrogenic zone with decreased numbers of glomeruli and proximal convoluted tubules. The resulting reduction in cortical thickness in *Krd* /+ mice correlates with a significant reduction in kidney weight compared to wild-type controls. Although other genes linked to *Pax-2* on mouse chromosome 19 are also encompassed in the 7-cM interval deleted in this mutant and may account for the early lethality of homozygous *Krd*/*Krd* embryos before E10.5, the observed heterozygous phenotype is very likely due to *Pax-2* haplo-insufficiency. Embryonic retina and kidney are major sites of *Pax-2* expression, and *Krd* mice also exhibit retinal malformations and cellular depletion and optic nerve abnormalities. The occurrence of a dominant phenotype resulting from a 50% reduction in gene dosage is an unusual genetic property shared by most *Pax* genes. In addition, a human pedigree dominantly expressing renal hypoplasia (requiring transplant in one case), proteinuria, vesicoureteral reflux, and optic nerve colobomas exhibits a single nucleotide deletion in exon 5 of human *PAX2*, causing a frameshift mutation (SANYANUSIN et al. 1995). Although renal histology was not examined, the overall phenotype appears very similar to that in the *Krd* mouse. Interestingly, mice doubly heterozygous for *Krd* and *Sd* may show an increased rate of renal agenesis (KELLER et al. 1994). Finally, mice carrying a null *Pax-2* allele have also been generated by homologous recombination using ES cells (TORRES et al. 1995). Heterozygotes exhibit a background-dependent reduction in kidney size similar to the reported *Krd* and human phenotypes. Homozygous *Pax*-2-deficient mice express bilateral renal and ureter agenesis and also lack genital tracts. The absence of these structures in *Pax-2* mutants suggests a role for *Pax-2* in the specification or differentiation of the intermediate mesoderm from which they are normally derived.

The expression pattern of the LIM homeodomain protein termed Lim-1 (or LH1) in the developing excretory system is very similar to that of *Pax-2* (FUJII et al. 1994). At E8.5, Liml expression appears most strongly in the nephrogenic cord and in a more anterior region consistent with expression in the pronephros. At E9.5, *Lim-1* expression is seen in the mesonephric duct and at E10.5 it is seen in both mesonephric tubules as well as mesonephric duct. Analogous to *Pax-2, Lim-1* expression at E11.5 is initially observed in the ureteric bud, but not in the mesenchymal tissue around the bud. At E13.5, however, *Lim-1* transcripts appear in both renal vesicles and S-shaped bodies. The knockout of *Lim-1* reveals renal agenesis as well as an absence of gonads and forebrain and midbrain tissue, although the renal agenesis phenotype is somewhat difficult to study, as most homozygotes fail to develop beyond E11.5 (SHAWLOT and BEHRINGER 1995). These results could suggest that Pax-2 and Lim-1 may interact directly or that one transcription factor may regulate the expression of the other. Significantly, *Pax-2* expression is preserved in the nephric system of Lim-1-deficient homozygotes (R. Behringer, personal communication). *Pax-2, Lim-1, ld*, and WT1 may thus function in a regulatory hierarchy controlling ureteric bud outgrowth and metanephric induction. The availability of mouse mutations in each of these genes should permit epistasis analyses to order them in a genetic pathway regulating early steps in nephrogenesis. Consistent with the view that *Pax-2* functions near the top of a pathway regulating metanephric induction, expression of *Pax-2* in *Wnt-4*-deficient embryos is unaffected, while that of *Pax-8* is eliminated (STARK et al. 1994).

Additional signals are evidently employed to control the execution of the induction process. Knockout mice carrying a null allele of the leucine zipper bHLH nuclear oncogene N-*myc* die at about E11.5, showing mesenchyme to epithelial transformation in the mesonephros but a reduced number of mesonephric tubules (STANTON et al. 1992; CHARRON et al. 1992). This may indicate that N-*myc* is required for proliferation or early steps of epithelial differentiation of the induced mesenchyme, a role consistent with its in vivo pattern of expression (MUGRAUER et al. 1988; MUGRAUER and EKBLOM 1991). A leaky mutation of N-myc generated by gene targeting results in a defect in growth and branching of distal bronchioles, but no kidney defect (MOENS et al. 1992). This approach enables a definition of later developmental effects of a gene which is an embryonic lethal mutation when null. Intermediate reduction in N-myc gene dosage by crossing the null and leaky heterozygotes may reveal other developmental roles.

The cytokine leukemia inhibitory factor (LIF) potently blocks nephrogenesis of cultured kidney rudiments, and LIF has been proposed to act downstream of the initial induction step, since LIF-inhibited rudiments are able to express syndecan-1, an early marker for differentiation (BARD and ROSS 1991; BARD et al. 1994; VAINIO et al. 1989). Since only trace amounts of LIF transcripts can be detected in the developing kidney (ROBERTSON et al. 1993), and LIF knockout mice have normal renal function (STEWART et al. 1992), it is unclear whether this finding is relevant to normal metanephric induction. Since LIF is a member of a larger gene family, however, including ciliary neurotrophic factor (CNTF), a functional role for other family members can be entertained. Interestingly, kidney rudiments cultured in the presence of LIF still develop a robust collecting system, implying that LIF does not impair the ability of the mesenchyme to reciprocally induce the ureter.

V. Establishment of Epithelial Polarity

The establishment of epithelial polarity is dependent on cell–cell and cell–matrix interactions. Attachment of the basolateral surface of epithelial cells to the underlying basement membrane is mediated by adhesion molecules, most notably integrins. Changes in expression patterns for extracellular matrix/adhesion molecules during the mesenchyme to epithelium transformation have been discussed above. It should be added that in the MDCK cell culture model, attachment to the substratum is necessary but not sufficient to establish apical–basolateral polarity (VEGAS-SALAS et al. 1987).

Cell–cell contacts are critical to the establishment of epithelial cell polarity. Using the Ca^{2+} switch model, GUMBINER et al. (1988) demonstrated that antibodies to E cadherin (uvomorulin) inhibited formation of all components of the junctional complex, the tight junction, zonula adherens, and desmosomes. These findings are consistent with the early expression of E cadherin in induced mesenchyme (VESTWEBER et al. 1985). It is unclear whether this molecule also participates in cell–cell recognition so that mesenchymal cells committed to an epithelial fate are thus distinguished from those mesenchymal cells destined to become interstitial cells or undergo apoptosis. Cells expressing cadherin at two different levels segregate into distinct populations when admixed in vitro, supporting a recognition function for these molecules (STEINBERG and TAKEICHI 1994). Cadherins cluster at sites of cell contact and interact with each other on adjacent cells through their extracellular domains. The cytoplasmic domains interact with catenins, which in turn link cadherins to the subcortical cytoskeleton and to signaling molecules, some of which have been implicated in nephrogenesis, such as the EGF RTK (HOSCHUETZKY et al. 1994). The interaction of cadherins with catenins is essential for full adhesion activity. Transfection of 3T3 cells with E cadherin results in a restriction of (Na^+-K^+)-ATPase to sites of cell–cell contact. This ability to polarize 3T3 cells depends upon the presence of the cytoplasmic cytoskeletal-binding domain (MCNEILL et al. 1990). As already mentioned, Wnt-4 may mediate mesenchymal condensation through modulation of catenins, thereby resulting in propagation of the inductive signal.

Tight junctions are established at the apical edges of cells and are thought to form a barrier separating apical and basolateral zones from mixing by lateral diffusion. In the *Drosophila* embryo, mutation of the fly homologue of mammalian *ZO-1*, a gene encoding a tight junction protein, leads to uncontrolled proliferation of imaginal disc epithelial cells and loss of their normal morphologic pattern (Woods and Bryant 1990).

Although much has been learned about general mechanisms of the biogenesis of epithelial polarity, much less is known about how epithelia acquire specific morphology and functions in different organs. HNF-1, a homeodomain-containing transcription factor, is initially expressed at E15 in S-shaped bodies and continues to be expressed postnatally in maturing proximal and distal tubules (Lazzaro et al. 1992). Knock out of HNF-1 produces a defect in glucose reabsorption in the proximal tubule and a syndrome of polyuria with profound glucose wasting (Pontoglio et al. 1996). The capacity of the proximal tubule to reabsorb glucose increases postnatally. HNF-1 is thus required for the development of this specific function of the mature proximal tubule.

VI. Tubule Formation

Transcripts for both HGF and its putative receptor, encoded by the c-*met* oncogene, are detectable in the embryonic kidney in both mesenchyme and ureteric bud at E11.5, and antibodies to HGF block tubulogenesis in vitro (Santos et al. 1994; Woolf et al. 1995). In MDCK cells, HGF has been identified as a fibroblast-derived morphogen capable of inducing tubule formation (Montesano et al. 1991a,b). In addition, the c-met gene product has been shown to be specifically capable of mediating mesenchymal to epithelial conversion (Tsarfaty et al. 1994; Sonnenberg et al. 1993). Woolf et al. (1995) examined the effect of an HGF-blocking antibody in organ culture. Their results suggest a dual role for HGF/c-met: to prevent apoptosis of mesenchymal cells, promoting their epithelial differentiation by an autocrine mechanism, and to enhance branching morphogenesis of the ureteric bud by a paracrine mechanism. HGF knockout mice, however, exhibit abnormalities of liver but not kidney development, indicating that HGF is not required for normal nephrogenesis (Schmidt et al. 1995). As discussed earlier, however, these results are not necessarily incompatible; other compensatory mechanisms may exist such that, while HGF may contribute to mesenchymal–epithelial transformation and branching tubular morphogenesis, it is not necessary. In mIMCD-3 cells, derived from the inner medullary collecting duct, TGF-α and EGF can induce the formation of branching tubular structures in collagen gels (Barros et al. 1995). Similar to results from organ culture, TGF-β inhibits tubulogenesis in this model. The well-recognized effects of TGF-β on matrix synthesis and degradation may mediate its effects on branching morphogenesis. Taub et al. (1990) obtained similar results with a tubulogenesis model using primary culture of 10- to 14-day-old mouse renal tubular cells in matrigel. HGF, TGF-α, and EGF also have differential effects on tubule elongation versus branching, with mIMCD-3 cells more likely to branch than MDCK cells. Thus, in addition to redundant functions, synergistic effects of growth factors are likely to be important in tubulogenesis and branching. The finding of abnormal cystic dilatation of collecting ducts in the EGFR knock out mice (THREADGILL et al. 1995) might be explained by a differential effect of EGF on elongation versus branching in the developing collecting duct.

As outlined earlier, the composition of extracellular matrix changes during nephrogenesis. Matrix components expressed in developing kidney in vivo when ureteric bud branching is occurring, such as laminin, type I collagen, fibronectin, and entactin, are also able to promote branching of MDCK cells in three-dimensional gels (SANTOS and NIGAM 1993). Type IV collagen, which is expressed after ureteric bud branching is largely completed, inhibits branching morphogenesis in this cell culture model. DAVIES et al. (1995) have demonstrated that the loss of sulfated proteoglycans (GAGS) by inhibition of sulfation by sodium chlorate or enzymatic degradation by heparatinase and chondroitinase inhibits ureteric bud growth and branching morphogenesis in organ culture. This results in a reduced number of nephrons, presumably reflecting fewer ureteric bud ramifications in metanephric mesenchyme. Differentiation of metanephric mesenchyme in regions which are induced by contact with the ureteric bud proceeds normally, indicating that mesenchyme-derived tubules do not depend on GAGS. It is possible that the effects of the stroma on ureteric bud branching demonstrated in BF-2 deficient mice could be the result of changes in the composition of the extracellular matrix.

Epithelial–mesenchymal interactions are associated with extensive tissue remodeling, requiring regulated degradation of extracellular matrix. Thus, differential expression of matrix proteases are likely also involved in the growth and branching of tubules through the extracellular matrix, although this has not been examined extensively in nephrogenesis. Stromelysin-1, a membrane-degrading matrix metalloproteinase (MMP), caused an increase in branching morphogenesis and increased proliferation and tumors of mammary duct epithelium when transgenically overexpressed under the control of a mammary gland-specific promoter (SYMPSON et al. 1994). A similar phenotype is obtained by overexpression of Wnt-1 (TSUKAMOTO et al. 1988). The modulation of extracellular matrix may increase the availability of growth factors, such as TGF and Wnt family members, which are normally sequestered in extracellular matrix. Thus both modulation of extracellular matrix or overexpression of Wnt may have the same effect of increasing the availability of Wnt to interact with its receptor on ductal epithelium. HGF has been shown to increase the expression of urokinase-type plasminogen activator and its receptor in MDCK cells (PEPPER et al. 1992). This finding directly links a factor known to promote branching morphogenesis in the kidney with the regulation of matrix proteases.

Epimorphin is a member of a novel class of mesenchymal proteins which is involved in cell contact-dependent epithelial–mesenchymal interaction. This protein is likely anchored to the cell surface by phosphatidylinositol and is expressed only in the mesenchyme of a variety of tissues, including the kidney (HIRAI et al. 1992). An antibody to epimorphin prevents tubulogenesis in lung explants and prevents establishment of apical–basolateral polarity. 3T3 cells transfected with the cDNA for epimorphin can substitute for endogenous mesenchyme in promoting lumen formation when cocultured with lung epithelial cells (HIRAI et al. 1992). Epimorphin is ubiquitously expressed, and therefore tissue-specific factors are likely also involved in mediating cell contact-dependent mesenchymal–epithelial signals in individual organs, including the kidney.

E. Glomerular Development

I. Glomerular Epithelium

Less is known about maturation of the glomerulus during kidney development than early events in mesenchyme to epithelium transformation and tubulogenesis. Nonetheless, a detailed understanding of glomerular development will be necessary for an understanding of human renal disease, since a large number of glomerular disorders, including those of the glomerular basement membrane (GBM), are important clinically. The lineage analyses by HERZLINGER et al. (1992) already discussed support earlier work by EKBLOM et al. (1980) indicating that there is a birth order of renal progenitor cells such that glomerular epithelial cells form first and differentiation further proceeds sequentially from proximal to distal tubule. Development of the specialized functions of glomerular components in particular have received little attention. For example, the glomerular visceral epithelial cell or podocyte contributes to the filtration barrier of the glomerulus with the formation of slit diaphragms. Electron microscopy and immunogold staining have identified the tight junction protein ZO-1 in this specialized structure, suggesting that it is a modified tight junction (SCHNABEL et al. 1990). However, little is known about the development of the podocyte, in spite of its central role in many glomerular diseases. J.A. Kreidberg recently reported that knock out of $\alpha 3$ integrin leads to an absence of foot process in podocytes (personal communication). This implies that interaction of laminin in the glomerular basement membrane with its integrin receptor on the developing podocyte is required for the formation of the foot processes.

Additional signals, not yet well characterized, are likely involved in differentiation of the podocyte. As indicated above, the fact that *WT1*-deficient mice express renal agenesis does not preclude a later role in glomerular development, as suggested by its late expression pattern in the podocyte. Histopathological analysis in patients with Denys-Drash syndrome reveals abnormal podocytes, consistent with this notion (HASTIE 1994). Interestingly, among the nine FGF family members identified to date, high levels of FGF2 (basic FGF) have been found in all epithelial cells of the developing chick kidney from the pronephric stage onwards (DONO and ZELLER 1994), and stable nuclear accumulation of FGF2 is observed in both differentiating and postmitotic podocytes of both meso- and metanephros. Since several high-affinity FGF receptors are also expressed during kidney morphogenesis (ORR-URTREGER et al. 1991; PETERS et al. 1992), a role in glomerulogenesis and other aspects of kidney morphogenesis is possible.

Notch-2, a transmembrane-signaling molecule which has been shown to specify neurogenic cell fate in *Drosophila* (reviewed by ARTAVANIS-TSAKONAS et al. 1995), has been knocked out in mouse. Interestingly, Notch-2-deficient mice have small kidneys with developmental arrest at the S-shaped stage, resulting in a lack of mature glomeruli in addition to defects in the developing distal tubule and collecting duct (T. Gridley, personal communication). In *Drosophila*, Notch is localized to apical membranes in polarized cells and thus may function in the establishment of apical–basolateral polarity. Genetic approaches in flies and worms have identified putative components of a Notch-signaling cascade which implicate it in relatively direct transmission of signals from the cell surface to the nucleus. The phenotype of the Notch-2 knockout suggests that vertebrate homologs of these genes are likely to be involved in glomerular development.

II. Mesangial Cells and Angiogenesis

Although early work suggested that the capacity for angiogenesis might reside within the kidney itself, grafting studies placing mouse metanephric mesenchyme onto quail chorioallantoic membrane (CAM) have shown that the glomerular endothelium and the blood vessels which grow into the developing kidney originate from outside of it (SARIOLA et al. 1983a). In addition, vascular ingrowth is specifically restricted to induced mesenchymes, occurring via capillary ingrowth (Fig. 6D–F), into the slit which forms at the proximal pole at the comma body stage of nephron development. Since uninduced mesenchyme does not support vascular ingrowth in these grafting experiments, it has been suggested that the differentiating epithelia secrete an angiogenic factor to attract endothelial cells. Work by RISAU and EKBLOM (1986) has suggested that a heparin-binding protein, possibly FGF, plays such a role in glomerular angiogenesis. Other factors, such as PDGF (see below) and renin, which is expressed in afferent arterioles of developing kidneys (GOMEZ et al. 1991), are likely involved in glomerular capillary formation. Since split products of collagen and fibronectin can exert an angiogenic effect, the extracellular matrix changes that occur after metanephric induction may also serve to stimulate angiogenic invasion of the glomerulus (SAXÉN 1987). When induced murine embryonic kidney is grafted onto CAM, vascular ingrowth follows a particular path, suggesting that differentiating nephric tissue provides guidance cues to invading capillaries. The vessels follow the border between early condensates and fibronectin-expressing stroma, but avoid areas of more developed epithelial tubules which express laminin on their basement membranes (SARIOLA et al. 1984). This finding is consistent with the purported role of fibronectin in mediating migratory events in development.

PDGF-β and its receptor (PDGF-βR) are examples of growth factors or their receptors which display a kidney phenotype when knocked out (LEVÉEN et al. 1994; SORIANO 1994). Deficient mice show a complete absence of mesangial cells from the glomerular tuft and a reduction in the glomerular capillary network. In early stages of glomerulogenesis, PDGF-βR is expressed in undifferentiated metanephric mesenchyme, and subsequently its expression is restricted primarily to mesangial cells, which also express PDGF (ALPERS et al. 1992). This has led to the suggestion that PDGF initially serves a paracrine role to recruit mesangial cells to the developing glomerulus and subsequently plays an autocrine role in stimulating proliferation of these cells. The reduced capillary network may be either a primary defect or suggests that mesangial cells provide a structural framework of elaborated matrix that permits capillary ingrowth into the glomerulus. However, quail–mouse chimeric experiments suggest that mesangial cells have their origin outside the developing kidney like the renal vasculature, and thus PDGF may be required for ingrowth of cells which can form both the capillary network and the supporting mesangial cells (SARIOLA et al. 1983a).

III. Glomerular Basement Membrane

An interesting finding from the PDGF knockout mice is that the GBM can form essentially normally in the absence of mesangial cells. This is consistent with previous work which suggested that the GBM is derived from the podocytes and endothelial cells and that the mature GBM results from fusion of epithelial and endothelial basal laminae (SARIOLA et al. 1983b; ABRAHAMSON 1985). Since the discovery that mutations in the noncollagenous domain of α_5-type IV collagen cause X-linked hereditary nephritis (BARKER et al. 1990), considerable effort has been made to understand the relationship of different type IV collagen chains in normal GBM development and as a potential cause of autosomal hereditary nephritis. Biochemical studies indicate that two populations of collagen heterodimers are formed, α_3- to α_5-type IV, synthesized by visceral epithelial cells, and α_1 -to α_2-type IV synthesized by mesangial and endothelial cells (KLEPPEL et al. 1992). Basement membrane of other tissues, such as lens and cornea, do not form disulphide bond-linked α-type IV dimers. In early stages of glomerular development (renal vesicle, comma, and S-shaped stages) α_1- and α_2 (IV)-collagens predominate in the basal lamina, along with laminin B1. As the GBM matures, α_3, α_4, and α_5(IV) and S -laminin replace α_1 and α_2(IV) and laminin B1 (MINER and SANES 1994). Double labeling experiments demonstrate colocalization of both early- and late-appearing α(IV)-collagens, indicating that this process represents a true replacement of the developing GBM rather than synthesis of a new GBM (MINER and SANES 1994). Mutations in both α_3- and α_4-type IV collagen have been shown to cause autosomal recessive hereditary nephritis (MOCHIZUKI et al. 1994). Interestingly, in Alport's syndrome where either α_3, α_4, or α_5 may be absent from the GBM, there is a significant increase in α_1 and α_2 (Kashtan and Kim 1992). Together these data indicate that αIV-collagens undergo a developmental switch during biogenesis of the GBM, similar to hemoglobin and myosins. Furthermore, there may be coordinate regulation of early (α_1 and α_2) and late chain (α_3, α_4, and α_5) expression. In S-laminin/laminin beta 2 deficient knock out mice laminin B1 is retained, suggesting that a feedback mechanism regulates maturation of the GBM. These mice develop congenital nephrotic syndrome in the early postnatal period with pathology that most closely resembles human minimal change disease (Noakes et al. 1995). Another well-described GBM defect is benign familial hematuria due to thin basement membrane disease. The gene has not been identified, but a GBM component is the most likely candidate.

F. Human Urogenital Anomalies and Malformation Syndromes

A wide variety of human renal congenital anomalies have been described, in some cases with simple Mendelian inheritance (reviewed in Crawfurd 1988). Others may be the result of a complex interaction between polygenic inheritance and/or environmental factors. A common renal anomaly is unilateral renal agenesis, thought to exist in about 0.1% of the population. Stochastic factors alone may also account for a certain incidence of renal anomalies. One interpretation of the large number of renal anomalies that exist is simply that the developing urogenital system is very sensitive to subtle perturbations in rate of cell growth or movement. This is exemplified by the presence of normal kidneys and of unilateral and bilateral renal agenesis in mice carrying the *limb deformity* (*ld*) mutation in a genetically inbred, isogenic strain of mice. Since the affected mice are identical, the variable penetrance appears to reflect environmental or stochastic factors. This is likely analogous to other human congenital disorders such as cleft palate/cleft lip (CL(P)) and congenital cardiac malformations such as ventricular septal defect (VSD), which doubtless have genetic and epigenetic components, but for which stochasm alone can also be invoked (Kurnit et al. 1987).

I. Renal Agenesis and Dysplasias

In this review we have already referred to several mouse mutations acting at different stages in kidney development which can result in renal agenesis (see Tables 1, 2). A survey by Holmes (1989) of infants born with bilateral renal agenesis confirms that this is a heterogeneous entity, not infrequently associated with anomalies of adjacent structures such as neural tube defects, imperforate anus, and genital tract anomalies, as well as distant structures (especially heart and cleft lip/palate). While the majority of the cases are thought to be sporadic, a familial tendency has been documented. Kallman's syndrome is an X-linked inherited disorder characterized by hypogonadism and anosmia which can be associated with unilateral or bilateral renal agenesis and adysplasia. The mutated gene in this defect encodes a protein (named ADMLX) with fibronectin type II repeats typical of adhesion molecules thought to be involved in cell migration, such as NCAM (Franco et al. 1991; Legouis et al. 1991). As already discussed, propagation of the inductive signal in the metanephric mesenchyme involves cell migration, a process that involves adhesion molecules.

Renal hypoplasia is usually bilateral and in most cases is thought to result from insufficient early branching of the ureteric bud, as suggested by a marked reduction in fetal lobes. The most common form is oligomeganephronia, in which nephron number is reduced to about 20% of normal and remaining nephrons display marked enlargement of glomeruli and proximal tubules. This disorder is not familial and, in contrast to small kidneys due to renal dysplasia, is rarely associated with other urinary tract or extrarenal anomalies (reviewed in Welling and Grantham 1991). Nonetheless, defects at multiple steps in kidney development could lead to a reduction in nephron number. Mice deficient for cyclooxygenase 2 (COX-2) have significantly reduced nephron number due to a defect in postnatal maturation of nephrons in the subcapsular nephrogenic zone (Morham et al. 1995).

The term renal dysplasia implies a defect in nephrogenesis after the inductive interaction between the mesenchyme and ureteric bud has taken place. Thus the spectrum of disorders given this designation is broad, reflecting arrested or defective metanephric differentiation at different stages and of variable severity. Association with an abnormally located ureteral orifice or with urinary tract anomalies, resulting in obstruction, is common. Cyst formation (multicystic dysplastic kidneys) and a variable degree of hypoplasia are not an infrequent association (reviewed in Welling and Grantham 1991). Thus agenesis/hypoplasia and dysplasia form a continuum. While most cases are sporadic, familial cases have been reported and dominant inheritance with variable penetrance has been suggested (Murugasu et al. 1991).

II. Renal Cystic Disease

Polycystic kidney disease (PKD) is a common group of hereditary diseases, affecting one in every 200 to 1000 individuals. There is a long list of causes of polycystic kidneys, in many cases occurring as part of multiple malformation syndromes, implying that multiple genetic pathways, in addition to environmental factors, can lead to cystogenesis. Mutations in at least 15 nonallelic loci in mouse (see Table 2 and Aziz 1995) and humans have been shown to cause PKD. Three general mechanisms have been proposed to lead to cyst formation: (1) increased cell proliferation, (2) increased secretion of fluid by cystic epithelium, and (3) abnormal cell–substratum interaction.

Adult-onset autosomal dominant type (ADPKD) is the most common form of PKD and is caused by mutations at one of three loci, two of which have been mapped to the short arms of chromosome 16 (PKD1) and chromosome 4 (PKD2) (PETERS et al. 1993). PKD1 accounts for about 90% of cases. The PKD1 gene has been identified and the predicted gene product is a large, 4034-amino acid novel protein with functional domains which suggest that it is involved in protein–protein and protein–carbohydrate interactions in extracellular matrix (EUROPEAN POLYCYSTIC KIDNEY DISEASE CONSORTIUM 1994; THE INTERNATIONAL POLYCYSTIC KIDNEY DISEASE CONSORTIUM 1995). The gene for PKD2 has also been identified and the predicted protein is 968 amino acids with homology to PKD1 and to the family of voltage-activated calcium and sodium channels (MOCHIZUKI et al. 1996).

The recessive form of PKD (ARPKD) is much less common than ADPKD, affecting one in 6000 to one in 14 000 individuals. In spite of the wide clinical spectrum of the disease, analysis of 60 families that cover the entire range of clinical phenotypes indicates that ARPKD is linked to a single genetic locus 6p21.1-p12 (GUAY-WOODFORD et al. 1995). Two mouse mutations in which the phenotypes are similar to the human disease, *cpk* and *Tg737,* are excluded as homologs of the human disease based on these linkage data. Nonetheless, mouse mutants which express the PKD phenotype will undoubtedly prove useful to understanding the pathogenesis of PKD, especially in connection with identified human PKD genes. These mouse mutations are included in Table 2.

Autosomal recessive medullary cystic disease (familial juvenile nephronophthisis) has been mapped in the region between D2S48 and D2S51 on chromosome 2p (ANTIGNAC et al. 1993). The early pathologic findings in this disease suggest a primary defect in formation of the tubular basement membrane (TBM) with later, secondary development of medullary cysts. Senior-Loken syndrome, which displays an identical renal phenotype but also includes ocular abnormalities, does not map to this locus.

A number of inherited systemic diseases are also associated with renal cysts. Both Von Hippel-Lindau and tuberous sclerosis are examples of the coexistence of renal tumors and renal cysts, supporting the notion that the unchecked proliferation that predisposes to neoplasia is also a mechanism of cystogenesis. Consistent with this idea, although biochemical confirmation has not yet been reported, the Von Hippel-Lindau gene has been identified and this gene is thought to encode a tumor suppressor gene. As anticipated, the gene is expressed in embryonic kidney (KESSLER et al. 1995). It is interesting that several oncogenes, including SV40 large T (KELLY et al. 1991) and c-*myc* (TRUDEL et al. 1991), when expressed transgenically result in formation of cystic kidneys. There is complementation between c-Myc and bcl-2 in oncogenesis (MCDONNELL and KORSMEYER 1991), and bcl-2 can inhibit c-Myc-induced apoptosis (BISSONETTE et al. 1992). It has been suggested that elimination of bcl-2 in *bcl-2*-deficient mice leads to unchecked endogenous c-Myc expression, cell proliferation, and cystic kidneys (VEIS et al. 1993). *Pax-2,* which may promote proliferation of mesenchymal cells but is turned off as epithelial differentiation proceeds, remains high in some cystic diseases (Eng et al. 1994). However, overexpression of *Pax-2* under the control of the cytomegalovirus (CMV) promoter leads to a disruption of kidney development similar to congenital nephrotic syndrome, without cystic disease (Dressler et al. 1993). The recent identification of the PKD1 gene, and the finding of a transgene-induced deletion of a putative cell cycle regulator in a mouse cystic kidney model (Moyer et al. 1994), should provide important clues to molecular mechanisms that mediate tubulogenesis.

III. Role of Environmental Factors

As already noted, renal agenesis is common in the general population, and environmental factors may well exert important effects on renal development. Intrauterine growth retardation in rats and possibly humans leads to a nephron deficit disproportionate to the reduction in body weight and probably to an increased susceptibility to hypertension (Merlet-Benichou et al. 1994). This likely reflects the potential for a variety of environmental factors to affect kidney growth and development. While the environmental factors affecting kidney development discussed below do not constitute a complete list, they were chosen to illustrate that a variety of factors, including drugs, congenital infections, and metabolic disturbances, can lead to malformations of the kidney and urinary tract.

A variety of drugs have been implicated in congenital renal anomalies. The interested reader is referred to a comprehensive review recently published by Lau and Kavlock (1994). Administration of angiotensin-converting enzyme (ACE) inhibitors to infant rats causes marked attenuation of glomerular growth, consistent with the proposed trophic effects in addition to the well-recognized hemodynamic effects of angiotensin II (Fogo et al. 1990; Brent and Beckman 1991). The effect of ACE inhibitors on kidney development is discussed in greater detail in Chap. 25 of this volume. Prolonged administration of gentamicin to pregnant rats causes a severe deficit in nephron number (Gilbert et al. 1990). Thalidomide embryopathy is associated with multiple urinary tract malformations, including unilateral renal agenesis, rotational anomalies, hydronephrosis, duplication defects, and horseshoe kidney (Warkany 1971). Maternal anticonvulsant therapy with trimethadione has been associated with unilateral renal agenesis (Zachai et al. 1975). The congenital rubella syndrome can involve the urinary tract, causing polycystic kidneys, unilateral renal agenesis, and duplication defects (Menser et al. 1967). Maternal diabetes mellitus has been suggested as a risk factor for bilateral renal agenesis (Holmes 1989).

In addition to causing reduced nephron numbers and hypoplasia or agenesis, environmental factors can also contribute to renal cyst formation. Butowski et al. (1985) have created a model of PKD by the administration of 2-amino-4,5-diphenylthiazol. Other chemicals have also been implicated in

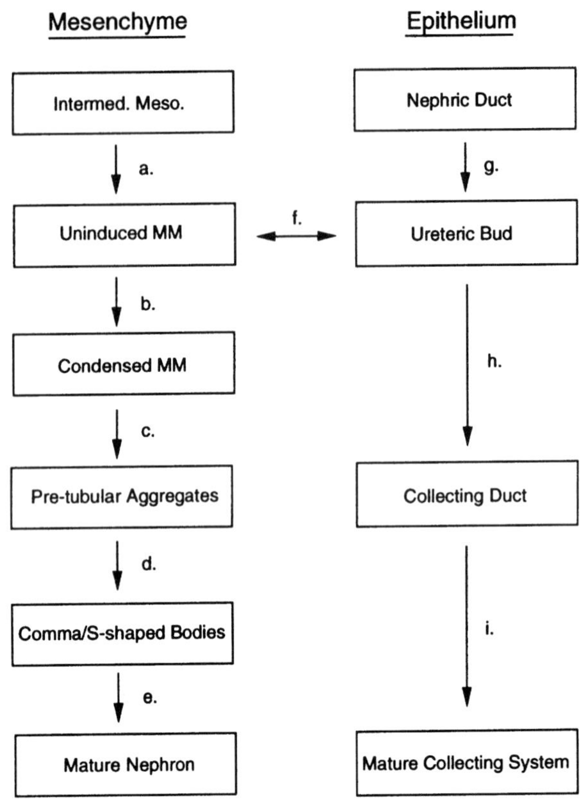

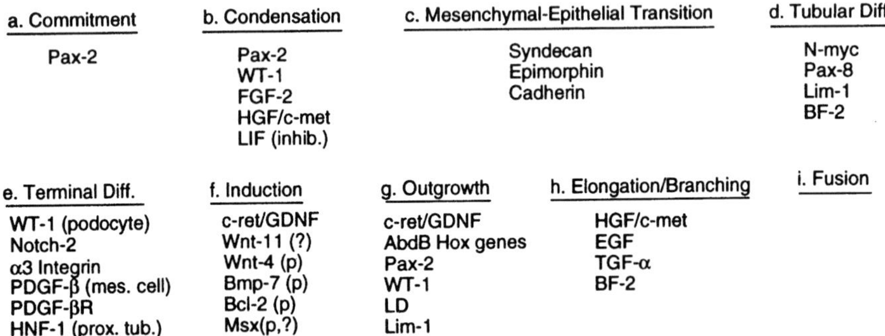

Fig. 7. Scheme tentatively assigning different gene products to discrete steps in kidney development. The separation of the mesenchymal and epithelial lineages except at the induction step (*step f*) is not intended to imply separate development; rather, reciprocal interactions are likely at many steps. *Question marks* denote uncertain assignment; *inhib* stands for inhibitory; *p* denotes gene products likely to participate in the propagation and maintenance steps of induction rather than the initial induction process. No genes have been directly implicated in collecting duct fusion

cystogenesis (RESNICK et al. 1976). A number of these chemicals may exert their toxic effects by generating oxidant stress. Since bcl-2 is thought to regulate an antioxidant pathway, there may be a biochemical link between some toxin-induced cystic disease and that seen in *bcl-2*-deficient mice (HOCKENBERY et al. 1993).

G. Conclusions

It has been estimated from the number of independent mutations isolated during saturation mutagenesis experiments in *Drosophila* that as many as 2500 genes may be involved in invertebrate eye development (HALDER et al. 1995). At the top, or near the top, of this developmental hierarchy, however, sit only one or at most a few genes. Because of the greater complexity of the vertebrate genome compared to the invertebrate genome, kidney development in mammals is likely to be even more complex. Nonetheless, the recent application of homologous recombination and embryonic stem cell technology to mammalian development has resulted in a surprisingly large number of genetic mouse mutants in which kidney development is disturbed. Thus, the genetic dissection of mammalian nephrogenesis, which only a few years ago seemed a daunting proposition, is now becoming a practical reality. Figure 7 is a summary of those genes which have been implicated in mediating discrete steps in nephrogenesis. Since many of these mutations appear to affect the early steps of ureteric bud outgrowth or metanephric induction, steps for which a powerful organ-based culture system is well established, considerable insight into the molecular pathways controlling kidney development is likely to come with the next several years.

Acknowledgments. The authors are indebted to the large number of colleagues who shared their results, especially work still in progress, so that this review might be as current as possible. In addition, the authors thank the large number of individuals who critiqued this manuscript while it was under preparation. R.M. is an Assistant Investigator of the Howard Hughes Medical Institute; M.R. is supported by a Clinician Scientist Award of the American Heart Association.

References

Abrahamson DR (1985) Origin of the glomerular basement membrane visualized after in vivo labeling of laminin in newborn rat kidneys. J Cell Biol 100: 1988–2000

Al-Awqati O (1992) Cellular and molecular mechanisms of renal development and tubulogenesis. Curr Opin Nephrol Hypertens 1: 53–58

Alpers CE, Seifert RA, Hudkins KL, Johnson RJ, Bowen-Pope DF (1992) Developmental patterns of the PDGF-β chain, PDGF receptor and alpha actin expression in human glomerulogenesis. Kidney Int 42: 390–399

Antignac C, Arduy CH, Beckmann JS, Benessy F, Gros F, Medhioub M, Hildebrandt F, Dufier J-L, Kleinknecht C, Broyer M, Weissenbach J, Habib R, Cohen D (1993) A gene for familial juvenile nephronophthisis (recessive medullary cystic kidney disease) maps to chromosome 2p. Nat Genet 3: 342–345

Argao EA, Kern MJ, Branford WW, Scott WJ, Potter SS (1995) Malformations of the heart, kidney, palate and skeleton in alpha-MHC-Hoxb7 transgenic mice. Mech Dev 52: 291–303

Artavanis-Tsakonas S, Matsuno K, Fortini M (1995) Notch signaling. Nature 268: 225–232

Avner ED, Sweeney WE Jr (1990) Polypeptide growth factors in metanephric growth and segmental nephron differentiation. Pediatr Nephrol 4: 372–377

Aziz N (1995) Animal models of polycystic kidney disease. BioEssays 17: 703–712

Bard JBL, Ross ASA (1991) LIF, the ES-cell inhibition factor, reversibly blocks nephrogenesis in cultured mouse kidney rudiments. Development 113: 193–198

Bard JBL, McConnell JE, Davies JA (1994) Towards a genetic basis for kidney development. Mech Dev 48: 3–11

Bard JBL, Davies JA, Karavanova I, Lehtonen E, Sariola H, Vainio S (1996) Kidney development: the inductive interactions. Sem Cell Dev Biol 7: 195–202

Barker DF, Hostikka SL, Zhou J, Chow LT, Oliphant AR, Gerken SC, Gregory MC, Skolnick MH, Atkin CL, Tryggvason K (1990) Identification of mutations in the COL4A5 collagen gene in Alport's syndrome. Science 248: 1224–1227

Barros EJG, Santos OFP, Matsmoto K, Nakamura T, Nigam SK (1995) Differential tubulogenic and branching morphogenetic activities of growth factors: implications for epithelial tissue development. Proc Natl Acad Sci USA 92: 4412–4491

Bernstein J, Cheng F, Roszka J (1981) Glomerular differentiation in metanephric culture. Lab Invest 450: 183–190

Bissonnette RP, Echeverri F, Mahboudi A, Green DR (1992) Apoptotic cell death induced by c-myc is inhibited by bcl-2. Nature 359: 552–553

Boulter CA, Aguzzi A, Evans MJ, Affara N (1992) A Chimeric Mouse Model for autosomal-dominant polycystic kidney disease. Contrib Nephrol 97: 60–70

Boyden EA (1927) Experimental obstruction of the mesonephric ducts. Proc Soc Exp Biol Med 24: 572–576

Bradley A (1987) Production and analysis of chimeric mice. In: Robertson EJ (ed) Teratocarcinomas and embryonic stem cells. IRL, Oxford, pp 113–151

Bradley RS, Cowin P, Brown AM (1993) Expression of Wnt-1 in PC12 cells results in modulation of plakoglobin and E-cadherin and increased cell adhesion. J Cell Biol 123: 1857–1865

Brent RL, Beckman DA (1991) Angiotensin-converting enzyme inhibitors, an embryopathic class of drugs with unique properties: information for clinical teratology counselors. Teratology 43: 543–546

Briand P, Kahn A, Vandewalle A (1995) Targeted oncogenesis: a powerful method to derive renal cell lines. Kidney Int 47: 388–394

Bruening W, Bardeesy N, Silverman BL, Cohn RA, Machin GA, Aronson J, Housman D, Pelletier J (1992) Germline intronic and exonic mutations in the Wilms' tumour gene (WT1) affecting urogenital development. Nat Genet 1: 144–148

Burrow CR, Wilson PD (1993) A putative Wilm's tumor-secreted growth factor activity required for primary culture of human nephroblasts. Proc Natl Acad Sci USA 90: 6066–6070

Butowski RJ, Carone FA, Grantham JJ, Hudson BG (1985) Tubular basement membrane changes in 2-amino-4,5-diphenylthiazol induced polycyctic kidney disease. Kidney Int 28: 744–751

Call KM, Glaser T, Ito CY, Buckler AJ, Pelletier J, Haber DA, Rose EA, Kral A, Yeager H, Lewis WH, Jones C, Housman D (1990) Isolation and characterization of a zinc finger polypeptide gene at the human chromosome 11 Wilms' tumor locus. Cell 60: 509–520

Cantley LG, Barros EJ, Gandhi M, Rauchman MI, Nigam SK (1994) Regulation of mitogenesis, motogenesis, and tubulogenesis by hepatocytes growth factor in renal collecting duct cells. Am J Physiol 267: F271–280

Capecchi MR (1989) The new mouse genetics: altering the genome by gene targeting. Trends Genet 5: 70–76

Capecchi MR (1994) Targeted gene replacement. Sci Am 270: 52–59

Charité J, de Graaf W, Sanbing S, Deschamps J (1994) Ectopic expression of Hoxb-8 causes duplication of the ZPA in the forelimb and homeotic transformation of axial structures. Cell 78: 589–601

Charron J, Malynn BA, Fisher P, Stewart V, Jeanotte L, Goff SP, Robertson EJ, Alt FW (1992) Embryonic lethality in mice homozygous for a targeted disruption of the N-*myc* gene. Genes Dev 6: 2248–2257

Chen SK, Jaffe L, Capovilla M, Botas J, Mann RS (1994) The DNA binding specificty of ultrabithorax is modulated by cooperative interactions with extradenticle, another homeoprotein. Cell 78: 603–615

Clapp WL, Abrahamson DR (1993) Regulation of kidney organogenesis: homeobox genes, growth factors, and Wilms tumor. Curr Opin Nephrol Hypertens 2: 419–429

Coles HSR, Burne JF, Raff MC (1993) Large-scale normal cell death in the developing rat kidney and its reduction by epidermal growth factor. Development: 118: 777–784

Crawfurd M d'A (1988) The genetics of renal tract disorders. Oxford University Press, Oxford

Davies J, Lyon M, Gallagher J, Garrod D (1995) Sulphated proteoglycan is required for collecting duct growth and branching but not nephron formation during development. Development 121: 1507–1517

Davis AP, Witte DP, Hseihli HM, Potter SS, Capecchi MR (1995) Absence of radius and ulna in mice lacking hoxa-11 and hoxd-11. Nature 375: 791–795

Deschamps J, de Jaaf R, Verrijzer P, de Gouw M, Destree O, Meijlink F (1987) The mouse Hox 2.3 homeobox-containing gene: regulation in differentiating pluripotent stem cells and expression pattern in embryos. Differentiation 35: 21–30

Dollé P, Izpisúa-Belmonte J-C, Brown JM, Tickle C, Duboule D (1991) *Hox-4* genes and the morphogenesis of mammalian genitalia. Genes Dev 5: 1767–1776

Dono R, Zeller R (1994) Cell-type-specific nuclear translocation of fibroblast growth factor-2 isoforms during chicken kidney and limb morphogenesis. Dev Biol 163: 316–330

Dressler GR, Deutsch U, Chowdhury K, Nornes HO, Gruss P (1992) Pax2, a new murine paired box containing gene and its expression in the developing excretory system. Development 109: 787–795

Dressler GR, Wilkinson JE, Rothenpieler UW, Patterson LT, Williams-Simons L, Westphal H (1993) Deregulation of Pax2 expression in transgenic mice causes severe kidney abnormalities. Nature 362: 65–67

Drummond IA, Madden SL, Rohwer-Nutter P, Bell GI, Sukhatme VP, Rauscher FJ (1992) III. Repression of the insulin like growth factor II gene by the Wilms tumor suppressor WT1. Science 257: 674–678

Dudley AT, Lyons KM, Robertson EJ (1995) A requirement for bone morphogenetic protein-7 during development of the mammalian kidney and eye. Genes Dev 9: 2795–2807

Durbec P, Marcos-Gutierrez CV, Kilkenny C, Grigoriou M, Warttiowaara K, Suvanto P, Smith D, Ponder B, Constantini F, Saarma M, Sariola H, Pachnis V (1996) GDNF signalling through the Ret receptor tyrosine kinase. Nature 381: 789–793

Durbeej M, Söderström S, Ebendal T, Birchmeier C, Ekblom P (1993) Differential expression of neutrophin receptors during renal development. Development 19: 977–989

Ekblom P, Weller A (1991) Ontogeny of tubulointerstitial cells. Kidney Int 39: 394–400

Ekblom P, Miettinen A, Saxén L (1980) Induction of brush border antigens of the proximal tubule in the developing kidney. Dev Biol 74: 263–274

Ekblom P, Lehtonen E, Saxén L, Timpl R (1981) Shift in collagen type as an early response to induction of the metanephric mesenchyme. J Cell Biol 89: 276–283

Ekblom P, Thesleff I, Miettinen A, Saxén L (1991) Organogenesis in a defined medium supplemented with transferrin. Cell Growth Differ 10: 281–288

Eng E, Wilson PD, Racusen LC, Burrow CR (1994) Aberrant Pax-2 expression in human autosomal dominant polycystic kidney disease. J Am Soc Nephrol 5: 621

Englert C, Hou X, Maheswaran S, Bennett P, Ngwu C, Re GG, Garvin AJ, Rosner

MR, Haber DA (1995) WT1 suppresses synthesis of the epidermal growth factor receptor and induces apoptosis. EMBO J 14: 4662–4675

European Polycystic Kidney Disease Consortium (1994) The polycystic kidney disease 1 gene encodes a 14 kb transcript and lies within a duplicated region on chromosome 16. Cell 77: 881–894

Flaherty L, Bryda EC, Collins D, Rudofsky U, Montgomery JC (1995) New mouse model for polycystic kidney disease with both recessive and dominant gene effects. Kidney Int 47: 552–558

Fogo A, Yoshida Y, Yared A, Ichikawa I (1990) Importance of angiogenic angiotensin II in the glomerular growth of maturing kidneys. Kidney Int 38: 1068

Franco B, Guioli S, Pragliola A, Incerti B, Bardoni B, Tonlorenzi R, Carrozzo R, Maestrini E, Pieretti M, Tailon-Miller P, Brown CJ, Willard HF, Lawrence C, Persico MG, Camerino G, Ballabio A (1991) A gene defect in Kallman's syndrome shares homology with neural cell adhesion and axonal path-finding molecules. Nature 353: 529–535

Fujii T, Pichel JG, Masanori T, Toyama R, Dawid IB, Westphal H (1994) Expression patterns of the murine LIM class homeobox gene *lim1* in the developing brain and excretory system. Dev Dyn 199: 73–83

Gilbert T, Lelievre-Pegorier M, Merlet-Benichou C (1990) Immediate and long-term renal effects of fetal exposure to gentamicin. Pediatr Nephrol 4: 445–450

Gluecksohn-Schoenheimer S (1943) The morphologic mutations of a dominant mutation in mice affecting tail and urinogenital system. Genetics 28: 340–348

Gluecksohn-Schoenheimer S (1945) The Sd-strain in mice. Genetics 30: 29–38

Gluecksohn-Waelsch S, Rota TR (1963) Development in organ culture of kidney rudiments from mouse mutant embryos. Dev Biol 7: 432–444

Godley LA, Kopp JB, Eckhaus M, Paglino JJ, Owens J, Varmus HE (1996) Wild-type p53 transgenic mice exhibit altered differentiation of the ureteric bud and possess small kidneys. Genes Dev 10: 836–850

Gomez RA, Pupilli C, Everett AD (1991) Molecular and cellular aspects of renin during kidney ontogeny. Pediatr Nephrol 5: 80–87

Gonzalez–Mariscal L, Chavez de Ramirez B, Cereijido M (1985) Tight junction formation in cultured epithelial cells (MDCK). J Membr Biol 86: 113–125

Grobstein C (1953a) Morphogenetic interaction between embryonic mouse tissues separated by a membrane filter. Nature 172: 869–871

Grobstein C (1953b) Inductive epithelial-mesenchymal interaction in cultured organ rudiments of the mouse. Science 118: 52–55

Grobstein C (1955) Inductive interactions in the development of the mouse metanephros. J Exp Zool 130: 319–340

Grobstein C (1956) Trans-filter induction of tubules in mouse metanephric mesenchyme. Exp Cell Res 10: 424–440

Grone H-J, Schwabe GC, Walther C, Gruss P (1993) Expression of developmental control genes pax-2 and pax-8 in adult kidney and during regeneration or renal tubular epithelia. J Am Soc Nephrol 4: 465 (abstr)

Gruenwald P (1952) Development of the excretory system. Ann N Y Acad Sci 55: 142–146

Gstraunthaler G (1988) Epithelial cells in culture. Renal Physiol Biochem 11: 1–42

Gu H, Marth JP, Orban PC, Mossman H, Rajewsky K (1994) Deletion of a DNA polymerase beta gene segment in T cells using cell type specific gene targeting. Science 265: 103–196

Guay-Woodford LM, Muecher G, Hopkins SD, Avner ED, Germino GG, Guillot AP, Herrin J, Hoooolleman R, Irons DA, Primack W, Thompson PD, Waldo FB, Lunt PW, Zerres K (1995) The severe perinatal form of autosomal recessive polycystic kidney disease maps to chromosome 6p21.1-p12: implications for genetic counseling. Am J Hum Genet 56: 1101–1107

Gumbiner B, Stevenson B, Grimaldi A (1988) The role of cell adhesion molecule uvomorulin in the formation and maintenance of epithelial junctional complexes. J Cell Biol 107: 1575–1587

Haber DA, Park S, Maheswaran S, Englert C, Re GG, Hazen-Martin DJ, Cens DA,

Garvin AJ (1993) WT-1 mediated growth suppression of Wilms tumor cells expressing a WT-1 splicing variant. Science 262: 2057–2059
Halder G, Callaerts P, Gehring W (1995) Induction of ectopic eyes by targeted expression of the eyeless gene in *Drosophila*. Science 267: 1788–1792
Hastie NP (1994) The genetics of Wilms' tumor: a case of disrupted development. Annu Rev Genet 28: 523–555
Hatini V, Huh SO, Herzlinger D, Soares VC, Lai E (1996) Essential role of stromal mesenchyme in kidney morphogenesis revealed by targeted disruption of Winged Helix transcription factor BF-2. Genes Dev 10: 1467–1478
Herzlinger D, Koseki C, Al-Awqati Q (1992) Metanephric mesenchyme contains multipotent stem cells whose fate is restricted after induction. Development 114: 565–572
Hink L, Nelson WJ, Papkoff J (1994) Wnt-1 modulates cell-cell adhesion in mammalian cells by stabilizing β-catenin binding to the cell adhesion protein cadherin. J Cell Biol 124: 729–741
Hirai Y, Takebe K, Takashina M, Kobayashi S, Takeichi M (1992) Epimorphin: a mesenchymal protein essential for epithelial morphogenesis. Cell: 69: 471–481
Hockenbery DM (1994) bcl-2 in cancer, development and apoptosis. J Cell Sci [Suppl] 18: 51–55
Hockenbery DM, Oltvai ZN, Yin X-M, Milliman CL, Korsmeyer SJ (1993) Bcl-2 functions in an antioxidant pathway to prevent apoptosis. Cell 75: 241–251
Hogan BLM, Costantini F, Lacy E (1986) Manipulating the mouse embryo. Cold Spring Harbor Laboratory, Cold Spring Harbour
Holmes LB (1989) Prevalence, phenotypic heterogeneity and familial aspects of bilateral renal agenesis/dysgenesis. Genetics of kidney disorders. Liss, New York, pp 1–11
Holst BD, Goomer RS, Wood IC, Edelman GM, Jones FS (1994) Binding and activation of the promoter for the neural cell adhesion molecule by Pax-8. J Biol Chem 269: 22245–22252
Hoschuetzky H, Aberle H, Kemler R (1994) Beta-catenin mediates the interaction of the cadherin-catenin complex with epidermal growth factor receptor. J Cell Biol 127: 1375–1380
Hsieh HM, Witte DP, Weinstein M, Branford W, Li H, Small K, Potter SS (1995) Hoxa11 structure, extensive antisense transcription, and function in male and female sterility. Development 121: 1373–1385
Hunt P, Krumlauf R (1991) Deciphering the Hox code: clues to patterning the brachial regions of the head. Cell 66: 1075–1078
Iakoubova OA, Dushkin H, Beier DR (1995) Localization of a murine recessive polycystic kidney disease mutation and modifying loci that affect disease severity. Genomics 26: 107–114
Jing S, Wen D, Yu Y, Holst PL, Luo Y, Fang M, Tamir R, Antonio L, Hu Z, Cupples R, Louis J-C, Hu S, Altrock BW, Fox GM (1996) GDNF-induced activation of the *ret* protein tyrosine kinase is mediated by GDNFR-alpha, a novel receptor for GDNF. Cell 85: 1113–1124
Karp SL, Ortizarduan A, Li SR, Neilson EG (1994) Epithelial differentiation of metanephric mesenchymal cells after stimulation with hepatocyte growth factor or embryonic spinal cord. Proc Natl Acad Sci USA 91: 5286–5290
Kashtan CE, Kim Y (1992) Distribution of the α1 and α2 chains of collagen IV and of collagens V and VI in Alport syndrome. Kidney Int 42: 115–126
Keller SA, Jones JM, Boyle A, Barrow LL, Killen PD, Green DG, Kapousta NV, Hitchcock PF, Swank PF, Meisler M (1994) Kidney and retinal defects (Krd), a transgene induced mutation with a deletion of mouse chromosome 19 that includes the *Pax-2* locus. Genomics 23: 309–320
Kelly KA, Agarwal N, Redders ST, Herrup KJ (1991) Renal cyst formation and multifocal neoplasia in transgenic mice carrying the simian virus 40 early region. J Am Soc Nephrol 2: 84–97
Kessel M, Gruss P (1991) Homeotic transformations of murine prevertebrae and concomitant alteration of the Hox code induced by retinoic acid. Cell 67: 89–104

Kessler PM, Vasvada SP, Rackley RR, Stackhouse T, Duh F-M, Latif F, Lerman MI, Zbar B, Williams BRG (1995) Expression of the Von Hippel-Lindau tumor suppressor gene, *VHL*, in human fetal kidney and during mouse embryogenesis. Mol Med 1: 457–466

Klein G, Langegger M, Timpl R, Ekblom P (1990) Role of laminin A chain in the development of epithelial cell polarity. Cell 55: 331–341

Kleppel MM, Fan WW, Cheong HI, Michael AF (1992) Evidence for separate networks of classical and novel basement membrane collagen. Characterization of α3(IV) Alport antigen heterodimer. J Biol Chem 267: 4137

Kreidberg JA, Sariola H, Loring JM, Maeda M, Pelletier J, Housman D, Jaenisch R (1993) WT-1 is required for early kidney development. Cell 74: 679–691

Kress C, Vogels R, De Graff W, Bonnerot C, Meijlink F, Nicolas J-F, Deschamps J (1990) Hox 2.3 upstream sequences mediate Lac Z expression in intermediate mesoderm derivatives in transgenic mice. Development 109: 775–786

Koseki C, Herzlinger D, Al-Awqati Q (1991) Integration of embryonic nephrogenic cells carrying a reporter gene into functioning nephrons. Am J Physiol 261: C550–C554

Kurnit DM, Layton WM, Matthysse S (1987) Genetics, chance and morphogenesis. Am J Hum Genet 41: 979–995

Larsson SH, Charlieu J-P, Miyagawa K, Engelkamp D, Rassoulzadegan M, Ross A, Cuzin F, van Heyningen V, Hastie ND (1995) Subnuclear localization of WT1 in splicing or transcription factor domains is regulated by alternative splicing. Cell 81: 391–401

Lau C, Kavlock RJ (1994) Functional toxicity in the developing heart, lung and kidney. In: Kimmel CA, Buelke-Sam J (eds) Development toxicology, 2nd edn. Raven, New York, pp 119–188

Lazzaro D, De Simone V, De Magistris L, Lehtonen E, Cortese R (1992) LFB1 and LFB3 homeoproteins are sequentially expressed during kidney development. Development 114: 469–479

LeBrun DP, Warnke RA, Cleary ML (1993) Expression of bcl-2 in fetal tissue suggests a role in morphogenesis. Am J Pathol 142: 743–753

Legouis R, Hardelin J-P, Levilliers J, Claverie J-M, Compain S, Wunderle V, Millasseau P, Le Paslier D, Cohen D, Caterina D, Bougeleret L, Delemarre-van de Waal H, Lutfalla G, Weissenbach J, Petit C (1991) The candidate gene for the X-linked Kallman syndrome encodes a protein related to adhesion molecules. Cell 67: 423–435

Levéen P, Pekny M, Gebre-Medhin S, Swolin B, Larsson E, Betsholtz C (1994) Mice deficient for PDGF-beta show renal, cardiovascular, and hematological abnormalities. Genes Dev 8: 1875–1887

Luo G, Hofmann C, Bronckers ALJJ, Sohocki M, Bradley A, Karsenty G (1995) BMP-7 is an inducer of morphogenesis, and is also required for eye development and skeletal patterning. Genes Dev 9: 2808–2820

Lyon MF, Searle AG (1989) Genetic variants and strains of the laboratory mouse. Fischer, Stuttgart

Lyon MF, Searle AG (1990) Genetic variants and strains of the laboratory mouse. Oxford University Press, Oxford

Lyons KM, Hogan BLM, Robertson EJ (1995) Colocalization of BMP7 and BMP2 RNAs suggests that these factors cooperatively mediate tissue interactions during murine development. Mech Dev 50: 71–83

Maas RL, Elfering SE, Glaser T, Jepeal L (1994) Deficient outgrowth of the ureteric bud underlies the renal agenesis phenotype in mice manifesting the *limb deformity* (*ld*) mutation. Dev Dyn 119: 214–228

McDonnell TJ, Korsmeyer SJ (1991) Progression from lymphoid hyperplasia to high grade malignant lymphoma in mice transgenic for the t(14;18). Nature 349: 254–256

McGinnis W, Garber RL, Wirz J, Kuroiwa A, Gehring WJ (1984) A homologous protein-coding sequence in drosophila homeotic genes and its conservation in other metazoans. Cell 37: 403–408

McGinnis M, Krumlauf R (1992) Homeobox genes and axial patterening. Cell 68: 283–302

McNeill H, Ozawa M, Kemler R, Nelson WJ (1990) Novel function of the cell adhesion molecule uvomorulin as an inducer of cell surface polarity. Cell 62: 309–316

Menser M, Robertson SEJ, Dorman DC, Gillespie AM, Murphy A (1967) Renal lesions in congenital rubella. Pediatrics 40: 901–904

Merlet-Benichou C, Gilbert T, Muffat-Joly M, Lelievre-Pegorier M, Levy B (1994) Intrauterine growth retardation leads to permanent nephron deficit in the rat. Pediatr Nephrol 8: 175–180

Miner JH, Sanes JR (1994) Collagen IV α3, α4, and α5 chains in rodent basal laminae; sequence, distribution, association with laminins, and developmental switches. J Cell Biol 127: 879–891

Mochizuki T, Wu G, Hayashi T, Xenophontos SL, Veldhuisen B, Saris JJ, Reynolds DM, Cai Y, Gabow PA, Pierides A, Kimberling WJ, Breuning MH, Deltas CC, Peters DJ, Somlo S (1996) PKD2, a gene for polycystic kidney disease that encodes an integral membrane protein. Science 272: 1339–1342

Mochizuki T, Lemmink HH, Mariyama M, Antignac C, Gubler M-C, Pirson Y, Verellen-Dumoulin C, Chan B, Schroder CH, Smets HJ, Reeders ST (1994) Identification of mutations in the α3(IV) and α4(IV) collagen genes in autosomal recessive Alport syndrome. Nat Genet 8: 77–82

Moens CB, Auerbach AA, Conlon RA, Joyner AL, Rossant JA (1992) A targeted mutation reveals a role for N-myc in branching morphogenesis in the embryonic mouse lung. Genes Dev 6: 691–704

Montesano R, Schaller G, Orci L (1991a) Induction of epithelial tubular morphogenesis in vitro by fibroblast-derived soluble factors. Cell 66: 697–711

Montesano R, Matsumotomo K, Nakamura T, Orci L (1991b) Identification of a fibroblast-derived epithelial growth factor as the hepatocyte growth factor. Cell 67: 901–908

Moore MW, Klein RD, Farinas I, Sauer H, Armanini M, Phillips H, Reichardt L, Ryan AM, Carver-Moore K, Rosenthal A (1996) Renal and neuronal abnormalities in mice lacking GDNF. Nature 382: 76–79

Morham SG, Langenbach R, Loftin CD, Tiano HF, Vouloumaanos N, Jennette JC, Mahleer JF, Kluckman KD, Ledford A, Lee CA, Smithies O (1995) Prostaglandin synthase 2 gene disruption causes severe renal pathology in the mouse. Cell 83: 473–482

Moyer JH, Lee-Tischler MJ, Kwon H-Y, Schrick JJ, Avner ED, Sweeney WE, Godfrey VL, Cacheiro NLA, Wilkinson JE, Woychik RP (1994) Candidate gene associated with a mutation causing recessive polycystic kidney disease in mice. Science 264: 1329–1333

Mugrauer G, Alt FW, Ekblom P (1988) N-*myc* proto-oncogene expression during organogenesis in the developing mouse as revealed by in situ hybridization. J Cell Biol 107: 1325–1335

Mugrauer G, Ekblom P (1991) Contrasting expression patterns of three members of the *myc* family of protooncogenes in the developing and adult mouse kidney. J Cell Biol 112: 13–25

Mundlos S, Pelletier J, Darveau A, Bachmann M, Winterpacht A, Zabel B (1993) Nuclear localization of the protein encoded by the Wilms tumor gene WT-1 in embryonic and adult tissues. Development 119: 1329–1341

Murugasu B, Cole BR, Hawkins EP, Blanton SH, Conley SB, Portman RJ (1991) Familial renal adysplasia. Am J Kidney Dis 28: 490–494

Nauta J, Ozawa Y, Sweeny WE, Rutledge JC, Avner ED (1993) Renal and biliary abnormalities in a new murine model of autosomal recessive polycystic kidney disease. Pediatr Nephrol 7: 163–172

Noakes PG, Miner JH, Gautam M, Cunningham JM, Sanes JR, Merlie JP (1995) The renal glomerulus of mice lacking s-laminin/laminin beta 2:nephrosis despite molecular compensation by laminin beta 1. Nature Gen 10: 400–406

Orr-Urtreger A, Givol D, Yayon A, Yarden Y, Lonai P (1991) Developmental expression of two murine fibroblast growth factor receptors *flg* and *bek*. Development 113: 1419–1434

Osathanondh V, Potter EL (1963a) Development of human kidney as shown by microdissection. II. Renal pelvis, calyces and papillae. Arch Pathol 76: 277–289

Osathanondh V, Potter EL (1963b) Development of human kidney as shown by microdissection. III. Formation and interrelationship of collecting tubules and nephron. Arch Pathol 76: 290–302

Pachnis V, Mankoo B, Costantini F (1993) Expression of the c-*ret* proto-oncogene during mouse embryogenesis. Development 119: 1005–1117

Pelletier J, Bruening W, Kashtan CD, Mauer SM, Maniel JC, Striegel JE, Houghton DC, Junien C, Habib R, Fouser L, Fine RN, Silverman BC, Haber DA, Housman D (1991) Germline mutations in the Wilm's tumor suppressor gene are associated with abnormal development and Denys-Darsh syndrome. Cell 67: 437–447

Pepper MS, Matsuomoto K, Nakamura T, Orci L, Montesano R (1992) HGF increases urokinase type plasminogen activator and u-PA receptor expression in MDCK cells. J Biol Chem 267: 20493–20496

Peratoni AP, Dove LF, Williams CL (1991) Induction of tubules in rat metanephrogenic mesenchyme in the absence of inductive tissue. Differentiation 48: 25–31

Peratoni AP, Dove LF, Karavanova I (1995) Basic fibroblast growth factor can mediate the early inductive events in renal development. Proc Natl Acad Sci USA 92: 4696–4700

Peters DJM, Spruit L, Saris JJ, Ravine D, Sandkuijl LA, Fossdal R, Boersma J, Vaneijk R, Norby S, Constantinoudeltas CD (1993) Chromosome 4 localization of a second gene for autosomal-dominant polycystic kidney disease. Nat Genet 5: 359–362

Peters KG, Werner S, Chen G, Williams LT (1992) Two FGF receptor genes are differentially expressed in epithelial and mesenchymal tissues during limb formation and organogenesis in the mouse. Development 114: 233–243

Phelps DE, Dressler GR (1993) Aberrant expression of Pax-2 in *Danforth's short tail* (*Sd*) mice. Dev Biol 157: 251–258

Pichel JG, Shen L, Sheng HZ, Granholm A-C, Drago J, Grinberg A, Lee EJ, Huang SP, Saarma M, Hoffer BJ, Sariola H, Westphal H (1996) Defects in enteric innervation and kidney development in mice lacking GDNF. Nature 382: 73–76

Plachov D, Chowdhury K, Walther C, Simon D, Guenet JL, Gruss P (1990) Pax8, a murine paired box gene expressed in the developing excretory system and thyroid gland. Development 110: 643–651

Pontoglio M, Barra J, Hadchouel M, Doyen A, Kress C, Bach JP, Babinet C, Yaniv M (1996) Hepatocyte nuclear factor 1 inactivation results in hepatic dysfunction, phenylketonuria and Fanconi syndrome. Cell 84: 575–585

Poole TJ, Steinberg MS (1981) Amphibian pronephric duct morphogenesis: segregation, cell rearrangement and directed migration of the Ambystoma duct rudiment. J Embryol Exp Morphol 63: 1–16

Poole TJ, Steinberg MS (1982) Evidence for the guidance of pronephric duct migration by a craniocaudally traveling adhesion gradient. Dev Biol 92: 144–158

Rauchman MI, Nigam SK, Delpire E, Gullans SR (1993) An osmotically tolerant inner medullary collecting duct cell line (mIMCD-3) from an SV40 transgenic mouse. Am J Physiol 265: F416–F424

Rauscher FJ, Morris JF, Tournay OE, Cook DM, Curran T (1990) Binding of the Wilms tumor locus zinc finger protein to the Egr-1 consensus sequence. Science 250: 1259–1262

Rauskolb M, Peifer M, Wieschaus E (1993) *extradenticle,* a regulator of homeotic gene activity, is a homolog of the homeobox-containing human proto-oncogene *pbxl*. Cell 74: 1101–12

Reeders ST (1992) Multilocus polycystic disease. Nat Genet 1: 235–237

Reeve AE, Eccles MR, Wilkins RJ, Bell GI, Millow IJ (1985) Expression of insulin-like growth factor II transcripts in Wilms tumor. Nature 317: 258–60

Resnick JS, Brown DM, Vernier RL (1976) Normal development and experimental models of cystic renal disease. In: Gradner KD (ed) Cystic diseases of the kidney. Wiley, New York

Riethmacher D, Brinkmann V, Birchmeier C (1995) A targeted mutation in the mouse E-cadherin gene results in defective preimplantation development. Proc Natl Acad Sci USA 92: 855–859

Risau W, Ekblom P (1986) Production of a heparin-binding angiogenesis factor by the developing kidney. J Cell Biol 103: 1101–1107

Ritvos O, Tuuri T, Eramma M, Sainio K, Hilden K, Saxen L, Gilbert SF (1995) Activin disrupts epithelial branching morphogenesis in developing glandular organs of the mouse. Mech Dev 50: 229–245

Robertson EJ (1987) Embryo-derived stem cells. In: Teratocarcinomas and embryonic stem cells. IRL, Oxford

Robertson M, Chambers I, Rathjen P, Nichols J, Smith A (1993) Expression of alternative forms of differentiation inhibiting activity (DIA/LIF) during mouse embryogenesis and in neonatal and adult tissues. Dev Genet 14: 165–173

Rogers SA, Ryan G, Hammerman MR (1991) Insulin-like growth factors I and II are produced in the metanephros and are required for growth and development in vitro. J Cell Biol 113: 1147–1453

Rothenpieler UW, Dressler GR (1993) *Pax-2* is required for mesenchyme to epithelium conversion during kidney development. Development 119: 711–720

Ryan G, Steele-Perkins V, Morris JF, Rauscher III FJ, Dressler GR (1995) Repression of Pax-2 by WT1 during normal kidney development. Development 121: 867–875

Sainio K, Gilbert SF, Lehtonen E, Nishi M, Kumar NM, Gilula NB, Saxén L (1992) Differential expression of gap junction mRNAs and proteins in the developing murine kidney and in experimentally induced nephric mesenchymes. Development 115: 827–837

Sanchez MP, Silos-Santiago I, Frisen J, He B, Lira SA, Barbacid M (1996) Renal agenesis and the absence of enteric neurons in mice lacking GDNF. Nature 382: 70–73

Santos OP, Nigam SK (1993) HGF-induced tubulogenesis and branching of epithelial cells is modulated by extracellular matrix and TGF-β. Dev Biol 160: 293–302

Santos OF, Barros EJ, Yang X-M, Matsumoto K, Nakamura T, Park M, Nigam SK (1994) Involvement of hepatocyte growth factor in kidney development. Dev Biol 163: 525–529

Sanyanusin P, Schimmenti LA, McNoe LA, Ward TA, Pierpont EM, Sullivan MJ, Dobyns WB, Eccles MR (1995) Mutation of the PAX 2 gene in a family with optic nerve colobomas, renal anomalies and vesicoureteral reflux. Nat Genet 9: 358–364

Sariola H, Ekblom P, Lehtonen E, Saxén L (1983a) Differentiation and vascularization of the metanephric kidney grafted on the chorioallantoic membrane. Dev Biol 96: 427–435

Sariola H, Timpl R, von der Mark K, Mayne R, Fitch JM, Linsenmeyer TR, Ekblom P (1983b) Dual origin of the glomerular basement membrane. Dev Biol 101: 86–96

Sariola H, Peault B, Le Douarin N, Buck C, Dieterlen F, Saxén L (1984) Extracellular matrix and capillary ingrowth in interspecies chimeric kidneys. Cell Growth Differ 15: 43–52

Sariola H, Aufderheide E, Bernhard H, Henke-Fahle S, Dippold W, Ekblom P (1988) Antibodies to cell surface ganglioside G_{D3} perturb inductive epithelial-mesenchymal interactions. Cell 54: 235–245

Sariola H, Saarma M, Sainio K, Arumae U, Palgi J, Vaahtokari A, Thesleff I, Karavanov A (1991) Dependence of kidney morphogenesis on the expression of nerve growth factor receptor. Science 254: 571–573

Satokata I, Benson G, Maas R (1995) Sexually dimorphic sterility in *Hoxa10*-deficient mice. Nature 374: 460–463

Saxén L (1984) Chimeric tissue combinations in the analysis of developmental mechanisms in the embryonic kidney. In: Le Douarin N, McLaren A (eds) Chimeras in developmental biology. Academic, London, pp 401–408

Saxén L (1987) Organogenesis of the kidney. Cambridge University Press, Cambridge

Saxén L, Lehtonen E (1978) Transfilter induction of kidney tubules is a function of the extent and duration of intercellular contact. J Embryol Exp Morphol 47: 97–109

Saxén L, Saksela E (1971) Transmission and spread of embryonic induction. Exp Cell Res 66: 369–377

Saxén L, Salonen J, Ekblom P, Nordling S (1983) DNA synthesis and cell generation cycle during determination and differentiation of the metanephric mesenchyme. Dev Biol 98: 130–138

Schmidt C, Bladt F, Goedecke S, Brinkman V, Zschiesche W, Sharpe M, Gherardi E, Birchmeier C (1995) Scatter factor/hepatocyte growth factor is essential for liver development. Nature 373: 699–702

Schnabel E, Anderson JM, Farquhar MG (1990) Diversity among tight junctions in rat kidney. J Cell Biol 111: 1255–1263

Schuchardt A, D'Agati V, Pachnis V, Constantini F (1996) Renal agenesis and hypodysplasia in ret-k-mutant mice result from defects in ureteric bud development. Development 122: 1919–1929

Schuchardt A, D'Agati V, Larsson-Blomberg L, Costantini F, Pachnis V (1994) Defects in the kidney and enteric nervous system of mice lacking the tyrosine kinase receptor Ret. Nature 367: 380–383

Schwarz-Sommer Z, Huijser P, Nacken W, Saedler H, Sommer H (1990) Genetic control of flower development by homeotic genes in *antirrhinum majus*. Science 250: 931–936

Scott MP, Weiner AJ (1984) Structural relationships among genes that control development: sequence homology between the *antennapedia, ultrabithorax*, and *fushi tarazu* loci of *drosophila*. Proc Natl Acad Sci USA 81: 4115–4119

Sharma PM, Bowman M, Yu BF, Sukumar S (1994) An rodent model for Wilms tumors: embryonal kidney neoplasms induced in rats by N-nitroso-N′-methylurea. PNAS USA 91: 9931–9935

Shawlot W, Berhringer R (1995) Requirement for Lim 1 in head-organizer function. Nature 374: 425–430

Shimamura K, Hirano S, Mcmahon AP, Takeichi M (1994) Wnt-1 dependent regulation of local E-cadherin and αN-catenin expression in the embryonic mouse brain. Development 120: 2225–2234

Sonnenberg E, Meyer D, Weidner KM, Birchmeier C (1993) HGF and its receptor, C-met, can mediate a signal exchange between mesenchyme and epithelia during mouse development. J Cell Biol 123: 223–236

Sonnenberg-Riethmacher E, Walter B, Riethmacher D, Godecke S, Birchmeier C (1996). The c-ros tyrosine kinase receptor controls regionalization and differentiation of epithelial cells in the epididymis. Genes Dev 10: 1184–1193

Sorenson CM, Rogers SA, Korsmeyer SJ, Hammerman MR (1995) Fulminant metanephric apoptosis and abnormal kidney development in bcl-2 deficient mice. Am J Physiol 37: F73–F78

Soriano P (1994) Abnormal kidney development and hematological disorders in PDGF-beta receptor mutant mice. Genes Dev 8: 1888–1896

Sorokin L, Sonnenberg A, Aumailley M, Timpl R, Ekblom P (1990) Recognition of the laminin E8 cell binding site by an integrin possessing the α6 subunit is essential for epithelial polarization in developing kidney tubules. J Cell Biol 111: 1265–1273

Stanton BR, Perkins AS, Tessarollo L, Sassoon DA, Parada L (1992) Loss of N-*myc* function results in embryonic lethality and failure of the epithelial components of the embryo to develop. Genes Dev 6: 2235–2247

Stark K, Vainio S, Vassileva G, McMahon AP (1994) Epithelial transformation of metanephric mesenchyme in the developing kidney regulated by Wnt-4. Nature 372: 679–683

Stein PL, Vogel H, Soriano P (1994) Combined deficiencies of Src, Fyn, and Yes tyrosine kinases in mutant mice. Genes Dev 8: 1999–2007

Steinberg MS, Takeichi M (1994) Experimental specification of cell sorting, tissue spreading, and specific spatial patterning by quantitative differences in cadherin expression. Proc Natl Acad Sci USA 91: 206–209

Stewart CL, Kasper P, Brunet LJ, Bhatt H, Gadi I, Köntgen F, Abbondanza SJ (1992) Blastocyst implantation depends on maternal expression of leukaemia inhibitory factor. Nature 359: 76–79

Strachan T, Read AP (1994) PAX genes. Curr Opin Genet Dev 4: 427–438

Stuart ET, Haffner R, Oren M, Gruss P (1995) Loss of p53 function through PAX-mediated transcriptional repression. EMBO J 14: 5638–5645

Sukhatme VP (1990) Early transcriptional events in cell growth: the Egr family. J Am Soc Nephrol 1: 859–866

Sukhatme VP (1992) The Egr transcription factor family: from signal transduction to kidney differentiation. Kidney Int 41: 550–553

Sympson CJ, Talhouk RS, Alexander CM, Chin JR, Clift CM, Bissell MJ, Werb Z (1994) Targeted expression of stromelysin-1 in mammary gland provides evidence for a role of proteinases in branching morphogenesis and the requirement for an intact basement membrane for tissue specific gene expression. J Cell Biol 125: 681–693

Takahashi H, Calvet JP, Dittemore-Hoover D, Yoshida K, Grantham JJ, Gattone VH (1991) A hereditary model of slowly progressive polycystic kidney disease in the mouse. J Am Soc Nephrol 1: 980–989

Taub M, Wang Y, Szezesny T M, Kleinman HK (1990) EGF or TGF-α is required for kidney tubulogenesis in matrigel cultures in scrum free medium. Proc Nat Acad Sci USA 87: 4002–4006

Tessarollo L, Nagarajan L, Parada LF (1992) c-ros: the vertebrate homolog of the sevenless tyrosine kinase receptor is tightly regulated during organogenesis in mouse embryonic development. Development 115: 11–20

The International Polycystic Kidney Disease Consortium (1995) Polycystic kidney disease: the complete structure of the PKD1 gene and its protein. Cell 81: 289–298

Theiler K, Gluecksohn-Waelsch S (1956) The morphological effects and the development of the *fused* mutation in the mouse. Anat Rec 125: 83–104

Threadgill DW, Dlugosz AA, Hansen L, Tennenbaum T, Lichti U, Yee D, LaMantia C, Mourton T, Herrup K, Harris R, Barnard JA, Yuspa SH, Coffey RJ, Magnuson T (1995) Targeted disruption of mouse EGF receptor: Effect of genetic background on mutant phenotype. Nature 269: 230–234

Tiedmann K, Egerer G (1984) Vascularization and glomerular ultrastructure in the pig mesonephros. Cell Tissue Res 238: 165–175

Torres M, Gomez-Pardo E, Dressler GR, Gruss P (1995) Pax-2 controls multiple steps of urogenital development. Development 121: 4057–4065

Torrey TW (1954) The early development of the human nephrons. Contrib Embryol, Carnegie Inst Wash 35: 75–97

Torrey TW (1965) Morphogenesis of the vertebrate kidney. In: DeHaan RL, Ursprung H (eds) Organogenesis. Holt, Rinehart and Winston, New York, pp 559–579

Treanor JJS, Goodman L, de Sauvage F, Stone DM, Poulsen KT, Beck CD, Gray C, Armanini MP, Pollock RA, Hefti F, Phillips HS, Goddard A, Moore MW, Buj-Bello A, Davies AM, Asai N, Takahashi M, Vandlen R, Henderson CE, Rosenthal A (1996) Charaterization of a multicomponent receptor for GDNF. Nature 382: 80–83

Trudel M, D'Agati V, Costantini F (1991) C-myc as an inducer of polycystic kidney disease in transgenic mice. Kidney Int 39: 665–671

Trumpp A, Blundell PA, de la Pompa JL, Zeller R (1992) The chicken *limb deformity* gene encodes nuclear proteins expressed in specific cell types during morphogenesis. Genes Dev 6: 14–28

Trupp M, Arenas A, Fainzilber M, Nilsson A-S, Sieber B-A, Grigoriou M, Kilkenny C, Salazar-Grueso E, Pachnis V, Arumae U, Sariola H, Saarma M, Ibanez CF (1996) Functional receptor for GDNF encoded by the *c-ret* proto-oncogene. Nature 381: 785–789

Tsarfaty I, Rong S, Resau JH, Rulong S, da Silva PP, Van de Woude G (1994) The met proto-oncogene mesenchymal to epithelial conversion. Science 263: 98–101

Tsukamoto AS, Grosschedl R, Guzman RC, Parslow T, Varmus HE (1988) Expression of the int-1 gene in transgenic mice is associated with mammary gland hyperplasia and adenocarcinomas in male and female mice. Cell 55: 619–625

Unsworth B, Grobstein C (1970) Induction of kidney tubules in mouse metanephrogenic mesenchyme by various embryonic mesenchymal tissues. Dev Biol 21: 547–556

Vainio S, Lehtonen E, Jalkaanen M, Bernfield M, Saxén L (1989) Epithelial-mesenchymal interactions regulate the stage specific expression of a cell surface proteoglycan, syndecan, in the developing kidney. Dev Biol 134: 382–392

van Dijk MA, Murre C (1994) *Extradenticle* raises the DNA binding specificity of homeotic selector genes products. Cell 78: 617–624

Vegas-Salas DE, Salas PJI, Gunderson D, Rodriguez-Boulan E (1987) Formation of the apical pole of epithelial (MDCK) cells: polarity of an apical protein is independent of tight junctions while segregation of a basolateral marker requires cell-cell interaction. J Cell Biol 104: 905–916

Veis DJ, Sorenson CM, Shutter JR, Korsmeyer S (1993) Bcl-2 deficient mice demonstrate fulminant lymphoid apoptosis, polycystic kidneys, and hypopigmented hair. Cell 75: 229–240

Vestweber D, Kemler R, Ekblom P (1985) Cell adhesion molecule uvomorulin during kidney development. Dev Biol 112: 213–221

Wang Z-Y, Madden SL, Deuel TF, Rauscher III FJ (1992) The human platelet-derived growth factor A-chain (PDGF-A) gene is a target for repression by the WT1 Wilms' tumor protein. J Biol Chem 267: 21999–22002

Warkany J (1971) Congenital malformations. Notes comments. Year Book Medical Publishers, Chicago, p 92

Wartiovaara J, Lehtonen E, Nordling S, Saxén L (1972) Do membrane filters prevent cell contacts? Nature 238: 407–408

Wartiovaara J, Nordling S, Lehtonen E, Saxén L (1974) Transfilter induction of kidney tubules: correlation with cytoplasmic penetration into nucleopore filters. J Embryol Exp Morph 31: 667–682

Welling LW, Grantham JJ (1991) Cystic and developmental diseases of the kidney. In: Brenner BM, Rector FC Jr (eds) The Kidney Saunders, Philadelphia

Werner H, Rauscher III FJ, Sukhatme VP, Drummond IA, Roberts CT Jr, LeRoith D (1994) Transcriptional repression of the insulin growth factor receptor (IGF-I-R) gene by the tumor suppressor WT1 involves binding to sequences upstream and downstream of the IGF-I-R gene transcription site. J Biol Chem 269: 12577–12582

Wolf G, Kuncio GS, Sun MJ, Neilson EG (1991) Expression of homeobox genes in a proximal tubular cell line derived from adult mice. Kidney Int 39: 1027–1033

Woolf AS, Joannou-Kolatsi M, Hardman P, Andermarcher E, Moorby C, Fine LG, Jat PS, Noble MD, Gherardi E (1995) Roles of HGF/scatter factor and the met receptor in the early development of the metanephros. J Cell Biol 128: 171–184

Woods DF, Bryant PJ (1990) The *discs-large* tumor suppressor gene of *Drosophila* encodes a guanylate kinase homolog located at septate junctions. Cell 66: 451–464

Yun K, Fidler AE, Eccles MR, Reeve AE (1993) Insulin-like growth factor II and WT-1 transcript localization in human fetal kidney and Wilms tumor. Cancer Res 53: 5166–5171

Zachai E, Mellman WJ, Neiderer B, Hanson J (1975) The fetal trimethadione syndrome. J Pediatr 87: 280–284

Zhang H, Bradley A. Mice deficient for BMP2 are non-viable and have defects in ammion/chorion and cardiac development. Development (submitted)

CHAPTER 7

Palate

E.F. Zimmerman

Development of the mammalian secondary palate is a complex series of events whose perturbation in humans is characterized by the birth defect cleft palate (CP). Human cleft palate is usually thought to be multifactorially inherited with a frequency of 4 per 10 000 live births and is genetically distinct from cleft lip (CL) with or without cleft palate, which has a frequency of 8 per 10 000 live births (Baird et al. 1994). A number of reviews have been written on palate development (Zimmerman 1984; Pisano and Greene 1986; Ferguson 1988).

A. Why Study the Palate?

Much research has focused on the palate by both teratologists and developmental biologists. Many teratogens which produce congenital malformations in particular perturb palate development, causing cleft palate. Striking examples are glucocorticoids, TCDD (dioxin), and retinoids, which may exert their teratogenic effects through endogenous physiological receptors (Pratt et al. 1984; Goldman 1984; Abbott et al. 1994a; Durand et al. 1992).

Further interest in studying the palate derives from its important developmental events: (a) the reorientation process, whereby the vertical palate shelves elevate to the horizontal position above the tongue, (b) the fusion process, whereby the two apposing shelves produce a loss of the medial epithelium as the shelves develop into a single palatal structure, and (c) mesenchymal–epithelial interactions.

Finally, the study of the developing palate has the relative advantage of ease compared with other developmental model systems. The developing palate occurs late in embryogenesis and thus is large and easily dissected, since it is exposed externally. Many useful in vitro systems have been developed. Palate epithelium and mesenchyme can be separated by proteolytic enzymes and chelating agents. Mesenchymal cells can be readily grown in culture, allowing biochemical and molecular biology experiments to be performed. Epithelium can be grown on suitable substrates to follow differentiation of oral, medial, and nasal epithelium to their natural phenotypic fate (Carette et al. 1991). Mesenchymal cells have been incubated in chemotactic systems and in collagen gels in order to study motility and traction (Zimmerman et al. 1983; Venkatasubramanian and Zimmerman 1983). Recombination of epithelium and mesenchyme from different species (murine, chicken, alligator) can be performed to study epithelial–mesenchymal interactions (Ferguson and Honig 1984). Furthermore, cultured mouse or human palates can be employed in serum-free medium to facilitate study of the role of growth factors or teratogens on growth and differentiation (Sharpe et al. 1992a; Abbott and Buckalew 1992). Finally, the reorientation process of palatal shelves and synthesis of extracellular matrix (ECM) can be studied in vitro by various techniques in which fetal heads are cultured with the tongue as well as cranium and mandible excised (Wee et al. 1976; Bulleit and Zimmerman 1985; Morris-Wiman and Brinkley 1993).

B. Morphogenesis of the Palate

The secondary palate forms from palatal shelves that grow out of the maxilla. These palatal shelves grow downward parallel to the lateral surfaces of the tongue in the mouse, rat, and human, in contrast to the chicken, in which they grow out of the maxilla horizontally towards each other. The mammalian vertical shelves reorient rapidly upward to the horizontal position above the tongue, a process which has been termed "flip-up". The horizontal shelves continue their growth towards each other and fuse with each other and with the inferior edge of the nasal septum. During the process of fusion, there is a "breakdown" of the medial edge epithelium. The loss of the epithelial seam is followed by mesenchymal replacement, which then undergoes marked differentiation. Mesenchyme of the anterior to midpalate undergoes intramembranous ossification (without chondrogenesis), forming the hard palate, which serves as the roof of the mouth. The posterior palate undergoes myogenesis, forming the soft palate, which is involved in speech. In addition, at the site where the maxilla is contiguous to the hard palate, tooth formation occurs. In the mandible, a similar process occurs, giving rise to the lower set of teeth.

Perturbation of these morphogenetic events by teratogens can lead to cleft palate, and the susceptible time period in which this occurs has been termed the critical period for cleft palate formation. Since palate formation is a late developmental event, the critical period for cleft palate is usually later than most malformations (see Schardein 1993). In the developing mouse, the critical period occurs at about gestational days 9–14. Surprisingly, two susceptible stages of development were observed in the hamster after retinoic acid administration (Shenefelt 1972). In addition to the normal critical period for palatogenesis, Rogers et al. (1993) observed that the period of greatest sensitivity for methanol-induced cleft palate was gestational days 6–7 in mice, with no cleft palate induced after day 10 of development.

C. Reorientation of the Palate

Much effort has been expended to elucidate the process whereby the vertical shelves reorient to a horizontal position. The term "flip-up" is used for the process by which the anterior shelves rapidly elevate within seconds to minutes, as observed when the tongue has been removed from the orofacial cavity (FERGUSON 1978). An intrinsic shelf force is progressively generated within the palatal shelves (WALKER and FRASER 1956). It is thought that, once this force is generated past a threshold to overcome the frictional resistance of the tongue, the shelves will reorient. The process is slower in vivo, taking place in the course of hours, with the shelves undergoing both growth and movement (MORRIS-WIMAN and BRINKLEY 1993). This reorientation process involves a flip-up in the anterior palate and an oozing remodeling mechanism of the midposterior palate as it flows over the back end of the tongue, which cannot move out the way (BRINKLEY and MORRIS-WIMAN 1987a). The intrinsic shelf force is multifactorial, arising from the activities and interplay of components in the shelves: the extracellular matrix, mesenchymal cells, and surrounding palatal epithelium.

I. Extracellular Matrix and Mesenchyme

The chief component generating the force appears to be regional accumulation in the extracellular matrix composed of glycosaminoglycans (GAG), predominantly hyaluronic acid (HA). PRATT et al. (1973) reported that approximately 65% of the total GAG synthesized in the rat palate at the time of reorientation is HA. KNUDSEN et al. (1985) identified HA as the predominant GAG in the mouse palate during its development and also showed the presence of both chondroitin-4- and -6-sulfate. The peak of HA synthesis has been shown to occur at day 13 of gestation, i.e., before shelf elevation (FOREMAN et al. 1991). HA is a highly electrostatically charged, open-coiled molecule, capable of binding up to ten times its weight in water (COMPER and LAURENT 1978). This hydration of water is thought to be responsible for the increase in swelling of the extracellular matrix that is observed during palate development (BRINKLEY and BOOKSTEIN 1986; FOREMAN et al. 1991). However, the increased hydration may be constrained by cell–cell, cell–matrix, or matrix–matrix interactions. TOOLE (1981) has postulated that, if these restraining interactions are disrupted, the tissue will experience a sudden expansion. This model could explain generation of an intrinsic shelf force during palate development. It has been observed that there is more HA in the anterior and midpalate than the posterior palate (KNUDSEN et al. 1985), which is consistent with a strong force to flip up the anterior end of the palate and with the posterior palate following the anterior palate by oozing or remodeling around the tongue. Further support for role of GAG in shelf reorientation comes from the observation that chlorcyclizine-treated palates are associated with degradation of HA and inhibition of palate shelf elevation (BRINKLEY and MORRIS-WIMAN 1987b).

The erectile shelf-elevating force is partly directed by large bundles of type I collagen which run down the center of the vertical shelf from its base to its tip (Ferguson 1988). These fibers serve to stiffen the palate and aid the shelf to reorient. Morris-Wiman and Brinkley (1992, 1993) investigated major components of the extracellular matrix infrastructure during reorientation and remodeling in vivo and in vitro. The major components of this infrastructure were collagen III, fibronectin, and HA. With remodeling, HA's domain within the mesenchyme was expanded. However, the distribution of fibronectin and collagen became more circumscribed, possibly due to degradation of these components to allow expansion of HA.

A number of studies have suggested that an area of the mesenchyme subjacent to the oral epithelium at the midpalate plays a significant role in shelf orientation. These midoral mesenchymal cells align themselves perpendicular to the basement membrane (Babiarz et al. 1979) and appear to act as a constraining force from the oral epithelium to the infrastructure of interior mesenchyme. After reorientation, the alignment of these cells disappears as if the constraining force is relieved (Wee et al. 1979). It has been considered that these midoral cells aid reorientation by a direct contractile effect or by producing a cellular tension to counteract the force of hyaluronate-dependent expansion of the extracellular space. Evidence to support the role of these mesenchymal cells in exerting a direct or indirect contractile influence on the reorientation process is provided by observations that (a) these midoral cells are enriched for actin-containing microfilaments (Kuhn et al. 1980); (b) palates in glycerinated heads in which adenosine triphosphate (ATP) can enter cells to condense actomyosin underwent increased reorientation, which was prevented by cytochalasin B (Wee and Zimmerman 1980); (c) mesenchymal cells are extremely migratory and exert strong tractional forces on appropriate substrates (Venkatasubramanian and Zimmerman 1983); and (d) serotonin stimulated the migration of these cells (Zimmerman et al. 1983; see below). Knudsen et al. (1985) considered the possibility that the role of these midoral mesenchymal cells modulate reorientation by exerting constraints on HA-dependent swelling of the extracellular space (ECS), since HA is lacking in this area. This notion was supported by Morris-Wiman and Brinkley (1992), who confirmed that HA is lacking in these midoral mesenchymal cells and that collagen III-, fibronectin-, and tenascin-positive fibrils are observed to radiate between these cells to the underlying basement membrane. This unique epithelial–mesenchymal region may play a role in shelf remodeling by anchoring the extracellular matrix infrastructure and preventing expansion in the oral direction.

II. Epithelium

The epithelium overlying the oral surface has also been implicated in directing reorientation. Epithelial cell shape, distribution, and/or layering changes, as well as rugae development and an increase in mitotic figures in the presumptive oral epithelium, correlate with the timing of reorientation (BRINKLEY 1984; LUKE 1984). Support for this hypothesis that the oral epithelium helps to direct reorientation comes from the observation that specific removal of the oral epithelium inhibits reorientation in vitro (BULLEIT and ZIMMERMAN 1985, 1986). A unifying hypothesis is that the oral epithelium and the associated underlying basement membrane of the palatal shelf exhibit traction, which serves to constrain and direct the swelling osmotic force generated by HA in the mesenchyme in much the same way that the material constrains an inflating balloon. The alignment and fibrillar attachment of the midoral mesenchymal cells from the underlying oral basement membrane and epithelium to the extracellular matrix infrastructure within the core of the palatal mesenchyme constraining HA may further serve to direct the elevating force away from the oral surface (ZIMMERMAN and WEE 1984; MORRIS-WIMAN and BRINKLEY 1992).

D. Fusion of the Palate

After reorientation of vertical shelves to positions above the tongue and growth of the horizontal shelves to apposing positions, the palate undergoes fusion, which encompasses the following stages:

1. Initial sloughing of the outer cells of the medial edge epithelium (MEE)
2. Epithelial contact and adherence
3. Thinning of the epithelial seam and disappearance of portions of the seam, leaving islands of epithelial clumps
4. Final mesodermal confluence

Peridermal cell death was observed in the MEE before shelf contact, and the basal epithelial cells were reported to possess carbohydrate-rich lateral surfaces, which presumably functioned to adhere the epithelial layers of the two apposing shelves (WATERMAN et al. 1973; WATERMAN and MELLER 1974). Programmed cell death has long been accepted as the mechanism that removes the peridermal cells and the basal layer prior to mesenchymal confluence (SHAPIRO and SWENEY 1969; PRATT and MARTIN 1975). Evidence to support this mechanism included: (a) cessation of DNA synthesis before palatal fusion (GREENE and PRATT 1976), (b) appearance of ultrastructural changes consistent with cell death (SHAPIRO and SWENEY 1969), and (c) increased lysosomal enzymes in the MEE (SMILEY 1972). Finally, when excised palatal shelves were cultured, it was observed that the complete MEE epithelium sloughed off, which was considered representative of what occurred in vivo. Moreover, addition of a glucocorticoid blocked the loss of MEE, which was considered evidence that this teratogen inhibited palatal fusion (GOLDMAN et al. 1981).

However, FITCHETT and HAY (1989) reinvestigated the mechanism of loss of epithelium during the fusion process. Analyzing palates temporally in vivo, they confirmed reports that the periderm of the two-layered MEE sloughs off after the shelves assume horizontal positions. They observed that, after the basal lamina disappeared, elongating basal MEE cells sent filopodial processes into adjacent connective tissue. These cells assumed progressive morphological characteristics of the mesenchyme, confirmed by the immunological marking of the mesenchymal cytoskeletal protein vimentin. They postulated that basal epithelial cells transform into the mesenchymal phenotype and proposed that the considerable cell death reported in vitro in the fused midline epithelium may be due to trapped peridermal cells.

An alternate fate of epithelial cells during palatal fusion was proposed by CARETTE and FERGUSON (1992). Using an in vitro mouse model, palates were labeled with DiI (1,1-dioctadecyl-3,3,3′,3′-tetramethylindocarbocyanine perchlorate) and the fate of the labeled MEE during palatal fusion was followed by confocal microscopy. These authors reported that MEE cells migrate nasally and orally out of the seam and are recruited into, and constitute epithelial triangles on, both the oral and nasal aspect of the palate. To eliminate potential culture artifacts, palates were labeled in utero by DiI (SHULER et al. 1992) and by carboxyfluorescein (GRIFFITH and HAY 1992), and the fate of basal epithelium was traced during subsequent fusion. Both studies support the concept of epithelial–mesenchyme transformation, since epithelium and not mesenchyme was labeled prior to fusion and only mesenchyme was labeled after fusion was complete. Interestingly, if labeled palatal shelves are not allowed to fuse in vitro, the basal epithelial cells do not form mesenchyme after sloughing (see above), indicating that formation of the epithelial midline seam is necessary to trigger its epithelial–mesenchyme transformation.

TCDD (dioxin) is a specific epithelial teratogen causing cleft palate by preventing loss of the epithelial seam during the fusion process (see Chap. 21).

E. Mesenchymal–Epithelial Interactions

The developing secondary palate, like developing limb and kidney, undergoes mesenchymal–epithelial interactions. Removal of the epithelium from the underlying mesenchyme has allowed recombination experiments employing homologous, heterologous, heterochronic, and isochronic combinations of mandibular, limb, and palatal tissues both within and between mouse, chick, and alligator embryos (FERGUSON and HONIG 1984). Morphological differences in palatal epithelium have been used as the endpoint to determine the role of mesenchyme in directing epithelial differentiation. It has been shown in alligator and chick that the mesenchyme signals nasal, medial, and oral epithelium of the palate in a species-specific manner, even to heterologous epithelia. Palatal epithelium plays a passive role, receiving instructions from the underlying mesenchyme. However, mouse palatal epithelium is biased to differentiate into nasal, medial, and oral phenotypes and will do so if the epithelium is placed on a neutral mesenchyme such as mandibular mesenchyme. In contrast, mouse palatal mesenchyme signals regionally specific epithelial differentiation to both heterologous epithelia or palatal epithelia from other species such as chick and alligator (FERGUSON 1988). Since nasal, medial, and oral epithelium is specified by the underlying mesenchyme, there has been a determined search for these signaling factors. Putative growth factors (epidermal growth factor, EGF; transforming growth factor (TGF) -α and -β; insulin-like growth factors, IGF) have been candidates that might instruct the epithelium directly or induce synthesis of extracellular matrix components which could carry out the signaling process.

F. Neurotransmitters

A considerable body of evidence indicates that neurotransmitters function in primitive and higher organisms to regulate morphogenesis, including proliferation, differentiation, cell motility, and metamorphosis (see review by LAUDER 1993). The role of neurotransmitters in regulating the process of palate reorientation has been investigated (ZIMMERMAN and WEE 1984). Employing an embryo culture system which permits palate elevation, various neurotransmitters and their antagonists were tested to observe their effects on this process. It was observed that the anterior and posterior palate respond independently to these neurotransmitters. Acetylcholine agonists stimulate posterior shelf reorientation, while serotonin (5-HT) stimulates the anterior palate to elevate. γ-Aminobutyric acid (GABA) produced an opposite effect to that of acetylcholine and 5-HT, inhibiting shelf movement at both the anterior and posterior ends.

I. Serotonin

While 5-HT stimulates the anterior end of the palate to elevate, its antagonist methysergide reverses the effect of 5-HT and inhibits anterior shelf reorientation by itself. The antagonist cyproheptidine also inhibits shelf rotation in vitro and delays the process of palate reorientation when injected into pregnant dams (WEE et al. 1979). Various aspects of cellular responsiveness and metabolism to 5-HT have been monitored. 5-HT was found to markedly stimulate cell motility in two in vitro systems. It markedly stimulated migration of mesenchymal cells out of palates explanted in a hydrated collagen matrix. The stimulation was agonist specific, since the antagonist methysergide blocked the response (VENKATASUBRAMANIAN and ZIMMERMAN 1983). In the other system employing chemotaxis, 5-HT markedly stimulated cell motility, a concentration of $10^{-5} M$ producing a 4.9-fold increase (ZIMMERMAN et al. 1983).

The effect of 5-HT on biochemical reactions associated with cell movement has also been observed. When palate mesenchymal cells were treated with 5-HT, protein carboxyl methylation, a process associated with motility, was stimulated by about 100% at concentrations ranging from 3×10^{-7} to 3×10^{-6} M. The effect of 5-HT on the second messengers cyclic adenosine monophosphate (AMP) and cyclic guanosine monophosphate (GMP) were also measured. At a concentration of 10^{-5} M, 5-HT markedly depressed cyclic AMP levels, while it increased cyclic GMP with a spike of stimulation (6.1-

fold) within 30s (ZIMMERMAN et al. 1983). These effects of 5-HT in stimulating cyclic GMP synthesis and protein carboxyl methylation and inhibiting cyclic AMP are consistent with neurotransmitter-stimulating proliferation as well as motility (LAUDER 1993).

5-HT is actively taken up by epithelium in early embryos (SHUEY et al. 1992) and in the palate (LAUDER and ZIMMERMAN 1988). In nonpalatal tissues, 5-HT-binding protein is observed in mesenchyme subjacent to epithelium, and 5-HT uptake inhibitors cause craniofacial defects in embryo culture in regions that actively take up 5-HT. Furthermore, about five to six cells deep in mesenchyme that is actively undergoing proliferation, 5-HT uptake inhibitors cause inhibition of proliferation. Thus it is likely that 5-HT plays a role in regulation of craniofacial growth, differentiation, and morphogenesis through epithelial–mesenchyme interactions (SHUEY et al. 1992).

Although specific 5-HT staining of the whole palate epithelium and other craniofacial epithelium was observed on day 12 of gestation, at day 14 staining for 5-HT was only observed in the invaginating epithelium of the midoral surfaces and developing tooth germs of the adjacent maxilla (LAUDER and ZIMMERMAN 1988). In addition, 5-HT staining has been localized in epithelium of the eye and in the nasal prominence at the point of invagination and fusion at day 11. These observations suggest that 5-HT may be involved in several types of morphogenetic movements. In addition, these studies have demonstrated that embryos at these stages have the capacity for 5-HT uptake and metabolism. Further, the embryonic heart, the placenta, and the maternal circulation may provide important sources of 5-HT during embryogenesis (SHUEY et al. 1990, 1993).

As indicated previously, midoral mesenchymal cells are aligned perpendicularly to the invaginated midoral epithelium. It is at this region that the midoral epithelium actively takes up 5-HT just prior to palate shelf reorientation. After reorientation, these midoral mesenchymal cells lose their orientation as the shelf swells, presumably due to deconstraining HA in the core mesenchyme. In addition, evidence was previously presented that this oral epithelium serves to direct the force of the shelf upwards to the nasal septum. It is of interest that the 5-HT antagonist methysergide in embryo culture blocks anterior to midshelf reorientation. The morphology of the shelf was abnormal, and although swelling on the nasal side mesenchyme occurred, swelling of midoral mesenchyme was markedly inhibited and orientation of these cells was abnormal (WEE et al. 1979). These observations are consistent with the hypothesis that the transient uptake of 5-HT in the invaginated epithelium at the midoral surface prior to the time of palate elevation serves to regulate the process. Possible mechanisms of 5-HT are: (a) increase of cell contractility of epithelium (and/or subjacent mesenchyme) and (b) degradation of extracellular matrix components of basement membrane and/or subjacent mesenchyme to deanchor the infrastructure constraining HA. As a consequence, HA may bind water and expand to reorient the palate. It has been shown that, during reorientation, fibronectin and collagen III was decreased, presumably by degradation (MORRIS-WIMAN and BRINKLEY 1992). One possibility is that 5-HT activates extracellular matrix proteases to allow HA expansion in the mesenchyme.

II. Catecholamines

The first report that catecholamines may play a role in palate development came from WATERMAN et al. (1976), who showed that these monamines (isoproterenol, epinephrine, norepinephrine, and dopamine) stimulated hamster adenylate cyclase, with maximal activity prior to and during palate fusion. Cultured murine palate mesenchymal cells have been shown to elevate cyclic AMP in a dose-dependent manner when exposed to increasing concentrations of various catecholamines (GABARINO and GREENE 1984; SCHREINER et al. 1986). β_2-Adrenergic receptors were characterized in cultured palate mesenchymal cells (GABARINO and GREENE 1984). Levels of catecholamines, including norepinephrine, epinephrine, and dopamine, have been measured in the embryonic palate and quantitative changes noted during development (ZIMMERMAN et al. 1981; PISANO and GREENE 1985). Although catecholamines are present in the palate, it is unlikely that they function in palate reorientation. Norepinephrine or dopamine at concentrations of 10^{-4} M produced no significant effect on either anterior or posterior palate reorientation in embryo culture (ZIMMERMAN and WEE 1984). However, the β-adrenergic catecholamine isoproterenol has been shown to inhibit proliferation of mouse embryonic palate mesenchymal cells in vitro. Evidence was presented that isoproterenol delayed initiation of cellular DNA synthesis via a receptor-mediated response. Since palatogenesis is characterized by differential growth of various cell types during development, it is possible that catecholamines may regulate growth locally (PISANO et al. 1986). Alternatively, the inhibition of growth may be a consequence of induced differentiation in the palate.

III. γ-Aminobutyric Acid and Diazepam

Since diazepam is teratogenic in experimental animals and functions, at least in part, through a GABAergic mechanism, the role of GABA in palate development and the teratogenic action of diazepam have been studied (ZIMMERMAN and WEE 1984). It is well established that benzodiazepines such as diazepam are teratogenic in experimental animals, causing cleft palate in mice (MILLER and BECKER 1975) and hamsters (SHAH et al. 1979), producing neural defects in mice (JURAND 1980) and hamsters (GURAM et al. 1982), and acting as a behavioral teratogen in rats (KELLOG et al. 1980). Diazepam may be teratogenic in humans (SAXÉN 1975; SAXÉN and SAXÉN 1975; AARSKOG 1975), although this conclusion has been disputed (SAFRA and OAKLEY 1976; ROSENBERG et al. 1983; SHIONO and MILLS 1984). In experimental animals, it was observed that diazepam produced a higher frequency of cleft palate in the SWV mouse strain than in the A/J strain (WEE and ZIMMERMAN 1983).

Furthermore, cleft palate sensitivity to diazepam was not associated with the H-2 locus (TOCCO et al. 1987), which is a characteristic of glucocorticoid and phenytoin teratogenesis (GOLDMAN 1984).

It was observed that tenfold more GABA is necessary to significantly inhibit palate reorientation in the diazepam-resistant A/J than the diazepam-sensitive SWV embryo in culture, and the effects were significantly reversed by the GABA antagonist picrotoxin. Diazepam at a concentration of 10^{-4} M inhibited palate reorientation in vitro, the effect being greater in SWV than in A/J mice, and picrotoxin partially but significantly reversed the inhibition produced by diazepam. These results were interpreted as indicating that diazepam inhibits shelf reorientation to produce cleft palate by interacting with a functional GABAergic system in the developing palate (WEE and ZIMMERMAN 1983).

Further attempts to characterize the putative GABAergic system in the palate have been carried out. GABA has been measured in the embryonic palate and found to undergo quantitative changes during development. GABA levels were significantly higher in palatal cells and in skin fibroblasts from the diazepam- and GABA-sensitive SWV than in those from the A/J strain (NORMAN et al. 1985; WEE et al. 1986). A GABA uptake mechanism has been sought using primary cultures of embryonic palate mesenchymal cells. It was shown that both palate cells and skin fibroblasts are capable of accumulating [^{3}H]GABA by saturable uptake mechanisms, characteristic of both high- and low-affinity transport systems. The V_{max} of the SWV high-affinity uptake system in palate cells and skin fibroblasts was 1.8-fold higher than that of comparable A/J cell cultures. Thus active GABA uptake and GABA levels appear to be genetically regulated in non-neural cells (WEE and ZIMMERMAN 1985).

In summary, evidence has been obtained that diazepam induces cleft palate by mimicking GABA in a putative GABAergic system in palate development. A significant finding is that genetic differences in both diazepam teratogenesis and in a GABAergic system have been observed. Comparing the SWV and A/J strains, the SWV mouse showed (a) a greater sensitivity to diazepam-induced cleft palate, (b) a greater sensitivity to GABA and diazepam inhibition of palate reorientation in embryo culture, and (c) a greater palatal GABA level and a more efficient GABA uptake system. Thus diazepam may be more teratogenic in mice (and possibly individuals in the human population) which possess a more efficient GABAergic system.

Nevertheless, these correlative studies do not prove that GABA regulates palate development. However, it was recently reported that a mouse mutation in the P locus which leads to cleft palate is associated with an alteration in the β_3-subunit of the type A GABA receptor. This genetic data provides strong evidence of an active role of GABA in palate development (CULIAT et al. 1993).

G. Growth Factors

It is commonly thought that development of the mammalian secondary palate is regulated, at least in part, by a complex interaction between extracellular matrix molecules and soluble agents such as families of growth factors. These growth factors, which include neurotransmitters functioning as hormones (as reviewed previously), may also be responsible for directing mesenchymal–epithelial interactions (FERGUSON 1988). Palatal mesenchymal cells are heterogeneous in the synthesis of various growth factors and extracellular matrix molecules. Synthesis of growth factors at specific locations in the palate may give rise to diffusion gradients, which would allow differing regional responses by epithelium and mesenchyme. It has been shown that the same growth factor may stimulate proliferation at one concentration and inhibit it at another (SPORN et al. 1987; MASSAGUE 1987). During palate development, gradients of growth factors may be set up by binding and sequestering them in particular areas by binding proteins or extracellular matrix molecules. In addition, growth factors may come from amniotic fluid and from the oronasal cavity. Active uptake of serotonin into palatal epithelium is a striking example. Gradients of growth factors, as well as gradients of specific receptors, may generate complex interactions. Finally, it has been shown that different growth factors interact with the same cells to either synergize or antagonize physiologic responses of the cells (SPORN et al. 1987; SHARPE et al. 1992a).

I. Glucocorticoids

Glucocorticoids and retinoids (see Chap. 21) were first shown to be teratogenic, causing cleft palate. Only later was it realized that these agents were endogenous to the embryo and functioned to regulate development, and the evidence that retinoids act as morphogens is now overwhelming. The teratogenic mechanisms of glucocorticoids in inhibiting proliferation of palatal mesenchymal cells (PRATT et al. 1984) and blocking palatal fusion through a prostaglandin pathway (GOLDMAN 1984) have been reviewed.

As part of these studies, it was shown that the amount of glucocorticoid receptor in palates varied with mouse strain, which in turn correlated with cleft palate sensitivity (SALOMON and PRATT 1976; GOLDMAN et al. 1977). Temporal and spatial expression of the glucocorticoid receptor (GR) and its mRNA have been analyzed. At day 12 of gestation, the GR mRNA was locally expressed in the region of palatal outgrowth in both mesenchymal and epithelial cells, and GR was expressed uniformly throughout the palate. By day 14, GR was expressed with higher levels in epithelial cells than in the mesenchyme. Abundant levels of mRNA of GR were observed during bone formation (ABBOTT et al. 1994b). Using a semiquantitative measure of GR mRNA in total shelves, LAU et al. (1993) reported that GR mRNA significantly decreased during the time of palate reorientation. Both studies showing the spatial and temporal expression of GR and its mRNA support the notion that glucocorticoid and its receptor function to regulate regional growth and differentiation during palatogenesis.

II. Transforming Growth Factor-α, Epidermal Growth Factor, and Epidermal Growth Factor Receptor

TGF-α binds to the EGF receptor (EGFR), and it is thought that TGF-α is the embryonic form of EGF. Since TGF-α and EGFR are localized to the palate, it is likely that they function in palate development. However, it is not clear whether EGF also plays a role in palate development, possibly deriving from the uterus (WERB 1990). In localization studies, TGF-α was present at day 12 throughout the palate mesenchyme, with little growth factor in the epithelia (Table 1). At day 13, histochemical staining increased in epithelia and mesenchyme at the tip of the palate. As palatal shelves fused together (day 14.5), intense staining was observed in the midline epithelial seam and the subjacent mesenchyme (DIXON et al. 1991). EGFR have been detected in the developing palate of mouse and rat embryos (NEXφ et al. 1980; ABBOTT et al. 1988; SHIOTA et al. 1990). Spatial differences in EGFR in palate epithelium have been observed. EGFR immunoreactivity is intense in the MEE. However, ABBOTT et al. (1988) observed equal immunostaining in nasal and oral epithelium, while SHIOTA et al. (1990) reported greater immunostaining on the oral surface than the nasal epithelium. Although mesenchymal cells in the midpalatal region exhibited apparently homogeneous staining of EGFR on day 12, it diminished in the mesenchyme beneath the MEE on day 13 (SHIOTA et al. 1990; BRUNET et al. 1993). This is the area of presumed epithelial–mesenchyme interactions. In this regard, HA, which is thought to generate the shelf force to reorient the palate, is synthesized in the mesenchyme and its synthesis is stimulated by EGF (TURLEY et al. 1985). Furthermore, EGFR immunoreactivity in the mesenchyme was more intense in the midpalate than posteriorly (SHIOTA et al. 1990). This is consistent with the notion that this receptor system is involved in regulation of HA and thus reorientation, since there is more HA in the anterior to midportion of the palate than at the posterior end (KNUDSEN et al. 1985).

In order to understand the function of these polypeptide growth factors, they have been incubated with various palate preparations in vitro. Proliferation of human embryonic palatal mesenchymal cells was stimulated by EGF in vitro (YONEDA and PRATT 1981). EGF inhibits degeneration of the MEE of the rat palate (Table 1) and promotes hypertrophy and keratinization of these cells (HASSELL 1975). Addition of TGF-α to cultured epithelial sheets stimulated the synthesis of tenascin and collagen IX in the MEE. Since TGF-α is known to localize to the MEE, this suggests a regulatory role of TGF-α for differentiation of the MEE during the fusion process (DIXON et al. 1993).

Families with genetically linked cleft palate have been analyzed for associations with alterations in a number of candidate genes thought to be involved in palatogenesis. It was observed that nonsyndromic cleft lip with or

Table 1. Peptide growth factors in palate development

Growth factor	Temporal and spatial localization	Effects	Possible role
TGF-α, EGF, EGFR	E12[a]: TGF-α throughout mesenchyme E13–E14.5: TGF-α increased at MEE and subjacent mesenchyme EGFR in medial, oral, and nasal (?) epithelium; in mesenchyme, which decreases subjacent to MEE later	EGF prevents degeneration of midline epithelial seam TGF-α stimulates synthesis of ECM components tenascin and collagen IX in MEE EGF stimulates synthesis of HA TGF-β, bFGF, IGF II decrease EGFR binding	Modulate differentiation of MEE, as well as HA synthesis in mesenchyme
TGF-β_1,-β_2,-β_3 and receptors	TGF-β_1 in mesenchyme, especially in developing tooth germ; in epithelium (?) TGF-β_2 in epithelium, not mesenchyme TGF-β_1 mRNA in epithelium, not mesenchyme TGF-β_2 mRNA in mesenchyme, not epithelium	TGF-β_1 inhibits proliferation of MEPM cells in vitro TGF-β_1 stimulates synthesis of fibronectin and collagen (I, III, V) in vitro	Epithelial–mesenchyme interactions: fusion at MEE, tooth germ differentiation, regulate mesenchyme
	TGF-β_3 mRNA same as TGF-β_1 but at higher levels, earlier in development, and particularly to MEE TGF-β receptor (I, II, III) present in MEPM cells	EGF downregulates TGF-β_3 mRNA levels in vitro Cyclic AMP, retinoids, TGF-β_1, TGF-β_2 increase TGF-β_3 mRNA levels TGF-β induce precocious differentiation of MEE	
IGF-I, -II	IGF II peptide and mRNA predominant over IGF I IGF II mRNA expressed later in development in horizontal palate, in mesenchyme and not epithelium IGF II peptide and binding protein in epithelium (nasal, medial)		Epithelium–mesenchyme interaction
aFGF, bFGF	Basement membrane and epithelium (nasal, medial), MEE at fusion	Stimulate GAG, especially HA, in vitro	Modulate mesenchymal ECM biosynthesis to facilitate epithelial seam degeneration

TGF, transforming growth factor; EGF, epidermal growth factor; EGFR, EGF receptor; IGH, immunoreactive growth hormone; aFGF, acidic fibroblast growth factor; bFGF, basic FGF; MEE, medial edge epithelium; MEPM, murine embryonic palate mesenchymal; ECM, extracellular matrix; HA, hyaluronic acid; IGF, insulin-like growth factor; AMP, adenosine monophosphate; GAG, glycosaminoglycans.
[a]Embryonic (gestational) day 12.

Table 2. Association of cleft palate in humans with candidate genes

Association observed	Study
Linkage of autosomal dominant clefting syndrome (Van der Woude syndrome) to loci on chromosome Iq	MURRAY et al. (1990)
Nonsyndromic cleft lip with or without cleft palate associated with genetic variation of TGF-α	ARDINGER et al. 1989
Cleft lip with or without cleft palate associated with TGF-α and retinoic acid receptor loci	CHENEVIX-TRENCH et al. 1992

without cleft palate was associated with a genetic variation of TGF-α (ARDINGER et al. 1989; CHENEVIX-TRENCH et al. 1992; Table 2). These studies provide strong evidence that TGF-α is involved in palatogenesis and that mutations in the TGF-α gene or other associated DNA sequences lead to cleft lip with or without cleft palate.

III. Transforming Growth Factors-β_1, -β_2, and -β_3, and Their Receptors

TGF-β, first discovered in an assay based on its ability to transform fibroblasts in culture, has been shown to have profound effects on nearly all cell types, influencing either proliferation, differentiation, or other aspects of their function. Three isoforms of TGF-β (TGF-β_1, -β_2, and β_3) have been identified in the palate which bind to multiple TGF receptors (TGF-βR I, II, and III). Although in the transforming assay TGF-β stimulated cell growth, it has been shown in other systems that it inhibits cell proliferation and stimulates differentiation. One of the most extensive roles of TGF-β is to enhance the formation of extracellular matrix, an action of extreme importance in embryogenesis (SPORN et al. 1987). The activity of TGF-β on extracellular matrix derives from its effects on both protein synthesis and degradation. TGF-β markedly increases mRNA levels for fibronectin and collagen I and III (IGNOTZ et al. 1987; RAGHOW et al. 1987; VARGA et al. 1987), acting on both transcription (ROSSI et al. 1988) and mRNA stability (PENTTINEN et al. 1988). TGF-β also inhibits breakdown of matrix proteins by suppressing the release of collagenase and plasminogen activator (EDWARDS et al. 1987; SAKSELA et al. 1987) and increasing release of protease inhibitors (LAIHO et al. 1987).

HEINE et al. (1987) reported immunochemical staining of TGF-β_1 in mesenchymal cells, especially in developing tooth buds in the contiguous maxilla, while not detecting peptide in the epithelium (Table 1). Employing in situ hybridization, LEHNERT and AKHURST (1988) detected TGF-β_1 mRNA in epithelium and not mesenchyme, implying synthesis of the TGF-β_1 in palatal epithelium and diffusion into the underlying mesenchyme to carry out an epithelial–mesenchymal interaction. Although GEHRIS et al. (1991) detected TGF-β_1 in both mesenchyme and epithelium during palate development, they indicated that their antibody recognized intracellular epithelial TGF-β_1 protein, results consistent with TGF-β_1 acting in a paracrine fashion in an epithelial–mesenchyme interaction.

In contrast, TGF-β_2 mRNA is localized to palatal mesenchyme and not epithelium (PELTON et al. 1990; FITZPATRICK et al. 1990), yet the growth factor was predominant in the epithelium (GEHRIS et al. 1991). These results support the hypothesis that TGF-β_2 also functions in a paracrine fashion; however, it is synthesized in the mesenchyme to exert its effect in the epithelium.

TGF-β_3 mRNA is expressed in the same areas as TGF-β_1, but with greater transcript prevalence, and is expressed earlier in development in the vertical palate epithelium. This expression continues in the horizontal palatal shelf, predominantly in the MEE, and is lost as the epithelial seam disrupts soon after palatal shelf fusion (FITZPATRICK et al. 1990; PELTON et al. 1990; GEHRIS et al. 1994). Thus, TGF-β_3 (and TGF-β_1) may function in the palatal fusion process.

Consistent with the role of the TGF-β family functioning in palatogenesis was the observation that the TGF-β receptor subtypes I, II, and III are present in mouse embryonic mesenchymal cells (LINASK et al. 1991).

Various studies have been carried out with the TGF-β isoforms with palate to determine their putative role in regulating its development. As indicated previously, at the time of palatal shelf elevation and fusion, a period of MEE differentiation, the TGF-β isoforms, particularly TGF-β_3, distribute predominantly to the MEE. It has been shown that TGF-β_1, TGF-β_2 or TGF-β_3 induce precocious differentiation of MEE (DIXON and FERGUSON 1992; GEHRIS and GREENE 1992). During the same developmental period, palate mesenchymal cells actively proliferate and synthesize extracellular matrix components (PRATT et al. 1973). Both of these processes, critical to normal palate development, are also capable of being modulated by the TGF-β growth factors (Table 1). TGF-β_1 has been shown to inhibit proliferation of murine embryonic palate mesenchymal (MEPM) cells in culture (LINASK et al. 1991; SHARPE et al. 1992a). TGF-β_1 stimulated synthesis of fibronectin and collagen (I, III, and V) in cultured palates or MEPM cells (DIXON and FERGUSON 1992; SHARPE et al. 1992a). The effects of TGF-β_1 in increasing collagen levels in MEPM cells were shown to be due to increased levels of mRNA and decreased collagen degradation (D'ANGELO et al. 1994).

IV. Insulin-Like Growth Factors I and II

Insulin-like growth factors (IGF) I and II and their mRNA levels were localized in the developing palate (FERGUSON et al. 1992). IGF II predominated over IGF I (Table 1). IGF II gene expression was developmentally regulated in the palate; it was absent in the vertical palate at days 12–13 of gestation and in the horizontal palate, and present at day 14 in mesenchyme but not epithelium. By contrast, IGF II peptide was predominantly localized to palate epithelium (particularly nasal and medial edge), but also to the mesenchyme at day 14. The distribution of the IGF-binding protein was similar to that of IGF

II peptide. These results suggest a paracrine function for IGF II whereby it is synthesized in the mesenchyme and diffuses into the epithelium to regulate epithelial differentiation, including fusion. This pattern of expression and distribution of peptide of IGF II is similar to that of TGF-β_2, described previously.

V. Acidic and Basic Fibroblast Growth Factor

Localization of acidic and basic fibroblast growth factor (aFGF, bFGF) during palate development was mostly in the basement membrane and epithelium rather than the mesenchyme (SHARPE et al. 1993). Immunochemical staining was predominantly in the nasal and medial aspects of the epithelium and was intense at the time of midline epithelial seam formation (Table 1). Effects of these growth factors on cultured MEPM cells indicated that responses were modulated by the culture substratum. FGF stimulated MEPM cell proliferation on plastic and on collagen, but inhibited it in collagen gels. Production of GAG, particularly HA and dermatan sulfate, was stimulated by FGF. These results have been interpreted as indicating that aFGF and bFGF modulate synthesis of extracellular matrix in the mesenchyme to facilitate seam degeneration of the MEE (SHARPE et al. 1993).

VI. Interactions

These studies suggest very strongly that growth factors play an important role in regulating palatogenesis. However, it is still difficult to understand the precise details of this process due to the number of growth factors and the different isoforms of each growth factor present in the palate. An appreciation of the complexity of this process derives from studies which show interactions of these various growth factors and their individual isoforms. For example, it has been shown that EGF downregulates TGF-β_3 mRNA levels in MEPM cells. Further, intracellular cyclic AMP, retinoids, and both TGF-β_1 and TGF-β_2 increase TGF-β_3 mRNA levels (GEHRIS et al. 1994). Another potential activity of these growth factors to produce interactions during palate development is to alter receptor binding, as was shown with TGF-β, IGF II, and bFGF on EGFR (SHARPE et al. 1992b; BRUNET et al. 1993).

It has been shown that latent, inactive TGF-β contains mannose-6-phosphate (M-6-P) residues and binds to the IGF II/M-6-P receptor. Activation of TGF-β requires the action of plasmin on the latent TGF-β on the IGF II/M-6-P receptor (DENNIS and RIFKIN 1991). Since TGF-β and IGF II co-localize in the midline epithelial seam, mesenchymal signaling of epithelium might be accomplished by passage of TGF-β_2 and IGF II (as well as TGF-α) from mesenchyme to epithelium at specific times of development. Thus these growth factor peptides may interact with cells in discrete regions of palate to regulate expression of each other to alter growth and differentiation.

H. Homeobox Genes

Homeobox genes regulate pattern development and have been shown to regulate the axial skeleton (see Chap. 2), including the limb (see Chap. 3). A number of studies have mapped expression of the murine homeobox genes during embryonic development. Surprisingly, homeobox gene expression has not been found in the developing palate, except for tooth germ development in the adjacent maxilla (see below). Instead it has been observed that homeobox genes are expressed in the developing branchial archs, including the first arch that leads to ultimate development of the secondary palate. Furthermore, gain of function and loss of function mutations in transgenic mice of various homeobox genes involved at the level of the first branchial arch produce cleft palate.

I. Patterns of Expression

A number of homeobox genes appear to be involved in craniofacial development, especially tooth germ formation. *Hox-7.1* is expressed at day 10 of gestation in the neural crest-derived mesenchyme of medial nasal, lateral nasal, maxillary, and mandibular processes. At day 12, expression is restricted to the mesenchyme immediately surrounding the developing tooth germs in the maxillary and mandibular processes (MacKenzie et al. 1991). Another homeobox gene, *Hox-8*, is expressed in the mesenchyme beneath sites of future tooth formation in a proximodistal gradient. It is later expressed in mouse oral epithelium of dental placodes and then in the buccal aspect of the invaginating dental lamina. Subsequently, expression switches from the differentiating mesenchymal ameloblasts to the differentiating epithelial odontoblasts. *Hox-7*, another homeobox gene, is expressed in the mesenchyme of the dental papilla and follicle at all stages. It has been suggested that the reciprocity of expression represents an interactive role between *Hox-7, Hox-8*, and other genes in regulating epithelial–mesenchymal interactions during dental differentiation (MacKenzie et al. 1992). *D1x-2* is expressed in the mesenchyme of the maxillary and mandibular branches of the first branchial arch and later in development extends into the ectoderm of the maxillary process and the olfactory pit and other aspects of the face and neck (Bulfone et al. 1993). In contrast, the *D1x-1* gene is expressed in the same tissues, although its onset of expression is about 2 days after *D1x-2* expression (Dollé et al. 1992). *T1x-1* is also expressed in the branchial arches. After expression in the first arch, expression is subsequently observed in the second, third, and fourth arches and then in developing tooth germ and in specific cell populations of the mandible (Raju et al. 1993). Employing in vitro recombination experiments, *Msx-1* and *Msx-2*, which are expressed in craniofacial development, were shown to undergo mesenchymal–epithelial interactions in the developing tooth (Jowett et al. 1993).

II. Homeobox Mutations and Cleft Palate

In order to determine the function of isolated homeobox genes during murine embryonic development, gain of function and loss of function mutations have been produced in transgenic mice. Various birth defects have been observed, affirming the role of homeobox genes in directing pattern formation. However, the number of homeotic transformations observed has been rather low, which has supported the notion that many of these homeobox genes function in a redundant manner. Thus if one homeobox gene is knocked out, its function will be replaced by *Hox* gene paralogues.

Among the malformations observed with ectopic expression and knockout of transgenes is cleft palate associated with other birth defects (Table 3). When a loss of mutation for the *Hoxa2* gene was created, homeotic transformations of skeletal elements were generated. *Hoxa2* is apparently the only *Hox* gene expressed at the level of rhombomere 2 (Wilkinson 1993); thus, no *Hox* gene paralogues exert redundant functions. In addition, mice were born with cleft palate. However, 18% of mutants lacked cleft palate, suggesting that the defect is not primary, but rather due to failure of palatal shelf reorientation. It was proposed that a mechanical stress to block reorientation was induced by a more caudal skeletal defect (Rijli et al. 1993).

The *Msx-1* homeobox gene is expressed at diverse sites of epithelial–mesenchymal interaction and has been implicated in signaling processes between tissue layers. Satokata and Maas (1994) reported that *Msx-1* deficient mice exhibit abnormal dental development, which would be expected, since *Msx-1* is expressed in tooth development. However, cleft palate is also associated with the mutation, which is surprising, since the gene is not expressed during palate development. However, in the *Msx-1* mutant, palatal shelves are developmentally delayed, although they do elevate. It is possible that the

Table 3. Homeobox mutations in transgenic mice lead to cleft palate

Type of mutation	Birth defects	References
Gain of function		
Hoxa7	Cleft palate, axial skeletal, open eye, nonfused pinnae	Balling et al. (1989)
Hoxb6	Cleft palate, open eye, micronathia, microtia, skeletal	Kaur et al. (1992)
Hoxb7	Cleft palate, open eye, VSD, skeletal	McLain et al. (1992)
Loss of function		
Hoxa2	Cleft palate, duplications of skeletal ear	Gendron-Maguire et al. (1993) Rijli et al. (1993)
Msx-1	Cleft palate, abnormal dental development	Satokata and Maas (1994)

VSD, ventrical septal defect.

dental mesenchyme cells contribute to palate formation or that disruption of dental development causes geometric alterations in the jaw relationships, resulting in cleft palate as a secondary consequence (FERGUSON 1994).

III. Signaling Relationships

It is not clear which signaling molecules direct homeobox gene expression during palatogenesis. Sonic hedgehog, which signals homeobox genes in various embryonic systems, has not been reported to be expressed in the developing palate. It is of interest that the peptide growth factors TGF-β and IGF II are expressed in the mesenchymal cells underlying the developing tooth germ epithelium, that 5-HT is actively taken up in this region, and that many of these homeobox genes are similarly expressed in tooth germ development (*Hox-7, Hox-7.1, Hox-8, D1x-2, Msx-2*). Thus these growth factors might play a role in regulating homeobox gene expression in the epithelial–mesenchymal interaction underlying dental development.

I. Association of Cleft Palate in Humans with Candidate Genes

As indicated previously, studies have been pursued in families with cleft palate to determine genes which are mutated that are responsible for this birth defect. Families with genetically linked cleft palate have been analyzed for associations with a number of candidate genes likely to be involved in palate formation, which have included homeobox, extracellular matrix, and growth factor genes. Most studies are carried out on tissue DNA disgested with restriction enzymes and subjected to restriction fragment length polymorphisms (RFLP) analysis with probes of various genes. Table 2 indicates the associations of cleft palate to genetic defects that have been observed.

References

Aarskog D (1975) Association between maternal intake of diazepam and oral clefts. Lancet 2: 219

Abbott BD, Buckalew AR (1992) Embryonic palatal responses to teratogens in serum-free organ culture. Teratology 45: 369–382

Abbott BD, Adamson ED, Pratt RM (1988) Retinoic acid alters EGF receptor expression during palatogenesis. Development 102: 853–867

Abbott BD, Perdew GH, Birnbaum LS (1994a) Ah receptor in embryonic mouse palate and effects of TCDD on receptor expression. Toxicol Appl Pharmacol 126: 16–25

Abbott BD, McNabb FMA, Lau C (1994b) Glucocorticoid receptor expression during the development of the embryonic mouse secondary palate. J Craniofac Genet Dev Biol 14: 87–96

Ardinger HH, Buetow LH, Bell GI, Bardach J, VanDemark DR, Murray JC (1989) Associations of genetic variation of the transforming growth factor-alpha gene with cleft lip and palate. Am J Hum Genet 45: 348–353

Babiarz BS, Wee EL, Zimmerman EF (1979) Palate morphogenesis. III. Changes in cell shape and orientation during shelf elevation. Teratology 20: 249–278

Balling R, Mutter G, Grus P, Kessel M (1989) Craniofacial abnormalities induced by ectopic expression of the homeobox I gene Hox 1.1 in transgenic mice. Cell 58: 337–347

Baird PA, Sadovnick AD, Yee IML (1994) Maternal age and oral cleft malformations: data from a population-based series of 576,815 consecutive livebirths. Teratology 49: 448–451

Brinkley LL (1984) Changes in cell distribution during mouse secondary palate closure in vivo and in vitro. I. Epithelial cells. Dev Biol 102: 216–227

Brinkley LL, Bookstein FL (1986) Cell distribution during mouse secondary palate closure. II. Mesenchymal cells. J Embryol Exp Morph 96: 111–130

Brinkley LL, Morris-Wiman J (1987a) Computer-assisted analysis of hyaluronate distribution during morphogenesis of the mouse secondary palate. Development 100: 629–636

Brinkley LL, Morris-Wiman J (1987b) Effects of chlorcyclizine-induced glycosaminoglycan alterations on patterns of hyaluronate distribution during morphogenesis of the mouse secondary palate. Development 100: 637–640

Brunet CL, Sharpe PM, Ferguson MWJ (1993) The distribution of epidermal growth factor binding sites in the developing mouse palate. Int J Dev Biol 37: 451–458

Bulfone A, Kim H-J, Puelles L, Porteus MH, Grippo JF, Rubenstein JLR (1993) The mouse *Dlx-2 (Tes-1)* gene is expressed in spatially restricted domains of the forebrain, face and limbs in midgestation mouse embryos. Mech Dev 40: 129–140

Bulleit RF, Zimmerman EF (1985) The influence of the epithelium on palate shelf reorientation. J Embryol Exp Morph 88: 265–279

Bulleit RF, Zimmerman EF (1986) The effect of reducing ATP levels on reorientation of the secondary palate. J Embryol Exp Morph 93: 73–84

Carette MJM, Lane EB, Ferguson MWJ (1991) Differentiation of mouse embryonic palatal epithelial sheets in culture: selective cytokeratin expression distinguishes between oral, medial edge and nasal epithelial cells. Differentiation 47: 149–161

Carette MJM, Ferguson MWJ (1992) The fate of medial edge epithelial cells during palatal fusion in vitro: an analysis by Dil labelling and confocal microscopy. Development 114: 379–388

Chenevix-Trench G, Jones K, Green AC, Duffy DL, Martin NG (1992) Cleft lip with or without cleft palate: associations with transforming growth factor alpha and retinoic acid receptor loci. Am J Hum Genet 51: 1377–1385

Comper WE, Laurent TC (1978) Physiological function of connective tissue polysaccharides. Physiol Rev 58: 255–315

Culiat CT, Stubbs L, Nicholls RD, Montgomery CS, Russell LB, Johnson DK, Rinchik EM (1993) Concordance between isolated cleft palate in mice and alterations within a region including the gene encoding the β3 subunit of the type A γ-aminobutyric acid receptor. Proc Natl Acad Sci USA 90: 5105–5109

D'Angelo M, Chen J-M, Ugen K, Greene RM (1994) TGFβ1 regulation of collagen metabolism by embryonic palate mesenchymal cells. J Exp Zool 270: 189–201

Dennis PA, Rifkin DB (1991) Cellular activation of latent transforming growth factor β requires binding to the cation-independent mannose-6-phosphate/insulin-like growth factor type II receptor. Proc Natl Acad Sci USA 88: 580–584

Dixon MJ, Ferguson MWJ (1992) The effects of epidermal growth factor, transforming growth factors alpha and beta and platelet-derived growth factor on murine palatal shelves in organ culture. Arch Oral Biol 37: 395–410

Dixon MJ, Garner J, Ferguson MWJ (1991) Immunolocalisation of epidermal growth factor (EGF), EGF receptor and transforming growth factor alpha (TGFα) during murine palatogenesis in vivo and in vitro. Anat Embryol (Berl) 184: 83–91

Dixon MJ, Carette MJM, Moser BB, Ferguson MWJ (1993) Differentiation of isolated murine embryonic palatal epithelium in culture: exogenous transforming growth factor alpha modulates matrix biosynthesis in defined experimental conditions. In Vitro Cell Dev Biol Anim 29A: 51–61

Dollé P, Price M, Duboule D (1992) Expression of the murine D1x-1 homeobox gene during facial, ocular and limb development. Differentiation 49: 93–99

Durand B, Saunders M, Leroy P, Leid M, Chambon P (1992) All-trans and 9-cis retinoic acid induction of CRABPII transcription is mediated by RAR-RXR heterodimers bound to DR1 and DR2 repeated motifs. Cell 71: 73–85

Edwards DR, Murphy G, Reynolds JJ, Whitham SE, Docherty AJ, Angel P, Heath JK (1987) Transforming growth factor beta modulates the expression of collagenase and metalloproteinase inhibitor. EMBO J 6: 1899–1904

Ferguson MWJ (1978) Palatal shelf elevation in the Wistar rat fetus. J Anat 125: 555-577

Ferguson MWJ (1988) Palate development. Development 103: 41–60

Ferguson MWJ (1994) Craniofacial malformations: towards a molecular understanding. Nat Genet 6: 329–330

Ferguson MWJ, Honig LS (1984) Epithelial-mesenchymal interactions during vertebrate palatogenesis. In: Zimmerman EF (ed) Palate development: normal and abnormal cellular and molecular aspects. Academic, New York, pp 137–164

Ferguson MWJ, Sharpe PM, Thomas BL, Beck F (1992) Differential expression of insulin-like growth factors I and II (IGF I and II), mRNA, peptide and binding protein 1 during mouse palate development: comparison with TGFβ peptide distribution. J Anat 181: 219–238

Fitchett JE, Hay ED (1989) Medial edge epithelium transforms to mesenchyme after embryonic palatal shelves fuse. Dev Biol 131: 455–474

Fitzpatrick DR, Denhez F, Kondaiah P, Akhurst RJ (1990) Differential expression of TGF beta isoforms in murine palatogenesis. Development 109: 585–595

Foreman DM, Sharpe PM, Ferguson MWJ (1991) Comparative biochemistry of mouse and chick secondary-palate development in vivo and in vitro with particular emphasis on extracellular matrix molecules and the effects of growth factors on their synthesis. Arch Oral Biol 36: 457–471

Garbarino MP, Greene RM (1984) Identification of adenylate cyclase-coupled β-adrenergic receptors in the developing mammalian palate. Biochem Biophys Res Comm 119: 193–202

Gehris AL, Greene RM (1992) Regulation of murine embryonic epithelial cell differentiation by transforming growth factors β. Differentiation 49: 167–173

Gehris AL, D'Angelo M, Greene RM (1991) Immunodetection of the transforming growth factors β1 and β2 in the developing murine palate. Int J Dev Biol 35: 17–24

Gehris AL, Pisano MM, Nugent P, Greene RM (1994) Regulation of TGFβ3 gene expression in embryonic palatal tissue. In Vitro Cell Dev Biol 30A: 671–679

Gendron-Maguire M, Mallo M, Zhang M, Gridley T (1993) Hoxa-2 mutant mice exhibit homeotic transformation of skeletal elements derived from cranial neural crest. Cell 75: 1317–1331

Goldman AS (1984) Biochemical mechanism of glucocorticoid- and phenytoin-induced cleft palate. In: Zimmerman EF (ed) Palate development: normal and abnormal cellular and molecular aspects. Academic, New York, pp 217–239

Goldman AS, Katsumata M, Yaffe SJ, Gasser DL (1977) Palatal cytosol cortisol-binding protein associated with cleft palate susceptibility and H-2 genotype. Nature 265: 643–644

Goldman AS, Herold R, Piddington R (1981) Inhibition of programmed cell death in the fetal palate by cortisol. Proc Soc Exp Biol Med 166: 418–424

Greene RM, Pratt R (1979) Developmental aspects of secondary palate formation. J Embryol Exp Morph 36: 225–245

Greene R, Pratt R (1979) Correlation between cAMP levels and cytochemical localization of adenylate cyclase during development of the secondary palate. J Histochem Cytochem 27: 924–931

Griffith CM, Hay ED (1992) Epithelial-mesenchymal transformation during palatal fusion: carboxyfluorescein traces cells at light and electron microscopic levels. Development 116: 1087–1099

Guram MS, Gill TS, Geber WF (1982) Comparative teratogenicity of chlordiazepoxide, amitriptyline, and a combination of the two compounds in the fetal hamster. Neurotoxicology 3: 83–90

Hassell JR (1975) The development of rat palatal shelves in vitro. An ultrastructural analysis of the inhibition of epithelial cell death and palate fusion by the epidermal growth factor. Dev Biol 45: 90–102

Heine UI, Munoz EF, Flanders KC, Ellingsworth LR, Lam Y-YP, Thompson NL, Roberts AB, Sporn MB (1987) Role of transforming growth factor-β in the development of the mouse embryo. J Cell Biol 105: 2861–2876

Ignotz RA, Endo T, Massague J (1987) Regulation of fibronectin and type I collagen mRNA levels by transforming growth factor-β. J Biol Chem 262: 6443–6446

Jowett AK, Vainio S, Ferguson MWJ, Sharpe PT, Thesleff I (1993) Epithelial-mesenchymal interactions are required for msx 1 and msx 2 gene expression in the developing murine molar tooth. Development 117: 464–470

Jurand A (1980) Malformations of the central nervous system induced by neurotropic drugs in mouse embryos. Dev Growth Differ 22: 61–78

Kaur S, Singh G, Stock JL, Schreiner CM, Kier AB, Yager KL, Mucenski ML, Scott WJ, Potter SS (1992) Dominant mutation of the murine Hox 2.2 gene results in developmental abnormalities. J Exp Zool 264: 323–326

Kellogg CD, Tervo D, Ison J, Parisi T, Miller RK (1980) Prenatal exposure to diazepam alters behavioral development in rats. Science 207: 205–207

Knudsen TB, Bulleit RF, Zimmerman EF (1985) Histochemical localization of glycosaminoglycans during morphogenesis of the secondary palate in mice. Anat Embryol (Berl) 173: 137–142

Kuhn EM, Babiarz BS, Lessard JL, Zimmerman EF (1980) Palate morphogenesis. I. Immunological and ultrastructural analyses of mouse palate. Teratology 21: 209–223

Laiho M, Saksela O, Keski-Oja J (1987) Transforming growth factor-β induction of type-I plasminogen activator inhibitor. J Biol Chem 262: 17467–17474

Lau EC, Li ZQ, Santon V, Slavkin HC (1993) Messenger RNA phenotyping for semiquantitative comparison of glucocorticoid receptor transcript levels in the developing embryonic mouse palate. J Steroid Biochem Mol Biol 46: 751–758

Lauder JM (1993) Neurotransmitters as growth regulatory signals: role of receptors and second messengers. Trends Neurosci 16: 233–240

Lauder JM, Zimmerman EF (1988) Sites of serotonin uptake in epithelia of the developing mouse palate, oral cavity, and face: possible role in morphogenesis. J Craniofac Genet Dev Biol 8: 265–276

Lehnert SA, Akhurst RJ (1988) Embryonic expression pattern of TGF beta type-1 RNA suggests both paracrine and autocrine mechanisms of action. Development 104: 263–273

Linask KK, D'Angelo M, Gehris AL, Greene RM (1991) Transforming growth factor-β receptor profiles of human and murine embryonic palate mesenchymal cells. Exp Cell Res 192: 1–9

Luke DA (1984) Epithelial proliferation and development of rugae in relation to palatal shelf elevation in the mouse. J Anat 138: 251–258

Mackenzie A, Leeming GL, Jowett AK, Ferguson MWJ, Sharpe PT (1991) The homeobox gene Hox 7.1 has specific regional and temporal expression patterns during early murine craniofacial embryogenesis, especially tooth development in vivo and in vitro. Development 111: 269–285

Mackenzie A, Ferguson MWJ, Sharpe PT (1992) Expression patterns of the homeobox gene, Hox-8, in the mouse embryo suggest a role in specifying tooth initiation and shape. Development 115: 403–420

Massague J (1987) The TGFβ family of growth and differentiation factors. Cell 49: 437–438

McLain K, Schreiner C, Yager KL, Stock JL, Potter SS (1992) Ectopic expression of Hox 2.3 induces craniofacial and skeletal malformations in transgenic mice. Mech Dev 39: 3–16

Miller RP, Becker BA (1975) Teratogenicity of oral diazepam and diphenylhydantoin in mice. Toxicol Appl Pharmacol 32: 53–61

Morris-Wiman J, Brinkley L (1992) An extracellular matrix infrastructure provides support for murine secondary palatal shelf remodelling. Anat Rec 234: 575–586

Morris-Wiman J, Brinkley L (1993) Rapid changes in the extracellular matrix accompany in vitro palatal shelf remodelling. Anat Embryol (Berl) 188: 75–85

Murray JC, Nishimura DY, Buetow KH, Ardinger HH, Spence MA, Sparkes RS, Falk RE, Falk PM, Gardner RJM, Harkness EM, Glinski LP, Pauli RM, Nakamura Y, Green PP, Schinzel A (1990) Linkage of an autosomal dominant clefting syndrome (Van der Woude) to loci on chromosome Iq. Am J Hum Genet 46: 486–491

Nexϕ E, Hollenberg MD, Figueroa A, Pratt RM (1980) Detection of epidermal growth factor-urogastrone and its receptor during fetal mouse development. Proc Natl Acad Sci USA 77: 2782–2785

Norman EJ, Wee EL, Berry HK, Zimmerman EF (1985) Rapid gas chromatographic-mass spectrometric quantitation of γ-aminobutyric acid in biological specimens. J Chromatogr 337: 21–27

Pelton RW, Hogan BLM, Miller DA, Moses HL (1990) Differential expression of genes encoding TGFs β1, β2, and β3 during murine palate formation. Dev Biol 141: 456–460

Penttinen RP, Kobayashi S, Bornstein P (1988) Transforming growth factor-β increases mRNA for matrix proteins both in the presence and in the absence of changes in mRNA stability. Proc Natl Acad Sci USA 85: 1105–1108

Pisano MM, Greene RM (1985) Quantitation of catecholamines in embryonic orofacial tissue. IRCS Med Sci 13: 900–901

Pisano MM, Greene RM (1986) Hormone and growth factor involvement in craniofacial development. IRCS Med Sci 14: 635–640

Pisano MM, Schneiderman MH, Greene RM (1986) Catecholamine modulation of embryonic palate mesenchymal cell DNA synthesis. J Cell Physiol 126: 84–92

Pratt RM, Martin GR (1975) Epithelial cell death and cyclic AMP increase during palatal development. Proc Natl Acad Sci USA 96: 874–877

Pratt RM, Goggins JF, Wilk AL, King CTG (1973) Acid mucopolysaccharide synthesis in the secondary palate of the developing rat at the time of rotation and fusion. Dev Biol 32: 230–237

Pratt RM, Kim CS, Grove RI (1984) Role of glucocorticoids and epidermal growth factor in normal and abnormal palatal development. In: Zimmerman EF (ed) Palate development: normal and abnormal cellular and molecular aspects. Academic, New York, pp 81–101

Raghow R, Postlethwaite AE, Keski-Oja J, Moses HL, Kang AH (1987) Transforming growth factor-β increases steady state levels of type I procollagen and fibronectin messenger RNAs posttranscriptionally in cultured human dermal fibroblasts. J Clin Invest 79: 1285–1288

Raju K, Tank S, Dubé ID, Kamel-Reid S, Bryce DM, Breitman ML (1993) Characterization and developmental expression of T1x-1, the murine homolog of HOX11. Mech Dev 44: 51–64

Rijli FM, Mark M, Lakkaraju S, Dierich A, Dollé P, Chambon P (1993) A homeotic transformation is generated in the rostral branchial region of the head by disruption of Hoxa-2, which acts as a selector gene. Cell 75: 1333–1349

Rogers JM, Barbee BD, Rehnberg BF (1993) Critical periods of sensitivity for the developmental toxicity of inhaled methanol. Teratology 47: 395
Rosenberg L, Mitchell AA, Parsells JL, Pashayan H, Louik C, Shapiro S (1983) Lack of relation of oral clefts to diazepam use during pregnancy. N Engl J Med 309: 1282–1285
Rossi P, Karsenty G, Roberts AB, Roche NS, Sporn MB, DeCrombrugghe B (1988) A nuclear factor I binding site mediates the transcriptional activation of a type I collagen promoter by transforming growth factor-β. Cell 52: 405–414
Safra MJ, Oakley GP (1976) Valium: an oral cleft teratogen? Cleft Palate. Craniofac J 13: 198–200
Saksela O, Moscatelli D, Rifkin DB (1987) The opposing effects of basic fibroblast growth factor and transforming growth factor beta on the regulation of plasminogen activator activity in capillary endothelial cells. J Cell Biol 105: 957–963
Salomon DS, Pratt RM (1976) Glucocorticoid receptors in murine embryonic facial mesenchyme cells. Nature 264: 174–177
Satokata I, Maas R (1994) Msx1 deficient mice exhibit cleft palate and abnormalities of craniofacial and tooth development. Nat Genet 6: 348–356
Saxén I (1975) Associations between oral clefts and drugs taken during pregnancy. Int J Epidemiol 4: 37–44
Saxén I, Saxén L (1975) Association between maternal intake of diazepam and oral clefts. Lancet 2: 498
Schardein JL (1993) Chemically induced birth defects. Dekker, New York
Schreiner CM, Zimmerman EF, Wee EL, Scott WJ (1986) Caffeine effects on cyclic AMP levels in the mouse embryonic limb and palate in vitro. Teratology 34: 21–27
Shah RM, Donaldson D, Burdett D (1979) Teratogenic effects of diazepam in the hamster. Can J Physiol Pharmacol 57: 556–561
Shapiro BL, Sweney L (1969) Electron microscopic and histochemical examination of oral epithelial-mesenchymal interaction (programmed cell death). J Dent Res 48: 652–660
Sharpe PM, Foreman DM, Carette MJM, Schor SL, Ferguson MWJ (1992a) The effects of transforming growth factor-β1 on protein production by mouse embryonic palate mesenchymal cells in the presence or absence of serum. Arch Oral Biol 37: 39–48
Sharpe PM, Brunet CL, Ferguson MWJ (1992b) Modulation of the epidermal growth factor receptor of mouse embryonic palatal mesenchyme cells in vitro by growth factors. Int J Dev Biol 36: 275–282
Sharpe PM, Brunet CL, Foreman DM, Ferguson MWJ (1993) Localisation of acidic and basic fibroblast growth factors during mouse palate development and their effects on mouse palate mesenchyme cells in vitro. Roux Arch Dev Biol 202: 132–143
Shenefelt RE (1972) Morphogenesis of malformations in hamsters caused by retinoic acid: relation to dose and stage at treatment. Teratology 5: 103–118
Shiono PH, Mills JL (1984) Oral clefts and diazepam use during pregnancy. N Engl J Med 311: 919–920
Shiota K, Fujita S, Akiyama T, Mori C (1990) Expression of the epidermal growth factor receptor in developing fetal mouse palates: an immunohistochemical study. Am J Anat 188: 401–408
Shuey DL, Yavarone M, Sadler TW, Lauder JM (1990) Serotonin and morphogenesis in the cultured mouse embryo. In: Lauder JM (ed) Molecular aspects of development and aging of the nervous system. Plenum, New York, pp 205–215
Shuey DL, Sadler TW, Lauder JM (1992) Serotonin as a regulator of craniofacial morphogenesis: site specific malformations following exposure to serotonin uptake inhibitors. Teratology 46: 367–378
Shuey DL, Sadler TW, Tamir H, Lauder JM (1993) Serotonin and morphogenesis. Anat Embryol (Berl) 187: 75–85

Shuler CF, Halpern DE, Guo Y, Sank AC (1992) Medial edge epithelium fate traced by cell lineage analysis during epithelial-mesenchymal transformation in vivo. Dev Biol 154: 318–330

Smiley GR (1972) An in vitro and in vivo study of single palatal processes. Anat Rec 173: 405–416

Sporn MB, Roberts AB, Wakefield LM, de Crombrugghe B (1987) Some recent advances in the chemistry and biology of transforming growth factor-beta. J Cell Biol 105: 1039–1045

Tocco DR, Renskers K, Zimmerman EF (1987) Diazepam-induced cleft palate and lack of correlation with the H-2 locus. Teratology 35: 439–445

Toole BP (1981) Glycosaminoglycans in morphogenesis. In: Hay E (ed) Cell biology of the extracellular matrix. Plenum, New York, pp 259–294

Turley EA, Hollenberg MD, Pratt RM (1985) Effects of epidermal growth factor/urogastrone on glycosaminoglycan synthesis and accumulation in vitro in the developing mouse palate. Differentiation 28: 279–285

Varga J, Rosenbloom J, Jimenez SA (1987) Transforming growth factor-β (TGF-β) causes a persistent increase in steady-state amounts of type I and type III collagen and fibronectin mRNAs in normal human dermal fibroblasts. Biochem J 247: 597–604

Venkatasubramanian K, Zimmerman EF (1983) Palate cell motility and substrate interaction. J Craniofac Genet Dev Biol 3: 143–157

Walker BE, Fraser FC (1956) Closure of the secondary palate in three strains of mice. J Embryol Exp Morph 4: 176–189

Waterman RE, Meller SM (1974) Alterations in the epithelial surfaces of human palatal shelves prior to and during fusion: a scanning electron microscopic study. Anat Rec 180: 111–136

Waterman RE, Ross LM, Meller SM (1973) Alterations in the epithelial surface of A/jax mouse palatal shelves prior to and during fusion: a scanning electron microscopic study. Anat Rec 176: 361–376

Waterman RE, Palmer GC, Palmer SJ, Palmer SM (1976) Catecholamine-sensitive adenylate cyclase in the developing golden hamster palate. Anat Rec 185: 125–138

Wee EL, Zimmerman EF (1980) Palate morphogenesis: II. Contraction of cytoplasmic processes in ATP-induced palate rotation in glycerinated mouse heads. Teratology 21: 15–27

Wee EL, Zimmerman EF (1983) Involvement of GABA in palate morphogenesis and its relation to diazepam teratogenesis in two mouse strains. Teratology 28: 15–22

Wee EL, Zimmerman EF (1985) GABA uptake in embryonic palate mesenchymal cells of two mouse strains. Neurochem Res 10: 1673–1688

Wee EL, Wolfson LG, Zimmerman EF (1976) Palate shelf movement in mouse embryo culture: evidence for skeletal and smooth muscle contractility. Dev Biol 48: 91–103

Wee EL, Babiarz BS, Zimmerman S, Zimmerman EF (1979) Palate morphogenesis. IV. Effects of serotonin and its antagonists on rotation in embryo culture. J Embryol Exp Morph 53: 75–90

Wee EL, Norman EJ, Zimmerman EF (1986) Presence of γ-aminobutyric acid in embryonic palates of AJ and SWV mouse strains. J Craniofac Genet Dev Biol 6: 53–61

Werb Z (1990) Expression of EGF and TGF-α genes in early mammalian development. Mol Reprod Dev 27: 10–15

Wilkinson DG (1993) Molecular mechanisms of segmental patterning in the vertebrate hindbrain and neural crest. Bioessays 15: 499–505

Yoneda T, Pratt RM (1981) Mesenchyme cells from the human embryonic palate are highly responsive to epidermal growth factor. Science 213: 563–565

Zimmerman EF (ed) (1984) Palate development: normal and abnormal cellular and molecular aspects. Academic, New York

Zimmerman EF, Wee EL (1984) Role of neurotransmitters in palate development. In: Zimmerman EF (ed) Palate development: normal and abnormal cellular and molecular aspects. Academic, New York, pp 37–63

Zimmerman EF, Wee EL, Phillips N, Roberts N (1981) Presence of serotonin in the palate just prior to shelf elevation. J Embryol Exp Morph 64: 233–250

Zimmerman EF, Clark RL, Ganguli S, Venkatasubramanian K (1983) Serotonin regulation of palatal cell motility and metabolism. J Craniofac Genet Dev Biol 3: 371–385

Section II
Common Biochemical, Metabolic, and Physiological Mechanisms of Abnormal Development

CHAPTER 8
Cell Death

T.B. KNUDSEN

A. Introduction

One of the major challenges facing teratologists is the elucidation of mechanisms by which birth defects occur. This requires an understanding of the primary targets of physiological and environmental agents that affect the health and well-being of the conceptus, the cellular events of pathogenesis, and the genes which place the embryo at risk. Cell death is gene-directed event in the developing embryo. Its' broad phylogenetic representation is consistent with a selective advantage in development for the involution of vestigial structures which have importance in phylogenetic but not ontogenetic development, reduction of superfluous cells, disposal of cells that have already completed their functions, elimination of cells which differentiate inappropriately, and suppression of potentially harmful or abnormal cells (SAUNDERS 1966; ELLIS et al. 1991).

Many drugs and chemicals induce excessive cell death in the developing embryo and also cause structural malformations in experimental animals; furthermore, the vulnerable territories of the embryo often embrace naturally occurring cell deaths at the time of exposure, and so the question arises as to whether induced cell death is a primary lesion in dysmorphogenesis or whether it merely reveals natural stress points in the embryo. With the recognition that most of these deaths represent the end stage of a regulated process known as apoptosis, the genetic circuitry of the cell death program provides an opportunity to explore functional relationships in issues such as why certain embryonic cells die in response to xenobiotic and metabolic toxins while others survive, which genes predispose the embryo to developmental abnormalities related to cell death, and how human teratogens interact with these genes. This chapter will review the historical concepts and recent advances in cell death research, giving special emphasis to mechanisms of drug toxicity in embryonic development.

B. Embryonic Cell Death

Glücksmann gave early recognition to the importance of cell death during morphogenesis and development. He wrote that cell degenerations can arise from the combination of "imprecise induction and proliferation stimuli with a limited adaptability of induced embryonic cells" (GLÜCKSMANN 1965). The basic tenet of this hypothesis is that morphogenetic cell death occurs not as a stochastic event, but as a purposeful cellular process whose incipience is integrated with the regulation of cell growth, migration, and differentiation. Only recently have the common signaling pathways become apparent. Drugs could directly interact with the pathways that regulate or implement cell death, or they might act indirectly by uncoupling the integration of cell death with other morphogenetic processes, to initiate an array of cell death lesions which may be considered "homotopic" or "heterotopic" modifications of the normal (orthotopic) pattern of cell death in the developing embryo.

I. Orthotopic Pattern

One of the earliest studies to demonstrate the embryonic control of cell death was published by SAUNDERS and colleagues (1962). These investigators examined the focal appearance of cell death in the posterior necrotic zone (PNZ) of the chick wing bud. Prospective PNZ tissue died on schedule even when explanted or transplanted hours or days earlier, suggesting that specific cells were programmed for deletion. The hypothetical "death sentence" could, however, be revoked if the graft was transplanted to a remote site of the wing bud, implicating local control by cell–cell signaling. An important concept shaped by these studies is thus that the pattern of cell death is programmed as a normal morphogenetic event in the embryo and that the mechanism involves an intrinsic sequence of steps initiated by an external stimulus to culminate, hours to days later, in the collapse of specific cells. This raises questions about how the cell death program is implemented and about the nature of the physiological signals which turn the process on and off in a dynamic, developing system.

One issue is whether programmed cell death is autonomous or whether it is facilitated by neighboring cells (à la Dr. Kevorkian). Here the nematode *Caenorhabditis elegans* has been a useful model because its amenability to cell lineage analysis. The ontogeny of cell death can be observed in developing nematodes with single-cell resolution. Of the 1090 somatic cells generated during development of an adult hermaphrodite nematode, 131 undergo programmed cell death (SULSTON and HORVITZ 1977; SULSTON et al. 1983). These deaths are rigidly controlled such that the same cells die in every animal and each at its own characteristic time. Random mutagenesis has yielded nematode strains which are defective in 14 genes required for signaling or execution of programmed cell death. Two genes, *ced-3* and *ced-4* (cell death abnormal), are necessary for programmed death. Loss of either function blocks the initiation of almost all of the 131 programmed cell deaths (ELLIS and HORVITZ 1986). A third gene, *ced-9*, is a negative regulator of cell death. When present in a mutated gain of function form, *ced-9* initiates excessive cell death in developing nematodes (HENGARTNER et al. 1992). An immediate question thus arises as to whether the normal products of these genes, *ced-3* and *ced-4* in particular, act in a cell-autonomous manner. In other words, do they function as mediators of cell death within prospective dying cells or do they send a Kevorkian signal to target cells? Genetic mosaic experiments using mutant and wild-type chimeras showed that the *ced-3* and *ced-4* gene products must function within dying cells or their close ancestors for competence to programmed cell death (YUAN and HORVITZ 1990). For example, in chimeric animals in which some cells contain a functional *ced-3* gene and others do not, programmed cell deaths proceed normally only in those cells in which *ced-3* is active. Since *ced-3* and *ced-4* are expressed during stages of nematode development when most of the programmed cell deaths occur, they are excellent candidate genes for primary cell death regulators (YUAN and HORVITZ 1992; YUAN et al. 1993). Cloning and sequencing of *ced-3* showed homology to the mammalian protease interleukin(IL)-1β-converting enzyme (ICE). This was an important breakthrough in cell death research and will be discussed later in the chapter.

II. Homotopic Patterns

In vertebrate embryos, dysregulation of the onset and expanse of programmed cell death zones is a very consistent feature of teratogenic treatment (MENKES et al. 1964; MENKES et al. 1970; MILAIRE and ROOZE 1983; SULIK et al. 1988). Menkes and coworkers observed that mechanistically diverse teratogenic agents may expand orthotopic cell death zones and that regions of cell death which were homologous in different species showed "phase specificity" by mapping to sites most vulnerable to induced abnormal development (MENKES et al. 1964, 1965). An example is the foyer preaxial primaire (Fpp) of the anterior mesoderm of the limb bud. Like the anterior necrotic zone in chick limbs, the fpp in rodents may play an important role in reduction of superfluous cells for the predigital limb field (MILAIRE 1963; ROOZE 1977). Anterior digital defects such as preaxial polydactylism, hyperphalangism of the first digit, ectrodactylism, and oligosyndactylism are foreshadowed by ectopic cell deaths in and around the fpp. Enlargement of the Fpp occurs in digital reduction deformities related to the *Oligosyndactylous* mutation (MILAIRE 1967) and ethylene glycol monomethyl ether (GREENE et al. 1987). Reduction of the Fpp has been observed in supernumerary digital deformities related to the *Hemimelia-extra toe* mutation (KNUDSEN and KOCHHAR 1981) and 5-bromo-2′-deoxyuridine (WISE and SCOTT 1982). Therefore, programmed cell death in the Fpp provides what might be considered a teratological scaffolding for diverse drugs, chemicals, and mutations which potentially disrupt development.

Sulik and coworkers proposed that the homotopic modification of cell death can explain complex teratological syndromes where the fetal malformations appear unrelated in terms of tissue type or location (SULIK et al. 1988). The evidence derives from comparative analysis of teratogenic treatment with ethanol, retinoic acid, radiation, antineoplastics, and hyperthermia. For example, retinoic acid induces cell death which has been regarded as phase specific during early pathogenesis of the cranial neural folds (SULIK et al. 1988), limb buds (KOCHHAR 1977; ALLES and SULIK 1989), caudal mesoderm (ALLES and SULIK 1990), and facial mesenchyme (OSUMI-YAMASHITA et al. 1990; MOTOYAMA et al. 1994). Limb reduction defects associated with retinoid excess have been correlated with precocious firing of physiological cell death in the apical ectodermal ridge (AER). These deaths could prematurely terminate the inductive capacity of the ridge ectoderm and consequently affect limb outgrowth (SULIK and DEHART 1988). Although it is not known whether retinoic acid initiates AER cell death by stimulating the cell death program directly or indirectly, whereby these cells are induced to a state of vulnerability to other physiological signals which then trigger cell death, an obvious question is which specific nuclear retinoic acid receptors (RAR, RXR) are involved. Several studies have examined the topographical relationship between the expression of these receptors and retinoid-induced cell death in susceptible and resistant territories of the mouse embryo. Teratogenic doses of retinoic acid clearly upregulate RAR-β2 expression in territories which exhibited induced cell death and were destined for malformation (SOPRANO et al. 1994; JIANG and KOCHHAR 1992; OSUMI-YAMASHITA et al. 1992; MOTOYAMA et al. 1994). While these observations may be interpreted as indicating that RAR-β2 are linked with physiological cell death, RAR-β2 expression is not limited to regions of programmed cell death (MENDELSOHN et al. 1991). Therefore, additional studies are needed to determine whether retinoid teratogenesis is directly related to cell death or to other aspects of pattern formation.

III. Heterotopic Patterns

Early limb development has been a popular model to explore the relationship between cell death and dysmorphogenesis, because the sensitive elements of the appendicular skeleton can be mapped to the territory of the limb bud from which these elements derive (SCOTT 1979; KNUDSEN 1990). For example, an early study compared the effect of two antineoplastic agents, cytosine arabinoside and hydroxyurea, which interfere with nucleic acid metabolism by different mechanisms (RITTER et al. 1973). Both drugs induced cell death in distal limb mesenchyme with different timing, and the abnormal limb phenotypes more closely reflected the time of cell death than drug exposure. Other studies confirmed the close correlation between cell death and malformation; furthermore, dose–response analysis linked limb malformations more closely with heterotopic cell death than growth arrest (KOCHHAR et al. 1978; MANSON and MILLER 1983).

One of the best-characterized models for teratogen-induced cell death studies is cyclophosphamide. An activated metabolite, phosphoramide mustard, is a mutagenic alkylating agent that produces a broad spectrum of malformations related to DNA damage (MIRKES 1985). MANSON et al. (1982) showed that cyclophosphamide-induced cell death in the mouse limb bud peaked 24 h after treatment; however, the role of cell death in the resulting abnormalities was questioned. When DNA integrity was monitored, the drug-related damage continued to mount between 24 and 72 h after treatment, during which time the limb buds became grossly abnormal but lacked heterotopic cell death. These findings were interpreted as indicating that a surviving population of mesenchymal cells carried DNA damage into the post-treatment recovery phase and that dysmorphogenesis was driven by the survival of abnormal cells rather than loss of cells per se (MANSON et al. 1982).

A direct link between teratogen-induced cell death and fetal malformation is often complicated by the ardent capacity of undifferentiated embryonic systems, particularly early embryos, to regulate cell number and compensate for cell loss (SNOW and TAM 1979; POWER and TAM 1993). Nevertheless, understanding why some but not all embryonic cells die after teratogen exposure is an important issue. A dramatic example of differential cytotoxicity is seen for cyclophosphamide (and other DNA-damaging agents) in early embryos, in which cell death is induced in the prosencephalic neuroepithelium, while the primitive heart is practically resistant. Total cell generation times differ between these regions by nearly 30% (9.45 h for the prosencephalic neuroepithelium, 13.37 h for the primitive heart); hence differential cytotoxicity may reflect an inherent difference in surveillance and repair capabilities related to the G_1 phase of the cell cycle (MIRKES et al. 1989). Genome surveillance systems which recognize and attempt to counter DNA damage have not been fully characterized in the developing embryo. Such systems, which play an important role in tumor suppression in differentiated tissues, are tightly integrated with embryonic growth control and might vary between different embryonic tissues and developmental stages. For example, higher cyclophosphamide exposure levels cause heterotopic cell death, whereas lower exposures primarily arrest cell growth (CHERNOFF et al. 1989; FRANCIS et al. 1990; LITTLE and MIRKES 1992). Surveillance and repair systems may be important determinants of differential cytotoxicity in the embryo and will be discussed later in the chapter.

C. Mechanisms of Cell Death

Early studies to investigate the control and function of cell death were directed at the morphological and subcellular changes that occur in a dying cell. The sequence of events fall into two categories: apoptosis (physiological cell death) and necrosis (lethal cell injury). These types of cell death differ fundamentally in their initiation and biological impact (KERR et al. 1972; WYLLIE et al. 1980; SEARLE et al. 1982) and thus have implications for understanding mechanisms of teratogenesis.

I. Necrosis

Necrosis or lethal cell injury results from injurious environmental stress such as high-level toxicant exposure, ischemia, complement attack, or direct cell trauma (LAIHO and TRUMP 1975; WYLLIE et al. 1980; SEARLE et al. 1982; KERR and HARMON 1991). Regions of contiguous cells are commonly struck en mass. The morphological changes are essentially the same in a wide variety of lethally injured cells and may be defined on the basis of early (reversible) and late (irreversible) changes. Early microscopic changes include clumping of loosely aggregated chromatin on the nuclear envelope, dilation of endoplasmic reticulum, partial dispersal of ribosomes, and condensation of the mitochondrial matrix (C phase). Mitochondrial respiration is believed to be intact during the C phase, and at this point the injured cell may recover. The mitochondrial matrix gradually swells as necrosis advances (S phase). Flocculent matrical densities form and are characteristic of an injured cell in which mitochondrial respiration is no longer intact; therefore, the transition from the C to S phase may signify the "point of no return" from which a cell can no longer be revived (LAIHO and TRUMP 1975). Rapid dissolution of most cellular organelles, including the nucleus (karyolysis) and plasma membrane (cytolysis), leads to eosinophilic "ghosting" under light microscopy.

A cell undergoing necrosis loses the properties of selective permeability and membrane ion-pumping activities due to membrane damage or disruption of energy metabolism. It is particularly the loss of calcium homeostasis which seems to deliver the lethal blow (SCHANNE et al. 1979; FARBER 1990). Disruption of plasma membrane integrity renders the cell permeable to vital dyes such as trypan blue which are excluded from a viable cell. In the same way, intracellular contents leak into the extracellular milieu. Sudden release of hydrolases, proteases, inflammatory cytokines, and other kinds of intracellular macromolecules initiates a harmful inflammatory reaction and leads to secondary scarring of the juxtaposed viable tissue; consequently, debris and dead cells from a necrotic area may persist for several days.

II. Apoptosis

In 1971, KERR described a different kind of cell death whereby the dying cell underwent condensation and fragmentation rather than swelling and dissolution. The phenomenon was originally referred to as "shrinkage necrosis," but was later named apoptosis (pronounced ap-o-to'sis, or ap-op-to'sis; FUNDER 1994) to give special recognition to its fundamental role in the steady state kinetics of healthy adult tissues, programmed cell deletion during normal embryonic development and metamorphosis, and a myriad of pathological processes including teratogenesis and cancer (KERR et al. 1972; KERR and HARMON 1991). Apoptosis or physiological cell death is an energy-dependent process which is engaged by cell–cell signaling or specific molecular damage (CORCORAN et al. 1994). It usually affects scattered single cells asynchronously rather than contiguous groups of cells en mass. The sequence of ultrastructural changes are essentially the same in a wide variety of apoptotic cells (KERR et al. 1972; WYLLIE et al. 1980; SEARLE et al. 1982). During the early phase, membrane-bound cell fragments known as apoptotic bodies are formed. Dense masses of chromatin aggregate beneath the nuclear envelope as the cell undergoes violent development of surface protuberances, which by time-lapse has been likened to "boiling" of the cytoplasm (SANDERSON 1976). The nucleus often breaks up at this stage to produce discrete fragments (karyorrhexis) which disperse into the corresponding cellular protuberances (Fig. 1A).

In the later phase, protuberances of varying size and shape separate from the cell. Surface proteins of an apoptotic cell are cross-linked by tissue transglutaminases preserving the integrity of apoptotic bodies (FESUS et al. 1987). Consequently, there is no leakage of intracellular contents into the extracellular milieu (KERR et al. 1972). The composition of a particular apoptotic body depends on the cellular constituents which happen to be present in the cytoplasmic protuberance that gave rise to it. Some bodies consist largely of condensed chromatin, whereas others may contain cytoplasmic elements only. However, the display of a well-preserved organelle structure in nascent apoptotic bodies has been interpreted as a sign of persisting metabolic activity (KERR et al. 1972). Membrane-bound apoptotic bodies may be extruded into a lumen, such as the amniotic cavity or gut lumen, but more often they are engulfed by neighboring tissue cells and macrophages. Herein they undergo a series of intracellular degenerative changes sometimes referred to as secondary necrosis. Since potentially toxic contents of dying cells are internally digested within phagosomes, a local inflammatory reaction is avoided. This is an important difference between apoptosis and necrosis (KERR et al. 1972).

Teratological cell deaths related to retinoic acid, cytosine arabinoside, cyclophosphamide, and purine nucleosides have been characterized by cellular condensation and fragmentation followed by phagocytosis (SCHWEICHEL and MERKER 1973; KOCHHAR et al. 1978; MANSON et al. 1982; YASUDA et al. 1987; JIANG and KOCHHAR 1992; CHEN et al. 1994; GAO et al. 1994). In an electron microscope study of spontaneous and experimentally induced cell deaths in rodent embryos, SCHWEICHEL and MERKER (1973) classified deaths in three categories: type 1 involved nuclear and cytoplasmic condensation, fragmentation, and phagocytosis; type 2 lacked apparent cellular condensation and fragmentation, but displayed cell inclusions interpreted as "autophagosomes"; and type 3 involved mitochondrial swelling and rapid disintegration of cell structures. Type-1 cell deaths occurred spontaneously in scattered single cells and were increased in embryos exposed to antineoplastic agents such as cyclophosphamide; type-2 cell deaths were evident spontaneously in territories of large-scale physiological cell death and experimentally in retinoid excess; and type-3 cell deaths were found in cartilage calcification zones but rarely in response to teratogens. On the basis of these descriptions, type-3 cell death closely resembles necrosis, and type 1 the first phase of apoptosis. However,

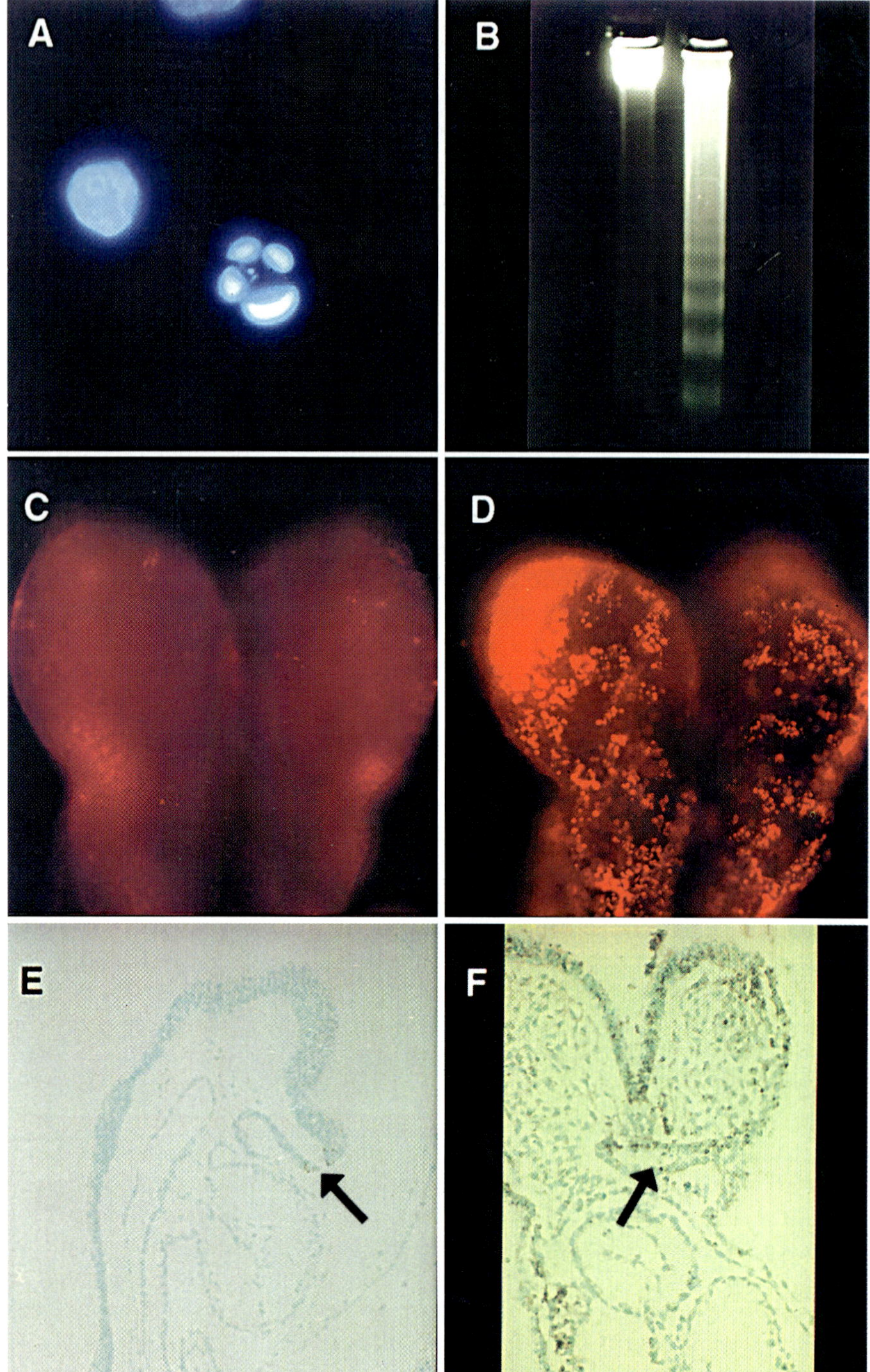
A
B
C
D
E
F

the morphological appearances of type-2 cell death may not have been correctly interpreted in that study. Autophagosomes, regarded by SCHWEICHEL and MERKER as an early sign of cellular degeneration, appear more typical of phagocytosed apoptotic bodies vis-à-vis the second phase of apoptosis (KERR et al. 1972). Thus the study by SCHWEICHEL and MERKER (1973) placed into proper context the fact that drugs and chemicals usually kill embryonic cells by apoptosis; however, when citing this paper the current terminology should be used (ALLES et al. 1991).

Apoptosis may be mapped in normal and abnormal embryos by supravital staining with Nile blue sulfate (SAUNDERS et al. 1962; PEXIEDER 1975; ALLES and SULIK 1989) or acridine orange (MENKES et al. 1979; GREGORY et al. 1991; ABRAMS et al. 1993; GAO et al. 1994). These basophilic dyes are trapped in an adenosine triphosphate (ATP)-dependent manner within free and phagocytosed apoptotic bodies and bind to the condensed nucleic acids contained within those structures (DARZYNKIEWICZ et al. 1992). Although acridine orange and Nile blue sulfate are not excluded from necrotic cells, the disruption of ATP metabolism precludes their active accumulation in a necrotic cell, and thus staining by these methods can be regarded as relatively specific for an apoptotic mode of cell death (ABRAMS et al. 1993). An example of acridine orange-positive staining is shown for the head fold of a normal day-8 mouse embryo and an abnormal embryo with widespread apoptosis related to 2′-deoxyadenosine intoxication (Fig. 1C,D). The quantitative power of supravital staining is limited, since apoptotic bodies from one dying cell may be engulfed by several phagocytes, or one phagocytic cell may contain debris from several apoptotic corpses. Either way, low incidences of apoptosis can have measurable impact on tissue mass. This is so because the morphological signs of apoptosis are brief, lasting perhaps only about 3 h, and so in a tissue such as rat liver as few as 0.5% hepatocytes might be visible as apoptotic per h, whereas the consequence several days later is the loss of 25% of hepatocytes (BURSCH et al. 1990). Likewise, a small amount of teratogen-induced apoptosis in a rapidly growing organ rudiment has the potential to manifest as major tissue deficiencies depending on compensatory growth. Intuitively, it is obvious that compensatory growth would be facilitated after teratogen-induced

◄

Fig. 1A–F. Techniques to demonstrate drug-induced apoptosis. **A** Human 697 pre-B leukemia cell exposed to 5 μ*M* 2-chloro-2′-deoxyadenosine (2-CdA) and stained with 4′,6-diamidino-2-phenylindole (DAPI). **B** Classical nucleosomal ladder in day-8 mouse embryos cultured for 3 h in the presence of 1 m*M* adenosine (*lane 1*) or 0.1 m*M* 2′-deoxyadenosine (*lane 2*) and pentostatin to block degradation. **C** Head fold of normal day-8 mouse embryo stained supravitally with acridine orange. **D** Heterotopic cell death induced by 2′-deoxyadenosine intoxication with 5 mg/kg pentostatin in vivo. **E** TUNEL assay of normal day-8 mouse embryo section counterstained with methyl green; *arrow* denotes programmed cell death in the prosencephalic neuroepithelium. **F** Heterotopic cell death (*arrow*) induced by 10 mg/kg 2-CdA on day 8 in the neuroepithelium but not the primitive heart, as revealed by TUNEL assay

apoptosis, as opposed to necrosis, where extensive collateral damage might be anticipated.

1. Chromatin Degradation

In 1980, Wyllie reported that the morphological changes in mouse thymocytes in which apoptosis were induced with glucocorticoids were accompanied by the degradation of DNA into oligonucleosome-length fragments. A direct correlation was observed between the proportion of nuclei with condensed chromatin and the proportion of DNA which did not pellet after centrifugation at 27 000 g. When the unsedimented DNA fraction was analyzed by agarose gel electrophoresis, a ladder was observed which consisted of DNA fragments whose molecular weights were integer multiples of 180–200 base pair (bp) subunits. In stark contrast, DNA from necrotic cells was randomly degraded to an electrophoretic smear. The unit size of the apoptotic DNA ladder corresponded to the length of DNA wrapped around the histone core of a nucleosome (185 bp). Thus Wyllie concluded that chromatin condensation in apoptosis was accompanied by cleavage of DNA in the linker region between nucleosomes (WYLLIE 1980; SCHWARTZMAN and CIDLOWSKI 1993). The regular DNA fragmentation pattern derived from integer multiples of 185 bp is commonly referred to as the nucleosomal ladder. Perhaps importantly, mitochondrial DNA remains intact during apoptosis (MURGIA et al. 1992), but is damaged to the same extent as nuclear DNA during necrosis (TEPPER and STUDZINSKI 1993). Several recent studies have implicated a decrease of mitochondrial transmembrane potential ($\Delta\psi m$) as an event in an apoptotic target cell; thus mitochondrial integrity and function may be needed to initiate changes leading to apoptotic cell death (COSSARIZZA et al. 1995; PETIT et al. 1995). DNA laddering has been observed in systems beyond thymocytes, including teratogen-induced cell death in early embryos for phosphoramide mustard (CHEN et al. 1994), 2′-deoxyadenosine (GAO et al. 1994; Fig. 1B), and *N*-acetoxy-2-acetylaminofluorene (THAYER and MIRKES 1995).

One advantage of the mouse thymocyte model is the ability to obtain relatively synchronized populations of apoptotic cells by glucocorticoid stimulation. This has facilitated the effort to purify and characterize the apoptotic nuclease (COHEN and DUKE 1984; COMPTON and CIDLOWSKI 1987). Although general consensus exists on calcium dependence and zinc inhibition, the apoptotic nuclease itself has eluded definitive identification. Leading candidates are NUC18 (GAIDO and CIDLOWSKI 1991), DNase I (UCKER et al. 1992; PEITSCH et al. 1993), and DNase II (BARRY and EASTMAN 1993). NUC18 is an 18-kDa endonuclease that was first isolated from glucocorticoid-treated thymocytes (GAIDO and CIDLOWSKI 1991; BORTNER et al. 1995). Partial sequencing identified it as being similar to cyclophilin A, a ubiquitous protein which avidly binds the immunosuppressant cyclosporine (MONTAGUE et al. 1994). Cyclophilin A is homologous to a heat shock-inducible enzyme, peptidyl prolyl *cis*-trans isomerase, which catalyzes the *cis*-trans isomerization of proline imide bonds to facilitate protein folding in vitro (FISCHER and SCHMID 1990; SYKES et al. 1993) and is developmentally regulated in first-trimester human decidua and placenta (MEIER et al. 1995). Thus nucleosomal laddering may involve a cyclophilin-like nuclear protein which plays a multifunctional role in developing and environmentally stressed tissues.

The premise that chromatin cleavage is a primary death lesion in apoptosis is supported largely by the fact that aurintricarboxylic acid, a general nuclease inhibitor, blocked DNA fragmentation without affecting calcium availability or protein synthesis (McCONKEY et al. 1989b; ARENDS et al. 1990; BATISTATOU and GREENE 1991). However, subsequent studies showed that apoptotic bodies may form independently of nucleosomal laddering, raising questions as to the specific relationship between the morphological and biochemical changes in chromatin structure as the cell death program unfolds. UCKER et al. (1992) studied a mutant fibroblast cell line which lacked the endonuclease responsible for nucleosomal laddering and found the cells to still display the characteristic nuclear morphology of apoptosis. Nucleosomal laddering may be dissociated from the nuclear condensation in normal cells by exposing them to various agents such as zinc (COHEN et al. 1992; X.M. SUN et al. 1994), 3-aminobenzamide (HOSHINO et al. 1993), and nicotinamide (GAO et al. 1994). However, the most definitive evidence for dissociation of nucleosomal laddering and chromatin condensation was provided by D.Y. SUN et al. (1994). Using a simple cell-free assay involving isolated nuclei, these investigators induced classical nucleosomal ladders with micrococcal nuclease, but chromatin condensation did not occur; conversely, the potassium ionophore valinomycin induced chromatin condensation without chromatin cleavage. Since characteristic morphological features of apoptosis may occur in the absence of classical nucleosomal laddering, and vice versa, internucleosomal DNA cleavage is neither necessary nor sufficient for apoptosis.

Classical nucleosomal laddering is invariably preceded by the degradation of chromatin into larger DNA fragments of 50 and 300 kilobase pairs (kbp). The higher-order chromatin degradation parallels chromatin condensation and now appears to be a necessary step in apoptosis (WALKER et al. 1991; BROWN et al. 1993; OBERHAMMER et al. 1993; BEERE et al. 1995). It is believed to involve cleavage of heterochromatin into single (50 kbp) and hexameric (300 kbp) loop structures which are released from their attachment sites to the nuclear scaffolding (FILIPISKI et al. 1990). Oligonucleosomes that form the characteristic DNA ladder appear when the 50-kbp fragments are subsequently degraded by an apoptotic nuclease (OBERHAMMER et al. 1993; ZHIVOTOVSKY et al. 1994); however, the latter change is not always seen, as evidenced by apoptosis in the embryonic C3H/10$T_{1/2}$ cell line (TOMEI et al. 1993) and insect metamorphosis (ZAKERI et al. 1993).

Detection of large (50 kbp) and small (185 bp) DNA fragments is possible by the in situ terminal transferase reaction, commonly known as TUNEL (terminal transferase dUTP nick end labeling). This technique is based on the specific binding of terminal deoxynucleotidyl transferase to free 3′-hydroxyl ends of DNA and the template-independent synthesis of a biotin- or digoxygenin-labeled deoxyuridine triphosphate (dUTP) deoxynucleotide heteropolymer (Gavrieli et al. 1992; Ansari et al. 1993; Wijsman et al. 1993). Due to the rapid nature of stepwise chromatin degradation in many cells, the transitional stages between cells with an unfragmented genome and cells with extensive DNA fragmentation are often not apparent. Chromatin is also degraded to kilobase-sized DNA fragments in necrosis (Bicknell and Cohen 1995), and Grasl-Kraupp et al. (1995) determined that TUNEL labeling is insufficient to discern between apoptosis, necrosis, and suboptimal tissue fixation; however, apoptotic cells label by TUNEL around the nuclear circumference, whereas in necrosis the cell labels evenly in the nucleus and cytoplasm and only after nuclear degradation (Kressel and Groscurth 1994). TUNEL-positive nuclei have clearly been mapped to classical sites of programmed cell death, such as interdigital cell death in the developing limb (Tone et al. 1994) and degeneration of the midline epithelial seam in the fusing secondary palate (Mori et al. 1994; Taniguchi et al. 1995). Correlations with spontaneous limb dysmorphogenesis have also been possible with this method (Zakeri et al. 1994), and TUNEL-positive cells appear as a consequence of exposure to drugs that initiate large-scale cell death and a classical nucleosomal ladder in the embryo (Thayer and Mirkes 1995; Wubah et al. 1996). An example of drug toxicity is shown in Fig. 1E,F for normal day-8 mouse embryos and after treatment with the potent 2′-deoxyadenosine analogue cladribine (2-chloro-2′-deoxyadenosine, 2-CdA). TUNEL-positive cells are widespread through the neuroepithelium of treated embryos, but are conspicuously absent from the primitive heart, as is seen for teratogenic treatment with other agents that cause DNA damage (Mirkes 1985; Gao et al. 1994; Thayer and Mirkes 1995).

2. Protease Involvement

Since cells may die without classical DNA laddering, a critical question arises as to the mechanism of initiation and control of high molecular weight DNA fragmentation. Recent studies have demonstrated that chromatin condensation and nuclear fragmentation is paralleled by the progressive degradation of the lamin A+B network in the nucleus of an apoptotic cell (Oberhammer et al. 1994). The filamentous lamin A+B network is depolymerized to 67- and 70-kDa subunits during chromatin condensation and then is cleaved to 46-kDa fragments in conjuction with nuclear fragmentation. Large (50–300 kbp) DNA fragments appear as the nuclear scaffolding is degraded in this manner, suggesting that release of chromatin loops from attachment to the nuclear matrix renders these sites of chromatin accessible to nuclease attack.

Proteolysis of specific nuclear proteins now appears to be a very early step in the initiation of apoptosis. Genetic evidence has identified *ced-3* as a death protease in nematodes. Normal function of the *ced-3* product is required for programmed cell death during nematode development, but not for normal cell viability (YUAN and HORVITZ 1990). The CED-3 protein is related to a mammalian cysteine protease known as ICE which cleaves the inactive precursor of IL-1β to its active cytokine (YUAN et al. 1993). Overexpression of murine ICE (mICE) caused Rat-1 fibroblast cells to undergo apoptosis. This required cysteine protease activity, because various mutant constructs of mICE did not cause cell death and death was preventable with the cowpox virus gene *crmA*, which encodes a specific ICE inhibitor (MIURA et al. 1993). Similar findings were made in dorsal root ganglion neurons cultured in the absence of nerve growth factor, a condition that would normally lead to cell death (GAGLIARDINI et al. 1994).

ICE-deficient mice generated by gene targeting are not critically impaired in spontaneous or induced apoptosis and are developmentally normal (KUIDA et al. 1995; LI et al. 1995). However, ICE is only one representative of a larger family of genes encoding apoptotic proteases, indicating that there may be functional redundancies or that other homologues are more important death proteases in developing mammalian tissues (KUMAR et al. 1994; WANG et al. 1994; KAMENS et al. 1995; FERNANDES-ALNEMRI et al. 1994, 1995). *Nedd-2* is one such homologue that is highly expressed in several tissues undergoing apoptosis during mouse embryonic development (KUMAR et al. 1994). Overexpression of *Nedd-2* mRNA has been shown to induce apoptosis in mammalian cells, and growth factor-dependent murine cells expressing *Nedd-2* antisense become significantly refractory to apoptosis upon cytokine removal (KUMAR 1995).

Protease involvement during apoptosis raises important questions as to the identity of substrates and products of these enzymes. One potential "death substrate" is poly(ADP-ribose) polymerase (PARP), a nuclear enzyme of DNA repair (MARKS and FOX 1991; NOSSERI et al. 1994). KAUFMAN (1989) reported that drug-induced nucleosomal laddering in human HL-60 cells is associated with degradation of PARP. Specific cleavage from 116 kDa to an 85-kDa fragment is an early marker for cell death induced by mechanistically diverse chemotherapeutic agents, including cytosine arabinoside and methotrexate (KAUFMAN et al. 1993). PARP cleavage occurs between Asp-216 and Gly-217 within a domain resembling one of the cleavage sites in pro-IL-1β. Properties of the PARP cleavage enzyme matched those of ICE to implicate a protease resembling ICE (prICE); however, PARP was not cleaved by ICE, nor was pro-IL-1β cleaved by prICE (LAZEBNIK et al. 1994).

Purification of prICE has been recently achieved (NICHOLSON et al. 1995; TEWARI et al. 1995). The protease is composed of 12-kDa and 17-kDa subunits, the smaller of which is identical to CPP32β, which is an ICE-CED-3 gene cloned from human Jurkat cells (FERNANDES-ALNEMRI et al. 1994). The names "apopain" (NICHOLSON et al. 1995), and "Yama", after the Hindu god of death (TEWARI et al. 1995), have been proposed. Inhibitor studies have shown that apopain is necessary but not sufficient for apoptosis (NICHOLSON et al. 1995). Like ICE, apopain is synthesized as an inactive proenzyme which must be activated by proteolytic cleavage; thus it appears that apoptosis involves a proteolytic cascade whereby apogenic signals activate a death protease which, in turn, cleaves PARP and other proteins. Many important questions remain as to the signal transduction pathways by which apogenic signals activate the proteolytic cascade which, to our knowledge, are yet to be explored within the field of developmental toxicology.

D. Implications for Drug Toxicity

I. Three Planes of Damage

Work in various tumor cell lines has shown that environmental stress conditions which induce cells to undergo necrosis may, under weaker exposures, initiate apoptosis. For example, murine tumor cells die exclusively by apoptosis when subjected to weak hyperthermia or oxidative stresses, whereas increasing the exposure enhances necrosis over apoptosis exclusively (HARMON et al. 1990; TAKANO et al. 1991; NOSSERI et al. 1994). Mild stress introduces "sublethal lesions" such as DNA damage, which can be recognized by cellular surveillance systems and transduced into a signal for apoptosis. As the stress reaches a critical load, loss of cellular homeostasis presides and necrosis occurs. Three successively higher planes of toxic damage can therefore be described: (1) sublethal damage below the threshold for apoptosis; (2) sublethal damage transduced to an apogenic signal; and (3) lethal cell injury inducing necrosis. Cellular pathways which transduce specific molecular damage into a signal that regulates changes in the rate of cell growth or initiates apoptosis are areas of active research (CORCORAN et al. 1994). In addition, cells may exist at a threshold for apoptosis and require tonic input from specific growth factors to stay alive (RAFF 1992). Therefore, signal transduction pathways that regulate cell growth and death may be important links to understanding how drugs influence the viability of embryonic cells.

II. Signal Transduction

One of the best-characterized second messenger systems in programmed cell death is cyclic adenosine monophosphate (AMP) (PRATT and MARTIN 1975). Pharmacological agents that elevate cyclic AMP levels also induce apoptosis, and endogenous cyclic AMP levels increase when apoptosis is induced (KIZAKI et al. 1990; MCCONKEY et al. 1990; MENTZ et al. 1995). In rat myelocytic leukemia cells, the apogenic effect of cyclic AMP can be reproduced by microinjecting the type I regulatory subunit of cyclic AMP-dependent protein kinase (PKA) and the catalytic subunit of PKA (LANOTTE et al. 1991; VINTERMYR et al. 1993). Okadaic acid or calyculin A, which are inhibitors of protein phosphatases-1 and -2A, can block the apogenic effect of ionizing radiation or hyperthermia, providing further evidence of specific protein phosphorylation at Ser/Thr residues of target proteins (BAXTER and LAVIN 1992; SONG and LAVIN 1993). Protein kinase C (PKC) may have an antagonistic effect. PKC activators, such as phorbol esters, prevented DNA fragmentation and cell death, and the PKC inhibitor staurosporine promoted apoptosis (McConkey et al. 1989a; Asakawa et al. 1993; Grant et al. 1994; Lucas et al. 1994). These results suggest that signal transduction events in apoptosis may involve the complex interplay of specific protein kinases and cellular phosphatases.

Downstream targets in the PKA/PKC cascade might include transcription factors such as AP-1 or p53. The AP-1 complex is a strong transcription factor formed from dimerization of c-*fos* and c-*jun* proto-oncogene products, and it is widely believed to be necessary, although not sufficient, for apoptosis. Transient expression of c-*fos* and c-*jun* is an early event in lymphoid and neuronal cells when these cells are induced to apoptosis by environmental stresses and growth factor deprivation (Colotta et al. 1990, 1992; Smeyne et al. 1993; Estus et al. 1994; Pandey and Wang 1995). Using *fos-lacZ* transgenic mice, Smeyne et al. (1993) showed that c-*fos* is chronically expressed in regions of programmed cell death such as the medial-edge epithelium of the palate prior to apoptosis.

III. Metabolic Imbalance

Many kinds of metabolic imbalances are transduced into apogenic signals. Vitamin deficiencies are associated with pathogenesis of malformations, and these conditions may lead to inappropriate cell death. For example, maternal zinc deficiency induces excessive cell death in rat embryos (Record et al. 1985; Jankowski et al. 1995). In studying the effects of zinc deficiency of human cell lines, Martin et al. (1991) reported that myeloid cells underwent necrosis primarily, whereas lymphoid cells died by apoptosis. Cell lineage is thus important. Other investigators reported differences in susceptibility to zinc deficiency related to the state of cell differentiation. Thymocytes become far more sensitive to zinc deficiency as they mature (McCabe et al. 1993). This is not necesssarily explained by inherent "apogenicity", since immature thymocytes are more sensitive to 2′-deoxyadenosine than are the more mature cells (A. Cohen et al. 1978; Seto et al. 1985).

Folate deficiency also induces apoptosis. Proerythroblasts from folate-deficient mice degenerate, in contrast to their normal counterparts, by a mechanism which is reversible with thymidine ($< 10\ \mu M$) (Koury and Horne 1994). It is well known that folate deprivation will affect cellular deoxynucleotide triphosphate (dNTP) pools because one-carbon groups are essential for the de novo synthesis of deoxyadenosine triphosphate (dATP), deoxyguanosine triphosphate (dGTP), and deoxythymidine triphosphate (dTTP) (St. James et al. 1994). Since fidelity of DNA replication and repair synthesis requires the correct balance of all four dNTP, and nuclear dNTP pools can only sustain DNA synthesis for a period of 30 s to 3 min, pool imbalances threaten genome integrity (Meuth 1989). In general, dNTP pool imbalances induce apoptosis, although it is not clear whether the death signal is initiated directly or indirectly through DNA damage (Yoshioko et al. 1987). For example, expansion of dATP pools by 2′-deoxyadenosine intoxication triggers apoptosis in cells of a variety of lineages and states of differentiation, including immature lymphoid cells (Seto et al. 1985; Kizaki et al. 1988; Carrera et al. 1990; Johnston et al. 1992; Robertson et al. 1993), early mouse embryos (Gao et al. 1994), and postmitotic sympathetic neurons (Wakade et al. 1995). During early pregnancy, 2′-deoxyadenosine is normally detoxified by adenosine deaminase in the decidua and placenta (Knudsen et al. 1991; Blackburn et al. 1992), but accumulates upon the inactivation of this enzyme pharmacologically (Knudsen et al. 1989, 1992) or genetically (Migchielsen et al. 1995; Wakamiya et al. 1995). Acute intoxication of day-8 mouse embryos with more than 50 μM 2′-deoxyadenosine severely amplifies dATP pools within 1.5 h and triggers large-scale apoptosis within 4.5 h (Gao et al. 1994), malformation by day 10, and death by day 12 (Airhart et al. 1993). Heterotopic cell deaths related to deoxyadenylate stress involve the neuroepithelium, but not the primitive heart (Gao et al. 1994). This closely resembles the cytotoxic pattern in cyclophosphamide (Mirkes 1985) and *N*-acetoxy-2-acetylaminofluorene (Thayer and Mirkes 1995) and indicates that different regions of the early embryo respond differently to DNA damaging agents.

E. Death Circuits

It has long been suspected that a state of cellular vulnerability exists in which cell death is induced readily by drugs (Kyprianou and Isaacs 1989). Movement of cells into and out of this state is probably influenced by oncogenes (Vaux et al. 1988) and tumor suppressor genes (Yonesch-Rouach et al. 1991). Several death-associated genes have been identified in hepatocytes (Fesus et al. 1987), hormone-dependent tissues (Buttyan et al. 1989), and thymocytes (Owens et al. 1991). Little is known concerning their roles in drug toxicity during embryonic development. RP-8, a gene specific to apoptosis in immature thymocytes (Owens et al. 1991), is expressed in the cerebellum of mutant *weaver* mice during spatially inappropriate granule cell death (Owens et al. 1995). Clusterin (TRPM-2, SGP-2, apolipoprotein J), a secreted protein specific to apoptosis in prostate and some but not all regions of physiological cell death (Buttyan et al. 1989), is upregulated in rat embryos following exposure to phosphoramide mustard (Chen et al. 1994). However, the strongest evidence for causality of cell death circuitry currently favors two genes, the *bcl-2* oncogene and the *p53* tumor suppressor gene.

I. B Cell Lymphoma/Leukemia-2

The *bcl-2* gene codes for a 26-kDa oncoprotein that functions as a negative regulator of apoptosis (Vaux 1993; Reed 1994). Deregulation of *bcl-2* expression in B lymphocytes is a critical event during the pathogenesis of follicular lymphoma (YUNIS et al. 1982; TSUJIMOTO and CROCE 1986). The first direct test of the biological impact of deregulated *bcl-2* expression was reported by VAUX et al. (1988). These investigators found that *bcl-2* cooperated with the powerful proto-oncogene c-*myc* to facilitate tumorigenesis. BCL-2 overproduction did not promote B cell proliferation, but instead conferred a growth factor-independent state of survival. In a broader sense, *bcl-2* participates in the molecular decision of whether a cell lives or dies. Numerous studies have examined the consequences of *bcl-2* misexpression in stable, transfected cell lines, transgenic mice, and antisense deletion. BCL-2 invariably hardens cell lines to oncogenic growth signals (SENTMAN et al. 1991; BISSONNETTE et al. 1992; GARCIA et al. 1992), to the killing effect of pleiotropic chemotherapeutic agents (FISHER et al. 1993; MANOME et al. 1993; MIYASHITA and REED 1992, 1993; OLIVER et al. 1993; WALTON et al. 1993; MIYASHITA et al. 1994; GAO et al. 1995), and to the cytotoxicity of priority environmental pollutants (ATEN et al. 1995). In fact, many toxicants antagonized by *bcl-2* (cytosine arabinoside, 5-fluoro-2′-deoxyuridine, methotrexate, nitrogen mustard, 2-CdA, and methyl mercury chloride) are known to be embryotoxic in laboratory animals. Mice deficient in BCL-2 exhibit a spectrum of cell death abnormalities, including polycystic kidney disease. This phenotype is apparently related to enhanced susceptibility of the metanephric rudiment to metabolic toxins (VEIS et al. 1993; NAKAYAMA et al. 1993). Embryos homozygous for the *bcl-2* null mutation display excessive mesenchymal cell death in the renal primordium after gestational day 12, leading to hypoplasia and a polycystic phenotype (SORENSON et al. 1995).

Several studies have investigated the tissue localization of BCL-2 during embryogenesis (HOCKENBERY et al. 1991; LEBRUN et al. 1993; LU et al. 1993; MERINO et al. 1994; VEIS NOVACK and KORSMEYER 1994). BCL-2 is present as early as day 8 of gestation in the mouse, first in the neuroepithelium and later in multiple differentiated tissues. The tissue localization was consistent with a role in protecting against apoptosis at developmental stages when programmed cell death would otherwise disrupt morphogenesis. Human *bcl-2* blocks programmed cell death when transfected into *C. elegans*, reflecting its ancestry as a functional homologue of *ced-9* (VAUX et al. 1992). However, a direct link between BCL-2 and teratogen-induced cell death has not been investigated.

Developmental toxicants may induce BCL-2-dependent apoptosis in cell lines. For example, the nucleoside analogue 2-CdA induces apoptosis in human leukemia cell lines and early mouse embryos (GAO et al. 1995; WUBAH et al. 1996). Cytotoxicity of 2-CdA is related to the Cl-dATP to dATP ratio achieved in the cell; higher ratios produce extensive DNA strand breakage (CHUNDURU et al. 1993; HENTOSH and GRIPPO 1994). A concentration of 1 μM 2-CdA produces physical strand breaks equivalent to 100 cGy (rads) ionizing radiation (GRIFFAG et al. 1989; HIROTA et al. 1989; CARRERA et al. 1990; CARSON et al. 1992). A strong correlation was found between AO-positive cell counts (apoptosis) in normal cells at 24 h and the Cl-dATP to dATP ratio, a dosimeter of biochemical stress, at 12 h (r^2, 0.95; $p=0.027$). Fourfold overproduction of BCL-2 (697/BCL2 cells) lowered the deoxyadenylate stress ratio from 7 to below 0.1, while fully protecting cell viability (GAO et al. 1995).

BCL-2 is an integral membrane protein in the rough endoplasmic reticulum, outer mitochondrial membrane, and nuclear envelope (CHEN-LEVY et al. 1989; MONAGHAN et al. 1992; KRAJEWSKI et al. 1993). Subcellular fractionation studies by HOCKENBERY et al. (1990) concluded that BCL-2 was an inner mitochondrial membrane protein and proposed a role in mitochondrial energy metabolism; however, JACOBSON et al. (1993) found that *bcl-2* transfection blocked staurosporine-induced apoptosis in a mutant fibroblast cell line which lacked an intact respiratory chain. Other studies have suggested that BCL-2 blocks downstream events such as calcium signaling (BAFFY et al. 1993; ZHONG et al. 1993; LAM et al. 1994) or superoxide radical damage (HOCKENBERY et al. 1993; KANE et al. 1993). The actions of BCL-2 can be opposed by other proteins that share significant homology with it. Recently, a 21-kDa homologue has been described, termed Bax (BCL-2-associated protein). BAX/BAX homodimers predispose cells to a state of vulnerability in which cell death is readily induced by a variety of unrelated agents; a less vulnerable state is conferred by BCL-2/BAX heterodimers (OLTVAI et al. 1993). Relative expression of *bcl-2* and *bax* genes in tissues may constitute an intrinsic determinant of apogenic potential (KRAJEWSKI et al. 1994); alternatively, other members of the *bcl-2* family may be involved (TAKAYAMA et al. 1995; YANG et al. 1995).

II. Tumor Suppressor Gene p53

DNA is a critical target for many natural and anthropogenic agents which perturb mammalian development. *A priori*, genes which recognize and attempt to counter DNA damage in the embryo might influence signal transduction events that suppress cell growth, stimulate repair systems, and initiate cell death (HOLBROOK and FORNACE 1991). One of the important positive regulators of apoptosis in mammalian tissues is the *p53* tumor suppressor gene. The *p53* gene encodes a transcription factor which is activated by certain kinds of environmental stress (KASTAN et al. 1991; KUERBITZ et al. 1992; LANE 1992; CLARKE et al. 1993; LOWE et al. 1993b) or oncogenic growth signals (DEBBAS and WHITE 1993; YONISH-ROUACH et al. 1991; HERMEKING and EICK 1994; SYMONDS et al. 1994; HOWES et al. 1994; PAN and GRIEP 1994; MORGANBESSER et al. 1994). p53 was initially discovered as an abundant 53-kDa cellular protein in early- to midgestation mouse embryos and in spontaneous and chemically induced tumors (LINZER and LEVINE 1979; MORA et al. 1980; OREN et al. 1982; ROGEL et al. 1985). Loss of *p53* function clearly contributes to neoplastic development, as evidenced by the detection of *p53* mutations in nearly half of human tumors of many tissue types (HOLLSTEIN et al. 1991; OLINER et al. 1992; HARRIS 1993; CHO et al. 1994) and the predisposition of *p53*-deficient mice to tumorigenesis (DONEHOWER et al. 1992; JACKS et al. 1994; PURDIE et al. 1994).

The *p53* gene participates in the cellular response to DNA damage through recognition of damaged DNA (KASTAN et al. 1991; FRITSCHE et al. 1993; LOWE et al. 1993a; NELSON and KASTAN 1994; COATES et al. 1995). Two landmark studies (CLARKE et al. 1993; LOWE et al. 1993b) showed that thymocytes from *p53*-deficient mice were refractory to the killing effect of ionizing radiation, whereas apoptosis was readily induced in their normal counterparts. Without *p53,* irradiated surviving thymocytes replicate through the DNA damage leading to genomic instability. *p53*-dependent apoptosis has since been documented in many cell types and following pleiotropic environmental stresses, particularly related to breaks in genomic DNA (CLARKE et al. 1993; LOWE et al. 1993a,b; MERRITT et al. 1994; COATES et al. 1995).

Genes acting downstream in the p53 cascade directly regulate cell cycle traverse at the G_1S restriction point to present the damaged cell with an opportunity for DNA repair (KASTAN et al. 1991). In circumstances involving more severe levels of DNA damage, p53 enables the cell death program to function (YONISH-ROUACH et al. 1993). This indicates an important role of p53 signaling in toxicant-tissue interactions which culminate in cell death. Candidate genes whose expression may be influenced by *p53* include *p21*$^{\mathrm{WAF1/Cip1}}$, a tight-binding inhibitor of the cyclin-dependent protein kinase that gates G_1/S phase transition (EL-DIERY et al. 1993; HARPER et al. 1993), *Gadd45,* a DNA damage-inducible gene which bears homology to regions of the 28S ribosomal RNA subunit (CARRIER et al. 1994; SMITH et al. 1994; YOSHIDA et al. 1994), *bcl-2,* which is negatively regulated by p53 (MIYASHITA et al. 1994), and *bax* which is positively regulated by p53 (OLTVAI et al. 1993; ZHAN et al. 1994; MIYASHITA and REED 1995).

Although *p53*-null mutant (*p53*$^{-/-}$) mouse embryos may develop normally to term, 8%–16% of them exhibit a severe anterior neural tube closure defect known as exencephaly (SAH et al. 1995). This abnormality is consistent with the fact that *p53* is normally expressed at high levels in the mouse embryo prior to day 13 of gestation (ROGEL et al. 1985; SCHMID et al. 1991). Another indication of the importance of *p53* function derives from recent data which has demonstrated increased incidences of diverse malformations, fetal resorptions, and postnatal deaths in a *p53*$^{+/-}$ heterozygous dam following a mildly teratogenic regimen of dioxin and benzo[a]pyrene as compared with a *p53*$^{+/+}$ dam (NICOL et al. 1995). These results implicate *p53* in normal development and suggest a role as a general teratological suppressor (NICOL et al. 1995).

The link between *p53* activity, apoptosis, and induced abnormal development has been recently investigated using 2-CdA as a model toxicant. Here, we observed that p53 may function as an initiator of chemical teratogenesis through induction of cell death (WUBAH et al. 1996). Treatment of pregnant mice with 5–10 mg 2-CdA/kg on day 8 of gestation resulted in large-scale apoptosis (Fig. 1E,F) and nuclear p53 accumulation. The p53 protein has a short half-life (less than 20 min) in normal cells, but it is stabilized for several hours by breaks in DNA. Physical strand breakage triggers hyperphosphorylation of the p53 protein (Kastan et al. 1991; Merritt et al. 1994). p53-dependent apoptosis was strong in teratogenic treatment of the neuroepithelium, but was conspicuously absent in the primitive heart (Wubah et al. 1996). Interestingly, *p53* transcripts are widely distributed in the embryo, but the heart was relatively weak (Schmid et al. 1991). Perhaps higher basal expression of the *p53* gene provides a selective advantage to some early embryonic tissues, such as the neuroepithelium, which repair DNA damage less efficiently (Mirkes et al. 1989). The advantage might be to move cells into a state of vulnerability whereby cell death is readily induced by genomic lesions at the expense of predisposing these tissues to drugs which directly or indirectly damage DNA.

To investigate the relationship between p53-dependent apoptosis and teratogenesis, we subjected day-8 mouse embryos on different *p53* gene backgrounds to 2-CdA. Abnormal development was manifested as eye defects by day 11, particularly lens agenesis (Wubah et al. 1996). Overall the incidences of these defects were 73.3% for *p53* wild-type fetuses, 52.5% for heterozygous mutants, and 2.2% for null mutants depending on the dosage (Fig. 2). Genetic models have produced apoptotic lens abnormalities which are blocked in the p53-deficient condition. Mouse embryos lacking the Rb (retinoblastoma) tumor suppressor exhibit excessive apoptosis in the lens ru-

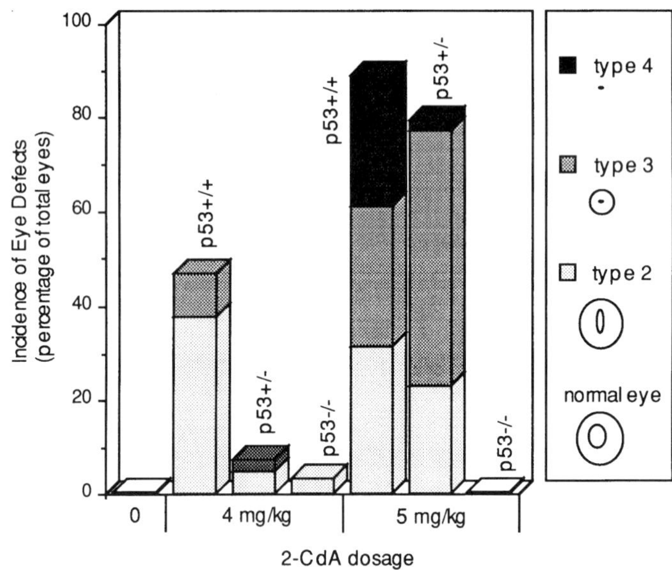

Fig. 2. Eye malformations induced by 2-CdA in mice. Pregnant TSG-p53 mice were treated with 2-chloro-2′-deoxyadenosine (2-CdA) on day 8 of gestation, and ocular phenotypes were scored on day 17. The morphologies are indicated on the *right*. (Data from Wubah et al. 1996)

diment on day 10, leading to eye reduction defects, whereas embryos doubly null for Rb and p53 show nearly complete suppression of the effect (MORGENBESSER et al. 1994). Another example is provided by transgenic mice which express the wild-type human *p53* gene under control of the crystallin gene promoter. These mice develop microphthalmia after birth as a result of a prenatal defect in lens fiber differentiation. In eyes of transgenic mice that express both wild-type and mutant p53 alleles, the mutant p53 can interfere with the apoptotic function of the wild-type product (NAKAMURA et al. 1995). Specific roles in lens morphogenesis and retinal differentiation have been suggested for Rb and p53 (HOWES et al. 1994; PAN and GRIEP 1994). Inactivation of Rb impairs retinal differentiation, leading to retinal degeneration by p53-dependent apoptosis at the normal time of photoreceptor differentiation. In the absence of p53, abnormal cells survive and eventually yield retinoblastomas (HOWES et al. 1994).

It is thus interesting to contrast these results with some of the earlier descriptions of cell death during eye development (GLÜCKSMANN 1965). Glücksmann proposed that the eye field initially possesses too many cells, formed in excess of the requirement for differentiation, and those cells which fail to properly adapt to the changing microenvironment die spontaneously. In lens development, the primary optic vesicle induces a wide ectodermal response which extends beyond the "real target" of the lens placode to form a surplus of peridermal cells and lens stalk cells; the extra cells become superfluous if they fail to enter the differentiation environment of the lens and subsequently degenerate. Thus while cell death may initially correct for an oversized rudimentary lens, another implication of controlled cell deletion is suppression of cells whose survival would be unproductive for embryonic development, as discovered almost 30 years later. The paradoxical role of p53 as a teratological suppressor (NICOL et al. 1995) and teratogenesis initiator (WUBAH et al. 1996) clearly calls for more information on its regulation and function during developmental toxicity. It would seem that p53 offers stressed embryos a choice between life and a "ride on the Pale Horse".

Acknowledgments. I am grateful to the National Institute of Child Health and Human Development for supporting this work through grant RO1 HD30302; I would also like to thank Ms. Judith Wubah and Dr. Monier Ibrahim, who were supported in part by grant T32 ES07282, for their assistance and Dr. Gao Xiang, Ms. Diep Nguyen, and Mr. Jeff Charlap for technical help.

References

Abrams JM, White K, Fessler LI, Steller H (1993) Programmed cell death during Drosophila embryogenesis. Development 117: 29–43

Airhart MJ, Robbins CM, Knudsen TB, Church JK, Skalko RG (1993) Occurrence of embryotoxicity in mouse embryos following in utero exposure to 2′-deoxycoformycin (pentostatin). Teratology 47: 17–27

Alles A, Alley K, Barrett JC, Buttyan R, Columbano A, Cope FO, Copelan EA, Duke RC, Farel PB, Gerhenson LE, Goldgaber D, Green DR, Honn KV, Hully J,

Isaacs JT, Kerr JFR, Krammer PH, Lockshin RA, Martin DP, McConkey DJ, Michaelson J, Schulte-Hermann R, Server AC, Szende B, Tomei LD, Tritton TR, Umansky SR, Valerie K, Warner HR (1991) Apoptosis: a general comment. FASEB J 5: 2127–2128

Alles AJ, Sulik KK (1989) Retinoic-acid-induced limb-reduction defects: perturbation of zones of programmed cell death as a pathogenetic mechanism. Teratology 40: 163–171

Alles AJ, Sulik KK (1990) Retinoic acid-induced spina bifida: evidence for a pathogenetic mechanism. Development 108: 73–81

Ansari B, Coates PJ, Greenstein BD, Hall PA (1993) In situ end-labelling detects DNA strand breaks in apoptosis and other physiological and pathological states. J Pathol 170: 1–8

Arends MJ, Morris RG, Wyllie AH (1990) Apoptosis. The role of the endonuclease. Am J Pathol 136: 593–608

Asakawa J, Tsiagbe VK, Thorbecke GJ (1993) Protection against apoptosis in chicken bursa cells by phosbol ester in vitro. Cell Immunol 147: 180–187

Aten J, Prigent P, Poncet P, Blanpied C, Claessen N, Druet P, Hirsch F (1995) Mercuric chloride-induced programmed cell death of a murine T cell hybridoma. I. Effect of the proto-oncogene Bcl-2. Cell Immunol 161: 98–106

Baffy G, Miyashita T, Williamson JR, Reed JC (1993) Apoptosis induced by withdrawal of interleukin-3 (IL-3) from an IL-3-dependent hematopoietic cell line is associated with repartitioning of intracellular calcium and is blocked by enforced Bcl-2 oncoprotein production. J Biol Chem 268: 6511–6519

Barry MA, Eastman A (1993) Identification of deoxyribonuclease II as an endonuclease involved in apoptosis. Arch Biochem Biophys 300: 440–450

Batistatou A, Greene LA (1991) Aurintricarboxylic acid rescues PC12 cells and sympathetic neurons from cell death caused by nerve growth factor deprivation: correlation with suppression of endonuclease activity. J Cell Biol 115: 461–471

Baxter GD, Lavin MF (1992) Specific protein dephosphorylation in apoptosis induced by ionizing radiation and heat shock in human lymphoid tumor lines. J Immunol 148: 1949–1954

Beere HM, Chresta CM, Alejo-Herberg A, Skladanowski A, Dive C, Larsen AK, Hickman JA (1995) Investigation of the mechanism of higher order chromatin fragmentation observed in drug-induced apoptosis. Mol Pharmacol 47: 986–996

Bicknell GR, Cohen GM (1995) Cleavage of DNA to large kilobase pair fragments occurs in some forms of necrosis as well as apoptosis. Biochem Biophys Res Commun 207: 40–47

Bissonnette RP, Echeverri F, Mahboubi A, Green DR (1992) Apoptotic cell death induced by c-myc is inhibited by *bcl-2*. Nature 359: 552–553

Blackburn MR, Gao X, Airhart MJ, Skalko RG, Thompson LF, Knudsen TB (1992) Adenosine in the early postimplantation mouse uterus: quantitation by HPLC-fluorometric detection and spatio-temporal regulation by 5'-nucleotidase and adenosine deaminase. Dev Dynam 194: 155–168

Bortner CD, Oldenburg NBE, Cidlowski JA (1995) The role of DNA fragmentation in apoptosis. Trends Cell Biol 5: 21–26

Brown DG, Sun X-M, Cohen GM (1993) Dexamethasone-induced apoptosis involves cleavage of DNA to large fragments prior to internucleosomal fragmentation. J Biol Chem 268: 3037–3039

Bursch W, Paffe S, Putz B, Barthel G, Schulte-Hermann R (1990) Determination of the length of the histological stages of apoptosis in normal liver and in altered hepatic foci of rats. Carcinogenesis 11: 847–853

Buttyan R, Olsson CA, Pintar J, Chang M, Bandyle M, Ng P, Sawczuk LS (1989) Induction of the TRPM-2 gene in cells undergoing programmed cell death. Mol Cell Biol 9: 3473–3481

Carrera CJ, Terai C, Lotz M, Curd JG, Piro LD, Beutler E, Carson DA (1990) Potent toxicity of 2-chlorodeoxyadenosine toward human monocytes in vitro and in vivo. J Clin Invest 86: 1480–1488

Carrier F, Smith ML, Bae I, Kilpatrick KE, Lansing TJ, Chen C-Y, Engelstein M, Friend SH, Henner WD, Gilmer TM, Kastan MB, Fornace AJ Jr (1994) Characterization of human Gadd45, a p53-regulated protein. J Biol Chem 269: 32672–32677

Carson DA, Wasson DB, Esparza LM, Carrera CJ, Kipps TJ, Cottam HB (1992) Oral antilymphocyte activity and induction of apoptosis by 2-chloro-2'-*arabino*-fluoro-2'-deoxyadenosine. Proc Natl Acad Sci USA 89: 2970–2974

Chen B, Cyr DG, Hales BF (1994) Role of apoptosis in mediating phosphoramide mustard-induced rat embryo malformations in vitro. Teratology 50: 1–12

Chen-Levy Z, Nourse J, Cleary ML (1989) The *bcl-2* candidate proto-oncogene product is a 24- kilodalton integral-membrane protein highly expressed in lymphoid cell lines and lymphomas carrying the t(14;18) translocation. Mol Cell Biol 9: 701–710

Chernoff N, Rogers JM, Alles AJ, Zucker RM, Elstein KH, Massaro EJ, Sulik KK (1989) Cell cycle alterations and cell death in cyclophosphamide teratogenesis. Teratogenesis Carcinog Mutagen 9: 199–209

Cho Y, Gorina S, Jeffrey PD, Pavletich NP (1994) Crystal structure of a p53 tumor suppressor-DNA complex: understanding tumorigenic mutations. Science 265: 346–355

Chunduru SK, Appleman JR, Blakley RL (1993) Activity of human DNA polymerases α and β with 2-chloro-2′-deoxyadenosine 5′-triphosphate as a substrate and quantitative effects of incorporation on chain extension. Arch Biochem Biophys 302: 19–30

Clarke AR, Purdie CA, Harrison DJ, Morris RG, Bird CC, Hooper ML, Wyllie AH (1993) Thymocyte apoptosis induced by p53-dependent and independent pathways. Nature 362: 849–852

Coates PJ, Save V, Ansari B, Hall PA (1995) Demonstration of DNA damage/repair in individual cells using in situ end labelling: association of p53 with sites of DNA damage. J Pathol 176: 19–26

Cohen A, Hirschhorn R, Horowitz SD, Rubinstein A, Palmar SH, Hong R, Martin DW Jr (1978) Deoxyadenosine triphosphate as a potentially toxic metabolite in adenosine deaminase deficiency. Proc Natl Acad Sci USA 75: 472–476

Cohen JJ, Duke RC (1984) Glucocorticoid activation of a calcium-dependent endonuclease in thymocyte nuclei leads to cell death. J Immunol 132: 38–42

Cohen GM, Sun X-M, Snowden RT, Dinsdale D, Skilleter DN (1992) Key morphological features of apoptosis may occur in the absence of internucleosomal DNA fragmentation. Biochem J 286: 331–334

Colotta F, Polentarutti N, Staffico M, Fincato G, Mantovani A (1990) Heat shock induces the transcriptional activation of c-fos protooncogene. Biochem Biophys Res Commun 168: 1013–1019

Colotta F, Polentarutti N, Sironi M, Mantovani A (1992) Expression and involvement of *c-fos* and *c-jun* protooncogenes in programmed cell death induced by growth factor deprivation in lymphoid cell lines. J Biol Chem 267: 18278–18283

Compton MM, Cidlowski JA (1987) Identification of a glucocorticoid-induced nuclease in thymocytes. A potential "lysis gene" product. J Biol Chem 262: 8288–8292

Corcoran GB, Fix L, Jones DP, Moslen MT, Nicotera P, Oberhammer FA, Buttyan R (1994) Apoptosis: molecular control point in toxicity. Toxicol Appl Pharmacol 128: 169–181

Cossarizza A, Franceschi C, Monti D, Salvioli S, Bellesia E, Rivabene R, Biondo L, Rainaldi G, Tinari A, Malorni W (1995) Protective effect of N-acetylcysteine in tumor necrosis factor-α-induced apoptosis in U937 cells: the role of mitochondria. Exp Cell Res 220: 232–240

Darzynkiewicz Z, Bruno S, Del Bino G, Gorczyca W, Hotz MA, Lassota P, Traganos F (1992) Features of apoptotic cells measured by flow cytometry. Cytometry 13: 795–808

Debbas M, White E (1993) Wild-type p53 mediates apoptosis by E1A, which is inhibited by Elb. Genes Dev 7: 546–554

Donehower LA, Harvey M, Slagle BL, McArthur MJ, Montgomery CA Jr, Butel JS, Bradley A (1992) Mice deficient for p53 are developmentally normal but susceptible to spontaneous tumours. Nature 356: 215–221

El-Deiry WS, Tokino T, Velculescu VE, Levy DB, Parsons R, Trent JM Lin D, Mercer WE, Kinzler KW, Vogelstein B (1993) WAF1, a potential mediator of p53 tumor suppressor. Cell 75: 817–825

Ellis HM, Horvitz HR (1986) Genetic control of programmed cell death in the nematode C. elegans. Cell 44: 817–829

Ellis RE, Yuan J, Horvitz HR (1991) Mechanisms and functions of cell death. Annu Rev Cell Biol 7: 663–698

Estus S, Zaks WJ, Freeman RS, Gruda M, Bravo R, Johnson EM Jr (1994) Altered gene expression in neurons during programmed cell death: identification of *c-jun* as necessary for neuronal apoptosis. J Cell Biol 127: 1717–1727

Farber JL (1990) The role of calcium in lethal cell injury. Chem Res Toxicol 3: 503–508

Fernandes-Alnemri T, Litwack G, Alnemri ES (1994) CPP32, a novel human apoptotic protein with homology to *Caenorhabditis elegans* cell death protein *Ced-3* and mammalian interleukin-1β-converting enzyme. J Biol Chem 269: 30761–30764

Fernandes-Alnemri T, Litwack G, Alnemri ES (1995) *Mch2*, a new member of the apoptotic *Ced-3/Ice* cysteine protease gene family. Cancer Res 55: 2737–2742

Fesus L, Thomazy V, Falus A (1987) Induction and activation of tissue transglutaminase during programmed cell death. FEBS Lett 224: 104–108

Filipiski J, Leblanc J, Youdale T, Sikorska M, Walker PR (1990) Periodicity of DNA folding in higher order chromatin structures. EMBO J 9: 1319–1327

Fischer G, Schmid FX (1990) The mechanism of protein folding. Implications of in vitro refolding models for de novo protein folding and translocation in the cell. Biochemistry 29: 2205–2212

Fisher TC, Milner AE, Gregory CD, Jackman AL, Aherne W, Hartley JA, Dive C, Hickman JA (1993) bcl-2 modulation of apoptosis induced by anticancer drugs: resistance to thymidylate stress is independent of classical resistance pathways. Cancer Res 53: 3321–3326

Francis BM, Rogers JM, Sulik KK, Alles AJ, Massaro EJ, Elstein KH, Zucker RM, Rosen MB, Chernoff N (1990) Cyclophosphamide teratogenesis: evidence for compensatory responses to induced cellular toxicity. Teratology 42: 473–482

Fritsche M, Haessler C, Brandner G (1993) Induction of nuclear accumulation of the tumor-suppressor protein p53 by DNA-damaging agents. Oncogene 8: 307–317

Funder J (1994) Apoptosis: two p or not p? Nature 371: 98

Gagliardini V, Fernandez P-A, Lee RKK, Drexler HCA, Rotello RJ, Fishman MC, Yuan J (1994) Prevention of vertebrate neuronal death by the *crmA* gene. Science 263: 826–828

Gaido ML, Cidlowski JA (1991) Identification, purification, and characterization of a calcium-dependent endonuclease (NUC18) from apoptotic rat thymocytes. NUC18 is not histone H2B. J Biol Chem 266: 18580–18585

Gao X, Blackburn MR, Knudsen TB (1994) Activation of apoptosis in early mouse embryos by 2′-deoxyadenosine exposure. Teratology 49: 1–12

Gao X, Knudsen TB, Ibrahim MM, Haldar S (1995) Bcl-2 relieves deoxyadenylate stress and suppresses apoptosis in pre-B leukemia cells. Cell Death Differ 2: 69–78

Garcia I, Martinou I, Tsujimoto Y, Martinou J-C (1992) Prevention of programmed cell death of sympathetic neurons by the *bcl-2* proto-oncogene. Science 258: 302–304

Gavrieli Y, Sherman Y, Ben-Sasson SA (1992) Identification of programmed cell death in situ via specific labeling of nuclear DNA fragmentation. J Cell Biol 119: 493–501

Glücksmann A (1965) Cell death in normal development. Arch Biol (Liege) 76: 419–437

Grant S, Tuner AJ, Bartimole TM, Nelms PA, Joe VC, Jarvis WD (1994) Modulation of 1-[β-D-arabinofuranosyl] cytosine-induced apoptosis in human leukemia cells by staurosporine and other pharmacological inhibitors of protein kinase C. Oncol Res 6: 87–99

Grasl-Kraupp B, Ruttkay-Nedecky B, Koudelka H, Bukowska K, Bursch W, Schulte-Hermann R (1995) In situ detection of fragmented DNA (TUNEL assay) fails to discriminate among apoptosis, necrosis, and autolytic cell death: a cautionary note. Hepatology 21: 1465–1468

Greene JA, Sleet RB, Morgan KT, Welsch F (1987) Cytotoxic effects of ethylene glycol monomethyl ether in the forelimb bud of the mouse embryo. Teratology 36: 23–34

Gregory CD, Dive C, Henderson S, Smith CA, Williams GT, Gordon J, Rickinson AB (1991) Activation of Epstein-Barr virus latent genes protects human B cells from death by apoptosis. Nature 349: 612–612

Griffag J, Koob R, Blakley RL (1989) Mechanisms of inhibition of DNA synthesis by 2-chlorodeoxyadenosine in human lymphoblastic cells. Cancer Res 49: 6923–6928

Harmon BV, Corder AM, Collins RJ, Gobe GC, Allen J, Allan DJ, Kerr JFR (1990) Cell death induced in a murine mastocytoma by 42–47 °C heating in vitro: evidence that the form of death changes from apoptosis to necrosis above a critical heat load. Int J Radiat Biol 58: 845–858

Harper JW, Adami GR, Wei N, Keyomarsi K, Elledge SJ (1993) The p21 Cdk-interacting protein Cip1 is a potent inhibitor of G1 cyclin-dependent kinases. Cell 75: 805–816

Harris CC (1993) p53: at the crossroads of molecular carcinogenesis and risk assessment. Science 262: 1980–1981

Hengartner MO, Ellis RE, Horvitz R (1992) *Caenorhabditis elegans* gene *ced-9* protects cells from programmed cell death. Nature 356: 494–499

Hentosh P, Grippo P (1994) Template 2-chloro-2′-deoxyadenosine monophosphate inhibits in vitro DNA synthesis. Mol Pharmacol 45: 955–961

Hermeking H, Eick D (1994) Mediation of c-myc-induced apoptosis by p53. Science 265: 2091–2093

Hirota Y, Yoshioka A, Tanaka S, Watanabe K, Otani T, Minowada J, Matsuda A, Ueda T, Wataya Y (1989) Imbalance of deoxyribonucleoside triphosphates, DNA double-strand breaks, and cell death caused by 2-chlorodeoxyadenosine in mouse FM3A cells. Cancer Res 49: 915–919

Hockenbery D, Nunez G, Milliman C, Schreiber RD, Korsmeyer SJ (1990) *Bcl-2* is an inner mitochondrial membrane protein that blocks programmed cell death. Nature 348: 334–336

Hockenbery D, Zutter M, Hickey W, Nahm M, Korsmeyer SJ (1991) BCL2 protein is topographically restricted in tissues characterized by apoptotic cell death. Proc Natl Acad Sci USA 88: 6961–6965

Hockenbery DM, Oltvai ZN, Yin X-M, Milliman CL, Korsmeyer SJ (1993) Bcl-2 functions in an antioxidant pathway to prevent apoptosis. Cell 75: 214–251

Holbrook NJ, Fornace AJ Jr (1991) Response to adversity: molecular control of gene activation following genotoxic stress. New Biol 3: 825–833

Hollstein M, Sidransky D, Vogelstein B, Harris CC (1991) p53 mutations in human cancers. Science 253: 49–53

Hoshino J, Beckmann G, Kroger H (1993) 3-Aminobenzamide protects the mouse thymocytes in vitro from dexamethasone-mediated apoptotic cell death and cytolysis without changing DNA strand breakage. J Steroid Biochem 44: 113–119

Howes KA, Ransom N, Papermaster DS, Lasudry JGH, Albert DM, Windle JJ (1994) Apoptosis or retinoblastoma: alternative fates of photoreceptors expressing the HPV-16 E7 gene in the presence or absence of p53. Genes Dev 8: 1300–1310

Jacks T, Remington L, Williams BO, Schmitt EM, Halachmi S, Bronson RT, Weinberg RA (1994) Tumor spectrum analysis in p53-mutant mice. Curr Biol 4: 1–7

Jacobson MD, Burne JF, King MP, Miyashita T, Reed JC, Raff MC (1993) Bcl-2 blocks apoptosis in cells lacking mitochondrial DNA. Nature 361: 365–369

Jankowski MA, Uriu-Hare JY, Rucker RB, Rogers JM, Keen CL (1995) Maternal zinc deficiency, but not copper deficiency or diabetes, results in increased cell death in the rat: implications for mechanisms underlying abnormal development. Teratology 51: 85–93

Jiang H, Kochhar DM (1992) Induction of tissue transglutaminase and apoptosis by retinoic acid in the limb bud. Teratology 46: 333–340

Johnston JB, Lee K, Verburg L, Blondal J, Mowat MRA, Israels LG, Begleiter A (1992) Induction of apoptosis in $CD4^+$ prolymphocytic leukemia by deoxyadenosine and 2′-deoxycoformycin. Leuk Res 16: 781–788

Kamens J, Paskind M, Hugunin M, Talanian RV, Allen H, Banach D, Bump N, Hackett M, Johnston CG, Li P, Mankovich JA, Terranova M, Ghayur T (1995) Identification and characterization of ICH-2, a novel member of the interleukin-1β-converting enzyme family of cysteine proteases. J Biol Chem 270: 15250–15256

Kane DJ, Sarafian TA, Anton R, Hahn H, Gralla EB, Valentine JS, Ord T, Bredesen DE (1993) Bcl-2 inhibition of neural death: decreased generation of reactive oxygen species. Science 262: 1274–1277

Kastan MB, Onyekwere O, Sidransky D, Vogelstein B, Craig RW (1991) Participation of p53 protein in the cellular response to DNA damage. Canc Res 51: 6304–6311

Kaufmann SH (1989) Induction of endonucleolytic DNA cleavage in human acute myelogenous leukemia cells by etoposide, camptothecin, and other cytotoxic anticancer agents: a cautionary note. Cancer Res 49: 5870–5878

Kaufmann SH, Desnoyers S, Ottaviano Y, Davidson NE, Poirier GG (1993) Specific proteolytic cleavage of poly(ADP-ribose) polymerase: an early marker of chemotherapy-induced apoptosis. Cancer Res 53: 3976–3985

Kerr JFR (1971) Shrinkage necrosis: a distinct mode of cellular death. J Pathol 105: 13–20

Kerr JFR, Harmon BV (1991) Definition and incidence of apoptosis: an historical perspective. In: Tomei LD, Cope FO (eds) Apoptosis: the molecular basis of cell death. Cold Spring Harbor Laboratory Press, Plainview, pp 5–29

Kerr JFR, Wyllie AH, Currie AR (1972) Apoptosis: a basic biological phenomenon with wide-ranging implications in tissue kinetics. Br J Cancer 26: 239–257

Kizaki H, Shimada H, Ohsaka F, Sakurada T (1988) Adenosine, deoxyadenosine, and deoxyguanosine induce DNA cleavage in mouse thymocytes. J Immunol 141: 1652–1657

Kizaki H, Suzuki K, Tadakuma T, Ishimura Y (1990) Adenosine receptor-mediated accumulation of cyclic AMP-induced T-lymphocyte death through internucleosomal DNA cleavage. J Biol Chem 265: 5280–5284

Knudsen TB (1990) In vitro approaches to the study of embryonic cell death in developmental toxicity. In: Kimmel GL, Kochhar DM (eds) In vitro methods in developmental toxicology: use in defining mechanisms and risk parameters. CRC Press, Boca Raton, pp 129–142

Knudsen TB, Kochhar DM (1981) The role of morphogenetic cell death during abnormal limb-bud outgrowth in mice heterozygous for the dominant mutation Hemimelia-extra toe (Hm^x). J Embryol Exp Morphol 65 [Suppl]: 289–307

Knudsen TB, Gray MK, Church JK, Blackburn MR, Airhart MJ, Kellems RE, Skalko RG (1989) Early postimplantation embryolethality in mice following in utero inhibition of adenosine deaminase with 2'-deoxycoformycin. Teratology 40: 615–626

Knudsen TB, Blackburn MR, Chinsky JM, Airhart MJ, Kellems RE (1991) Ontogeny of adenosine deaminase in the mouse decidua and placenta: immunolocalization and embryo transfer studies. Biol Reprod 44: 171–184

Knudsen TB, Winters RS, Otey SK, Blackburn MR, Airhart MJ, Church JK, Skalko RG (1992) Effects of (R)-deoxycoformycin (pentostatin) on intrauterine nucleoside catabolism and embryo viability in the pregnant mouse. Teratology 45: 91–103

Kochhar DM (1977) Cellular basis of congenital limb deformity induced by vitamin A. Birth Defects 13: 111–154

Kochhar DM, Penner JD, McDay JA (1978) Limb development in mouse embryos. II. Reduction defects, cytotoxicity and inhibition of DNA synthesis produced by cytosine arabinoside. Teratology 18: 71–92

Koury M, Horne DW (1994) Apoptosis mediates and thymidine prevents erythroblast destruction in folate deficiency anemia. Proc Natl Acad Sci USA 91: 4067–4071

Krajewski S, Tanaka S, Takayama S, Schibler MJ, Fenton W, Reed JC (1993) Investigation of the subcellular distribution of the bcl-2 oncoprotein: residence in the nuclear envelope, endoplasmic reticulum, and outer mitochondrial membrane. Cancer Res 53: 4701–4714

Krajewski S, Krajewski M, Shabaik A, Miyashita T, Wang HG, Reed JC (1994) Immunohistochemical determination of in vivo distribution of Bax, a dominant inhibitor of Bcl-2. Am J Pathol 145: 1323–1336

Kressel M, Groscurth P (1994) Distinction of apoptotic and necrotic cell death by in situ labelling of fragmented DNA. Cell Tissue Res 278: 549–556

Kuerbitz SJ, Plunkett BS, Walsh WV, Kastan MB (1992) *p53* is cell cycle checkpoint determinant following irradiation. Proc Natl Acad Sci USA 89: 7491–7495

Kuida K, Lippke JA, Ku G, Harding MW, Livingston DJ, Su MS-S, Flavell RA (1995) Altered cytokine export and apoptosis in mice deficient in interleukin-1β converting enzyme. Science 267: 2000–2003

Kumar S (1995) Inhibition of apoptosis by the expression of antisense *Nedd2*. FEBS Lett 368: 69–72

Kumar S, Kinoshita M, Noda M, Copeland NG, Jenkins NA (1994) Induction of apoptosis by the mouse *Nedd2* gene, which encodes a protein similar to the product of the *Caenorhabditis elegans* cell death gene *ced-3* and the mammalian IL-1β-converting enzyme. Genes Dev 8: 1613–1626

Kyprianou N, Isaacs JT (1989) "Thymineless" cell death in androgen-independent prostatic cancer cells. Biochem Biophys Res Commun 165: 73–81

Laiho KU, Trump BF (1975) Studies on the pathogenesis of cell injury. Effects of inhibitors of metabolism and membrane function on the mitochondria of Ehrlich ascites tumor cells. Lab Invest 32: 163–183

Lam M, Dubyak G, Chen L, Nunez G, Miesfeld RL, Distelhorst CW (1994) Evidence that BCL-2 represses apoptosis by regulating endoplasmic reticulum-associated Ca^{2+} fluxes. Proc Natl Acad Sci USA 91: 6569–6573

Lane DP (1992) p53, guardian of the genome. Nature 358: 15–16

Lanotte M, Riviere JB, Hermouet S, Houge G, Vintermyr OK, Gjertsen BT, Doskeland SO (1991) Programmed cell death (apoptosis) is induced rapidly and with positive cooperativity by activation of cyclic adenosine monophosphate-kinase I in a myeloid leukemia cell line. J Cell Physiol 146: 73–80

Lazebnik YA, Kaufman SH, Desnoyers S, Poirier GG, Earnshaw WC (1994) Cleavage of poly(ADP-ribose) polymerase by a proteinase with properties like ICE. Nature 371: 346–347

LeBrun DP, Warnke RA, Cleary ML (1993) Expression of bcl-2 in fetal tissues suggests a role in morphogenesis. Am J Pathol 142: 743–753

Li P, Allen H, Banerjee S, Franklin S, Herzog L, Johnston C, McDowell J, Paskind M, Rodman L, Salfeld J, Towne E, Tracey D, Wardwell S, Wei F-Y, Wong W, Kamen R, Seshadri T (1995) Mice deficient in IL-1β-converting enzyme are de-

fective in production of mature IL-1β and resistant to endotoxic shock. Cell 80: 401–411
Linzer DIH, Levine AJ (1979) Characterization of a 54K dalton cellular SV40 tumor antigen present in SV40 transformed cells and uninfected embryonal carcinoma cells. Cell 17: 43–52
Little SA, Mirkes PE (1992) Effects of 4-hydroperoxycyclophosphamide (4-OOH-CP) and 4-hydroperoxydechlorocyclophosphamide (4-OOH-deClCP) on the cell cycle of post implantation rat embryos. Teratology 45: 163–173
Lowe SW, Ruley HE, Jacks T, Housman DE (1993a) p53-dependent apoptosis modulates the cytotoxicity of anticancer agents. Cell 74: 957–967
Lowe SW, Schmitt EM, Smith SW, Osborne BA, Jacks T (1993b) p53 is required for radiation-induced apoptosis in mouse thymocytes. Nature 362: 847–849
Lu Q-L, Poulsom R, Wong L, Hanby A (1993) BCL-2 expression in adult and embryonic non-haematopoietic tissues. J Pathol 169: 431–437
Lucas M, Sanchez-Margalet V, Sanz A, Solano F (1994) Protein kinase C activation promotes cell survival in mature lymphocytes prone to apoptosis. Biochem Pharmacol 47: 667–672
Manome Y, Weichselbaum RR, Kufe DW, Fine HA (1993) Effect of bcl-2 on ionizing radiation and 1-β-D-arabinofuranosylcytosine-induced internucleosomal DNA fragmentation and cell survival in human myeloid leukemia cells. Oncol Res 5: 139–144
Manson JM, Miller ML (1983) Contribution of mesenchymal cell death and mitotic alteration to asymmetric limb malformations induced by MNNG. Teratogenesis Carcinog Mutagen 3: 335–353
Manson JM, Papa L, Miller ML, Boyd C (1982) Studies of DNA damage and cell death in embryonic limb buds induced by teratogenic exposure to cyclophosphamide. Teratogenesis Carcinog Mutagen 2: 47–59
Marks DI, Fox RM (1991) DNA damage, poly(ADP-ribosyl)ation and apoptotic cell death as a potential common pathway of cytotoxic drug action. Biochem Pharmacol 42: 1859–1867
Martin SJ, Mazdai G, Strain JJ, Cotter TG, Hannigan BM (1991) Programmed cell death (apoptosis) in lymphoid and myeloid cell lines during zin deficiency. Clin Exp Immunol 83: 338–343
McCabe MJ Jr, Jiang SA, Orrenius S (1993) Chelation of intracellular zinc triggers apoptosis in mature thymocytes. Lab Invest 69: 101–110
McConkey DJ, Hartzell P, Jondal M, Orrenius S (1989a) Inhibition of DNA fragmentation in thymocytes and isolated thymocyte nuclei by agents that stimulate protein kinase C. J Biol Chem 264: 13399–13402
McConkey DJ, Hartzell P, Nicotera P, Orrenius S (1989b) Calcium-activated DNA fragmentation kills immature thymocytes. FASEB J 3: 1843–1849
McConkey DJ, Orrenius S, Jondal M (1990) Agents that elevate cAMP stimulate DNA fragmentation in thymocytes. J Immunol 145: 1227–1230
Meier U, Beier-Hellwig K, Klug J, Linder D, Beier HM (1995) Identification of cyclophilin A from human decidual and placental tissue in the first trimester of pregnancy. Mol Hum Reprod 1 (Hum Reprod 10): 1305–1310
Mendelsohn C, Ruberte E, LeMeur M, Morriss-Kay G, Chambon P (1991) Developmental analysis of the retinoic acid-inducible RAR-β2 promoter in transgenic animals. Development 113: 723–734
Menkes B, Litvac B, Ilies A (1964) Spontaneous and induced cell degeneration in relation to teratogenesis. Rev Roum Embryol Cytol (Ser Embryol) 1: 47–60
Menkes B, Deleanu M, Ilies A (1965) Comparative study of some areas of physiological necrosis in the embryo of man, some laboratory mammals and fowl. Rev Roum Embryol Cytol (Ser Embryol) 2: 161–171
Menkes B, Sandor S, Ilies A (1970) Cell death in teratogenesis. In: Woollam DH (ed) Advances in teratology. Academic, New York, pp 169–215
Menkes B, Prelipceanu O, Caplanasan I (1979) Vital fluorochroming as a tool for

embryonic cell death research. In: Persaud TVN (ed) Abnormal embryogenesis. University Park Press, Baltimore, pp 219–241 (Advances in the study of birth defects, vol 3)

Mentz F, Mossalayi MD, Ouaaz F, Debre P (1995) Involvement of cAMP in CD3 T cell receptor complex- and CD2-mediated apoptosis of human thymocytes. Eur J Immunol 25: 1798–1801

Merino R, Ding L, Veis DJ, Korsmeyer SJ, Nunez G (1994) Developmental regulation of the bcl-2 protein and susceptibility to cell death in B lymphocytes. EMBO J 13: 683–691

Merritt AJ, Potten CS, Kemp CJ, Hickman JA, Balmain A, Lane DP, Hall PA (1994) The role of p53 in spontaneous and radiation-induced apoptosis in the gastrointestinal tract of normal and p53-deficient mice. Cancer Res 54: 614–617

Meuth M (1989) The molecular basis of mutations induced by deoxynucleotide triphosphate pool imbalances in mammalian cells. Exp Cell Res 181: 305–316

Migchielsen AAJ, Breuer ML, van Roon MA, te Riele H, Zurcher C, Ossendorp F, Toutain S, Hershfield MS, Berns A, Valerio D (1995) Adenosine-deaminase-deficient mice die perinatally and exhibit liver-cell degeneration, atelctasis and small intestinal cell death. Nature Gen 10: 279–287

Milaire J (1963) Etude morphologique et cytochimique du developpment des membres chez la souris et chez la taupe. Arch Biol (Liege) 74: 129–317

Milaire J (1967) Histochemical observations on the developing foot of normal, oligosyndactylous (Os/+) and syndactylous (sm/sm) mouse embyros. Arch Biol (Liege) 78: 223–288

Milaire J, Rooze M (1983) Hereditary and induced modifications of the normal necrotic patterns in the developing limb buds of the rat and mouse: facts and hypotheses. Arch Biol (Bruxelles) 94: 459–490

Mirkes PE (1985) Cyclophosphamide teratogenesis: a review. Teratogenesis Carcinog Mutagen 5: 75–88

Mirkes PE, Ricks JL, Pascoe-Mason JM (1989) Cell cycle analysis in the cardiac and neuroepithelial tissues of day 10 rat embryos and the effects of phosphoramide mustard, the major teratogenic metabolite of cyclophosphamide. Teratology 39: 115–120

Miura M, Zhu H, Rotello R, Hartwieg EA, Yuan J (1993) Induction of apoptosis in fibroblasts by IL-1β-converting enzyme, a mammalian homolog of the *C. elegans* cell death gene *ced-3*. Cell 75: 653–660

Miyashita T, Reed JC (1992) *bcl-2* gene transfer increases relative resistance of S49.1 and WEH17.2 lymphoid cells to cell death and DNA fragmentation induced by glucocorticoids and multiple chemotherapeutic drugs. Cancer Res 52: 5407–5411

Miyashita T, Reed JC (1993) Bcl-2 oncoprotein blocks chemotherapy-induced apoptosis in a human leukemia cell line. Blood 81: 151–157

Miyashita T, Reed JC (1995) Tumor suppressor p53 is a direct transcriptional activator of the human bax gene. Cell 80: 293–299

Miyashita T, Krajewski S, Krajewski M, Wang HG, Lin HK, Liebermann DA, Hoffman B, Reed JC (1994) Tumor suppressor p53 is a regulator of bcl-2 and bax gene expression in vitro and in vivo. Oncogene 9: 1799–1805

Monaghan P, Robertson D, Amos TAS, Dyer MJS, Mason DY, Greaves MF (1992) Ultrastructural localization of BCL-2 protein. J Histochem Cytochem 40: 1819–1825

Montague JW, Gaido ML, Frye C, Cidlowski JA (1994) A calcium-dependent nuclease from apoptotic rat thymocytes is homologous with cyclophilin. Recombinant cyclophilins A, B, and C have nuclease activity. J Biol Chem 269: 18877–18880

Mora PT, Chandrasekaran K, McFarland VW (1980) An embryo protein induced by SV40 virus transformation of mouse cells. Nature 288: 722–724

Morgenbesser SD, Williams BO, Jacks T, DePinho RA (1994) p53-dependent apoptosis produced by Rb-deficiency in the developing mouse lens. Nature 371: 72–74

Mori C, Nakamura N, Okamoto Y, Osawa M, Shiota K (1994) Cytochemical identification of programmed cell death in the fusing fetal mouse palate by specific labelling of DNA fragmentation. Anat Embryol (Berl) 190: 21–28

Motoyama J, Taki K, Osumi-Yamashita N, Eto K (1994) Retinoic acid treatment induces cell death and the protein expression of retinoic acid receptor β in the mesenchymal cells of mouse facial primordia in vitro. Dev Growth Differ 36: 281–288

Murgia M, Pizzo P, Sandona D, Zanovello P, Rizzuto R, Di Virgilio F (1992) Mitochondrial DNA is not fragmented during apoptosis. J Biol Chem 267: 10939–10941

Nakamura T, Pichel JG, Williams-Simons L, Westphal H (1995) An apoptotic defect in lens differentiation caused by human p53 is rescued by a mutant allele. Proc Natl Acad Sci USA 92: 6142–6146

Nakayama K-I, Nakayama K, Negishi I, Kuida K, Shinkai Y, Louie MC, Fields LE, Lucas PJ, Stewart V, Alt FW, Loh DY (1993) Disappearance of the lymphoid system in Bcl-2 homozygous mutant chimeric mice. Science 261: 1584–1588

Nelson WG, Kastan MB (1994) DNA strand breaks: the DNA template alterations that trigger p53-dependent DNA damage response pathways. Mol Cell Biol 14: 1815–1823

Nicholson DW, Ali A, Thornberry NA, Vaillancourt JP, Ding CK, Gallant M, Gareau Y, Griffin PR, Labelle M, Lazebnik YA, Munday NA, Raju SM, Smulson ME, Yamin T-T, Yu VL, Miller DK (1995) Identification and inhibition of the ICE/CED-3 protease necessary for mammalian apoptosis. Nature 376: 37–43

Nicol CJ, Harrison ML, Laposa RR, Gimelshtein IL, Wells PG (1995) A teratologic suppressor role for *p53* in benzo[a]pyrene-treated transgenic *p53*-deficient mice. Nature Gen 10: 181–187

Nosseri C, Coppola S, Ghibelli L (1994) Possible involvement of poly(ADP-ribosyl) polymerase in triggering stress-induced apoptosis. Exp Cell Res 212: 367–373

Oberhammer FA, Wilson JW, Dive C, Morris ID, Hickman JA, Wakeling AE, Walker PR, Sikorska M (1993) Apoptotic death in epithelial cells: cleavage of DNA to 300 and/or 50 kb fragments prior to or in the absence of internucleosomal fragmentation. EMBO J 12: 3679–3684

Oberhammer FA, Hochegger K, Froschl G, Tiefenbacher R, Pavelka M (1994) Chromatin condensation during apoptosis is accompanied by degradation of lamin A + B, without enhanced activation of cdc2 kinase. J Cell Biol 126: 827–837

Oliner JD, Kinzer KW, Meltzer PS, George DL, Vogelstein B (1992) Amplification of a gene encoding a p53- associated protein in human sarcomas. Nature 358: 80–86

Oliver FJ, Marvel J, Collins MKL, Lopez-Rivas A (1993) Bcl-2 oncogene protects a bone marrow-derived pre-B cell line from 5-fluoro, 2'-deoxyuridine-induced apoptosis. Biochem Biophys Res Commun 194: 126–132

Oltvai ZN, Milliman CL, Korsmeyer SJ (1993) Bcl-2 heterodimerizes in vivo with a conserved homolog, Bax, that accelerates programed cell death. Cell 74: 609–619

Oren M, Reich NC, Levine AJ (1982) Regulation of the cellular p53 tumour antigen in teratocarcinoma cells and their differentiated progeny. Mol Cell Biol 2: 443–449

Osumi-Yamashita N, Noji S, Nohno T, Koyama E, Doi H, Eto K, Taniguchi S (1990) Expression of retinoic acid genes in the neural crest derived cells during mouse facial development. FEBS Lett 264: 71–74

Osumi-Yamashita N, Iseki S, Noji S, Nohno T, Koyama E, Taniguchi S, Doi H, Eto K (1992) Retinoic acid treatment induces the ectopic expression of retinoic acid receptor β gene and excessive cell death in the embryonic mouse face. Dev Growth Differ 34: 199–209

Owens GP, Hahn WE, Cohen JJ (1991) Identification of mRNAs associated with programmed cell death in immature thymocytes. Mol Cell Biol 11: 4177–4188

Owens GP, Mahalik TJ, Hahn WE (1995) Expression of the death-associated gene RP-8 in granule cell neurons undergoing postnatal cell death in the cerebellum of *weaver* mice. Dev Brain Res 86: 35–47

Pan H, Griep AE (1994) Altered cell cycle regulation in the lens of HPV-16 E6 or E7 transgenic mice: implications for tumor suppressor gene function in development. Genes Dev 8: 1285–1299

Pandey S, Wang E (1995) Cells en route to apoptosis are characterized by the upregulation of *c-fos*, *c-myc*, *c-jun*, *cdc2*, and RB phosphorylation, resembling events or early cell cycle traverse. J Cell Biochem 58: 135–150

Peitsch MA, Polzar B, Stephan H, Crompton T, MacDonald HR, Mannherz HG, Tschopp J (1993) Characterization of the endogenous deoxyribonuclease involved in nuclear DNA degradation during apoptosis (programmed cell death). EMBO J 12: 371–377

Petit PX, Lecoeur H, Zorn E, Dauguet C, Mignotte B, Gougeon M-L (1995) Alterations in mitochondrial structure and fucntion are early events of dexamthasone-induced thymocyte apoptosis. J Cell Biol 130: 157–167

Pexieder T (1975) Cell death in the morphogenesis and teratogenesis of the heart. Adv Anat Embryol Cell Biol 51: 1–100

Power M-A, Tam PPL (1993) Onset of gastrulation, morphogenesis and somitogenesis in mouse embryos displaying compensatory growth. Anat Embryol (Berl) 187: 493–504

Pratt RM, Martin GR (1975) Epithelial cell death and cyclic AMP increases during palatal development. Proc Natl Acad Sci USA 72: 874–879

Purdie CA, Harrison DJ, Peter A, Dobbie L, White S, Howie SEM, Salter DM, Bird CC, Wyllie AH, Hooper ML, Clarke AR (1994) Tumor incidence, spectrum and ploidy in mice with a large deletion in the p53 gene. Oncogene 9: 603–609

Raff MC (1992) Social controls on cell survival and cell death. Nature 356: 397–399

Record IR, Tulsi RS, Dreosti IE, Fraser FJ (1985) Cellular necrosis in zinc-deficient rat embryos. Teratology 32: 389–405

Reed JC (1994) Bcl-2 and the regulation of programmed cell death. J Cell Biol 124: 1–6

Ritter EJ, Scott WJ, Wilson JG (1973) Relationship of temporal patterns of cell death and development to malformations in the rat limb. Possible mechanisms of teratogenesis with inhibitors of DNA synthesis. Teratology 7: 219–226

Robertson LE, Chubb S, Meyn RE, Stroy M, Ford R, Hittelman WN, Plunkett W (1993) Induction of apoptotic cell death in chronic lymphocytic leukemia by 2-chloro-2'-deoxyadenosine and 9-β-D-arabinosyl-2-fluoradenine. Blood 81: 143–150

Rogel A, Popliker M, Webb CG, Oren M (1985) p53 cellular tumor antigen: analysis of mRNA levels in normal adult tissues, embryos, and tumors. Mol Cell Biol 5: 2851–2855

Rooze MA (1977) The effects of the Dh gene on limb morphogenesis in the mouse. Birth Defects (Orig Art Ser) 13/1: 69–95

Sah VP, Attardi LD, Mulligan GJ, Williams BO, Bronson RT, Jacks T (1995) A subset of *p53*-deficient embryos exhibit exencephaly. Nature Gen 10: 175–180

Sanderson CJ (1976) The mechanism of T cell mediated cytotoxicity. II. Morphological studies of cell death of time-lapse microcinematography. Proc R Soc Lond [B] 192: 241–255

Saunders JW Jr (1966) Death in embryonic systems. Science 154: 604–612

Saunders JW, Gasseling MT, Saunders LC (1962) Cellular death in morphogenesis of the avian wing. Dev Biol 5: 146–178

Schanne FAX, Kane AB, Young EE, Farber JL (1979) Calcium dependence of toxic cell death: a final common pathway. Science 206: 700–702

Schmid P, Lorenz A, Hameister H, Montenarh M (1991) Expression of p53 during mouse embryogenesis. Development 113: 857–865

Schwartzman RA, Cidlowski JA (1993) Apoptosis: the biochemistry and molecular biology of programmed cell death. Endocr Rev 14: 133–151

Schweichel JU, Merker HJ (1973) The morphology of various types of cell death in prenatal tissues. Teratology 7: 253–266

Scott WJ (1979) Physiological cell death in normal and abnormal rodent limb development. Adv Study Birth Defects III: 135–142

Searle J, Kerr JFR, Bishop CJ (1982) Necrosis and apoptosis: distinct modes of cell death with fundamentally different significance. Pathol Ann 2: 229–259

Sentman CL, Shutter JR, Hockenbery D, Kanagawa O, Korsmeyer SJ (1991) bcl-2 inhibits multiple forms of apoptosis but not negative selection in thymocytes. Cell 67: 879–888

Seto S, Carrera CJ, Kubota M, Wasson DB, Carson DA (1985) Mechanism of deoxyadenosine and 2-chlorodeoxyadenosine toxicity to nondividing human lymphocytes. J Clin Invest 75: 377–383

Smeyne RJ, Vendrell M, Hayward M, Baker SJ, Miao GG, Schilling K, Robertson LM, Curran T, Morgan JI (1993) Continuous *c-fos* expression precedes programmed cell death in vivo. Nature 363: 166–169

Smith ML, Chen I-T, Zhan Q, Bae I, Chen C-Y, Gilmer TM, Kastan MB, O'Connor PM, Fornace AJ Jr (1994) Interaction of the p53-regulated protein Gadd45 with proliferating cell nuclear antigen. Science 266: 1376–1380

Snow MHL, Tam PPL (1979) Is compensatory growth a complicating factor in mouse teratology? Nature 279: 555–557

Song Q, Lavin MF (1993) Calyculin A, a potent inhibitor of phosphatases-1 and -2A, prevents apoptosis. Biochem Biophys Res Commun 190: 47–55

Soprano DR, Gyda M III, Jiang H, Harnish DC, Ugen K, Satre M, Chen L, Soprano KJ, Kochhar DM (1994) A sustained elevation in retinoic acid receptor-β2 mRNA and protein occurs during retinoic acid-induced fetal dysmorphogenesis. Mech Dev 45: 243–253

Sorenson CM, Rogers SA, Korsmeyer SJ, Hammerman MR (1995) Fulminant metanephric apoptosis and abnormal kidney development in bcl-2-deficient mice. Am J Physiol 268: F73–F81

St James SJ, Basnakian AG, Miller BJ (1994) In vitro folate deficiency induces deoxynucleotide pool imbalance, apoptosis, and mutagenesis in chinese hamster ovary cells. Cancer Res 54: 5075–5080

Sulik KK, Dehart DB (1988) Retinoic-acid-induced limb malformations resulting from apical ectodermal ridge cell death. Teratology 37: 527–537

Sulik KK, Cook CS, Webster WS (1988) Teratogens and craniofacial malformations: relationships to cell death. Development 103 [Suppl]: 213–232

Sulston JE, Horvitz HR (1977) Post-embryonic cell lineages of the nematode *Caenorhabditis elegans*. Dev Biol 82: 110–156

Sulston JE, Schierenberg E, White JG, Thompson N (1983) The embryonic cell lineage of the nematode *Caenorhabditis elegans*. Dev Biol 100: 64–119

Sun DY, Jiang S, Zheng L-M, Ojcius DM, Young JD-E (1994) Separate metabolic pathways leading to DNA fragmentation and apoptotic chromatin condensation. J Exp Med 179: 559–568

Sun X-M, Snowden RT, Dinsdale D, Ormerod MG, Cohen GM (1994) Changes in nuclear chromatin precede internucleosomal DNA cleavage in the induction of apoptosis by etoposide. Biochem Pharmacol 47: 187–195

Sykes K, Gething MJ, Sambrook J (1993) Proline isomerases function during heat shock. Proc Natl Acad Sci USA 90: 5853–5857

Symonds H, Krall L, Remington L, Robles MS, Lowe S, Jacks T, Van Dyke T (1994) p53-dependent apoptosis suppresses tumor growth and progression in vivo. Cell 78: 703–711

Takayama S, Sato T, Krajewski S, Kochel K, Irie S, Millan JA, Reed JC (1995) Cloning and functional analysis of BAG-1: a novel Bcl-2 binding protein with anticell death activity. Cell 80: 279–284

Takano YS, Harmon BV, Kerr JFR (1991) Apoptosis induced by mild hyperthermia in human and murine tumour cell lines: a study using electron microscopy and DNA gel electrophoresis. J Pathol 163: 329–336

Taniguchi K, Sato N, Uchiyama Y (1995) Apoptosis and heterophagy of medial edge epithelial cells of the secondary palatine shelves during fusion. Arch Histol Cytol 58: 191–203

Tepper CG, Studzinski GP (1993) Resistance of mitochondrial DNA to degradation characterizes the apoptotic but not the necrotic mode of human leukemia cell death. J Cell Biochem 52: 352–361

Tewari M, Quan LT, O'Rourke K, Desnoyers S, Zeng Z, Beidler DR, Poirier GG, Salvesen GS, Dixit VM (1995) Yama/CPP32β, a mammalian homolgue of CED-3, is a *CrmA*-inhibitable protease that cleaves the death substrate poly(ADP-ribose) polymerase. Cell 81: 801–809

Thayer JM, Mirkes PM (1995) Programmed cell death and N-acetoxy-2--acetylaminofluorene-induced apoptosis in the rat embryo. Teratology 51: 418–429

Tomei LD, Shapiro JP, Cope FO (1993) Apoptosis in C3H/10T½ mouse embryonic cells: evidence for internucleosomal DNA modification in the absence of double-stranded cleavage. Proc Natl Acad Sci USA 90: 853–857

Tone S, Tanaka S, Minatogawa Y, Kido R (1994) DNA fragmentation during programmed cell death in the chick limb buds. Exp Cell Res 215: 234–236

Tsujimoto Y, Croce CM (1986) Analysis of the structure, transcripts, and protein products of bcl-2, the gene involved in human follicular lymphoma. Proc Natl Acad Sci USA 83: 5214–5218

Ucker DS, Obermiller PS, Eckhart W, Apgar JR, Berger NA, Meyers J (1992) Genome digestion is a dispensable consequence of physiological cell death mediated by cytotoxic T lymphocytes. Mol Cell Biol 12: 3060–3069

Vaux DL (1993) Toward an understanding of the molecular mechanisms of physiological cell death. Proc Natl Acad Sci USA 90: 786–789

Vaux DL, Cory S, Adams JM (1988) *Bcl-2* gene promotes haemopoietic cell survival and cooperates with c-myc in immortalized pre-B cells. Nature 335: 440–442

Vaux DL, Weissman IL, Kim SK (1992) Prevention of programmed cell death in *Caenorhabditis elegans* by human *bcl-2*. Science 258: 1955–1957

Veis DJ, Sorenson CM, Shutter JR, Korsmeyer SJ (1993) Bcl-2-deficient mice demonstrate fulminant lymphoid apoptosis, polycystic kidneys, and hypopigmented hair. Cell 75: 229–240

Veis Novack D, Korsmeyer SJ (1994) *Bcl-2* protein expression during murine development. Am J Pathol 145: 61–73

Vintermyr OK, Gjertsen BT, Lanotte M, Doskeland SO (1993) Microinjected catalytic subunit of cAMP-dependent protein kinase induces apoptosis in myeloid leukemia (IPC-81) cells. Exp Cell Res 206: 157–161

Wakade AR, Przywara DA, Palmer KC, Kulkarni JS, Wakade TD (1995) Deoxynucleoside induces neuronal apoptosis independent of neurotrophic factors. J Biol Chem 270: 17986–17992

Wakamiya M, Blackburn MR, Jurecic R, McArthur MJ, Geske RS, Cartwright J Jr, Mitani K, Vaishnav S, Belmont JW, Kellems RE, Finegold MJ, Montgomery CA Jr, Bradley A, Caskey CT (1995) Disruption of the adenosine deaminase gene causes hepatocellular impairment and perinatal lethality in mice. Proc Natl Acad Sci USA 92: 3673–3677

Walker PR, Smith C, Youdale T, Leblanc J, Whitfield JF, Siikorska M (1991) Topoisomerase II-reactive chemotherapeutic drugs induce apoptosis in thymocytes. Cancer Res 51: 1078–1085

Walton MI, Whysong D, O'Connor PM, Hockenbery D, Korsmeyer SJ, Kohn KW (1993) Constitutive expression of human Bcl-2 modulates nitrogen mustard and camptothcin induced apoptosis. Cancer Res 53: 1853–1861

Wang L, Miura M, Bergeron L, Zhu H, Yuan J (1994) *Ich-1*, an *Ice/ced-3*-related gene, encodes both positive and negative regulators of programmed cell death. Cell 78: 739–750

Wijsman JH, Jonker RR, Keijzer R, van de Velde CJH, Cornlisse CJ, van Dierendonck JH (1993) A new method to detect apoptosis in paraffin sections: in situ end-labeling of fragmented DNA. J Histochem Cytochem 41: 7–12

Wise LD, Scott WJ Jr (1982) Incorporation of 5-bromo-2'-deoxyuridine into mesenchymal limb-bud cells destined to die: relationship to polydactyly induction in rats. J Embryol Exp Morph 72: 125–141

Wubah J, Ibrahim M, Gao X, Ngyuen D, Pisano MM, Knudsen TB (1996) Teratogen-induced eye defects mediated by *p53*-dependent apoptosis. Curr Biol 6: 60–69

Wyllie AH (1980) Glucocorticoid-induced thymocyte apoptosis is associated with endogenous endonuclease activation. Nature 284: 555–556

Wyllie AH, Kerr JFR, Currie AR (1980) Cell death: the significance of apoptosis. Int Rev Cytol 68: 251–306

Yang E, Zha J, Jockel J, Boise LH, Thompson CB, Korsmeyer SJ (1996) Bad, a heterodimeric partner for Bcl-XL and Bcl-2, displaces Bax and promotes cell death. Cell 80: 285–291

Yasuda Y, Konishi H, Kihara T, Tanimura T (1987) Developmental anomalies induced by all-trans-retinoic acid in fetal mice. II. induction of abnormal neuroepithelium. Teratology 35: 355–366

Yonish-Rouach E, Resnitzky D, Lotem J, Sachs L, Kimchi A, Oren M (1991) Wild-type p53 induces apoptosis of myeloid leukaemic cells that is inhibited by interleukin-6. Nature 352: 345–347

Yoshida T, Schneider EL, Mori N (1994) Cloning of the rat Gadd45 cDNA and its mRNA expression in the brain. Gene 151: 253–255

Yoshioka A, Tanaka S, Hiraoka O, Koyama Y, Hirota Y, Ayusawa D, Seno T, Garrett C, Wataya Y (1987) Deoxyribonucleoside triphosphate imbalance. 5-fluorodeoxyuridine-induced DNA double strand breaks in mouse FM3A cells and the mechanism of cell death. J Biol Chem 262: 8235–8241

Yuan J, Horvitz HR (1990) The *Caenorhabditis elegans* genes *ced-3* and *ced-4* act cell autonomously to cause programmed cell death. Dev Biol 138: 33–41

Yuan J, Horvitz HR (1992) The *Caenorhabditis elegans* cell death gene *ced-4* encodes a novel protein and is expressed during the period of extensive programmed cell death. Development 116: 309–320

Yuan J, Shaham S, Ledoux S, Ellis HM, Horvitz HR (1993) The C. elegans cell death gene *ced-3* encodes a protein similar to mammalian interleukin-1β-converting enzyme. Cell 75: 641–652

Yunis JJ, Oken MM, Kaplan ME, Ensrud KM, Howe RR, Theologides (1982) Distinctive chromosomal abnormalities in histologic subtypes of non-hodgkin's lymphoma. N Engl J Med 307: 1231–1236

Zakeri ZF, Quaglino D, Latham T, Lockshin RA (1993) Delayed internucleosomal DNA fragmentation in programmed cell death. FASEB J 7: 470–478

Zakeri ZF, Quaglino D, Ahuja HS (1994) Apoptotic cell death in the mouse limb and its suppression in the Hammertoe mutant. Dev Biol 165: 294–297

Zhivotovsky B, Cedervall B, Jiang S, Nicotera, Orrenius S (1994) Involvement of Ca^{2+} in the formation of high molecular weight DNA fragments in thymocyte apoptosis. Biochem Biophys Res Commun 202: 120–127

Zhan Q, Fan S, Bae I, Guillof C, Liebermann DA, O'Connor PM, Fornace AJ Jr (1994) Induction of *bax* by genotoxic stress in human cells correlates with normal p53 status and apoptosis. Oncogene 9: 3743–3711

Zhong L-T, Sarafian T, Kane DJ, Charles AC, Mah SP, Edwards RH, Bredesen DE (1993) bcl-2 inhibits death of central neural cells induced by multiple agents. Proc Natl Acad Sci USA 90: 4533–4537

CHAPTER 9
Cellular Responses to Stress

P.E. MIRKES

A. Introduction

Since the evolution of the cellular phenotype, prokaryotic and eukaryotic cells have had to cope with adverse changes in their environment. Although cells have evolved many distinct stress responses, this chapter will focus on three major, highly conserved, response systems, i.e., the genotoxic response system, which is activated by DNA damage; the oxidative stress response system, which is activated by excess reactive oxygen species (ROS) and imbalances in the oxidant/antioxidant status within cells; and the heat shock response, which is activated by exposure to heat and other agents that adversely affect protein folding (Fig. 1). The sections dealing with each of the stress response systems begin with a description of the prokaryotic stress response because, in most instances, the prokaryotic systems are the best understood. This is followed by a discussion of the eukaryotic stress response systems, focusing on yeast and mammals. Finally, each section concludes with a discussion about what is known concerning the induction of these stress response systems in mammalian embryos, particularly postimplantation mammalian embryos. Normal embryonic development requires a precisely orchestrated chain of temporal and spatial events, and any alterations in this chain could lead to altered development and subsequent pathogenesis. Although the mammalian embryo develops within the protective environment of the uterus, this protection is not absolute and we now know that mammalian development can be perturbed by a wide variety of chemical and physical agents, many of which are known to induce one or more of these stress systems in nonembryonic systems. Thus understanding the embryo's stress response capabilities is essential to the understanding of how developmental toxicants exert their toxicity.

B. Cellular Responses to Stress

I. Genotoxic Stress Response

1. Introduction

Genotoxic stress, i.e., DNA damage, poses a significant challenge to a cell because, if left unrepaired, it could lead to mutations, altered cell function, or

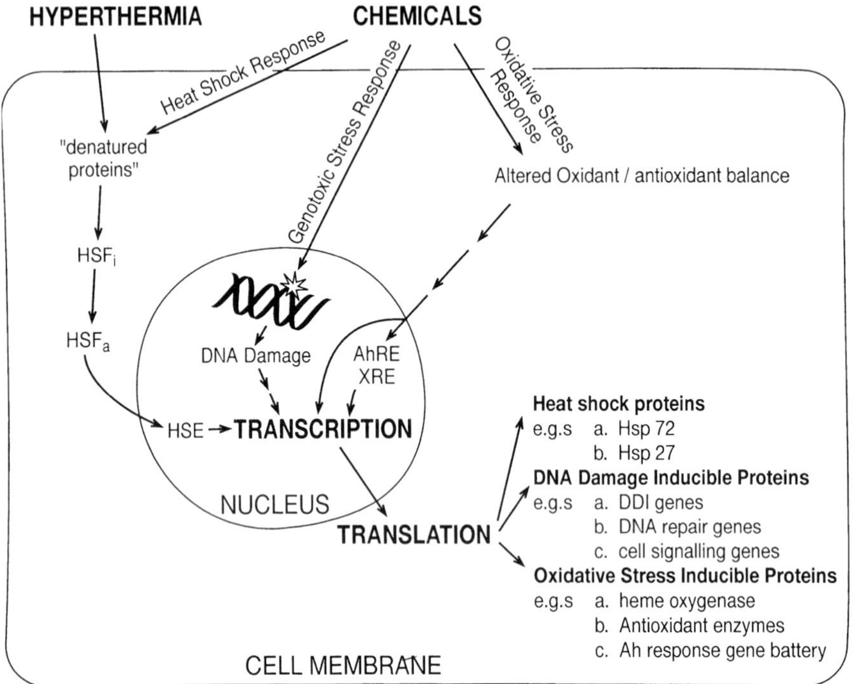

Fig. 1. Cellular response pathways to agents that induce protein denaturation, DNA damage, or altered oxidant/antioxidant balance. All three pathways involve mechanisms for sensing stress, alterations in gene transcription, and the production of specific stress proteins. *XRE*, xenobiotic-responsive element; *HSE*, heat shock element; *HSF*, heat shock transcription factor; *AhRE*, aromatic hydrocarbon-responsive element; *DDI*, DNA damage inducible

cell death. It is not surprising, therefore, that cells from bacteria to humans have evolved responses to DNA damage that involve, in part, mechanisms to repair damaged DNA. This section will provide an overview of the genotoxic stress response, focusing first on prokaryotes, then on lower eukaryotes, and finally on the mammalian genotoxic stress response.

2. Prokaryotes

Although our understanding of cellular responses to DNA damage is still incomplete, the nature and regulation of the genotoxic stress response in bacteria, specifically *Escherichia coli* (E. coli), are known in the greatest detail. Several excellent reviews of the bacterial response to genotoxic stress, known as the SOS response, are available (LITTLE and MOUNT 1982; WALKER 1984, 1985), and the reader is referred to them for a detailed description. This section is intended to highlight the major aspects of the genotoxic stress response in bacteria as a basis for comparison to the mammalian genotoxic stress response (see below).

E. coli respond to DNA damage by inducing two major regulatory networks, i.e., the SOS network and the adaptive response network (WALKER 1985). The SOS network was the first to be characterized, and it was shown that more than 17 genes were induced in response to DNA damage or interference with DNA replication (KENYON and WALKER 1980; WALKER 1984). The best characterized of the 17 genes (SOS regulon) induced by DNA damage are the *UvrA,-B,-C,* and *-D* genes. The gene products of the *UvrA,-B,* and *-C* genes interact to form the UvrABC endonuclease, an enzyme complex that is able to recognize a wide range of DNA lesions, such as ultraviolet (UV) induced pyrimidine dimers, UV-induced pyrimidine-pyrimidone photoproducts, guanosines modified with platinum compounds, and psoralen adducts, and to remove them by excision repair (SANCAR and RUPP 1983; FRANKLIN and HASELTINE 1984). Some of the other genes induced by DNA damage include *recA, recN,* and *ruv*, genes thought to play a role in daughter-strand gap and double-strand breaks repair (LLOYD et al. 1984; WEST et al. 1981, 1982; WEST and HOWARD-FLANDERS 1984). Although the functions of some of the SOS genes are unknown, one gene, *sulA*, is not a DNA repair gene, but rather functions in growth arrest (PAPATHANASIOU and FORNACE 1991). In response to this signal, a specific protease function of the RecA protein is reversibly activated. Activated RecA, in turn, cleaves the LexA protein, a repressor of SOS regulon. Proteolytic cleavage of LexA relieves this repression and leads to the coordinate induction of the SOS genes (WALKER 1984).

In addition to the SOS response, DNA damage also induces an adaptive response reminiscent of the thermotolerance response induced by heat stress (see Sect. B.III.4). The adaptive response is induced by methylating and ethylating agents in the following manner. If *E. coli* are exposed to a low concentration of a methylating agent, e.g., *N*-methyl-*N*'-nitro-*N*-nitrosoguanidine (MNNG) they become resistant to the mutagenic and lethal affects of a subsequent higher dose of the same agent (SAMSON and CAIRNS 1977). This induced resistance, termed the adaptive response, involves at least four genes, *alkA, alkB, aidB,* and *ada. AlkA* encodes a DNA glycosylase (3-methyladenine-DNA glycosylase II), an enzyme with broad substrate specificity capable of removing 3-methyladenine, 3-methylguanine, O^2-methylcytosine, and O^2-methylthymine (KARRAN et al. 1982; EVENSEN and SEEBERG 1982; YAMAMOTO et al. 1983; MCCARTHY et al. 1984). The *aidB* gene encodes a 60-kDa protein that is homologous to several mammalian acyl coenzyme A dehydrogenases (LANDINI et al. 1994). The *ada* gene encodes a 39-kDa protein that serves at least two function; it is a positive regulator of the adaptive response (SEDGWICK 1983; LEMOTTE and WALKER 1985) and an alkyltransferase capable of repairing O^6-methylguanine, O^4-methylthymine, and methylphosphotriester lesions (TEO et al. 1984). The *alkB* gene product plays some as yet undefined role in the repair of alkylated DNA (YAMAMOTO et al. 1983; KATAOKA et al. 1983; KATAOKA and SEKIGUCHI 1985).

3. Eukaryotes

a) Yeast

In yeast more than 30 genes have been identified that function in DNA repair, i.e., the *RAD* genes (Friedberg 1988). These genes, belonging to three epistasis groups, function in nucleotide excision repair (*RAD*3 group), recombinational repair (*RAD52* group), and mutagenic processes (*RAD6* group). Like the *sulA* gene in *E. coli*, the *RAD9* gene in yeast encodes a growth arrest function. In addition, several of these DNA repair genes have been shown to be inducible by DNA damage. Likewise, DNA ligase (Johnson et al. 1986), DNA polymerase (Johnston et al. 1987), and ribonucleotide reductase (Elledge and Davis 1987) are also induced by DNA damage. In addition to these genes, McClanahan and McEntee (1984) have isolated cDNA clones for four genes that are induced by DNA damage, i.e., the *DDR* (DNA damage responsive) genes. Finally, six *DIN* (damage-inducible) genes of unknown function have been isolated by Ruby and Szostak (1985). These authors estimate that there may be over 80 *DDI* genes in yeast, representing nearly 1% of the total genes in yeast. Although the functions of yeast *DDI* genes are less well understood compared to the situation in bacteria, it is clear that yeast dedicate a significant part of their genome to their genotoxic stress response. Given the potentially large number of genes involved in that response, it is also clear that the yeast genotoxic stress response is a complex phenomenon.

b) Mammals

α) DNA Damage-Inducible Genes. Because of the large number of genes induced by DNA damage in bacteria and yeast and the complexity of the genotoxic stress response that this suggests, one would predict that there is also a complex genotoxic stress response in mammals. Even though research concerning the mammalian genotoxic stress response has a relatively short history, this prediction has already been verified and more than 50 genes (see Table 1) have been identified whose mRNA are induced by DNA damage. Several excellent reviews (Herrlich et al. 1994; Gewirtz 1993; Fornace et al. 1992; Fornace 1992) have recently appeared that provide a detailed analysis of the mammalian genotoxic stress response, and the goal of the present review is to highlight the major aspects of this response. A quick review of Table 1 reveals the complexity of the mammalian genotoxic stress response in that genes with very diverse functions, such as DNA repair, transcriptional regulation, cell signalling, growth control, and responses to injury, and many with still unknown functions, are induced by DNA damage.

β) DNA Repair Genes. As is the case in bacteria and yeast, some of the mammalian *DDI* genes encode for proteins involved in DNA repair, although, as has been pointed out, DNA repair genes constitute a small fraction

Table 1. Mammalian DNA-damage inducible genes (DDI)

Gene	Inducing agents				
	UV	X-ray	Alk	H_2O_2	Go
DNA repair					
β-Polymerase	–	–	+	+	–
O^6-MT	+	+	+		
MAG		+	+		
Cell Signaling					
c-*fos*	+	+	+	+	–
c-*jun*	+	+	+	+	+
jun-B	+	+		+	+
jun-D	+		+	+	+
Protein kinase C		+			
EGR1		+			+
c-myc	+			+	
TNF	+	+			
IL-1	+	+			
Interferon-α	+	+			
bFGF	+	+			
EGF receptor	+				
DNA damage-inducible (*DDI*)					
DDIA class I clones	+	–	–	–	–
DDIA18	+	–	–	–	–
DDIA class II clones	+		+		
DDIA9	+	–	+	+	–
DDIA15	+	–	+	+	–
DDIU cDNA	+				
DDIU1	+	–	+	–	
DDIU2	+	–	+	–	
DDIU3	+	–	+	–	
Growth arrest and DNA Damage (gadd)					
gadd45	+	+	+	+	+
gadd153	+	–	+	+	+
gadd34	+	–	+	+	+
gadd33	+	–	+	+	+
gadd7	+	–	+	+	+
sprI	+				+
sprII	+				+
spr2–1	+				+
RP-2, RP-8		+			+
MyD118		–	+		+
Others					
NFκB		+			
C-H-*ras*	+				
C-5(*trk-2* homologue)	+				
HIV-1 LTR	+	–	–		
Mo-MuSV LTR	+	+			
RaLV	+				
SV40 promoter	+		+		
Virus-like 30s element		+			
Heme oxygenase	+	+		+	
Nmo-1		+	+		

Table 1. (*Contd.*)

Gene	Inducing agents				
	UV	X-ray	Alk	H_2O_2	Go
Metallothioneins	+				
Collagenase	+	+	+		
Uroplasminogen activator	+				
tis-Plasminogen activator	+	+			
Ornithine decarboxylose	+	+			
Ubiquitin	–	–	+	–	–
hsp70	–	–	+	–	–
hsp27	+	–	+	–	–
hsp89 α and-β	–	–	+	–	–
grp78	–		+		

For references to specific genes see Fornace (1992), Papathanasiou and Fornace (1991), Fornace et al. (1992) and Keyse (1993). O^6-MT, O^6-methyltransferase; MAG, N^3-methyladenine glycosylase; TNF, tumor necrosis factor; IL, interleukin; bFGF, basic fibroblast growth factor; EGF, epidermal growth factor; NF, nuclear factor.

of *DDI* genes induced. Given the complexity of mammalian DNA repair (Sancar 1994; Hanawalt 1994; Modrich 1994), it seems likely that other inducible DNA repair genes remain to be discovered. To date, three DNA repair genes are known to be induced by genotoxic stress, i.e., β-polymerase, O^6-methyltransferase (O^6-MT), and N^3-methyladenine glycosylase (MAG). All three of these DNA repair genes function in the repair of non-UV-type base damage.

γ) Cell-Signaling Genes. Among other genes that are induced by DNA damage are those involved with cell signaling. These include certain immediate-early or early response genes such as c-*fos, jun-B, jun-D,* and *EGR1* that are known to play a role in growth control, differentiation, and response to injury. Given that these genes, in turn, can control a variety of downstream genes, it is not surprising that these genes are activated by a variety of stresses including growth arrest. One of the challenges will be to understand the relationships between the initial DNA damage, the activation of these genes, and the subsequent responses that a cell activates in an attempt to restore homeostasis. Other cell-signaling genes induced by DNA damage are certain cytokines such as tumor necrosis factor (TNF), interleukin (IL)-1, interferon-α and basic fibroblast growth factor (bFGF). These cytokines are known to play a role in inflammation and response to tissue injury so that, again, it is not surprising that these factors are induced by DNA damage. It will also not be surprising if other cytokines are shown to be inducible by DNA damage.

δ) Other DNA Damage-Inducible Genes. Recent efforts, using subtractive hybridization strategies, have identified a large number of *DDI* genes (For-

NACE et al. 1989). These can be divided into two basic groups, those that are inducible by some form of DNA damage, i.e., the *DDIA* and *DDIU* genes, and those induced by DNA damage and growth arrest, i.e., the *gadd* genes. Within the group of genes induced by DNA damage but not growth arrest, two subgroups are apparent, i.e., those that are only induced by UV-type DNA damage (class I) and those induced by multiple DNA damaging agents (class II and *DDIU*). The functional significance of these subgroups is presently unclear; however, as more of these genes are characterized, new insights into their function will be forthcoming. The *gadd* genes, designated *gadd7, -33, -34, -45,* and *-153*, constitute an interesting group of genes that is coordinately regulated by agents that induce DNA damage and at the same time force cells out of the cell cycle (PAPATHANASIOU and FORNACE 1991; FORNACE et al. 1989). Recent work has shown that induction of the *gadd* genes involves both transcriptional and post-transcriptional control mechanisms (JACKMAN et al. 1994). Inducing agents include methymethansulfonate (MMS), nitrogen mustard, mephalan, MNNG, UV radiation, *N*-acetoxy-2-acetylaminofluorine (AAF), cis-Pt (II) diaminedichloride, cadmium salts, and hydrogen peroxide, with MMS being the most consistent and strongest inducer. Although little is known about the function of the *gadd* gene products, it is now known that *gadd153* encodes a nuclear protein, also called CHOP-10, which is homologous to CAAT enhancer-binding proteins (JACKMAN et al. 1994; RON and HABENER 1992). Thus *gadd153* may serve as a negative regulator of CAAT enhancer-binding protein. Further, *gadd45* is known to be a p53-responsive gene, thereby relating at least one of the *gadd* genes to pathways leading to growth control and apoptosis.

Two other groups of growth arrest and DNA damage-related genes, the *spr* and *MyD*, have also been identified. The *spr* (small proline rich) genes are a group of *DDI* genes isolated from human keratinocytes (KARTASOVA and VAN DE PUTTE 1988), and one of them, *sprI*, has strongly homology to *gadd33* (FORNACE 1992). These genes encode 8-to 10-kDa proteins of unknown functions that are rich in proline, cysteine, and glutamine. *MyD* genes were isolated from murine M1 myeloblastic leukemia cell line stimulated to undergo terminal differentiation by IL-6 (LORD et al. 1990). This group of genes includes five transcription factors, four of which are known to be inducible by DNA damage. In addition, two genes from this group, *MyD116* and *MyD118*, share homology with *gadd34* and *gadd45,* respectively. Once again, the function of these *gadd*-related genes is not known, although it has been suggested that *gadd34*/*MyD116* and *gadd45*/*MyD118* may be involved in apoptosis, because cell death often follows growth arrest accompanying differentiation or DNA damage (FORNACE 1992). This could be a significant suggestion to follow up, because it is well known that cell death is a common consequence of exposure to most, if not all, developmental toxicants.

ε) Regulation of the Genotoxic Response. Information just reviewed clearly indicates the potential complexity of the cellular response to DNA damage in higher eukaryotes, but we now need to consider what is known about the regulation of this response, i.e., what the sensor is and how DNA damage is transduced into the activation of DNA repair, growth control, and other pathways related to coping with genotoxic stress. Although many of the factors involved in the regulation of the genotoxic response are yet to be discovered, two candidates for the sensor of DNA damage are already known. One is poly (adenosine diphosphate-ribose) polymerase (PARP) (De Murcia et al. 1994). PARP is a multifunctional protein containing two fragments involved in automodification (ribosylation and nicotinamide adenine dinucleotide (NAD) binding and another fragment that includes a DNA-binding domain. The DNA-binding domain contains two zinc-containing mini-domains that recognize and bind to a single-stranded DNA break. As result of binding to damaged DNA, PARP covalently modifies a family of nuclear proteins as well as itself. Clearly PARP can "sense" single-standed DNA breaks; however, the relationship to the subsequent induction of the genotoxic stress response is unknown. The second candidate for sensor of DNA damage is DNA-activated protein kinase (DNA-PK) (Anderson 1993). In vitro, DNA-PK binds to and is activated by DNA containing a free end. As a result of activation, DNA-PK has been shown to phosphorylate a variety of proteins including the transcription factors c-jun, c-fos, and c-myc that are also known to be inducible by DNA damage. Thus, DNA-PK may play a pivotal role in activating a cascade of gene expression in response to genotoxic stress. In addition to the phosphorylation of transcription factors, DNA-PK also phosphorylates p53. Given that p53 is a key factor in regulating the G_1/S transition of the cell cycle, it is tempting to speculate that DNA-PK may also be a link between DNA damage and subsequent alterations in the cell cycle mediated by p53.

ξ) Mammalian Embryonic DNA Damage-Inducible Genes. The majority of genes induced by DNA damage have been identified within the last 6 years, which explains the paucity of information regarding the function of these genes. Nonetheless, a wealth of information is available concerning the induction of these genes in a wide variety of cell types. In contrast, virtually nothing is known about the induction of any of the *DDI* genes in mammalian embryos. However, a knowledge of the mechanisms available to mammalian embryos to cope with DNA damage is crucial to our understanding of how developmental toxicants, many of which directly or indirectly damage DNA, induce abnormal sequelae. To gain such knowledge, my laboratory has begun to assess the capacity of mammalian postimplantation embryos to respond to DNA damage induced by selected embryotoxins known to induce various types of DNA damage (Little and Mirkes 1994). Our approach is to expose embryos cultured in vitro to selected developmental toxicants and then assess the expression of *DDI* genes. As our embryotoxins we have used MMS, which induces primarily DNA single-strand breaks, 4'-(9-acridinylamino) methan-m-anisidide (mAMSA), a topoisomerase inhibitor that produces protein-associated DNA strand breaks, 4-hydroperoxycyclophosphamide (CP), which produces primarily DNA crosslinks, and AAF, which produces DNA crosslinks and strand breaks. Using these embryotoxins we have shown that, of the four, only MMS induces the expression of *gadd153* above the basal constitutive level. Using in situ hybridization, we show that, although *gadd153* is expressed in all tissues of the embryo, the heart exhibits higher constitutive levels of *gadd153* and, furthermore, this gene is induced to higher levels in the heart than other embryonic tissues. These findings demonstrate several important points. First and not unexpectedly, mammalian embryos can mount a response to DNA damage, although we have not formally ruled out the possibility that induction of *gadd153* is related to damage-induced growth arrest. Second, DNA damage itself is not sufficient to induce *gadd153* expression, as evidenced by the negative results with AAF, CP and mAMSA. The fact that single-strand DNA appears to be the signal for the bacterial SOS response (Papathanasiou and Fornace 1991), the positive induction by MMS, which produces DNA single-strand breaks, and the lack of induction by the other embryotoxins, which do not, may indicate that mammalian embryonic cells are responding to single-stranded DNA as well. Third, responses to genotoxic stress may be cell specific, as shown by the higher constitutive and induced levels of *gadd153* in embryonic heart cells. Why this is so is unclear, but may be related to state of differentiation or proliferation, to name only the most obvious possibilities. Whatever the underlying reasons, cell-specific responses to genotoxic stress could play a major role in determining organ-specific developmental toxicity.

II. Oxidative Stress Response

1. Introduction

Although the use of oxygen provided a clear evolutionary advantage for aerobic organisms, this advantage came with a price. Aerobic organisms must defend themselves from the toxic byproducts of oxygen metabolism, i.e., ROS. In response to these toxic metabolities of oxygen, prokaryotic and eukaryotic organisms have evolved elaborate defense mechanisms, including enzymatic and nonenzymatic antioxidants. Although these defenses are normally adequate to detoxify ROS generated by cellular metabolic processes, these frontline defenses may not be sufficient under conditions of oxidative stress. Thus, prokaryotes and eukaryotes have evolved oxidative stress-inducible strategies to deal with excess ROS. This selection will begin with an overview of the oxidative stress response in bacteria, in which we have the most complete understanding, and then moves on to the mammalian response.

2. Prokaryotes

a) The oxyR Regulon Gene

Two major oxidative stress response systems have been identified in bacteria: the hydrogen peroxide and superoxide responses. When *E. coli* and *Salmonella typhimurium* are exposed to hydrogen peroxide, they rapidly begin to synthesize approximately 30 proteins, as detected on two-dimensional gels (CHRISTMAN et al. 1985; GREENBERG and DEMPLE 1989). Of these 30 proteins, 12 are maximally induced within the first 10 min after treatment, whereas an additional 18 proteins are induced within the first hour. Although many of these oxidative stress-inducible proteins have not been identified, three enzymes are known which are important in oxidant detoxification. These include catalase, glutathione reductase, and alkyl hydroperoxide reductase. Catalase converts hydrogen peroxide to oxygen and water and thereby directly removes this potentially toxic metabolite. Glutathione reductase plays a role in maintaining the pool of reduced glutathione, which in turn plays a role in the detoxification of hydrogen peroxide and in maintaining the reduced state of cellular proteins. Alkyl hydroperoxide reductase converts physiological hydroperoxides generated by oxidative stress to their corresponding alcohols.

These three enzymes and six other early hydrogen peroxide-inducible proteins are coordinately regulated in bacteria under the control of the *oxyR* gene (CHRISTMAN et al. 1989). These coordinately regulated genes are thus called *oxyR* regulon or stimulon. The *oxyR* gene product is a 34-kDa protein homologous to a family of bacterial proteins known to regulate downstream genes. Interestingly, neither the synthesis nor the levels of the OxyR protein increase after exposure to hydrogen peroxide, suggesting that activation of the *oxyR* regulon is achieved by activating the endogeneous OxyR protein. Subsequent work showed that, under oxidant stress, the OxyR protein becomes oxidized. This oxidation, in turn, induces a conformational change by which OxyR is activated and thereby able to transduce the oxidative stress signal directly to RNA polymerase. Thus in the hydrogen peroxide response system, the protein that regulates key response genes also serves as the sensor of oxidative stress. This is one of the few instances in which the stress sensor has been identified and it mechanism of action elucidated.

b) The SoxRS Regulon Gene

The second major oxidative stress response system in bacteria is the response to excess superoxide, typically induced experimentally by paraquat. Two-dimensional gels revealed that approximately 40 proteins are induced by redox cycling agents that generate superoxide (GREENBERG and DEMPLE 1989; WALKUP and KOGOMA 1989). Again, many of these inducible proteins have not been identified; however, several enzymes have been identified, including superoxide dismutase (Mn SOD), endonuclease IV (Endo IV), glucose-6-phosphate dehydrogenase (G6PD), and fumarase C. All of these inducible proteins have demonstrated roles in antioxidant defense. Mn SOD directly converts superoxide to hydrogen peroxide; Endo IV is involved in repair of oxidative damage to DNA; G6PD is a key enzyme in the maintenance of reduced nicotinamide adenine dinucleotide phosphate (NADPH), an essential cofactor for enzymes, e.g., glutathione reductase; and fumerase is a TCA cycle enzyme.

Like the proteins belonging to *oxyR* regulon, these enzymes and at least six other proteins are coordinately regulated by the soxRS locus. The soxRS locus is composed of two genes, *soxR* and *soxS*, which encode two proteins, a 17-kDa (SoxR) and a 12.9-kDa (SoxS) one, related to known prokaryotic transcription regulators. Regulation of *soxRS* regulon is more complicated than regulation of *oxyR* regulon. In the induction of the *soxRS* regulon, the SoxR protein is a redox-sensitive protein (sensor) that may be activated directly by superoxide through an iron–sulfur center in its C-terminal cysteine cluster (HIDALGO and DEMPLE 1994). Once activated, SoxR binds to the soxS promoter, thereby activating transcription of the *soxS* gene. The resultant SoxS protein in turn binds to downstream promoters and activates transcription of genes belonging to *soxRS* regulon (LI and DEMPLE 1994). Thus, both the OxyR and the SoxR transcription factors appear to be redox-sensitive proteins that may act both as molecular sensors of oxidative stress and transcription activators of genes involved in antioxidant defense.

3. Eukaryotes

a) Stress-Inducible Genes

Although the oxidative stress response in bacteria is reasonably well understood, our understanding of how mammalian cells respond to oxidative stress is much less complete. Nonetheless, studies using mammalian cells have now shown that a variety of proteins are induced by oxidative stress, and those identified in a literature search are presented in Table 2. These proteins range in function from proto-oncogene early response gene products such as c-fos and c-myc, cytokines such as IL-1, antioxidant enzymes such as superoxide dismutase (SOD), proteases such as collagenase to the metal-binding proteins such as metallothionein. With the exception of the antioxidants and perhaps metallothionein, the exact role of many of these proteins in the oxidative stress response remains to be elucidated. In addition, *gadd* genes and *DDI* genes are also inducible by oxidative stress. Whether these genes are induced in response to oxidative stress per se, i.e., ROS, or the ROS-induced DNA damage and/or growth arrest that ensues is often unclear. The induction of *gadd* and *DDI* genes by oxidative stress and DNA damage serves to make the point that the same stress protein can be induced by different stresses.

b) Heme Oxygenase

Although the proteins described above have been shown to be induced either directly or indirectly by oxidative stress, the protein most consistently induced

Table 2. Mammalian oxidative stress-inducible genes

Oxidative stress-inducible genes	Inducers	References
Proto-oncogenes c-*fos*, c-*myc*, *jun-B*, *jun-D*, c-*jun*	UV, X-ray, H_2O_2	Fornace (1992)
Transcription factors NFkB	UV	Brach et al. (1991); Fornace (1992)
Gadd genes *gadd7*, *-33*, *-34*, *-45*, *-153*	UV, H_2O_2	Fornace (1992)
DNA damage-inducible genes (DDI) *DDIA9* *DDIA15* β-Polymerase	UV, H_2O_2	Fornace (1992)
Metallothionein (MT) MT I, MT II	UV, TPA	Bauman et al. (1991); Herrlich et al. (1996)
Proteases Collagenase Plasminogen activator	UV, TPA	Miskin et al. (1981); Aggeler and Murnane (1990)
Antioxidants		
heme oxygenase	UV, X-rays, H_2O_2 Sodium arsenite, cadmium chloride, menadione	Applegate et al. (1991); Janssen et al. (1992, 1993); Maines (1984); Keyse (1990)
Antioxidant enzymes SOD, catalase, GPX	Asbestos	Holley et al. (1992); Jornot et al. (1992); Janssen et al. (1992)
Cytokines Interleukin-1 TNF	UV, ionizing radiation	Hallahan et al. (1989); Woloschak et al. (1990)
Aromatic hydrocarbon (Ah) response gene battery Cytochrome P1–450 Cytochrome P3–450 NAD(P)H: menadione oxidoreductase (*Nmo-1*) Aldehyde dehydrogenase (*Aldh-1*) UDP-glucuronosyltransferase (*Ugt-1*) Glutathione transferase (*G4-1*)	TCDD, phytoalexins, menadione, MMS, butylated hydroxytoluene	Nebert et al. (1990)

SOD, superoxidase dismutase; TNF, tumor necrosis factor; NADPH, reduced nicotinamide adenine dinucleotide phosphate; UDP, uridine diphosphate; UV, ultraviolet; TPA, 12-*o*-tetradecanoylphorbol-13-acetate; TCDD, 2, 3, 7, 8-tetrachlorodibenzo-*p*-dioxin; MMS, methylmethansulfonate; NF, nuclear factor; GPX, glutathione peroxidase.

by oxidative stress in a variety of mammalian cells is the heme-degrading enzyme heme oxygenase (HO). HO exists as two isoforms, HO-1, which is inducible by a variety of agents, and HO-2, which appears not to be inducible.

HO plays an essential role in heme metabolism by cleaving heme to biliverdin, which is subsequently converted to bilirubin. HO-1 can be induced by UVA radiation (320–380 nm), X-ray irradiation, therapeutic agents (e.g., cimetidine), organic solvents (e.g., carbon tetrachloride), hormones (e.g., insulin), bacterial toxins, sulfhydryl reagent depleters (e.g., diethyl maleate), halogenated hydrocarbons (e.g., Arochlor, Monsanto), alkylating agents (e.g., ethionine), and organic metal complexes (e.g., organotins) (MAINES 1984), as well as hydrogen peroxide, cadmium chloride, sodium arsenite, and menadione (JANSSEN et al. 1993).

Although the function of HO-1 induced by oxidative stress is not completely understood, recent work suggests the following scenario. HO-1, induced by oxidative stress, converts heme (hemoglobin/cytochrome P450) to biliverdin and in the process releases free iron. This release of free iron leads to the induction of ferritin (VILE and TYRRELL 1993), which then binds intracellular free iron. Finally, ferritin acts as an antioxidant by sequestering free iron required for the production of the highly toxic hydroxyl radical via the Fenton reaction. In addition to the antioxidant effects of increased ferritin, it has also been proposed that the increased production of bilirubin, associated with HO-1-mediated metabolism of heme, serves an antioxidant function (STOCKER 1990), because it is known that bilirubin is an effective scavenger of singlet oxygen as well as the superoxide and peroxyl radicals (JANSSEN et al. 1993). Thus the induction of HO-1 may enhance the antioxidant status of cells both through an increase in ferritin and in bilirubin and thereby ameliorate the toxic effects of oxidative stress.

c) Aromatic Hydrocarbon-Responsive Gene Battery

As outlined previously, some of the genes induced by oxidative stress in bacteria are coordinately regulated, i.e., *oxyR* and *soxRS* regulons. The question now arises of whether mammalian cells have evolved a similar strategy. Recent work involving the aromatic hydrocarbon-responsive (Ah) gene battery indicates that the Ah gene battery may represent a major, coordinated response to oxidative stress in mammalian cells (NEBERT et al. 1990; NEBERT 1994). In the mouse, the Ah gene battery comprises at least six genes. Two of these genes, *CYP1A1* and *CYP1A2*, encode two phase I drug-metabolizing enzymes, cytochrome P1–450 and P3–450, respectively. These enzymes function by inserting one atom of oxygen into endogenous or xenobiotic substrates and in the process often produce reactive/toxic products. In addition, four phase II genes, *Nmo-1* (NAD(P)H:menadione oxidoreductase), *Aldh-1* (aldehyde dehydrogenase), *Ugt-1* (uridine diphosphate-glucuronosyltransferase), and Gt-1 (glutathione transferase) are also part of this battery. These four enzymes act on oxygenated intermediates formed by phase I enzymes to produce hydrophilic products that are readily excreted from the cell. Two major regulatory elements have been discovered in these Ah genes. The upstream regulatory regions of *Cyp1a1* and *Cyp1a2* genes contain multiple Ah elements (AhRE, also called dioxin-responsive elements, DRE, and xenobiotic-responsive elements, XRE). The four phase II genes also contain at least one AhRE and in addition one antioxidant response element (ARE or EpRE). Recent work has shown that this ARE is activated in cells exposed to hydrogen peroxide and a variety of redox cyclers (RUSHMORE et al. 1991). Thus all six Ah genes can be upregulated by the activated Ah receptor and, in addition, the four phase II genes can also be upregulated independently of the Ah receptor through a putative transcription factor that binds to EpRE/ARE. The activity of this putative transcription factor, and thereby the four phase II genes, is negatively regulated by the gene product of another gene, *Nmo-1n*, located on chromosome 7. In response to oxidative stress, the *Nmo-1n* gene is turned off and phase II genes turned on, leading to the production of at least four proteins which function to combat oxidative stress.

d) Embryonic Oxidative Stress-Inducible Genes

Although no systematic analysis of oxidative stress-inducible genes in mammalian embryos has been reported, several related papers have been published. Studies in postimplantation rat and mouse embryos have shown that SOD, catalase, and glutathione peroxidase mRNA and enzyme activities are detectable from the early somite stage on (REIMER and SINGH 1990; EL-HAGE and SINGH 1990). In addition, other studies have shown that lead 2,3,7,8-tetrachlorodibenzo-*P*-dioxin (TCDD), and excess glucose can lead to alterations in either antioxidant enzyme activity or Ah receptor expression (SOMASHEKARAIAH et al. 1992; ERIKSSON and BORG 1993; ABBOTT et al. 1994). Thus, early postimplantation mammalian embryos have at least some components of their antioxidant defense system in place and can respond to developmental toxicants that have the potential to upset the oxidant–antioxidant balance. Nonetheless, our understanding of how the mammalian embryo responds to oxidative stress is rudimentary at this time. Clearly, additional research in this area is needed.

III. Heat Shock Response

1. Introduction

Cells from bacteria to humans respond to elevated temperature by activating the heat shock response, which is characterized by activation of heat shock gene transcription and the synthesis of heat shock proteins. In addition, many of the heat shock genes and their corresponding proteins exhibit remarkable homology between bacteria and humans. For example, members of the heat shock protein 70 (Hsp-70) family share a 50% homology at the amino acid level between bacteria and humans. The evolutionary conservation of the heat shock response and heat shock proteins suggest that heat shock proteins play an indispensable role in cell survival. Although the initial discovery of the heat shock response in *Drosophila* involved induction by heat, subsequent studies by RITOSSA (1962) and others have shown that the so-called heat shock response can also be induced by a variety of stresses, including amino acid analogues, heavy metals, uncouplers of oxidative phosphorylation, inhibitors of electron transport, inhibitors of protein synthesis (e.g., puromycin), steroid hormones, antineoplastic drugs, and certain teratogens, carcinogens, and mutagens (NOVER 1991). Because the heat shock response can be induced by a variety of agents in addition to heat, some have suggested that this response be renamed the stress response and the proteins induced be designated stress proteins. Nonetheless, to distinguish this stress response system from the oxidative and genotoxic stress response systems, the heat shock response designation will be used in this chapter.

The fact that heat shock genes can be activated by a variety of stresses initially led to confusion concerning the mechanism by which cells recognize an adverse change in their environment, i.e., what stress-induced alteration signals the induction of the heat shock response. Based on the fact that many inducers of the heat shock response also adversely affect protein conformation, HIGHTOWER (1980) initially hypothesized that the accumulation of abnormally folded proteins was the trigger that activated the heat shock response. Support for this hypothesis came from studies showing that cells grown in the presence of amino acid analogues, which lead to abnormally folded proteins, undergo a heat shock response (GOFF and GOLDBERG 1985; KELLEY and SCHLESSINGER 1978). Definitive supporting evidence came from studies showing that the microinjection of denatured proteins into frog oocytes was sufficient to activate the heat shock response. The knowledge that abnormal protein folding triggers the induction of heat shock proteins suggested also that at least one of the functions of heat shock proteins was to facilitate normal protein folding. This suggestion has been confirmed by more recent data showing that many heat shock proteins are constitutively expressed in the absence of stress and that they function as molecular chaperones in facilitating protein folding and assembly (see below).

Although stress-induced alterations in protein folding appear to be the intracellular signal that a cell is experiencing stress, how this signal is transduced into the nucleus, with the concomitant activation of heat shock genes, has also been recently addressed. The key to this signal transduction pathway is the heat shock transcription factor (HSF). A current model (MORIMOTO 1993) of how HSF regulates inducible heat shock gene expression involves the interaction of HSF with one of the major heat shock proteins, HSP-70, which is constitutively synthesized in the absence of stress. In the absence of stress HSF is present as a monomer in the cytosol, where it transiently interacts with HSP-70. This interaction apparently stabilizes the conformation of HSF in a non-DNA-binding form. During stress, the accumulation of misfolded or aggregated proteins competes with HSF for available HSP-70. The result is that HSP-70 is titrated from its association with HSF. The dissociation of HSP-70 from HSF removes the negative regulatory influence of HSP-70 and allows HSF to assemble into trimers. This trimeric form of HSF then migrates into the nucleus, binds to consensus elements (HSE) in the promoter region of all heat shock genes, and activates heat shock gene transcription, which culminates in the synthesis and accumulation of heat shock proteins. The accumulation of newly synthesized HSP-70 to levels that exceed the demand for interaction with misfolded or aggregated proteins allows HSP-70 to again interact and thereby inactivate HSF.

As is the case for many key regulatory molecules, it is now known that multiple HSFs exist, and to date three HSFs, HSF1, -2, and -3, have been identified (Nakai and Morimoto 1993). Interestingly, individual HSFs can be differentially regulated. For example, HSF1, but not HSF2, is activated by elevated temperatures, heavy metals, amino acid analogues and oxidative stress, whereas HSF2, but not HSF1, is activated during differentiation of erythroleukemia cells (Sistonen et al. 1994) and during mouse spermatogenesis (Morimoto 1993). This differential regulation may be related to the fact that HSF1 and HSF2 exist in the cytoplasm of unstressed cells as monomers and dimers, respectively. This finding implies that the stress-induced oligomerization to the trimeric DNA-binding state may occur via two distinct pathways. In addition to the two activation pathways involving HSF1 and HSF2, it is now known that HSF3 is not activated by conditions that activate either HSF1 or HSF2, suggesting that HSF3 may respond to yet "undiscovered cellular stresses" (Nakai and Morimoto 1993).

Thus, prokaryotic and eukaryotic cells have evolved an intricately organized process for sensing and responding to a variety of stressful situations that adversely affect cellular protein folding, assembly, and presumably function. With the description of the heat shock response as background, the following sections will first summarize an extensive literature concerning the heat shock proteins and their putative functions and then conclude with a discussion on concerning the heat shock response, heat shock proteins, and normal/abnormal development.

2. Heat Shock Proteins

In the major reference source entitled *Heat Shock Response*, Nover (1991) lists more than 50 heat shock proteins that have been identified in prokaryotes and eukaryotes based upon differences in molecular weight. The actual number of distinct heat shock proteins that can be induced is difficult to ascertain because of species differences and differences in the methods used to analyze heat shock proteins. Currently, heat shock proteins are characterized primarily by their molecular weights and for convenience are grouped together into families based upon size. To provide a useful overview, the major bacterial, yeast, and mammalian heat shock proteins and their cellular location and function are listed in Table 3. In all three groups, five major families of heat shock proteins are recognized, i.e., the HSP-100, -90, -70, -60 families and the so-called low molecular weight heat shock protein family. The roles of the major heat shock proteins are discussed in the following sections.

Table 3. Major prokaryotic and eukaryotic heat shock proteins (Hps)

Organism	Hsp	Cellular location	Function
Prokaryotes (*E. Coli*)	Hsp-90		Protease, ATPase, protein translocation/export
	Hsp-94		
	Hsp-70 (htpg)		
	Hsp-70 (dnak)		ATPase; 5′-nucleotidase; replication; thermotolerance
	Hsp-60 (groEL)		Protein folding, phage assembly, ATPase
	Hsp-40 (dnaJ)		Protein translocation, λDNA replication
	Hsp-20 (grpE)		DNA replication
	Hsp-10 (groES)		Protein folding, phage assembly, ATPase; DNA replication
Eukaryotes			
Yeast	Hsp-104	Cytoplasm/nucleus	Thermotolerance, ethanol tolerance, spore viability
	Hsp-90	Cytoplasm/nucleus	
	Hsc-82 (constitutive)		
	Hsp-82 (constitutive/inducible)		
	Hsp-70		
	Ssal-4 (constitutive + /or inducible)	Cytoplasm	Protein folding/translocation
	Ssb 1/2 (hs-repressible)	Cytoplasm	Folding of nascent polypeptides
	Ssc 1 (transiently inducible)	Mitochondria	Protein import
	Kar2 (qrp78)	ER	
	Hsp-60		
	mif4	Mitochondria	Protein folding/assembly
	Tcp1	Cytoplasm	Protein folding/assembly
	Hsp-26	Cytoplasm/nucleus	Structurally related to crystallin
Mammals	Hsp-105/110	Cytoplasm/nucleus/nucleolus	
	Hsp-90		
	Hsp-90	Cytoplasm/nucleus	Hormone function; protein chaperone
	GRP-94	ER	
	Hsp-70		
	Hsp-72 (Hsp-70)	Cytoplasm/nucleus	Protein folding/translocation, thermotolerance
	Hsp-73 (Hsc-70)	Cytoplasm/nucleus	
	GRP-78 (Bip)	ER	Protein folding/import
	mtp-70	Mitochondria	
	Hsp-60/58	Mitochondria	Protein folding/assembly of oligomeric protein complexes
	Hsp-32	Cytoplasm	heme oxygenase
	Hsp-28	Cytoplasm/nucleus	Structurally related to crystallin

ER, endoplasmic reticulum.

3. Heat Shock Proteins as Chaperones

Although many mechanistic details remain to be elucidated, it is now known that heat shock proteins function in the absence of stress as molecular chaperones, i.e., they interact with other proteins to insure that they are properly folded during or immediately after translation, translocated to correct final destination, or maintained in a state conducive to appropriate biological activity. To illustrate the role of heat shock proteins as chaperons, two examples will be discussed: (1) the role of *E. coli* heat shock proteins Hsp-10 (groES), -20 (grpE), -40 (dnaJ), -60 (groEL), and -70 (dnaK) in facilitating protein folding and (2) the role of Hsp-60, -70, and -90 in steroid receptor function.

The role of heat shock proteins in protein folding and assembly has been most intensively studied in *E. coli*, and a current model based upon accumulated data from many laboratories (Georgopoulos and Welch 1993) has been developed. In this model, three putative protein-folding pathways are outlined. In pathway 1, Hsp-70 (dnaK), a protein with ATPase activity, interacts with nascent, unfolded polypeptides as they emerge from the ribosome/polysome. Presumably, this interaction prevents subsequent inappropriate inter- and intramolecular interactions that could result in improper folding of the mature protein. The final release of a properly folded protein is regulated by two other proteins, Hsp-40 (dnaJ) and Hsp-20 (grpE), and requires adenosine triphosphate (ATP) hydrolysis. In pathway 2, the newly synthesized polypeptide is transferred, with the participation of Hsp-40, -20 and perhaps -70, to another chaperone machine consisting of Hsp-60 (groEL) and Hsp-10 (groES). As part of the chaperone machine, Hsp-60 exists as a complex consisting of 14 Hsp-60 molecules organized into two double rings of seven subunits each. At the electron microscopic level, this structure appears as a stack of four donuts with a central core. The transferred protein is wholly or partially bound within the central core of the Hsp-60 chaperone machine. Binding of the unfolded polypeptide triggers the association of Hsp-10 to the opposite side of the heat shock protein chaperone machine. Hydrolysis of ATP then results in the release of a fully folded protein and the dissociation of Hsp-60 from the Hsp-10, although multiple rounds of polypeptide binding to and release from the Hsp-60 chaperone machine may be required to achieve final folding of the protein. Pathway 3 may be involved in the folding and/or assembly of proteins that, as functional proteins, exist as homo- or hetero-oligomers. Thus, in *E. coli* at least five proteins play a role in the folding and assembly of functional proteins. Although not as well understood in mammalian cells, protein folding is also achieved in conjunction with heat shock proteins (Welch 1992), and homologues of several *E. coli* chaperones have been identified.

More recent work has also revealed that heat shock proteins play an important role in steroid receptor function is mammals (Smith and Toft 1993). Similar to the situation for protein folding in *E. coli*, multiple proteins are apparently involved in regulating steroid receptor function. To date, seven proteins have been consistently observed in association with various steroid receptors, the so-called receptor-associated proteins (RAP). These include two heat shock proteins (Hsp-70 and -90), two other heat-inducible proteins (p60 and p59), and three other proteins (p54, p50, and p23). Although specific functions for each of the RAP is unknown, in general these proteins function to maintain the receptor in a stable conformation specifically receptive to ligand binding. Presumably ligand binding causes changes (possibly conformational ones) that lead to the dissociation of the RAP and activation of the receptor. It is important to point out that different receptors may differ in their requirements for specific RAPs. For example, the stable avian progesterone avian receptor complex includes Hsp-70, but not p59, whereas the glucocorticoid receptor includes p59, but not Hsp-70. In addition, it is possible that receptor–RAP complexes may vary quantitatively between different cell types within a tissue or organism (SMITH and TOFT 1993). If so, this could have important implications for developmental toxicants that directly or indirectly affect hormone-signaling pathways and may explain, in part, some examples of tissue-specific responses to developmental toxicants. This possibility becomes more intriguing with more recent data showing that other key regulatory molecules, such as the heme-regulated eIF-2 alpha kinase (MATTS et al. 1992), the Ah receptor (PONGRATZ et al. 1992), and Myo-D1 (SHAKNOVICH et al. 1992), have been shown to be associated with specific RAPs. Thus heat shock proteins may play a central role in modulating the activity of a whole range of key regulatory molecules.

4. Heat Shock Proteins and Thermotolerance

One of the most intriguing aspects of heat shock protein biology relates to the phenomenon of thermotolerance. Thermotolerance is induced when cells are subjected to a mild heat shock followed some time later by an acute heat shock. Under these conditions, cells exposed to the double heat shock are protected, at least partially, from the toxic effects of heat-induced stress compared to cells that only receive the acute exposure. Although most of the research concerning thermotolerance has been conducted using cultured cells, thermotolerance can also be induced at the tissue/organ (MARQUEZ et al. 1994) and the whole organism level (WHITE et al. 1994). In addition, work from several laboratories has shown that thermotolerance can be induced in postimplantation mammalian embryos (see below).

Because exposure to elevated temperatures consistently leads to the induction of heat shock proteins and because most heat shock proteins are constitutively expressed in the absence of stress, it was logical to speculate that one or more of these heat shock proteins plays a role in thermotolerance. At this writing, three heat shock proteins have been implicated in thermotolerance, i.e., Hsp-27, -70, and -104. In vertebrates, Hsp-70 appears to play a major role in thermotolerance, and support for this association is based upon the following lines of evidence: (a) in a variety of cell types (LI and LASZLO

1985; Li and Werb 1982), accumulation of Hsp-70 correlates closely with the induction of thermotolerance; (b) heat-resistant cell lines overexpress Hsp-70 (Laszlo and Li 1985; Anderson et al. 1989); (c) thermotolerance is abolished by the microinjection of antibodies against Hsp-70 into fibroblasts (Riabowol et al. 1988); (d) thermotolerance is enhanced after the microinjection of Hsp-70 mRNA into mouse oocytes (Hendry and Kola 1991); (e) transformation of cells with Hsp-70 genes leads to enhanced thermotolerance (Li et al. 1991; Angelidis et al. 1991; Williams et al. 1993), and (f) cells transfected with plasmids containing multiple copies of the 5′ control region of Hsp-70, which bind available HSF, were unable to mount a thermotolerance response (Johnston and Kucey 1988). Although these extensive data strongly support a role for Hsp-70 in thermotolerance, it is important to point out that some studies fail to implicate Hsp-70 in thermotolerance (Carper et al. 1987). Moreover, the critical biological process(es) protected by Hsp-70 at elevated temperatures remain to be elucidated. Thus, one of the remaining challenges is to elucidate the mechanism(s) by which Hsp-70 induces thermotolerance.

Although less extensively studied, Hsp-27 has also been implicated in the induction of thermotolerance in vertebrate cells. This association is based upon two lines of evidence. First, the production of thermotolerant Chinese hamster ovary cells by hyperthermic selection of mutagenized cells was associated specifically with the overexpression of Hsp-27 (Chrétien and Landry 1988). Second, transfection of Chinese hamster ovary cells with plasmids containing Hsp-27 conferred thermotolerance (Landry et al. 1989; Lavoie et al. 1993). To probe the nature of Hsp-27-related thermotolerance, various cellular processes were compared in thermotolerant and control cells. Protein synthesis, RNA synthesis, rRNA processing, and protein degradation were disrupted by elevated temperature to a similar extent in thermotolerant and control cells. However, in thermotolerant cells the microfilament (actin) network was dramatically protected from the disruptive effects of elevated temperature (Lavoie et al. 1993, 1995). In addition, Hsp-27 expression in thermotolerant cells is associated with reduced aggregation of nuclear proteins and an enhanced rate of recovery from aggregation after heat shock (Kampinga et al. 1994). Thus Hsp-27 may protect cells from heat-induced cell death by preventing the disruption of key cytoplasmic and nuclear proteins. One major unanswered question is the relative roles of Hsp-27 and Hsp-70 in the induction of thermotolerance.

Although Hsp-27 and -70 play the dominant role in the induction of thermotolerance in vertebrates, Hsp-104 appears to play a dominant role in yeast, particularly at extreme temperatures. Cells carrying mutations in the *HSP104* gene die at 100–1000 times the rate of wild-type cells under thermotolerance conditions (Sanchez and Lindquist 1990). In addition, Hsp-104 plays a role in the tolerance of yeast cells to ethanol, particularly under extreme conditions. Thus, cells carrying Hsp-104 mutations are 1000-fold less tolerant to 20% ethanol compared to wild-type cells (Sanchez et al. 1992). Interestingly, Hsp-104 apparently does not play a major role in tolerance to heavy metals or arsenite, because mutations in HSP-104 have little or no effect on survival of cells exposed to these agents (SANCHEZ et al. 1992). These results suggest that the damage induced by heavy metals and arsenite is different from the damage induced by heat and ethanol and that some protein other than Hsp-104 plays a role in repairing the damage induced by heavy metals and arsenite. It will be interesting to determine whether these findings in yeast also apply to vertebrate systems.

The results obtained with heavy metals and arsenite also suggest that different heat shock proteins cooperate to minimize damage induced by various toxic stresses. Evidence to support such cooperation exists for Hsp-70 and Hsp-104 in yeast. Yeast cells contain three heat-inducible Hsp-70 genes, *SSA1, SSA3,* and *SSA4.* Mutations that eliminate the functions of all three of these genes simultaneously have no effect on survival at extreme temperatures (50 °C), but prevent growth at moderately high temperatures (37 °C) (WERNER-WASHBURNE et al. 1987). In addition, although mutations in Hsp-104 eliminate survival at extreme temperatures, cells carrying these mutations do exhibit a transient thermotolerance (i.e., for some minutes). However, in cells with null alleles for all three *HSP-70* genes and Hsp-104, both the transient and long-term thermotolerance are blocked. Although details are lacking, these results suggest that, in yeast, Hsp-70 may function to prevent or repair damage induced by moderate stress, whereas Hsp-104 takes over under conditions of extreme stress. Again, it will be interesting to determine whether such cooperation exists among vertebrate heat shock proteins, for example, Hsp-27 and -70, both of which have been shown to play a role in thermotolerance.

5. Heat Shock Proteins and Mammalian Development

HSP-70 is one of the first genes activated at the one- to two-cell stage in the mammalian embryo (BENSAUDE et al. 1983; HEIKKILA 1993; CHRISTIANS et al. 1995). Apparently, the activation of Hsp-70 at this early stage of development is independent of the HSE, suggesting that regulation of transcription involves some mechanism other than the binding of HSF (BEVILACQUA and MANGIA 1993). In addition, Hsp-70 cannot be induced by heat shock at this stage of development (BEVILACQUA and MANGIA 1993) and continues to be refractory until the blastocyst stage (WITTIG et al. 1983; HAHNEL et al. 1986; MORANGE et al. 1984). The molecular mechanism underlying the inability of preimplantation mammalian embryos to mount a heat shock response is unknown. It is interesting to speculate that the inability to mount a heat shock response, generally believed to be a protective response to stress, may be related to the heightened sensitivity of the preimplantation embryo to perturbation by developmental toxicants.

Once the blastocyst stage is reached, embryos are capable of responding to heat stress with the induction of several heat shock proteins (MIRKES 1987; WALSH et al. 1987; HIGO et al. 1989; BENNETT et al. 1990; HONDA et al. 1991), the most prominent being Hsp-72. Hsp-72 can be induced in day-10 rat embryos by temperatures above 40 °C. At these temperatures, synthesis of Hsp-72 can be detected within 30–60 min (MIRKES 1987) after exposure, and accumulation of Hsp-72 protein can be detected within 2.5 h after exposure (MIRKES and DOGGETT 1992). Once synthesized, Hsp-72 protein can be detected in the embryo for up to 24 h. Thus, temperatures that exceed the normal growth temperatures (37 °–38 °C) by more than 3 °C rapidly induce the synthesis and accumulation of specific heat shock proteins. Furthermore, the ability of embryos to mount a heat shock response changes dramatically during early postimplantation development (MIRKES et al. 1991). A 43 °C heat shock for 30 min induces the synthesis of several heat shock proteins, most prominently Hsp-70, in gestational day 9 and 10 rat embryos. However, embryos exposed on day 11 of gestation exhibit an attenuated heat shock response, while day-12 embryos do not mount a heat shock response when exposed to an identical heat shock exposure. Thus, the ability to mount a heat shock response varies considerably as a function of developmental stage. Although the mechanisms underlying this variability are presently unknown, an understanding of the factors that control the embryos capacity to respond to heat stress could lead to significant insights into the factors that control the sensitivity of mammalian embryos to developmental toxicants.

Exposure to temperatures that induce the heat shock response also induce abnormal development. Gestational day-10 rat embryos exposed in vitro to hyperthermia exhibit increased cell death, growth retardation, and abnormal somitogenesis (G.L. KIMMEL et al. 1993, 1995; MIRKES 1985; WALSH et al. 1987). Similar exposures in vivo resulted primarily in abnormalities of the ribs and vertebral elements (C.A. KIMMEL et al. 1993a). In addition, a good correlation exists between exposures that induce abnormal development and exposures that induce the expression of heat shock proteins, particularly HSP-70 and -90 (C.A. KIMMEL et al. 1991, 1994). Moreover, the induction of heat shock protein synthesis has been shown to precede abnormal somitogenesis by several hours (FISHER et al. 1991). Thus, heat shock proteins may serve as biomarkers of heat effects on development. Although heat shock proteins may serve as biomarkers, they appear not to be causally related to heat-induced abnormal development, because heat shock proteins can be induced by exposures that do not disrupt development (MIRKES and DOGGETT 1992).

As discussed previously, a variety of agents, some of them known teratogens, can induce the synthesis of heat shock proteins in nonembryonic systems. This knowledge led GERMAN (1984) to hypothesize that the induction of the heat shock response provides a common pathway by which diverse environmental agents induce developmental abnormalities, i.e., that a teratogen is an agent capable of inducing a heat shock response. He further proposed that it was not the induction of heat shock genes and proteins that caused the abnormal development, but rather the stress-induced alteration of the normal developmental program of gene regulation. An early test of this hypothesis indicated that the teratogen sodium arsenite (AS) did indeed induce a heat shock response in postimplantation mouse embryos (GERMAN et al. 1986). In a subsequent study, (AS) was shown to induce the same set of heat shock proteins as induced by hyperthermia (MIRKES and CORNEL 1992). In addition, the teratogens phenytoin (HANSEN et al. 1988) and retinoic acid (ANSON et al. 1987, 1991) also induce the synthesis of heat shock proteins. In a more extensive study, postimplantation rat embryos were exposed in vitro to embryotoxic concentrations of five selected chemical teratogens, AAF, cadmium chloride (CAD), cyclophosphamide (CP), AS, and sodium salicylate (SAL). Although all five agents induced abnormal development and growth retardation, only AS and SAL induced a heat shock response, as judged by the accumulation of Hsp-72 (MIRKES et al. 1994). Thus not all chemical teratogens apparently induce a heat shock response; however, it should be pointed out that, in the latter study, only Hsp-72 was used as a marker for the heat shock response. Other stress proteins related to the heat shock, genotoxic, and oxidative stress responses may have been induced by AAF, CP, and CAD. More research is needed to determine not only those stress genes that are induced by teratogens, but also those developmentally important genes that are affected by teratogens. The relationships between chemical teratogens, induction of stress responses, alterations in developmental gene regulation, and teratogenesis remain unclear, and further research in these areas should yield key insights into the mechanisms of chemical teratogenesis.

A discussion of heat shock proteins and mammalian development would not be complete without some discussion concerning the development of thermotolerance in mammalian embryos. Several laboratories have demonstrated that thermotolerance can be induced in postimplantation mouse (KAPRON-BRAS and HALES 1991) and rat embryos (MIRKES 1987; WALSH et al. 1987). Moreover, KAPRON-BRAS and HALES (1991) have shown that mild hyperthermia (5 min at 43 °C) can confer protection from a subsequent embryotoxic exposure to cadmium. Likewise, FINNELL et al. (1993) have reported that mild hyperthermia protects fetuses from a subsequent teratogenic exposure to valproic acid. Conversely, they also reported that valproic acid was effective in reducing the risk of hyperthermia-induced neural tube defects. These cumulative results clearly demonstrate that early postimplantation mammalian embryos are capable of mounting a thermotolerance response, but do not elucidate the underlying mechanism. Based upon the work with cells in culture reviewed previously, one of the obvious potential factors playing a role in embryonic thermotolerance is Hsp-72. Although considerable evidence can be marshaled to support the hypothesis that Hsp-72 plays an important role in cellular thermotolerance (see above), confirmatory evidence in mammalian embryos has not been forthcoming. Using different approaches, three groups have come to a similar conclusion, i.e., induction of Hsp-72 is not sufficient to explain the induction of thermotolerance (FINNELL et al. 1993; KAPRON-BRAS and HALES 1992; HARRIS et al. 1991) Although current evidence indicates that Hsp-72 is not sufficient to explain the induction of thermotolerance in embryos, it is also too early to eliminate Hsp-72 from further consideration. What is needed now is a more direct approach to the role of Hsp-72, as well as other heat shock proteins, in which the potentially confounding effects of heat on other cellular processes are reduced or eliminated. Perhaps appropriate transgenic animals can be constructed to help elucidate the role of potential proteins in thermotolerance.

C. Summary and Future Directions

Data reviewed in this chapter document that cells have evolved, over evolutionary time, complex responses to various type of stress that have the potential of damaging key cellular macromolecules. Research, particularly in prokaryotes, has identified many of the key elements involved in cellular responses to genotoxic, oxidative, and heat stress and, more importantly, the molecular mechanisms that underlie stress protein functions. Nowhere is this clearer than our current understanding of the role of bacterial heat shock proteins in protein folding and assembly, processes essential for cell function. We now know how heat shock proteins are organized as chaperonin machines and some of the details concerning how these complexes facilitate protein folding and assembly. Although our understanding of prokaryotic stress proteins is often quite sophisticated, much less is known about eukaryotic stress proteins. With respect to mammalian embryonic stress proteins, there is a dearth of information, making embryonic stress protein biology a potentially fruitful area of inquiry. Stress proteins are the cell's attempt to survive when exposed to potentially toxic conditions, and clearly these proteins must play a critical role in the developing embryo, in which the complex program of development is under very stringent temporal and spatial constraints.

Given the obvious importance of embryonic stress response systems, there are some unresolved issues that demand more study. First, much more data are needed describing the embryo's stress response repertoire. As discussed in this chapter, some of the basics of the embryonic heat stress response have been documented; however, virtually nothing is known concerning the embryos ability to cope with oxidative and genotoxic stress. However, if animal studies are informative, failure to cope with oxidative/genotoxic stress may underlie many instances of teratogen-induced maldevelopment. Thus we need to discover what capabilities the embryo possesses and how the embryonic stress response systems are regulated. Second, a systematic analysis of developmental toxicant-induced stress response systems is also needed to determine how universal the induction of a stress response by developmental toxicants is and what signal transduction pathways couple the interaction of developmental toxicants at the cell membrane with the induction of stress response gene transcription. Third, data are needed describing the developmental acquisition of various stress response systems, i.e., to establish when during development embryos are capable of inducing stress responses. As described earlier, the heat shock response cannot be induced prior to the blastocyst stage of development. The consequence of this for the preimplantation embryo needs to be investigated, as does the possible existence of alternative stress response systems in the preimplantation embryo. Finally, little, if any, data exist concerning the ability of different tissues within the embryo to respond to stress with the induction of one or more of the stress response systems. We have noted that cells of the embryonic heart constitutively express several stress response proteins, e.g., Hsp-27 and *gadd153*. We need to establish whether the constitutive expression of these and other stress proteins might explain the resistance of these cells to the cytotoxic effects of several developmental toxicants and whether the constitutive levels of stress proteins and/or the ability to synthesize stress proteins after exposure to developmental toxicants explains, at least in part, tissue sensitivities to developmental toxicity. These are just some of the many questions that require answers. Answers to these questions will provide not only key insights into normal and abnormal development, but also clues to ways in which embryonic stress response systems can be therapeutically manipulated.

Acknowledgements. The author's research summarized in this chapter was supported by grants from the National Institutes of Health and the March of Dimes Birth Defects Foundation. I wish to thank the members of my research laboratory for their camaraderie and productivity and John Rajan for secretarial assistance. This chapter is dedicated to my mentor and colleague, Dr. Thomas H. Shepard, who initially supported my entry into the field of Developmental Toxicology.

References

Abbott BD, Perdew GH, Birnbaume LS (1994) Ah receptor in embryonic mouse palate and effects of TCDD on receptor expression. Toxicol Appl Pharmacol 126: 16–25

Aggeler J, Murnane JP (1990) Enhanced expression of procollagenase in ataxia-telangiectasia and xeroderma pigmentosum fibroblasts. In vitro Cell Dev Biol 26: 915–922

Anderson CW (1993) DNA damage and the DNA-activated protein kinase. Trends Biochem Sci 18: 433–437

Anderson RL, Van-Kersen I, Kraft PE, Hahn GM (1989) Biochemical analysis of heat resistant mouse tumor cell strains: a new member of the HSP70 family. Mol Cell Biol 9: 3509–3516

Angelidis CE, Lazaridis I, Pagoulatos GN (1991) Constitutive expression of heat shock protein 70 in mammalian cells confers thermoresistance. Eur J Biochem 199: 35–39

Anson JF, Hinson WG, Pikin JL, Kwarta RF, Hansen DK, Young JF, Burns ER, Casciano DA (1987) Retinoic acid induction of stress proteins in fetal mouse limb buds. Dev Biol 121: 542–547

Anson JF, Laborde JB, Pipkin JL, Hinson WJ, Hansen DK, Sheehan DM, Young JF (1991) Target tissue specificity of retinoic acid-induced stress proteins and malformations in mice (see comments). Teratology 44: 19–28

Applegate LA, Luscher P, Tyrrell RM (1991) Induction of heme oxygenase: a general response to oxidant stress in cultured mammalian cells. Cancer Res 51: 974–978

Bauman JW, Liu J, Liu YP, Klaassen CD (1991) Increase in metallothionein produced by chemicals that induce oxidative stress. Tox Appl Pharmacol 110: 347–354

Bennett GD, Mohl VK, Finnell RH (1990) Embryonic and maternal heat shock responses to a teratogenic hyperthermic insult. Reprod Toxicol 4: 113–119

Bensaude O, Babinet C, Morange M, Jacob F (1983) Heat shock proteins, first major products of zygotic gene activity in mouse embryo. Nature 305: 331–333

Bevilacqua A, Mangia P (1993) Activity of a microinjected inducible murine hsp68 gene promoter depends on plasmid configuration and the presence of heat shock elements in mouse dictyate oocytes but not in two-cell embryos. Dev Genet 14: 92–102

Brach MA, Hass R, Sherman ML, Gunji H, Weichselbaum R, Kufe D (1991) Ionizing radiation induces expression and binding activity of the nuclear factor kappa B. J Clin Invest 88: 691–695

Carper SW, Duffy JJ, Gerner EW (1987) Heat shock proteins in thermotolerance and other cellular processes. Cancer Res 47: 5249–5255

Chrétien P, Landry J (1988) Enhanced heat constitutive expression of the 27 Kda heat shock proteins in heat resistant variants from Chinese hamster cells. J Cell Physiol 137: 157–166

Christians E, Campion E, Thompson EM, Renard JP (1995) Expression of the HSP 70.1 gene, a landmark of early zygotic activity in the mouse embryo, is restricted to the first burst of transcription. Development 121: 113–122

Christman MF, Morgan RW, Jacobson FS, Ames BN (1985) Positive control of a regulon for defense against oxidative stress and some heat shock proteins in Salmonella typhimurium. Cell 41: 753–762

Christman MF, Storz G, Ames BN (1989) OxyR, a positive regulator of hydrogen peroxide-inducible genes in Escherichia coli and Salmonella typhimurium, is homologous to a family of bacterial regulatory proteins. Proc Natl Acad Sci USA 86: 3484–3488

de Murcia G, Menissier-de-Murcia J, Schreiber V (1994) Poly(ADP-ribose) polymerase: a molecular nick-sensor [published erratum appears in Trends Biochem Sci 19: 250]. Trends Biochem Sci 19: 172–176

el Hage S, Singh SM (1990) Temporal expression of genes encoding free radical-metabolizing enzymes is associated with higher mRNA levels during in utero development in mice. Dev Genet 11: 149–159

Elledge SJ, Davis RW (1987) Identification and isolation of the gene encoding the small subunit of ribonucleotide reductase from Saccharomyces cerevisiae: DNA damage-inducible gene required for mitotic viability. Mol Cell Biol 7: 2783–2793

Erickson UJ, Borg LA (1993) Diabetes and embryonic malformations. Role of substrate-induced free-oxygen radical production for dysmorphogenesis in cultured rat embryos. Diabetes 42: 411–419

Evensen G, Seeberg E (1982) Adaptation to alkylation resistance involves the induction of a DNA glycosylase. Nature 296: 773–775

Finnell RH, Van-Waes M, Bennett GD, Eberwine JH (1993) Lack of concordance between heat shock proteins and the development of tolerance to teratogen-induced neural tube defects. Dev Genet 14: 137–147

Fisher BR, Kimmel GL, Kimmel CA, Heredia DJ (1991) The association of heat-induced alterations in protein synthesis with somite defects in rat embryos. Teratology 43: 465

Fornace AJ Jr (1992) Mammalian genes induced by radiation; activation of genes associated with growth control. Annu Rev Genet 26: 507–526

Fornace AJ Jr, Nebert DW, Hollander MC, Luethy JD, Papathanasiou M, Fargnoli J, Holbrook NJ (1989) Mammalian genes coordinately regulated by growth arrest signals and DNA-damaging agents. Mol Cell Biol 9: 4196–4203

Fornace AJ Jr, Jackman J, Hollander MC, Hoffman-Liebermann B, Liebermann DA (1992) Genotoxic-stress-response genes and growth-arrest genes. gadd, MyD, and other genes induced by treatments eliciting growth arrest. Ann NY Acad Sci 663: 139–153

Franklin WA, Haseltine WA (1984) Removal of UV light-induced pyrimidine-pyrimidone (6–4) products from Escherichia coli DNA requires the uvrA, uvrB, and uvrC gene products. Proc Natl Acad Sci USA 81: 3821–3824

Friedberg EC (1988) Deoxyribonucleic acid repair in the yeast Saccharomyces cerevisiae. Microbiol Rev 52: 70–102

Georgopoulos C, Welch WJ (1993) Role of the major heat shock proteins as molecular chaperones. Annu Rev Cell Biol 9: 601–634

German J (1984) Embryonic stress hypothesis of teratogenesis. Am J Med 76: 293–301

German J, Louie E, Banerjee D (1986) The heat shock response in vivo: experimental induction during mammalian organogenesis. Teratog Carcinog Mutagen 6: 555–562

Gewirtz DA (1993) DNA damage, gene expression, growth arrest and cell death. Oncol Res 5: 397–408

Goff SA, Goldberg AL (1985) Production of abnormal proteins in E. coli stimulates transcription of lon and other heat shock genes. Cell 41: 587–595

Greenberg JT, Demple B (1989) A global response induced in Escherichia coli by redox cycling agents overlaps with that induced by peroxide stress. J Bacteriol 171: 3933–3939

Hahnel AC, Gifford DJ, Heikkila JJ, Schultz GA (1986) Expression of the major heat shock protein (hsp 70) family during early mouse embryo development. Teratog Carcinog Mutagen 6: 493–510

Hallahan DE, Spriggs DR, Bockett MA, Kufe DW, Weichselbaum RR (1989) Increased tumor necrosis factor alpha mRNA after cellular exposure to ionizing radiation. Proc Natl Acad Sci USA 86: 10104–10107

Hanawalt PC (1994) Transcription-coupled repair and human disease. Science 266: 1957–1958

Hansen DK, Anson JF, Hinson WG, Pipkin JL Jr (1988) Phenytoin-induced stress protein synthesis in mouse embryonic tissue. Proc Soc Exp Biol Med 189: 136–140

Harris C, Juchau MR, Mirkes PE (1991) Role of glutathione and hsp 70 in the acquisition of thermotolerance in postimplantation rat embryos. Teratology 43: 229–239

Heikkila JJ (1993) Heat shock gene expression and development. II. An overview of mammalian and avian developmental systems. Dev Genet 14: 87–91

Hendrey J, Kola I (1991) Thermolability of mouse oocytes is due to the lack of expression and/or inducibility of Hsp 70. Mol Reprod Dev 28: 1–8

Herrlich P, Angel P, Rahmsdorf HJ, Mallick U, Poting A, Hieber L, Lucke-Huhle C, Schorpp M (1986) The mammalian genetic stress response. Adv Enzyme Regul 25: 485–504

Herrlich P, Sachsenmaier C, Radler-Pohl A, Gebel S, Bluttner C, Rahmsdorf HJ (1994) The mammalian UV response: mechanism of DNA damage induced gene expression. Adv Enzyme Regul 34: 381–395

Hidalgo E, Demple B (1994) An iron-sulfur center essential for transcriptional activation by the redox-sensing SoxR protein. EMBO J 13: 138–146

Hightower LE (1980) Cultured animal cells exposed to amino acid analogues or puromycin rapidly synthesize several polypeptides. J Cell Physiol 102: 407–427

Higo H, Lee JY, Satow Y, Higo K (1989) Elevated expression of protooncogenes accompany enhanced induction of heat-shock genes after exposure of rat embryos in utero to ionizing radiation. Teratogen Carcinogen Mutagen 9: 191–198

Holley JA, Janssen YM, Mossman BT, Taatjes DJ (1992) Increased manganese superoxide dismutase protein in type II epithelial cells of rat lungs after inhalation of crocidolite asbestos or cristobalite silica. Am J Pathol 141: 475–485

Honda K, Hatayama T, Takahashi K, Yukioka M (1991) Heat shock proteins in human and mouse embryonic cells after exposure to heat shock or teratogenic agents. Teratogen Carcinogen Mutagen 11: 235–244

Jackman J, Alamo I Jr, Fornace AJ Jr (1994) Genotoxic stress confers preferential and coordinate messenger RNA stability on the five gadd genes. Cancer Res 54: 5656–5662

Janssen YM, Marsh JP, Absher MP, Hemenway D, Vacek PM, Leslie KO, Born PJ, Mossman BT (1992) Expression of antioxidant enzymes in rat lungs after inhalation of asbestos or silica. J Biol Chem 267: 10625–10630

Janssen YM, van-Houten B, Born PJ, Mossman BT (1993) Cell and tissue responses to oxidative damage. Lab Invest 69: 261–274

Johnson AL, Barker DG, Johnston LH (1986) Induction of yeast DNA ligase genes in exponential and stationary phase cultures in response to DNA damaging agents. Curr Genet 11: 107–112

Johnston LH, White JH, Johnson AL, Lucchini G, Plevani P (1987) The yeast DNA polymerase I transcript is regulated in both the mitotic cell cycle and in meiosis and is also induced after DNA damage. Nucleic Acids Res 15: 5017–5030

Johnston RN, Kucey BL (1988) Competitive inhibition of hsp 70 gene expression causes thermosensitivity. Science 242: 1551–1554

Jornot L, Junod AF (1992) Response for human endothelial cell antioxidant enzymes to hyperoxia. Am J Respir Cell Mol Biol 6: 107–115

Kampinga HH, Brunsting JF, Stege GJ, Konings AW, Landry J (1994) Cells overexpressing Hsp 27 show accelerated recovery from heat induced nuclear protein aggregation. Biochem Biophys Res Commun 204: 1170–1177

Kapron-Brás CM, Hales BF (1991) Heat-shock induced tolerance to the embryotoxic effects of hyperthermia and cadmium in mouse embryos in vitro. Teratology 43: 83–94

Kapron-Brás CM, Hales BF (1992) Genetic differences in heat-induced tolerance to cadmium in cultured mouse embryos are not correlated with changes in a 68-kD heat shock protein. Teratology 46: 191–200

Karran P, Hjelmgren T, Lindahl T (1982) Induction of a DNA glycosylase for N-methylated purines is part of the adaptive response to alkylating agents. Nature 296: 770–773

Kartasova T, van de Putte P (1988) Isolation, characterization, and UV-stimulated expression of two families of genes encoding polypeptides of related structure in human epidermal keratinocytes. Mol Cell Biol 8: 2195–2203

Kataoka H, Sekiguchi M (1985) Molecular cloning and characterization of the alkB gene of Escherichia coli. Mol Gen Genet 198: 263–269

Kataoka H, Yamamoto Y, Sekiguchi M (1983) A new gene (alkB) of Escherichia coli that controls sensitivity to methyl methane sulfonate. J Bacteriol 153: 1301–1307

Kelley PM, Schlesinger MJ (1978) The effect of amino acid analogues and heat shock on gene expression in chicken embryo fibroblasts. Cell 15: 1277–1286

Kenyon CJ, Walker GC (1980) DNA-damaging agents stimulate gene expression at specific loci in Escherichia coli. Proc Natl Acad Sci USA 77: 2819–2823

Keyse SM (1990) Oxidant stress lead to transcriptional activation of the human heme oxygenase gene in cultured cells. Mol Cell Biol 10: 4967–4969

Keyse SM (1993) The induction of gene expression in mammalian cells by radiation. Cancer Biol 4: 119–128

Kimmel CA, Kimmel GL, Lu C, Heredia DJ, Fisher BR, Brown NT (1991) Stress protein synthesis as a potential biomarker for heat-induced developmental toxicity. Teratology 43: 465

Kimmel CA, Cuff JM, Kimmel GL, Heredia DJ, Tudor N, Silverman PM (1993) Skeletal development following heat exposure in the rat. Teratology 47: 229–242

Kimmel CA, Claggett TW, Kimmel GL, Tudor N, Hogan KA (1994) Segmentation anomalies and stress proteins induced by heat on gestation days (GD) 10, 11 or 12 in the rat. Teratology 49: 409

Kimmel GL, Cuff JM, Kimmel CA, Heredia DJ, Tudor N, Silverman PM (1993) Embryonic development in vitro following short-duration exposure to heat. Teratology 47: 243–251

Kimmel GL, Claggett TW, Kimmel CA, Bolou B, Todor N (1995) The progression of effects of heat stress on somitogenesis in the rat. Teratology 51: 157–158

Landini P, Hajec LI, Volkert MR (1994) Structure and transcriptional regulation of the Escherichia coli adaptive response gene aidB. J Bacteriol 176: 6583–6589

Landry J, Chrétien P, Lambert H, Hickey E, Weber LA (1989) Heat shock resistance conferred by expression of the human HSP27 gene in rodent cells. J Cell Biol 109: 7–15

Laszlo A, Li GC (1985) Heat resistant variants of Chinese hamster fibroblasts altered in expression of heat shock protein. Proc Natl Acad Sci USA 82: 8029–8033

Lavoie JN, Gingras-Breton G, Tanguay RM, Landry J (1993) Induction of Chinese hamster HSP27 gene expression in mouse cells confers resistance to heat shock. HSP27 stabilization of the microfilament organization. J Biol Chem 268: 3420–3429

Lavoie JN, Lambert H, Hickey E, Weber LA, Landry J (1995) Modulation of cellular thermoresistance and actin filament stability accompanies phosphorylation-induced changes in the oligomeric structure of heat shock protein 27. Mol Cell Biol 15: 505–516

Lemotte PK, Walker GC (1985) Induction and autoregulation of ada, a positively acting element regulating the response of Escherichia coli K-12 to methylating agents. J Bacteriol 161: 888–895

Li GC, Laszlo A (1985) Thermotolerance in mammalian cells: a possible role for heat shock proteins. In: Atkinson BG, Walden DB (eds) Changes in eukaryotic gene expression in response to environmental stress. Academic, Orlando, p 227

Li GC, Werb Z (1982) Correlation between synthesis of heat shock proteins and development of thermotolerance in Chinese hamster fibroblasts. Proc Natl Acad Sci USA 79: 3218–3222

Li Z, Demple B (1994) SoxS, an activator of superoxide stress genes in Escherichia coli. Purification and interaction with DNA. J Biol Chem 269: 18371–18377

Li GC, Li LG, Liu YK et al. (1991) Thermal response of rat fibroblasts stably transfected with the human 70-kDa heat shock protein-encoding gene. Proc Natl Acad Sci USA 88: 1681–1685

Little JW, Mount DW (1982) The SOS regulatory system of Escherichia coli. Cell 29: 11–22

Little SA, Mirkes PE (1994) Induction of the growth arrest-DNA damage induced gene (gadd 153) in rat embryos in response to genotoxic agents. Teratology 49: 400

Lloyd RG, Benson FE, Shurvinton CE (1984) Effect of ruv mutations on recombination and DNA repair in Escherichia coli K12. Mol Gen Genet 194: 303–309

Lord KA, Hoffman-Liebermann B, Liebermann DA (1990) Complexity of the immediate early response of myeloid cells to terminal differentiation and growth arrest includes ICAM-1, Jun-B and histone variants. Oncogene 5: 387–396

Maines MD (1984) New Developments in the regulation of heme metabolism and their implications. Crit Rev Toxicol 12: 241–314

Marquez CM, Sneed PK, Li GC, Mak JY, Phillips TL (1994) HSP 70 synthesis in clinical hyperthermia patients: preliminary results of a new technique. Int J Radiat Oncol Biol Phys 28: 425–430

Matts RL, Xu Z, Pal JK, Chen JJ (1992) Interactions of the heme-regulated eIF-2 alpha kinase with heat shock proteins in rabbit reticulocyte lysates. J Biol Chem 267: 18160–18167

McCarthy TV, Karran P, Lindahl T (1984) Inducible repair of O-alkylated DNA pyrimidines in Escherichia coli. EMBO J 3: 545–550

McClanahan T, McEntee K (1984) Specific transcripts are elevated in Saccharomyces cerevisiae in response to DNA damage. Mol Cell Biol 4: 2356–2363

Mirkes PE (1985) Hyperthermia-induced heat shock response and thermotolerance in postimplantation rat embryos. Dev Biol 119: 115–122

Mirkes PE, Cornel L (1992) A comparison of sodium arsenite- and hyperthermia-induced stress responses and abnormal development in cultured postimplantation rat embryos. Teratology 46: 251–259

Mirkes PE, Doggett B (1992) Accumulation of heat shock protein 72 (hsp 72) in postimplantation rat embryos after exposure to various periods of hyperthermia (40–43 °C) in vitro: evidence that heat shock protein 72 is a biomarker of heat-induced embryotoxicity. Teratology 46: 301–309

Mirkes PE, Grace RH, Little SA (1991) Developmental regulation of heat shock protein synthesis and HSP 70 RNA accumulation during postimplantation rat embryogenesis. Teratology 44: 77–89

Mirkes PE, Doggett B, Cornel L (1994) Induction of heat shock response (HSP 72) in rat embryos exposed to selected chemical teratogens. Teratology 49: 135–142

Miskin R, Ben-Ishai R (1981) Induction of plasminogen activator by UV light in normal and xeroderma pigmentosum fibroblasts. Proc Natl Acad Sci USA 78: 6236–6240

Modrich P (1994) Mismatch repair, genetic stability and cancer. Science 266: 1959–1960

Morange M, Diu A, Bensaude O, Babinet C (1984) Altered expression of heat shock proteins in embryonal carcinoma and mouse early embryonic cells. Mol Cell Biol 4: 730–735

Morimoto RI (1993) Cells in stress: transcriptional activation of heat shock genes. Science 259: 1409–1410

Nakai A, Morimoto RI (1993) Characterization of a novel chicken heat shock transcription factor, heat shock factor 3, suggests a new regulatory pathway. Mol Cell Biol 13: 1983–1997

Nebert DW (1994) Drug-metabolizing enzymes in ligand-modulated transcription. Biochem Pharmacol 47: 25–37

Nebert DW, Petersen DD, Fornace AJ Jr (1990) Cellular responses to oxidative stress: the [Ah] gene battery as a paradigm. Environ Health Perspect 88: 13–25

Nover L (1991) Heat shock response. CRC Press, Boca Raton

Papathanasiou MA, Fornace AJ Jr (1991) DNA-damage inducible genes. Cancer Treat Res 57: 13–36

Pongratz I, Mason GG, Poellinger L (1992) Dual roles of the 90-kDa heat shock protein hsp90 in modulating functional activities of the dioxin receptor. Evidence that the dioxin receptor functionally belongs to a subclass of nuclear receptors which require hsp 90 both for ligand binding activity and repression of intrinsic DNA binding activity. J Biol Chem 267: 13728–13734

Reimer DL, Singh SM (1990) In situ hybridization studies on murine catalase mRNA expression during embryonic development. Dev Genet 11: 318–325

Riabowol KT, Mizzen LA, Welch WJ (1988) Heat shock is lethal to fibroblasts microinjected with antibodies against hsp 70. Science 242: 433–436

Ritossa F (1962) A new puffing pattern induced by temperature shock and DNP in Drosophila. Experientia 18: 571–573

Ron D, Habener JF (1992) CHOP, a novel developmentally regulated nuclear protein that dimerizes with transcription factors C/EBP and LAP and functions as a dominant negative inhibitor of gene transcription. Genes Dev 6: 439–453

Ruby SW, Szostak JW (1985) Specific Saccharomyces cerevisiae genes are expressed in response to DNA-damaging agents. Mol Cell Biol 5: 75–84

Rushmore TH, Morton MR, Pickett CB (1991) The antioxidant responsive element. Activation by oxidative stress and identification of the DNA consensus sequence required for functional activity. J Biol Chem 266: 11632–11639

Samson L, Cairns J (1977) A new pathway for DNA repair in Escherichia coli. Nature 267: 281–283

Sancar A (1994) Mechanisms of DNA excision repair. Science 266: 1954–1956

Sancar A, Rupp WD (1983) A novel repair enzyme: UVRABC excision nuclease of Escherichia coli cuts a DNA strand on both sides of the damaged region. Cell 33: 249–260

Sanchez Y, Lindquist SL (1990) HSP104 required for induced thermotolerance. Science 248: 1112–1115

Sanchez Y, Taulien J, Borkovich KA, Lindquist S (1992) HSP104 is required for tolerance to many forms of stress. EMBO J 11: 2357–2364

Sedgwick B (1983) Molecular cloning of a gene which regulates the adaptive response to alkylating agents in Escherichia coli. Mol Gen Genet 191: 466–472

Shaknovich R, Shue G, Kohtz DS (1992) Conformational activation of a basic helix-loop-helix protein (MyoD1) by the C-terminal region of murine HSP90 (HSP84). Mol Cell Biol 12: 5059–5068

Sistonen L, Sarge KD, Morimoto RI (1994) Human heat shock factors 1 and 2 are differentially activated and can synergistically induce hsp 70 gene transcription. Mol Cell Biol 14: 2087–2099

Smith DF, Toft DO (1993) Steroid receptors and their associated proteins. Mol Endocrinol 7: 4–11

Somashekaraiah BV, Padmaja K, Prasad AR (1992) Lead-induced lipid peroxidation and antioxidant defense components of developing chick embryos. Free Radic Biol Med 13: 107–114

Stocker R (1990) Induction of haem oxygenase as a defence against oxidative stress. Free Radic Res Commun 9: 101–112

Teo I, Sedgwick B, Demple B, Li B, Lindahl T (1984) Induction of resistance to alkylating agents in E. Coli: the ada+ gene product serves both as a regulatory protein and as an enzyme for repair or mutagenic damage. EMBO J 3: 2151–2157

Vile GF, Tyrrell RM (1993) Oxidative stress resulting from ultraviolet A irradiation of human skin fibroblasts leads to a heme oxygenase-dependent increase in ferritin. J Biol Chem 268: 14678–14681

Walker GC (1984) Mutagenesis and inducible responses to deoxyribonucleic acid damage in Escherichia coli. Microbiol Rev 48: 60–93

Walker GC (1985) Inducible DNA repair systems. Annu Rev Biochem 54: 425–457

Walkup LK, Kogoma I (1989) Escherichia coli proteins inducible by oxidative stress mediated by the superoxide radical. J Bacteriol 171: 1476–1484

Walsh DA, Klein NW, Hightower LE, Edwards MJ (1987) Heat shock and thermotolerance during early rat embryo development. Teratology 36: 181–191

Welch WJ (1992) Mammalian stress response: cell physiology, structure/function of stress proteins, and implications for medicine and disease. Physiol Rev 72: 1063–1081

Werner-Washburne M, Stone DE, Craig EA (1987) Complex interactions among members of an essential subfamily of hsp 70 genes in Saccharomyces cerevisiae. Mol Cell Biol 7: 2568–2577

West SC, Howard-Flanders P (1984) Duplex-duplex interactions catalyzed by RecA protein allow strand exchange to pass double-strand breaks in DNA. Cell 37: 683–691

West SC, Cassuto E, Howard-Flanders P (1981) Mechanism of E. coli RecA protein directed strand exchanges in post-replication repair of DNA. Nature 294: 659–662

West SC, Cassuto E, Howard-Flanders P (1982) Postreplication repair in E. coli: strand exchange reactions of gapped DNA by RecA protein. Mol Gen Genet 187: 209–217

White CN, Hightower LE, Schultz RJ (1994) Variation in heat-shock proteins among species of desert fishes (Poeciliidae, Poeciliopsis). Mol Biol Evol 11: 106–119

Williams RS, Thomas JA, Fina M, German Z (1993) Human heat shock protein 70 (hsp 70) protects murine cells from injury during metabolic stress. J Clin Invest 92: 503–508

Wittig S, Hensse S, Keitel C, Elsner C, Wittig B (1983) Heat shock gene expression is regulated during teratocarcinoma cell differentiation and early embryonic development. Dev Biol 96: 507–514

Woloschak GE, Chang-Liu CM, Jones PS, Jones CA (1990) Modulation of gene expression in Syrian hamster embryo cells following ionizing radiation. Cancer Res 50: 339–344

Yamamoto Y, Kataoka H, Nakabeppu Y, Tsuzuki T, Sekiguchi M (1983) The genes involved in the repair of alkylated DNA in Escherichia coli K-12. In: Friedberg EG, Bridges BA (eds) Cellular responses to DNA damage. Liss, New York, pp 271–278

CHAPTER 10

Cell–Cell Interactions

P.J. LINSER

A. Introduction

My initial contact with teratology occurred in 1974 while I was in graduate school studying developmental biology and teratology at the University of Cincinnati. At that time my teachers included such world-renowned teratologists as James G. Wilson, Josef Warkany, Ernest Zimmerman, William Scott, and Harold Kalter. Also at that time, the University of Cincinnati was one of only a very few institutions that actually granted a Ph.D. degree in "Developmental Biology." This interdisciplinary degree was based firmly on the study of abnormal development as a window into the regulation of normal embryogenesis.

In the ensuing 20 years, developmental biology has expanded geometrically as a specific scientific discipline. The explosion of new technologies has provided for direct analyses of questions that could only be inferred from teratological studies in the past. In 1974, molecular genetics was still in its infancy and dependent on such nearly forgotten methodologies as the Cot curve (LEWIN 1994). Students in awe of the complexities of normal development as evidenced by the results of teratologic errors had never heard of cDNA, polymerase chain reaction (PCR), or synthetic oligonucleotides, nor would they have guessed that the points of the compass would be used one day to describe powerful scientific techniques that would influence our directions in other ways.

As a new student of normal and abnormal development, I naturally explored numerous conceptual vantage points in my quest for a research direction that could stimulate my interests for the duration of a career. One of the central themes in developmental research has always been the question of how complex systems emerge from simpler beginnings. Does the orderly progression from undifferentiated egg to a vast array of tissues and cell types occur by time- and cell division-dependent lineal steps? Or is differentiation a malleable process that responds to some set of specific cues? Classical studies had shown that the development of higher organisms was characterized by instructive events, events that seemed to trigger changes in cells and tissues as in the complex morphogenesis of the vertebrate eye. Such apparent instructive events led to the now common term "embryonic induction" (SPEMANN and MANGOLD 1964). Implicit in this concept was the notion that cells have the capacity to send and receive signals of some sort that trigger developmental responses in the participating cells, such that changes in gene expression and morphogenic reshaping events can be measured. The studies of GURDON (1963) and his colleagues showed that individual cells retain the genetic potential to fulfill the entire spectrum of cell differentiation and that the environmental cues which a cell encounters have the potential to mold the character of that cell and its daughter cells.

Although the hypothesis that cells differentiate in response to an unfolding cascade of microenvironmental cues was not new in 1974, the now accepted dogma that cell to cell interactions provide many of the regulatory stimuli that control normal development was just then achieving wide acceptance. As a student, the implications of this concept seemed staggering. If indeed a cell became what it was "told" to become, might we not learn enough to finally be able to "tell" cells to stop being cancerous or to "tell" cells to fill in a missing functionality that produced a specific birth defect? Thus it appeared to me as a young (and naive) student that a true understanding of developmental regulation could only come from studying cell to cell interactions.

Cell–cell interactions is a phrase that can be applied to many distinguishable phenomena. In other words, there are many types of cell–cell interactions recognized in the literature today. The types of cell interactions can be categorized on several bases. Perhaps the simplest way to distinguish types of interaction is by examining the distance covered by the signaling molecule(s).

A large number of cell interactions or signaling processes are mediated by effectors that operate over a relatively long distance. By this I simply mean that many communication events occur as one cell synthesizes a signal molecule and then releases or deposits the molecule in the local microenvironment. Reception and response to the signal substance is then based on interaction of that substance with specific molecular "machinery" in or on another cell. Examples of such cell–cell interaction effector molecules include hormones, growth factors, metabolites, and components of the extracellular matrix (ECM).

A second general category of cell–cell interactions operates over short intercellular distances and specifically involves the contact of two cells through their plasma membranes. This type of interaction has been assigned a number of designations in recent years and will be the main focus of this review. Perhaps the most widely used phrase that encompasses studies on such phenomena is "contact-mediated cell–cell recognition" (ALBERTS et al. 1994; MONROY and MOSCONA 1979) Another less inclusive, although widely employed term for this area of study is the term "cell adhesion."

B. Cell–Cell Recognition and Cell Adhesion

The conceptual basis for the plethora of investigations underway which focus on cell–cell recognition was birthed by the pioneering studies of such individuals as WILSON (1907) and GALTSOFF (1925). Using the very primitive multicellular organisms (sponges), these early investigators demonstrated that individual cells have the capacity to identify appropriate partners (recognition) and to associate specifically with those partners (adhesion) in a tissue-specific fashion. During the ensuing three quarters of a century, many hundreds of investigators have amassed data which show that contact-mediated cell–cell recognition is fundamental to the evolution and embryonic development of all multicellular organisms. The literature is fraught with hypothetical arguments that the emergence of membrane-bound recognition machinery is one of the primary enabling steps in the evolution of complex organisms (reviewed in EDELMAN and CROSSIN 1991). Among the documented developmental phenomena that rely on cell–cell recognition are cell migration (ALBERTS et al. 1994; BRONNER-FRASER 1993; ONO et al. 1994; RANSCHT and BRONNER-FRASER 1991), epithelial–mesenchymal interactions and transitions (BERNFIELD et al. 1984; YANG et al. 1995; FADOOL and LINSER 1994), neurite outgrowth and path finding (GOODMAN and SHATZ 1993; ONO et al. 1994), synaptogenesis (TRISLER 1987), activation and inactivation of specific genes (MORRIS and MOSCONA 1970; LINSER and MOSCONA 1984; MOSCONA et al. 1979; MOSCONA and LINSER 1983), immune surveillance (JANEWAY and TRAVERS 1994), and cell differentiation in a very general sense (GILBERT 1994; BAUER et al. 1992). From this partial list, it is evident that cell–cell recognition influences most or all aspects of embryonic development. Therefore, it is intuitively obvious that perturbations of cell–cell recognition phenomena could lead to teratological defects. Furthermore, since nearly 70% of all human birth defects remain unexplained in terms of causation (SCHARDEIN 1993), cell–cell recognition as a regulatory process seems a likely source of some of these defects. This hypothesis has been suggested by many investigators, but to date surprisingly few true examples of cell–cell recognition failures resulting in human or animal birth defects have been documented. The fact that cell–cell recognition failure as part of the etiology of specific defects is uncommon in the literature may only be a reflection of our inability to make the appropriate investigative connections. What is currently known about the large numbers of cell surface molecules that act in these processes, and their modes of action, may give us insight into why it has been difficult to ascribe certain teratological events to this area of developmental biology.

I. Cell–Cell Recognition and/or Adhesion Molecules

The earliest studies which demonstrated that cells of even primitive organisms possessed the intrinsic capacity to identify appropriate partners and to adhere to those partners led to similar observations of cognitive and adhesive properties in cells of higher organisms (MONROY and MOSCONA 1979). Functional assays were devised by a number of investigators which made it possible to measure such phenomena in vitro (MOSCONA 1961; STEINBERG 1964; MCCLAY et al. 1981; EDELMAN 1983; ROTH and WESTON 1967). For the most part, these assays were designed to detect and even quantitate adhesive interactions. Once it was possible to assay some cellular or developmental event that was dependent on interactions, it became possible to search for the specific cell surface molecules that participated in the defined event. Examples of the methods for identifying specific cell–cell recognition molecules will be discussed in more detail later. During the past 15 years this multistep approach has led to the identification of several "families" of cell–cell recognition or adhesion molecules (ALBELDA and BUCK 1990; EDELMAN 1985; FADOOL and LINSER 1993b; HAUSMAN and MOSCONA 1976; GRENNINGLOH et al. 1990; TAKEICHI 1990).

Table 1 presents a list of the families of cell adhesion molecules and several members of each type. The assignment of a given molecule to a family is based on a number of criteria, including Ca^{2+} dependence, ligand binding, and molecular sequence (ALBERTS et al. 1994). For the purpose of this review, I have listed five such families: members of the immunoglobulin (Ig) super gene family, cadherins, integrins, selectins, and others. For extensive reviews of cell–cell recognition molecules, see EDELMAN and CROSSIN (1991), ALBELDA and BUCK (1990), HYNES (1992), TAKEICHI (1990) or ALBERTS et al. (1994).

Table 1. Families of cell–cell adhesion molecules

Family	Examples	Binding characteristics	Cytoskeletal association	Cell junctions
Cadherins	E, N, P, R	Ca^{2+} dependent, homophylic	Actin via catenins	Adhesion belts and desmosomes
Ig superfamily	NCAM, L1, CD8, 5A11, MHcs	Ca^{2+} independent, homophylic and heterophylic	Various	No
Integrins	Many heterodimeric combinations subunits	Ca^{2+} dependent, heterophylic	Actin via talin	Focal contacts
Selectins	P, L, E	Ca^{2+} dependent, heterophylic	Unknown	No
Others	Glycosyl transferases, proteoglyans, cognin	Various	Unknown	No

Ig, immunoglobulin; NCAM, neural cell adhesion molecule; MHcs, major histocompatibility complex antigens.

C. How Could Normal Functioning Be Disrupted (Teratogenesis)?

A basic tenet of teratology is that viable birth defects result from specific and limited perturbations. Agents that have very broad reaching effects on embryonic events tend to be lethal, whereas agents that influence only a few functionalities can induce viable defects. The processes that cell–cell recognition influence are many and varied, and it seems obvious that any very generalized inhibition of these events would be lethal. Indeed, it is clear that wholesale disruption of the cell–cell relationships of an embryonic tissue leads to impairment or even termination of developmental progress (McKEEHAN 1957). The current state of our knowledge shows that there are only a few examples in which specific recognition molecules only influence one or a few developmental events. In general, each type of recognition molecule is expressed in numerous cell types and exhibits up- and down-modulation as a function of development. In order to examine the potential role of these molecules in normal and abnormal development, we must first make some deductions concerning the possible ways in which their actions could be perturbed.

The first possible means of disrupting the function of a cell–cell recognition molecule relates to the earliest events in the processes these molecules mediate, i.e., ligand–receptor binding interactions. As listed in table 1, the known cell–cell recognition or adhesion molecules on a given cell all have (by definition) a receptor–ligand-type association with molecules on the surface of an interactive cell. Interference with such specific molecular interactions could, therefore, block all subsequent events that are influenced by a particular molecule. Indeed, much of our knowledge of cell–cell recognition molecules comes from the use of various probes that can specifically block and hence identify these interactions. Thus chemical agents or abnormal circumstances that can prohibit the interaction of a recognition molecule with its ligand could lead to failure of interdependent developmental events and teratogenesis.

The second obvious method of perturbing cell–cell recognition phenomena is to inhibit the timely expression of the receptors and/or their ligands. In other words, agents or circumstances that block the expression of cell–cell recognition molecules could lead to failure of events dependent on their function. As the literature shows that most recognition molecules are in themselves developmentally regulated, it seems likely that inappropriately high or low levels of the molecules and/or their receptors could lead to defects.

The final general way of perturbing the regulatory influence of cell–cell recognition molecules is to block events subsequent to the initial recognition step. The most advanced studies of cell–cell recognition have revealed many links to cascading intracellular signaling processes. The intracellular signaling processes then lead to changes in cell physiology, shape, or gene expression such that cell phenotype and, hence, development is affected. Therefore, any agents or events that impinge directly on the cascading intracellular processes triggered by cell–cell recognition could also generate defects.

D. In Vitro and In Situ Analyses

I. Ligand–Receptor Interaction Blockade

As mentioned earlier, one of the most productive approaches to the molecular analyses of cell–cell recognition has been a three-step approach: first, to identify a developmental event that is measurable and dependent on cell–cell interactions; second, to devise an assay of the event; finally, to perturb the process with probes that can potentially shed some light on specific molecules that are involved and then use the probes to aid in isolation. This stepwise analysis has been used by many investigators, and it is safe to say that all of the major families of cell–cell recognition molecules were first identified in this manner (EDELMAN 1983; ALBELDA and BUCK 1990; SPRINGER 1990; TAKEICHI 1990; LASKY 1992; MAILLET and SHUR 1993).

Conceptually, the most obvious cell–cell recognition-dependent event that might be used as a focus of study is the actual reassociation of cells into appropriate groupings. With the advent of methods for dissociating embryonic tissues into single cell suspensions (MOSCONA 1952; TOWNES and HOLTFRETER 1955), it became possible to devise simple assays for cell reassociation, more frequently called aggregation. One of the most notable contributions of MOSCONA (1952) to the field of experimental developmental biology was the first use of proteolytic enzymes for the purpose of generating single-cell suspensions from embryonic organ rudiments. This advance made the analysis of cell–cell recognition in higher vertebrate systems practical. Numerous investigators devised ways to assay the potential of dissociated cells to reaggregate in a quantifiable manner (MOSCONA 1961; STEINBERG 1964; EDELMAN 1983; MCCLAY et al. 1981; ROTH and WESTON 1967). One of the simplest assay systems has also provided many of the landmark breakthroughs.

Rotation of suspension cultures of dissociated single cells in physiological buffers leads to the gradual formation of multicellular aggregates. Counting the disappearance of single cells or the appearance and size of multicellular aggregates as a function of time in rotary culture was one of the first quantitative assays of cell–cell recognition (MOSCONA 1952, 1961). Using this very simple assay, various investigators showed that a large number of molecules and conditions influence cell reaggregation.

The aggregation assay and culture system provided a simple means of examining the earliest events in cell–cell interactions. It also was useful in demonstrating that many of the mature characteristics of a complex tissue could manifest under the apparent control of tissue-autonomous control (see, for example, LINSER and MOSCONA 1979). In other words, undifferentiated tissues could be shown to undergo normal cell differentiation and functional maturation in such an aggregate culture, isolated from embryonic factors

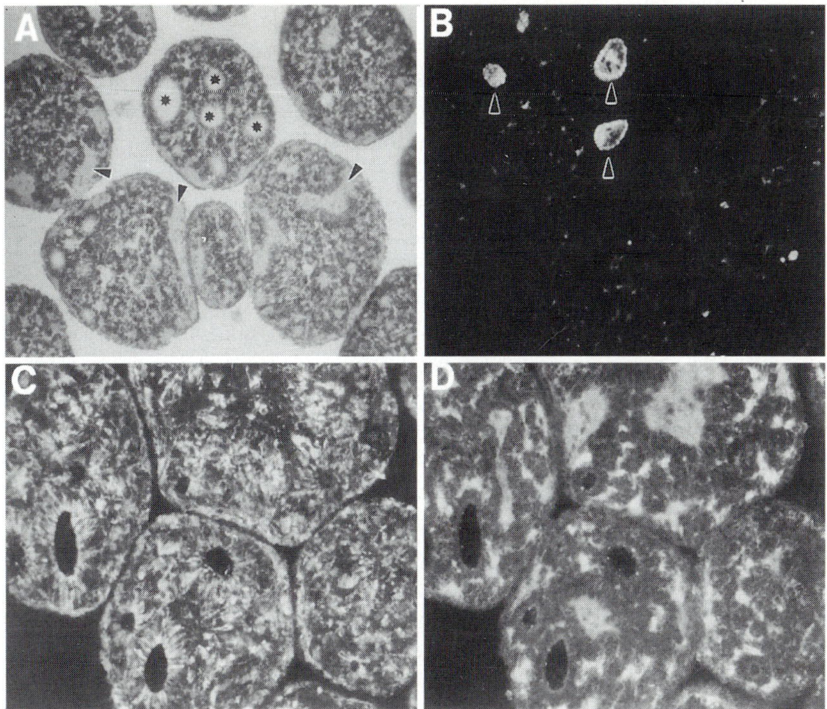

Fig. 1A–D. The intrinsic capacity of embryonic retina cells to differentiate in aggregate cultures. Undifferentiated embryonic chick retina (at day 7) was dissociated with trypsin into a suspension of single cells. The suspension was rotated in growth medium (199 + 10% fetal bovine serum) for 7 days with medium changes every 48 h. Hydrocortisone was added after 5 days (LINSER and MOSCONA 1979). At the end of the 7 days, the aggregates were prepared for histological analyses. **A** Standard hematoxylin and eosin (H&F) staining of aggregate sections. The aggregates are characterized by structural features of mature retina including defined areas of neuropil (plexiform layers, *arrowheads*) and rosettes of organized photoreceptor cells (*asterisks*). **B** The photoreceptor rosettes stain with fluorescent peanut agglutinin (*arrowheads*), which is a definitive marker of photoreceptor and retina maturation (JOHNSON et al. 1986). **C** Indirect immunofluorescence for glutamine synthetase, an enzyme that is expressed only by Müllerian glial cells under the control of neuronal cell–glial cell interactions (LINSER and MOSCONA 1979, 1984). **D** Immunofluorescence of the HNK-1 carbohydrate antigen found on cell adhesion molecules such as neural cell adhesion molecule (NCAM) (MEHRKE et al. 1987). In retina aggregates, HNK-1 staining defines the regions of neuron–neuron interaction (neuropil/plexiform layer)

(MOSCONA and LINSER 1983). Thus, regulatory events that occur after initial contact and recognition between cells has taken place also can be probed in vitro. Figure 1 shows a structural and immunohistochemical analysis of aggregates of embryonic retina cells as an example. In this case, undifferentiated retina tissue was dissociated into single cells, reaggregated, and then maintained in rotary suspension cultures for 7 days. The aggregates were prepared for histological analyses, and sections were then either stained with standard hematoxylin and eosin (H&E) or with antibodies which identified developmentally regulated molecules. Such aggregates exhibit structural features typical of a normal intact neural retina and also show cell differentiation and the expression of specific markers as would occur in situ. One of the important observations of the 1970s was that certain examples of cell differentiation and maturation showed a marked dependence on the three-dimensional relationship of the cells. In the case of the embryonic retina, maturation of the neuroglial cells was shown to be dependent on contact-mediated cell–cell interactions between the developing glial cells and retinal neurons (LINSER and MOSCONA 1979, 1984). Examples of this type of interaction-dependent development have been described in numerous systems and afford an approach to investigating complex developmental regulation and the experimental assessment of teratogenic influences.

Earlier studies also found that certain pharmacological agents and drugs influenced cell aggregation (see, for example, HAUSMAN and MOSCONA 1973). Some of these, such as 5-bromodeoxyuridine, cytosine arabinoside, retinoic acid, ethanol, and hormones are known to be teratogenic under certain circumstances (SHEPARD 1983). Most of these molecules probably had their influence on reaggregation indirectly, i.e., they affected macromolecular synthesis and processing in general or interfered with events linked to cell recognition other than the primary events of cell contact and adhesion.

Recently, DASTON et al. (1991) showed that a number of classical teratological agents have demonstrable effects on cell–cell association-dependent phenomena in the aggregation assay using embryonic chick retina cells. The first insight into the identity of specific cell–cell recognition molecules came from the use of probes that bind to cell surface components in a more or less specific manner, blocking the association of molecules that normally act as receptor–ligand pairs.

The first type of molecular probe that will be discussed that was used to block ligand–receptor interactions in cell–cell recognition was plant lectin (SACHS 1974). The sensitivity of cell associations to proteases and the need for protein synthesis for some aspects of cell–cell recognition (MOSCONA and MOSCONA 1963) indicated that at least some of the specific molecules involved were cell surface proteins. Further, the inhibitory capacity of certain lectins showed that not just proteins but more specifically glycoproteins were important (SACHS 1974). Lectins did not prove to be very useful in early studies for identifying individual components of the recognition system, since most lectins recognize carbohydrate moieties that are common to several glycoconjugates. The analyses with lectins did show, however, that cell–cell recognition phenomena could be perturbed by soluble molecules that blocked normal cell adhesion through specific receptors.

Figure 2 summarizes selected studies which showed that heterotypic cell–cell interactions could be effectively blocked with soluble molecules such as plant lectins. In this case, the succinylated derivative of concanavalin- A (S-ConA) was added to reaggregation cultures of embryonic retina cells. After an appropriate time to allow for "normal" cell differentiation in controls, the disposition and maturation of neuroglial cells and neurons was assessed biochemically and by immunohistochemistry. In the absence of the lectin, as already shown in Fig. 1, neurons and glial cells intermingle as in the intact tissue. Glial maturation occurs, as defined by glutamine synthetase (GS) expression (LINSER and MOSCONA 1979, 1984), and antibodies to this enzyme can be used to define the distribution of the glia. Earlier studies demonstrated that maturation of the glia and GS expression is dependent on three-dimensional cell–cell interactions. In other words, in a monolayer environment that minimizes cell–cell contact, GS expression is reduced manyfold relative to a reaggregated control (MORRIS and MOSCONA 1970; LINSER and MOSCONA 1984). In the aggregates formed in the presence of S-ConA, the normal intermingling of neurons and glial cells was prevented (LINSER 1987; LINSER and PERKINS 1987). Instead, the glial cells formed a central core and neurons were distributed as an enveloping layer around the glial core. GS expression in these aggregates was severely attenuated, presumably due to the inhibition of normal neuronal–glial relationships.

Biochemical analyses of plasma membrane glycoproteins that interact with the S-ConA revealed a set of approximately ten glycopeptides that were detectable in high concentrations on the surface of the retina cells when dissociated and challenged to reform tissue. Later stages of retina progressively lose the capacity to reassociate in a histotypic fashion and also lose the ability to produce the high levels of this set of glycoconjugates that seem to correlate with S-ConA effects. The results are consistent with the hypothesis that all or some of these lectin-binding glycopeptides are intimately involved in neuronal–glial interactions and the regulation of glial maturation in retina. Ten is a large number of individual proteins to go after at one time, so like many others we focused our attention on a newly emerging technology for the purpose of identifying individual molecules that play important roles in cell–cell interaction-dependent phenomena.

It should also be noted here that the capacity of cells in vitro to adhere to lectin-coated surfaces was explored as a screen for teratogenic substances (BRAUN et al. 1979, 1982). These investigations seemed to find that exposing established cell lines to a specific subgrouping of traditional teratogens led to reductions in the capacity of the cells to adhere to lectin-coated surfaces. The implications of these studies have never been explored in greater detail, but it seems quite interesting that cell adhesiveness might be directly affected by certain types of teratogens.

One of the most significant advances in the quest to identify the molecular basis of cell–cell recognition was the addition of an immunological approach. In 1962, MOSCONA and MOSCONA showed that antiserum raised against suspensions of embryonic cells could inhibit reaggregation. The nearly absolute specificity of antigen–antibody interactions provided a potential tool for finally identifying specific molecules. However, for immunology to succeed in this vein, a way of producing a monospecific antiserum was required. Several

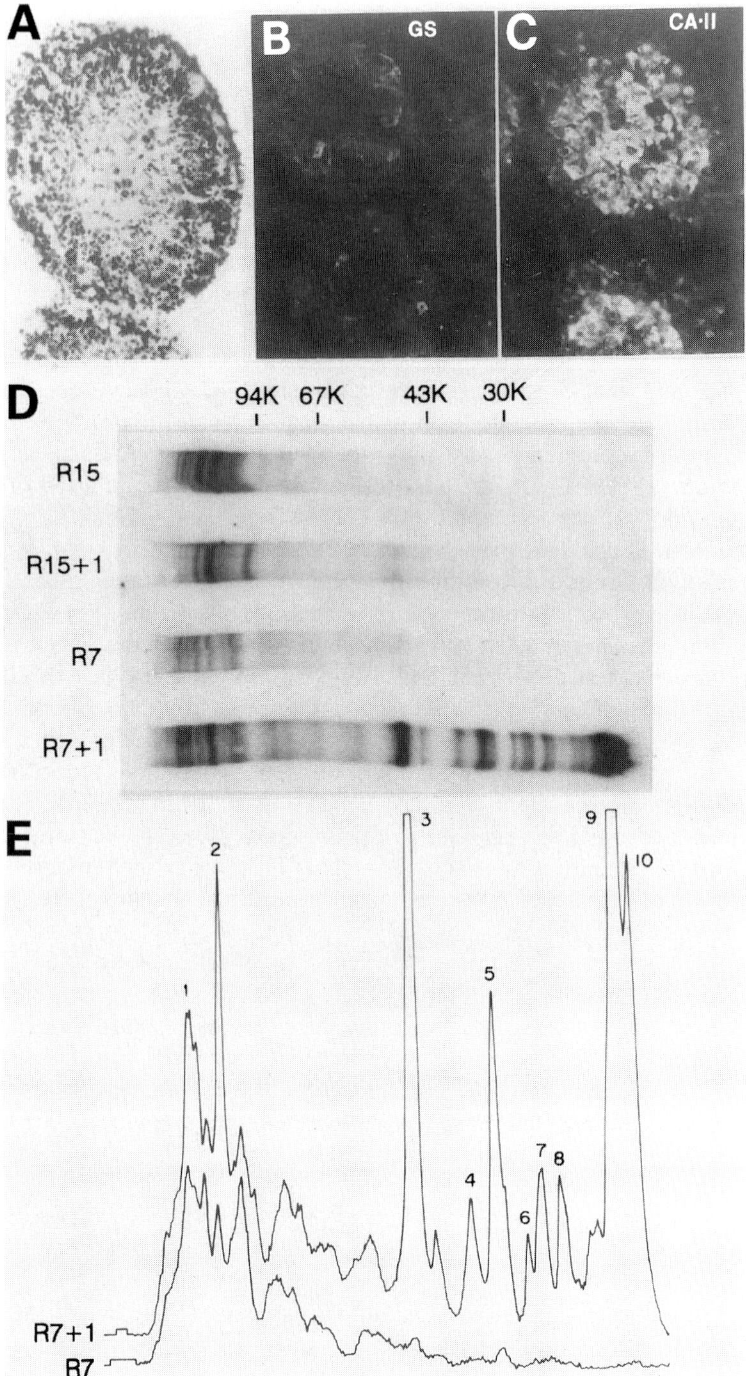
A
B
GS
C
CA·II
D
94K
67K
43K
30K
R15
R15+1
R7
R7+1
E
1
2
3
4
5
6
7
8
9
10
R7+1
R7

laboratories pursued this goal simultaneously. In some instances, purely biochemical methods were used in attempts to purify cell membrane components that influenced cell association (Lilien and Moscona 1967; Hausman and Moscona 1973). Once candidate molecules with expected characteristics were isolated, they were then used as immunogens to produce antisera with inhibitory properties (Hausman and Moscona 1973, 1979). In a combined immunological–biochemical approach, the research group in G. Edelman's laboratory developed what they termed the "iterative" procedure for generating a monospecific polyclonal rabbit antiserum to a cell–cell recognition molecule. From their efforts, neural- cell adhesion molecule (N-CAM) was first identified (Edelman 1983).

The "iterative" immunization procedure was truly a breakthrough in the field of cell–cell recognition, but it was very lengthy and laborious. Hence, few laboratories attempted this approach. The landmark work of Kohler and Milstein in 1975, which made the production of monoclonal antibodies (mAb) practical, influenced the progress of research in the field of cell–cell recognition probably more than any other single event. Many laboratories around the world had developed assays for cell–cell interaction-dependent events and were seeking ways to analyze the molecular bases of those interactions. mAb technology made it fairly easy to produce hundreds to thousands of unique antibodies to cell surface components. With functional assays in hand, investigators produced many individual mAb and screened them for

Fig. 2A–E. Usefulness of retina aggregate cultures in studying cell–cell interactions and the molecular basis of those interactions. Aggregates were prepared as described in the legend to Fig. 1, except that succinylated concanavalin A was added to the culture medium initially. After 48 h, the medium was replaced with lectin-free growth medium. After 7 days, the aggregates were analyzed for structural and biochemical development. **A** Hematoxylin and eosin (H&E) staining of sectioned aggregates. The aggregates lack photoreceptor cell rosettes and are characterized by a central core of lightly stained glial cells surrounded by an envelope of darker neurons. **B** The central core cells fail to immunostain for glutamine synthetase, indicating failed neuronal–glial interaction. **C** A marker of the glia, carbonic anhydrase-II (CA-II), which is not regulated by neuronal–glial interactions, defines the central core glia. Hence, the lectin prohibited interaction of neuroblast and glioblast cells, which produced a failure of histogenesis and specific aspects of glial maturation (Linser 1987). **D** Sodium dodecyl sulfate polyacrylamide gel electrophoresis (SDS-PAGE) analysis of retina membrane glycopeptides that react with the lectin. Plasma membranes were isolated from early and late retina (*R7* and *R15*, respectively) and from the same age retinas after trypsinization and overnight recovery (*R7+1* and *R15+1*). After separation on gradient gels and electroblotting to nitrocellulose, the blots were probed with I^{125}-labeled succinylated concanavalin A. Molecular mass markers are depicted across the *top* of the gel autoradiogram. Note that the early retina, when challenged by trypsinization (R7+1) overproduces several lectin-binding components and that this capacity is lost in the older retina (R15+1). **E** Comparison of densitometric scans of the R7 and R7+1 samples. Note that approximately ten different glycopeptides (*1–10*) show exaggerated expression in response to trypsin dissociation and the challenge to reaggregate and continue tissue differentiation

measurable effects in their particular assay system. Once an individual antibody was characterized that could perturb a particular interaction-dependent event, the antibody could be used to identify and purify specific molecules or macromolecules. In this manner, representatives of all of the major families of cell–cell recognition (adhesion) molecules were identified (ALBELDA and BUCK 1990; FADOOL and LINSER 1993b; LASKY 1992; RISAU et al 1986; TAKEICHI 1990).

As described above, retina cell cultures were used to define certain developmental phenomena that were dependent on cell–cell interactions. Furthermore, soluble molecules such as lectins could be used to perturb such interactions and help limit the range of potential macromolecular components of the interactive process to pursue biochemically (LINSER 1987; SACHS 1974). By immunizing mice with live, dissociated embryonic retina cells from chick, a panel of mAb was produced. When these antibodies were screened for the capacity to interfere with glial maturation and GS expression in cultured, interactive retina cells, a particular mAb was identified (LINSER and PERKINS 1987; FADOOL and LINSER 1993b). The mAb (called 5A11) was used to immunopurify a glycoprotein from retina plasma membranes, and peptide sequence analyses revealed that the 5A11 antigen was a member of the Ig super gene family (FADOOL and LINSER 1993b). Furthermore, several other laboratories around the world were actively studying this or homologous proteins from several vertebrates, including man (MIYAUCHI et al. 1991), rodents (FOSSUM et al. 1991), and chickens (RISAU et al. 1986).

The structural characteristics of 5A11 antigen and its homologues (HT7, neurothelin, basigin, OX47) are as follows:

- 32-kDa polypeptide typical of a member of the Ig superfamily
- 13- to 23- kDa additional carbohydrate
- Variable carbohydrate structure (differential lectin affinity)
- Unique transmembrane domain
- Homodimer
- Cytoskeletal association

The distribution of 5A11 and its homologues is as follows:

- Epithelial structures and blood cells
- High levels associated with periods of embryonic induction
- Change in distribution during development from generalized to cell- specific

Its functional characteristics are the following:

- Antibodies inhibit reaggregation of retina cells
- Antibodies inhibit neuronal–glial interactions in retina
- Antibodies inhibit neurite outgrowth over a glial substrate
- Inducible at the blood–brain barrier

The power of the mAb approach for studying the molecular basis of cell–cell interactions soon led to the isolation, characterization, and even cloning of many hundreds of specific cell- surface molecules. This, of course, led to structural and functional analyses that would further substantiate the potential of cell–cell interactions as a primary target of teratogenesis. Table 1 also lists some of the features of the various cell–cell recognition molecules with respect to their known functions. In general, cell recognition molecules operate through a specific receptor–ligand interaction. In some cases the receptor and the ligand are the same molecule on apposed cells (homotypic interaction), and in other cases they are distinct gene products (heterotypic interaction). The receptor–ligand interactions provide specificity to the "recognition" phenomenon, but the ensuing events which influence the interacting cells in terms of physiology (HYNES 1992), shape (GUMBINER 1993), motility (BRONNER-FRASER 1993), or gene expression (LINSER 1987) are additional areas which may be susceptible to teratogenic influences.

As already discussed, specific cell adhesion or cell–cell recognition interactions can be inhibited by molecules such as lectins or antibodies that bind at or near the ligand-binding region of a specific recognition molecule. Therefore, similar molecules with this type of binding specificity might have potential as teratogens. In some instances, immunoglobulins have been shown to cross the placental barrier (HUNZIKER and WEGMANN 1986), so it is at least possible that interference with cell–cell recognition during embryonic development could occur from maternally delivered immunoglobulin or lectin-like molecules. It is interesting to note that JORDAN and KARWOSKI (1994) have recently reported that orally administered mAb to cell surface antigens of the neural retina could later be detected in the retina after apparently crossing the blood–retinal barrier. Since some antibodies have already been noted to cross the placental barrier, damage to the embryo or fetus might actually take place should a pregnant female ingest or receive by injection antibodies with specificity for cell–cell recognition molecules. This seems admittedly far fetched at this time, but raises questions that should be investigated.

Investigations into the molecular structure and its relationship to function of cell–cell recognition molecules have provided additional possibilities for perturbation and teratogenesis. Several laboratories have attempted to identify structural domains within given macromolecules that control specific functionalities (BOUCAUT et al. 1984; NAIDET et al. 1987; RUOSLAHTI and PIERSCHBACHER 1987). The best known example of this was generated by several laboratories working on the ECM adhesion molecules called integrins. One particularly informative set of studies tested the capacity of synthetic peptides to block interactions between the cell surface integrin complex and its ligands. Specific proteolytic fragments of ECM molecules such as fibronectin were tested for their capacity to interfere with cell–ECM adhesion (reviewed in YAMADA 1991). Subsequently, smaller peptide sequences were produced synthetically based on the sequences of the active fragments. In this manner, a minimal binding domain of three amino acids was identified. The sequence of the tripeptide is Arg-Gly-Asp (RGD) (RUOSLAHTI and PIERSCHBACHER 1987). This tripeptide has been reported to interfere with a plethora of developmental processes when introduced artificially into the environment of developing embryonic tissues (RUOSLAHTI and PIERSCHBACHER 1987). Among functionalities that have been shown to depend on RGD-based interactions are neural crest cell migration (BRONNER-FRASER 1993), vertebrate eye morphogenesis (SVENNEVIK and LINSER 1993), gastrulation (BOUCAT et al. 1984), oligodendrocyte maturation (CARDWELL and ROME 1988), and even development of invertebrates such as *Drosophila* (NAIDET et al. 1987). Larger synthetic peptides with RGD as the core have been useful in identifying the ligand domains of a number of ECM and cell surface molecules (YAMADA 1991). RGD as the core of developmentally important cell–cell and cell–ECM recognition systems has been reported in evolutionarily diverse animal models including vertebrates and invertebrates (BOUCAUT et al. 1984; NAIDET et al. 1987). In our own laboratory, we showed that the RGD tripeptide and antibodies to the integrin complex were mutually effective in dramatically disrupting vertebrate eye morphogenesis (SVENNEVIK and LINSER 1993).

The importance of cell–cell interactions in regulating normal development is no longer in doubt. Analyses performed both in vitro and in situ have demonstrated that soluble molecules such as lectins, antibodies, or binding site-mimicking peptides can generate developmental failures. Such failures can be very specific or very general in nature. The potential for teratogenesis from such molecules seems to exist, but the very nature of these molecules does not support any high probability of embryonic exposure. Unless we identify an as yet unknown class of molecules that can act as antagonists to cell–cell recognition phenomena, it seems that blockage of receptor–ligand interactions is an unlikely source of developmental defects in situ. Many (although not all) of the known cell adhesion molecules depend on Ca^{2+} for binding activity. Therefore, it is possible that Ca^{2+} -chelating agents or molecules such as lead which compete with Ca^{2+} might be teratogenic due to just such an antagonistic inhibition of cell–cell recognition. A number of cell–cell recognition molecules have been identified in soluble forms circulating in blood or other fluids (ALBERTS et al 1994; DURBEC et al 1992; JANSSEN et al. 1994; KROG et al. 1992; PARADIES and GRUNWALD 1993; JANEWAY and TRAVERS 1994). It is possible that such molecules could act as antagonists or inappropriate stimulators of events should the soluble forms be supplied at sensitive stages of development. In a few instances, soluble forms are associated with disease states (JANSSEN et al. 1994). If the soluble recognition molecules can cross the placenta, then they might provide a source of teratogenic inhibition of cell–cell recognition phenomena.

II. Availability (Expression and Functional Regulation)

The vast majority of cell–cell recognition molecules that have been studied show dynamic patterns of developmental expression (ALBERTS et al. 1994). In other words, the concentration and distribution of specific molecules changes dramatically with the development and maturation of every tissue. In addition, several cell–cell recognition molecules exist in a variety of forms which can be influenced by differential RNA splicing or post-translational modifications (EDELMAN and CROSSIN 1991; FADOOL and LINSER 1993a; HYNES 1992; RUTISHAUSER 1993). The importance of cell–cell recognition in development seems evident from the studies outlined above using specific inhibitors to block their adhesive or recognition functions. The fact that their expression and even molecular form is developmentally regulated suggests that their proper functioning is influenced by temporal and spatial expression control. Thus, chemical agents or circumstances which might influence the expression, translation, or modification of specific recognition molecules might effect additional secondary events that are triggered by cell–cell recognition processes. Several lines of investigation support the hypothesis that, without the right amount of the right cell–cell recognition molecule in the right form on the right cells at the right time, developmental defects are produced.

Perhaps the most widely studied of the cell–cell recognition molecules is the neural cell adhesion molecule (NCAM; see Table 1). Figure 3 shows a diagrammatic representation of the four major proteins generated from splice variants of the single gene (EDELMAN and CROSSIN 1991). Appearance of the different polypeptides is developmentally regulated, and each has potentially different specific roles. One of the intriguing features of the NCAM molecule is the presence of varying quantities of a peculiar carbohydrate group, α-2,8-linked polysialic acid (PSA). As cell surface proteins, most if not all cell–cell recognition molecules are glycosylated to widely varying degrees. In many instances, the carbohydrate groups linked to the polypeptide are sufficiently unique that much speculation has developed concerning their intrinsic importance to the recognition process. In some cases, specific molecular roles have been implicated by inhibition of synthesis or enzymatic cleavage of the carbohydrate group. The PSA moiety of NCAM attracted experimental attention early for two reasons. First, the specific structure of the PSA (α-2,8 linkage) is typical in certain bacteria, but is only rarely found on glycoproteins of higher vertebrates (RUTISHAUSER 1993). Second, the amount of this sugar polymer which is attached to the NCAM polypeptide varies widely with development. The current model for the role of PSA in NCAM-mediated cell–cell adhesion is that high levels of the carbohydrate produce an organized aqueous environment or shell around the adhesive domains of the protein. This reduces or modifies the strength of adhesive interactions and produces very weak associations between apposed cells. NCAM that contains little or no PSA can form tighter NCAM–NCAM bonds on adjacent cells. It is also produced at times in development when the specific cells associate very tightly with one another (EDELMAN 1985; RUTISHAUSER 1993).

The post-translational addition of PSA to NCAM and other specific carbohydrate structures to cell–cell recognition molecules in general may be an area of concern for teratogenesis. The importance of the NCAM PSA has been demonstrated by the injection of a specific endoglycosidase that only cleaves PSA (endo-n; RUTISHAUSER 1993) into embryos at various stages of development (ONO et al. 1994). The results showed that neural birth defects that

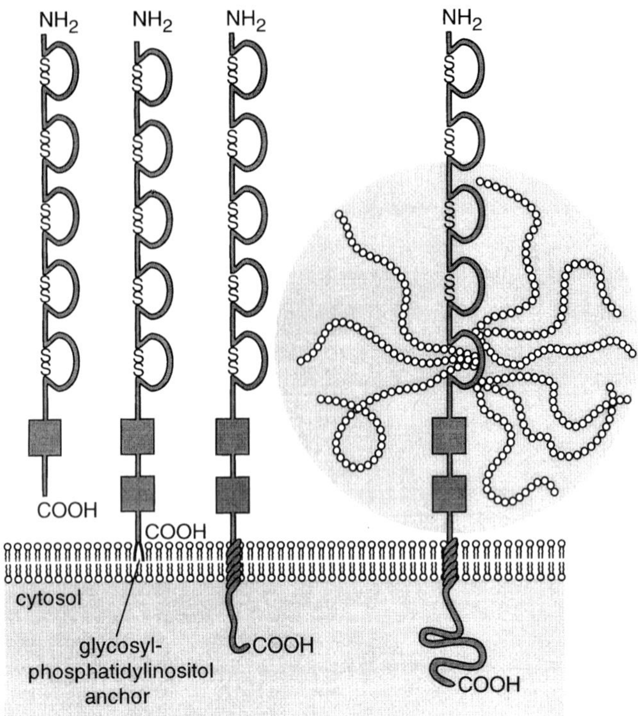

Fig. 3. Four major forms of the immunoglobulin (Ig) superfamily adhesion molecule NCAM (neural cell adhesion molecule). RNA splicing produces several translational products that vary primarily in the carboxyl terminus. This affects the type of molecular association with the plasma membrane. The drawing furthest to the *right* also depicts long chains of polysialic acid and a region of "structured" water (*gray, shaded sphere*) that encapsulates a portion of the polypeptide core. All forms of NCAM can be produced with or without polysialic acid side chains of variable length, but it was drawn only on one molecule here for the purpose of clarity. (After ALBERTS et al. 1994; EDELMAN and CROSSIN 1991; RUTISHAUSER 1993)

result from a failure of neuronal migration can be produced by such modifications of NCAM (ONO et al. 1994). Hence, pharmacological agents that might influence the processing pathways for production and membrane-trafficking cell membrane molecules could be teratogenic due to secondary effects on cell–cell recognition phenomena. It is also interesting to note that the splotch (*Sp*) mutation in mice involves inappropriately timed addition of PSA to NCAM during neural tube development (NEALE and TRASLER 1994). The *Sp* mutation produces neural tube defects, including spina bifida and exencephaly. Recently, NEALE and TRASLER (1994) have shown that high concentrations of PSA are evident on NCAM at a time when control animals show very little. This suggests that NCAM-bearing cells may be less mutually adhesive in the mutant, resulting in failures of tissue morphogenesis. We will return to the issue of genetic failures in NCAM expression later in this review.

As has already been stated, cell–cell recognition molecules exhibit dynamic patterns of expression that reflect modulation of the absolute concentrations of individual molecules on the surface of cells. An obvious hypothesis that can be derived from this observation is that events that can lead to altered levels of cell–cell recognition molecules at the cell surface might also lead to developmental defects. A number of exciting molecular approaches have been used to increase or decrease the expression of specific recognition molecules both in vitro and in situ. One particularly striking example of developmental aberration as a result of overexpression of a cell–cell recognition molecule was described by DETRICK et al. (1990) regarding neural development in *Xenopus* embryos. In this particular study, the investigators injected RNA for N cadherin (see Table 1) into *Xenopus* embryos prior to the stage in development when it first appears in the embryo. The RNA injection led to the ectopic expression of N cadherin at the cell surface. The inappropriate appearance of the protein at the cell surface was then correlated with subsequent severe developmental errors in organogenesis. In contrast, other embryos that were similarly treated with NCAM RNA showed no anomalous development (DETRICK et al. 1990).

A central feature of brain development is the formation of ordered laminations within specific brain regions. The development of the CNS is typically described as having three key phases: cell mitosis, cell migration, and cell maturation. All three of these phases are influenced by cell–cell interactions. The cell migration process is critical to establishing functional neural networks, and failure of migration has been identified as the underlying problem in Kallman's syndrome, a human example of developmental brain malformation (FRANCO et al. 1991; LEGOUIS et al. 1991). Recent studies again implicate additional specific cell–cell recognition molecules in this critical aspect of neurogenesis. GALILEO et al. (1992) produced retroviral vectors designed to interfere with integrin production in the infected cells. The strategy of their approach was to produce a replication-defective retrovirus that carried a marker gene (β-galactosidase) and a sequence that, when transcribed, would produce an antisense copy of a portion of the β-subunit of the integrin heterodimer mRNA. Antisense inhibition of translation is currently being pursued widely as a powerful tool for delineating the role of specific molecules in specific cells. In Galileo's experiment, the retrovirus was used to infect embryonic chick optic tectum in situ. The potential of the infected cells to achieve normal mature positions within the laminations of the tectum was then evaluated in comparison to controls infected with virus that only carried the marker β-galactosidase gene. The results showed that the antisense-integrin virus did indeed reduce the levels of integrin at the surface of the infected cells. As an apparent result of this, the infected cells failed to migrate normally, generating a situation that could lead to impaired brain function (GALILEO et al. 1992). These experiments again demonstrate that the level of specific cell–cell recognition molecules is an additional factor in achieving normal embryogenesis.

One of the most important new approaches for understanding a specific gene's developmental significance is to produce a genetic "knockout" (EVANS et al. 1993; KWEE et al. 1995; MAYADAS et al. 1993; TOMASIEWICZ et al. 1993). In general this is accomplished either by chemical or radiation-based mutagenesis and selection for fortuitous loss of function mutations or by homologous recombination of DNA sequences into specifically targeted genes. To date only a few nonlethal knockouts of major cell–cell recognition molecules have been accomplished. Of particular interest to this discussion is the recent production of a mouse line which does not produce the major (180-kDa) translational product of the NCAM gene (TOMASIEWICZ et al. 1993). Perhaps the most surprising finding in the analysis of these mice is that the homozygous NCAM(–) mice survive to adulthood and show a nearly normal phenotype. A number of subtle structural anomalies in the nervous system of these animals have been reported (TOMASIEWICZ et al. 1993). The most pronounced deficit is a failure of neuronal migration in the brain, which leads to incomplete maturation of the olfactory region and a concomitant loss of olfactory input and olfaction-dependent behaviors. Although other behavioral deficits may be detected in these animals which indicate additional developmental errors, it seems remarkable that loss of the function of this molecule has so little obvious effect on embryogenesis. Other gene knockouts have been generated in what seem to be important functions with little obvious effects. The reason for this probably relates to the evolution of redundancy in many biological systems that allows for a certain amount of developmental error and recovery. This will be discussed again in the conclusion.

III. Downstream Signaling Cascade

Cell–cell recognition, as defined for the purpose of this review, operates by way of specific molecular interactions between ligands and receptors at the cell surface. Interference with the receptor–ligand interaction by several means can lead to developmental failure or errors in vitro. Interference of this sort might come from several types of potentially teratogenic substances or genetic mutations as discussed above.

Equally as important as the actual interactions that occur between cell–cell recognition molecules on the communicating cells are the intracellular events triggered by those interactions. In some instances we have no information concerning the mechanistic link between the actions of specific cell–cell recognition molecules and resulting changes in cell character. However, many investigations have demonstrated that, in general, cell–cell recognition molecules and cell adhesion molecules are associated with second messenger systems (ALBERTS et al. 1994; ALBELDA and BUCK 1990). Indeed, nearly every type of second messenger system that has been identified to date has a reported connection to cell–cell interaction phenomena in some way. Furthermore, many cell adhesion molecules are also linked directly or indirectly to structural components of cells, including the cytoskeleton and junctional structures or junction formation (GUMBINER 1993). Therefore, any genetic defect or chemical teratogen that influences membrane ion traffic, G protein systems, cytoskeletal integrity, gap junctional condutance, protein kinase, phospholipase activities, etc. has the potential to interfere directly or indirectly with the process of cell–cell recognition. Many of the second messenger cascades and their involvement in teratologic events are discussed by other authors elsewhere in this volume. At the time of this writing, our knowledge of developmental malformations caused by blockade of second messenger cascades linked specifically to cell adhesion molecules or cell–cell recognition molecules is only inferential. Similarly, gap junctional communication between embryonic cells clearly can influence subsequent developmental decisions (WARNER et al. 1984). The definitive experiments in this regard were performed by injecting antibodies to gap junctional subunits directly into the cytoplasm of embryonic cells. This is not an event that is likely to be mimicked by external environmental molecules. A number of externally supplied substances such as retinoic acid and H^+ ion also influence gap junctional conductance and are teratological (MEHTA and LOEWENSTEIN 1991). However, such toxicants influence numerous cellular processes and internal signaling cascades, so it is exceeedingly difficult to ascribe primary significance to gap junctions for any developmental perturbations.

E. Conclusions

The last 20 years have yielded an explosion of information concerning the roles and mechanisms of cell–cell interactions in normal embryonic development. The percentage of human birth defects that remain unexplained still hovers near 70%. We still do not know whether some substantial portion of that 70% may be linked to failures of cell–cell interactions, but the knowledge that we have gained in the last two decades can certainly help generate some educated hypotheses.

First, it is possible that perturbation of cell–cell recognition phenomena is not a source of many nonlethal developmental defects. Although it is quite a simple matter to interfere with cell–cell recognition in vitro with blockade of specific receptor–ligand interactions, this may be very unlikely to occur in utero. The types of molecules that we know will prevent such interactions (i.e., lectins, antibodies, peptides, soluble forms of cell adhesion molecules, etc.) have very limited access to the mammalian embryo or fetus.

Another reason why perturbation of cell–cell recognition processes may not be a substantial source of developmental defects is the apparent redundancy of these systems. The last two decades have witnessed the discovery of hundreds of cell adhesion molecules and other cell–cell recognition molecules. The molecular genetic analyses of many of these have revealed multigene families whose proportions are currently a matter of speculation. In many well-documented cases, it appears that there are overlapping patterns of expression and functional significance for many of the known molecules. Thus it may be possible that this complicated system for cell–cell signaling has evolved with considerable flexibility. Indeed, in the few instances in which specific genes have been "knocked out," the effects on embryogenesis have been surprisingly minimal (MAYADAS et al. 1993; TOMASIEWICZ et al. 1993). It should be noted here, however, that what appears as a "minimal" effect in a mutated mouse might actually manifest as a severe syndrome of defects in a human. It is exceptionally difficult to bridge the gap between what a developmental abnormality produces in the character and behavior of a mouse or frog and what a similar defect might cause in a human.

The foundations of the arguments above concerning why cell–cell recognition might not be involved in generating birth defects can serve equally well as foundations for the opposite argument. The apparent redundancy of molecular mechanisms for cell–cell recognition may not be redundancy at all but rather "fine-tuning." The plethora of individual genes, splice variants, heteromeric combinations, and differentially modified translational products that occur make it possible for each cell to possess a unique array of cell–cell recognition molecules at every stage of development. The complexity of recognition and signaling systems surely defines the limits of complexity that can be achieved in tissue and organogenesis. Thus, the more complex an organism, the more complex are the requirements placed on cell–cell recognition and signaling processes. Human development, being arguably the most complicated such process on earth, must require exquisitely complicated regulation. Specific failures in any phase of cell–cell recognition would likely lead to anomalous development or perhaps a loss in complexity as defined by a reduction in function.

Cell–cell recognition and communication is a multistep process, and it is feasible that any one of the steps could be limiting for the final dependent regulatory events. Mutations or chemical insults that target any step may be teratogenic. Therefore, it seems more likely that the reason we have not attributed more of the unexplained human birth defects to problems of cell–cell recognition is because, even after 20 years of extensive research in this area, we still do not know how to recognize the appropriate correlations.

References

Alberts B, Bray D, Lewis J, Raff M, Roberts K, Watson JD (1994) Cell junctions, cell adhesion, and the extracellular matrix. In: Robertson M (ed) Molecular biology of the cell. Garland, New York, pp 949–1011

Albelda SM, Buck C (1990) Integrins and other cell adhesion molecules. FASEB J 4: 2868–2880

Bauer GE, Balsamo J, Lilien J (1992) Cadherin-mediated adhesion in pancreatic islet cells is modulated by a cell surface N-acetylgalactosaminylphosphotransferase. J Cell Sci 103: 1235–1241

Bernfield M, Banarjee SD, Koda JE, Rapraegar AC (1984) Remodeling of basement membranes as a mechanism of morphogenetic tissue interaction. In: Trelstad RL (ed) The role of the extracellular matrix in development. Liss, New York, p 545

Boucaut J-C, Darribere T, Poole TJ, Aoyama H, Yamada KM, Thiery JP (1984) Biologically active synthetic peptides as probes of embryonic development: a competitive inhibitor of fibronectin function inhibits gastrulation in amphibian embryos and neural crest cell migration in avian embryos. J Cell Biol 99: 1822–1830

Braun AG, Emerson DJ, Nichinson BB (1979) Teratogenic drugs inhibit tumor cell attachment to lectin-coated surfaces. Nature 282: 507–509

Braun AG, Buckner CA, Emerson DJ, Nichinson BB (1982) Quantitative correspondence between in vivo and in vitro activity of teratogenic agents. Proc Natl Acad Sci USA 79: 2056–2060

Bronner-Fraser M (1993) Mechanisms of neural crest migration. Bioessays 15: 221–230

Cardwell MC, Rome LH (1988) Evidence that an RGD-dependent receptor mediates the binding of oligodendrocytes to a novel ligand in a glial-derived matrix. J Cell Biol 107: 1541–1549

Daston GP, Baines D, Yonker (1991) Chick embryo neural retina cell culture as a screen for developmental toxicity. Toxicol Appl Pharmacol 109: 352–366

Detrick RJ, Dickey D, Kintner CR (1990) The effects of N-cadherin misexpression on morphogenesis in xenopus embryos. Neuron 4: 493–506

Durbec P, Gennarini G, Goridis C, Rougan G (1992) A soluble form of the F3 neuronal cell adhesion molecule promotes neurite outgrowth. J Cell Biol 117: 877–887

Edelman G (1985) The cell in contact: adhesions and junctions as morphogenetic determinants. Wiley, New York

Edelman GM (1983) Cell adhesion molecules. Science 219: 450–457

Edelman GM, Crossin KL (1991) Cell adhesion molecules: implications for a molecular histology. Annu Rev Biochem 60: 155–190

Evans SC, Lopez LC, Shur BD (1993) Dominant negative mutation in cell surface beta-1,4-galactosyltransferase inhibits cell-cell and cell-matrix interactions. J Cell Biol 120: 1045–1057

Fadool JM, Linser PJ (1993a) Differential glycosylation of the 5A11/HT7 antigen by neural retina and epithelial tissues in the chicken. J Neurochem 60: 1354–1364

Fadool JM, Linser PJ (1993b) 5A11 Antigen is a cell recognition molecule which is involved in neuronal-glial interactions in avian neural retina. Dev Dynam 196: 252–262

Fadool JM, Linser PJ (1994) Spacial and temporal expression of the 5A11/HT7 antigen in the chick embryo: association with morphogenetic events and tissue maturation. Rouxs Arch Dev Biol 203: 328–339

Fossum S, Mallett S, Barclay AV (1991) The MRC OX-47 antigen is a member of the immunoglobulin superfamily with an unusual transmembrane sequence. Eur J Immunol 21: 671–679

Franco B, Guioli S, Pragliola A, Incerti A, Bardoni B, Tonlorenzi R, Carrozzo R, Maestrini E, Pieretti M, Tailon-Miller P, Brown CJ, Willard HF, Lawrence C, Persico MG, Camerino G, Ballabio A (1991) A gene deleted in Kallmann's syndrome shares homology with neural cell adhesion and axonal-pathfinding molecules. Nature 353: 529–535

Galileo DS, Majors J, Horwitz AF, Sanes JR (1992) Retrovirally introduced antisense integrin RNA inhibits neuroblast migration in vivo. Neuron 9: 1117–1131

Galstoff PS (1925) Regeneration after dissociation: an experimental study on sponges. J Exp Zool 42: 223–255

Gilbert S (1994) Developmental biology, 4th edn. Sinauer, Sunderland, MA

Goodman CS and Shatz CJ (1993) Developmental mechanisms that generate precise patterns of neuronal connectivity. Neuron 10 [Suppl]: 77–98

Grenningloh G, Bieber AJ, Rehm EJ, Show PM, Traquina ZR, Hortsch M, Patel NH, Goodman CS (1990) Molecular genetics of neuronal recognition in Drosphila: evolution and function of immunoglobulin superfamily cell adhesion molecules. Cold Spring Harb Symp Quant Biol 55: 327–340

Gumbiner BM (1993) Proteins associated with the cytoplasmic surface of adhesion molecules. Neuron 11: 551–564

Gurdon JB (1963) Nuclear transplantation in amphibia and the importance of stable nuclear changes in promoting cellular differentiation. Q Rev Biol 38: 54

Hausman RE, Moscona AA (1973) Cell surface interactions: differential inhibition by proflavine of embryonic cell aggregation and production of specific cell-aggregating factor. Proc Natl Acad Sci USA 70: 3111–3114

Hausman RE, Moscona AA (1976) Isolation of retina-specific cell-aggregating factor from membranes of embryonic neural retina. Proc Natl Acad Sci USA 73: 3594–3598

Hausman RE, Moscona AA (1979) Immunologic detection of retina cognin on the surface of embryonic cells. Exp Cell Res 119: 191–204

Hunziker RD, Wegmann TG (1986) Placental immunoregulation. Crit Rev Immunol 6: 245–285

Hynes RO (1992) Integrins: versatility, modulation and signalling in cell adhesion. Cell 69: 11–25

Janeway C, Travers P (1994) Immunology. Current Science and Garland, London

Janssen BA, Lugmani RA, Gordon C, Hemingway (1994) Correlation of blood levels of soluble vascular cell adhesion molecule-1 with disease activity in systemic lupus erythematosus and vasculitis. Br J Rheumatol 33: 1112-1116

Johnson LV, Hageman GS, Blanks JC (1986) Interphotoreceptor matrix domains ensheathe vertebrate cone photoreceptor cells. Invest Ophthal Vis Sci 27: 129–135

Jordan BL, Karwoski CJ (1994) In vivo delivery of monoclonal antibodies to retinal Müller cells. Invest Ophthal Vis Sci 34(4): 1366

Kohler G, Milstein C (1975) Continuous cultures of fused cells secreting antibody of predefined specificity. Nature 256: 495–497

Krog L, Olsen M, Dalseg AM, Roth J, Bock E (1992) Characterization of soluble neural cell adhesion molecule in rat brain, CSF, and plasma. J Neurochem 59: 838–847

Kwee L, Baldwin HS, Shen HM, Stewart CL, Buck C, Buck CA, Labow M (1995) Defective development of the embryonic and extraembryonic circulatory systems in vascular cell adhesion molecule (VCAM-1) deficient mice. Development 121: 489–503

Lasky LA (1992) Selectins: interpreter of cell-specific carbohydrate information during inflammation. Science 258: 964–969

Legouis R, Hardelin J-P, Levilliers J, Claverie J-m, Compain S, Wunderle V, Millasseau P, Le Paslier D, Cohen D, Caterina D, Bougueleret L, Delemarre-Van de Wall H, Lutfaila G, Weissenbach J, Petit C (1991) The candidate gene for the X-linked Kallman syndrome encodes a protein related to adhesion molecules. Cell 67: 423–435

Lewin B (1994) Genes V. Oxford University Press, Oxford

Lilien JE and Moscona AA (1967) Cell aggregation: its enhancement by a supernatant from cultures of homologous cells. Science 157: 70–72

Linser P, Moscona AA (1979) Induction of glutamine synthetase in embryonic retina: localization in Müller fibres and dependence on cell interactions. Proc Natl Acad Sci USA 76: 6476–6480

Linser P, Moscona AA (1984) The influence of neuronal-glial interactions on glia-specific gene expression in embryonic retina. In: Lauder JM, Nelson PG (eds) Gene expression and cell-cell interactions in the developing nervous system. Plenum, New York, pp 185–202

Linser PJ (1987) Neuronal-glial interactions in retina development. Am Zool 27: 161–169

Linser PJ, Perkins MS (1987) Regulatory aspects of the in vitro development of retinal Müller glial cells. Cell Differ 20: 189–196

Maillet CM, Shur BD (1993) Uvomorulin, LAMP-1, and laminin are substrates for cell surface beta-1,4-galactosyltransferase on F9 embryonal carcinoma cells: comparisons between wild type and mutant 5.51 att (–) cells. Exp Cell Res 208: 282–295

Mayadas TN, Johnson RC, Rayburn H, Hynes RO, Wagner DD (1993) Leukocyte rolling and extravasation are severely compromised in selectin-deficient mice. Cell 74: 541–554

McClay DR, Wessel GM, Marchase RB (1981) Intercellular recognition: quantitation of initial binding events. Proc Natl Acad Sci USA 78: 4975–4979

McKeehan MS (1957) Cytological aspects of embryonic lens induction in the chick. J Exp Zool 117: 31–64

Mehrke G, Jockusch H, Schachner M (1987) Development of mammalian nerve-muscle synapses in culture: lack interference by antibodies to the neural cell adhesion molecule N-CAM and its L2/HNK-1 carbohydrate epitope. Neurosci Lett 78: 247–252

Mehta PP, Lowenstein WR (1991) Differential regulation of communication by retinoic acid in homologous and heterologous junctions between normal and transformed cells. J Cell Biol 113: 371–379

Miyauchi T, Masuzawa Y, Muramatsu T (1991) The basigin group of the immunoglobulin superfamily: complete conservation of a segment in and around transmembrane domains of human and mouse basigin and chicken HT7 antigen. J Biochem (Tokyo) 110: 770–774

Monroy A, Moscona AA (1979) Cell aggregation: Construction of multicellular systems from cell suspensions. In: Monroy A, Moscona AA (eds) Introductory concepts in developmental biology. University of Chicago Press, Chicago, pp 155–205

Morris JE, Moscona AA (1970) Induction of glutamine synthetase in embryonic retina: its dependence on cell interactions. Science 167: 1736–1738

Moscona AA (1952) Cell suspensions from organ rudiments of chick embryos. Exp Cell Res 3: 535–539

Moscona AA (1961) Rotation-mediated histogenetic aggregation of dissociated cells: a quantifiable approach to cell interactions in vitro. Exp Cell Res 22: 455–475

Moscona AA, Linser PJ (1983) Developmental and experimental changes in retinal glia cells: cell interactions and control of phenotype expression and stability. In: Moscona AA, Monroy A (eds) Current topics in developmental biology, vol 18. Academic, New York, pp 155–188

Moscona AA, Moscona MH (1962) Specific inhibition of cell aggregation by antiserum to suspensions of embryonic cells. Anat Rec 142: 319

Moscona AA, Mayerson P, Linser PJ, Moscona M (1979) Induction of glutamine synthetase in the neural retina of the chick embryo: localization of the enzyme in Müller fibers and effects of BrdU and cell separation. In: Giacobini E, Vernadakis A, Shahar A (eds) Tissue culture in neurobiology. Academic, New York, pp 111-127

Moscona MH, Moscona AA (1963) Inhibition of adhesiveness and aggregation of dissociated cells by inhibitors of protein and RNA synthesis. Science 142: 1070–1071

Naidet C, Semeriva M, Yamada KM, Thiery JP (1987) Peptides containing the cell-attachment recognition signal arg-gly-asp prevent gastrulation in Drosophila embryos. Nature 325: 348–350

Neale SA, Trasler DG (1994) Early sialylation on N-CAM in Splotch neural tube defect mouse embryos. Teratology 50: 118–124

Ono K, Tomasiewicz, Magnuson T, Rutishauser U (1994) N-CAM mutation inhibits tangential neuronal migration and is phenocopied by enzymatic removal of polysialic acid. Neuron 13: 595–609

Paradies NE, Grunwald GB (1993) Purification and characterization of NCAD90, a soluble endogenous form of N-cadherin, which is generated by proteolysis during

retinal development and retains adhesive and neurite promoting function. J Neurosci Res 36: 33–45
Ranscht B, Bronner-Fraser M (1991) T-cadherin expression alternates with migrating neural crest cells in the trunk of the avian embryo. Development 111: 15–22
Risau W, Hallman R, Albrect U, Henke-Fahle S (1986) Brain induces the expression of an early cell surface marker for blood-brain barrier-specific endothelium. EMBO J 5: 3179–3183
Roth S, Weston JA (1967) The measurement of intercellular adhesion. Proc Natl Acad Sci USA 58: 974–980
Rutishauser U (1993) NCAM and its polysialic acid moiety: a mechanism for pull/push regulation of cell interactions during development? Development [Suppl] 1992: 99–104
Ruoslahti E, Pierschbacher (1987) New perspectives in cell adhesion: RGD and integrins. Science 238: 492–497
Sachs L (1974) Lectins as probes for changes in membrane dynamics in malignancy and cell differentiation. In: Moscona AA (ed) The cell surface in development. Wiley, New York, pp 127–139
Schardein JL (1993) Chemically induced birth defects, 2nd edn. Dekker, New York
Shepard TH (1993) Catalog of teratogenic agents, 4th edn. Johns Hopkins University Press, Baltimore
Spemann H, Mangold H (1964) Induction of embryonic primordia by implantation of organizers from different species. In: Willer BH, Oppenheimer JM (eds) Foundations of experimental embryology. Prentice-Hall, Englewood, p 144
Springer TA (1990) Adhesion receptors of the immune system. Nature 346: 425–433
Steinberg MS (1964) The problem of adhesive selectivity in cellular interactions. In: Locke M (ed) Cellular membranes in development. Academic, New York, pp 321–366
Svennevik E, Linser PJ (1993) The inhibitory effects of integrin antibodies and the RGD tripeptide on early eye development. Invest Ophthal Vis Sci 34: 1774–1784
Takeichi M (1990) Cadherins: a molecular family important in selective cell–cell adhesion. Annu Rev Biochem 59: 237–252
Tomasiewicz H, Ono K, Yee D, Thompson C, Goridis C, Rutishauser U, Magnuson T (1993) Genetic deletion of a neural cell adhesion molecule variant (N-CAM-180) produces distinct defects in the central nervous system. Neuron 11: 1163–1174
Townes PL, Holtfreter J (1955) Directed movements and selective adhesion of embryonic amphibian cells. J Exp Zool 128: 53–120
Trisler D (1987) Molecular markers of cell positions in avian retina are responsible for synapse formation. Am Zool 27: 189–206
Warner AE, Guthrie SC, Gilula NB (1984) Antibodies to gap junctional protein selectively disrupt junctional communication in the early amphibian embryo. Nature 311: 127–131
Wilson HV (1907) On some phenomena of coalescence and regeneration in sponges. J Exp Zool 5: 245–258
Yamada KM (1991) Fibronectin and other cell interactive glycoproteins. In: Hay ED (ed) Cell biology of extracellular matrix, 2nd edn. Plenum, New York, pp 111–148
Yang JT, Rayburn H, Hynes RO (1995) Cell adhesion events mediated by $\alpha4$ integrins are essential in placental and cardiac development. Development 121: 549–560

CHAPTER 11
Growth Factor Disturbance

G.T. O'NEILL and R.J. AKHURST

A. Introduction

Growth factors are extracellular polypeptides that, by interaction with their corresponding transmembrane receptors, induce a cascade of intracellular signalling responses, which culminate in changes to the phenotype of the target cell (NILSEN-HAMILTON 1994). These phenotypic changes include alterations in cellular proliferation or differentiation or alterations in extracellular matrix deposition, which when observed as effects on tissues, can result in altered tissue integrity or morphogenesis.

The study of growth factors is highly pertinent to teratology, as many xenobiotic agents, including 2,3,7,8-tetrachlorodibenzo-*p*-dioxin (TCDD), glucocorticoids and retinoids, are known to affect growth factor expression or modify the effects of growth factors both in vitro and in vivo (MAHMOOD et al. 1992; SLAGER et al. 1993; DOHR et al. 1994; SLAVIN et al. 1994; POCKWINSE et al. 1995; VOGEL and ABEL 1995; WANG et al. 1995). In particular, studies on the mechanisms whereby TCDD, glucocorticoids and retinoids result in clefting of the mammalian secondary palate have implicated a number of growth factor pathways (ABBOTT et al. 1992; ABBOTT and BUCKALEW 1992). As will be seen in this review, perturbation of growth factor expression during embryogenesis can result in prenatal death or congenital malformation. It is therefore quite plausible that some of the alterations in growth factor action which occur in response to teratogens actually mediate the changes which results in congenital malformation. An understanding of the exact function played by growth factors in the embryo, fetus *and* adult is therefore of central importance in elucidating the molecular pathogenesis of birth defects and the mode of action of teratogens.

There is a vast body of literature describing the sites of expression of various growth factors and their receptors during embryogenesis, and a great deal of effort has been expended in examining the action of growth factors on cells in vitro (NILSEN-HAMILTON 1994). In this review, however, we will concentrate on the attempts that have been made to understand the role of growth factors in embryogenesis by perturbing growth factor function, in vivo, in whole-embryo culture or in embryonic explants in culture. We will first overview the technological approaches to perturbation of growth factors in vivo and then go on to review studies on five families of "classical" growth factors, with an emphasis on genetic dissection of their function utilising transgenic animals, gene knockouts and the occurrence of natural mutants in mice and humans. In a short review of this nature, we have had to restrict ourselves to the classical growth factors and have not attempted to review studies on other secretory molecules, such as the activins, bone morphogenetic proteins (KINGSLEY 1994) and *Sonic hedgehog* (LAUFER et al. 1994), which are all keys players in body pattern determination.

B. Technological Approaches

Two broad experimental approaches have been taken to address the exact function of individual growth factors and their receptors in embryogenesis, namely the expression of ectopic growth factor/receptor genes and the ablation of growth factor/receptor function. The technological approach used is influenced both by the nature of the experimental aims and the animal system being investigated.

Examination of the effects of applying ectopic growth factor to a cell type or organ system in vivo has been approached using (a) transgenic mice with cell type-targeted expression of individual growth factors (PIERCE et al. 1989; CUI et al. 1995), (b) implantation of growth factor-saturated heparin-acrylic beads into chick embryos (NISWANDER et al. 1993; COHN et al. 1995), (c) transfection of organ or embryo cultures with retroviral-driven growth factor expression systems (MIMA et al. 1995) or (d) injection of growth factor/receptor-encoding mRNAs into *Xenopus* embryos (SMITH and HARLAN 1991).

The neutralisation of growth factor/receptor activity has been attempted by the use of specific antibodies and antisense oligodeoxynucleotides. Antisense oligodeoxynucleotides designed to hybridise with, and thus knock out expression of specific growth factor mRNA have been successfully used in tissue culture and in some organ culture systems (POTTS et al. 1991; SOUZA et al. 1994; BRUNET et al. 1995). Antisense oligodeoxynucleotides are very small molecules, consisting of 16–21 nucleotides, that rapidly penetrate an individual cell. However, the inefficient penetration of the oligodeoxynucleotide into whole tissues has pre-empted most studies in vivo. The short half-life of the oligodeoxynucleotides once inside the cell has limited this approach to the study of very specific developmental events occurring over a time scale of hours rather than days (LALLIER and BRONNER-FRASER 1993). Finally, the very small size of the oligodeoxynucleotides has raised the question of the specificity of RNA targeting (WOOLF et al. 1992). Blocking antibody studies have been used to complement an antisense deoxyoligonucleotide approach (BRUNET et al. 1995). Provided that targeted knockout of both the growth factor and its encoding mRNA result in the same phenotype, this approach can be very informative.

An alternative approach which has been used to specifically interfere with growth factor or receptor gene function, and which has been used in studies on

Xenopus, chick and mice, is the generation of dominant negative mutant genes (MERCOLA et al. 1990; HEMMATI-BRIVANLOU and MELTON 1992; PETERS et al. 1994). The bioactive forms of most growth factors and their receptors are homo-or heterodimeric. Mutation of growth factor genes can be designed to generate dysfunctional ligands which, if present in excess over the wild-type form, can block normal activity (MERCOLA et al. 1990; Fig. 1). Dominant negative receptors have also successfully been used to ablate receptor function (BRAND et al. 1993; FILVAROFF et al. 1994). The consensus strategy for generating dominant negative growth factor receptors has been the use of truncated forms of the receptor that lack the intracellular signalling domain but possess the extracellular and transmembrane domains required for ligand binding and receptor dimerisation (Fig. 1). Dominant negative receptors have the advantage that, in contrast to ligands, the wild-type receptors are normally expressed at very low levels and their activity is therefore easily titrated against. Furthermore, ablation of the receptor would have a cell-autonomous effect, making the interpretation of resultant phenotype more easy to analyse.

Dominant negative mutant DNA constructs can be delivered to the embryo via transgenesis in mice, using retroviral vectors for chick or direct injection of RNA into *Xenopus* embryos. One disadvantage of the dominant negative receptor approach is that the overexpression of a receptor, be it a wild

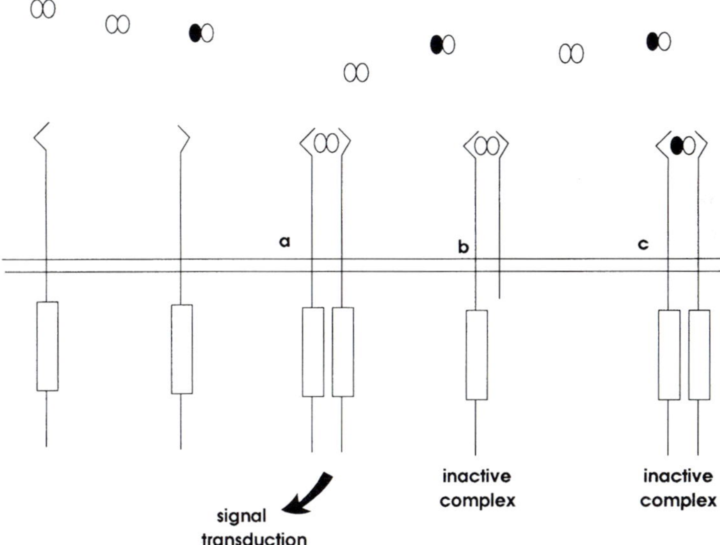

Fig. 1. Dominant negative growth factor ligands and receptors. *Open elipses* are wild-type ligand monomers, *filled-in* elipses are dominant negative ligands, and *open oblongs* depict the kinase domains of transmembrane receptors. *a* denotes homodimeric growth factor–receptor interaction; *b* denotes a heterodimeric dominant negative receptor, binding to a wild-type homodimeric ligand; *c* denotes a heterodimeric wild-type dominant ligand binding to a wild-type homodimeric receptor

type or dominant negative, can force a non-physiological association with distantly related receptors with which interaction would not normally occur. Once again, therefore, one must be cautious in the interpretation of results. For example, utilising a dominant negative transforming growth factor (TGF)-β receptor one might incidentally knock out activin signalling in addition to TGF-β signalling (Ten Dijke et al. 1994).

An exceedingly powerful tool with which to address the function of any specific gene in vivo is the use of embryonic stem (ES) cell gene targeting technology to generate mice in which the expression of individual genetic elements has been ablated or "knocked out" (Mansour et al. 1988; Thomas and Capecchi 1986). Homologous recombination between a DNA-targeting vector and a specific chromosomal site can be utilised to mutate genes in mouse ES cells in situ (Fig. 2). These cells are derived from mouse blastocysts and have the unique property of retaining their totipotency even after prolonged in vitro culture. If injected into host blastocysts, they can contribute to all tissues of the embryo, including the germ line. Thus "designer" lines of mice can be generated with heterozygous or homozygous mutations of any gene in which a partial genomic sequence has been characterised. This approach has been utilised to knock out genes encoding ligands and receptors of several growth factor families. One significant drawback of this approach is that *all* tissues of the knockout animal will lack the gene. If the gene knockout is embryolethal, it restricts analysis of growth factor ligand or receptor function up to the stage when developmental arrest occurs. Ideally, the exploitation of inducible or "switch on/off" gene expression systems, in which mutant gene expression could be regulated or modified in a stage or tissue-specific manner, would circumvent these problems. This technology is being developed (Lakso et al. 1992; Furth et al. 1994), but has not as yet been applied for studies on growth factors.

One of the problems in the interpretation of all knockout studies in transgenic mice, whether using dominant negative or gene knockout technology, is the possibility of functional redundancy of growth factor ligands and receptors. Additionally, in a genetically manipulated organism, other growth factors that exhibit similar bioactivity may be "recruited" to act as a substitute and compensate for the effects of total gene ablation. At present there is no solid evidence that alternative growth factor recruitment occurs, although this argument has been put forward for the relative lack of congenital abnormalities in TGF-α and TGF-β_1 knockout mice (Luetteke et al. 1993; Shull et al. 1992).

A final resource, not to be overlooked in a review of this nature, is the existence of spontaneously occurring growth factor ligand and receptor mutants of mouse and humans. With the mapping of the mouse and human genomes, and the analysis of mouse developmental mutants and hereditary congenital malformations in humans, a number of growth factors or growth factor receptors have been found to be responsible for causing birth defects or embryolethality in these organisms.

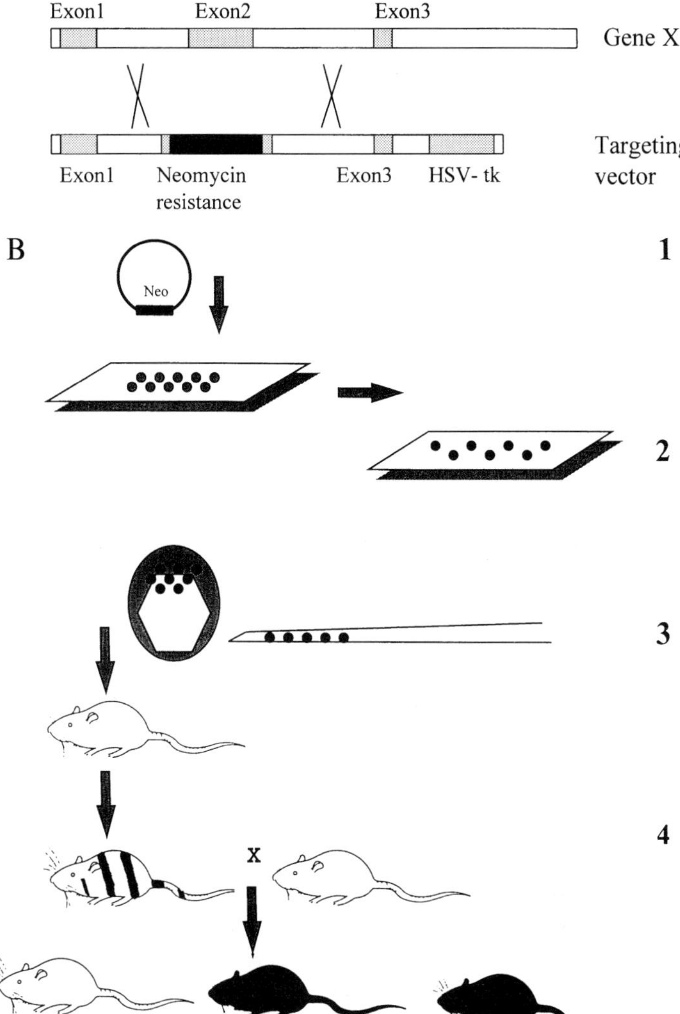

Fig. 2A,B. Generation of gene "knockout" mice. **A** Positive/negative selection of targeted embryonic stem (ES) cells. The targeting vector carries a region of homology with gene X, a neomycin resistance gene (*Neo*^{r}) sited within an exon of gene X and a copy of the herpes simplex virus (*HSV*)-*tk* gene. Homologous recombination between gene X sequences in the targeting vector and the endogenous gene results in the insertion of the neomycin gene into gene X and the disruption of its coding sequence. ES cells with homologous targeting will be resistant to G418 and gancyclovir. Cells in which random non-homologous recombination has occurred with the linearised plasmid will be resistant to G418, but sensitive to gancyclovir. **B** ES cells transfected with the targeting vector are cultured in media supplemented with G418 and gancyclovir to select putative ES cell clones that carry disrupted copies of gene X (*1*, *2*). Chimeric offspring are produced following the microinjection of ES cells into blastocyst-stage embryos (*3*) and their transfer to pseudopregnant recipient mice (*4*). First-generation offspring will be chimeric. A proportion of the second-generation offspring will be heterozygous for the ES cell-derived knockout

C. Perturbation Studies on Growth Factor Families

I. Transforming Growth Factor-β

TGF-βs are a closely related family of three, mainly homodimeric ligands, TGF-β_1, -β_2 and -β_3 (Table 1). They have been reported to be mainly growth inhibitory in vitro, but can also indirectly induce cell proliferation. They can modulate cell differentiation positively or negatively and have a major impact on extracellular matrix production. The biology and biochemistry of TGF-β has been recently reviewed (ROBERTS and SPORN 1990; SPORN and ROBERTS 1992), as has the possible function(s) of this family in vertebrate embryogenesis (AKHURST 1994). TGF-β are members of a much larger superfamily of molecules, including the activins, inhibins, Müllerian inhibitory substance (MIS) and the decapentaplegic Vg-1-related (DVR) group of molecules (KINGSLEY 1994). All three TGF-βs interact with a hetero-oligomeric receptor complex containing two transmembrane serine/threonine kinase receptors, TβRI and TβRII (WRANA et al. 1994), which initiates the TGF-β biological response. In addition, there are a number of other TGF-β-binding cell surface proteins, notably β-glycan (or the type III TGF-β receptor, TβRIII) (LOPEZ-CASILLAS et al. 1993) and its endothelial homologue, endoglin (CHEIFETZ et al. 1992), which can modulate the ability of each TGF-β isoform to bind to the receptor complex (LOPEZ-CASILLAS et al. 1993).

A number of approaches have been taken to address the function of TGF-β in embryogenesis. From descriptive studies, all three TGF-β have been implicated in different aspects of cardiogenesis (POTTS and RUNYAN 1989; AKHURST et al. 1990; DICKSON et al. 1993) and palatogenesis (FITZPATRICK et al. 1990; PELTON et al. 1990). To address the question of which isoform or isoforms play the most critical role in these processes, two independent groups utilised antisense oligonucleotides to knock out specific TGF-β isoform expression in organ or tissue explants of the developing secondary palate (BRUNET et al. 1995) and heart (POTTS et al. 1991). POTTS et al. (1991) demonstrated that antisense oligonucleotides against TGF-β_3, but not TGFβ_1 or -β_2, could block the differentiation of cardiac mesenchymal cushion tissue from endocardial cells in the chick. Similarly, BRUNET et al. (1995) found that antisense oligonucleotides and antibodies against TGF-β_3, but not TGF-β_1 or TGF-β_2, could block mouse palatal shelf fusion in organ culture. However, much controversy still surrounds the deoxyoligonucleotide approach to growth factor ablation (WOOLF et al. 1992). A more comprehensive analysis of the role of TGF-β_3 in cardiogenesis and palatogenesis awaits the characterisation of the prospective developmental anomalies in TGF-β_3 knockout mice.

A number of groups have attempted to generate transgenic mice which over express TGF-β_1 in vivo, but were unsuccessful either because the transgene was not expressed or gave a lethal phenotype (SELLHEYER et al. 1993). However, transgenic mice were successfully generated by tissue-targeting

Table 1. Growth factors and their receptors

Growth factor family	Members	Synonym	Molecular mass (kDa)	Receptor/binding proteins Signal transduction
Transforming growth factor (TGF)-β	TGF-β_1 TGF-β_2 TGF-β_3		25	Type I/type II heterodimeric receptor[a] All ligands bind via type I and II receptor complex
Epidermal growth factor (EGF)/ TGF-α	TGF-α EGF Vaccinia growth factor (VGF)		5.6	EGR receptor (EGFR)[b]
Insulin-like growth factor (IGF)	IGF I		21	IGF I receptor (IGF1R)[c]
	IGF II		22	IGF II receptor (IGF2R)/mannose-6-phosphate, (Man-6-P) receptor[d]
Fibroblast growth factor (FGF)	FGF-1	aFGF	20–30kDa	FGFR1 (*flgI*/ *cek*-1)[b,e]
	FGF-2	bFGF		
	FGF-3	*int-2*		FGFR2 (*Bek; kgfr*)[b,e]
	FGF-4	*hst-1*/kFGF		
	FGF-5			FGFR3[b,e]
	FGF-6			
	FGF-7	Keratinocyte growth factor (KGF)		FGFR4[b,e]
	FGF-8			
	FGF-9			
Platelet-derived growth factor	PDGF AA PDGF AB		28–35kDa	PDGF-α[b] (binding PDGF A and B chains)
	PDGF BB			PDGF-β[b] (binding PDGF B chain only)

aFGF, acidic FGF; bFGF, basic FGF.
[a]Serine/threonine kinase.
[b]Tyrosine kinase.
[c]Signal transducing receptor for both IGF I and IGF II.
[d]IGF2R man-6-P receptor is a transmembrane glycoprotein without kinase activity.
[e]FGFR1-4 isoforms, produced by differential splicing of the extracellular immunoglobulin-like domains, have variable ligand-binding affinities.

TGF-β_1 expression either to mammary gland epithelium (PIERCE et al. 1993) or to epidermis (CUI et al. 1995) using tissue-specific gene promoters. Both studies gave somewhat unexpected results based on predictions from in vitro studies but, in doing so, have made a further contribution to an understanding of TGF-β function in vivo.

Although TGF-β is a potent epithelial growth inhibitor in vitro, and gene expression studies have implicated this growth factor as an endogenous growth regulator within the epidermis, overexpression of TGF-β_1 in keratinocytes, using keratin-promoter transgenic mice, did not lead to hypoplasia but *stimulated* keratinocyte proliferation (either directly or indirectly). Indeed, utilising this transgenic model, CUI et al. (1995) demonstrated that epidermal keratinocytes are normally refractile to negative growth regulation by TGF-β in vivo, since they express only very low levels of TβRII. The widespread reports of TGF-β as a potent negative growth regulator (in vitro) are due to the induction of its receptor when cells are placed in culture (CUI et al. 1995). In the skin, the growth-regulatory function of TGF-β_1 is only seen when epidermal homeostasis is perturbed. Both TGF-β_1 and TβRII expression levels are elevated in response to hyperplasia, thus acting in concert to restore epidermal homeostasis.

There have been similar findings on transgenic studies of the mammary gland epithelia. Two reports, using independent transgenic models which targeted TGF-β_1 expression to different mammary epithelial cell types, confirmed the general inhibitory effects of TGF-β on mammary gland development in vivo (JHAPPAN et al. 1993; PIERCE et al. 1993). However, in mouse mammary tumor virus (MMTV) TGF-β_1 transgenic mice, with targeted expression to all mammary epithelia, there were cell type-specific differences in response to the ectopic TGF-β_1 gene expression. Although recombinant TGF-β_1 led to ductal epithelial cell hypoplasia, there was no effect of the transgene on epithelial alveolar outgrowths which occur during pregnancy. Thus the MMTV TGF-β_1 transgenic model demonstrated the differential sensitivity to TGF-β of two epithelial cell types within the mammary gland in vivo, a result which could not have been obtained from in vitro studies on mammary epithelial cells (PIERCE et al. 1993).

TGF-β gene knockout mice have now been generated for TGF-β_1 (SHULL et al. 1992; KULKARNI et al. 1993) and, at the time of writing, lines of mice that carry TGF-β_2 and -β_3 knockouts have been established (T. Doetschmann, personal communication). An in depth analysis of the TGF-β_1-null phenotype has provided invaluable insight into the numerous functions of TGF-β_1 in vivo (KULKARNI and KARLSSON 1993; LETTERIO et al. 1994; GEISER et al. 1993; SHULL and DOETSCHMAN 1994; HINES et al. 1994; DICKSON et al. 1995)

In the original reports, it was clear that lack of TGF-β_1 does not necessarily lead to embryolethality (SHULL et al. 1992; KULKARNI et al. 1993), since approximately 50% of homozygous TGF-β_1 null animals were born and appeared developmentally normal. However, these juvenile mice developed a fatal wasting syndrome, commencing 2 weeks postnatally, caused by multifocal inflammation. The nature of the pathology of this postnatal lethal phenotype has now been studied in detail. Inflammation was most severe in the lungs and heart, and death was most likely due to cardiopulmonary failure (KULKARNI et al. 1994, 1995). The first histopathological lesions appeared around 8 days post-partum (p.p), when increased adhesion of leukocytes to vascular endothelium was observed. This increased cell adhesion to endothelial cells and their extracellular matrix appeared to be due to elevated expression of certain integrin molecules on the leukocyte cell surface, namely lymphocyte function-associated gene (LFA)-1 (integrin α_L β_2) and very late activation antigen (VLA)-4 (integrin α_4 β_2) (HINES et al. 1994). GEISER et al. (1993) also showed that there was elevated expression of major histocompatibility (MHC) class I and II molecules in multiple tissues of the TGF-β_1 null animals.

The fact that developmentally normal TGF-β_1 null pups were born to heterozygous TGF-β_1 null females raised the question of whether TGF-β_1 is indeed at all necessary for normal mammalian development. However, LETTERIO et al. (1994) demonstrated that TGF-β_1 can cross the placenta intact and suggested that some null fetuses might receive adequate levels of maternal TGF-β_1 (from their heterozygous mothers) to rescue them from intrauterine death. DICKSON et al. (1995) examined the cause of intrauterine loss of the TGF-β_1 null conceptuses which did not reach parturition. They found that the cause of death, which occurred specifically at 9.5–10.5 days, appeared to have been restricted to the extraembryonic yolk sac and involved defects in vasculogenesis and haematopoiesis. The embryos had thus been starved of oxygen and nutrients, prior to the establishment of the chorioallantoic placental circulation. The specific cellular defects in the null mice appeared to be lack of endothelial tube formation and the establishment of weak endothelial connections, probably due to inadequate endothelial differentiation. This would be consistent with TGF-β_1 acting as an inducer of endothelial differentiation (MADRI et al. 1988) and with the fact that TGF-β can increase expression of cell adhesion molecules, extracellular matrix synthesis and tight junction formation by endothelial cells (MADRI et al. 1992). It was interesting that, although TGF-β_1 is a potent endothelial cell growth inhibitor, there appeared to be no endothelial cell hyperplasia in the null animals. Defective haematopoiesis was also apparent in the yolk sac, despite the fact that TGF-β_1 has been reported to be a potent inhibitor of haematopoiesis in vitro (OHTA et al. 1987). Thus in these early embryos it seems likely that the primary function of TGF-β_1 in vivo is as a modulator of cellular differentiation and phenotype and not, as might have been predicted from in vitro studies, as a potent negative growth regulator.

The finding of yolk sac endothelial defects in TGF-β_1 null animals is intriguing, since the first human hereditary disorder reported to be associated with the regulation of TGF-β bioactivity is also characterised by vascular dysplasia. Hereditary haemorrhagic telangiectasia type I is caused by lesions in the endoglin gene (MCALLISTER et al. 1994), the endothelial equivalent of

TβRIII. However, endoglin might have other, TGF-β-independent, functions, and it remains to be seen whether the genetic lesions identified in the haemorrhagic telangiectasia type I patients affect TGF-β metabolism.

II. Fibroblast Growth Factors

The fibroblast growth factor (FGF) family consists of nine structurally related proteins FGF-1 (acidic FGF), FGF-2 (basic FGF), FGF-3 (*int*-2), FGF-4 (*hst-1*/k-FGF), FGF-5, FGF-6, FGF-7 (keratinocyte growth factor, KGF), FGF-8, and FGF-9. Each has a molecular mass of 20–30 kDa (Table 1). FGF-1, FGF-2 and FGF-9 are unusual in lacking a conventional hydrophobic signal sequence for secretion from the cell. They are probably secreted from the cell by a novel, as yet uncharacterised, secretory pathway. The nine FGFs signal via a family of four genetically distinct transmembrane tyrosine kinase receptors, FGFR1 (*cek-1*/*flgI*), FGFR2 (BEK and KGFR), FGFR3 and FGFR4 (Miller and Rizzino 1994). The nine FGFs have differential binding affinities for each of the FGFR, and receptor and ligand diversity are further amplified by differential splicing of both FGF and FGFR transcripts. FGFR2, for example, has two alternatively spliced forms, BEK, which has preferential binding for FGF-1 and -2, and KGFR, which binds with highest affinity to FGF-7 (Werner et al. 1992; Miki et al. 1992).

The FGFs were originally characterised as eliciting proliferative responses in mesodermal and neuroectodermal derivatives, though they have also been found to stimulate epithelial proliferation and alter cell motility, differentiation and cell survival (Miller and Rizzino 1994). For the developmental biologist, however, one of the most important functions of this family of growth factors is in mesoderm induction and determination of spatial patterning (Slack et al. 1987).

During vertebrate embryogenesis, the earliest patterning events are the induction of mesoderm from ectoderm, by endodermally derived factors, and the determination of dorsoventral and anteroposterior axes. Most studies to determine the candidate molecules involved in these processes have utilised *Xenopus* as a model system. Several groups have found that culturing explanted *Xenopus* ectoderm (animal caps) in the presence of members of the FGF and activin families could induce mesoderm formation from ectodermal explants (Kimelman and Kirschner 1987; Rosa et al. 1988; Slack and Isaacs 1989; Slack et al. 1989; Dawid 1994). It was also suggested that FGF tended to induce more posterior/ventral-type mesoderm, whereas activin induced more dorsal/anterior mesoderm, though the exact fate of the induced mesoderm depended on the concentration of activin employed (Green and Smith 1990). The identification of which specific molecules and receptors are involved has been highly controversial. Indeed there must be a complex interplay of different molecules resulting in mesoderm formation and determination of cell type. The use of dominant negative receptors has, however, assisted in the dissection of these complex processes.

Overexpression of a truncated FGF receptor in *Xenopus* by injection of the encoding mRNA into oocytes and embryos abolished mesoderm formation in ectodermal explants of injected embryos (AMAYA et al. 1991). Furthermore, in whole embryos, the dominant negative receptor led specifically to a deficit of trunk (posterior) mesoderm structures, thus supporting the view that FGF molecules are responsible for inducing mesoderm of a more posterior/ventral phenotype (AMAYA et al. 1991). More recently, detailed studies using the dominant negative FGFR have helped dissect the complex interplay between activin- and FGF-mediated effects. CORNELL and KIMELMAN (1994) and SCHLUTE-MERKER et al. (1994) independently demonstrated that the dominant negative FGFR could abolish mesoderm induction by activin and related molecules, although the intracellular signalling pathways for FGF and activins are quite distinct. The induction by activin of posterior/dorsal-expressed genes, such as *Xbra* and cardiac actin, was highly sensitive to the dominant negative FGFR effects, whereas the induction of more anterior structures, characterised by head organiser-specific genes, was not. These results demonstrated that activin signalling alone is not sufficient for induction of certain mesodermal structures, but that induction of trunk mesoderm requires some synergystic interaction between FGF and activin-like molecules.

A role for FGF in mammalian mesoderm induction has not yet been directly demonstrated. Descriptive studies of FGF ligand and receptor expression in mouse embryos have demonstrated that most are expressed during very early post-implantation development and throughout fetal development (ORR-URTREGER et al. 1991; WILKINSON et al. 1988; WILKINSON et al. 1989; HEBERT et al. 1991; NISWANDER and MARTIN 1992). Despite widespread temporal and spatial expression patterns from the initial stages of post-implantation development, ablation of several FGF genes by gene knockout appeared not to induce severe developmental anomalies, indicating that other ligand receptor systems with similar activities are recruited or compensate.

Expression of FGF-3 has been localised to the embryonic mesoderm during gastrulation and early somite stages, where it has been proposed to exert a role in cell migration. At later stages of development, it is localised to the sensory and support cells of the ear, the retina and the cerebellum (WILKINSON et al. 1988; WILKINSON et al. 1989). Null FGF-3 homozygous mice were found to be viable and fertile, and the developmental anomalies were identified as inner ear defects and a lesser number of vertebrae in the tail (MANSOUR et al. 1993). The pattern of embryonic and fetal FGF-5 expression has indicated that it may have a role in gastrulation and myogenesis (HEBERT et al. 1991). However, analysis of the phenotype of FGF-5 knockout mice showed an identical phenotype to that of the classical mouse mutant *angorra*. The only anomaly was sustained hair growth in FGF-5 null homozygotes, indicating that FGF-5 regulates the transition between the proliferative phase of the hair growth cycle (anagen) and follicle regression (categen) (HEBERT et al. 1994).

In contrast to FGF-3 and FGF-5, the ablation of FGF-4 expression by gene knockout technology did result in severe developmental anomalies as early as the egg cylinder stage (5.5 days post-coitum). Null homozygotes initiated a maternal decidual response in the uterus, but exhibited disorganised or absent embryonic structures. The early embryolethality associated with ablation of FGF-4 demonstrates that this particular ligand must play a primary role either in the regulation of mesoderm formation, gastrulation or embryo–uterine interactions (Feldman et al. 1995).

Disabling receptor function can interfere with the signalling response of several FGF ligands. Ablation of FGFR1 function in mouse embryos induced early growth defects, but the embryos retained the ability to gastrulate. The anomalies associated with these embryos indicated that FGFR1 may transduce signals that specify mesodermal cell fates and regional patterning early after implantation (Yamaguichi et al. 1994). It would also suggest that FGF-4 must act via another FGFR to mediate its effects prior to gastrulation (Feldman et al. 1995).

In addition to mesoderm induction and determination of body axes, FGF have also been shown to be critical in both initiation of limb bud outgrowth and tissue patterning during limb development. The apical ectodermal ridge (AER) at the outer rim of the developing limb bud provides inductive signals to the underlying mesoderm to maintain proximodistal outgrowth. In chick, removal of the AER results in a failure of limb bud outgrowth. FGF-4 (Niswander et al. 1993) and FGF-8 (Crossley and Martin 1995) are expressed in the AER, and FGF-4-impregnated beads can compensate for removal of the AER (Niswander et al. 1993). When beads impregnated with FGF-1, FGF-2 or FGF-4 are placed into the interlimbal flank of a developing chick embryo, a complete new limb is induced at that site (Cohn et al. 1995). This ectopic delivery of FGF-4 regulates the expression of key inductive signalling genes, such as *Sonic hedgehog*, which are required to establish pattern formation and growth of the developing limb (Cohn et al. 1995). FGF-8 is the most likely candidate as the endogenous inducer of AER activity, since this FGF is expressed in the ectoderm as the AER becomes a morphological entity, and its expression is maintained until the AER regresses (Crossley and Martin 1995). FGFR1 and FGFR2 are likely candidates for the responding receptor, as they are expressed in limb bud mesenchyme and ectoderm, respectively (Orr-Urtreger et al. 1991). In chimeric mice which transgenically overexpress FGF4, generated using ES cell technology, multiple small limb bud structures develop in the interlimbal flank (Cohn et al. 1995), suggesting that FGF-4 can induce a limb bud field de novo from lateral plate mesoderm in both avian and mammalian species.

Over the past year, a number of genetic mutations have been identified in genes encoding the FGFR in human familial and spontaneous disorders which affect skeletal development. Mutations in the FGFR have so far all been dominantly acting and due to missense mutations. Achondroplasia (ACH), the most common form of dwarfism, is an autosomal dominant disorder characterised by short limbs and macrocephaly and has been found to be due to a missense mutation in *FGFR3*. Of 26 sporadic and familial cases examined, all had a glycine to arginine substitution (G380R) in the transmembrane domain of *FGFR3* (Shiang et al. 1994; Rousseau et al. 1994). Thanatophoric dysplasia (TD), which is a sporadic neonatal lethal disorder similar in phenotype to homozygous ACH, has also been shown to have various but distinct missense mutations in *FGFR3* (Tavormina et al. 1995). This disorder is characterised by severe micromelic shortening of the limbs, macrocephaly, reduced vertebral body height and poor cellular proliferation and column formation in the cartilaginous growth plates of the long bones. All 16 of the mutations in type I TD were found in the kinase domain of the receptor, whereas mutations causing another type of TD were located in the extracellular domain of the receptor.

Several genetically determined forms of craniosynostosis (premature fusion of the skull bones) have been reported, and some of these, namely Pfeiffer, Crouzon, Jackson-Weiss and Apert syndromes, have been shown to have mutations in *FGFR2* (Rutland et al. 1995; Reardon et al. 1995). Crouzon and Pfeiffer syndromes are both characterised by craniosynostosis and flattening of the mid-face, but Pfeiffer syndrome, in addition, has digital abnormalities. Apert syndrome has craniosynostosis associated with severe syndactyly, as well as various abnormalities of the skin, skeleton, brain and internal organs (Wilkie et al. 1995). Wilkie et al. (1995) speculated that syndactyly in Apert syndrome is caused by a reduced anteroposterior length or thickening of the AER during the earliest stages of limb bud development.

Like ACH and TD, the mutations which cause Pfeiffer, Crouzon, Jackson-Weiss and Apert syndromes are all missense mutations in differing regions of the *FGFR2* gene. The localisation of *FGFR2* mutations in Apert syndrome predicts that both the KGFR and BEK isoforms of FGFR2 would be affected. High-level BEK expression in the developing skull and long bones would explain the craniosynostosis. Similarly, the syndactyly in Apert syndrome would be explained by KGFR expression in the AER of the limb bud. *FGFR2* mutations giving rise to Crouzon syndrome are predominantly in exon 9, which, due to differential splicing, only affects the BEK isoform of the receptor. This might explain the absence of a limb phenotype in Crouzon patients, since BEK is only expressed diffusely in limb bud mesenchyme. However, there are exceptions to this rule. Some Crouzon patients have been identified with a mutation in exon 7, which *should* affect both KGFR and BEK, but these patients do not have limb abnormalities (Wilkie et al. 1995). Furthermore, identical *FGFR2* mutations occur in both Pfeiffer and Crouzon syndromes, though phenotypically these syndromes are quite distinct, especially with respect to presence or absence of syndactyly (Rutland et al. 1995). Conversely, Pfeiffer syndrome has been shown to be genetically heterogeneous, with some families having point mutations in *FGFR1*. These observations remain an enigma. Presumably there must be other genetic polymorphisms, probably within *FGFR2* itself, which modify the phenotypic effect of each mutation (Rutland et al. 1995).

III. Platelet-Derived Growth Factors

Platelet-derived growth factors (PDGF) are a family of three homo- and heterodimeric ligands, PDGF AA, BB and AB, composed of two distinct polypeptides, PDGFA and PDGFB. There are two tryosine kinase receptors, PDGFR-α and PDGFR-β. PDGFR-α binds both the A and B chains of PDGF with high affinity, whereas PDGFR-β binds only the B chain with high affinity. PDGFR dimerisation is a consequence of ligand binding and is essential for signal transduction (Bowen-Pope and Seifert 1994).

The PDGF are mitogenic for many mesenchymal cell types; there are also chemotactic and can induce cellular differentiation and extracellular matrix modification. In the adult, PDGF has a major role to play in inflammation and wound healing. However, an imbalance in its control has also been implicated in a number of pathological disorders, including atherosclerosis, fibrosis and glumerulonephritis (Bowen-Pope and Seifert 1994). An understanding of the role of PDGF in normal development, has depended, until now, on descriptive studies on sites of protein and gene expression. However, the recent identification of PDGFR-α gene deletion as the molecular defect in *Patch* (*Ph*) mice (Stephensen et al. 1991) and the generation of homozygous null PDGFB and PDGFR-β (Leveen et al. 1994; Soriano 1994) mice has provided more definitive data as to their exact function in vivo.

Patch (PDGFR-α) is a recessive lethal mutation, causing pigmentation defects in the heterozygous state and leading to prenatal death with craniofacial abnormalities in the homozygous state. As might have been predicted, the developmental phenotype of *Ph*/*Ph* fetuses is cell autonomous, affecting cell types which express the PDGFR-α gene (Orr-Urtreger et al. 1992; Morrison-Graham et al. 1993). *Ph*/*Ph* embryos have defects which are mainly associated with non-neuronal neural crest-derived tissues, indicating that PDGFR-α might affect these tissues both directly, by modulating cell growth and survival, and/or indirectly, via modulation of the surrounding extracellular matrix. Embryonic day (E) 9.0 *Ph*/*Ph* mutant embryos have neural tube deformities. Later, by E12, they have craniofacial abnormalities, cleft palate, a shortened neck and spina bifida from the cervical level down. The craniofacial abnormalities are related to defects in neural crest-derived tissue.

In later *Ph*/*Ph* embryos (E15-E18), neural crest-derived tissue of the heart, namely the aorticopulmonary septae, is also defective, resulting in one large, single vessel emanating from the heart rather than two independent vessels. The enamel organ of the tooth and the dental papilla, both neural crest-derived, are also absent. Finally, there are defects in corneal development, with a reduction in the number of corneal fibroblasts, also related to defects in neural crest. Thus the major function of PDGFR-α in fetal development appears to be control of migration and phenotype of non-neuronal neural crest (Morrison-Graham et al. 1993).

More recently, PDGFB and PDGFR-β knockout mice have been generated. In the homozygous null state, both are perinatally lethal and have quite similar phenotypes (LEVEEN et al. 1994; SORIANO 1994). The phenotypes of the PDGFB and PDGFR-β null animals are virtually identical except in possession of cardiac abnormalities, which are specific for the PDGFB knockouts. At E16, the PDGFB null animals have enlarged hearts with very thin ventricular walls and dilated thoracic arteries. In the most severe cases, this results in intrauterine death, associated with generalised blood vessel dilation and oedema. It is noticeable that similar heart defects are present in both *Ph*/*Ph* and homozygous null PDGFB fetuses, but not in homozygous PDFGR-β animals. Together with the fact that the defects in the *Ph*/*Ph* fetuses are more severe than the PDGFB knockout fetuses, this might suggest that the cardiac defect in the PDGFB null animals is due to a partial failure in PDFGR-α signalling, which might normally be initiated by PDGF AB or BB. In contrast, PDGFR-β does not appear to play a role in cardiogenesis (LEVEEN et al. 1994; SORIANO 1994).

The phenotypes shared by PDGFB and PDGFR-β null mice include abnormal kidney glomeruli, anaemia (erythroblastosis) and thrombocytopaenia. The abnormal glomeruli are caused by complete absence of mesangial cells. This is consistent with the fact that these cells normally express both PDGFB and PDGFR-β and that mesangial cells are mitogenically activated by PDGF. Thus the mutant phenotypes might represent defects in mesangial progenitor migration, cell proliferation, differentiation or survival (LEVEEN et al. 1994; SORIANO 1994).

IV. Transforming Growth Factor-α

TGF-α is a small growth factor (50 amino acids) related to epidermal growth factor (EGF) which shares the same transmembrane tyrosine kinase receptor as EGF, the EGF receptor (EGFR) (Table 1). TGF-α is unusual in that unprocessed membrane-bound forms (approximately 160 amino acids) of the growth factor are biologically active. The action of TGF-α and EGF is mainly mitogenic, though in synergy with other growth factors they can also modulate cellular phenotype (DERYNCK 1992). TGF-α has been implicated as playing an important role in embryogenesis, since there is widespread expression of this growth factor in mouse embryos (WILCOX and DERYNCK 1988). It has also been implicated in epithelial–mesenchymal interactions during palatogenesis, primarily due to its localisation within the medical epithelial seam of the secondary palate, where it may regulate that composition of the extracellular matrix (DIXON et al. 1993), and also by association of a genetic polymorphism linked to the human TGF-α gene in individuals with clefting of the secondary palate (FENG et al. 1994). However, the absence of palatal defects in TGF-α knockout mice would indicate that other growth factors can compensate for the unspecified role of TGF-α during palate formation (LUETTEKE et al. 1993; MANN et al. 1993).

Two independent groups generated TGF-α gene knockout mice (LUETTEKE et al. 1993; MANN et al. 1993). Rather surprisingly, the homozygous

TGF-α null animals were healthy and fertile, thus lines of breeding homozygous animals were established. This observation would strongly argue against TGF-α playing a significant primary role any developmental process, unless we were to infer that EGF, or another related ligand, can substitute for lack of TGF-α. The major "defect" in the homozygous TGF-α null lines was pronounced waviness of the coat and variable eye defects. Surprisingly, although TGF-α is known to be expressed in interfollicular and follicular keratinocytes of the skin, interfollicular epidermal morphology and histology were normal. Furthermore, the TGF-α mice did not have a deficit in wound healing, as might otherwise have been predicted (MANN et al. 1993). The dishevelled appearance of the coat in the TGF-α knockout mice was due to abnormalities in the hair shaft, with the development of a curved hair follicle. All follicular types were affected, including whisker follicles. It is known that TGF-α is expressed in the differentiating keratinocytes of the follicle (and interfollicular epidermis) and that the EGFR is expressed in the proliferative compartment (basal keratinocytes of the epidermis and hair bulb), but the mechanism for establishment of wavy hair follicles in the null mice remains unknown. It might reflect abnormalities in keratinocyte proliferation, differentiation or tissue morphogenesis.

The TGF-α null mice also had eye abnormalities, though the penetrance of this phenotype, unlike wavy hair, was not complete. Eye abnormalities were often unilateral. The affected mice showed open eyes at birth (which later closed due to scarring), superficial opacity, microphthalmia, retinal defects and corneal neovascularization and scarring. The phenotype was the consequence of both primary and secondary effects.

Interestingly, the TGF-α null phenotype is very similar to two classical mouse mutations, waved-1 (*wa-1*) and waved-2 (*wa-2*) (LUETTEKE et al. 1994; FOWLER et al. 1995). It has now become evident that *wa-1* is caused by a mutation in the TGF-α gene and that *wa-2* is due to a point mutation in the EGFR (*egfr*). The generation of EGFR gene knockout mice has not yet been achieved, but one might predict a more severe phenotype than the TGF-α knockouts *wa-1* or *wa-2*, since complete absence of the EGFR (unlike a point mutation) would knockout both TGF-α and EGF-mediated effects.

V. Insulin-Like Growth Factors

The insulin-like growth factor (IGF) family is comprised of insulin and IGF I and II. Insulin is primarily involved in the regulation glucose metabolism, whereas IGF I and IGF II induce mitogenic responses in target tissues and can induce the differentiation of cultured cells. IGF I is a 70-amino acid polypeptide that is expressed during both embryogenesis, to promote fetal growth, and in the adult, where it mediates many of the effects of growth hormone. IGF II is only expressed during embryogenesis, where its primary action is to promote fetal growth. The IGF, unlike the other growth factors mentioned in this review, have predominantly endocrine modes of delivery to target tissues.

The actions of IGF I and IGF II are mediated predominantly by the IGF 1 receptor, IGF1R, a conventional transmembrane tyrosine kinase. IGF2R, is an unusual monomeric transmembrane glycoprotein that is a multifunctional receptor which also acts as the mannose-6-phosphate receptor. It plays a role in lysosomal targeting and signal transduction (WARD et al. 1994). Blocking ligand binding to IGF2R with specific antibodies was not found to interfere with the mitogenic effects of IGF II. Thus IGF II binding to IGF2R does not directly alter growth, but might regulate circulating levels of the IGF II by processes such as internalisation or degradation (WANG et al. 1994; see below).

Disruption of the IGF II gene by knockout techniques resulted in fetal growth retardation, the onset of which was noticeable from 16 days post-coitum. The resulting offspring, although viable, were found to have only 60% of normal body weight, and the dwarf phenotype was maintained throughout neonatal and adult life (DECHIARA et al. 1990). IGF II is an imprinted locus in which only the paternally derived allele is expressed during embryogenesis, that derived from the mother being transcriptionally silent. Thus heterozygous null IGF II offspring in which the null allele was paternally derived exhibited the same phenotype as homozygous null animals (DECHIARA et al. 1990).

Targeted knockout of the IGF I gene also produced significant growth retardation in homozygous null offspring, which had approximately 60% of normal body weight at birth. There was also genetic variability in the expressivity of the IGF I homozygous null phenotype. On C57B16 or 129/Sv inbred backgrounds, animals with no IGF I exhibited neonatal lethality, but IGF I null animals survived to maturity when these strains were crossed onto a MF1 genetic background (LIU et al. 1993). In contrast to IGF II, no heterozygote offspring expressed the defective phenotype, indicating that the IGF I gene is not genomically imprinted (LIU et al. 1993; BAKER et al. 1993).

Targeted knockout of the IGF1R gene, which is the main receptor through which IGF I and IGF II signal growth responses, was found to induce a neonatal lethal phenotype in the homozygous null state, with a reduction in birth weight (LIU et al. 1993). Neonatal death was attributed to an unspecified respiratory failure. The null phenotype included skin, muscle and organ hypoplasia and retarded craniofacial ossification equivalent to a period of 2 developmental days. Like IGF I, the IGF1R heterozygotes were normal, demonstrating that *IGF1R* is not an imprinted gene (LIU et al. 1993).

The cross-breeding of mice with knockouts of IGF I and IGF1R was not found to result in additive phenotypic effects, verifying that IGF I signals mainly, if not solely, through the IGF1R receptor. In contrast, the doubly homozygous offspring that were derived from crosses between IGF II and IGF1R knockouts exhibited a further reduction in size, approximating to only 35% of normal birthweight. These results demonstrate that IGF II must signal through another receptor (denoted XR) in addition to IGF1R (LIU et al. 1993; BAKER et al. 1993).

The exact function of IGF2R is still under investigation. As eluded to earlier, this receptor does not mediate the growth effects of IGF II. The encoding gene, *igf2/mpr*, is paternally imprinted. Homozygous null *igf2/mpr* animals or heterozygotes in which the disrupted allele was maternally derived were in fact larger rather than smaller than wild-type litter mates at birth (WANG et al. 1994). In addition to effects on fetal growth, the IGF2R null phenotype also included missorting of mannose-6-phosphate-tagged proteins.

The phenotype of IGF2R null animals was neonatally lethal, but was rescued by crossbreeding to an IGF II null background, giving further support to the proposition that one function of IGF2R is to regulate the levels of circulating IGF II by its internalisation or degradation. The neonatal overgrowth of the IGF2R knockout mice was attributed to excess levels of circulating IGF II (WANG et al. 1994).

1. Human Disorders Associated with Insulin-Like Growth Factor II Gene Dysfunction

Beckwith-Wiedemann Syndrome (BWS) is a congenital overgrowth syndrome characterised by gigantism, macroglossia, renal and abdominal wall defects and predisposition to embryonal tumours. It is a genetically diverse condition in which a proportion of the affected individuals exhibit cytogenetic lesions at 11p15.5, i.e. involving the IGF II gene (SLATTER et al. 1994). The cytogenetic lesions are either paternally derived duplications or maternally derived transclocations or inversions, invoking the involvement of a maternally imprinted gene. The subset of BWS individuals with lesions at 11p15.5 are predisposed to Wilms tumour. One can conclude that BWS is at least partially caused by overexpression of the paternally derived IGF II gene. In this respect, it is interesting that IGF II is overexpressed in Wilms tumours (REEVE et al. 1985; SCOTT et al. 1985). Disruption of the mechanisms that regulate imprinted gene function, such as uniparental disomy, where two copies of one parental chromosome region are present (HENRY et al. 1991), or loss of maternal imprinting (by translocation or inversion) and subsequent biallelic expression, implicate IGF II as the candidate gene in BWS and as a predisposing tumour gene in Wilm's tumour (STEENMAN et al. 1994; MOULTON et al. 1994).

D. Conclusions

In conclusion, it is certain that modulation of growth factor expression or of growth factor activity is mechanistic in the pathogenesis of both experimentally induced and spontaneously occurring congenital malformations. Important information concerning growth factor function in vivo has been gained taking a molecular genetic approach to dissect the complex network of growth factor interactions. External agents certainly modulate growth factor action, and a molecular approach will be needed to identify which growth factor pathways are perturbed by individual environmental toxins.

References

Abbott BD, Buckalew AR (1992) Embryonic palatal responses to teratogens in serum-free organ culture. Teratology 45: 369–382

Abbott BD, Harris MW, Birnbaum LS (1992) Comparisons of the effects of TCDD and hydrocortisone on growth factor expression provide insight into their interaction in the embryonic mouse palate. Teratology 45: 35–53

Akhurst RJ (1994) The transforming growth factor β family in vertebrate embryogenesis. In: Nilsen-Hamilton M (ed) Growth factors and signal transduction in development. Wiley-Liss, New York, pp 97–122

Akhurst RJ, Fitzpatrick DR, Gatherer D, Lehnert SA, Millan FA (1990) The role of TGFβs in mammalian embryogenesis. Prog Growth Factor Res 2: 153–168

Amaya E, Musci TJ, Kirschener MW (1991) Expression of a dominant negative mutant of the FGF receptor disrupts mesoderm formation in Xenopus embryos. Cell 66: 257–270

Baker J, Liu J, Robertson EJ, Efstradiatis A (1993) Role of insulin-like growth factors in embryonic and postnatal growth. Cell 75: 73–82

Bowen-Pope DF, Seifert RA (1994) Platelet-derived growth factors in development. In: Nilsen-Hamilton M (ed) Growth factors and signal transduction in development. Wiley-Liss, New York, pp 51–73

Brand T, MacLellan WR, Schneider MD (1993) A dominant-negative receptor for type β transforming growth factors created by deletion of the kinase domain. J Biol Chem 268: 11500–11503

Brunet CL, Sharpe PM, Ferguson MWJ (1995) Inhibition of TGFβ3 (but not TGFβ1 or TGFβ2) activity prevents normal mouse embryonic palate fusion. Int J Dev Biol 39: 345–355

Cheifetz S, Bellon T, Cales C, Vera S, Bernabeu C, Massague J, Letarte M (1992) Endoglin is a component of the transforming growth factor-β receptor system in human endothelial cells. J Biol Chem 267: 19027–19030

Cohn MJ, Izpisua-Belmonte JC, Abud H, Heath JK, Tickle C (1995) Fibroblast growth factors induce additional limb development from the flank of chick embryos. Cell 80: 739–746

Cornell RA, Kimelman D (1994) Activin-mediated mesoderm induction requires FGF. Development 120: 453–462

Crossley PH, Martin GR (1995) The mouse FGF 8 gene encodes a family of polypeptides and is expressed in regions that direct outgrowth and patterning in the developing embryo. Development 121: 439–451

Cui W, Fowlis DJ, Cousins FM, Duffie E, Bryson S, Balmain A, Akhurst RJ (1995) Concerted action of TGFβ1 and its type II receptor in control of epidermal homeostasis in transgenic mice. Genes Dev 9: 945–955

Dawid IB (1994) Embryonic induction and axis formation in Xenopus laevis: growth factor action and early response genes. In: Nilsen-Hamilton M (ed) Growth factors and signal transduction in development. Wiley-Liss, New York, pp 199–227

DeChiara TM, Efstradiatis A, Robertson EJ (1990) A growth-deficient phenotype in heterozygous mice carrying an insulin-like growth factor II gene disrupted by targeting. Nature 345: 78–80

Derynck R (1992) The physiology of transforming growth factor α. Adv Cancer Res 58: 27–52

Dickson MC, Slager HG, Duffie E, Mummery CL, Akhurst RJ (1993) TGFβ2 RNA and protein localisations in the early embryo suggest a role in cardiac development. Development 117: 625–639

Dickson MC, Martins JS, Cousins FM, Kulkarni AB, Karlsson S, Akhurst RJ (1995) Transforming growth factor-beta 1 is essential for hematopoiesis and endothelial differentiation in vivo. Development 121: 1845–1854

Dixon MJ, Foreman D, Schor S, Ferguson MWJ (1993) Epidermal growth factor and transforming growth factor alpha regulate extracellular matrix production by embryonic mouse palatal mesenchymal cells cultured in a variety of strata. Roux Arch Dev Biol 203: 140–150

Dohr O, Vogel C, Abel J (1994) Modulation of growth factor expression by 2,3,7,8-tetrachlorodibenzo-p-dioxin. Exp Clin Immunogenet 11: 142–148

Feldman B, Poueymirou W, Papaioannou VE, DeChiara TM, Goldfarb M (1995) Requirement of FGF-4 for postimplantation mouse development. Science 267: 246–249

Feng H, Sassani R, Bartlett SP, Lee A, Hecht JT, Malcolm S, Winter RM, Vintiner GM, Buetow KH, Gasser DL (1994) Evidence from family studies for linkage disequilibrium between TGFα and a gene for nonsyndromic cleft lip with or without cleft palate. Am J Hum Genet 55: 932–936

Filvaroff EH, Ebner R, Derynck R (1994) Inhibition of myogenic differentiation in myoblasts expressing a truncated type II TGF-β receptor. Development 120: 1085–1095

Fitzpatrick DR, Denhez F, Kondaiah P, Akhurst RJ (1990) Differential expression of TGF beta isoforms in murine palatogenesis. Development 109: 585–595

Fowler KJ, Walker F, Alexander W, Hibbs ML, Nice EC, Bohmer RM, Mann GB, Thumwood C, Maglitto R, Danks JA et al. (1995) A mutation in the epidermal growth factor receptor in waved-2 mice has a profound effect on receptor biochemistry that results in impaired location. Proc Natl Acad Sci USA 92: 1465–1469

Furth PA, Onge ST, Boger H, Gruss H, Gossen M, Kistner A, Bujard H, Hennighausen L (1994) Temporal control of gene expression in transgenic mice by a tetracycline-responsive promoter. Proc Natl Acad Sci USA 91: 9302–9306

Geiser A, Letterio JJ, Kulkarni AB, Karlsson S, Roberts AB, Sporn MB (1993) TGF-β1 controls expression of major histocompatibility genes in the postnatal mouse: aberrant histocompatibility antigen expression in the pathogenesis of the TGF-β1 null phenotype. Proc Natl Acad Sci USA 90: 9944–9948

Green JBA, Smith JC (1990) Graded changes in dose of a *Xenopus* activin A homologue elicit stepwise transitions in embryonic cell fate. Nature 347: 391–394

Hebert JM, Boyle M, Martin GR (1991) mRNA localization studies suggest that murine FGF5 plays a role in gastrulation. Development 112: 407–418

Hebert JM, Rosenquist T, Gotz J, Martin GR (1994) FGF5 as a regulator of the hair growth cycle: evidence from targeted and spontaneous mutations. Cell 78: 1017–1025

Hemmati-Brivanlou A, Melton DA (1992) A truncated activin receptor inhibits mesoderm induction and formation of axial structures in *Xenopus* embryos. Nature 359: 609–614

Henry I, Bonaiti-Pellie C, Chehensse V, Beldjord C, Schwartz C, Utermann G, Junien C (1991) Uniparental paternal disomy in a genetic cancer-predisposing syndrome. Nature 351: 665 667

Hines KL, Kulkarni AB, McCarthy JB, Tian H, Ward JM, Christ M, McCartney-Francis L, Furcht LT, Karlsson S, Wahl SM (1994) Synthetic fibronectin peptides interrupt inflammatory cell infiltration in transforming growth factor β1 knockout mice. Proc Natl Acad Sci USA 91: 5187–5191

Jhappan C, Geiser AG, Kordon EC, Bagheri D, Hennighausen L, Roberts AB, Smith GH, Merlino G (1993) Targeting expression of a transforming growth factor β1 transgene to the pregnant mammary gland inhibits alveolar development and lactation. EMBO J 12: 1835–1845

Kimelman D, Kirschner M (1987) Synergistic induction of mesoderm by FGF and TGF-beta and the identification of an mRNA coding for FGF in the early Xenopus embryo. Cell 51: 869–877

Kingsley DM (1994) The TGF-β superfamily: new members, new receptors, and new genetic tests of function in different organisms. Genes Dev 8: 133–146

Kulkarni AB, Karlsson S (1993) Transforming growth factor-β1 (TGF-β1) knockout mice: a mutation in one cytokine causes dramatic inflammatory disease. Am J Pathol 143: 3–9

Kulkarni AB, Huh C, Becker D, Geiser A, Lyght M, Flanders KC, Roberts AB, Sporn MB, Ward JM, Karlsson S (1993) Transforming growth factor-β1 null mutation in mice causes excessive inflammatory response and early death. Proc Natl Acad Sci USA 90: 770–774

Kulkarni AB, Ward JM, Geiser AG et al. (1994) TGF-β1 knockout mice: immune dysregulation and pathology. In: Abrahm NG, Shadduck RK, Levine AS et al (eds) Molecular biology of haematopoiesis. Intercept, Andover, pp 749–757

Kulkarni AB, Ward JM, Yaswen L, Mackall CL, Bauer SR, Huh CG, Gress RE, Karisson S (1995) Transforming growth factor-β1 null mice: an animal model for inflammatory disorders. Am J Pathl 146: 1–12

Lakso M, Sauer B, Mosinger B, Lee EJ, Manning RW, Yu S, Mulder KL, Westphal H (1992) Targeted oncogene activation by site-specific recombination in transgenic mice. Proc Natl Acad Sci USA 89: 6232–6236

Lallier T, Bronner-Fraser M (1993) Inhibition of neural crest cell attachment by integrin antisense oligonucleotides. Science 259: 692–695

Laufer E, Nelson CE, Johnson RL, Morgan BA, Tabin C (1994) *Sonic hedgehog* and FGF-4 act through a signalling cascade and feedback loop to integrate growth and patterning of the developing limb bud. Cell 79: 993–1003

Letterio JJ, Gieser AG, Kulkarni AB, Roche NS, Sporn MB, Roberts AB (1994) Maternal rescue of the transforming growth factor β knockout. Science 264: 1936–1938

Leveen P, Pekny M, Gebre-Medhin S, Swolin B, Larsson E, Betsholtz C (1994) Mice deficient for PDGF B show renal, cardiovascular, and hematological abnormalities. Genes Dev 16: 1875–1887

Liu J, Baker J, Perkins AS, Robertson EJ, Efstradiatis A (1993) Mice carrying null mutations of the genes encoding insulin-like growth factor I (*Igf-1*) and type 1 IGF receptor (Igf1r). Cell 75: 59–72

Lopez-Casillas F, Wrana JL, Massague J (1993) Betaglycan presents ligand to the TGFβ signaling receptor. Cell 73: 1435–1444

Luetteke NC, Qiu TH, Peiffer RL, Oliver P, Smithies O, Lee DC (1993) TGFα deficiency results in hair follicle and eye abnormalities in targeted and waved-1 mice. Cell 73: 263–278

Luetteke NC, Philips HK, Qui TH, Copeland NG, Earp HS, Jenkins NA, Lee DC (1994) The mouse waved-2 phenotype results from a point mutation in the EGF receptor kinase. Genes Dev 8: 399–413

Madri JA, Pratt BM, Tucker AM (1988) Phenotypic modulation of endothelial cells by transforming growth factor-beta depends upon the composition and organization of the extracellular matrix. J Cell Biol 106: 1375–1384

Madri JA, Bell L, Merwin JR (1992) Modulation of vascular cell behaviour by transforming growth factor β. Mol Reprod Dev 32: 121–126

Mahmood R, Flanders KC, Morriss-Kay GM (1992) Interactions between retinoids and TGFβs in mouse morphogenesis. Development 115: 67–74

Mann BG, Fowler KJ, Gabriel A, Nice EC, Williams RL, Dunn AR (1993) Mice with a null mutation of the TGFα gene have abnormal skin architecture, wavy hair, and curly whiskers and often develop corneal inflammation. Cell 73: 249–261

Mansour SL, Thomas KR, Capecchi MR (1988) Disruption of the proto-oncogene int-2 in mouse embryo-derived stem cells: a general strategy for targeting mutations to non-selectable genes. Nature 336: 348–352

Mansour SL, Goddard D, Capecchi MR (1993) Mice homozygous for a targeted disruption of the proto-oncogene *int-2* have developmental defects in the tail and the inner ear. Development 117: 13–28

McAllister KA, Grogg DW, Johnson DW et al (1994) Endoglin, a TGF-β binding protein of endothelial cells, is the gene for hereditary haemorrhagic telangiectasia type 1. Nat Genet 8: 345–351

Mercola M, Deininger PL, Shamah SM, Porter J, Wang C, Stiles CD (1990) Dominant-negative mutants of a platelet-derived growth factor gene. Genes Dev 4: 2333–2341

Miki T, Bottaro DP, Fleming TP, Smith CL, Burgess WH, Chan AM, Aaronson SA (1992) Determination of ligand-binding specificity by alternate splicing: two distinct growth factor receptors encoded by a single gene. Proc Natl Acad Sci USA 89: 246–250

Mima T, Ueno H, Fischman DA, Willimas LT, Mikawa T (1995) FGFR is required for in vivo mycocyte proliferation at early embryonic stages of heart development. Proc Natl Acad Sci USA 92: 467–71

Miller K, Rizzino A (1994) Developmental regulation and signal transduction pathways of fibroblast growth factors and their receptors. In: Nilsen-Hamilton M (ed) Growth factors and signal transduction in development. Wiley-Liss, New York, pp 19–50

Morrison-Graham K, Schatteman GC, Bork T, Bowen-Pope DF, Weston JA (1993) A PDGF receptor mutation in the mouse (Patch) perturbs the development of a non-neuronal subset of neural crest-derived cells. Development 115: 133–142

Moulton T, Crenshaw T, Hao Y et al (1994) Epigenetic lesions at the H19 locus in Wilm's tumour patients. Nat Genet 7: 440–447

Nilsen-Hamilton M (1994) Growth factors and signal transduction in development. Wiley-Liss, New York

Niswander L, Martin G (1992) FGF-4 expression during gastrulation, myogenesis, limb and tooth development in the mouse. Development 114: 755–768

Niswander L, Tickle C, Vogel A, Booth I, Martin GR (1993) FGF-4 replaces the apical ectodermal ridge and directs outgrowth and patterning of the limb. Cell 75: 579–587

Ohta M, Greenberger JS, Anklesaria P, Bassols A, Massague J (1987) Two forms of transforming growth factor-beta distinguished by multipotential haematopoietic progenitor cells. Nature 329: 539–541

Orr-Urtreger A, Givol D, Yayon A, Yarden Y, Lonai P (1991) Developmental expression of two murine fibroblast growth factor receptors, *flg* and *bek*. Development 113: 1419–1434

Orr-Urtreger A, Bedford MT, Do M, Eisenbach L, Lonai P (1992) Developmental expression of the α receptor for platelet-derived growth factor, which is deleted in the embryonic lethal *patch* mutation. Development 115: 289–303

Pelton RW, Dickinson ME, Moses HL, Hogen BLM (1990) In situ hybridisation analysis of TGFβ3 RNA expression during mouse development: comparative study with TGFβ1 and TGFβ2. Development 110: 609–620

Peters K, Werner S, Liao X, Wert S, Whitsett J, Williams L (1994) Targeted expression of a dominant negative FGF receptor blocks branching morphogenesis and epithelial differentiation of the mouse lung. EMBO J 13: 3296–3301

Pierce DFJ, Johnson MD, Matsui Y, Robinson SD, Gold LI, Purchio AF, Daniel CW, Hogen BLM, Moses HL (1993) Inhibition of mammary duct development but not alveolar outgrowth during pregnancy in transgenic mice expressing active TGF-β1. Genes Dev 7: 2308–2317

Pierce GF, Mustoe TA, Lingelbach J, Masakowski VR, Griffin GL, Senior RM, Deuel TF (1989) Platelet-derived growth factor and transforming growth factor-beta enhance tissue repair activities by unique mechanisms. J Cell Biol 109: 429–440

Pockwinse SM, Stein JL, Lian JB, Stein GS (1995) Developmental and stage-specific cellular responses to vitamin D and glucocorticoids during differentiation in the odontoblast phenotype: interrelationship of morphology and gene expression by in situ hybridisation. Exp Cell Res 216: 244–260

Potts JD, Runyan RB (1989) Epithelial-mesenchymal cell transformation in the embryonic heart can be mediated, in part, by transforming growth factor beta. Dev Biol 134: 392–401

Potts JD, Dagle JM, Walder JA, Weeks DL, Runyan RB (1991) Epithelial-mesenchymal transformation of embryonic cardiac endothelial cells is inhibited by a modified antisense oligodeoxynucleotide to transforming growth factor β3. Proc Natl Acad Sci USA 88: 1516–1520

Reardon W, Winter RM, Rutland P, Pulleyn LJ, Jones BM, Malcolm S (1995) Mutations in the fibroblast growth factor receptor 2 gene cause Crouzen syndrome. Nat Genet 9: 98–103

Reeve AE, Eccles MR, Wilkins RJ, Bell GI, Millow LJ (1985) Expression of insulin-like growth factor-II transcripts in Wilm's tumour. Nature 317: 258–260

Roberts AB, Sporn MB (1990) The transforming growth factor-betas. In: Sporn MB, Roberts AB (eds) Peptide growth factors and their receptors. Springer, Berlin Heidelberg New York, pp 419–472

Rosa F, Roberts AB, Danielpour D, Dart LL, Sporn MB, Dawid IB (1988) Mesoderm induction in amphibians: the role of TGF-beta 2-like factors. Science 239: 783–785

Rousseau F, Bonaventure J, Legeal-Mallet L, Pelet A, Rozet J, Maroteaux P, Le Merrer M, Munnich A (1994) Mutations in the gene encoding fibroblast growth factor receptor-3 in achondroplasia. Nature 371: 252–254

Rutland P, Pulleyn LJ, Reardon W et al (1995) Identical mutations in the FGFR2 gene cause both Pfeiffer and Crouzon syndrome phenotypes. Nat Genet 9: 173–176

Schulte-Merker S, Smith JC, Dale L (1994) Effects of truncated activin and FGF receptors and of follistatin on the inducing activities of BVg1 and activin: does activin play a role in mesoderm induction. EMBO J 13: 3533–3541

Scott J, Cowell J, Robertson ME et al (1985) Insulin-like growth factor II gene expression in Wilm's tumour and embryonic tissues. Nature 317: 260–261

Sellheyer K, Bickenbach JR, Rothnagel JA, Bundman D, Longley MA, Krieg T, Roche NS, Roberts AB, Roop DR (1993) Inhibition of skin development by overexpression of transforming growth factor β1 in the epidermis of transgenic mice. Proc Natl Acad Sci USA 90: 5237–5241

Shiang R, Thompson LM, Zhu Y, Church DM, Fielder TJ, Bocian M, Winokur ST, Wasmuth JJ (1994) Mutations in the transmembrane domain of FGFR3 cause the most common form of dwarfism, achondoplasia. Cell 78: 335–342

Shull MM, Doetschman T (1994) Transforming growth factor-β1 in reproduction and development. Mol Reprod Dev 39: 239–246

Shull MM, Ormsby I, Kier AB et al (1992) Targeted disruption of the mouse transforming growth factor-β1 gene results in multifocal inflammatory disease. Nature 359: 693–699

Slack JMW, Isaacs HV (1989) Presence of basic fibroblast growth factor in the early Xenopus embryo. Development 105: 147–153

Slack JM, Darlington BG, Heath JK, Godsave SF (1987) Mesoderm induction in early Xenopus embryos by heparin-binding growth factors. Nature 326: 197–200

Slack JMW, Darlington BG, Gillespie LL, Godsave SF, Isaacs HV, Paterno GD (1989) The role of fibroblast growth factor in early Xenopus development. Development 107 [Suppl]: 141–148

Slager HG, Van Inzen W, Freund E, Van den Eijnden-van Raaij AJ, Mummery CL (1993) TGFβ in the early mouse embryo-implications for regulation of muscle formation and implantation. Dev Genet 14: 212–224

Slavin J, Unemori E, Hunt TK, Amento E (1994) TGFβ and dexamethasone have opposing effects on collagen metabolism in low passage human dermal fibroblasts in vitro. Growth Factors 11: 205–213

Slatter RE, Elliott M, Welham K, Carrera PN, Schofield PN, Maher ER (1994) Mosaic uniparental disomy in Beckwith-Wiedmann syndrome. J Med Genet 31: 749–753

Smith WC, Harland RM (1991) Injected Xwnt-8 RNA acts early in Xenopus embryos to promote formation of a vegetal dorsalizing center. Cell 67: 753–765

Soriano P (1994) Abnormal kidney development and hematological disorders in PDGF beta-receptor mutant mice. Genes Dev 8: 1888–1896

Souza P, Sedlackova L, Kuliszewski M, Wang J, Liu J, Tseu I, Liu M, Tansell AK, Post M (1994) Antisense oligogeoxynucleotides targeting PDGF-B mRNA inhibit cell proliferation during embryonic rat lung development. Development 120: 2163–2173

Sporn MB, Roberts AB (1992) Transforming growth factor β: recent progress and new challenges. J Cell Biol 119: 1017–1021

Steenman MJC, Rainier S, Dobry CJ, Grundy P, Horon IL, Feinberg AP (1994) Loss of imprinting of IGF2 is linked to reduced expression and abnormal methylation of H19 in Wilm's tumour. Nat Genet 7: 433–439

Stephensen DA, Mercola M, Anderson E, Wang C, Stiles CD, Bowen-Pope DF, Chapman VM (1991) The platelet-derived growth factor receptor alpha subunit is deleted in the mouse mutant patch (*Ph*). Proc Natl Acd Sci USA 88: 6–10

Tavormina PL, Shiang R, Thompson LM, Zhu Y, Wilkin DJ, Lachman RS, Wilcox WR, Rimoin DL, Cohn DH, Wasmuth JJ (1995) Thanatophoric dysplasia (types I and II) caused by distinct mutations in fibroblast growth factor receptor 3. Nat Genet 9: 321–328

Ten Dijke P, Yamashiata H, Ichijo H, Franzen P, Laiho M, Miyazono K, Heldin C (1994) Characterisation of type I receptors for transforming growth factor-β and activin. Science 264: 101–104

Thomas KR, Capecchi MR (1986) Introduction of homologous DNA sequences into mammalian cells induces mutations in the cognate gene. Nature 324: 34–28

Vogel C, Abel J (1995) Effect of 2,3,7,8-tetrachlorodibenzo-p-dioxin on growth factor expression in the human breast cancer cell line, MCF-7. Arch Toxicol 69: 259–265

Wang J, Kuliszewski M, Yee W, Sedlackova L, Xu J, Tseu I, Post M (1995) Cloning and expression of glucocorticoid-induced genes in fetal rat lung fibroblasts: transforming growth factor β3. J Biol Chem 270: 2722–2728

Wang Z, Fung MR, Barlow DP, Wagner EF (1994) Regulation of embryonic growth and lysosomal targeting by the imprinted *Igf2/Mpr gene*. Nature 372: 464–467

Ward CR, Garside RN, Dhir RN et al (1994) The insulin family of growth factors and signal transduction during development. In: Nilsen-Hamilton M (ed) Growth factors and signal transduction in development. Wiley-Liss, New York, pp 1–18

Werner S, Duan DR, de Vries C, Peters KP, Johnson D, Williams LT (1992) Differential splicing in the extracellular region of the fibroblast receptor 1 generates receptor variants with different ligand binding specificities. Mol Cell Biol 12: 82–88

Wilcox JN, Derynck R (1988) Developmental expression of transforming growth factors alpha and beta in the mouse fetus. Mol Cell Biol 8: 3415–3422

Wilkie AOM, Slaney SF, Oldridge M et al (1995) Apert syndrome results from localised mutations of FGFR2 and is allelic with Crouzen syndrome. Nat Genet 9: 165 172

Wilkinson DG, Peters G, Dickson C, McMahon AP (1988) Expression of the FGF-related proto-oncogene int-2 during gastrulation and neurulation in the mouse. EMBO J 7: 691–695

Wilkinson DG, Bhatt S, McMahon AP (1989) Expression pattern of the FGF-related proto-oncogene int-2 suggests muliple roles in fetal development. Development 105: 131–136

Woolf TM, Melton DA, Jennings CGB (1992) Specificity of antisense oligonucleotides in vivo. Proc Natl Acad Sci USA 89: 7305–7309

Wrana JL, Attisano L, Weiser R, Ventura F, Massague J (1994) Mechanism of activation of the TGF-β receptor. Nature 370: 341–347

Yamaguichi TP, Harpal K, Henkeyer M, Rossant J (1994) FGFR-1 is required for embryonic growth and mesodermal patterning during mouse gastrulation. Genes Dev 8: 3032–3038

CHAPTER 12

Targeted Gene Disruptions as Models of Abnormal Development

T.W. SADLER, E.T. LIU, and K.A. AUGUSTINE

A. Introduction

For years teratologists have had access to mice with spontaneously occurring mutations resulting in unique phenotypes. Animals with defects in many organ systems are available and provide valuable information about the origin of specific congenital malformations. However, most of these mutations involve recessive genes, and the genes themselves are unknown (KALTER 1980). Consequently, there are no markers available to identify mutant embryos before they overtly exhibit an altered phenotype. Thus it is difficult to investigate the earliest cellular events and mechanisms responsible for abnormal phenotypes, since only 25% of all embryos in a litter are affected and there is no way to identify which embryos they are before they exhibit grossly abnormal morphology. As a result, the amount of information provided by these models has been limited, and interpretations of the cellular mechanisms of teratogenesis somewhat speculative.

With the advent of molecular biology, new genetic models of abnormal development are now available that provide markers for early identification of affected embryos and, in some cases, the potential to target specific genes for study. These popular approaches include the following: (a) insertional mutations, (b) null mutant mice created by homologous recombination in embryonic stem cells (knockout mice), and (c) disruption of gene expression using antisense oligonucleotides. Of the three, insertional mutations are the least specific, since individual genes are not targeted as they are in the other two approaches. All three techniques permit molecular identification of affected embryos and, therefore, the opportunity to study the earliest possible alterations leading to abnormal phenotypes. They also afford the opportunity to study the effects of gene–teratogen interactions.

B. Insertional Mutants

Transgenic mice with insertional mutations are generated by injecting a DNA construct into zygotes. The foreign DNA occasionally integrates in a host chromosome in such a way that it disrupts an endogenous gene, resulting in a mutation (insertional mutagenesis). A well-studied example of this approach was generated by injecting zygotes with a DNA construct containing a *Drosophila* heat shock gene (*hsp-70*) and a herpesvirus thymidine kinase gene (McNeish et al. 1988). Progeny from a transgenic mouse resulting from the procedure were interbred and produced offspring with limb and craniofacial defects. Hindlimbs were affected more than forelimbs and were often severely truncated, giving rise to the term *legless* (*lgl*) for the mutation. Only pups homozygous for the transgene were affected, suggesting that an autosomal recessive mutation had occurred (McNeish et al. 1988, 1990).

The advantage of this model was the fact that the foreign DNA serves as a molecular tag to identify mutant embryos early in development and to characterize the mutation. In this regard, subsequent studies of this model showed that there were two tightly linked insertion sites that mapped to chromosome 12 near the situs inversus viscerum locus (*iv*) (Singh et al. 1991; Schreiner et al. 1993). Complementation tests showed that *lgl* and *iv* were allelic and suggested that the insertion had altered more than one gene. Unfortunately, further analysis has not resulted in the identification of potential candidate genes that might have been disrupted to produce the abnormalities.

However, characterization of the phenotypes produced by the mutation has been completed. Homozygous animals have truncated hindlimbs with absence of structures distal to the femur, preaxial forelimb defects, absence of the olfactory bulbs, malformations of the cerebral hemispheres, facial clefts, and situs inversus in 50% of the mutants (McNeish et al. 1990; Singh et al. 1991). The value of the model for teratology has been illustrated by additional studies that analyzed pathogenic events leading to the limb abnormalities. These analyses showed that an episode of cell death occurs in mutant hindlimbs at the 34- to 35-somite stage. Initially, this death is located in the dorsal, preaxial mesoderm, but later becomes more encompassing and extends throughout the mesoderm. There is also an episode of cell death in the apical ectodermal ridge, such that by the 45-somite stage no thickened ectoderm characteristic of the ridge is present (Singh et al. 1991).

Another advantage of the model is the potential to study gene–teratogen interactions. In this regard, retinoic acid was shown to exacerbate the malformations observed in the homozygote mutant embryos, leading to the hypothesis that the mutation had altered a retinoid-responsive gene (Scott et al. 1994). Interestingly, a peak period of sensitivity to the effects of retinoic acid occurred 48 h prior to the first appearance of the limb buds. Thus positional information for prospective limb bud cells may have been altered by the interaction of the mutation and the teratogen.

Although the insertional mutation approach has produced some models that have enhanced our insight into the pathogenesis of certain birth defects and the interactions of genes and teratogens, the disadvantages of the approach make it less than optimal for such studies. The main problem is the lack of specificity. For example, effects might result from the transgene itself or from disruptions in genes where the transgene inserts. Then, if multiple insertion sites occur, multiple genes might be affected, making interpretation of the genetic origin of abnormal phenotypes difficult. In addition, if the mutation is recessive, only 25% of the offspring will be affected, so that studying the earliest occurring mechanisms of teratogenesis will be difficult, since the majority of embryos in a litter will not be abnormal.

C. Knockout Mice

Targeted mutations (site-directed mutagenesis; THOMAS and CAPECCHI 1987) by homologous recombination in embryonic stem cells have resulted in the capacity to create transgenic mice lacking expression of specific genes. The gene to be targeted is knocked out in embryonic stem cells, and these cells are injected into blastocysts to generate germ line chimeras. In turn, these mice are used to generate lines that can be bred to homozygosity, followed by examination of null mutant fetuses lacking expression of the gene. This approach has resulted in large numbers of mice lacking expression of genes that appear to be important in development. In fact, the number of phenotypes produced has flooded the field of developmental biology with models that have yet to be well characterized (Table 1).

Examples of the success of the knockout approach abound. Two interesting reports that appeared almost simultaneously described knockouts of the *Wnt-1* proto-oncogene (THOMAS and CAPECCHI 1990; MCMAHON and BRADLEY 1990), which is expressed in the midbrain, hindbrain, and spinal cord during head fold formation and neurulation (WILKINSON et al. 1987; MCMAHON et al. 1992). Both groups achieved similar results, showing that null mutant mice were missing large regions of the midbrain and an absence of the cerebellum, defects consistent with the expression pattern of the gene. However, some variability in the phenotype occurred in the study by THOMAS and CAPECCHI (1990), as illustrated by the fact that one mouse homozygous for the null alleles survived to adulthood and developed the posterior portion of the cerebellum. This result was not readily explained, but it was hypothesized that, since *Wnt-1* is a secreted protein (VAN DEN HEUVAL et al. 1989), its inductive influence might exist across a gradient that could produce differences in the penetrance of the *Wnt-1* mutation. In this manner, some mice might be missing only the anterior region of the cerebellum (THOMAS and CAPECCHI 1990).

Another interesting feature of these *Wnt-1* knockout mice was the absence of defects in other tissues that express the gene, such as the spinal cord and neural crest cells. Both groups attributed this result to redundancy (THOMAS and CAPECCHI 1990; MCMAHON and BRADLEY 1990) and suggested that other members of the *Wnt* family may overcome the loss of *Wnt-1* (ROELINK and NUSSE 1991; MCMAHON et al. 1992). A candidate for this function is *Wnt-3a*, which is coexpressed in the neural tube with *Wnt-1* and shares approximately 60% homology with this gene.

Redundancy is a confounding issue in the targeted mutation approach, and many examples exist where genes were knocked out and the resultant

Table 1. Examples of targeted mutations to create transgenic "knockout mice"

Targeted gene	Affected organs and tissues	Reference
int-1 (*Wnt-1*)	Absent midbrain, partial or complete absence of the cerebellum	THOMAS and CAPECCHI (1990); MCMAHON and BRADLEY (1990)
int-2	Inner ear, tail, and vertebral anomalies	MANSOUR et al. (1993)
N-myc	Limb, lung, stomach, liver, heart, and CNS	SARVAI et al. (1993)
engrailed-1	Absent colliculi and cerebellum, absent third and fourth cranial nerves, limb, sternum	WURST et al. (1994)
HNF-3β	Node, notochord	ANG and ROSSANT (1994)
Hoxa1 (*Hox-1.6*)	Delayed hindbrain neural tube closure, absent cranial nerves, skull, inner ear	LUFKIN et al. (1991)
Hoxa3 (*Hox-1.5*)	Absent thymus and parathyroid glands, reduced thyroid and submaxillary glands, heart, major arteries, face	CHISAKA and CAPECCHI (1991)
Apob	Exencephaly, hydrocephaly	HOMANICS et al. (1995)
CRABP I and II	Outgrowth of extra tissue on digit 5; otherwise mice were normal	LAMPRON et al. (1995)
goosecoid	Aplastic nasal cavities, defective tongue muscles, middle ear anomalies	YAMADA et al. (1995); RIVERA-PEREZ et al. (1995)
p53	Exencephaly in a subset of homozygous embryos	SAH et al. (1995)

For a complete current listing of knockout mice, see BRANDON et al. (1995). CNS, central nervous system.

phenotypes did not include abnormalities in all regions of gene expression, suggesting that redundancy occurred. Such results might be expected, since many genes have been found to be members of large families with overlapping expression patterns. Thus, if a gene is knocked out and does not result in abnormalities in tissues in which it is expressed the question arises of whether this result indicates that the gene is expendable and not important for that developmental event or whether redundant genes made up for the loss of the targeted gene. There is also the problem of targeted knockouts of genes whose expression patterns suggest that they play roles at critical stages of embryogenesis, but their loss results in no abnormal phenotypes or inconsistent results. For example, knockouts of c-*src* (SORIANO et al. 1991), CRABP I and CRABP II (LAMPRON et al. 1995), and others have produced no severely altered phenotypes, yet these genes are expressed at key times in development. In other cases, only a subset of homozygous knockout embryos show an abnormal phenotype, such as *p53*-deficient mice, where 5%–7% of homozygotes are exencephalic (SAH et al. 1995). The question to be answered is whether these results document the lack of importance of these genes, the phenomenon of functional redundancy, compensation, or the inability to reveal function due to the protected environment of laboratory animals and their facilities (Chambon 1994; Lampron et al. 1995).

Another phenomenon clouding the interpretation of results from the targeted mutation approach is compensation. For example, when a gene is absent from the time of fertilization onward, is it the loss of function of that gene that is being investigated, or the ability of the organism to regulate its development in the absence of that gene, or a complex interplay of the two processes. Logic dictates that a complex model exists in which the organism attempts to find ways to compensate for the loss of the gene, either by substituting redundant genes or other forms of regulation (Routtenberg 1995). Excellent examples of this complexity and the capacity for compensation are given in two studies reported by Capecchi and coworkers. In the first, 80% of *fgf*-3 homozygotes were observed to have functional inner ear structures on one side, while the other side was completely defective (Mansour et al. 1993). In the second, null mutant fetuses for *hoxb*5 exhibited differences in the severity of alterations in skeletal development on one side of the body versus the other (Rancourt et al. 1995). In both cases, differences were random with respect to sidedness, suggesting that even within the same embryo compensation can occur and that it may occur unequally, as cells destined to form bilateral structures independently adapt to circumvent the loss of a gene product.

In addition to issues involving redundancy and compensation that complicate interpretations of results from the site-directed mutagenesis approach, there are other difficulties with this model. First, it is expensive and time consuming. Housing costs for mice in a transgenic facility are high, in part because it often takes 1 or 2 years to generate null mutant fetuses. Second, targeting genes that are expressed early in development may result in lethality at the earliest stages of organogenesis. In this situation, an analysis of function is complicated unless embryos can be collected and analyzed prior to their death, a difficult task if they are dying at early stages of development. In addition, if the gene is expressed at different times during development, but its loss leads to early embryonic death, then study of the gene's function at later stages will be impossible. A third difficulty with this approach is the inherent problem of assessing redundancy between different genes and interactions of genes involved in signaling pathways. To conduct such studies requires the creation of multiple knockouts. Creation of such models is possible, but usually requires breeding two lines of knockout mice to homozygosity for both genes. Such an approach is time consuming, costly, and increasingly difficult if interactions between more than two genes are to be investigated. Finally, the targeted mutation approach usually results in recessive mutations that must be bred to homozygosity to demonstrate a phenotype. Since in many cases homozygotes do not survive or are incapable of breeding, heterozygous matings must be employed to generate the null phenotype. Thus, in any given litter, only 25% of the embryos will exhibit the abnormal phenotype to be studied. While this result is satisfactory for descriptions of developmental defects, studies designed to elucidate the earliest cellular alterations that cause the abnormalities are less feasible. For example, during early stages of development, the absence of a tissue might be due to decreased cell proliferation, increased cell death, altered cell migration, or other factors. These processes can be studied by a variety of techniques, but the feasibility of the approaches becomes questionable if only 25% of the embryos in a litter are destined to exhibit some type of cellular disruption. Thus, unless the cellular events leading to an abnormal phenotype were dramatic, it might be impossible to adequately define the underlying mechanisms responsible for the morphological alterations.

In summary, the use of site-directed mutagenesis to create gene knockouts in transgenic mice has resulted in the establishment of complex models with phenotypes generated by an interplay of the loss of a gene product and the organism's attempt to compensate for that loss. Furthermore, investigations into the cellular mechanisms responsible for the phenotypes are also complicated. The net result has been the creation of large numbers of animals with developmental defects that have been well described morphologically, but that have not been well characterized biologically. Thus, before the scientific community becomes completely overwhelmed with these animals at a tremendous cost in research funds, alternative models designed to delineate mechanisms responsible for the abnormalities should be developed.

D. Antisense

Another means of "knocking out" genes during development involves the use of antisense technology and mouse (or rat) whole embryo culture. With this approach, antisense oligonucleotides are microinjected into the amniotic cavity of mouse embryos at early stages of neurulation (Fig. 1A,B). These probes are cDNA of approximately 20 bases that are directed to a specific sequence in the mRNA of the targeted gene (FAKLER et al. 1994). The cDNA is rapidly endocytosed (LOKE et al. 1989; YAKUBOV et al. 1989) by most if not all cells of the embryo (AUGUSTINE et al. 1993). Once inside the cell, small quantities escape from the endosomes (LAKE et al. 1989), reach the targeted mRNA, and bind to it. The message is then cleaved at the binding site by the enzyme RNase H that recognizes double-stranded RNA (HELENE and TOULME 1990; WAGNER 1994) and is ubiquitous in cells (AGRAWAL et al. 1990). The antisense probes may also function by preventing ribosomal attachment to mRNA and by binding to DNA in the nucleus to inhibit transcription (HELENE and TOULME 1990). The net result is loss of gene function and reduction in the targeted protein. The antisense technique also affords the opportunity to study gene–teratogen interactions with relative ease. Thus specific genes can be disrupted by antisense targeting followed by exposure of the embryo to a developmental toxicant. Ongoing studies in our laboratory

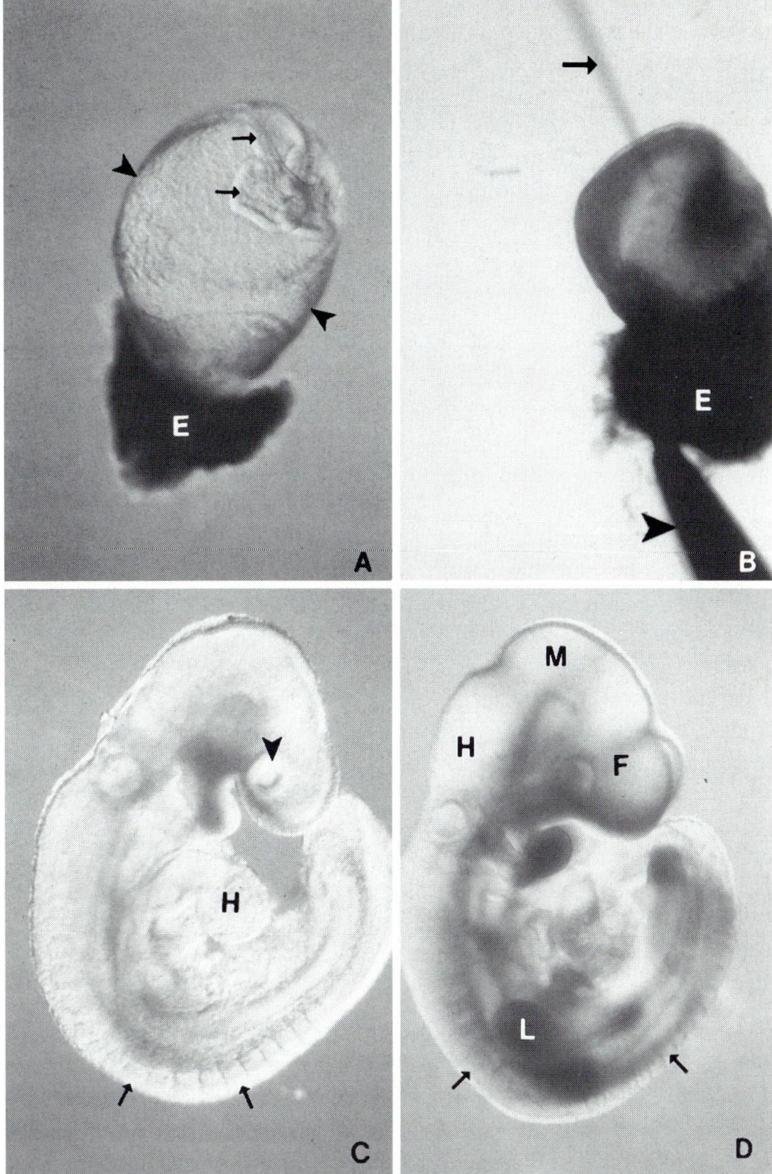

Fig. 1. **A** Day-9 mouse embryo (three- to five-somite stage) as prepared for culture showing an intact visceral yolk sac (*arrowheads*), ectoplacental cone (*E*), and embryo (*arrows*). **B** Day-9 mouse embryo prepared for culture and being microinjected with oligodeoxynucleotides. The injection needle (*arrow*) is positioned to pierce the visceral yolk sac in a region devoid of embryonic tissue, while forceps (*arrowhead*) hold the ectoplacental cone (*E*) for stability. **C** Embryo after 24 h of culture showing normal development. The optic vesicle (*arrowhead*), heart (*H*), and 19 pairs of somites (*arrows*) are present. **D** Embryo after 48 h of culture showing normal development. By this time, 30–32 pairs of somites are present (*arrows*), forelimb buds (*L*) have appeared, and forebrain (*F*), midbrain (*M*), and hindbrain (*H*) regions are clearly distinguishable

are using such an approach to target genes that appear to play protective roles against oxidative stress, such as catalase and superoxide dismutase. Antisense-treated embryos are then challenged with toxicants that induce a variety of reactive species to test the protective role of these enzymes. Following injection, embryos are cultured for 24–48 h and developmental defects are characterized. Since the culture system supports normal embryonic growth and development throughout the period of neurulation and limb bud formation, numerous morphological parameters are available for study (Fig. 1C,D; New 1978; Sadler 1979).

This combined use of antisense and whole embryo culture is obviously limited by the stages during which embryos will grow in culture and the genes that are expressed during those stages. However, since embryos can be grown from early stages of gastrulation and neurulation, numerous genes involved in axis formation and patterning of the central nervous system are available for study. In addition, genes regulating neural crest migration, cardiac morphogenesis, and early stages of limb bud formation can be targeted. Thus the system can be employed to study genetic regulation of events occurring at initial stages of organogenesis.

Using the antisense approach, genes involved in formation of the forebrain, cerebellum, facial development, and axis formation have been studied (Augustine et al. 1993, 1995a,b; Sadler et al. 1995). Targeting of *Wnt-1*, the homologue of the *Drosophila* gene *wingless* (*wg*), which is expressed in the mesencephalon and myelencephalon (Wilkinson et al. 1987; McMahon et al. 1992), documented its role in formation of the cerebellar region (Fig. 2A,B), a result consistent with that obtained using null mutant mice (Thomas and Capecchi 1990; McMahon and Bradley 1990). However, additional defects were observed using the antisense approach to target *Wnt-1* that were not present in null mutant fetuses (Augustine et al. 1993, 1995a). These abnormalities included facial and cardiac defects that were attributed to effects on neural crest cells that also express the gene (McMahon et al. 1992). Results showing that antisense studies sometimes produce phenotypes that are not identical to those obtained in null mutant animals have also been obtained

→

Fig. 2. A Embryo injected with *Wnt-1* sense oligonucleotides (40 μ*M*) at the three- to five-somite stage and cultured for 48 h showing normal development, including formation of the forebrain (*F*), midbrain (*M*), and hindbrain (*H*) regions. **B** Embryo injected with *Wnt-1* antisense oligonucleotides (40 μ*M*) at the three- to five-somite stage and cultured for 48 h. Both the forebrain (*F*) and hindbrain (*H*) regions are underdeveloped, while the spinal cord, somites (*arrows*), and forelimb buds (*L*) are unaffected. **C** Embryo injected with a mixture of antisense oligonucleotides to *Wnt-1* (20 μ*M*) and *Wnt-3a* (20 μ*M*) at the three- to five-somite stage and cultured for 48 h. The cranial region is underdeveloped, especially in the area of the hindbrain (*H*); kinks appear in the neural tube (*arrows*), and somites are indistinct. **D** Embryo exposed to *engrailed-1* antisense oligonucleotides (60 μ*M*) at the three- to five-somite stage and cultured for 48 h showing hindbrain defects (*H*), kinks in the neural tube (*arrows*) small forelimb buds (*L*), and shortening of the embryonic axis (caudal dysgenesis)

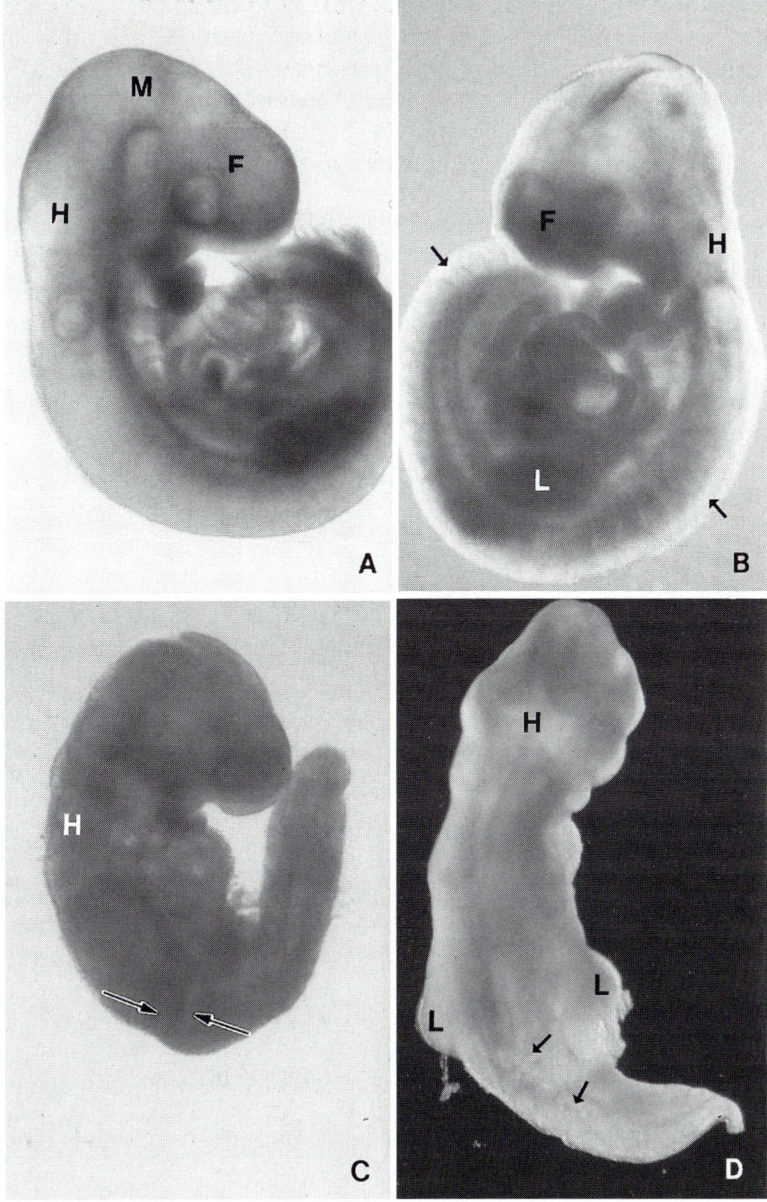

following targeted disruption of *engrailed-1* (WURST et al. 1994; AUGUSTINE et al. 1995b; SADLER et al. 1995) and *Wnt-3a* (TAKADA et al. 1994; AUGUSTINE et al. 1995) and may reflect inherent differences in the two techniques. For example, as stated previously, null mutant embryos represent a complex model in which a resultant phenotype may occur through the interaction of loss of a gene and attempts to regulate development in its absence. On the other hand, targeted disruptions of genes in culture occur acutely in the middle of organogenesis and leave little time for the embryo to compensate for the loss of a specific protein. Furthermore, the culture method allows little time for recovery from that loss, since embryos do not survive past the early limb bud stage of development. Therefore, different results with the two systems might be anticipated.

One advantage of the antisense method was recently illustrated by targeting *Wnt-1* and *Wnt-3a* simultaneously (AUGUSTINE et al. 1993, 1995a). These two genes are family members with approximately 60% homology and are coexpressed in the spinal cord region of neurulating embryos. Since *Wnt-1* knockout mice showed no spinal cord abnormalities, it was postulated that *Wnt-3a* protected against the loss of *Wnt-1* (ROELINKE and NUSSE 1991; MCMAHON et al. 1992). By targeting both genes in embryo culture, the redundancy hypothesis could be tested. Interestingly, brain, neural tube, and axis defects occurred when both genes were targeted with antisense oligonucleotides (AUGUSTINE et al. 1993, 1995a; Fig. 2C). Thus *Wnt-3a* can serve as a redundant gene for *Wnt-1*, and the culture system can disclose these types of interactions between two or more genes much more efficiently than trying to generate double knockout mice.

Use of the antisense approach can also reveal new expression patterns and roles for genes that have not been discovered in knockout mice. For example, antisense disruption of *engrailed-1* produced abnormalities in areas that normally express the gene, including the mesencephalon and rhombencephalon, and limb defects, as had been shown in knockout mice (WURST et al. 1994; SADLER et al. 1995). However, embryos exposed to antisense also exhibited caudal dysgenesis, represented by a failure of caudal segments to develop (Fig. 2D) (SADLER et al. 1995). Since this abnormality has been ascribed to disruption of mesoderm formation by the process of gastrulation (DUHAMEL 1961; ALLES and SULIK 1993), it was hypothesized that *engrailed-1* must be expressed in the primitive streak and play a role in this process. In situ hybridization studies had failed to demonstrate *engrailed-1* expression in this region (DAVIDSON et al. 1988; DAVIS et al. 1991), and no disruption of axis formation was present in null mutant mice (WURST et al. 1994). However, subsequent polymerase chain reaction (PCR) analysis revealed *engrailed-1* expression in the tail region of neurulating mouse embryos and, together with the phenotypic data on caudal dysgenesis in embryos exposed to antisense probes to *engrailed-1*, showed that the gene plays a role in this morphogenetic process, a role that had not been discovered using transgenic mouse models (SADLER et al. 1995).

Virtually any gene can be targeted using the antisense approach, as can multiple genes. Specificity of the probes can be achieved with 18–20 bases, and probes of longer length may in fact have a greater chance of producing nonspecific effects (DASH et al. 1987). The oligos are usually uniformly substituted with sulfur on the phosphate groups (phosphorothioated probes) to protect against nuclease degradation and increase their half-life (CAMPBELL et al. 1990; HELENE 1991). Recent evidence suggests that chimeric probes in which only the first six bases at each end are modified, leaving the middle six to eight bases as unmodified phosphodiesters, improves RNase H activity while maintaining nuclease resistance (HELENE 1991). Further, coupling the probes to fusogenic proteins (BONGARTZ et al. 1994) or combining them with liposomes (ZELPHATI et al. 1994) may improve their uptake and delivery to the targeted mRNA. Uptake with standard unmodified phosphodiester probes occurs rapidly in injected mouse embryos, usually within 10 min, but their half-life is only 10–20 min (AUGUSTINE et al. 1993). Uniformly modified probes are taken up by cells within 30 min, and intact probes can still be detected at 24 h (AUGUSTINE et al. 1993). Thus the oligonucleotides are like chemical agents whose pharmacokinetics must be factored into the experimental design. The pharmacokinetic characteristics of the targeted gene and its protein must also be considered and will influence results. For example, since it is the mRNA (and perhaps also transcription) that is targeted, any existing protein present at the time that embryos are exposed to oligonucleotides will not be affected. Thus there may be some delay before an effect on a protein is observed, and this delay will depend on the rate of turnover of existing protein present at the time of injection.

As mentioned previously, specificity of the oligonucleotides and their effects is an essential consideration (STEIN and KRIEG 1994; WAGNER 1994). Appropriate sequences must be selected that are specific for the mRNA to be targeted. Sites to be targeted within the mRNA do not appear to be critical, although sites somewhere near the AUG initiation codon or in the 3′ untranslated region are good starting points (DEAN et al. 1994). However, several oligonucleotides targeted to different regions of the mRNA may need to be tried before an effective site will be discovered (FAKLER et al. 1994). The secondary structure of the mRNA and the ability of RNase H to act at the targeted site presumably influence the effectiveness of the oligonucleotides, but criteria for selecting optimum sites have not yet been established (BENNETT et al. 1994; DEAN et al. 1994; FAKLER et al. 1994). In addition to selecting appropriate antisense oligonucleotides, proper controls must be employed and the following criteria met:

1. Studies should use two or more control oligonucleotides that document specificity. These may incorporate the use of sense, mismatched (probes in which two or four bases are altered in the original antisense sequence), or scrambled (random reshuffling of the original antisense sequence) oligonucleotides. Another control experiment involves targeting a different region of the same mRNA with another antisense oligonucleotide and producing the same phenotype as the original probe. Each of these potential controls is not without caveats, since sense probes at high concentrations may affect transcription and specifically disrupt the targeted gene (AUGUSTINE et al. 1993); mismatch probes may have specific effects, since only eight bases are required to specifically target mRNA (FAKLER

et al. 1994); and as mentioned previously, not all antisense probes will work effectively to target a message.

2. An effect on the targeted mRNA or protein, by PCR, in situ hybridization, immunohistochemistry, etc., should be documented.

If these two conditions are met, the antisense approach in embryo culture can provide stage-specific information about gene function during development. Furthermore, since the antisense technique results in a high percentage of affected embryos (in most cases approximately 70% show a phenotypic effect), studies directed toward elucidating the underlying biological phenomena responsible for altered phenotypes, e.g., effects on cell death, cell migration, and cell proliferation, can be investigated much more readily than in knockout mice, where usually only the homozygotes exhibit a phenotype and thus only 25% of the animals are affected.

Disadvantages of the antisense approach can be deduced from the prior discussion, including the following:

1. Studies are restricted by the stages at which embryos can be cultured.
2. Specificity of the probes must be documented, since these are highly negatively charged molecules that can have nonspecific effects (WAGNER 1994).
3. Kinetics involved in uptake and distribution of the oligonucleotides and the half-life of the proteins must be considered and may preclude effective targeting of some genes.

Despite these concerns, when used appropriately the antisense approach is a powerful one and provides a complementary technique to site-directed mutagenesis as a means of delineating genetic regulation of development.

E. Gene–Teratogen Interactions

All three models that have been discussed can be used to study gene–teratogen interactions. Thus a gene of interest could be targeted and then genetically defective embryos could be exposed to a toxicant to determine the potential for interactions between the missing or defective gene and the insult. Such an approach has shown that the insertional mutation *legless* is more sensitive to retinoic acid-induced teratogenesis (SCOTT et al. 1994), but the mechanism for this effect is unknown. As mentioned previously, the lack of specificity of the insertional mutation approach makes planning gene–teratogen studies less feasible with this model.

Targeting specific genes with the other techniques of knockout mice and antisense technology should be more successful for gene–teratogen studies. Knockout mice have already been employed in such studies to test for increased sensitivity to the retinoids (LAMPRON et al. 1995), and our laboratory has initiated studies using antisense probes to target catalase, an antioxidant enzyme (MICHAELS et al. 1994). Previous studies have shown that this enzyme may be important in protecting the embryo from oxidative stress (JENKINSON et al. 1986), and preliminary data suggest that embryos with antisense attenuation of this gene product are severely growth retarded and malformed (T.W. Sadler, unpublished data). Thus this enzyme should be a candidate for investigating gene–teratogen interactions using the antisense approach.

Transgenic *p53*-deficient mice have also been used to study gene–teratogen interactions. This model is particularly interesting because on the one hand *p53* appears to be an important teratological suppressor gene, but on the other it may increase sensitivity to certain teratogens. For example, *p53*-deficient mice exposed to benzo[a]pyrene exhibited a two- to four fold higher incidence of embryotoxicity and teratogenicity than normal *p53* controls (Nicol et al. 1995), thus indicating a potential protective effect of the gene. In contrast, *p53*-deficient mice exposed to chlorodeoxyadenosine are less sensitive to this chemical's toxic effects, illustrating that in some cases the gene may enhance a compound's embryotoxicity (T.B. Knudsen, unpublished data; see Chap. 13 for more details).

It should be noted that these types of studies target specific genes and then determine the effects of a teratogen, the assumption being that, if there is an increased sensitivity to a toxicant, then it is probably due to an interaction between that gene and the toxicant, either directly or indirectly through a signaling pathway controlled by the targeted gene. Such an approach is in contrast to that taken by administering a toxicant and then monitoring changes in gene expression using subtractive hybridization or differential display. In these approaches, changes in gene expression following exposure are catalogued, and attempts are then made to link the changes to toxicant-induced abnormalities. The problem with these approaches is their lack of specificity, the numbers of genes that show altered expression patterns, and difficulties in establishing primary versus secondary effects on gene expression i.e., it is very difficult to know whether genes involved in production of the defects or those involved in repair and recovery are being examined. Any toxicant that causes abnormal phenotypes can be expected to disrupt the expression of many genes. Therefore, planning and interpreting gene–teratogen interaction studies with these types of approaches should be done judiciously and with caution.

Acknowledgement. This work was supported by NIH grant HD29495.

References

Agrawal S, Mayrand SH, Zamecnik PC, Pederson T (1990) Site-specific excision from RNA by RNase H and mixed-phosphate-backbone oligodeoxynucleotides. Proc Natl Acad Sci USA 87: 1401–1405

Alles AJ, Sulik KK (1993) A review of caudal dysgenesis and its pathogenesis in an animal model. Birth Defects (Orig Art Ser) 29: 83–102

Ang SL, Rossant J (1994) *HNF-3β* is essential for node and notochord formation in mouse development. Cell 78: 561–574

Augustine K, Liu ET, Sadler TW (1993) Antisense attenuation of *Wnt*-1 and *Wnt*-3a expression in whole embryo culture reveals roles for these genes in craniofacial, spinal cord, and cardiac morphogenesis. Dev Genet 14: 500–520

Augustine KA, Liu ET, Sadler TW (1995a) Interactions of *Wnt*-1 and *Wnt*-3a are essential for neural tube patterning. Teratology 51: 107–119

Augustine KA, Liu ET, Sadler TW (1995b) Antisense inhibition of *Engrailed* genes in mouse embryos reveals roles for these genes in craniofacial and neural tube development. Teratology 51: 292–299

Bennett CF, Condon TP, Grimm S, Chan H, Chiang MY (1994) Inhibition of endothelial cell adhesion molecule expression with antisense oligonucleotides. J Immunol 152: 3530–3540

Bongartz JP, Aubertin AM, Milhaud PG, Lebleu B (1994) Improved biological activity of antisense oligonucleotides conjugated to a fusogenic peptide. Nucleic Acids Res 22: 4681–4688

Brandon EP, Idzerda RL, McKnight GS (1995) Targeting the mouse genome: a compendium of knockouts. Curr Biol 5: 625–634, 758–765, 873–881

Chambon P (1994) The retinoid signaling pathway: molecular and genetic analyses. Semin Cell Biol 5: 115–125

Campbell JM, Bacon TA, Wickstrom E (1990) Oligodeoxynucleoside phosphorothioate stability in subcellular extracts, culture media, sera, and cerebrospinal fluid. J Biochem Biophys Methods 20: 259

Chisaka O, Capecchi MR (1991) Regionally restricted developmental defects resulting from targeted disruption of the mouse homeobox gene *hox* 1.5. Nature 350: 473–479

Dash P, Lotan I, Knapp M, Kaudel ER, Goelet P (1987) Selective elimination of mRNAs in vivo: complementary oligodeoxynucleotides promote RNA degradation by an RNase H-like activity. Proc Natl Acad Sci USA 84: 7896–7900

Davidson D, Grahm E, Sime C, Hill R (1988) A gene with sequence similarity to *Drosophila engrailed* is expressed during the development of the neural tube and vertebrae in the mouse. Development 104: 305–316

Davis CA, Holmyard DP, Millen KJ, Joyner AL (1991) Examining pattern formation in mouse chicken and frog embryos with an *En*-specific antiserum. Development 111: 287–298

Dean NM, McKay R, Condon TP, Bennett CF (1994) Inhibition of protein kinase C-α expression in human A549 cells by antisense oligonucleotides inhibits induction of intercellular adhesion molecule 1 (ICAM-1) mRNA by phorbol esters. J Biol Chem 269: 16416–16424

Duhamel B (1961) From mermaid to anal imperforation: the syndrome of caudal regression. Arch Dis Child 36: 152–155

Fakler B, Herlitze S, Amthor B, Zenner HP, Ruppersberg JP (1994) Short antisense oligonucleotide-mediated inhibition is strongly dependent on oligo length and concentration, but almost independent of location of the target sequence. J Biol Chem 269: 16187–16194

Helene C (1991) Rational design of sequence-specific oncogene inhibitors based on antisense and antigene oligonucleotides. Eur J Cancer 27: 1466–1471

Helene C, Toulme JJ (1990) Specific regulation of gene expression by antisense, sense, and antigene nucleic acids. Biochem Biophys Acta 1049: 99–125

Homanics GE, Maeda N, Traber MG, Kayden HJ, Dehart DB, Sulik KK (1995) Exencephaly and hydrocephaly in mice with targeted modification of the apolipoprotein B (Apob) gene. Teratology 51: 1–10

Jenkinson PC, Anderson D, Gangolli SD (1986) Malformations induced in cultured rat embryos by enzymatically generated active oxygen species. Teratogenesis Carcinog Mutagen 6: 547–554

Kalter H (1980) A compendium of the genetically induced congenital malformations of the house mouse. Teratology 21: 397–429

Lampron C, Rochette-Egly C, Gorry P, Dolle P, Mark M, Lufkin T, LeMeur M, Chambon P (1995) Mice deficient in cellular retinoic acid binding protein II (CRABP II) or in both CRABP I and CRABP II are essentially normal. Development 121: 539–548

Loke SL, Stein CA, Zhang XH, Mori K, Nakarishi M, Subasinghe C, Cohen JS, Neckers LM (1989) Characterization of oligonucleotide transport into living cells. Proc Natl Acad Sci USA 86: 3474–3478

Lufkin T, Dierich A, LeMeur M, Mark M, Chambon P (1991) Disruption of the *Hox-1.6* homeobox gene results in defects in a region corresponding to its rostral domain of expression. Cell 66: 1105–1119

Mansour SL, Goddard JM, Capecchi MR (1993) Mice homozygous for a targeted disruption of the protooncogene *int*-2 have developmental defects in the tail and inner ear. Development 117: 13–28

McMahon AP, Bradley A (1990) The *Wnt*-1 (*int*-1) proto-oncogene is required for development of a large region of the mouse brain. Cell 62: 1073–1085

McMahon AP, Joyner AL, Bradley A, McMahon JA (1992) The midbrain-hindbrain phenotype of *Wnt*-1/*Wnt*-1 mice results from stepwise deletion of engrailed-expressing cells by 9.5 days postcoitum. Cell 69: 581–595

McNeish JP, Scott WJ, Potter SS (1988) *Legless,* a novel mutation found in PHT-1 transgenic mice. Science 241: 837–839

McNeish JD, Thayer J, Walling K, Sulik KK, Potter SS, Scott WJ (1990) Phenotypic characterization of the transgenic mouse insertional mutation, *legless*. J Exp Zool 253: 151–162

Michaels C, Raes M, Toussaint O, Remacle J (1994) Importance of Se-glutathione peroxidase, catalase, and Cu/Zn-SOD for cell survival against oxidative stress. Free Radiat Biol Med 17: 235–248

New DAT (1978) Whole embryo culture and the study of mammalian embryos during organogenesis. Biol Rev 53: 81–122

Nicol CJ, Harrison ML, Laposa RR, Gimelshtein IL, Wells PG (1995) A teratologic suppressor role for *p53* in benzo [a] pyrene-treated transgenic *p53*-deficient mice. Nature Genet 10: 181–187

Rancourt DE, Tsuzuki T, Capecchi MR (1995) Genetic interaction between *hoxb*-5 and *hoxb*-6 is revealed by nonallelic noncomplementation. Genes Dev 9: 108–122

Rivera-Perez JA, Mallo M, Gendron-Maguire M, Gridley T, Behringer RR (1995) *Goosecoid* is not an essential component of the mouse gastrula organizer, but is required for craniofacial and rib development. Development 121: 3005–3012

Roelink H, Nusse R (1991) Expression of two members of the *Wnt* family during mouse development-restricted temporal and spatial patterns in the developing neural tube. Genes Dev 5: 381–388

Routtenberg A (1995) Knockout mouse fault lines. Nature 374: 314–315

Sadler TW (1979) Culture of early somite mouse embryos during organogenesis. J Embryol Exp Morphol 49: 17–25

Sadler TW (1995) Langman's medical embryology, 7th edn. Williams and Wilkins, Baltimore

Sadler TW, Liu ET, Augustine KA (1995) Antisense targeting of *Engrailed*-1 causes abnormal axis formation in mouse embryos. Teratology 51: 300–310

Sah VP, Attardi LD, Mulligan GJ, Williams BO, Bronson RT, Jacks T (1995) A subset of *p-53*-deficient embryos exhibit exencephaly. Nature Genet 10: 175–180

Sarvai S, Shimono A, Wakamatsu Y, Palmes C, Hanaoka K, Kondoh H (1993) Defects of embryonic organogenesis resulting from targeted disruption of the *N-myc* gene in the mouse. Development 117: 1445–1455

Schreiner CM, Scott WJ, Supp DM, Potter SS (1993) Correlation of forelimb malformation asymmetries with visceral organ situs in the transgenic mouse insertional mutation, *legless*. Dev Biol 158: 560–562

Scott WJ, Collins MD, Ernst AN, Supp DM, Potter SS (1994) Enhanced expression of limb malformations and axial skeleton alterations in *legless* mutants by transplacental exposure to retinoic acid. Dev Biol 164: 277–289

Singh G, Supp DM, Schreiner C, McNeish JD, Merker HJ, Copeland NG, Jenkins NA, Potter SS, Scott WJ (1991) *Legless* insectional mutation: morphological, molecular, and genetic characterization. Genes Dev 5: 2245–2255

Soriano P, Montgomery C, Gesk R, Bradley A (1991) Targeted disruption of the *c-src* proto-oncogene leads to osteopetrosis in mice. Cell 64: 693–702

Stein CA, Krieg AM (1994) Problems in interpretation of data derived from in vitro and in vivo use of antisense oligodeoxynucleotides. Antisense Res Dev 4: 67–69

Takada S, Stark KL, Vassileva G, McMahon J, McMahon AP (1994) *Wnt*-3a regulates somite and tailbud formation in the mouse embryo. Genes Dev 8: 174–189

Thomas KR, Capecchi MR (1987) Site-directed mutagenesis by gene targeting in mouse embryo-derived stem cells. Cell 51: 503–512

Thomas KR, Cappechi MR (1990) Targeted disruption of the murine *int*-1 proto-oncogene resulting in severe abnormalities in midbrain and cerebellar development. Nature 346: 847–850

van den Heuval M, Nusse R, Johnston P, Lawrence PA (1989) Distribution of the *wingless* gene product in *Drosophila* embryos: a protein involved in cell-cell communication. Cell 59: 739–749

Wagner RW (1994) Gene inhibition using antisense oligodeoxynucleotides. Nature 372: 333–335

Wilkinson DG, Bailes JA, McMahon AP (1987) Expression of the proto-oncogene *int*-1 is restricted to specific neural cells in the developing mouse embryo. Cell 50: 79–88

Wurst W, Auerbach AB, Joyner AL (1994) Multiple developmental defects in *Engrailed*-1 mutant mice: an early mid-hindbrain deletion and patterning defects in forelimb and sternum. Development 120: 2065–2075

Yakubov LA, Deeva EA, Zarytova VF, Ivanova EM, Ryte AS, Yurchenko LV, Vlassov VV (1989) Mechanism of oligonucleotide uptake by cells: involvement of specific receptors? Proc Natl Acad Sci USA 86: 6454–6458

Yamada G, Mansouri A, Torres M, Stuart ET, Blum M, Schultz M, De Robertis EM, Gruss P (1995) Targeted mutation of the murine *goosecoid* gene results in craniofacial defects and neonatal death. Development 121: 2917–2922

Zelphati O, Wagner E, Leserman L (1994) Synthesis and anti-HIV activity of thiocholesteryl-coupled phosphodiester antisense oglionucleotides incorporated into immunoliposomes. Antiviral Res 25: 13–25

CHAPTER 13

Nucleotide Pool Imbalance

C. Lau

A. Introduction

Ribonucleotides (NTP) and deoxyribonucleotides (dNTP) are the basic building blocks of RNA and DNA. While the major nucleotides are found in all mammalian cells, their relative pool sizes are quite different depending on the cell type and species. These unique nucleotide pool patterns are maintained, to a large extent, by the tightly regulated biosynthetic and metabolic pathways through a number of feedback controls. Disruption of these balances have been shown to have profound effects on DNA synthesis and replication in cultured cells (Ashman and Davidson 1981; Topal and Baker 1982; Das et al. 1983; Newman and Miller 1983; Meuth 1984; Nicander and Reichard 1985; Yoshioka et al. 1987; Phear and Meuth 1989; Kohalmi et al. 1991; Wataya et al. 1993). In fact, this biological feature has become a fundamental tenet for the design of chemotherapeutic agents in the past half century. These drugs typically inhibit the biosynthesis of purine and pyrimidine nucleotides, block the reduction of NTP, or behave as analogue substitutes for the natural precursors to be incorporated into nucleic acids. Because tumor cells usually divide at a rapid rate, upsetting the supply and balance of nucleotide pools invariably will bring a halt to DNA synthesis and an arrest of further cell proliferation. Indeed, recent successes with chemotherapy against a substantial number of neoplastic diseases attest to the validity of this strategy.

Rapid cell proliferation is also a hallmark characteristic of an embryo. Since the discovery of antimetabolites as chemotherapeutic agents, teratologists and obstetricians have been concerned with the potential of these compounds to cause birth defects in humans. Indeed, numerous studies have been conducted with animal models to characterize the embryotoxic and teratogenic properties of these drugs. While each of these antimetabolites exhibits a unique profile of toxicity, they invariably cause cellular damages in the embryos, leading to a variety of structural abnormalities and an overall growth retardation. Based on these observations, Wilson (1973) has proposed altered nucleic acid integrity and function as one of the major mechanisms of teratogenesis. This chapter will review some of the recent findings with drugs that alter the intracellular pool of nucleotides and their disruptive effects in embryonic development.

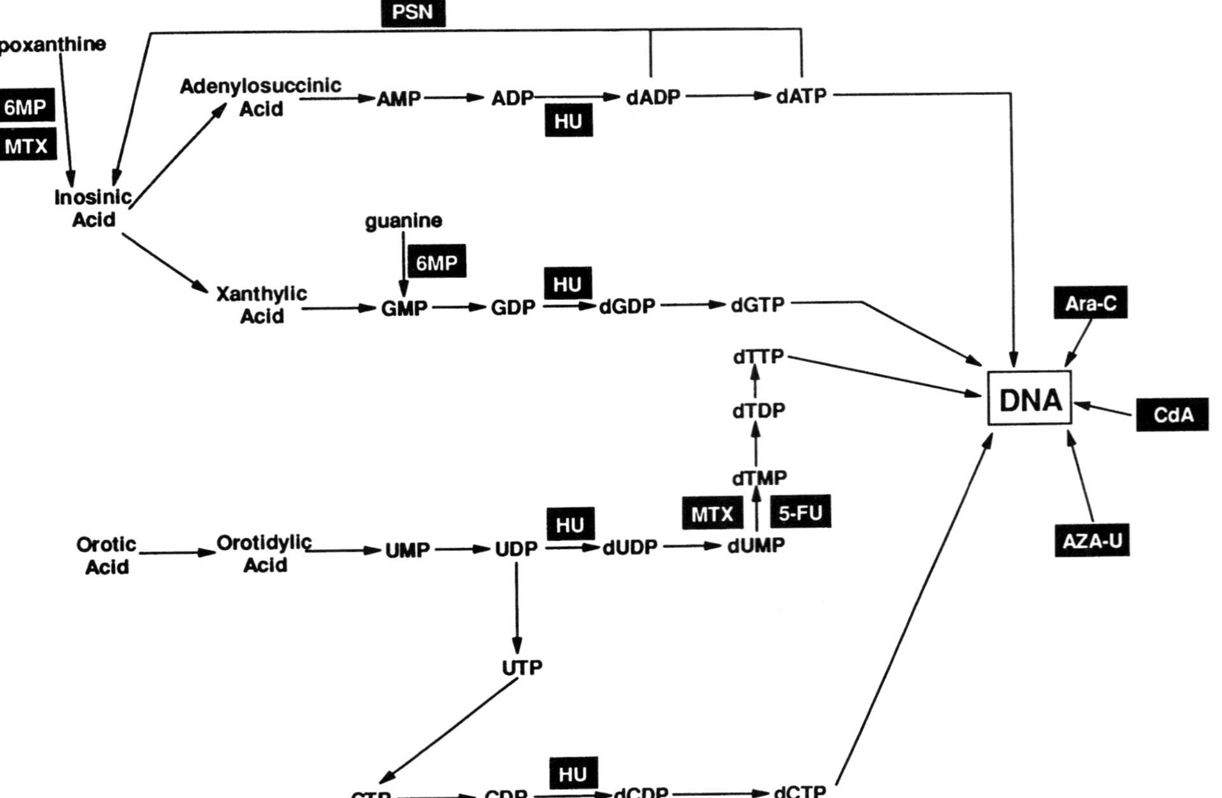

Fig. 1. Biosynthesis and metabolism of purine and pyrimidine nucleotides and sites of actions for various antimetabolites. See text for detailed description. *Ara-C*, cytosine arabinoside; *AZA-U*, azauridine; *CdA*, chlorodeoxyadenosine; *5-FU*, 5-fluorouracil; *HU*, hydroxyurea; *6MP*, 6-mercaptopurine; *MTX*, methotrexate; *PSN*, pentostatin

While it is beyond the scope of this chapter to review the biosynthesis and metabolism of nucleotides in details, Fig. 1 provides a simplified scheme of these reaction pathways and steps that drugs are thought to interfere with. De novo synthesis of purine nucleotides starts out with the sugar moiety ribose 5-phosphate as substrate and involves formation of the intermediate inosine 5′-phosphate through a series of open-chain and then ring-closure reactions. Inosinic acid is then converted to adenylic acid (adenosine monophosphate, AMP) through adenylosuccinic acid, and to guanylic acid (guanosine monophosphate, GMP) through xanthylic acid. Phosphorylation of AMP and GMP yields adenosine di- (ADP) and triphosphate (ATP) and guanosine di- (GDP) and triphosphate (GTP), respectively. Purine nucleotide synthesis is primarily regulated by end product feedback mechanisms at several key enzymatic steps. In addition to de novo synthesis, AMP and GMP can be formed from free purines (adenine and guanine) and purine nucleosides via the salvage pathway. On the other hand, pyrimidine nucleotide synthesis starts with the carbamylation of aspartic acid to form the intermediate orotic acid through a series of ring-building and -closure reactions. The pyrimidine ring structure is then attached to ribose phosphate to yield orotidylic acid, which is then decarboxylated to form uridylic acid (uridine monophosphate, UMP). UMP is phosphorylated to uridine di- (UDP) and triphosphate (UTP) and converted to cytidine 5′-phosphate (CTP) with the amino group donated from glutamine. Biosynthesis of pyrimidine nucleotides is regulated by the reaction products CTP and UTP at the initial enzymatic step of the pathway (carbamylation of aspartic acid). Like purine nucleotides, pyrimidine nucleotides can also be derived from free pyrimidines (uracil, uridine, and cytidine) and pyrimidine nucleosides by salvage pathways.

Purine and pyrimidine dNTP are generated directly by reduction of the corresponding NTP at the 2′ carbon atom. This process is catalyzed by a single enzyme ribonucleotide reductase using all four NTP diphosphates (ADP, GDP, CDP, and UDP) as substrates. The enzyme has a catalytic subunit and a regulatory subunit, and the latter determines its overall activity and substrate specificity; more importantly, it is regulated allosterically by a complex system of positive and negative feedback effects (Eriksson and Thelander 1978; Hunting and Henderson 1982; Reichard 1985; Cory 1987; Jackson 1982). The major feedback mechanisms are depicted in Fig. 2. Thus, if ATP is bound to the regulatory subunit, the enzyme will reduce CDP or UDP; this reaction is inhibited by deoxy-ATP (dATP), thymidine triphosphate (dTTP), and deoxy-GTP (dGTP). If dTTP binds to the regulatory site, the enzyme becomes GDP reductase, and the reaction is inhibited by dATP and dGTP. If dGTP binds to the site, the enzyme acts as an ADP reductase, and the reaction is inhibited by dATP and dTTP. The net effect of these stimuli and inhibitors therefore permits a balanced supply of the reduced diphosphates and hence of the triphosphates, the immediate precursors for DNA synthesis.

dNTP pools are generally very small – on the order of 1%–10% of the size of corresponding NTP pools and 1000-fold smaller than the total amount

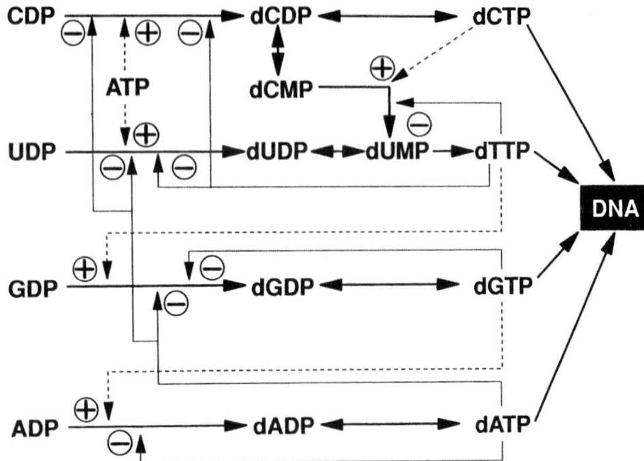

Fig. 2. Regulation of nucleotide pools by the ribonucleoside diphosphate reductase system. Reaction pathways are shown as *heavy solid lines*, positive feedback effects as *dashed lines*, and negative feedback effects as *thin solid lines*

of dNTP residues in DNA; however, significant amounts of all four dNTP are detectable even in cells that are not undergoing active division, suggesting that the absence of DNA synthesis in these cells does not simply result from a limited supply of precursors (SKOOG et al. 1974; TOPAL and BAKER 1982). When the cell traverses through S phase of the cell cycle, where DNA synthesis takes place, all dNTP pools expand dramatically; indeed, fluctuation of the pool levels tends to correlate with the demands of DNA synthesis. Yet, despite these expanded pool sizes, steady state supplies of dNTP within the cells are very limited and estimated to be enough to sustain DNA synthesis for only 30 s to 5 min (SKOOG et al. 1973, 1974; PRESCOTT 1976). Considering that DNA synthesis in a dividing cell typically lasts for several hours, rapid replenishment to maintain a balanced pool of these nucleic acid substrates is essential for cell replication to proceed. Alterations of one or more specific pools of dNTP have thus been shown to increase the frequency of mutation in cell cultures, and even a brief depletion of one or more dNTP pools will precipitate cell death (MEUTH 1984; YOSHIOKA et al. 1987; HIROTA et al. 1989; JAMES et al. 1994).

The consequences of excessive cell death and reduced cell proliferation caused by nucleotide pool imbalance during embryogenesis have drawn considerable attention from embryologists and teratologists and have been frequent topics of discussion (CHAUBE and MURPHY 1968; RITTER et al. 1973; W.J. SCOTT 1977; KNUDSEN and KOCHHAR 1981; SNOW 1983, 1986; AIRHART et al. 1993; ABBOTT et al. 1993; SHUEY et al. 1994a,b,c; Chap. 8, this volume). Intuitively, the hypothesis seems fairly straightforward. Because of the rapid turnover of dNTP, even a brief interruption of their synthesis will alter the pool levels, restrict DNA synthesis, and perturb cell cycle and division, resulting in cell death and diminished cell population required for tissue growth and differentiation. These cellular deficits in turn may lead to malformation. To a certain extent, some of the embryotoxic characteristics of antiproliferative agents support this contention. When these drugs were given to humans or laboratory animals at high doses during the early stages of pregnancy, embryolethality and resorption were invariably observed. In fact, some of these compounds (most notably methotrexate, MTX) were designed and are still used as abortifacient in humans (STOVALL et al. 1990; CREININ 1993). Even when the embryo survived the insults of these agents, hypoplasia in a variety of organs and overall growth retardation were found. Furthermore, a salient example linking nucleotide imbalance to teratogenesis was provided by the elegant work of NELSON (1960), who examined the nature of congenital defects in the rats produced by transient (2-day) maternal folic acid deficiency during the post-implantation period. Folates and folic acid are converted to methylenetetrahydrofolate, which is a key methyl donor in purine biosynthesis and conversion of dUMP to dTMP. It is not surprising that altered folate metabolism during pregnancy has been associated with developmental abnormalities in the embryo and fetus. (For a detailed review and discussion of this topic, see Chap. 15 of this volume). NELSON's findings illustrated that when the deficiency period occurred early (on day 7 of gestation), defects in cerebrum, ventral abdominal wall, and heart were seen; when the deficiency began later (on day 11), a different pattern of deformities was observed which included the lens, kidney, skeleton, and palate. The sequence of organs adversely affected by folate deficiency roughly parallels the order of normal organogenesis. These results are in excellent agreement with the basic principle of teratology, which states that the stage of embryonic development during which a teratogen acts will determine the nature of abnormalities produced. Thus findings from this study and numerous others which associated nucleotide pool imbalance during embryogenesis with congenital defects further implicate this metabolic disorder as a key mechanism of teratogenesis.

Nucleotide pool imbalance in cells can be produced experimentally by three approaches: (1) treatment with a number of antimetabolite drugs which block biosynthesis and/or metabolism; (2) addition of excess natural precursors which upset the pool balance or introduction of analogue substitutes to be incorporated erroneously into RNA and DNA; and (3) use of somatic cell mutants with altered synthesis of DNA precursors. In cancer research, all three approaches are routinely used. However, in embryological and teratological studies, most of the investigators prefer antimetabolites as a means to upset the nucleotide pool balance. This chapter will therefore focus on this area of research and provide an overview of the mechanisms of actions for these agents and their respective embryotoxicity. The material presented here will be limited primarily to mammalian species and in vivo studies. (For further information, readers should consult other recent works on this topic using the chick embryo model as well as cultured embryo and cultured organ models; GRUBB and MONTIEGEL 1975; MANSON et al. 1977; NEUBERT et al. 1977; SCHMID 1984; GRAFTON et al. 1987; ABBOTT et al. 1993; SHUEY et al. 1994c).

B. Determination of Nucleotide Pools

Both NTP and dNTP can be separated by means of high-performance liquid chromatography (HPLC) and detected by ultraviolet spectrophotometry. SIMPSON and BROWN (1986) have provided an excellent review of sample preparations, HPLC methods and applications in determining the nucleotide pool profiles in a variety of cell extracts and physiological samples. AREZZO (1987) described a simple and sensitive method for the measurement of NTP and dNTP by anionic exchange HPLC under isocratic conditions, while CROSS et al. (1993) reported a reverse-phase HPLC method under gradient conditions. More recently, the advances of capillary electrophoresis methodology have enabled even faster analytical times and higher resolving power to be achieved, in comparison to conventional HPLC, for the separation and quantification of nucleotides. Indeed, a typical retention time of 20 min and a detection level in the micromolar range of NTP and dNTP have been demonstrated using these techniques (TAKIGIKU and SCHNEIDER 1991; O'NEILL et al. 1994).

Although it is feasible to separate and quantify all the NTP and dNTP in a single chromatographic assay, simultaneous measurements of both nucleotide species in mammalian tissues prove to be much more difficult, because the cells typically contain NTP in amounts several orders of magnitude greater than those of dNTP. Thus the dNTP peaks in the chromatograms are completely masked by the NTP peaks. To circumvent this problem, two laboratories (GARRETT and SANTI 1979; RITTER and BRUCE 1979) described procedures to destroy and remove the NTP selectively by periodate oxidation; the remaining dNTP could then be readily measured by HPLC. Indeed, improvements of this technique have been elaborated by RITTER et al. (1980) and TANAKA et al. (1984). In particular, RITTER and coworkers (1980) indicated that rapid fixation of tissue was required to avoid the artificially high dNTP values. In addition, these investigators were among the first to evaluate the dNTP levels of mammalian embryos, although pooling of all the embryos from a single litter was necessary for the measurement. Recent modifications of these procedures in our laboratories have further allowed reliable assessment of the dNTP profiles in an individual rat embryo at gestational day 14 (L. Mole, personal communication).

dNTP can also be measured by an indirect enzymatic method utilizing DNA polymerase (SOLTER and HANDSCHUMACHER 1969; LINDBERG and SKOOG 1970; SKOOG 1970; HUNTING and HENDERSON 1981). The method is based upon the ability of the enzyme to incorporate dNTP, in the presence of an appropriate primer template, into an acid-insoluble DNA product. The reaction is carried out with the DNA polymerase and three of the dNTP in excess; the limiting factor is the dNTP of interest. If one of the dNTP present in excess is radioactive, the amount of the limiting dNTP can be determined by precipitation of the DNA product and assay for radioactivity. One advantage of this method is its sensitivity, and detection of picomole amounts of dNTP have been reported (LINDBERG and SKOOG 1970; SKOOG 1970). However, this approach is limited to evaluation of one dNTP at a time, and changes in the dNTP pool within the cells in response to experimental manipulation may alter the specific activity of the labeled dNTP added, thereby introducing inaccuracies to the estimation. In view of the limitations associated with the enzymatic method and the advances made in HPLC, the latter analytical technique has gained wider acceptance among researchers interested in quantifying the nucleotide pool sizes.

C. Interruption of Pyrimidine Nucleotide Pools

I. Fluoropyrimidines

5-Fluorouracil (5-FU) and its deoxynucleoside analogue 5-fluoro-2′-deoxyuridine (FUdR) are widely used in the treatment of solid tumors of the gastrointestinal tract, breast, head, and neck. These drugs have short half-lives, with plasma clearance ranging from 10 to 20 min. In a pregnant rat model, 5-FU exhibits non-linear pharmacokinetics, crosses the placenta readily, and the relative fetal exposure increases in a dose-dependent fashion (BOIKE et al. 1989). In order to exert their cytotoxicity, these compounds must be converted enzymatically to the 5′-monophosphate nucleotides (F-UMP and F-dUMP). Their cellular metabolism and mechanisms of action have been described (PINENDO and PETERS 1988; PARKER and CHENG 1990). In brief, these drugs act in the following ways: (a) inhibition of thymidylate synthetase (TS), the enzyme that catalyzes the conversion of uridylate to thymidylate (SANTI et al. 1974; DANENBERG and LOCKSHIN 1981), and to a lesser extent thymidine kinase, the enzyme in the salvage pathway (NORD and MARTIN 1991; TSUKAMOTO and KOJO 1991; STOLFI et al. 1992; BAGRIJ et al. 1993), resulting in depletion of cellular thymidylate pool and cessation of DNA synthesis; (b) incorporation of F-UTP into RNA, causing miscoding during translation and production of nonfunctional proteins (NAKADA and MAGASANIK 1964; ROSEN et al. 1969; DOLNICK and PINK 1983, 1985; AKAZAWA et al. 1986; DOONG and DOLNICK 1988); and/or (c) incorporation of F-dUTP into DNA, leading to strand breaks and damages (MAJOR et al. 1982; CHENG and NAKAYAMA 1983; SCHUETZ et al. 1984). The relative contribution of each of these mechanisms in determining fluoropyrimidine cytotoxicity (and developmental toxicity) remains uncertain and may vary among tumor types and cell lines (Valeriote and Santelli 1984; Takimoto et al. 1993), although TS inhibition generally requires lower drug concentrations than those necessary for direct RNA and DNA effects.

In rat embryos, recent data from our laboratory have shown that TS activity is highly sensitive to 5-FU. Strong enzyme inhibition was produced at a dose as low as 5 mg/kg given to the pregnant dam. Maximal enzyme inhibition (> 75%) was reached between 20 and 40 mg/kg, and this was achieved rapidly within 2-4 h after treatment. Consistent with the depression of TS activity, the dTTP pool in the embryo was greatly depleted (by up to 60%; Ritter et al. 1980; observation in our laboratory), although significant changes were observed only when the rats received 20 mg/kg or higher doses of 5-FU. Interestingly, the dGTP pool was also reduced dramatically, by as much as 50%, while a concomitant four- to five-fold elevation of the dCTP pool was observed, perhaps resulting from feedback disinhibition of the ribonucleotide reductase system. Incorporation of labeled 5-FU and FUdR into mouse embryonic RNA and, to a lesser extent, DNA has been reported previously (Dagg et al. 1966; Ohmori 1972). However, results from our recent study with rat embryos indicated that the frequency of 5-FU nucleotide substitution for the natural precursors in embryonic RNA and DNA was in fact rather low, being estimated at 0.1% for RNA and 0.04% for DNA (compared to 2% substitution in total RNA of human KB cells where miscoding of mRNA was observed; Dolnick and Pink 1985). Furthermore, examination of rat embryonic proteins by two-dimensional gel electrophoresis after exposure to 5-FU did not reveal any significant shift in the protein migration pattern, suggesting that direct incorporation of fluoronucleotides into nucleic acids perhaps plays only a minor role in this agent's embryotoxic actions. Regardless of its precise mechanism of action, 5-FU markedly interfered with DNA synthesis in the mouse embryos, leading to cell death, fetal malformation, and lethality (Skalko and Jacobs 1978). Similar antiproliferative effects of 5-FU in rat embryos have also been reported (Lau et al. 1992; Abbott et al. 1993; Shuey et al. 1994a,b); DNA and protein levels in embryonic palate, limb buds, and liver as well as overall fetal weight gains were depressed in a dose-dependent manner. In addition to its direct actions on embryonic cell cycle progression (Elstein et al. 1993a,b), 5-FU also impedes fetal growth by interfering with erythropoiesis; the resultant anemic conditions thus may deprive the fetus of proper oxygenation and nutrients (Shuey et al. 1994b; Zucker et al. 1995).

In laboratory animal models, including rat, mouse, hamster, and monkey, numerous studies have shown fluorophyrimidines to be teratogenic (Karnofsky et al. 1958; Dagg 1960; Chaube and Murphy 1968; Wilson 1971; Shah and MacKay 1978; Shuey et al. 1994a). The most commonly observed abnormalities are cleft palate and hindlimb and tail defects. A comprehensive evaluation of 5-FU-induced external and skeletal malformations was conducted recently by Shuey and coworkers (1994a), who noted that the incidence and severity of these anomalies increased in a dose-dependent fashion; additionally, there was a significant association of fetal weight deficits and malformation status. These findings were consistent with those reported by GRAFTON and coworkers (1987) using an in vitro cultured embryo model. Detailed morphological characterization of the palatal and hindlimb defects produced by 5-FU has been provided by SHAH and colleagues (SHAH and BURDETT 1978; SHAH and MACKAY 1978; SHAH and WONG 1980; SHAH et al. 1984, 1989; BEN-KHAIAL and SHAH 1994), KATAGIRI (1983), ABBOTT et al. (1993), and SHUEY et al. (1994a,c).

A single clinical case of multiple congenital anomalies in a fetus exposed to 5-FU during the first trimester was reported by STEPHENS et al. (1980). The lesions described by these investigators (bilateral radial aplasia and absent thumb) were generally consistent with those seen in the animal models. However, as the authors themselves pointed out, their case most likely involved a basic genetic abnormality, and 5-FU may have exacerbated the ongoing maldevelopment of some structures. Indeed, in other cases involving intrauterine exposure to 5-FU, neither morbidity, mortality, nor any significant ill effects were noted in the newborns (STADLER and KNOWLES 1971; ODOM et al. 1990; LE et al. 1991).

If 5-FU-induced cytotoxicity is caused by depletion of thymidylate, the question arises of whether replenishment of the nucleotide by exogenous administration of natural pyrimidines might reverse the toxicity. Differing results from several studies on this issue have been reported. SANTELLI and VALERIOTE (1980) described potentiative effects of 5-FU cytotoxicity against leukemia by purines and pyrimidines, and NORD et al. (1992) showed that uridine co-treatment was correlated with an enhanced incorporation of 5-FU into cellular RNA; however, BAGRIJ et al. (1993) and KRALOVANSZKY et al. (1993) recently reported that uridine provided protection against gastrointestinal toxicity caused by 5-FU. At present, it is not clear whether these disagreements reflect differences in cell types or other unknown factors. In a teratological study, DAGG and KALLIO (1962) coadministered fluorodeoxyuridine with various doses of thymidine and reported mixed ameliorative effects depending on the structure examined and susceptibility of the structure to FUdR alone. Hence the frequency of tail abnormalities, which was highly sensitive to FUdR, was effectively decreased by thymidine at all doses. The hindlimb, which is less sensitive to FUdR than the tail, was protected by thymidine at low doses; however, the incidence of malformation in this structure actually increased when high doses of thymidine were given along with FUdR. In contrast, no ameliorative effect of thymidine was seen in the palate, which was least sensitive to FUdR (among the structures examined); instead, thymidine co-treatment readily increased the incidence of cleft palate. WILSON et al. (1969) and SCHUMACHER et al. (1969) evaluated various natural pyrimidines (uracil, uridine, thymine, and thymidine) and reported that all of them potentiated the teratogenic effects of 5-FU. These investigators suggested that the enhanced toxicity might be associated with altered metabolism of 5-FU, as non-specific pyrimidine-degrading enzymes in maternal liver were temporarily saturated by the combined concentrations of 5-FU and the exogenously administered pyrimidines. The net effect was tantamount to increasing the dose of 5-FU or diminished hepatic function. Indeed, when carbon tetrachloride was co-administered with 5-FU to compromise the maternal hepatic function, a striking increase in 5-FU-induced teratogenic effects was noted, indicating that the intact liver served to protect the embryo from 5-FU toxicity by catabolizing the drug. In a subsequent study, FORSTHOEFEL and WILLIAMS (1975) confirmed these findings and further suggested that the teratogenic effects might be mediated by cell death. Results from studies which attempt to use pyrimidine replenishment to reverse the 5-FU-induced embryotoxicity are therefore rather difficult to interpret. They may represent the net effect of direct ameliorative action at the cellular level and indirect enhancement of toxicity by altering the pharmacokinetics of 5-FU. The final outcomes may also depend on specific cell types and on how the tissue structure as a whole might have been affected by the 5-FU-evoked cell cycle disruption. In this regard, in vitro studies using organ culture (such as palate and limb bud) may offer a simpler and more direct approach to addressing this particular research question.

II. Other Halogenated Pyrimidines

Although primarily employed as experimental tools to investigate developmental processes, other halogenated pyrimidines such as 5-bromodeoxyuridine (BUdR), iododeoxyuridine, and chlorodeoxyuridine have also been studied in a variety of biological systems (RUFFOLO and FERM 1965; SKALKO et al. 1971; SKALKO and PACKARD 1973, 1975; PERCY 1975; SHAH and MCKAY 1978; BANNIGAN and LANGMAN 1979; SCOTT 1981; WISE and SCOTT 1982; FRANZ and KLEINEBRECHT 1982; NAKASHIMA et al. 1984; BANNIGAN 1985; BANNIGAN et al. 1990). The teratogenic effects of BUdR, in particular, have been extensively evaluated. Depending on the stage of embryonic development at which BUdR was given, polydactyly, neural tube defects, and cleft palate have been observed. In the case of polydactyly, occurrence of normal cell necrosis in a mesodermal zone which had a role in determining digital number was prevented (SCOTT 1981; WISE and SCOTT 1982). In contrast, BUdR-induced exencephaly involved elimination of a critical number of neuroepithelial cells prior to neural fold fusion (NAKASHIMA et al. 1984; BANNIGAN 1985). For induction of cleft palate, BANNIGAN and coworkers (1990) suggested that BUdR interfered with formation of the anterior end of Meckel's cartilage, initiating a chain of events leading to the failure of palatal shelf reorientation. The primary mechanisms of embryotoxic action of these halogenated pyrimidines appear to involve direct incorporation into DNA (SKALKO and PACKARD 1975); indeed, the severity of BUdR toxicity has been shown to be directly related to the proliferative rates of embryonic tissues (BANNIGAN et al. 1981).

III. Cytosine Arabinoside

Cytosine arabinoside (Ara-C, cytarabine) is an important antimetabolite used in the therapy of acute myelocytic leukemia and breast cancer. The deoxycytidine analogue is converted to Ara-CMP, Ara-CDP, and Ara-CTP by the appropriate nucleotide kinases. Ara-CTP is then incorporated into newly synthesized DNA instead of dCTP, causing abrupt interruption of DNA synthesis in many cells. The precise mechanism of Ara-C cytotoxicity is still not well understood. Previously, Ara-C was thought to be a competitive inhibitor of DNA polymerase. However, studies now indicate that inhibition of DNA synthesis in mammalian cells occurs at Ara-CTP concentrations 1/100 or less of those required for inhibition of DNA polymerase. Recent work by Ross et al. (1992) with HL-60 human leukemic cells showed that Ara-C caused the formation of small fragments of nascent DNA of less than 100 base pairs and a loss of the normally preponderant larger fragments of elongating DNA. These data thus lend support to the hypothesis that Ara-C does not act on the initiation of DNA synthesis, but rather interferes with the elongation of DNA strands (Woodcock 1987). Additionally, increasing evidence indicates that various transcription factors and apoptosis (programmed cell death) can be activated by Ara-C and that these processes may be involved in the common pathways of Ara-C cytotoxicity (Bhalla et al. 1992; Brach et al. 1992a,b; Datta et al. 1992). In mouse embryo cells exposed to Ara-C, the pools of dATP and dTTP were expanded threefold, that of dGTP was transiently elevated, and that of dCTP was reduced by more than 50% (Skoog and Nordenskjold 1971). Concomitantly, DNA synthesis was greatly reduced. Removal of Ara-C from the incubation medium normalized the nucleotide pools and slowly restored DNA synthesis.

The teratogenic potential of Ara-C has been evaluated in several studies with laboratory rodent models. In general, Ara-C acutely inhibits DNA synthesis in embryos, retards embryonic and fetal growth, and induces malformations. The adverse outcomes are dose dependent (Ritter et al. 1971; Ortega et al. 1991), with resorption commonly observed at high doses, and are related to the stage of organogenesis at which the embryos were exposed. Thus fused ribs and heart defects were seen in rats given Ara-C on day 10 of gestation (Ritter 1984), cleft palate and lips in mice of the same age (Marcickiewicz et al. 1986), while digital anomalies and kinky tails were found in rats and mice exposed at later gestational stages (Ritter et al. 1971; Scott et al. 1975; Goto and Endo 1987). Even within a particular defect, the pattern of limb bone deficiencies was specific for each developmental stage at which Ara-C was administered; the site of defect moved distally along the limb as development of the structure advanced (Kochhar et al. 1978). It is of interest to note that a preponderance of oligodactyly was found in the forelimbs and polydactyly in the hindlimbs of mice given Ara-C (Endo et al. 1987), although Scott et al. (1975) reported that polydactyly could be produced in both limbs of rats when the antimetabolite was administered at the appropriate times of development. Several investigators (RITTER et al. 1973; SCOTT et al. 1975; KOCHHAR et al. 1978; NAGOMI and OOHIRA 1980) further reported that the limb defects were associated with changes or absence of synchronized proliferation between mesodermal and ectodermal cells in various regions of the limb bud (perhaps due to differential susceptibility of these cells to Ara-C), as well as altered patterns of cell death in the preaxial ectoderm. Abnormalities in the brain, including segmental cerebellar hypoplasia, microcephalus, hydrocephalus, and dilation of lateral ventricles, retinal dysplasia, and focal microcytic renal cortical dysplasia leading to dilated tubules, have also been described in rats and mice treated with Ara-C during late gestation or postnatally (ADLARD et al. 1975; PERCY 1975; KASUBUCHI et al. 1977; OHNO 1984), while the central nervous system effects were further correlated with deficits in behavioral development (GRAY et al. 1986). Unlike 5-FU, the teratogenic effects of Ara-C were completely eliminated when large doses of deoxycytidine were administered simultaneously (CHAUBE et al. 1968; KOCHHAR et al. 1978). Indeed, deoxycytidine was also found to prevent cell death in the limbs of Ara-C-treated embryos.

A clinical case study describing major congenital abnormalities in a baby born to an Ara-C-treated mother was reported by WAGNER and colleagues (1980). Cytarabine was given to a woman to maintain remission from acute lymphocytic leukemia when she became pregnant. A baby was born with ear deformities and limb defects which included a lobster claw deformity with missing digits in the arm, shortened and bowed femur, each leg containing a single bone, and each foot composed of an os calcis and only two lateral metatarsals. In another case, cytarabine and thioguanine (a purine antimetabolite, see below) were given to the same mother during the first trimester of two separate pregnancies; congenital malformations were observed in only one of the babies (SCHAFER 1981). Similar to the previous report, distal limb defects which included absent medial digits of both feet and missing distal phalanges of both thumbs were described. These findings are thus comparable with those observed in animal models and suggest a teratogenic potential of Ara-C in humans.

IV. Azauridine

Azaribine, the prodrug form of azauridine, is effective in treating psoriasis, mycosis fungoides, and polycythemia vera. 6-Azauridine interrupts de novo synthesis of the pyrimidine base by inhibiting orotidine 5-phosphate decarboxylase (HANDSHUMACHER et al. 1962; SKODA 1975). Embryotoxic and teratological effects of azauridine have been described in rats (SAUNDERS et al. 1961; GUTOVA et al. 1971), mice (YOSHIHARA and DAGG 1967; VORHERR and WELCH 1970; DOSTAL and JELINEK 1979), rabbits (SAKSENA and CHAUDHURY 1970), and monkeys (VAN WAGENEN et al. 1970). Embryolethality was produced by high doses and exposure during early stages of pregnancy (days 5–10 for rats, days 21–30 for monkeys). Anatomical defects observed include cleft palate and lip, limb deformities, vertebral abnormalities, and hypoplasia of the caudal part of the trunk. In a clinical study, 6-azauridine was given to pregnant women as an abortifacient; although changes of trophoblast were noted, no spontaneous abortions occurred (VOJTA and JIRASEK 1966).

D. Interruption of Purine Nucleotide Pools

I. 6-Mercaptopurine and 6-Thioguanine

6-Mercaptopurine (6-MP) and 6-thioguanine (6-TG) are two of the oldest antineoplastic agents in current clinical use and are effective in the treatment of several types of leukemia. Despite almost half of a century of active research, the underlying mechanisms of action of these drugs remain to be clearly defined. Interference of de novo purine synthesis through inhibition of various enzymes resulting in depletion of purine nucleotide pools, direct incorporation of 6-MP and 6-TG into either RNA or DNA, disruption of cellular membrane glycoproteins, and induction of cellular differentiation have all been proposed as potential mechanisms of action (for reviews, see LENNARD 1992; BOSTROM and ERDMANN 1993). In fact, the cytotoxic effects of 6-MP and 6-TG are likely multifactorial.

Congenital malformations produced by a single dose of 6-MP in rats have been described by KARNOFSKY (1960), BRAGONIER and CARVER (1968), and KURY et al. (1968). MERKER et al. (1975) and SCOTT et al. (1980) further elaborated on the 6-MP-induced limb defects which involved aberrant ectodermal and mesodermal interactions in the rat limb bud, while SHAH and coworkers (SHAH and BURDETT 1978; BURDETT and SHAH 1988; BURDETT et al. 1988) described the aspects of palatal dysmorphogenesis following exposure to 6-MP in hamsters. Abnormal brain development has also been reported by ARAKAWA et al. (1967) and ADHAMI (1979). In addition, REIMERS and SLUSS (1978) showed that the deleterious effects of 6-MP can persist through second and third generations in mice. Interestingly, in a series of studies, HURLEY and colleagues (HIRSCH and HURLEY 1978; AMEMIYA et al. 1986, 1989) demonstrated that pregnant rats fed diets containing high concentrations of zinc and subsequently exposed to 6-MP had a reduction of drug-induced embryotoxicity. Mineral analysis of maternal and fetal tissues revealed pronounced effects of 6-MP on metabolism of zinc, copper, iron, calcium, and magnesium. These investigators postulated that the 6-MP teratogenesis might be associated with altered disposition and metabolism of minerals due to induction of maternal metallothionein synthesis, and high levels of dietary zinc might be able to ameliorate some of the deleterious effects of the drug.

Developmental toxicity of azathioprine, an immunosuppressive agent which is chemically related to 6-MP, has also been assessed. While teratogenic effects of this drug were reported in mice (NEUBERT et al. 1977), only a high incidence of embryolethality and fetal growth retardation were observed in rats (J.R. SCOTT 1977), and standard doses of azathioprine in the treatment of rheumatic diseases have not been found to increase the risk of congenital anomalies in humans (OSTENSEN 1992). On the other hand, ribavirin, an antiviral agent which acts by inhibiting inosine monophosphate dehydrogenase and blocks the biosynthesis of guanine nucleotides, has been reported to be teratogenic in mice (KOCHHAR et al. 1980).

II. Deoxycoformycin and Chlorodeoxyadenosine

Deoxycoformycin (pentostatin) is most effective in the treatment of hairy cell leukemia, although its potential uses against immunological and cardiovascular diseases are under active exploration (for a review, see KLOHS and KRAKER 1992). Pentostatin acts by inhibiting the enzyme adenosine deaminase (ADA), which converts adenosine to inosine and 2′-deoxyadenosine to 2′-deoxyinosine, thereby blocking the purine salvage metabolic pathways. ADA is expressed at low levels in most mammalian tissues, but is highly expressed in some differentiating tissues, including the uteroplacental unit (ARONOW et al. 1989; KNUDSEN et al. 1988, 1991; CHINSKY et al. 1990). The role of ADA at this maternal–embryonal interface remains to be determined. In view of this preferential expression of ADA in embryonic and differentiating tissues, KNUDSEN and coworkers (1989, 1992; AIRHART et al. 1993) examined the consequence of inhibiting this enzyme by pentostatin during various stages of embryonic development. In response to inhibition of ADA by pentostatin, adenosine and deoxyadenosine levels in the mouse embryos were elevated dramatically and abruptly within 30 min in a dose-dependent manner, ranging from a five-fold (0.05 mg/kg) to a 40-fold (5 mg/kg) increase for adenosine, and a 200-fold (0.5 mg/kg) to 600-fold (5 mg/kg) increase for deoxyadenosine. Indeed, a general correlation was indicated between embryolethality and the length of adenine nucleoside pool expansion. In addition, after a single treatment (5 mg/kg) of pentostatin on gestational day 7, severe abnormalities in neural tube closure and in craniofacial and limb development were detected which were followed by embryonic death and resorption by gestational day 12. Although the specific cause of these effects is not clear, results from a recent study suggested that the expanded pool of dATP might activate the apoptotic pathways and alter the patterns of cell death in the embryos (GAO et al. 1994).

2-Chloro-2′-deoxyadenosine (2-CdA) is a purine nucleoside analogue that is resistant to ADA and is used clinically for the treatment of low-grade lymphomas (KAY et al. 1992). It is readily phosphorylated within cells and causes rapid inhibition of DNA synthesis. Previously, this effect was shown to be related to inhibition of ribonucleotide reductase, leading to a depletion of intracellular pools of dNTP (PLUNKETT and SAUNDERS 1991). However, recent studies have indicated that inhibition of DNA synthesis occurs more rapidly than the decline of dNTP pools (CHUNDURU et al. 1993). Rather, inhibition of

DNA chain extension by incorporation of 2-CdATP into DNA has been suggested as an alternative mechanism. In addition to its direct DNA-inhibitory effects, 2-CdA has been associated with induction of apoptosis and programmed cell death (CARSON et al. 1992; ROBERTSON et al. 1993). In light of these findings, WUBAH et al. (1996) evaluated the embryotoxicity of 2-CdA in mice; preliminary data indicate that this agent is embryotoxic and teratogenic, and its actions may be mediated by excessive cell death. Indeed, recent reports which linked apoptosis to teratogen-induced cell death further lend credence to this possibility (CHEN et al. 1994; GAO et al. 1994).

III. Hydroxyurea

The primary role of hydroxyurea (HU) in chemotherapy is in the management of myeloproliferative disorders. This structurally simple compound is well absorbed and diffused into cells, where it quenches the tyrosyl free radical at the active site of the M_2 protein subunit of ribonucleotide reductase, inactivating the enzyme (YARBRO 1992). Because ribonucleotide reductase constitutes a focal point of control for dNTP synthesis, it is hardly surprising that inhibition of this enzyme upsets the intracellular dNTP pools. Indeed, SKOOG and NORDENSKJOLD (1971) reported that addition of HU to mouse embryo cells led to an almost instantaneous decrease in the dGTP pool (to less than 10% of basal level) and a diminished size of the dATP pool (to less than 25%); in contrast, the dTTP pool was expanded two-fold, while that of dCTP was relatively unaltered. Corresponding to the imbalanced nucleotide pools, DNA synthesis was inhibited, producing cell death in S phase of the cell cycle and synchronization of the fraction of cells that survive. Repair of DNA damaged by other chemicals or irradiation was also impeded by HU. Additionally, inhibition of RNA synthesis has been suggested as a possible mode of action for HU (KROWKE and BOCHERT 1975).

Embryotoxicity and teratogenicity of HU have been reported in several animal species including rats, mice, rabbits, cats, and monkeys (CHAUBE and MURPHY 1973; WILSON et al. 1975; DESESSO and JORDAN 1977; HERKEN et al. 1978; KHERA 1979). Cleft palate, hindlimb abnormalities, and necrosis in the brain and spinal cord are the major defects described. MILLICOVSKY and DESESSO (1980, 1981) reported a plethora of cardiovascular abnormalities in rabbit embryos, but these effects were likely related to a maternally mediated mechanism involving uterine ischemia. Amelioration of HU-induced congenital malformations has drawn considerable interest from several laboratories. While HU has been shown to elevate the intracellular pools of pyrimidines (SKOOG and NORDENSKJOLD 1971), administration of dCMP was reported to protect the embryos from HU toxicity (CHAUBE and MURPHY 1973; HERKEN 1984). An explanation for these paradoxical findings is presently not available, although HERKEN (1984) suggested that regenerative processes directed by dCMP might be involved in the protective effect. DESESSO et al. (1994) and DESESSO and GOERINGER (1990a,b) postulated that the rapid embryonic cell death induced by HU might be mediated through formation of hydrogen peroxide and the extremely reactive hydroxyl free radical; thus administration of free radical scavengers should reduce the developmental toxicity of HU. Indeed, antioxidants such as propyl gallate, ethoxyquin, and nordihydroguaiaretic acid, as well as D-mannitol have been shown to be effective in reducing the embryotoxic effects of HU.

IV. Methotrexate

The antifolate methotrexate (MTX) is an effective drug in the management of a variety of neoplastic diseases and psoriasis. It is also widely used as an immunosuppressive agent against host reactions resulting from marrow transplantation as well as in the therapy of rheumatoid arthritis. To understand the mechanism of action of MTX, one must appreciate the complexities of folate metabolism and their multiplicity of function within the cell. The topics of folate metabolism and antifolate actions have been reviewed extensively (JOHNSON and CHEPENIK 1981; OLSEN 1991; Chap. 15, this volume), and discussion here will be brief. Folate functions as a cofactor in a variety of one-carbon transfer reactions, including those involved in the synthesis of amino acids, purines, and thymidylate. To be active, dihydrofolate must be first reduced to tetrahydrofolate by the enzyme dihydrofolate reductase (DHFR). MTX is a structural analogue of folic acid; it binds tightly, but reversibly, to DHFR and inhibits the enzyme activity, thereby depleting the intracellular reserves of reduced folate cofactor and leading to alterations of nucleotide pools and interruption of DNA synthesis. Although it is well established that inhibition of folate-dependent enzymes contributes to the cytotoxic effects of MTX, the precise mechanism or mechanisms underlying these cellular effects are still under active and intense investigation. Several laboratories have shown that the level of reduced folates in the cell is only partially depleted (50%–60%), and the responses of reduced folates to DHFR inhibition by MTX vary considerably among different cell types (RHEE et al. 1992). In addition, while inhibition of de novo synthesis of purines and thymidylate is a major mechanism for MTX, depletion of nucleotides alone may not account for all of its cytotoxic actions. In a recent study, KWOK and TATTERSALL (1992) reported that hypoxanthine potentiated the effect of MTX in producing DNA fragmentation in murine leukemic cells, an event accompanied by a significant increase of intracellular levels of dATP. These authors postulated that the resultant elevation of dATP pools during DNA synthesis served as a signal for the cells to commit themselves to the apoptotic pathway and programmed cell death. Furthermore, a direct inhibitory effect of MTX on thymidine kinase has been indicated (ABONYI et al. 1992), suggesting yet another possible site of action.

The adverse consequence of folate deficiency during embryonic and fetal development has long been recognized (NELSON 1960; WARKANY 1978, 1986; JOHNSON and CHEPENIK 1981; see also Chap. 15, this volume). Numerous studies with animal models (rats, mice, cats, rabbits, monkeys) have illustrated the embryotoxic and teratogenic effects of MTX (SKALKO and GOLD 1974; KHERA 1976; JORDAN et al. 1977; WILSON et al. 1979; DARAB et al. 1987). Consistent with other pharmacologic agents of this class, the types of malformations seen varied with the time of treatment during gestation; these include cleft palate and lip, hydrocephaly, microphalmia, vertebral anomalies, and limb and digital defects. While the teratologic profile of MTX is not unique among the other drugs discussed thus far, its embryotoxicity in humans is perhaps the most profound and best established. In fact, MTX is still in use to date as an abortifacient in clinics (STOVALL et al. 1990; CREININ 1993). The teratogenic effects of the antifolate aminopterin (a structural analogue of MTX) were reported as early as 1952 by THIERSCH in unsuccessfully attempted clinical abortion. Subsequently, these findings were confirmed by numerous case reports and have been summarized periodically in reviews (WARKANY 1978, 1986; FELDKAMP and CAREY 1993). By and large, the defects observed, which include cleft palate and lip and cranial and limb malformations, are consistent with those noted in animal models.

In cancer treatment, the efficacy of folate antagonists was greatly enhanced when leucovorin (folinic acid) was found to be effective in the "rescue" of host cells from toxicity. In a similar fashion, DESESSO and GOERINGER (1991, 1992) recently noted that the developmental toxicity of MTX can also be decreased by this compound and its functional analogue. These results thus confirm that impaired one-carbon metabolism is indeed the process underlying MTX developmental toxicity and underscores the potential usefulness of folinic acid in antifolate therapy during pregnancy.

E. Conclusion

A review of the literature thus reveals a rich history of research on developmental biology and toxicology with antimetabolites, and these investigations continue to be actively pursued. While much exploration undoubtedly still lies ahead, the molecular mechanisms of action of the majority of these drugs are now better characterized. With various means of enzyme inhibition to upset the balance of dNTP pools, erroneous incorporation into nucleic acids, and interference with DNA chain elongation and repairs, these compounds invariably bring DNA synthesis to a halt and produce untimely cell death. At high doses and during early stages of gestation, it is hardly surprising that these embryotoxic effects translate into embryolethality and resorption. In this regard, the adverse developmental outcomes of these agents resemble those of X-ray irradiation. At lower doses, a common profile of major malformations can be identified across the board with this class of agents, including cleft palate, various degrees of limb defects, and some forms of cranial abnormalities. The preponderance of these three types of malformation might have resulted from a single exposure or a brief period of exposure during gestation when ontogeny of these structures was most active and susceptible to perturbation. However, longer treatment periods in some studies and clinical reports have indicated a very similar profile of defects. On the other hand, these defects represent three of the most commonly observed frank terata, and the high incidence associated with antimetabolite treatment may simply reflect the particular vulnerability of these embryonic structures.

Based on cell culture studies, the specificity of various antimetabolites is typically conferred at the stage of cell cycle (primarily S phase) rather than among cell types. Hence, unless the cells develop resistance to a particular agent (an intriguing process presently under intense investigation in cancer research), the cytotoxic effects are quite uniform. However, a much more confounding picture is produced by these chemical insults within the embryonic structures. While cell death is still a prevalent feature of these perturbations, some cells within an embryonic tissue may die, but the neighboring cells survive (W.J. Scott 1977). The embryonic heart cells in fact appear to be largely resistant to the cytotoxicity of various drugs (Mirkes and Greenway 1985; Mirkes et al. 1991; Knudsen and Ibrahim 1995). The underlying mechanism for this tissue-specific cytotoxicity is unclear. In the latter case, the mouse embryonic heart has been shown to lack or display only low levels of p53 gene expression (Schmid et al. 1991). Because p53 is known to be involved in the cell death program, differences in the rates of its gene expression may preferentially confer resistance of the embryonic heart to chemical insults. Alternatively, the malformations produced by antimetabolites may reflect the vulnerability of these embryonic structures to the absence of a critical cell mass or excessive cell death. Work from Scott's and Shah's laboratories which focused on limb and palate development, respectively, tends to support this contention. For instance, Scott and colleagues (Scott 1981; Scott et al. 1975, 1980) have indicated that uncoordinated proliferation and death between the ectodermal and mesodermal cells might be involved in the expression of limb dysmorphogenesis induced by various antimetabolites. On the other hand, the mechanistic linkage between the anti-proliferative effect of teratogens and the pathogenesis of anatomical defects is far from being simple and straightforward. As W.J. Scott (1977) pointed out, while it was easy to attribute the absence of a critical cell mass to the cause of a defect, there were examples of agents which reduced cell number but produced tissue-excess deformities such as polydactyly (Scott et al. 1975). Furthermore, Snow and colleagues have demonstrated restorative growth in the embryonic tissues following cytotoxic damage, particularly when this damage occurred during early embryonic stages (Snow and Tam 1979; Snow 1983, 1986). For instance, when mouse embryos were exposed to mitomycin C at presomite stages (days 6.5 or 7 postcoitum), cellular reduction and morphological retardation were initially observed, but the embryos apparently recovered from these deficits within days. Thus reduction of cell population alone is not likely to account for the induction of defects, and other cellular mechanisms must also have participated. These may entail failure of the tissues to respond to injury, altered programs of natural cell death, aberrant cell–cell interactions, and altered gene expression. Some of these possibilities are focal points of discussion in other chapters of this volume (see Chaps. 8–12). Hence, while research in the past several decades has presented us with detailed descriptions of congenital malformations produced by different antimetabolites which act by upsetting the intracellular nucleotide balance, a formidable gap of information still remains concerning the cellular mechanisms of teratogenesis.

One approach to dissecting the anatomy of birth defects is to use these antimetabolites as experimental probes and evaluate the resultant cellular aberrations. Indeed, work from Scott's and Shah's laboratories typifies these efforts and has provided a wealth of information on limb and palate development. More recently, KNUDSEN and coworkers (1995) have begun to use these agents as research tools to explore the molecular and genetic events involved in congenital malformations. Other investigators have taken a different approach and have constructed mathematical models to describe the cellular actions of these antimetabolites based on enzyme kinetics. For instance, MORRISON and ALLEGRA (1989) formulated a mathematical model of folate cycle in human breast cancer cells and were able to describe the inhibitory effects of MTX on purine and pyrimidine biosynthesis. In his recent volume, JACKSON (1992) provided an excellent discussion on computer simulation models of multienzyme systems in describing cytokinetics, pharmacokinetics of anticancer drugs, their inhibition kinetics, and the value of these models in designing and predicting their therapeutic uses. In an ongoing study, Setzer and coworkers in our laboratories have adopted these computer models for the prediction of 5-FU toxicity (R.W. SETZER, personal communication). SHUEY et al. (1994a) have taken these efforts further and provided a biologically based dose–response model for 5-FU developmental toxicity. These investigators evaluated the critical biological events which might intervene between the administration of 5-FU and the expression of adverse developmental outcomes. This model is illustrated in Fig. 3. Mathematical relationships between the dose–response profiles of these events were constructed (Fig. 4) and integrated into an empirical model to assess the risk potentials of 5-FU in producing limb defects. The predicted risk index for 5-FU based on this model was in reasonable agreement with the experimental data (Fig. 5). These studies thus may spearhead the next phase of efforts in search of a better understanding of congenital birth defects.

Concern about the use of these antimetabolites in the management of cancer during pregnancy has been addressed by several reviews (BLATT et al. 1980; DOLL et al. 1988; CALIGIURI and MAYER 1989; FELDKAMP and CAREY 1993). Isolated case reports of congenital malformations associated with in utero exposure to this class of drugs are available and have been described above, although the value of these individual case reports in defining the teratogenic properties of a drug remains controversial. Detection of potential teratogenicity of chemotherapeutics is also complicated by the facts that few patients are pregnant during treatment, and those who conceive may choose to

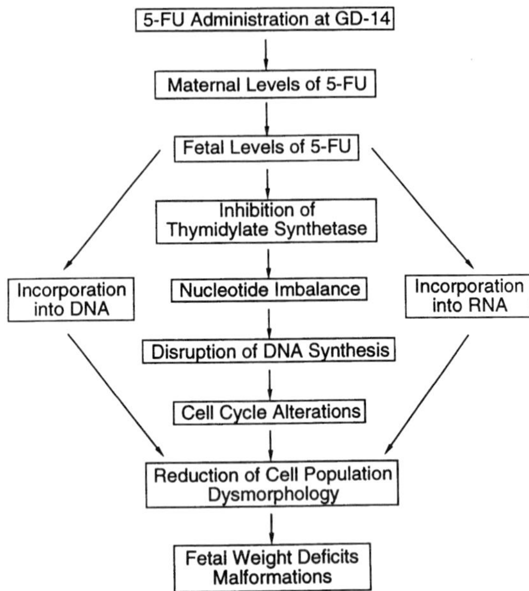

Fig. 3. Proposed critical events in the biologically based dose–response model for developmental toxicity of 5-fluorouracil (*5-FU*). *GD*, gestational day

have an abortion. Nevertheless, a survey of clinical literature that focuses primarily on anatomical abnormalities suggests that, with the possible exceptions of MTX and aminopterin, treatment with antimetabolites prior to conception or after the first trimester may not be as hazardous as previously anticipated. On the other hand, it must be noted that, while an agent may be used alone with relative safety, the same agent used in combination with others can become teratogenic (BLATT et al. 1980). Moreover, the possibility that prenatal exposure to these antimetabolites may alter development of embryonic and fetal physiological functions remains largely unexplored. Indeed, recent work by SHUEY et al. (1994b) and ZUCKER et al. (1995) which reported fetal anemia following maternal exposure to 5-FU illustrates this particular need for future research.

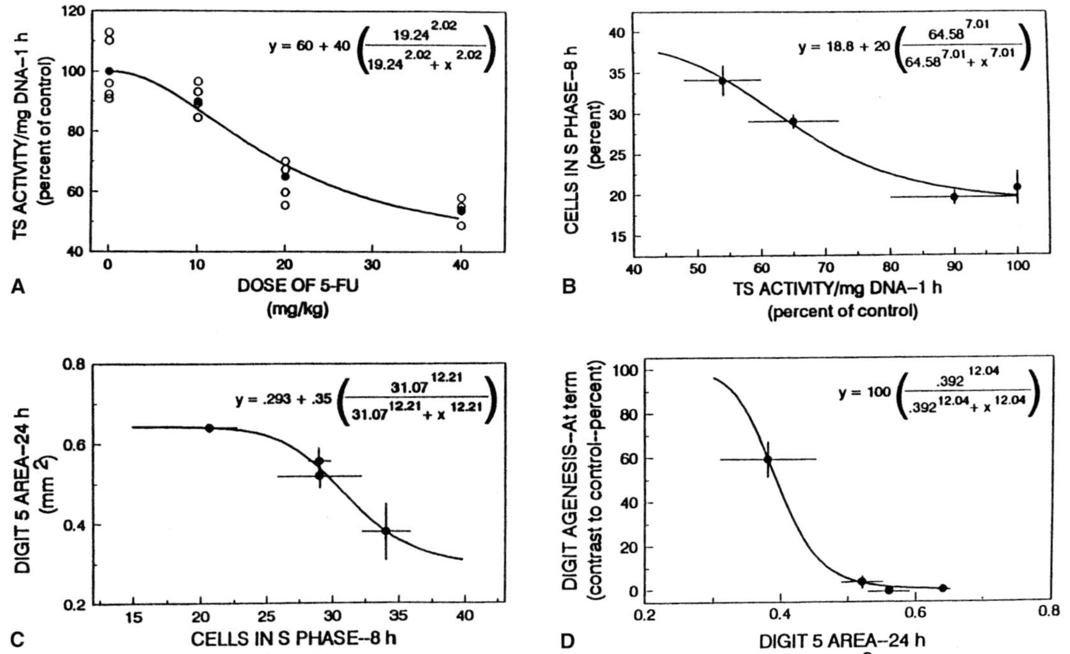

Fig. 4A–D. Mathematical relationships between successive biochemical and cellular events involved in the expression of hindlimb defects after exposure to 5-fluorouracil (5-FU). **A** Administered dose – thymidylate synthetase (*TS*) inhibition. **B** TS inhibition – altered cell cycle. **C** Altered cell cycle – growth reduction and dysmorphogenesis. **D** Growth reduction and dysmorphogenesis – malformations. (For a detailed description and discussion, see SHUEY et al. 1994a)

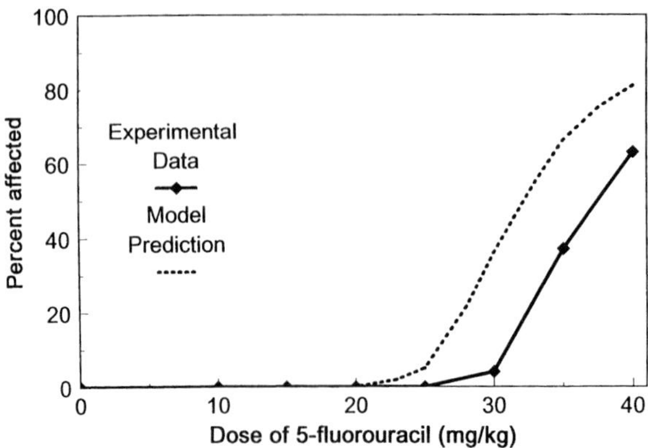

Fig. 5. Comparison between an integrated empirical model for the developmental toxicity of 5-fluorouracil based on inhibition of thymidine synthetase and actual experimental findings. (Modified from Shuey et al. 1994a)

References

Abbott BD, Lau C, Buckalew AR, Logsdon TR, Setzer W, Zucker RM, Elstein KH, Kavlock RJ (1993) Effects of 5-fluorouracil on embryonic rat palate in vitro: fusion in the absence of proliferation. Teratology 47: 541–554

Abonyi M, Prajda N, Hata Y, Nakamura H, Weber G (1992) Methotrexate decreases thymidine kinase activity. Biochem Biophys Res Commun 187: 522–528

Adhami H (1979) Influence of 6-mercaptopurine on the prenatal development of the rat cortex. Z Mikrosk Anat Forsch 93: 21–32

Adlard BPF, Dobbing J, Sands J (1975) A comparison of the effects of cytosine arabinoside and adenine arabinoside on some aspects of brain growth and development in the rat. Br J Pharmacol 54: 33–39

Akazawa S, Kumai R, Yoshida K, Ayusawa D, Shimizu K, Seno T (1986) The cytotoxicity of 5-fluorouracil is due to its incorporation into RNA and not its inhibition of thymidylate synthase as evidenced by the use of a mouse cell mutant deficient in thymidylate synthase. Jpn J Cancer Res 77: 620–624

Airhart MJ, Robbins CM, Knudsen TB, Church JK, Skalko RG (1993) Occurrence of embryotoxicity in mouse embryos following in utero exposure to 2′-deoxycoformycin (pentostatin). Teratology 47: 17–27

Amemiya K, Keen CL, Hurley LS (1986) 6-Mercaptopurine-induced alterations in mineral metabolism and teratogenesis in the rat. Teratology 34: 321–334

Amemiya K, Hurely LS, Keen CL (1989) Effect of 6-mercaptopurine on ^{65}Zn distribution in the pregnant rat. Teratology 39: 387–393

Arakawa T, Fujii M, Hayashi T (1967) Dilation of cerebral ventricles of rat offsprings induced by 6-mercaptopurine administration to dams. Tohoku J Exp Med 91: 143–148

Arezzo F (1987) Determination of ribonucleoside triphosphates and deoxyribonucleoside triphosphate in Novikoff hepatoma cells by high-performance liquid chromatography. Anal Biochem 160: 57–64

Aronow B, Lattier D, Silbiger R, Dusing M, Hutton J, Jones G, Stock J, McNeish J, Potter S, Witte D, Wiginton D (1989) Evidence for a complex regulatory array in the first intron of the human adenosine deaminase gene. Genes Dev 3: 1384–1400

Ashman CR, Davidson RL (1981) Bromodeoxyuridine mutagenesis in mammalian cells is related to deoxyribonucleotide pool imbalance. Mol Cell Biol 1: 254–260

Bagrij T, Kralovanszky J, Gyergyay F, Kiss E, Peters GJ (1993) Influences of uridine treatment in mice on the protection of gastrointestinal toxicity caused by 5-fluorouracil. Anticancer Res 13: 789–793

Bannigan JG (1985) The effects of 5-bromodeoxyuridine on fusion of the cranial neural folds in the mouse embryos. Teratology 32: 229–239

Bannigan JG, Langman J (1979) The cellular effect of 5-bromodeoxyuridine on the mammalian embryo. J Embryol Exp Morphol 36: 623–638

Bannigan JG, Langman J, van Breda A (1981) The uptake of 5-bromodeoxyuridine by the chicken embryo and its effects upon growth. Anat Embryol Bere 162: 425–434

Bannigan JG, Cottell DC, Morris A (1990) Study of the mechanisms of BUdR-induced cleft palate in the mouse. Teratology 42: 79–89

Ben-Khaial GS, Shah RM (1994) Effects of 5-fluorouracil on collagen synthesis in the developing palate of hamster. Anticancer Drugs 5: 99–104

Bhalla K, Tang C, Ibrado AM, Grant S, Tourkina E, Holladay C, Hughes M, Mahoney ME, Huang Y (1992) Granulocyte-macrophage colony-stimulating factor/interleukin-3 fusion protein (pIXY 321) enhances high dose ara-C-induced programmed cell death or apoptosis in human myeloid leukemia cells. Blood 80: 2883–2890

Blatt J, Mulvihill JJ, Ziegler JL, Young RC, Poplack DG (1980) Pregnancy outcome following cancer chemotherapy. Am J Med 69: 828–832

Boike GM, Deppe G, Young JD, Malone JM, Malviya VK, Sokol RJ (1989) Chemotherapy in a pregnant rat model. 2. 5-Fluorouracil: nonlinear kinetics and placental transfer. Gynecol Oncol 34: 191–194

Bostrom B, Erdmann G (1993) Cellular pharmacology of 6-mercaptopurine in acute lymphoblastic leukemia. Am J Pediatr Hematol Oncol 15: 80–86

Brach MA, Hermann F, Kufe DW (1992a) Activation of the AP-1 transcription factor by arabinofuranosylcytosine in myeloid leukemia cell. Blood 79: 728–734

Brach MA, Kharbanda SM, Hermann F, Kufe NW (1992b) Activation of the transcription factor kB in human KG-1 myeloid leukemia cells treated with 1-β-D-arabinofuranosylcytosine. Mol Pharmacol 41: 60–63

Bragonier JR, Carver MJ (1968) The influence of 6-mercaptopurine on rat placenta and fetus. Biochem Pharmacol 17: 1689–1697

Burdett DN, Shah RM (1988) Gross and cellular analysis of 6-mercaptopurine-induced cleft palate in hamster. Am J Anat 181: 179–194

Burdett DN, Waterfield JD, Shah RM (1988) Vertical development of the secondary palate in hamster embryos following exposure to 6-mercaptopurine. Teratology 37: 591–597

Caligiuri MA, Mayer RJ (1989) Pregnancy and leukemia. Semin Oncol 16: 388–396

Carson DA, Watson DB, Esparza LM, Carrera CJ, Kipps TJ, Cottam HB (1992) Oral antilymphocytic activity and induction of apoptosis by 2-chloro-2′-arabino-fluoro-2′-deoxyadenosine. Proc Natl Acad Sci USA 89: 2970–2974

Chaube S, Murphy ML (1968) The teratogenic effects of the recent drugs active in cancer chemotherapy. Adv Teratol 3: 181–237

Chaube S, Murphy ML (1973) Protective effect of deoxycytidylic acid (CdMP) on hydroxyurea-induced malformations in rats. Teratology 7: 79–88

Chaube S, Kreis W, Uchida K, Murphy ML (1968) The teratogenic effect of 1-β-D-arabinofuranosylcytosine in the rat. Protection by deoxycytidine. Biochem Pharmacol 17: 1213–1226

Chen B, Cyr DG, Hales BF (1994) Role of apoptosis in mediating phosphoramide mustard-induced rat embryo malformations in vitro. Teratology 50: 1–12

Cheng YC, Nakayama K (1983) Effects of 5-fluoro-2′-deoxyuridine metabolism in HeLa cells. Mol Pharmacol 23: 171–174

Chinsky JM, Ramamurphy V, Fanslow WC, Ingolia DE, Blackburn MR, Shaffer KT, Higley HR, Trentin JJ, Rudolph FB, Knudsen TB, Kellems RE (1990) Developmental expression of adenosine deaminase in the upper alimentary tract of mice. Differentiation 42: 172–183

Chunduru SK, Appleman JR, Blakley RL (1993) Activity of human DNA polymerases α and β with 2-chloro-2′-deoxyadenosine 5′-triphosphate as a substrate and quantitative effects of incorporation on chain extension. Arch Biochem Biophys 302: 19–30

Cory JG (1987) Unresolved issues in the study of mammalian ribonucleotide reductase. Adv Enzyme Regul 26: 287–299

Creinin MD (1993) Methotrexate for abortion at ≤ 42 days gestation. Contraception 48: 519–525

Cross DR, Miller BJ, James SJ (1993) A simplified HPLC method for simultaneously quantifying ribonucleotides and deoxyribonucleotides in cell extracts or frozen tissues. Cell Proliferation 26: 327–336

Dagg CP (1960) Sensitive stages for the production of developmental abnormalities in mice with 5-fluorouracil. Am J Anat 106: 89–96

Dagg CP, Kallio E (1962) Teratogenic interaction of fluorodeoxyuridine and thymidine. Anat Rec 142: 301–302

Dagg CP, Doerr A, Offutt C (1966) Incorporation of 5-fluorouracil-2-C^{14} by mouse embryos. Biol Neonate 10: 32–46

Danenberg PV, Lockshin A (1981) Florinated pyrimidines as tight-binding inhibitors of thymidylate synthetase. Pharmacol Ther 13: 69–90

Darab DJ, Minkoff R, Sciote J, Sulik KK (1987) Pathogenesis of median facial clefts in mice treated with methotrexate. Teratology 36: 77–86

Das SK, Benditt EP, Loeb LA (1983) Rapid changes in deoxynucleoside triphosphate pools in mammalian cells treated with mutagens. Biochem Biophys Res Commun 114: 458–464

Datta R, Kharbanda S, Kufe DW (1992) Transcriptional and post-transcriptional regulation of H1 histone gene expression by 1-β-D-arabinofuranosylcytosine. Mol Pharmacol 41: 64–68

DeSesso JM, Goeringer GC (1990a) The nature of the embryo-protective interaction of propyl gallate with hydroxyurea. Reprod Toxicol 4: 145–152

DeSesso JM, Goeringer GC (1990b) Ethoxyquin and nordihydroguaiaretic acid reduce hydroxyurea developmental toxicity. Reprod Toxicol 4: 267–275

DeSesso JM, Goeringer GC (1991) Amelioration by leucovorin of methotrexate developmental toxicity in rabbits. Teratology 43: 201–215

DeSesso JM, Goeringer GC (1992) Methotrexate-induced developmental toxicity in rabbits is ameliorated by 1-(*p*-tosyl)-3,4,4-trimethylimidazolidine, a functional analog for tetrahydrofolate-mediated one-carbon transfer. Teratology 45: 271–283

DeSesso JM, Jordan RL (1977) Drug-induced limb dysplaisas in fetal rabbits. Teratology 15: 199–211

DeSesso JM, Scialli AR, Goeringer GC (1994) D-mannitol, a specific hydroxyl free radical scavenger, reduces the developmental toxicity of hydroxyurea in rabbits. Teratology 49: 248–259

Doll DC, Ringenberg QS, Yarbro JW (1988) Management of cancer during pregnancy. Arch Intern Med 148: 2058–2064

Dolnick BJ, Pink JJ (1983) 5-Fluorouracil modulation of dihydrofolate reductase RNA levels in methotrexate-resistant KB cells. J Biol Chem 258: 13299–13306

Dolnick BJ, Pink JJ (1985) Effects of 5-fluorouracil on dihydrofolate reductase and dihydrofolate reductase mRNA from methotrexate-resistant KB cells. J Biol Chem 260: 3006–3014

Doong S-L, Dolnick BJ (1988) 5-Fluorouracil substitution alters pre-mRNA splicing in vitro. J Biol Chem 263: 4467–4473

Dostal M, Jelinek R (1979) Embryotoxicity of transplacentally and intraamniotically administered 6-azauridine in mice. Teratology 19: 143–148

Elstein KH, Zucker RM, Andrews JE, Ebron-McCoy M, Shuey DL, Rogers JM (1993a) Effects of developmental stage and tissue type on embryo/fetal DNA distributions and 5-fluorouracil-induced cell cycle perturbations. Teratology 48: 335–364

Elstein KH, Zucker RM, Shuey DL, Lau C, Chernoff N, Rogers JM (1993b) Utility of murine erythroleukemic cell (MELC) in assessing mechanisms of action of DNA-active developmental toxicants: application to 5-fluorouracil. Teratology 48: 75–87

Endo A, Sakai N, Ohwada K (1987) Analysis of diurnal difference in teratogen (Ara-C) susceptibility in mouse embryos by a progressive phase-shift method. Teratogenesis Carcinog Mutagen 7: 475–482

Eriksson S, Thelander L (1978) Allosteric regulation of calf thymus ribonucleotide reductase. Ciba Found Symp 68: 165–175

Feldkamp M, Carey JC (1993) Clinical teratology counseling and consultation case report: low dose methotrexate exposure in the early weeks of pregnancy. Teratology 47: 533–539

Forsthoefel PF, Williams ML (1975) The effects of 5-fluorouracil and 5-fluorodeoxyuridine used alone and in combination with normal nucleic acid precursors on development of mice in lines selected for low and high expression of Strong's luxoid gene. Teratology 11: 1–20

Franz J, Kleinebrecht J (1982) Teratogenic and clastogenic effects of BUdR in mice. Teratology 26: 195–202

Gao X, Blackburn MR, Knudsen TB (1994) Activation of apoptosis in early mouse embryos by 2′-deoxyadenosine exposure. Teratology 49: 1–12

Garrett C, Santi DV (1979) A rapid and sensitive high pressure liquid chromatography assay for deoxyribonucleoside triphosphates in cell extracts. Anal Biochem 99: 268–273

Goto T, Endo A (1987) Dose- and stage-related sex differences in the incidence of cytosine arabinoside induced digit anomalies in the mouse fetus. Teratology 35: 35–40

Grafton TF, Bazare JJ, Hansen DK, Sheehan DM (1987) The in vitro embryotoxicity of 5-fluorouracil in rat embryos. Teratology 36: 371–377

Gray LE, Kavlock RJ, Ostby J, Ferell J, Rogers J, Gray K (1986) An evaluation of figure-eight maze activity and general behavioral development following prenatal exposure to forty chemicals: effects of cytosine arabinoside, dinocap, nitrofen, and vitamin A. Neurotoxicology 7: 449–462

Grubb RB, Montiegel EC (1975) The teratogenic effects of 6-mercaptopurine on chick embryos in ovo. Teratology 11: 179–185

Gutova M, Elis J, Raskova H (1971) Teratogenic effect of 6-azauridine in rats. Teratology 4: 287–294

Handshumacher RE, Calabresi P, Welch AD, Bono VH, Fallon HJ, Frei E (1962) Summary of current information on 6-azauridine. Cancer Chemother Rep 21: 1–18

Herken R (1984) The influence of deoxycytidine monophosphate (dCMP) on the cytotoxicity of hydroxyurea in the embryonic spinal cord of the mouse. Teratology 30: 83–90

Herken R, Merker HJ, Krowke R (1978) Investigation of the effect of hydroxyurea on the cell cycle and the development of necrosis in the embryonic CNS of mice. Teratology 18: 103–118

Hirota Y, Yoshioka A, Tanaka S, Watanabe K, Otani T, Minowada J, Matsuda A, Ueda T, Wataya Y (1989) Imbalance of deoxyribonucleoside triphosphates, DNA double-strand breaks, and cell death caused by 2-chlorodeoxyadenosine in mouse FM3A cells. Cancer Res 49: 915–919

Hirsch KS, Hurley LS (1978) Relationship of dietary zinc to 6-mercaptopurine teratogenesis and DNA metabolism in the rat. Teratology 17: 303–314

Hunting D, Henderson JF (1981) Determination of deoxyribonucleoside triphosphates using DNA polymerase: a critical evaluation. Can J Biochem 59: 723–727

Hunting D, Henderson JF (1982) Models of the regulation of ribonucleotide reductase and their evaluation in intact mammalian cells. CRC Crit Rev Biochem 13: 325–348

Jackson RC (1992) The theoretical foundations of cancer chemotherapy introduced by computer models. Academic, San Diego

James SJ, Basnakian AG, Miller BJ (1994) In vitro folate deficiency induces deoxynucleotide pool imbalance, apoptosis, and mutagenesis in Chinese hamster ovary cells. Cancer Res 54: 5075–5080

Johnson EM, Chepenik KP (1981) Teratogenicity of folate antagonist. In: Juchau MR (ed) The biochemical basis of chemical teratogenesis. Elsevier/North-Holland, Amsterdam, pp 137–158

Jordan RL, Wilson JG, Schumacher HJ (1977) Embryotoxicity of the folate antagonist methotrexate in rats and rabbits. Teratology 15: 73–80

Karnofsky DA (1960) Influence of anti-metabolites inhibiting nucleic acid metabolism on embryonic development. Trans Assoc Am Physicians 73: 334–347

Karnofsky DA, Murphy ML, Lacon CR (1958) Comparative toxicologic and teratologic effects of 5-fluoro-substituted pyrimidines in the chick embryos and pregnant rat. Proc Am Assoc Cancer Res 2: 312–313

Kasubuchi Y, Wakaizumi S, Shimada M, Kusunoki T (1977) Cytosine arabinoside-induced transplacental dysgenetic hydrocephalus in mice. Teratology 16: 63–70

Katagiri N (1983) Vascular pattern and limb development. I. Normal development of the limbbud vasculature in the mouse and its aberrations induced by 5-fluorouracil. Hiroshima J Med Sci 32: 485–500

Kay AC, Saven A, Carrera CJ, Carson DA, Thurston D, Beutler E, Piro LD (1992) 2-Chlorodeoxyadenosine treatment of low grade lymphomas. J Clin Oncol 10: 371–377

Khera KS (1976) Teratogenicity studies with methotrexate, aminopterin, and acetylsalicylic acid in domestic cats. Teratology 14: 21–28

Khera KS (1979) A teratogenicity study on hydroxyurea and diphenylhydantoin in cats. Teratology 20: 447–452

Klohs WD, Kraker AJ (1992) Pentostatin: future directions. Pharmacol Rev 44: 459–477

Knudsen TB, Ibrahim MM (1995) Co-specificity of nuclear p53 accumulation and apoptotic cell death in early mouse embryos following DNA damage. Teratology 51: 173

Knudsen TB, Kochhar DM (1981) Limb development in mouse embryo. III. Cellular events underlying the determination of altered skeletal patterns following treatment with 5′-fluoro-2 deoxyuridine. Teratology 23: 241–251

Knudsen TB, Green JD, Airhart MJ, Higley HR, Chinsky JM, Kellems RE (1988) Developmental expression of adenosine deaminase in placental tissues of the early postimplantation mouse embryo and uterine stroma. Biol Reprod 39: 937–951

Knudsen TB, Gray MK, Church MR, Blackburn MR, Airhart MJ, Kellems RE, Skalko RG (1989) Early postimplantation embryolethality in mice following in utero inhibition of adenosine deaminase with 2′-deoxycoformycin. Teratology 40: 615–626

Knudsen TB, Blackburn MR, Chinsky JM, Airhart MJ, Kellems RE (1991) Ontogeny of adenosine deaminase in the mouse decidua and placenta: immunolocalization and embryo transfer studies. Biol Reprod 44: 171–184

Knudsen TB, Winters RS, Obey SK, Blackburn MR, Airhart MJ, Church JK, Skalko RG (1992) Effects of (R)-deoxycoformycin (pentostatin) on intrauterine nucleoside catabolism and embryo viability in the pregnant mouse. Teratology 45: 91–103

Kochhar DM, Penner JD, McDay JA (1978) Limb development in mouse embryos. II. Reduction defects, cytotoxicity and inhibition of DNA synthesis produced by cytosine arabinoside. Teratology 18: 71–92

Kochhar DM, Penner JD, Knudsen TB (1980) Embryotoxic, teratogenic, and metabolic effects of ribavirin in mice. Toxicol Appl Pharmacol 52: 99–112

Kohalmi SE, Glattke M, McIntosh EM, Kunz BA (1991) Mutational specificity of DNA precursor pool imbalance in yeast arising from deoxycytidine deaminase deficiency or treatment with thymidylate. J Mol Biol 220: 933–946
Kralovanszky J, Prajda N, Kerper-Fronius S, Bagrij T, Kiss E, Peters GJ (1993) Biochemical consequences of 5-fluorouracil gastrointestinal toxicity in rats: effect of high-dose uridine. Cancer Chemother Pharmacol 32: 243–248
Krowke R, Bochert G (1975) Inhibition of RNA synthesis, a possible mode of the embryotoxic action of hydroxyurea. Naunyn Schmiedebergs Arch Pharmacol 288: 7–16
Kury G, Chaube S, Murphy ML (1968) Teratogenic effect of some purine analogues on fetal rats. Arch Pathol 86: 395–402
Kwok JBJ, Tattersall MHN (1992) DNA fragmentation, dATP pool elevation and potentiation of antifolate cytotoxicity in L1210 cells by hypoxanthine. Br J Cancer 65: 503–508
Lau C, Cameron AM, Rogers JM, Shuey DL, Kavlock RJ (1992) Development of biologically-based dose-response models: correlations between developmental toxicity of 5-fluorouracil (5-FU) and its inhibition of thymidylate synthetase (TS) activity in the rat embryo. Teratology 45: 457
Le LV, Pizzuti DJ, Grenberg M, Reid R (1991) Accidental use of low-dose of 5-fluorouracil in pregnancy. J Reprod Med 36: 872–874
Lennard L (1992) The clinical pharmacology of 6-mercaptopurine. Eur J Clin Pharmacol 43: 329–339
Lindberg U, Skoog L (1970) A method for the determination of dATP and dTTP in picomole amounts. Anal Biochem 34: 152–160
Major PP, Egan E, Herrick D, Kufe DW (1982) 5-Fluorouracil incorporation into DNA of human breast carcinoma cells. Cancer Res 42: 3005–3009
Manson JM, Dourson ML, Smith CC (1977) Effects of cytosine arabinoside on in vivo and in vitro mouse limb development. In Vitro 13: 434–442
Marcickiewicz J, Chazan B, Niemiec T, Sokolska G, Troszynski M, Luczak M, Szmigielski S (1986) Microwave radiation enhances teratogenic effect of cytosine arabinoside in mice. Biol Neonate 50: 75–82
Merker HJ, Pospisil M, Mewes P (1975) Cytotoxic effects of 6-mercaptopurine on the limb-bud blastemal cells of rat embryos. Teratology 11: 199–218
Meuth M (1984) The genetic consequences of nucleotide precursor pool imbalance in mammalian cells. Mutat Res 126: 107–112
Millicovsky G, DeSesso JM (1980) Cardiovascular alterations in rabbit embryos in situ after a teratogenic dose of hydroxyurea: an in vivo microscopic study. Teratology 22: 115–124
Millicovsky G, DeSesso JM (1981) Effects of hydroxyurea on hemodynamics of pregnant rabbits: a maternally mediated mechanism of embryotoxicity. Am J Obstet Gynecol 140: 747–752
Mirkes PE, Greenway JC (1985) Uptake and binding of tritium from (chloroethyl ^{3}H) cyclophosphamide by rat embryos in vitro. Teratology 31: 373–380
Mirkes PE, Ellison A, Little SA (1991) Resistance of rat embryonic heart cells to the cytotoxic effects of cyclophosphamide does not involve aldehyde dehydrogenase-mediated metabolism. Teratology 43: 307–318
Morrison PF, Allegra CJ (1989) Folate cycle kinetics in human breast cancer cells. J Biol Chem 264: 10552–10566
Nagomi H, Oohira A (1980) Experimental study on pathogenesis of polydactyly of the thumb. J Hand Surg [Am] 5: 443–450
Nakada D Magasanik B (1964) The roles of inducer and catabolite repressor in the synthesis of β-galactosidase by *Escherichia coli*. J Mol Biol 8: 105–127
Nakashima K, Ninomiya H, Fujiki Y (1984) The effect of 5-bromodeoxyuridine on mouse embryos during neurulation in vitro. Experientia 40: 924–929

Nelson MM (1960) Teratogenic effects of pteroylglutamic acid deficiency in the rat. In: Wolstenholme GEW, O'Connor CM (eds) Ciba Foundation symposium on congenital malformations. Little and Brown, Boston, pp 134–157

Neubert D, Lessmollmann U, Hinz N, Dillmann I, Fuchs G (1977) Interference of 6-mercaptopurine riboside, 6-methyl-mercaptopurine riboside and azathioprine with the morphogenetic differentiation of mouse extremities in vivo and in organ culture. Naunyn Schmiedebergs Arch Pharmacol 298: 93–105

Newman CN, Miller JH (1983) Mutagen-induced changes in cellular deoxycytidine triphosphate and thymidine triphosphate in Chinese hamster ovary cells. Biochem Biophys Res Commun 114: 34–40

Nicander B, Reichard P (1985) Relations between synthesis of deoxyribonucleotides and DNA replication in 3T6 fibroblasts. J Biol Chem 260: 5376–5381

Nord LD, Martin DS (1991) Loss of murine tumor thymidine kinase activity in vivo following 5-fluorouracil (FUra) treatment by incorporation of FUra into RNA. Biochem Pharmacol 42: 2369–2375

Nord LD, Stolfi RL, Martin DS (1992) Biochemical modulation of 5-fluorouracil with leucovorin or delayed uridine rescue: correlation of antitumor activity with dosage and FUra incorporation into RNA. Biochem Pharmacol 43: 2543–2549

Odom LD, Plouffe L, Butler WJ (1990) 5-Fluorouracil exposure during the period of conception: report on two cases. Am J Obstet Gynecol 163: 76–77

Ohmori K (1972) Teratogenic effects of 5-fluoro-2′-deoxyuridine in pregnant mice. Teratology 5: 71–80

Ohno M (1984) Neuroanatomical study of somatomotor cortex in microcephalic mice induced by cytosine arabinoside. Brain Dev 6: 528–538

Olsen EA (1991) The pharmacology of methotrexate. J Am Acad Dermatol 25: 306–318

O'Neill KO, Shao X, Zhao Z, Malik A, Lee ML (1994) Capillary electrophoresis of nucleotides on Ucon-coated fused silica columns. Anal Biochem 222: 185–189

Ortega A, Puig M, Domingo JL (1991) Maternal and developmental toxicity of low doses of cytosine arabinoside in mice. Teratology 44: 379–384

Ostensen M (1992) Treatment with immunosuppressive and disease modifying drugs during pregnancy and lactation. Am J Reprod Immunol 28: 148–152

Parker WB, Cheng YC (1990) Metabolism and mechanism of action of 5-fluorouracil. Pharmacol Ther 48: 381–395

Percy DH (1975) Teratogenic effects of the pyrimidine analogues 5-iododeoxyuridine and cytosine arabinoside in late fetal mice and rats. Teratology 11: 103–118

Phear G, Meuth M (1989) The genetic consequences of DNA precursor pool imbalance: sequence analysis of mutation induced by excess thymidine at the hamster aprt locus. Mutat Res 214: 201–206

Pinendo HM, Peters GFJ (1988) Fluorouracil: biochemistry and pharmacology. J Clin Oncol 6: 1653–1664

Plunkett W, Saunders PP (1991) Metabolism and action of purine nucleoside analogs. Pharmacol Ther 49: 239–268

Prescott DM (1976) Reproduction of eukaryotic cells. Academic, New York

Reichard P (1985) Ribonucleotide reductase and deoxyribonucleotide pools. Basic Life Sci 31: 33–45

Reimers TJ, Sluss PM (1978) 6-Mercaptopurine treatment of pregnant mice: effects on second and third generations. Science 201: 65–67

Rhee MS, Coward KJ, Galivan J (1992) Depletion of 5,10-methylenetetrahydrofolate and 10-formyltetrahydrofolate by methotrexate in cultured hepatoma cells. Mol Pharmacol 42: 909–916

Ritter EJ (1984) Potentiation of teratogenesis. Fund Appl Toxicol 4: 352–359

Ritter EJ, Bruce LM (1979) The quantitative determination of deoxyribonucleoside triphosphates using high performance liquid chromatography. Biochem Med 21: 16–21

Ritter EJ, Scott WJ, Wilson JG (1971))Teratogenesis and inhibition of DNA synthesis induced in rat embryos by cytosine arabinoside. Teratology 4: 7–14

Ritter EJ, Scott WJ, Wilson JG (1973) Relationship of temporal patterns of cell death and development to malformations in the rat limb. Possible mechanisms of teratogenesis with inhibitors of DNA synthesis. Teratology 7: 219–226

Ritter EJ, Scott WJ, Wilson JG, Lampkin BC, Neely JE (1980) Effect of 5-fluoro-2′-deoxyuridine on deoxyribonucleotide pools in vivo. J Natl Cancer Inst 65: 603–605

Robertson LE, Chubb S, Meyn RE, Story M, Ford R, Hittleman WN, Plunkett W (1993) Induction of apoptotic cell death in chronic lymphocytic leukemia by 2-chloro-2′-deoxyadenosine and 9-β-D-arabinosyl-2-fluoroadenine. Blood 81: 143–150

Rosen R, Rothman F, Weigert MG (1969) Miscoding caused by 5-fluorouracil. J Mol Biol 44: 363–375

Ross DD, Cuddy DP, Cohen N, Hensley DR (1992) Mechanistic implications of alterations in HL-60 cell nascent DNA after exposure to 1-β-D-arabinofuranosylcytosine. Cancer Chemother Pharmacol 31: 61–70

Ruffolo P, Ferm V (1965) The embryocidal and teratogenic effects of 5-bromodeoxyuridine in the pregnant hamster. Lab Invest 14: 1547–1553

Saksena SK, Chaudhury RR (1970) The antifertility effect of 2′,3′,5′,-tri-O-acetyl-6-azauridine. Part II. In rabbits. Indian J Med Res 58: 374–376

Santelli G, Valeriote F (1980) In vivo potentiation of 5-fluorouracil cytotoxicity against AKR leukemia by purine, pyrimidines, and their nucleosides and deoxynucleosides. J Natl Cancer Inst 64: 69–72

Santi DV, McHenry CS, Sommer S (1974) Mechanism of interaction of thymidylate synthetase with 5-fluorouridylate. Biochemistry 13: 471–480

Saunders MA, Wiesner BP, Yudkin J (1961) Control of fertility by 6-azauridine. Nature 189: 1015–1016

Schafer AI (1981) Teratogenic effects of antileukemic chemotherapy. Arch Intern Med 141: 514–515

Schmid BP (1984) Monitoring of organ formation in rat embryos after in vitro exposure to azathioprine, mercaptopurine, methotrexate or cyclosporin A. Toxicology 31: 9–21

Schmid P, Lorenz A, Hameister H, Montenarh M (1991) Expression of p53 during mouse embryogenesis. Development 113: 857–865

Schuetz JD, Wallace HJ, Diasio RB (1984) 5-Fluorouracil incorporation into DNA of CF-1 mouse bone marrow cells as a possible mechanism of toxicity. Cancer Res 44: 1358–1363

Schumacher HJ, Wilson JG, Jordan RL (1969) Potentiation of the teragenic effects of 5-fluorouracil by natural pyrimidines. II. Biochemical aspects. Teratology 2: 99–106

Scott JR (1977) Fetal growth retardation associated with maternal administration of immunosuppressive drugs. Am J Obstet Gynecol 128: 668–674

Scott WJ (1977) Cell death and reduced proliferative rate. In: Wilson JG, Fraser FC (eds) Mechanisms and pathogenesis. Plenum, New York, pp 81–98 (Handbook of teratology, vol 2)

Scott WJ (1981) The pathogenesis of bromodeoxyuridine induced polydactyly. Teratology 23: 383—389

Scott WJ, Ritter EJ, Wilson JG (1975) Studies on induction of polydactyly in rats with cytosine arabinoside. Dev Biol 45: 103–111

Scott WJ, Ritter EJ, Wilson JG (1980) Ectodermal and mesodermal cell death patterns in 6-mercaptopurine riboside-induced digital deformities. Teratology 21: 271–279

Shah RM, Burdett DN (1978) Developmental abnormalities induced by 6-mercaptopurine in the hamster. Can J Physiol Pharmacol 57: 53–58

Shah RM, MacKay RA (1978) Teratological evaluation of 5-fluorouracil and 5-bromo-2-deoxyuridine on hamster fetuses. J Embryol Exp Morphol 43: 47–54

Shah RM, Wong DTW (1980) Morphological study of cleft palate development in 5-fluorouracil-treated hamster fetuses. J Embryol Exp Morphol 57: 119–128
Shah RM, Wong DT, Suen RS (1984) Ultrastructural and cytochemical observations on 5-fluorouracil-induced cleft-palate development in hamster. Am J Anat 170: 567–580
Shah RM, Chen YP, Burdett DN (1989) Palatal shelf reorientation in hamster embryos following treatment with 5-fluorouracil. Histol Histopathol 4: 449–456
Shuey DL, Lau C, Logsdon TR, Zucker RM, Elstein KH, Narotsky MG, Setzer W, Kavlock RJ, Rogers JM (1994a) Biologically based dose-response modeling in developmental toxicology: biochemical and cellular sequelae of 5-fluorouracil exposure in the developing rat. Toxicol Appl Pharmacol 126: 129–144
Shuey DL, Zucker ZM, Elstein KH, Rogers JM (1994b) Fetal anemia following maternal exposure to 5-fluorouracil in the rat. Teratology 49: 311–319
Shuey DL, Buckalew AR, Wilke TS, Rogers JM, Abbott BD (1994c) Early events following maternal exposure to 5-fluorouracil lead to dysmorphology in cultured embryonic tissues. Teratology 50: 379–386
Simpson RC, Brown PR (1986) High-performance liquid chromatographic profiling of nucleic acid components in physiological samples. J Chromatogr 379: 269–311
Skalko RG, Gold MP (1974) Teratogenicity of methotrexate in mice. Teratology 9: 159–164
Skalko RG, Jacobs DM (1978) The effects of 5-fluorouracil on [^{3}H]nucleoside incorporation into DNA of mouse embryos and maternal tissues. Exp Mol Pathol 29: 303–315
Skalko RG, Packard DS (1973) The teratogenic response of the mouse embryo to 5-iododeoxyuridine. Experientia 29: 198–200
Skalko RG, Packard DS (1975) Mechanisms of halogenated nucleoside embryotoxicity. Ann NY Acad Sci 255: 552–558
Skalko RG, Packard DS, Schwendimann RN, Raggio JF (1971) The teratogenic response of mouse embryos to 5-bromodeoxyuridine. Teratology 4: 87–93
Skoda J (1975) Azapyrimidine nucleotides. In: Eichler O, Farah H, Herken H, Welch AD (eds) Handbook of experimental pharmacology, vol 38. Springer, Berlin Heidelberg New York, pp 348–372
Skoog L (1970) An enzymatic method for the determination of dCTP and dGTP in picomole amounts. Eur J Biochem 17: 202–208
Skoog L, Nordenskjold B (1971) Effects of hydroxyurea and 1-β-D-arabinofuranosylcytosine on deoxyribonucleotide pools in mouse embryo cells. Eur J Biochem 19: 81–91
Skoog KL, Nordenskjold BA, Bjursell KG (1973) Deoxyribonucleoside-triphosphate pools and DNA synthesis in synchronized hamster cells. Eur J Biochem 33: 428–432
Skoog KL, Bjursell KG, Nordenskjold BA (1974) Cellular deoxyribonucleotide triphosphate pools levels and DNA synthesis. Adv Enzyme Regul 12: 345–354
Snow MHL (1983) Restorative growth in mammalian embryos. In: Kalter H (ed) Issues and reviews in teratology, vol 1. Plenum, New York, pp 251–284
Snow MHL (1986) Uncoordinated development of embryonic tissue following cytotoxic damage. In: Welsch F (ed) Approaches to elucidate mechanisms in teratogenesis. Hemisphere, Washington, pp 83–98
Snow MHL, Tam PPL (1979) Is compensatory growth a complicating factor in mouse teratology? Nature 279: 555–557
Solter AW, Handschumacher RE (1969) A rapid quantitative determination of deoxyribonucleoside triphosphates based on the enzymatic synthesis of DNA. Biochim Biophys Acta 174: 585–590
Stadler HE, Knowles J (1971) Fluorouracil in pregnancy: effect on the neonate. JAMA 217: 214–215

Stephens JD, Golbus MS, Miller TR, Wilber RR, Epstein CJ (1980) Multiple congenital anomalies in a fetus exposed to 5-fluorouracil during the first trimester. Am J Obstet Gynecol 137: 747–749

Stolfi RL, Colofiore JR, Nord LD, Koutcher JA, Martin DS (1992) Biochemical modulation of tumor cell energy: regression of advanced spontaneous murine breast tumors with a 5-fluorouracil-containing drug combination. Cancer Res 52: 4074–4081

Stovall TG, Ling FW, Buster JE (1990) Reproductive performance after methotrexate treatment of ectopic pregnancy. Am J Obstet Gynecol 162: 1620–1624

Takigiku R, Schneider RE (1991) Reproducibility and quantitation of separation for ribonucleoside triphosphates and deoxyribonucleoside triphosphates by capillary zone electrophoresis. J Chromatogr 559: 247–256

Takimoto CH, Voeller DB, Strong JM, Anderson L, Chu E, Allegra CJ (1993) Effects of 5-fluorouracil substitution on the RNA conformation and in vitro translation of thymidylate synthase messenger RNA. J Biol Chem 268: 21438–21442

Tanaka K, Yoshioka A, Tanaka S, Wataya Y (1984) An improved method for the quantitative determination of deoxyribonucleoside triphosphates in cell extracts. Anal Biochem 139: 35–41

Thiersch JB (1952) Therapeutic abortions with a folic acid antagonist, 4-aminoteroylglutamic acid (4-amino P.G.A.). Am J Obstet Gynecol 63: 1298–1304

Topal MD, Baker MS (1982) DNA precursor pool: a significant target for N-methyl-N-nitrosourea in C3H/10T1/2 clone 8 cells. Proc Natl Acad Sci USA 79: 2211–2215

Tsukamoto I, Kojo S (1991) The effects of fluorouracil on thymidylate synthase and thymidine kinase in regenerating rat liver after partial hepatectomy. Biochim Biophys Acta 1074: 52–55

Valeriote F, Santelli G (1984) 5-Fluorouracil (5-FUra). Pharmacol Ther 24: 107–132

Van Wagenen G, DeConti RC, Handschumacher RE, Wade ME (1970) Abortifacient and teratogenic effects of triacetyl-6-azauridine in the monkey. Am J Obstet Gynecol 108: 272–281

Vorherr H, Welch AD (1970) The mode of interruption of pregnancy by 6-azauridine in mice and rats. Biochem Pharmacol 19: 1001–1006

Vojta M, Jirasek J (1966) 6-Azauridine-induced changes of the trophoblast in early human pregnancy. Clin Pharmacol Ther 7: 162–165

Wagner VM, Hill JS, Weaver D, Baehner RL (1980) Congenital abnormalities in baby born to cytarabine treated mother. Lancet 2: 98–99

Warkany J (1978) Aminopterin and methotrexate: folic acid deficiency. Teratology 17: 353–358

Warkany J (1986) Aminopterin and methotrexate: folic acid deficiency. In: Sever JL, Brent RL (eds) Teratology update: environmentally induced birth defect risks. Liss, New York, pp 39–43

Wataya Y, Hwang H, Nakazawa T, Takahashi K, Otani M, Igaki T (1993) Molecular mechanisms of cell death induced dNTP pool imbalance. Nucleic Acids Symp Ser 29: 109–110

Wilson JG (1971) Use of rhesus monkeys in teratological studies. Fed Proc 30: 104–109

Wilson JG (1973) Environment and birth defects. Academic, New York

Wilson JG, Jordan RL, Schumacher H (1969) Potentiation of the teratogenic effects of 5-fluorouracil by natural pyrimidines. I. Biological aspects. Teratology 2: 91–98

Wilson JG, Scott WJ, Ritter EJ, Fradkin R (1975) Comparative distribution and embryotoxicity of hydroxyurea in pregnant rats and rhesus monkeys. Teratology 11: 169–178

Wilson JG, Scott WJ, Ritter EJ, Fradkin R (1979) Comparative distribution and embryotoxicity of methotrexate in pregnant rats and rhesus monkeys. Teratology 19: 71–79

Wise LD, Scott WJ (1982) Incorporation of 5-bromo-2-deoxyuridine into mesenchymal limb-bud cells destined to die: relationship to polydactyly induction in rats. J Embryol Exp Morphol 72: 125–141

Woodcock DM (1987) Cytosine arabinoside toxicity: molecular events, biological consequences, and their implications. Semin Oncol 2 Suppl 1: 251–256

Wubah JA, Ibrahim MM, Gao X, Nguyen D, Pisano MM, Knudsen TB (1996) Teratogen-induced eye defects mediated by p53-dependent apoptosis. Curr Biol 6: 60–69

Yarbro JW (1992) Mechanism of action of hydroxyurea. Semin Oncol 19 Suppl 9: 1–10

Yoshihara H, Dagg CP (1967) Teratogenicity of 6-azauridine in inbred mice. Anat Rec 157: 345

Yoshioka A, Tanaka S, Hiraoka O, Koyama Y, Hirota Y, Ayusawa D, Seno T, Garrett C, Wataya Y (1987) Deoxyribonucleoside triphosphate imbalance: 5-fluorodeoxyuridine-induced DNA double strand breaks in mouse FM3A cells and the mechanism of cell death. J Biol Chem 262: 8235–8241

Zucker RM, Elstein KH, Shuey DL, Rogers JM (1995) Flow cytometric detection of abnormal fetal erythropoiesis: application to 5-fluorouracil-induced anemia. Teratology 51: 37–44

CHAPTER 14

Interference with Embryonic Intermediary Metabolism

E.S. HUNTER, III

A. Introduction

The purpose of this chapter is to describe normal intermediary metabolism during development and, where possible, to discuss how metabolic perturbation leads to dysmorphogenesis. Embryonic metabolism is dynamic, and what is true about substrate utilization at any specific stage of development may not be true at other stages. For this reason, this review will focus on metabolism during the period of organogenesis.

Most of the information we have about normal embryonic intermediary metabolism during organogenesis has focused on glucose as the substrate. Based on many different studies, it has been established that glucose is an important, if not essential, substrate during organogenesis. Glucose metabolism can be perturbed by decreased availability of substrate, competition for transport, or inhibition of enzymes involved in glucose metabolism, and all of these adversely affect development. Furthermore, increased glucose concentrations have also been shown to alter development and may be associated with malformations in children born to insulin-dependent diabetic women. Thus the regulation of glucose availability and utilization is critical for normal embryogenesis.

B. Normal Glucose Metabolism

Glucose metabolism, in simplest terms, can be described as the degradation of glucose to pyruvate and the subsequent metabolism of pyruvate to CO_2. The metabolism of glucose to pyruvate is defined as glycolysis or the Embden-Meyerhoff pathway and occurs in the cell's cytosol. Pyruvate metabolism to CO_2 in the mitochondria is known as the tricarboxylic acid cycle or the Krebs cycle (Fig. 1). Glucose is transported into cells by facilitated diffusion using specific transport proteins. Once glucose enters the cell, the first steps in metabolism include 2 adenosine triphosphate (ATP)-dependent phosphorylations. The phosphorylated hexose is metabolized to two glyceraldehyde-3-phosphate triose molecules. The metabolism of glyceraldehyde-3-phosphate to pyruvate includes the production of two ATP molecules. Pyruvate is then transported across the mitochondrial membranes and is metabolized to ace-

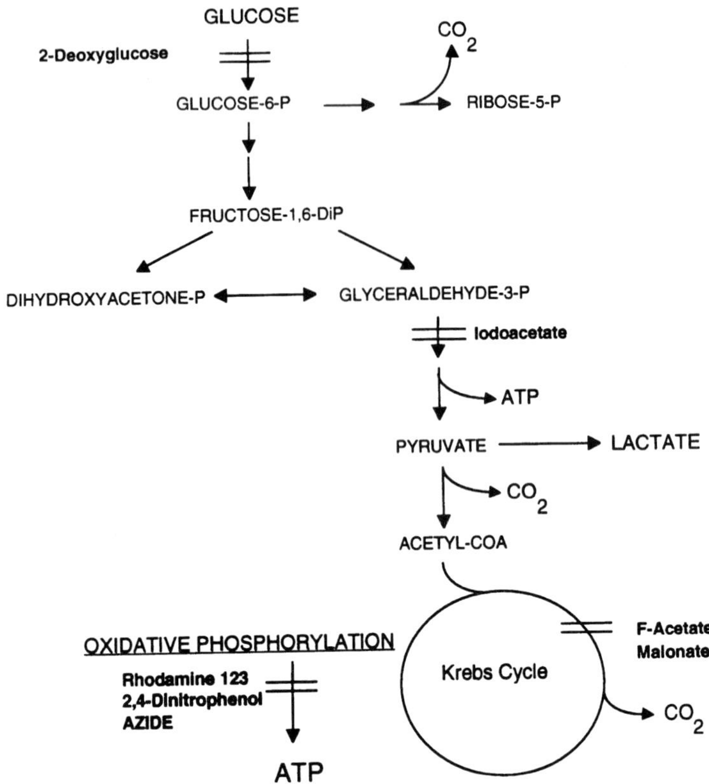

Fig. 1. Glucose metabolism by glycolysis and the Krebs cycle. The sites of action of inhibitors discussed in this chapter are indicated. *ATP*, adenosine triphosphate; *CoA*, coenzyme A

tylcoenzyme A (acetyl-CoA). Acetyl-CoA then enters the Krebs cycle by condensing with oxaloacetic acid to form citric acid. Citrate is then metabolized back to oxaloacetate (OAA) through a number of steps generating CO_2 and guanosine triphosphate (GTP).

In addition to glycolysis and Krebs cycle metabolism, intermediates of these pathways are used by other pathways. For example, the pentose phosphate pathway (PPP) uses glucose-6-phosphate to generate ribose moieties, dihydroxyacetone-phosphate is used for the production of glycerol, or pyruvate can be metabolized to lactic acid as a terminal metabolic step. Acetyl-CoA can be used for many metabolic processes, e.g., lipogenesis or ketone body formation, in addition to entering the Krebs cycle. Similarly, intermediates of the Krebs cycle are used for amino acid synthesis. For example, α-ketoglutarate is used for glutamate and glutamine synthesis.

One important result of glucose metabolism is the generation of energy stored as ATP. Glycolysis yields two ATP molecules for each molecule of glucose metabolized, but the Krebs cycle does not produce any ATP by its actions. Instead, reduced nicotinamide adenine dinucleotide (NADH) and reduced flavin adenine dinucleotide ($FADH_2$) generated from Krebs cycle metabolism are used for ATP production by oxidative phosphorylation (Ox-Phos). Ox-Phos transfers electrons from NADH and $FADH_2$ through the electron transport chain and pumps protons into the intermitochondrial space. This proton gradient across the inner mitochondrial membrane is then used by ATPase to generate ATP in the mitochondria. Since each acetyl-CoA entering the mitochondria drives the Krebs cycle for two cycles and because there are three sites for proton pumping in the electron transport chain, each molecule of glucose metabolized by the Krebs cycle generates 30 ATP molecules.

The enzymatic pathways of glucose utilization by embryos appear to be the same as those used in adult tissues. However, differences in enzymatic activities in adult and embryonic tissues, produced by substrate or cofactor regulation or different isozymes, result in different rates of glucose metabolism by specific pathways. Not only are there differences in the rates of specific substrate metabolism, but there are also differences in the relative utilization of pathways. Adult glucose metabolism is characterized by a high rate of Krebs cycle metabolism and Ox-Phos, with a low rate of lactate production under aerobic conditions. However, the pattern of metabolism used during development is stage dependent. Immediately, following fertilization, conceptuses are dependent upon Krebs cycle metabolism, but by the blastocyst stage glycolysis is the major pathway. Throughout early postimplantation development, embryonic metabolism is characterized by a tremendously high rate of glucose metabolism, with lactate as the predominant metabolic product. Once a vascular system has been established in the embryo and adnexa, an increasing activity of the Krebs cycle is observed. Thus during the period of gastrulation and organogenesis, the embryo is largely dependent upon glycolytic metabolism as the predominant pathway for glucose utilization. During the fetal period as organ development occurs, metabolism becomes stage and organ specific. Generally, with increased gestational age there is an increasing activity of the Krebs cycle. This general pattern of change is depicted in Fig. 2.

C. Preimplantation Pattern of Glucose Metabolism

Although the focus of this chapter is organogenesis, it is important to briefly outline metabolism during the preimplantation period. For a complete discussion of preimplantation-stage metabolism, there are a number of very good reviews (e.g., BIGGERS and BORLAND 1976; BIGGERS et al. 1989; LEESE 1990). In simplest terms, during the preimplantation stage rodent conceptuses undergo a shift from dependence on Krebs cycle and pyruvate metabolism to requiring glucose for a predominately glycolytic pattern of utilization.

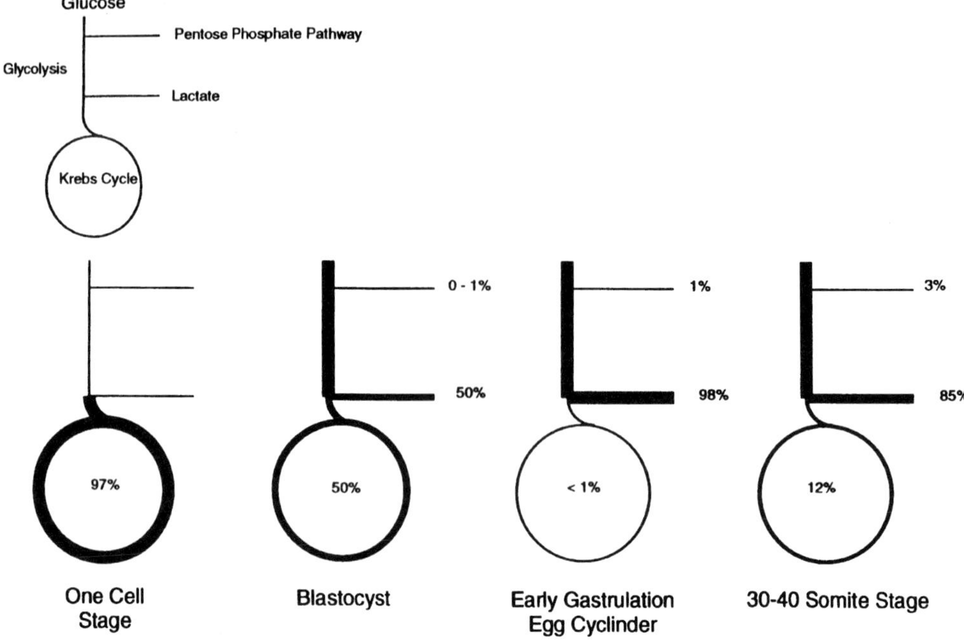

Fig. 2. Relative utilization of glycolysis and the Krebs cycle during early rodent development. (Adapted from Clough 1985)

Much of the research on substrate utilization by preimplantation-staged conceptuses has arisen from attempts to define the optimum culture medium to support development following in vitro fertilization. At the earliest time points, pyruvate or OAA is necessary to support development. However, by the two-cell stage there is an increasing number of pyruvate precursors (lactate, phospho-enol-pyruvate) that can support development, but it is not until the eight-cell stage that glucose is sufficient (BIGGERS 1967). Many studies have also measured the utilization of substrate from culture medium. The rate of pyruvate uptake by mouse conceptuses was not changed from the unfertilized oocyte through the eight- to 16-cell stage and then decreased to low levels in blastocysts. At the early stages, there is low utilization of glucose up through the morula stage, when an eight- to tenfold increase in glucose utilization occurs (LESSE and BARTON 1984). Qualitatively, similar changes in substrate utilization have been reported by several groups (e.g., BRINSTER 1973, and WALES 1986).

One underlying question concerns the cause of the increase in glucose utilization and the shift away from the efficiency of the Krebs cycle for energy production. There is probably no single reason for this shift, but instead a series of events that cumulatively result in an increased requirement for glucose, including an increasing demand for macromolecular synthesis.

Barbehenn et al. (1974) provide evidence that glucose utilization is limited by a low activity of phosphofructokinase (PFK) up through the morula stage of development. The regulatory mechanism of PFK in the embryo is not entirely clear, but Barbehenn and coworkers propose that high intracellular concentrations of citrate may produce the inhibitory effect. However, high concentrations of ATP in the embryo at earlier stages of development may also serve to downregulate PFK activity, while, conversely, decreasing ATP to adenosine diphosphate (ADP) ratios, observed with increased gestational age, may serve to upregulate PFK activity at later stages.

PFK activity may also be limited by the fact that there are only low concentrations of fructose-6-phosphate (F6P) due to low levels of hexokinase activity. With increased gestational age, there is a fourfold increase in hexose-6-phosphate (H6P), which parallels the increase in PFK activity. The increase in H6P likely results from the increased activity of hexokinase, which doubles from the one-cell stage to the eight-cell stage and is nearly eightfold higher in blastocysts than in one-cell-stage embryos (Ayabe et al. 1994).

Since the goal of assessing the effects of toxicants on rodent embryos is to understand their potential effects on human development, it is crucial to determine whether human embryonic metabolism is similar or dissimilar to that of other mammals. In this regard, the preimplantation stage offers an important window for comparing human and rodent embryos. For example, the human embryo follows a pattern of increasing glucose metabolism up to blastocyst stage similar to that seen in other mammals. The rate of lactate production from glucose increases 200- to 300-fold from the two-cell stage to the blastocyst (Wales et al. 1987). However, unlike the rodent, humans maintain their utilization of pyruvate for oxidative metabolism. Additionally, there are subtle differences in pathway regulation. For example, hexokinase appears to be the rate-determining step of glucose utilization in human conceptuses, since glucose utilization follows hexokinase activity (Leese et al. 1993). Qualitatively, metabolism of human preimplantation embryos follows patterns similar to those of other mammals, but quantitatively there are differences. (For a review of human preimplantation embryonic metabolism e.g. see Wales et al. 1987; Leese et al. 1993).

These studies indicate that the basic pattern of metabolism found in preimplantation human embryos is similar to the pattern of metabolism in other mammals and that agents that perturb rodent embryogenesis through disruption of metabolism would be expected to also affect human development. However, there are important differences in metabolism between rodents and humans that may modify the human response to metabolic perturbation. For example, the ability of human conceptuses to utilize pyruvate suggests that humans may be able to compensate for a change in nutrient availability better than other mammals during the preimplantation stage.

D. Glucose Metabolism During the Post-implantation Stage

During the period immediately following implantation, mouse conceptus metabolism appears to continue the pattern present in the blastocyst of low Krebs cycle and high glycolytic metabolism. Thus during the post implantation period and through the period of neurulation, lactate accounts for more than 91% of the catabolic output from glucose in mouse embryos (CLOUGH and WHITTINGHAM 1983).

During the period of organogenesis, the rodent embryo utilizes glucose at a rate greater than any other tissue (TANIMURA and SHEPARD 1970a). Furthermore, glucose appears to be an essential medium factor, i.e., normal growth and development of embryos in vitro does not occur if glucose is excluded from the culture medium (GUNBERG 1976; COCKCROFT 1979). The high rate of lactate formation by the rodent embryo during organogenesis in vitro was initially reported by NEGELEIN (1925). Later studies corroborated this result (KLEIBER et al. 1943) and were extended to embryos of different gestational ages (NEUBERT 1970; NEUBERT et al. 1971; TANIMURA and SHEPARD 1970a; SHEPARD et al. 1970). At the early stages of organogenesis, studies compared rates of ^{14}C-lactate production from U-^{14}C-glucose by mouse embryos on days 8 and 11 of gestation and a decreasing rate of lactate production with increasing gestational age was described (HORTON et al. 1985; HUNTER and SADLER 1987). These data also indicated that lactate accounted for 94% of the catabolic utilization of glucose by day-9 mouse conceptuses in vitro. NEUBERT et al. (1971) reported that the percentage of glucose utilized for lactate synthesis by rat embryos in vitro remained constant (80%–90%) throughout the period of days 12–14, i.e., during the later stage of organogenesis. In contrast, TANIMURA and SHEPARD (1970a) reported that the synthesis of ^{14}C-lactate from ^{14}C-glucose accounted for 90% of the glucose utilized by the day-11 rat embryo, but decreased to 80% and 68% for day-12 and -13 rat embryos in vitro, respectively. Although the later study corroborated NEUBERT's (1970) assessment of lactate as the primary metabolic product of embryonic glucose metabolism, TANIMURA and SHEPARD (1970a) described the rate of embryonic lactate production as decreasing with increasing gestational age. Thus during the period of organogenesis, the rodent embryo utilizes glucose primarily by the glycolytic pathway, with a subsequent synthesis of lactate that can account for as much as 95% of the catabolic output from glucose. With increased gestational age, the rate of lactate synthesis from glucose substrate decreases, with a concomitant increase in Krebs cycle metabolism of glucose-generated intermediates. Thus the conceptus shifts toward an increasingly efficient system for energy production. There is no information explaining why this metabolic shift occurs. One explanation may be that there is a move away from a period of rapid growth and high anabolic demands to a period of energy requirement for differentiation.

A recent report took the additional step of assessing glucose metabolism in the embryo proper and the visceral yolk sac (VYS) as separate tissues (AKA-

ZAWA et al. 1994). In this study, results showed that day-10 and -11 embryos metabolized glucose to lactate at a rate 25%–33% lower than that of the membranes. These studies also confirmed a decrease of approximately 60% in lactate production by the conceptus over days 10–11. When day-10 embryos were compared to day-11 embryos, there was a 14-fold increase in the rate of CO_2 produced from 6-^{14}C-glucose, indicating an increased rate of glucose metabolism by the Krebs cycle. There was no apparent shift in the rate of CO_2 produced from 1-^{14}C-glucose substrate by the PPP. In the membranes there was a six fold increase in CO_2 production from Krebs cycle metabolism. These studies clearly demonstrate metabolic differences in the embryo and membranes as well as the magnitude of change during this stage of development.

Using a slightly different approach, HUNTER and SADLER (1989) calculated the rates of CO_2 and lactate production by the embryo within the VYS by subtracting the rate of product formation by the VYS and ectoplacental cone (YSP) from that of the whole conceptus. Results showed that the rate of total CO_2 production by the embryo and adnexa was similar. However, the rate of lactate production by the embryo was 1.5 times greater than the surrounding tissues. Differences in these results from AKAZAWA and coworkers (1994) may be due to the inclusion of the ectoplacental cone in the HUNTER and SADLER study or to differences in the types of substrates or the concentrations of substrates available to the embryo within the VYS compared to an embryo placed directly in culture medium. Despite these differences, both studies confirm that the embryo and the surrounding membranes exhibit different rates of substrate utilization during the period of organogenesis, thereby indicating the requirement for future studies to consider the embryo and membranes separately.

I. The Krebs Cycle and the Pentose Phosphate Pathway

In addition to utilization of the glycolytic pathway by rodent embryos, activities of other metabolic pathways such as the Krebs cycle and the PPP have been determined. Initial investigations into the relative utilization of these pathways by day-13 to -15 rat embryos were reported by DEMEYER and DE PLAEN (1964). The ratio of radioactive CO_2 produced from 1-^{14}C-glucose to that produced from 6-^{14}C-glucose metabolism (C1 to C6 ratio) ranged from 25 to 28. This ratio was ten fold higher than that observed for maternal liver. The high rate of utilization of the PPP compared to the Krebs cycle was confirmed (KOHLER and PETERS 1970) in day-12 rat embryos, in which 1.5% of the glucose consumed was utilized by the PPP, while less than 0.2% was metabolized by the Krebs cycle. These results agree with those reported by NEUBERT (1970), in which 2.9% of the glucose utilized was metabolized by the PPP, while only 0.1% was metabolized by the Krebs cycle in day-13 rat embryos in vitro.

These studies were extended (TANIMURA and SHEPARD 1970a; SHEPARD et al. 1970) to compare different gestationally aged rat embryos (days 11–13) in vitro. Although these studies supported evidence for the high C1 to C6 ratios observed previously, they additionally served to document changes in the relative utilization of the PPP and Krebs cycle pathways. The C1 to C6 ratio decreased with increased gestational age from 11.3 on day 11 and 3.6 and 4.6 on days 12 and 13, respectively. These data suggested a high utilization of glucose by the PPP early in the embryonic period, followed by a relative increase in the utilization by the Krebs cycle during later developmental stages. During the period immediately following implantation in mouse embryos, i.e., days 6.5–8.5, the C1 to C6 ratio increased from 3.4 to 25.1 (CLOUGH and WHITTINGHAM 1983). Additionally, evidence for high utilization of the PPP compared to the Krebs cycle during early organogenesis in mouse conceptuses in vitro shows a C1 to C6 ratio of 48 in day-8 neurulation-staged conceptuses. This ratio decreases with increased gestational age to 23, 5.7, and 3.6 in day -9,-10, and -11 conceptuses, respectively (HUNTER and SADLER 1988). Thus the relative utilization of glucose by the PPP is much greater than for Krebs cycle metabolism during early organogenesis in rodent embryos.

1. The Pentose Phosphate Pathway

Because the rate of glucose metabolism by the Krebs cycle is very low, the rate of CO_2 produced from 1-^{14}C-glucose metabolism can be used to measure the rate of metabolism by the oxidative branch of the PPP. This pathway uses NADP as a cofactor and results in the generation of ribose-5-phosphate (R5P), which is subsequently used for de novo purine and pyrimidine synthesis. KOHLER and BRAND (1970) reported that the rate of R5P synthesis via the oxidative portion of the PPP was 2.2–5.7 times greater than that required for nucleic acid synthesis in the day-12 and -14 rat embryo. There is a more than threefold increase in the V_{max} of two key enzymes in the PPP, glucose-6-phosphate dehydrogenase (G-6-PD) and 6-phospho-gluconate dehydrogenase (6-P-GD), occurring from day 12 to 15 in rat embryos. Because of a high rate of nucleic acid synthesis occurring during early organogenesis, the critical need for R5P synthesis for nucleotide synthesis, and dramatic changes in the activities of G-6-PH and 6-P-GD, these authors proposed that modulation of these enzymes during organogenesis is a potential site for a teratogenic insult.

In addition to the oxidative branch of the PPP, there is a nonoxidative pathway that also results in the generation of R5P. The nonoxidative branch metabolizes glyceraldehyde-3-phosphate and F6P through the transaldolase and transketolase enzymes. To determine whether the nonoxidative portion of the PPP is functional in embryonic tissues, the rates of 1-, 2-, and 6-^{14}C-glucose incorporation into the ribose moiety of RNA and DNA were evaluated. These studies showed a substantial incorporation of 1-^{14}C-glucose into the ribose moiety, thus indicating that the transaldolase–transketolase branch of the pathway is functional during organogenesis (KOHLER and BRAND 1970). The authors further established that these different branches of the PPP do not function as a "cycle," but rather that each branch functions as a "half-cycle," each generating R5P. One unanswered question is whether either half-cycle can be upregulated to compensate for a decreased metabolism by the other in order to maintain R5P concentrations.

2. The Krebs Cycle

The importance of the Krebs cycle for glucose metabolism is poorly characterized in neurulation-staged embryos, and most studies have simply evaluated the generation of $^{14}CO_2$ from 6-^{14}C-glucose. As described previously, the decreased C1 to C6 ratio observed with advancing gestational age is, in part, the result of an increased rate of "glucose" utilization by Krebs cycle metabolism during organogenesis. When the rates of CO_2 production are compared during this stage, there is a 100-fold increase from day 8 to day 11 in mouse conceptuses (HUNTER and SADLER 1988). Furthermore, this increase in Krebs cycle activity is coupled to an increase in Ox-Phos during organogenesis, as evidenced by a higher rate of oxygen utilization with advancing gestational age (SPIELMANN and LUCKE 1973). Although these studies indicate that there is a profound shift in the pattern of metabolism to an increasing utilization of Krebs cycle and Ox-Phos, the specific enzymatic basis for this change has not been fully described.

Metabolism of pyruvate to acetyl-CoA is the first step for preparing glycolytic intermediates to enter the Krebs cycle. By monitoring 3,4-^{14}C-glucose metabolism, NEUBERT (1970) reported that 12% of metabolized glucose enters the acetyl-CoA pool in day-12 rat embryos. Since parallel experiments showed that only 0.1% of 6-^{14}C-glucose was metabolized by the Krebs cycle, this result indicated that a large proportion of glucose substrate was being used for other purposes, such as macromolecular synthesis. More recent evidence confirms the high rates of pyruvate metabolism to acetyl-CoA during early organogenesis (days 8–11) in mouse conceptuses in vitro (E.S. Hunter, unpublished result). When the rate of pyruvate metabolism by the VYS and ectoplacental cone was measured separately, there was a very high rate of pyruvate metabolism by the extraembryonic tissues on day 9. The differences between embryonic and extraembryonic pyruvate metabolism suggest that there may be substantial differences in the energy-producing pathways used by the embryo and the adnexa, but this remains to be established.

The activity of Krebs cycle enzymes has been evaluated in whole-embryo homogenates as well as in isolated intact mitochondria. BASS (1970) reported a marked increase in the specific activity of succinate and α-ketoglutarate dehydrogenases in isolated embryonic mitochondria between days 12 and 14 in rat embryos. In contrast, there was no change in specific cytochrome oxidase activity over the same time period. Similarly, OERTER and BASS (1975) reported that there was a 45-fold increase in total cytochrome oxidase activity between days 11 and 14, but that the change was proportional to the increase in DNA in the embryo. MACKLER et al. (1971) measured the activities of succinate dehydrogenase, ATPase, cytochrome oxidase, and NADH oxidase (antimycin sensitive) in rat embryo homogenates from day 10 to 14. These studies indicated that the activities of all of these enzymes increases with increased gestational age, except NADH oxidase, which plateaus at day 12. It is interesting to note that the level of activity of each enzyme increased in a parallel fashion and to a similar extent, which may suggest a coordinated regulation of enzymatic activity. Further evaluations of NADH oxidase (antimycin A sensitive) indicated an increase in activity between days 11 and 12 in the embryo. However, marked differences between the right and left sides of the embryo were found on day 11. There was a trend towards a difference between sides on days 12 and 13 (FANTEL et al. 1991). These studies indicate that increasing enzymatic activity is observed between the early somite stage and establishment of the vascular system. However, there appear to be differences in the time at which the peak activities are reached, e.g., cytochrome oxidase activity peaks between days 12 and 14, whereas ATPase and NADH oxidase peak on day 12. Although these differences remain to be resolved, the overall trend towards increasing activity with increasing gestational age during organogenesis appears to be the most important point.

HOMMES et al. 1971 evaluated the activity of several Krebs cycle enzymes in fetal liver and found that aconitase and fumarase had lower activities than those observed in adult liver mitochondria. When pyruvate metabolism was monitored, there was a high rate of acetyl-CoA production, suggesting that mitochondria adapt to an anabolic pattern of metabolism and that high rates of lipogenesis would be expected. Whether this same metabolic pattern exists in the embryo remains to be established, but this may provide important clues as to potential regulatory sites of metabolic maturation.

In addition to the metabolism of carbohydrate-generated intermediates, there is also evidence for Krebs cycle metabolism of noncarbohydrate substrate during organogenesis. For example, day -8 to -11 mouse conceptuses metabolize β-hydroxybutyrate (HUNTER and SADLER 1988), and preliminary studies indicate that glutamine is also metabolized by the Krebs cycle in day-8 mouse conceptuses (E.S. Hunter, unpublished data). Additionally, REECE et al. (1985) have reported that early head fold-staged rat conceptuses use 300 μg lipids/ml culture medium over a 24- to 48-h culture period. Whether these lipids are simply incorporated into the embryo or whether they are metabolized and then used for energy production or for de novo lipid synthesis remains to be answered. In this regard, day-13 rat embryos have been shown to consume oxygen in the absence of glucose substrate, suggesting that lipid metabolism has occurred (DE PLAEN 1970).

Although the changes in Krebs cycle activity have been linked to an increased dependence on Ox-Phos, it is important to remember that cytosolic NADH (generated by glycolysis) can be transferred to the mitochondria for energy production. There is no evidence for presence of the cytosolic α-glycerol phosphate dehydrogenase that would be involved in NADH shuttle across mitochondrial membrane (NEUBERT 1970), but there is evidence for activity of the intramitochondrial form that metabolizes glycerol back to dihydroxyacetone-phosphate by FAD-dependent metabolism. There is no evidence either for or against a functional malate-aspartate shuttle system. Thus whether reducing equivalents produced by glycolysis are being used for the generation of ATP during organogenesis remains an unanswered question. Recent data indicating that uncouplers and inhibitors of Ox-Phos induce malformations in early somite mouse embryos in culture (Hunter and Parker 1996) provide evidence that Ox-Phos may have an important function earlier in development than previously thought.

The changes observed in the utilization of the glycolytic and Krebs cycle pathways during gestation have been interpreted to reflect changes in the dependence of the embryo on the Krebs cycle for energy production during development. Interestingly, this change parallels an increased O_2 tension required to maintain normal embryonic growth and development in vitro (New 1978; Sadler 1979), an observation which is consistent with an increased activity of oxidative phosphorylation in the "older" embryo. Although specific increases in several mitochondrial enzymes have been reported, a clear and concise explanation for this metabolic shift has not been established.

One complication with much of the data that has been discussed is that it relies on whole embryo or whole conceptus metabolism. This type of analysis has the disadvantage that it might mask tissue- or organ-specific differences in metabolism. To address this complication, comparisons of embryonic and extraembryonic glucose metabolism have been reported (Akazawa et al. 1994; Hunter and Sadler 1989). In an elegant series of studies, Miki and coworkers (1988) evaluated the activities of enzymes involved in energy metabolism (lactate and succinate dehydrogenases and cytochrome oxidase) using histochemical methods in rat embryos from days 9.5 to 12.5 of gestation. Throughout the embryonic tissues evaluated (neural tube, somites, myocardium, and mesoderm), the activity of lactate dehydrogenase was evenly distributed and the activity increased through to day 12, suggesting a similar use of glycolysis by all tissues. However, when mitochondrial succinate dehydrogenase and cytochrome oxidase were monitored, the activities increased earlier in the myocardium compared to the other tissues. Even on day 12 there was only low to moderate enzymatic activity in the neural tube. Thus the ability of tissues to use the Krebs cycle and Ox-Phos is different within the embryo and appear to be under distinct and unique molecular regulation.

II. Anabolic Uses

In addition to the rates of CO_2 and lactate produced by a metabolic pathway, the flux of glucose C into macromolecules has been evaluated. The ability of embryos to accumulate glucose C in macromolecules is well established (Tanimura and Shepard 1970a; Hunter and Sadler 1988), and, in fact, carbons from glucose are found in most macromolecules. For example, the use of glucose for macromolecular synthesis was about five times greater in day-11 rat embryos than day-13 embryos (KOHLER and BRAND 1970). When the incorporation of glucose C into lipids, RNA, DNA, and proteins was compared, lipids accumulated twice as much glucose as other macromolecules. Fatty acids (free and esterified) and phospholipids were highly labeled and accounted for 85% of the accumulation in lipids.

KROWKE et al. (1971) have also evaluated the incorporation of glucose into macromolecules following administration of glucose substrates in vivo. In day-11 rat embryos, these authors reported similar levels of accumulation in lipid, RNA, and protein fractions. DNA incorporation was 50% of that of RNA, and low levels of accumulation were present in glycosaminoglycans and carbohydrates.

One important use of glucose is the generation of R5P by the PPP. Glucose incorporation in RNA and DNA has been described by several groups. Using techniques similar to KROWKE et al. (1971), BOCHERT et al. (1973), evaluated the incorporation of glucose and phosphate in RNA and DNA of day-12 rat embryos. These studies indicated that there was a high activity of the PPP and that 70%–85% of the glucose label of nucleotides was in the ribose moiety. Additionally, glucose C was found in purine and pyrimidine bases, indicating that glucose metabolism to the appropriate intermediate occurred to support de novo synthesis of these bases in the embryo. ROWE and MCEWEN (1983) and ROWE and KALAIZIS (1985) further evaluated de novo purine synthesis from glycine and serine, respectively. These authors documented that, in addition to serine, there is an important unknown source of one-carbon units used in purine biosynthesis.

These studies document the ability of the embryo to synthesize macromolecules from a variety of substrates, including glucose and amino acids. Although the importance of macromolecular synthesis to normal development seems self-apparent, relatively little work has been carried out to understand the enzymes and or key substrates involved, such as glucose.

E. Perturbation of Glucose Metabolism

The importance of providing glucose as a substrate to the embryo during development is best documented by the adverse effects of an improper supply. To this end, exposure of conceptuses to either insufficient concentrations (hypoglycemia) or excess (hyperglycemia) have been shown to induce dysmorphogenesis during the early stages of organogenesis.

I. Hypoglycemia

Administration of hypoglycemia-inducing agents produces a variety of malformations, including exencephaly and skeletal defects, in vivo. For example, administration of insulin has been reported to induce malformations (LICHTENSTEIN et al. 1951; CHOMETTE 1955; BRINSMADE et al. 1956; SMITHBERG and

RUNNER 1963; HANNAH and MOORE 1971; BUCHANAN et al. 1986; TANIGAWA et al. 1991). Initially, it was not clear whether the adverse effects were produced by insulin itself or the resulting hypoglycemia (see for example, LANDAUER 1972). Evidence now indicates that hypoglycemia is responsible for the teratogenic effects of insulin administration (BUCHANAN et al. 1986 and references in the following paragraphs).

Evidence that glucose is required for embryonic growth is derived partly from studies carried out to define and refine the nutrients and supplements for embryo culture medium. Because of the complex interaction among nutrients in vivo, it has been difficult, if not impossible, to make this assessment without using the in vitro whole embryo culture, where precise changes can be independently regulated. One of the earliest studies to substantiate the requirement for glucose was reported by GUNBERG (1976) using early somite-staged rat embryos. When whole serum was dialyzed and used as culture medium, there was little or no embryonic development. However, when glucose was added to dialyzed serum, growth and development was similar to that obtained in whole serum. Fructose and pyruvate were unable to support development, whereas mannose supported normal heart contractions and closure of the anterior neural tube, although growth was depressed relative to embryos grown in whole serum or dialyzed serum plus glucose.

COCKCROFT (1979), using head fold-staged rat embryos, evaluated the requirements of the embryo for glucose, vitamins, and amino acids by using dialyzed rat serum as the culture medium. Without glucose, embryonic growth and development were severely depressed compared to growth in whole serum. Neither pyruvate nor a mixture of amino acids and vitamins was able to support normal embryogenesis. However, with glucose, amino acids, and vitamins embryonic growth (as monitored by protein content) and morphogenesis were comparable to that achieved in vivo over the same time frame.

Although these studies indicated that the absence of glucose was detrimental to embryonic growth and development, the question of how much glucose was required to support normal development remained unanswered. Using serum from insulin-induced hypoglycemic rats, SADLER and HUNTER (1987) showed that, when the initial glucose concentration (normal glucose concentrations, approximately 150 mg/dl; 8.3 m*M*) was reduced to 60 mg/dl (3.3 m*M*) or less, high rates of malformations were produced in embryos placed into culture at the early somite stage (three to five somites) and that levels of 40 mg/dl (2.2 m*M*) were not only dysmorphogenic, but also reduced embryonic growth as monitored by protein content. Confirmation of the dysmorphogenic effects of insulin-induced hypoglycemic medium has been reported for rat conceptuses (AKAZAWA et al. 1987; ELLINGTON 1987). Duration of exposure to hypoglycemia is also an important consideration. Relatively brief exposures (1–4 h) to glucose concentrations of 20–40 mg/dl (1.1–2.2 m*M*) disrupted normal development in neurulation-staged mouse embryos (three to five somites) despite a return to normal glucose concentrations after exposure to hypoglycemia (SMOAK and SADLER 1990). Similarly, AKAZAWA et al. (1989) reported that a 1-h exposure to 2.2–2.5 m*M* glucose induced dysmorphogenesis in head fold-staged rat conceptuses in vitro.

The question arose as to how hypoglycemia affects glucose utilization by the conceptus and whether this effect explains its teratogenicity. For these studies, HUNTER and SADLER (1989) evaluated the production of lactate from glycolysis as well as the rates of CO_2 formation by the PPP and Krebs cycle. Throughout the 24-h exposure to 40 mg/dl glucose, there was a decrease in lactate production by neurulating mouse conceptuses. At glucose concentrations of 40 and 80 mg/dl, there was a decrease in the rate of CO_2 production by both the Krebs cycle and PPP following an 8- to 12-h exposure. Since glycolysis had been proposed as the primary route of energy production from glucose, these results suggested that hypoglycemia-induced dysmorphogenesis was produced by a decrease in energy production that was accompanied by decreased rates of metabolism by the PPP.

In addition to an effect on glucose metabolism, these results established that anabolic utilization of glucose for DNA and protein synthesis was compromised by exposure to hypoglycemic medium. I.W. Smoak and T.W. Sadler (unpublished results) further evaluated the effect of hypoglycemia of embryonic energy production by measuring embryonic ATP content. Their studies showed that a 12-h exposure to 20 mg/dl glucose reduced embryonic ATP content, thereby supporting the hypothesis that glycolysis is important for energy production and that perturbed energy production contributed to hypoglycemia-induced dys-morphogenesis. However, it is important to recall that short exposures to hypoglycemia (1–4 h), which did not reduce embryonic ATP content, were sufficient to affect embryogenesis, indicating that the mechanism of hypo-glycemia-induced defects may not be due solely to perturbations in energy production.

II. Hyperglycemia

In addition to toxic effects produced by a lack of glucose, an overabundance of this substrate has also been shown to induce malformations in vitro and may be associated with malformations associated with maternal insulin-dependent diabetes. Although maternal diabetes is characterized by a disruption in carbohydrate homeostasis, there are many metabolic changes that occur as a consequence of glucose imbalance, and the most recent information suggests that diabetes-induced malformations are multifactorial, resulting from exposure to many agents (FREINKEL et al. 1986; SADLER et al. 1988, 1989).

The dysmorphogenic effects of hyperglycemia have been reported by many groups (COCKCROFT and COPPOLA 1977; SADLER 1980; COCKCROFT 1984; KATASE et al. 1992). In general, glucose concentrations four-to-six fold greater than the normoglycemic levels (approximately 8 m*M*) are required to induce dysmorphogenesis during neurulation in rodent embryos. Additional work has described a variety of histological effects produced by exposure to hyperglycemic medium, including the following: effects on mitochondrial morphology (FINLEY and NORTON 1991); advanced cellular maturation with differentiation and decreased numbers of mitotic figures in the neuroepithelium (REECE et al. 1985); a reduction in the rough endoplasmic reticulum, decreased number of lipid droplets, and an increase in lysosomal-like structures in the VYS cell (PINTER et al. 1986); and a lengthening of neuroectodermal microvilli (SHEPARD et al. 1993). Although the effects of exposure to hyperglycemia have been well described, there is no definitive explanation as to the mechanism responsible for its toxicity.

A variety of mechanistic studies regarding glucose-induced dysmorphogenesis have been carried out, but no definitive answer obtained. Several studies have focused on a true or functional deficiency of arachidonic acid and prostaglandins as mediating the toxic effects (GOLDMAN et al. 1985; PINTER et al. 1986, 1988; BAKER et al. 1990; GOTO et al. 1992). These studies showed that addition of arachidonate or prostaglandin E_2 to culture medium reduced the incidence of neural tube closure defects induced by glucose. However, ENGSTRÖM et al. (1991) reported that hyperglycemia increased arachidonate uptake compared to control embryos. PINTER et al. (1988) analyzed the fatty acid content in the yolk sac and embryo following exposure to hyperglycemia and reported that the percentage of arachidonate in phospholipids and in nonesterified fatty acids was either unchanged or tended to be increased following exposure to hyperglycemia. These studies indicate that there is no apparent deficiency in arachidonate levels, but do not substantiate changes in its utilization for prostaglandin synthesis.

Hyperglycemia-induced alterations in *myo*-inositol transport have also been reported for a variety of tissues. Several studies have now shown that supplementation of culture medium with *myo*-inositol ameliorates the induction of neural tube closure defects produced by excess glucose and that hyperglycemia leads to a decrement in embryonic *myo*-inositol concentration (SUSSMAN and MATSCHINSKY 1988; BAKER et al. 1990; HASHIMOTO et al. 1990; HOD et al. 1986, 1990; WEIGENSBERG et al. 1990). The decreased accumulation of *myo*-inositol by hyperglycemia or by scylloinositol, a nonmetabolized analogue, has been shown to decrease phosphoinositide hydrolysis in response to endothelin-1 stimulation (STRIELEMAN et al. 1992; STRIELEMAN and METZGER 1993). The authors proposed that hyperglycemia resulted in a decreased ability of the embryo to respond to signals directing embryogenesis, which ultimately led to dysmorphogenesis (STRIELEMAN and METZGER 1993).

The metabolism of glucose to sorbitol by aldose reductase and the subsequent accumulation of sorbitol have also been proposed to mediate glucose-induced dysmorphogenesis (ERIKSSON et al. 1986; HOD et al. 1986; SUSSMAN and MATSCHINSKY 1988). However, aldose reductase inhibitors did not reduce the incidence of glucose-induced malformations (HOD et al. 1986; ERIKSSON et al. 1989). Additionally, ERIKSSON et al. (1989) reported that embryonic accumulation of sorbitol, produced by addition of fructose to the medium, was not associated with dysmorphology, thus providing further evidence that sorbitol was not responsible for the induction of glucose-induced malformations.

Another alternative that has been proposed for glucose-induced defects is that the effects are mediated through nonenzymatic glycation of embryonic proteins (KUBOW et al. 1993). When culture medium is supplemented with acetylsalicylic acid, there is a decrease in the incidence of hyperglycemia-induced malformations and in the embryonic concentration of glycated proteins. Studies to confirm and extend this observation are needed to substantiate the role of glycated proteins as responsible for induction of malformations.

ERIKSSON and BORG (1991, 1993) have reported that the dysmorphogenic effects of glucose can be ameliorated through the addition of free radical scavengers (superoxide dismutase, catalase, or glutathione peroxidase) to the culture medium. These experiments suggest that addition of excess substrate generates free oxygen radicals, which in turn are responsible for the dysmorphogenic effects. These authors have additionally shown that inhibiting the transport of a glucose-derived pyruvate to the mitochondria also reduces glucose-induced malformations. These studies suggest that the dysmorphogenic effects of glucose are at least in part the result of too much oxidative substrate entering the mitochondria and that the subsequent metabolic derangement results in the generation of free oxygen radicals.

III. Other Substrates

In addition to increased levels of carbohydrates inducing malformations, alternative metabolic substrates have also been shown to produce malformations. Pyruvate is an essential substrate for early preimplantation embryos; however, during the postimplantation stage, excess pyruvate induces dysmorphogenesis. A number of studies have indicated that addition of pyruvate to the culture medium is not beneficial (SADLER and NEW 1980), but the concentration dependence for the induction of malformations has only recently been evaluated (E.S. Hunter, unpublished results). These experiments indicated that relatively high concentrations of pyruvate (≥5 m*M*) cause malformations in neurulating mouse embryos.

ERIKSSON and BORG (1993) reported that superoxide dismutase would prevent the induction of malformations by 3 m*M* pyruvate in rat embryos. However, in neurulating mouse embryos, it was only able to decrease the incidence of malformations at 5 m*M* pyruvate, but not at higher pyruvate concentrations (E.S. Hunter, unpublished). Additional experiments focused on the possible prevention of pyruvate-induced malformations by inhibiting the transport of pyruvate into mitochondria using α-cyano-4-hydroxycinnamic acid (CHC). CHC failed to reduce the incidence of pyruvate-induced defects. This result suggested that the toxic effects of pyruvate are not mediated solely through its mitochondrial metabolism. One hypothesis for this effect is that pyruvate is metabolized to lactate and that lactate is responsible for the dysmorphogenic effects.

Since a high rate of glucose is metabolized to acetyl-CoA by the day-12 rat embryo, the rate of pyruvate metabolism by the mouse conceptus on days 8–

11 using ^{14}C-pyruvate has also been evaluated (E.S. Hunter, unpublished). These experiments show that the rate of pyruvate utilization remains relatively constant over this stage of development. However, based on the metabolism of glucose, the metabolic fate of acetyl-CoA shifts from an anabolic to a catabolic pattern. In order to determine whether pyruvate metabolism is important at the neurulation stage, embryos were exposed to the mitochondrial pyruvate transport inhibitor CHC. CHC produced malformations in a concentration-dependent fashion with concentrations of 250 μM or more inducing dysmorphogenesis. Although these experiments support the hypothesis that pyruvate metabolism is important during this stage of development there is no information about how much of an inhibition in pyruvate transport was induced by the toxic levels of CHC.

IV. Glycolytic Inhibitors

In addition to changes in the availability of glucose as a substrate, perturbation of glucose utilization through metabolic inhibitors has also been shown to alter development. In chick embryos, pioneer work into the metabolic requirements during the period of neural tube formation and closure was performed by Spratt (1950). Perturbation of glycolytic metabolism by iodoacetate or sodium fluoride (inhibitors of glyceraldehyde-3-phosphate dehydrogenase and enolase, respectively) produced degeneration of the neural tube. However, normal morphogenesis occurred when pyruvate was added to iodoacetate-containing medium. Thus a dependence upon the metabolism of glucose by the glycolytic pathway and subsequent utilization of the Krebs cycle to maintain morphogenesis during the period of neurulation was proposed for the chick embryo. This pattern of glucose utilization is in marked contrast to the lack of glucose metabolism by the Krebs cycle in neurulation staged rodent conceptuses. There is no current explanation as to why chick embryonic glucose metabolism would be different from rodent metabolism at the same gestational stage.

De Meyer and De Plaen (1964) reported that administration of 2-deoxyglucose (2DG), which inhibits glucose uptake and phosphorylation, to pregnant rats during the early stage of organogenesis, i.e., days 9 and 10, produced a high incidence of resorptions, i.e., embryonic or fetal death. Furthermore, anomalies such as cleft palate were observed among 45% of the surviving fetuses. Since 2DG also affects maternal carbohydrate metabolism, as do all metabolic inhibitors, a change in maternal physiology may have been responsible for the developmental effects and not the effect of the xenobiotic on the embryo.

Following the development of the whole embryo culture system (New and Stein 1964; New 1966, 1967), investigators were able to determine the direct effects of metabolic inhibitors on embryonic metabolism and morphogenesis and compare these results to the growth and development of embryos not exposed to the compounds. Tanimura and Shepard (1970b) reported that

10 m*M* 2-DG severely inhibited somitogenesis, yet was not lethal in day-12 rat embryos grown for a 22-h period. Additionally, following a 3-h exposure, 2DG induced a 60% decrease in U-^{14}C-glucose metabolism. These studies documented the direct inhibitory effects of 2DG on embryonic glucose metabolism and the correlation between this effect and alterations in morphogenesis.

The inhibitory effects of 2DG on embryonic glucose metabolism assessed by TANIMURA and SHEPARD (1970b) correlates well with the inhibitory effects observed by DE MEYER and DE PLAEN (1964) on the rate of CO_2 production from 1- and 6-^{14}C-glucose metabolism. SPIELMANN et al. (1973) further observed an inhibition of O_2 utilization following addition of 2DG to culture medium. However, the latter two groups did not correlate the observed biochemical alteration to an effect on morphogenesis.

GUNBERG (1976) further assessed the effects of an inhibition of the glycolytic pathway on morphogenesis by adding 0.1 m*M* iodoacetate to culture medium. Day-10 rat embryos exposed to iodoacetate (an inhibitor of glyceraldehyde-3-phosphate dehydrogenase) exhibited necrosis or no growth. However, unlike the chick embryo (SPRATT 1950) and the preimplantation rodent embryo (GARDNER and LEESE 1988), the addition of pyruvate to iodoacetate-containing medium did not reverse the embryolethal effect. Thus it was concluded that the day-10 rat embryo was not capable of utilizing pyruvate for Krebs cycle metabolism to circumvent an inhibition of the glycolytic pathway, a result which is consistent with the low activity of the Krebs cycle system at this stage of development (SHEPARD et al. 1970).

To better understand the relationship between inhibition of glucose utilization and dysmorphogenesis, HUNTER and TUGMAN (1996) evaluated the morphological and biochemical effects of 2DG and iodoacetic acid exposure on three- to six-somite mouse embryos in vitro. Both of these inhibitors produced dysmorphogenic and growth-inhibitory effects. Additionally, there was only a small reduction (approximately 20%) in lactate production (glycolysis) following a 24-h exposure period to concentrations of these agents that induced 100% malformations. These results suggests that even relatively small changes in glycolytic metabolism are detrimental to the neurulation-staged embryo.

Since production of ATP is likely to be perturbed following an inhibition of glucose metabolism, embryonic ATP content following exposure to 2DG was determined. Following a 24-h exposure to the toxicant, there was no reduction in ATP content in the whole embryo or in the VYS. This was an unexpected result, since I.W. Smoak and T.W. Sadler (unpublished results) had shown a reduction in ATP content following a 12-h exposure to hypoglycemic medium. However, MCCANDLES and SCOTT (1981) reported decreased ATP levels in the neuroepithelium following exposure to 6-aminonicotinamide, while demonstrating no changes in whole embryo ATP. Therefore, the ATP content in different regions of the embryo (head, midpiece, tail, and heart) were evaluated separately following exposure to 2DG. There was a specific reduction in ATP content in both the head and midpiece regions after a 24-h exposure, but no reduction following a 12-h exposure. Additional studies demonstrated that the regional specificity of 2DG-induced malformations was not due simply to differences in 2DG distribution in the embryo (HUNTER and TUGMAN 1995). These experiments suggest that dysmorphogenesis produced by 2DG is not due to a reduction in embryonic ATP content.

Mannose, an epimer of glucose, does not directly inhibit glycolytic metabolism, but rather, by competing with glucose for phosphorylation, inhibits glucose utilization. Mannose was shown to produce dysmorphogenesis in day-10.5 rat embryos (FREINKEL et al. 1983, 1984). Following exposure to 1.5 mg/ml (8.3 m*M*) mannose, embryos exhibited high rates of neural (65.2%) and extraneural (91.3%) defects. Utilization of glucose was decreased in the presence of a teratogenic concentration of mannose, although total hexose uptake (i.e., glucose plus mannose) was not affected. Furthermore, in the presence of mannose, the production of lactate by the embryo was significantly reduced. These data suggest that the decreased rate of embryonic glycolysis produced by exposure to mannose produced the dysmorphogenic response. Supporting evidence for this hypothesis was derived from the amelioration of mannose-induced neural tube defects by hyperglycemia. Under these conditions, i.e., 7.2 mg/ml (40 m*M*) glucose plus 1.5 mg/ml mannose, the amount of hexose metabolized by glycolysis (lactate production) was the same as that in control embryos, i.e., 1.2 mg/ml (6.7 m*M*). These data again suggest that the rodent embryo in vitro is dependent upon the glycolytic pathway as an essential pathway during the early period of organogenesis. Furthermore, these results support the hypothesis that an inhibition of glucose metabolism can induce embryonic malformations.

To further evaluate the relative utilization of glycolysis and the Krebs cycle, FREINKEL and colleagues (1983) compared the toxicity of mannose using rat conceptuses grown in the presence of mannose plus either 5% O_2 (normoxia) or 20% O_2 (hyperoxia). These experiments showed a decrease in the incidence of mannose-induced malformations at the higher oxygen concentration. These studies strongly suggest that oxidative phosphorylation may be induced as an alternative energy source during early organogenesis. In support of this hypothesis, MACKLER et al. (1973) reported that exposure of dams to 85% O_2 on days 8–11 increased mitochondrial NADH oxidase activity in day-11 rat embryos.

There are also animal models with genetic deficiencies of glycolytic enzymes. For example, the importance of intact glucose-metabolizing pathways is fully established by the embryolethal effects of deletion of either glucose phosphate isomerase (MERKLE and PRETSCH 1992) or triose phosphate isomerase (MERKLE and PRETSCH 1989) in mutant mice. The homozygous embryos of both mutants die during the early postimplantation stage, but the heterozygotes appear to be unaffected by the decreased metabolic capabilities. Interestingly, by day 9.5 of gestation in conceptuses that were homozygous for the glucose phosphate isomerase deficiency, only extraembryonic tissues were present (WEST 1993). This observation indicates that extraembryonic tissues have a differential ability to compensate for loss of glycolytic activity through an increased dependence on non-glycolytic pathways and substrates for anabolic and catabolic processes.

V. Pentose Phosphate Pathway Inhibitors

In addition to the embryonic utilization of the glycolytic and Krebs cycle pathways for energy production, the rates of glucose metabolism by the PPP and the utilization of intermediates generated by this pathway for nucleic acid synthesis have also been assessed. As described above, the rate of glucose metabolism by the PPP appears to reach a maximum early in the period of organogenesis and can account for as much as 2.0% of the total utilization of glucose. During this period of gestation, a high rate of nucleic acid synthesis occurs, suggesting an important function for this pathway.

The teratogenic effects of 6-aminonicotinamide (6AN), an inhibitor of the PPP, were reported as early as 1959 (PINSKY and FRASER 1959). Since this initial study, 6AN has been reported to induce multiple malformations, including neural tube defects (CHAMBERLAIN 1963), cleft palate (CHAMBERLAIN 1965; DIEWERT 1979; BIDDLE and FRASER 1979), cleft lip (TRASLER 1978), abnormal limb development (MCLACHLAN 1980), and lumbosacral defects (TANAKA 1981).

The mechanism by which 6AN inhibits the PPP involves the replacement of the nicotinamide moiety of pyridine nucleotides with the 6-amino analogue (JOHNSON and MCMALL 1956), forming 6-aminonicotinamide adenine dinucleotide (phosphate) [6-ANAD(P)]. In embryonic rat tissue, 6-ANADP competitively inhibits the enzymatic activity of G-6-PD, and 6-P-GD (BARRACH 1970). However, 6-P-GD is approximately 20 times more sensitive to this inhibition than G-6-PD.

Exposure of embryos to 6AN in vivo completely blocked the embryonic utilization of the oxidative portion of the PPP (KOHLER and NEUBERT 1968), a result supported by an accumulation of 6-phosphogluconate in embryonic tissue following administration of 6AN in vivo (BARRACH 1970). Utilizing radiolabeled glucose as substrate, the effects of 6AN on the metabolism and incorporation of intermediates into the ribose moieties of DNA and RNA have been reported (BARRACH 1970). Following administration of 6AN, the rate of ^{14}C-CO_2 generated from 1-^{14}C-glucose was decreased, and the incorporation of 6-^{14}C-glucose into RNA and DNA decreased to less than 40% of control values. Thus a decreased rate of embryonic nucleic acid synthesis was observed following exposure to 6AN in vivo. These effects have been corroborated in later studies in which the incorporation of ^{3}H-thymidine into DNA was inhibited by 50%–70% (RITTER et al. 1972), the flux of ^{14}C-orotic acid into DNA was inhibited (BARRACH 1970), and a quantitative decrease in RNA and DNA was observed (SCHEIL et al. 1977) after administration of 6AN.

The ketone body β-hydroxybutyrate (BOHB) has been reported to inhibit glucose utilization in both adult and fetal tissues. In adult tissue, this effect is in part mediated by an inhibition of the enzyme phosphofructokinase (Newsholme et al. 1962). However, in fetal brain the mechanism or site of this effect has not been elucidated (Shambaugh et al. 1977).

The dysmorphogenic effects of exposure to BOHB in vitro have been described by a number of investigators (Horton and Sadler 1983; Lewis et al. 1983; Sheehan et al. 1985; Hunter and Sadler 1987; Moore et al. 1989). Concentrations as low as 8 m*M* DL-BOHB induced malformations among three- to four- somite mouse embryos, while 32 m*M* induced growth retardation and neural tube defects (Horton and Sadler 1983). BOHB did not induce cell death (Horton and Sadler 1983), but reduced the mitotic index in the neurectoderm of the mesencephalon and rhombencephalon (Horton and Sadler 1985). Ultrastructurally, the only alteration observed was progressive mitochondrial swelling, which occurred in all embryonic cell types.

Since BOHB inhibits glucose metabolism in adult and fetal tissues (Newsholme et al. 1962; Shambaugh et al. 1977), the effects of BOHB on embryonic glucose metabolism were also evaluated. Horton et al. (1985) observed a 33% inhibition in the rate of lactate production from glucose by day-11 mouse embryos exposed to 32 m*M* DL-BOHB in vitro. However, when embryos were evaluated during the critical period of neural tube closure, no effect on glycolysis was observed following a 4-h exposure to 32 m*M* DL-BOHB (Horton et al. 1985). Thus inhibition of glycolytic metabolism was not responsible for the dysmorphogenic effects of BOHB.

Further studies of the effects of BOHB on fetal brain indicated that DL-BOHB inhibits the de novo synthesis of pyrimidines (Bhasin and Shambaugh 1982) and purines (Shambaugh et al. 1984) in vitro. The inhibition of pyrimidine synthesis has been shown to occur at the site of orotic acid synthesis and was proposed to be mediated by an allosteric effect of BOHB on carbamyl-phosphate synthetase II. Both BOHB and acetoacetate (ACAC) inhibit the synthesis of adenine. However, they do not affect purine salvage pathways in fetal brain in vitro. During the period of neurulation, the rodent embryo in vitro utilizes de novo pathways for purine nucleosides synthesis (Rowe and McEwen 1983). The enzymes of the purine salvage pathway, hypoxanthine phosphoribosyl-transferase and adenine phosphoribosyl-transferase, are present in rat embryos during embryogenesis (Rowe and McEwen 1983). However, there is no available information on the ability of embryonic tissue to compensate for decreased de novo synthesis by salvage pathways during the period of neural tube closure.

Hunter et al. (1987) further pursued the effects of BOHB on embryonic glucose utilization during the period of neurulation in mouse embryos. There was no effect of BOHB on glycolytic or Krebs cycle metabolism during a 24-h exposure period to 32 m*M* BOHB. However, there was a specific reduction in the rate of glucose metabolism by the PPP between 8 and 24 h of culture. The effects of BOHB on the PPP did not appear to be the result of direct inhibition, but rather result from a change in the redox potential (NADH to NAD ratio) produced by metabolism of the ketone body.

As mentioned previously, exposure to BOHB had been shown to inhibit de novo purine and pyrimidine synthesis at the site of orotic acid synthesis in fetal tissues. Although there was no reduction in embryonic orotate synthesis produced by exposure to BOHB, there was a reduction in the synthesis of uridine monophosphate (UMP) (HUNTER et al. 1987). Since UMP is produced from orotate and R5P, this observation supported the hypothesis that BOHB's reduction in glucose metabolism by the PPP resulted in a decrement in R5P.

Ribose at both 5 and 7.5 m*M* in combination with BOHB reduced the incidence of neural tube defects compared to BOHB alone. This result provided evidence that the mechanism of BOHB-induced neural tube defects was due in large part to a decreased de novo nucleoside synthesis, DNA synthesis, and hence mitotic index (as reported by HORTON and SADLER 1983) in the neuroepithelium.

In contrast to the effects of BOHB on glucose metabolism in five- to six-somite embryos (HUNTER et al. 1987), SIIUM and SADLER (1990) observed that in gestationally younger conceptuses (two- to three-somite stage) the predominant effect of BOHB was not on glucose utilization, but was a 34.3% decrease in de novo pyrimidine synthesis similar to the effect of BOHB in fetal tissue (BHASIN and SHAMBAUGH 1982). The reason for the differences in metabolic effect may be due to differences in the ability of the conceptus to metabolize the ketone body at these different gestational ages.

More recent evidence has indicated that the dysmorphogenic effects of BOHB may also result from generation of reactive oxygen species (ERIKSSON and BORG 1993). Although the genesis of reactive oxygen species (ROS) has not been described following exposure to BOHB, these factors may result from disruption of normal mitochondrial metabolism.

The studies of the effects of 6AN and BOHB further support the important role of the PPP in the generation of intermediates required for macromolecular synthesis during organogenesis and suggest that there is a critical role of the redox potential in regulating metabolism during this stage of development.

VI. Krebs Cycle Inhibitors

As described above, the rate of glucose metabolism by the Krebs cycle is low and accounts for less than 0.2% of the glucose utilized by the conceptus during the neurulation stage of rodent development. However, there is a high flux of glucose C into acetyl-CoA, as monitored by pyruvate metabolism on days 8–11 (E.S. Hunter, unpublished result). There is also a growing body of evidence to indicate that the flux of pyruvate across the mitochondrial membrane and Krebs cycle metabolism are important metabolic events during early organogenesis. For example, the adverse developmental effects of fluoroacetate, a well-characterized Krebs cycle inhibitor, on neurulating mouse embryos have been reported using whole embryo culture (HUNTER and PARKER submitted). Preliminary studies on the ATP content in different regions of the conceptus following a 24-h exposure to fluoroacetate showed a 35%–40% reduction in the tail and head regions, but no concomitant effects in the heart, midpiece, or VYS (HUNTER and PARKER 1996). There were no changes in ATP content following a 6-h exposure to fluoroacetate, suggesting that an effect on energy production is not solely responsible for the induction of malformations following an inhibition of the Krebs cycle.

Confirmatory studies of the dysmorphogenic effects of inhibition of Krebs cycle metabolism have also been conducted using malonate (HUNTER and PARKER 1996), a specific competitive inhibitor of succinate dehydrogenase (WHITE et al. 1978). However, whether this disruption of Krebs cycle activity affects development because of a primary effect on energy production or macromolecular synthesis or some other mechanism is unknown. Since Krebs cycle intermediates are used for macromolecular synthesis, inhibition of Krebs cycle metabolism may alter development via this mechanism (Fig. 3). As described previously, NEUBERT (1970) reported that 12% of the glucose used by day-12 rat embryos was metabolized to acetyl-CoA. Acetyl-CoA is condensed with OAA to form citrate, which can be further metabolized by the Krebs cycle or can cross the mitochondrial membrane to the cytosol to be used for de novo lipogenesis through its degradation to acetyl-CoA by ATP citrate lyase. (-)-Hydroxycitrate, a specific inhibitor of citrate metabolism, produced high rates of malformations in early somite-staged mouse embryos in vitro (E.S. Hunter and T.W. Sadler, unpublished results). However, there has been no description of how much of an inhibition of ATP citrate lyase was associated with the induction of dysmorphogenesis. Nonetheless, perturbation in Krebs cycle metabolism appears not only to inhibit the generation of reducing equivalents by the cycle, but also may decrease the flux of substrate to the cytosol for macromolecular synthesis.

Metabolic inhibitors have also been used to assess the metabolic capabilities of isolated organs in culture. Thus inhibition of glycolysis by iodoacetate produced a complete depression of the heart rate from day-12 and -13 rat embryos (COX and GUNBERG 1972). However, pyruvate was able to circumvent the blockage and provide sufficient energy to maintain contractions (but is unable to prevent dysmorphogenesis in the whole embryo, GUNBERG 1976). Inhibition of Krebs cycle metabolism (by malonate) or Ox-Phos (by 2,4-dinitrophenol) differentially affected heart contractions in tissue of these gestational ages. These data suggest that during an early stage of cardiogenesis, i.e., day 12, energy production occurs primarily from the metabolism of glucose by the glycolytic pathway. However, the data also show that either some dependence upon Ox-Phos for energy production by embryonic cardiac tissue is present or that the heart has the capacity to compensate for a decrease in glycolytic flux by using the Krebs cycle. In contrast, hearts from day-13 embryos are more dependent on the utilization of the Krebs cycle and the

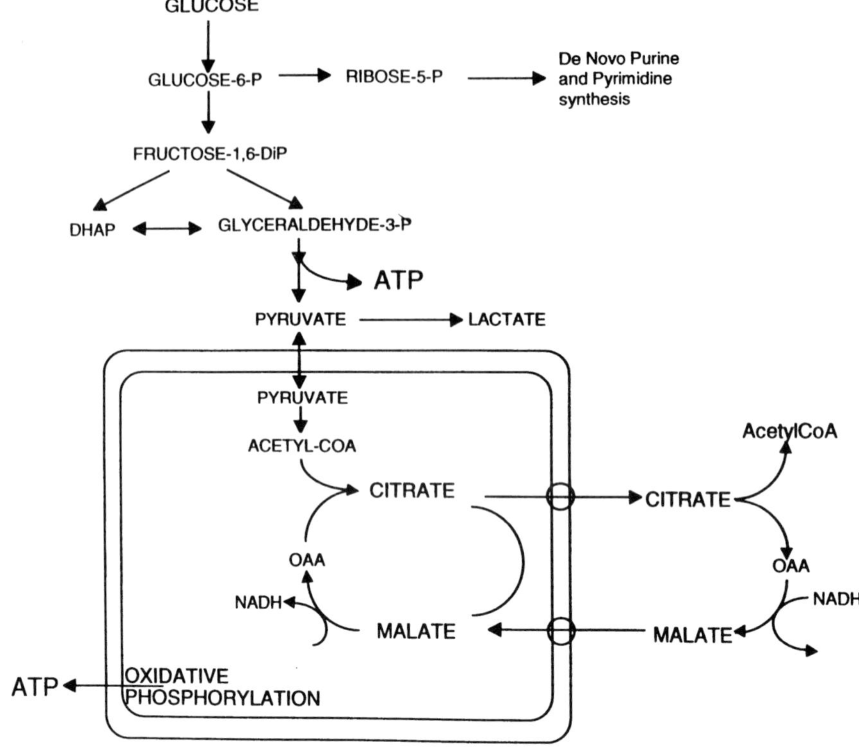

Fig. 3. Proposed model of citrate/malate shuttle for embryonic metabolism during early organogenesis. *DHAP*, dihydroxyacetone phosphate; *ATP*, adenosine triphosphate; *CoA*, coenzyme A; *NADH*, reduced nicotinamide adenine dinucleotide; *OAA*, oxaloacetate

electron transport system for energy production than hearts from younger embryos. Thus the differential effects of metabolic inhibitors have been substantiated by the enzymatic changes in the heart reported by MIKI et al. (1988).

VII. Oxidative Phosphorylation Inhibitors

Ox-Phos is the pathway used to transport electrons from cofactors (NADH and $FADH_2$) generated from glycolysis and the Krebs cycle and functions to generate ATP. The embryotoxicity and growth-inhibitory effects of thiamphenicol and chloramphenicol have been reported by several groups (MACKLER et al. 1975; BASS 1975; OERTER and BASS 1972, 1975). These agents were used as tools to study the role of mitochondrial metabolism during embryogenesis because of the proposed selective inhibition of mitochondrial protein synthesis. Administration of these xenobiotics on different days of gestation produced differential effects, indicating that the embryo's dependence on mitochondrial metabolism changed with gestational age. For example, BASS and

OERTER (1977) reported that the day-9 rat embryo appeared to be insensitive to thiamphenicol administration compared to the day-10 or -11 embryo. When administration of either agent occurred on days 10 and 11, there was a marked decrease (50%) in cytochrome oxidase activity in day-12 mitochondria (BASS 1975; BASS et al. 1978). Chloramphenicol completely blocked cytochrome oxidase synthesis during the first 24 h after administration in day-12 embryos (OERTER and BASS 1975). Following administration of an embryolethal dose (100 mg/kg) of thiamphenicol, embryonic ATP content was decreased by 50%. However, the relationship between inhibition of mitochondrial activity and ATP content could not be substantiated, although the observation that less than a 50% decrease in ATP was associated with embryolethality was important.

Further studies on the effects of perturbation in mitochondrial energy production showed that chloramphenicol induced growth retardation and high rates of resorptions, but was not teratogenic, when rats were exposed during organogenesis (MACKLER et al. 1975). Administration of chloramphenicol reduced mitochondrial NADH oxidase and cytochrome oxidase activity in day-14 fetuses. In contrast, malonate, which is a potent inhibitor of succinate oxidase, did not alter mitochondrial metabolism and was neither teratogenic nor embryotoxic. Although these data indicate that agents that perturb mitochondrial metabolism have the capacity to adversely affect development, there was no correlation between their potency as metabolic inhibitors and the developmental toxicities induced. One explanation for this discrepancy may have been that maternal metabolism and distribution of these compounds may have limited embryonic exposure and prevented their developmental toxicity. This difference likely explains why malonate induced dysmorphogenesis in vitro (HUNTER and PARKER 1996), but not in vivo.

The importance of Ox-Phos during neurulation has also been studied. As described above, the embryo uses glycolysis as the predominant route of glucose metabolism with little (less than 0.2%) of the glucose C completes the Krebs cycle metabolism. Based on this pattern of glucose metabolism, it has been proposed that Ox-Phos is not an important process during early organogenesis. Direct evidence of the rates of enzymatic activity, as described previously, also suggest that this pathway may not be an important source of energy. In an effort to compare the function of Krebs cycle metabolism for catabolic versus anabolic processes, the direct effects of uncouplers and inhibitors of Ox-Phos were assessed in neurulation-staged mouse conceptuses using whole embryo culture. Despite the proposed unimportance of Ox-Phos, rhodamine-123 (R123), an uncoupler of Ox-Phos, produced high rates of malformations, including neural tube closure defects in neurulation mouse embryos (HUNTER and PARKER 1996). Additionally, relatively brief exposure periods of 3 and 6 h were sufficient to induce neural tube defects, further demonstrating the sensitivity of embryos to this biochemical perturbation. Time-course studies of ATP content in the head, heart, and VYS demonstrated that exposure to R123 decreased ATP content in the head and heart and did so in a time frame that suggested a cause and effect relationship with the induction of dysmorphogenesis (HUNTER and PARKER 1996). Since the serum half-life of R123 is approximately 15–60 min (SWEATMAN et al. 1990), the relatively brief embryonic exposure following maternal administration of R123 may explain the lack of developmental effects of this agent in vivo (HOOD et al. 1988, 1989).

Further evidence to support the importance of Ox-Phos during neurulation comes from the dysmorphogenic effects of perturbing Ox-Phos by 2,4,-dinitrophenol or sodium azide (HUNTER and PARKER 1996). 2,4-Dinitrophenol is a well-characterized uncoupler of Ox-Phos. In contrast, sodium azide perturbs Ox-Phos by inhibiting electron transport by cytochrome oxidase.

Further evidence to support the importance of Ox-Phos during organogenesis comes from studies documenting the teratogenic and embryotoxic effects of cyanide and azide (DOHERTY et al. 1982; SANA et al. 1990). Cyanide administered to golden hamsters on days 6–9 by slow infusion resulted in neural tube defects. Coadministration of thiosulfate, a cyanide antagonist, ameliorated the teratogenic effects of cyanide. The authors further indicated that these studies support the hypothesis that the toxicity of some aliphatic nitriles are the result of a metabolic production of cyanide. In contrast, administration of azide to Syrian hamsters using an osmotic minipump on days 7–9 did not result in teratogenicity. However, there was an increase in embryonic resorptions compared to controls.

One remaining challenge with regard to the function of Ox-Phos will be to determine whether the reducing equivalents (NADH or $FADH_2$) originate from cytosolic metabolism or from the Krebs cycle. Hunter and Tugman have proposed that NADH generated from glycolysis is used by Ox-Phos. This working hypothesis proposes that there is a coordinated transport of malate into the mitochondria and citrate into the cytosol. In the cytosol, citrate is metabolized to acetyl-CoA and OAA by ATP citrate lyase. Acetyl-CoA is used for lipogenesis, and OAA is metabolized to malate via an NADH-dependent enzyme. Malate is then transported back into the mitochondria and metabolized to OAA to generate NADH in the mitochondria. OAA condenses with newly generated acetyl-CoA forming citrate, and the "cycle" begins again. There is no direct evidence to support this hypothesis, but it is consistent with the toxicity produced by inhibition of ATP citrate lyase and uncouplers and inhibitors of Ox-Phos. It also explains why there is a high rate of pyruvate metabolism without Krebs cycle metabolism of glucose.

F. Future Research

There are a number of important questions that remain to be solved with regard to intermediary metabolism during the period of organogenesis. First, there is little if any information about the regulatory steps in carbohydrate metabolism during organogenesis. Therefore, mechanisms responsible for the shift from a glycolytic to a Krebs cycle pattern of metabolism are unknown. Since this shift is a fundamental change in carbohydrate metabolism, the signals and enzymes involved may be important targets for toxicants.

Second, recent information has suggested that Ox-Phos may be an important pathway for energy production, even during early stages of organogenesis. Therefore, although the predominate route of glucose metabolism is glycolysis, is this pathway the primary route of energy production? If Ox-Phos is in fact an important energy source, is NADH from glycolytic metabolism used by Ox-Phos or are NADH and $FADH_2$ only generated by the Krebs cycle. These questions are important in determining and comparing the importance of toxicants that perturb metabolism by these different pathways during development.

Another important series of questions relate to the transport schemes and mechanisms used by the conceptus. At preimplantation stages, glucose is transported directly to all cells of the conceptus. However, there is little information concerning postimplantation stages before a definitive placenta is established and when neurulation and other morphogenetic events occur.

In addition to understanding the relationships between transport of nutrients, the importance of glucose in macromolecular synthesis has long been overlooked and has not been fully appreciated. Although the utilization of glucose for macromolecular synthesis has been established, further information regarding the regulation of the pathways involved and their sensitivity to toxicants is essential.

Additionally, although a number of metabolic changes induced by exposure to inhibitors and toxicants have been described, cause and effect relationships between these changes and the resulting malformations have not been well documented. For example, many papers have reported that a change in glycolytic metabolism resulted in a change in ATP content, which in turn resulted in abnormal development, yet how changes in ATP affect the embryo remains to be answered. Thus what are the sensitive ATP-dependent processes in the embryo? How much of a change in ATP content is required to affect a given process, i.e., is a 10% decrease in ATP important? What about non-ATP sources of high-energy phosphates such as GTP and phosphocreatine. These and other questions remain to be addressed with regard to how changes in intermediary metabolism ultimately result in dysmorphogenesis.

We have a rather complete understanding of the rates of glucose metabolism by the rodent conceptus during the period of organogenesis and, based on these rates of product formation, we have a good understanding of the relative utilization of this sugar by different pathways and the changes in the utilization by these pathways at different gestational ages. However, if we critically look at what we know, it is clear that we only have a rudimentary knowledge of intermediary metabolism. Unlike both earlier and later stages of gestation, there is little information regarding key regulatory enzymes of the pathways of glucose metabolism, and there is little information about changes in enzymatic activity that explain the relative shift to increasing Krebs cycle metabolism of glucose. Additionally, we know little about what alternative pathways are available to the embryo and the embryo's ability to compensate for a perturbation in metabolism. Furthermore, not only do these studies need to be performed for the whole conceptus, but there is also a critical need to perform these experiments in specific tissues and organ anlagen during early organogenesis. It may be that differences in metabolism or the ability of tissues to compensate for perturbation in substrate flux may help explain the tissue specificity for a variety of toxicants. Thus the study of normal intermediary metabolism should be the focus of future investigations.

References

Akazawa S, Akazawa M, Hashimoto M, Yamaguchi Y, Kuriya N, Toyama K, Ueda Y, Nakanishi T, Mori T, Miyake S, Nagataki S (1987) Effects of hypoglycemia on early embryogenesis in rat embryo organ culture. Diabetologia 30: 791–796

Akazawa M, Akazawa S, Hashimoto M, Akashi M, Yamazaki H, Tahara D, Yamamoto H, Yamaguchi Y, Nakanishi T, Nagataki S (1989) Effects of brief exposure to insulin-induced hypoglycemic serum during organogenesis in rat embryo culture. Diabetes 38: 1573–1578

Akazawa S, Unterman T, Metzger BE (1994) Glucose metabolism in separated embryos and investing membranes during organogenesis in the rat. Metabolism 43: 830–835

Ayabe T, Tsutsumi O, Taketani Y (1994) Hexokinase activity in mouse embryos developed in vivo and in vitro. Hum Reprod 9: 347–351

Baker L, Piddington R, Goldman A, Eller J, Moehring J (1990) Myo-inositol and prostaglandins reverse the glucose inhibition of neural tube fusion in cultured mouse embryos. Diabetologia 33: 593–596

Barbehenn EK, Wales RG, Lowry OH (1974) The explanation for the blockade of glycolysis in early mouse embryos. Proc Natl Acad Sci USA 71: 1056–1060

Barrach HJ (1970) Effects of 6-aminonicotinamide on the glucose metabolism of mammalian embryonic tissue. In: Bass R, Beck F, Merker HJ, Newbert D, Randhahn (eds) Metabolic pathways in mammalian embryos during organogenesis and its modification by drugs. Free University, Berlin, pp 365–384

Bass R (1970) Respiration and oxidative phosphorylation of mitochondrial fractions isolated from rat embryos. In: Bass R, Beck F, Merker H-J, Neubert D, Randhahn B (eds) Metabolic pathways in mammalian embryos during organogenesis and its modification by drugs. Freie Universität, Berlin, pp 309–319

Bass R (1975) Significance of mitochondrial function for embryonic development: proposal of a new mechanism for the induction of embryolethal effects. In: Neubert D, Merker HJ (eds) New approaches to the evaluation of abnormal embryonic development. Thieme, Stuttgart, pp 524–541

Bass R, Oerter D (1977) Embryonic development and mitochondrial function. 2. Thiamphenicol induced embryotoxicity. Naunyn–Schmiedebergs Arch Pharmacol 296: 191–197

Bass R, Schmidt C (1971) Respiration and oxidative phosphorylation in embryo-mitochondria (rats). Naunyn Schmeidebergs Arch Pharmacol 270 [Suppl]: R6

Bass R, Oerter D Krowke R, Spielmann H (1978) Embryonic development and mitochondrial function. III. Inhibition of respiration and ATP generation in rat embryos by thiamphenicol. Teratology 18: 93–102

Bhasin S, Shambaugh GE III (1982) Fetal fuels. V. Ketone bodies inhibit pyrimidine biosynthesis in fetal rat brain. Am J Physiol 243: E234–E239

Biddle FG, Fraser FC (1979) Genetic independence of the embryonic reactivity differences to cortisone and 6-AN-induced cleft palate in the mouse. Anat Embryol (Berl) 154: 49–54

Biggers JD (1971) New observations on the nutrition of the mammalian oocyte and the preimplantation embryo. In: Blandan RJ (ed) The biology of the blastocyst. University of Chicago Press, Chicago, pp 319–329

Biggers JD, Borland RM (1976) Physiological aspects of growth and development of the preimplantation mammalian embryo. Annu Rev Physiol 38: 95–119

Biggers JD, Whittingham DG, Donahue RP (1967) The pattern of energy metabolism in the mouse oocyte and zygote. Proc Natl Acad Sci USA 58: 560

Biggers JD, Gardner DK, Leese HJ (1989) Control of carbohydrate metabolism in preimplantation mammalian embryos. In: Rosenblum IY, Heyner S (eds) Regulation of growth in development. CRC Press, Boca Raton, pp 19–32

Bochert G, Mewes P, Krowke R (1973) Labelling of RNA and DNA moieties of mammalian embryos in-vivo by 14C-U-Glucose and 32P-orthophosphate. Naunyn Schmiedebergs Arch Pharmacol 277: 413–428

Brinsmade AF, Büchner F, Rübsaamen H (1956) Missbildungen am Kaninchenembryi durch Insulininjektion beim Muttertier. Naturwissenschaften 43: 259

Brinster RL (1973) Nutrition and metabolism of the ovum, zygote and blastocyst. In: Green RO, Astwood EB (eds) Handbook of Physiology, sect 7, vol II, Part 2. American Physiology Society, Washington, p 165

Buchanan TA, Schemmer JK, Freinkel (1986) Embryotoxic effects of brief maternal insulin-hypoglycemia during organogenesis in the rat. J Clin Invest 78: 643-649

Chamberlain JG (1963) Deleterious effects of 6-aminonicotinamide on implantation and early embryonic development in the Long Evans rat. Anat Rec 145: 312

Chamberlain JG (1965) Development of cleft palate induced by 6-aminonicotinamide late in gestation. Anat Rec 156: 31–40

Chomette G (1955) Entwicklungstörungen nach Insulinschock beim trächtigen Kaninchen. Beitr Pathol Anat 115: 439–451

Clough JR (1985) Energy metabolism during mammalian embryogenesis. Biochem Soc Trans 13: 77–79

Clough JR, Whittingham DG (1983) Metabolism of ^{14}C-glucose by postimplantation mouse embryos in vitro. J Embryol Exp Morphol 74: 133–142

Cockcroft DL (1979) Nutrient requirements of rat embryos undergoing organogenesis in vitro. J Reprod Fertil 57: 505–510

Cockcroft DL (1984) Abnormalities induced in cultured rat embryos by hyperglycemia. Br J Exp Pathol 65: 625–636

Cockcroft DL, Coppola PT (1977) Teratogenic effects of excess glucose on head-fold rat embryos in culture. Teratology 16: 141–146

Cox SJ, Gunderg DL (1972) Metabolic utilization by isolated embryonic rat hearts in vitro. J Embryol Exp Morphol 28: 235–245

De Meyer R, De Plaen J (1964) An approach to the biochemical study of teratogenic substances on isolated rat embryo. Life Sci 3: 709–713

de Plaen JL (1970) Respiration of rat embryos in vitro. In: Bass R, Beck F, Merker H-J, Neubert D, Randhahn B (eds) Metabolic pathways in mammalian embryos during organogenesis and its modification by drugs. Freie Universität, Berlin, pp 261–269

Diewert VM (1979) Correlation between mandibular retrognathia and induction of cleft palate with 6-AN in the rat. Teratology 19: 213–227

Doherty PA, Ferm VH, Smith RP (1982) Congenital malformations induced by infusion of sodium cyanide in the golden hamster. Toxicol Appl Pharmacol 64: 456–464

Ellington SKL (1987) Development of rat embryos cultures in glucose-deficient media. Diabetes 36: 1372–1378

Engström E, Haglund A, Eriksson UJ (1991) Effects of maternal diabetes or in vitro hyperglycemia on uptake of palmitic and arachidonic acid by rat embryos. Pediatr Res 30: 150–153

Eriksson UJ, Borg LAH (1991) Protection by free radical scavenging enzymes against glucose-induced embryonic malformations in vitro. Diabetologia 34: 325–331
Eriksson UJ, Borg LAH (1993) Diabetes and embryonic malformations. Role of substrate-induced free radical production for dysmorphogenesis in cultured rat embryos. Diabetes 42: 411–419
Eriksson UJ, Naeser P, Brolin SE (1986) Increased accumulation of sorbitol in offspring of manifest diabetic rats. Diabetes 35: 1356–1363
Eriksson UJ, Broin SE, Naeser P (1989) Influence of sorbitol accumulation on growth and development of embryos cultured in elevated levels of glucose and fructose. Diabetes Res 11: 27–32
Fantel AG, Person RE, Burroughs-Glein C, Shepard TH, Juchau MR, Mackler B (1991) Asymmetric development of mitochondrial activity in rat embryos as a determinant of the defect patterns induced by exposure to hypoxia, hyperoxia and redox cycles in vitro. Teratology 44: 355–362
Finley BE, Norton S (1991) Effects of hyperglycemia on mitochondrial morphology in the region of the anterior neuropore in the explanted rat embryo model: evidence for a modified Reid hypothesis as a mechanism for diabetic teratogenesis. Am J Obstet Gynecol 165: 1661–1666
Freinkel N, Lewis NJ, Akazawa S, Gorman L, Potaczek M (1983) The honeybee syndrome: teratogenic effects of mannose during organogenesis in rat embryo culture. Trans Assoc Am Physicians 96: 44–45
Freinkel N, Lewis NJ, Akazawa S, Roth SI, Gorman L (1984) The honeybee syndrome – implications of the teratogenicity of mannose in rat-embryo culture. N Engl J Med 310: 223–230
Freinkel N, Cockcroft DL, Lewis NJ, Gorman L, Akazawa S, Phillips LS, Shambaugh GE III (1986) The 1986 McCollum award lecture. Fuel-mediated teratogenesis during early organogenesis: the effects of glucose, ketones, or somatomedin inhibitor during rat embryo culture. Am J Clin Nutr 44: 986–95
Gardner DK, Leese HJ (1988) The role of glucose and pyruvate transport in regulating nutrient utilization by preimplantation mouse embryos. Development 104: 423–429
Goldman AS, Baker L, Piddington R, Mary B, Herold R, Elgler J (1985) Hyperglycemia-induced teratogenesis is mediated by a functional deficiency of arachidonic acid. PNAS USA 82: 8227–8231
Goto MP, Goldman AS, Uhing MR (1992) PGE2 prevents anomalies induced by hyperglycemia or diabetic serum in mouse embryos. Diabetes 41: 1644–1650
Gunberg DL (1976) In vitro development of postimplantation rat embryos cultured on dialyzed rat serum. Teratology 14: 65–70
Hannah RS, Moore KL (1971) Effects of fasting and insulin on skeletal development in rats. Teratology 4: 135–140
Hashimoto M, Akazawa S, Akazawa M, Akashi M, Yamamoto H, Maeda Y, Yamaguchi Y, Yamasaki H, Tahara D, Nakanishi T, Nagataki S (1990) Effects of hyperglycemia on sorbitol and myo-inositol contents of cultured embryos: treatment with aldose reductase inhibition and myo-inositol supplementation. Diabetologia 33: 597–602
Hod M, Star S, Passonneau JV, Unterman TG, Freinkel N (1986) Effect of hyperglycemia on sorbitol and myo-inositol content of cultured rat conceptuses: failure of aldose reductase inhibitors to modify myo-inositol depletion and dysmorphogenesis. Biochem Biophys Res Commun 140: 974–980
Hod M, Star S, Passonneau JV, Unterman TG, Freinkel N (1990) Glucose-induced dysmorphogenesis in the cultures rat conceptus: prevention by supplementation with myo-inositol. Isr J Med Sci 26: 541–544
Hommes FA, Luit-De Haan G, Richters AR (1971) The development of some Krebs cycle enzymes in rat liver mitochondria. Biol Neonate 17: 15–23
Hood RD, Ranganathan S, Jones CL, Ranganathan PN (1988) Teratogenic effects of a

lipophilic cationic dye rhodamine 123, alone and in combination with 2-deoxyglucose. Drug Chem Toxicol 11: 261–274

Hood RD, Jones CL, Ranganathan S (1989) Comparative developmental toxicity of cationic and neutral rhodamines in mice. Teratology 40: 143–150

Horton WE Jr, Sadler TW (1983) Effects of maternal diabetes on early embryogenesis. Alterations in morphogenesis produced by the ketone body, β-hydroxybutyrate. Diabetes 32: 610–616

Horton WE Jr, Sadler TW (1985) Mitochondrial alterations in embryos exposed to β-hydroxybutyrate in whole embryo culture. Anat Rec 213: 94–101

Horton WE Jr, Sadler TW, Hunter ES III (1985) Effects of hyperketonemia on mouse embryonic and fetal glucose metabolism in vitro. Teratology 31: 227–233

Hunter ES III, Parker JT (1996) Redefining embryonic intermediary metabolism: Oxidative phosphorylation is important during early organogenesis. Teratology (submitted)

Hunter ES III, Sadler TW, Wynn RE (1987) D(-)-betahydroxybutyrate-induced effects on mouse embryos in vitro. Teratology 36: 259–264

Hunter ES III, Sadler TW (1988) Embryonic metabolism of foetal fuels in whole-embryo culture. Toxicol In Vitro 2: 163–167

Hunter ES III, Sadler TW (1989) Fuel-mediated teratogenesis: biochemical effects of hypoglycemia during neurulation in mouse embryos in vitro. Am J Physiol 257: E269–E276

Hunter ES III, Tugman JA (1996) Inhibitors of glycolytic metabolism affect neurulation staged mouse conceptuses in vitro. Teratology (in press)

Hunter ES III, Sadler TW, Wynn RE (1987) A potential mechanism of DL-beta-hydroxybutyrate-induced malformations in mouse embryos. Am J Physiol 253: E72–E80

Johnson W, McMall JD (1956) Antimetabolite activity of 6-aminonicotinamide. Fed Proc 15: 284

Katase T, Tetsuo M, Hamada T, Yakushiji M (1992) The effects of glucose concentration on early embryogenesis using the whole embryo culture system on rats. Asia Ocean J Obstet Gynaecol 18: 363–369

Kleiber M, Cole HH, Smith AH (1943) Metabolic rates of rat fetuses in vitro. J Cell Comp Physiol 22: 167–176

Kohler E, Brand K (1970) Studies on the pentose phosphate pathway in rat embryos. In: Bass R, Beck F, Merker H-J, Neubert D, Randhahn B (eds) Metabolic pathways in mammalian embryos during organogenesis and its modification by drugs. Free University, Berlin, pp 361–365

Kohler E, Neubert D (1968) Aktivität NADP-bedürftiger Oxydoreduktasen in embryonalem Gewebe der Ratte. Naunyn Schmiedebergs Arch Pharmacol Exp Pathol 260: 154

Kohler E, Peters H (1970) Studies on the metabolism of glucose in mammalian embryonic tissue. Naunyn Schmiedebergs Arch Pharmacol 266: 371–372

Krowke R, Siebert G, Neubert D (1971) Biochemical screening test with 14C-glucose and 32P-phosphate for the evaluation of embryotoxic effects in vivo. Naunyn Schmiedebergs Arch Pharmacol 271: 274–278

Kubow S, Yaylayan V, Mandeville S (1993) Protection by acetylsalicylic acid against hyperglycemia-induced glycation and neural tube defects in cultured early somite mouse embryos. Diabetes Res 22: 145–158

Landauer W (1972) Is insulin a teratogen? Teratology 5: 129–136

Leese HJ (1990) The energy metabolism of the preimplantation embryo. In: Wiley LM, Heyner S (eds) Early embryo development and paracrine relationships. Wiley-Liss, New York, pp 67–78 (UCLA symposia on molecular and cellular biology, new series vol 117)

Leese HJ, Barton AM (1985) Production of pyruvate by isolated mouse cumulus cells. J Exp Zool 234: 241

Leese HJ, Conafhan J, Martin KL, Hardy K (1993) Early human embryo metabolism. Bioessays 15: 259–264

Lewis NJ, Akazawa S, Freinkel N (1983) Teratogenesis from β-hydroxybutyrate during organogenesis in rat embryo organ culture and enhancement by subteratogenic glucose. Diabetes 32 [Suppl 1]: 11A

Lichtenstein H, Guest GM, Warkany J (1951) Abnormalities of offspring of white rats given protamine zinc insulin during pregnancy. Proc Soc Exp Biol Med 78: 398–402

Mackler B, Grace R, Duncan HM (1971) Studies on mitochondrial development during embryogenesis in the rat. Arch Biochem Biophys 144: 603–610

Mackler B, Grace R, Bargman GJ, Shepard TH (1973) Studies of mitochondrial energy systems during embryogenesis in the rat. Arch Biochem Biophys 158: 662–666

Mackler B, Grace R, Tippit DF, Lemire RJ, Shepard TH, Kelley VC (1975) Studies of the development of congenital anomalies in rats. III. Effects of inhibition of mitochondrial energy systems on embryonic development. Teratology 12: 291–296

McCandles DW, Scott WJ (1981) The effect of 6-aminonicotinamide on energy metabolism in rat embryo neural tube. Teratology 23: 391–395

McLachlan JC (1980) The effects of 6-AN on limb development. J Embryol Exp Morphol 55: 307–318

Merkle S, Pretsch W (1989) Characterisation of triosephosphate isomerase mutants with reduced enzyme activity in mus musculus. Genetics 123: 837–844

Merkle S, Pretsch W (1992) A glucosephosphate isomerase (GPI) null mutation in mus musculus: evidence that anaerobic glycolysis is the predominate energy delivering pathway in early post-implantation embryos. Comp Biochem Physiol 101B: 309–314

Miki A, Mizoguchi A, Mizoguti H (1988) Histochemical studies of enzymes of the energy metabolism of postimplantation rat embryos. Histochemistry 88: 489–495

Moore DCP, Stanisstreet M, Clarke CA (1989) Morphological and physiological effects of β-hydroxybutyrate on rat embryos grown in vitro at different stages. Teratology 40: 237–251

Negelein E (1925) Über die glykolytische Wirkung des embryonalem Gewebes. Biochem Z 165: 122

Neubert D (1970) Aerobic glycolysis in mammalian embryos. In: Bass R, Beck F, Merker HZ, Neubert D, Randhahn B (eds) Metabolic pathways in mammalian embryos during organogenesis and its modification by drugs. Free University, Berlin, pp 225–246

Neubert D, Peters H, Teske S, Kohler E, Barach H-J (1971) Studies on the problem of "aerobic glycolysis" occurring in mammalian embryos. Naunyn Schmiedebergs Arch Pharmacol 268: 234–241

New DAT (1966) Development of rat embryos cultured in blood sera. J Reprod Fertil 12: 509–524

New DAT (1967) Development of explanted rat embryos in circulating medium. J Embryol Exp Morphol 17: 513–525

New DAT (1978) Whole-embryo culture and the study of mammalian embryos during organogenesis. Biol Rev 53: 81–122

New DAT, Stein KF (1964) Cultivation of post-implantation mouse and rat embryos on plasma clots. J Embryol Exp Morphol 12: 101–111

Newsholme EA, Randle PJ, Manchester KL (1962) Inhibition of the phosphofructokinase reaction in perfused rat heart by respiration of ketone bodies, fatty acids and pyruvate. Nature (Lond) 193: 270–271

Oerter D, Bass R (1975) Embryonic development and mitochondrial function. 1. Effects of chloramphenicol infusion on the synthesis of cytochrome oxidase and DNA in rat embryos during late organogenesis. Naunyn-Schmiedebergs Arch Pharmacol 290: 175–189

Pinsky L, Fraser FC (1959) Production of skeletal malformations in the offspring of pregnant mice treated with 6-aminonicotinamide. Biol Neonate 1: 106–112

Pinter E, Reece EA, Leranth CZ, Garcia-Segure M, Sanyal MK, Hobbins JC, Mahoney MJ, Naftlin F (1986) Arachidonic acid prevents hyperglycemia-associated yolk sac damage and embryopathy. Am J Obstet Gynecol 155: 691–702

Pinter E, Reece EA, Ogburn PL Jr, Turner S, Hobbins JC, Mahoney MJ, Naftolin F (1988) Fatty acid content of yolk sac and embryo in hyperglycemia-induced embryopathy and effect of arachidonic acid supplementation. Am J Obstet Gynecol 159: 1484–1490

Reece EA, Pinter E, Leranth CZ, Garcia-Segure M, Sanyal MK, Hobbins JC, Mahoney MJ, Naftlin F (1985) Ultrastructural analysis of malformations of the embryonic neural axis induced by in vitro hyperglycemic conditions. Teratology 32: 363–373

Ritter EJ, Scott WJ, Wilson JG (1972) DNA synthesis inhibition associated with 6-aminonicotinamide (6-AN) teratogenesis in the rat embryo. Teratology 5: 265–266

Rowe PB, Kalaizis A (1985) Serine metabolism in rat embryos undergoing organogenesis. J Embryol Exp Morphol

Rowe PB, McEwen SE (1983) De novo purine synthesis in cultured rat embryos undergoing organogenesis. Proc Natl Acad Sci USA 80: 7333–7336

Sadler TW (1979) Culture of early somite embryos during organogenesis. J Embryol Exp Morphol 49: 17–25

Sadler TW (1980) Effects of maternal diabetes on early embryogenesis.II. Hyperglycemia-induced exencephaly. Teratology 21: 349–356

Sadler TW, Hunter ES III (1987) Hypoglycemia: how much is too little for the embryo. Am J Obstet Gynecol 157: 190–193

Sadler TW, New DAT (1980) Comparison of headfold stage mouse embryos cultured in rat serum versus a partially defined medium. J Embryol Exp Morphol 66: 109–116

Sadler TW, Hunter ES III, Balkan W, Horton WE Jr (1988) Effects of maternal diabetes on embryogenesis. Am J Perinatol 5: 319–326

Sadler TW, Hunter ES III, Wynn RE, Phillips LS (1989) Evidence for multifactorial origin of diabetes-induced embryopathies. Diabetes 38: 70–74

Sana TR, Ferm VH, Smith RP, Kruszyna R, Kruszyna H, Wilcox DE (1990) Embryotoxic effects of sodium azide infusions in the Syrian hamster. Fund Appl Toxicol 15: 754–759

Scheil H-G, Claussen U, Grote W (1977) Effect of 6-amino-nicotinamide in teratogenic doses on the DNA and RNA concentration in rabbit embryos. Teratology 16: 297–300

Shambaugh GE III, Kohler RA, Freinkel N (1977) Fetal fuels. II. Contribution of selected carbon fuels to oxidative metabolism in rat conceptus. Am J Physiol 233: E457–E461

Shambaugh GE III, Angulo MC, Kohler RA (1984) Fetal fuels. VII. Ketone bodies inhibit synthesis of purines in fetal rat brain. Am J Physiol 247: E111–E117

Sheehan EA, Beck F, Clarke CA, Stanisstreet M (1985) Effects of β-hydroxybutyrate on rat embryos grown in culture. Experientia 41: 273–275

Shepard TH, Tanimura T, Robkin MA (1970) Energy metabolism in early mammalian embryos. Dev Biol [Suppl] 4: 42–58

Shepard TH, Park HW, Pascoe-Mason J (1993) Glucose causes lengthening of the microvilli of the neural plate of the rat embryo and produces a helical pattern on their surface. Teratology 48: 65–74

Shum L, Sadler TW (1990) Biochemical basis for D,L-betahydroxy-butyrate-induced teratogenesis. Teratology 42: 553–563

Smithberg M, Runner MN (1963) Teratogenic effects of hypoglycemia treatments in inbred strains of mice. Am J Anat 113: 479–489

Smoak IW, Sadler TW (1990) Embryopathic effects of short-term exposure to hypoglycemia in mouse embryos in vitro. Am J Obstet Gynecol 163: 619–624

Spielmann H, Lucke I (1973) Changes in the respiratory activity of different tissues of rat and mouse embryos during development. Naunyn-Schmiedeberg Arch Pharmacol 278: 151–164

Spielmann H, Meyer-Wendecker R, Spielmann F (1973) Influence of 2-deoxy-D-glucose and sodium fluoroacetate on respiratory metabolism of rat embryos during organogenesis. Teratology 7: 127–134

Spratt NT (1950) Nutritional requirements of the early chick embryo. III. The metabolic basis of morphogenesis and differentiation as revealed by the use of inhibitors. Biol Bull Mar Biol Lab Woods Hole 99: 120–135

Strieleman PJ, Metzger BE (1993) Glucose and scyllo-inositol impair phosphoinositide hydrolysis in the 10.5-day cultured rat conceptus: A role in dysmorphogenesis? Teratology 48: 267–278

Strieleman PJ, Connors MA, Metzger BE (1992) Phosphoinositide metabolism in the developing conceptus. Effects of hyperglycemia and scyllo-inositol in rat embryo culture. Diabetes 41: 989–992

Sussman I, Matschinsky FM (1988) Diabetes affects sorbitol and myo-inositol levels of neuro-ectodermal tissue during embryogenesis in rat. Diabetes 37: 974–981

Sweatman TW, Seshadri R, Israel M (1990) Metabolism and elimination of rhodamine 123 in the rat. Cancer Chemother Pharmacol 27: 205–210

Tanaka O (1981) Comparison of teratogenic effects of administration by 6-aminonicotinamide on days 7 and 9 of gestation in mice. Shimane J Med Sci 5: 109–122

Tanigawa K, Kazaguchi M, Tanaka O, Kato Y (1991) Skeletal malformations in rat offspring. Long-term effect of maternal insulin-induced hypoglycemia during organogenesis. Diabetes 40: 1115–1121

Tanimura T, Shepard TH (1970a) Glucose metabolism by rat embryos in vitro. Proc Soc Exp Biol Med 135: 51–53

Tanimura T, Shepard TH (1970b) Utilization of glucose carbon by rat embryos in vitro and in vivo and effects of 2-deoxy-D-glucose. Teratology 3: 210

Trasler DG (1978) A selection experiment for distinct types of 6-AN-induced cleft lip in mice. Teratology 18: 49–54

Wales RG (1986) Measurement of metabolic turnover in single mouse embryos. J Reprod Fertil 76: 717

Wales RG, Whittingham DG (1973) The metabolism of specifically labeled lactate and pyruvate by two-cell mouse embryos. J Reprod Fertil 33: 207–222

Wales RG, Whittingham DG, Hardy K, Graft IL (1987) Metabolism of glucose by human embryos. J Reprod Fertil 79: 289–297

Weigensberg MJ, Garcia-Palmer F-J, Freinkel N (1990) Uptake of myo-inositol by early-somite rat conceptuses. Transport kinetics and effects of hyperglycemia. Diabetes 39: 575–582

West JD (1993) A genetically defined animal model of an embryonic pregnancy. Hum Reprod 8: 1316–1323

White A, Handler P, Smith EL, Hill RL, Lehman IR (1978) Principles of biochemistry, 6th edn. McGraw-Hill, New York

CHAPTER 15

Alterations in Folate Metabolism as a Possible Mechanism of Embryotoxicity

D.K. HANSEN

A. Introduction

Interest in folic acid has increased recently as a result of reports of a protective effect of dietary supplementation on the incidence of neural tube defects (NTD) in humans. This chapter will describe folic acid, assays for this vitamin, and characteristics of the deficient state. Biochemical reactions involving folic acid will be reviewed, as will studies which report a protective effect with respect to NTD. A number of compounds which adversely affect folic acid concentrations will be reviewed along with any evidence of a role for folic acid in the developmental toxicity of these compounds.

Pteroylglutamic acid, or folic acid, is a member of the B group of vitamins. The structure of folic acid monoglutamate, which consists of a pteridine ring, *p*-aminobenzoic acid, and a single glutamic acid molecule, is shown in Fig. 1.

The pteridine ring structure is often reduced, and many folates are present as dihydro- or tetrahydro- derivatives (i.e., 5, 6, 7, 8-tetrahydrofolate). Substitutions occur primarily at the N-5 and N-10 positions, and hydrogen atoms can be replaced by formyl, methyl, formimino, or two hydrogen atoms in the case of methylene or methenyl derivatives. Reduced folic acid may be conjugated with one or more molecules of glutamic acid; in the liver, most of the reduced folic acid is conjugated with an average of five to eight glutamate residues in peptide linkage (COSSINS 1984). The polyglutamate chain appears to aid in cellular retention of folates. Additionally, polyglutamated folates are better substrates for various folate-requiring enzymes than are monoglutamate derivatives. The term folate will be used to refer to this group of compounds of different reduction and substitution states as well as polyglutamate status.

I. Dietary Sources

Humans cannot synthesize folic acid and are therefore dependent on dietary sources. Such sources include liver, green leafy vegetables, legumes, citrus fruits and juices, and fortified breads and cereals. The amount of folate in these foods is variable and may be altered by processing and cooking.

Dietary folates are usually polyglutamated derivatives, and the glutamate molecules are removed by a folylpolyglutamate hydrolase (conjugase) enzyme

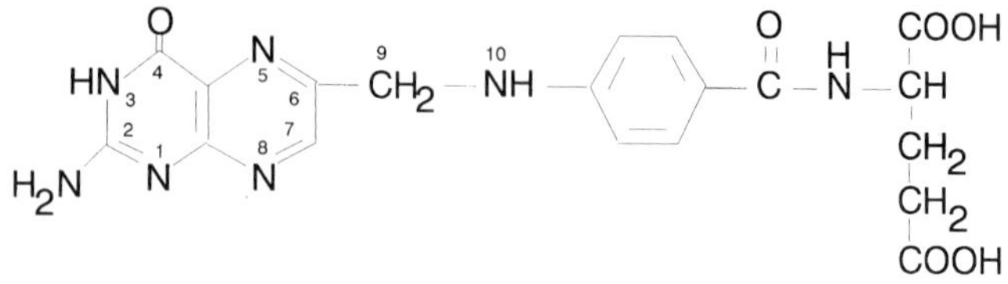

Fig. 1. Structure of folic acid monoglutamate

in the intestine. Folate is absorbed primarily as monoglutamate derivatives. Most of the folate is reduced and methylated in the intestinal cells and enters the circulation as 5-methyltetrahydrofolate monoglutamate (Rosenberg and Selhub 1986).

II. Recommended Dietary Allowances

The U.S. Recommended Dietary Allowance (RDA) is currently set at 0.18 mg/day for adult men and nonpregnant women and 0.4 mg/day for pregnant women (Food And Nutrition Board 1989). This represented a decrease of approximately 50% from the earlier RDA of 0.4 mg/day for adult men and nonpregnant women and 0.8 mg/day for pregnant women (Food And Nutrition Board 1980). Dietary surveys such as the Nutrition Examination Survey (NHANES II) and the Continuing Surveys of Food Intakes by Individuals (CSFII) have indicated that the daily intake for U.S. women is around 0.20 mg/day (Bailey 1992). Daily intake is also around 0.20 mg/day for women in the United Kingdom, but is lower (0.15 mg/day) for Canadian women (Bailey 1995). Block and Abrams (1993) found that about 50% of women take in less than 100% of the RDA.

III. Assay Methods

One of the more common assays for folate is based on the growth of various folate-requiring bacteria (Bird et al. 1969). The bacteria usually used for this assay are *Lactobacillus casei* (ATCC 7469), *Streptococcus faecalis* (ATCC 8043), and *Pediococcus cerevisiae* (ATCC 8081) (Cossins 1984). The ability of these micro-organisms to grow on various folate derivatives forms the basis for determination of the types of derivatives present in a sample. None of these bacteria grow well on polyglutamated folates, and glutamate molecules must be removed by treatment with a conjugase enzyme prior to assay. The original assay was recently successfully modified using a microplate reader (Newman and Tsai 1986; Horne and Patterson 1988), making it easier and less expensive. This assay is sensitive to 10 fmol folate.

However, the microbiological assay is time-consuming (requiring an overnight incubation for sufficient growth of the bacteria) and expensive for clinical laboratories to use. Therefore, radioassay methods were developed which are based on the competition of labeled and unlabeled folates for a limited number of binding sites on a folate-binding protein (reviewed in COSSINS 1984). Radioassays are generally used in the clinical setting to quantitate serum and erythrocyte folate levels. A new nonisotopic method, cloned enzyme donor immunoassay, has been described recently (VAN DER WEIDE et al. 1992). This assay was comparable in sensitivity and reliability to the radioassay for determining plasma folate concentrations. This assay system may become more common in the clinical setting in the future.

These methods are useful in quantitating total folate, but are less useful in quantitating individual folate derivatives. Several gel filtration and ion-exchange assays have been developed and were reviewed by COSSINS (1984). Various high-performance liquid chromatography (HPLC) techniques have also been developed. Detection and quantitation systems vary and include growth of *L. casei* (MCMARTIN et al. 1981; MCMARTIN 1984), ultraviolet (UV) absorbance (DUCH et al. 1983; WEGNER et al. 1986), or electrochemical detection (NATSUHORI et al. 1991). Some of these methods have only been tested with a few folate derivatives; others are time-consuming and/or may require a preliminary clean-up step prior to HPLC analysis. Additionally, these methods will only detect monoglutamated folates, and so samples must be treated with a conjugase enzyme prior to analysis. Methods have been described which are capable of detecting and quantitating folate polyglutamates. SELHUB (1989) has described a method which utilizes milk folate-binding protein for affinity purification prior to HPLC analysis; a diode array detection system was used for more precise identification of the pteridine ring structure. Another method is based on entrapment of 5,10-methylenetetrahydrofolate into a stable ternary complex with thymidylate synthase and [^{3}H] fluorodeoxyuridine monophosphate followed by gel electrophoresis to separate the polyglutamates (PRIEST and DOIG 1986). All other folate derivatives are converted to 5,10-methylenetetrahydrofolate by a series of enzymatic steps. This assay compares favorably with an HPLC method, but requires the use of several enzymes that are not available commercially (PRIEST et al. 1992).

IV. Characteristics of Folate Deficiency

The folate present in serum or plasma is often quantitated to determine whether a deficiency state is present. These folate levels are extremely variable and are dependent on recent dietary folate intake (COSSINS 1984). Folate concentrations in red cells are higher and somewhat less variable than those in plasma or serum. Because the lifespan of red blood cells is at least 120 days, measurements of folate in erythrocytes is generally considered to be indicative of folate status over the prior 3–4 months.

Although a low serum folate concentration is the first sign of folate deficiency, it is not always an ideal screen for a deficiency state. It may be within the normal range even though the patient has megaloblastic anemia due to folate deficiency. Serum folate levels below 2 ng/ml or erythrocyte folate concentrations less than 150 ng/ml are considered to be low. As folate deficiency progresses, the erythrocyte folate level decreases and morphological changes appear in bone marrow cells. Changes in the peripheral blood then appear, and the mean corpuscular volume increases. Eventually anemia develops, and as it becomes more severe the bone marrow becomes megaloblastic (LINDENBAUM and ALLEN 1995). Since megaloblastic anemia can also be due to cobalamin deficiency, this possibility must be ruled out by determination of serum cobalamin level and normal absorption of oral cobalamin (CHANARIN 1986).

B. Biochemical Pathways Involving Folates

I. One-Carbon Metabolism

Folates are involved in the synthesis of purines, thymidine, and amino acids. This occurs primarily through a series of enzymatic steps in which one-carbon units at different oxidation states are transferred in a series of metabolic steps. These steps are outlined in Fig. 2.

Compounds can enter the one-carbon pool at various sites. The major entry point appears to be through serine; however, formate, formiminoglutamic acid (FIGLU), dimethylglycine, and sarcosine can also supply one-carbon units at various levels of oxidation (WAGNER 1995). Dimethylglycine and sarcosine are products of choline metabolism and enter the pool only in hepatic mitochondria. FIGLU is a product of histidine catabolism and in the liver is metabolized to glutamic acid by the transfer of the formimino group to tetrahydrofolate (SHANE and STOKSTAD 1984). Increased amounts of FIGLU are excreted following a histidine load and suggest a folate deficiency. Serine and glycine are interconvertible by the activity of serine hydroxymethyltransferase, and this pathway is quantitatively the most important source of one-carbon units (MACKENZIE 1984).

II. Involvement in Methionine Metabolism

The methyl group from 5-methyltetrahydrofolate is transferred to homocysteine to form methionine. This amino acid or its derivatives can be used in protein synthesis, in a variety of transmethylation reactions, *trans*-sulfuration reactions, or in polyamine synthesis (FINKELSTEIN 1990). Methionine is metabolized to *S*-adenosylmethionine (SAM), the major methyl donor for hundreds of methylation reactions involving proteins, DNA, lipids, and a variety of other compounds (e.g., epinephrine; WAGNER 1995).

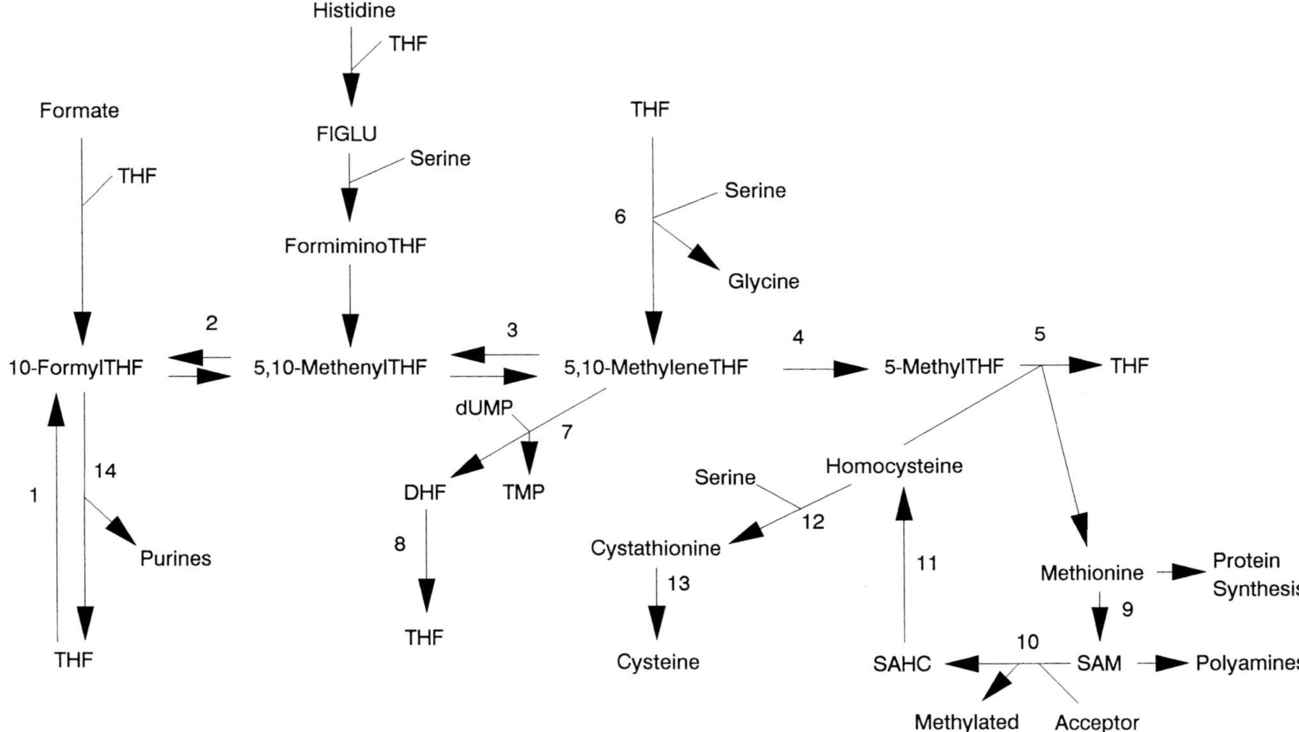

Fig. 2. Several of the metabolic steps in one-carbon metabolism and methionine metabolism. *THF*, tetrahydrofolate; *SAM*, *S*-adenosylmethionine; *SAHC*, *S*-adenosylhomocysteine; *FIGLU*, formiminoglutamic acid; *dUMP*, deoxyuridine monophosphate; *DHF*, dihydrofolate reductase. The enzymes depicted are *1*, 10-formyltetrahydrofolate synthase; *2*, 5,10-methenyltetrahydrofolate cyclohydrolase; *3*, 5,10-methylenetetrahydrofolate dehydrogenase; *4*, 5,10-methylenetetrahydrofolate reductase; *5*, methionine synthase; *6*, serine hydroxymethyl transferase; *7*, thymidylate synthase; *8*, dihydrofolate reductase; *9*, methionine adenosyltransferase; *10*, various methyl transferases; *11*, *S*-adenosylhomocysteinase; *12*, cystathionine β-synthase; *13*, cystathionase; *14*, phosphoribosylglycinamide transformylase or aminocarboxamide ribotide transformylase

C. Embryotoxicity of Folate Deficiency

I. Human Studies

Some of the earliest studies in humans which suggested that alterations in folate metabolism might be associated with adverse pregnancy outcomes came from Great Britain in the mid-1960s. HIBBARD et al. (1965) observed an increase in FIGLU excretion during pregnancy in mothers of malformed infants; the incidence of 3.6% was more than double the 1.5% found in mothers of normal infants. Later in that same year, HIBBARD and SMITHELLS (1965) reported a higher frequency of women with an increased FIGLU excretion among those giving birth to infants with severe malformations, particularly malformations of the central nervous system. Smithells' group hypothesized that defective maternal folate metabolism might be associated with fetal malformations, especially malformations of the central nervous system. To investigate this possibility, levels of several vitamins were assayed during the first 13 weeks of pregnancy (SMITHELLS et al. 1976). Six mothers of infants with central nervous system defects (three cases of anencephaly and one case each of myelomeningocele with hydrocephalus, meningocele, and microcephaly) had significantly lower red cell folate levels than did controls; serum folate levels were not different (SMITHELLS et al. 1976).

These results suggested that folate supplementation might be able to decrease the incidence of central nervous system defects. In an intervention trial, it was observed that periconceptional supplementation with multiple vitamins significantly decreased the incidence of NTD in subsequent pregnancies in individuals who had previously had a child with a NTD (SMITHELLS et al. 1980). A second study done in the same manner also found a significant decrease in recurrence of NTD among offspring of supplemented versus unsupplemented women (SMITHELLS et al. 1983). The vitamins used in these studies contained 0.36 mg folic acid. Although the authors attributed the protective effect to folic acid, the presence of other vitamins in the tablets may have confounded that conclusion.

Several other studies examined the role of diet, and especially folic acid, in the etiology of NTD. Most studies showed some protection (LAURENCE et al. 1981; MULINARE et al. 1988; MILUNSKY et al. 1989); however, there were studies that demonstrated no beneficial effect of supplemental folate (MILLS et al. 1989). A landmark study was published in 1991 (MRC VITAMIN STUDY RESEARCH GROUP 1991). This was a randomized trial involving women from several medical centers in various countries. Women who had already had a child with a NTD were recruited into the study, which was designed to specifically examine the role of folic acid in protection. In the folic acid-supplemented groups ($n=593$) there were six NTD (1.0%), and in the non-folate-supplemented groups ($n=602$) there were 21 NTD (3.5%). The trial was originally designed to include over 2000 women, but was stopped about half way through. Due to the clear protective effect of folic acid (4 mg/day), it was considered unethical to keep this beneficial treatment from the placebo group. The folic acid supplement reduced the risk for another NTD child by 72%.

The MRC study clearly showed a protective effect of folic acid supplementation on the recurrence of NTD, but it was not clear whether the vitamin would be effective in occurrent cases, which make up about 95% of all cases. Czeizel and Dudas (1992) conducted a randomized controlled trial which included 4156 women who had not previously delivered a child with a NTD. Women were randomly assigned to the multiple vitamin and mineral or to the trace element groups. There were no NTD among 2104 pregnancies in the multiple vitamin group and six cases among 2052 pregnancies in the control group which received a tablet containing various trace elements. The multivitamin–mineral supplement contained 0.8 mg folic acid. Although it has been interpreted that folic acid decreased the occurrence of NTD, the study was not designed in such a way that would unambiguously permit analysis of the effect of folic acid not in combination with other vitamins and minerals.

The role of dietary folate alone has also been investigated (reviewed in Bower et al. 1993). Bower and Stanley (1989) found a decreasing risk for NTD with increasing levels of dietary and supplemental folic acid during the first 6 weeks of pregnancy. Werler et al. (1993) found a 60% reduction in the risk of occurrence of NTD among women taking daily vitamin supplements containing 0.4 mg folic acid. They also observed a trend toward a decreased relative risk among women who were not taking supplements but who were getting a relatively high dietary intake of folic acid. Yates et al. (1987) and Wild et al. (1994) found no differences in dietary folate intake between women who had two or more NTD pregnancies and control women.

The preponderance of evidence suggests that periconceptional supplementation with folic acid decreases the risk of having a child with a NTD (reviewed in Czeizel 1995). Therefore, the U.S. Public Health Service has recommended that women increase their daily intake of folic acid (Morbidity and Mortality Weekly Report 1992). Since the neural tube is completely closed by postconception day 28 (O'Rahilly and Muller 1994), folate supplementation must occur very early to prevent NTD. Since many pregnancies are unplanned, the U.S. Public Health Service has recommended that all women who are capable of becoming pregnant, not just those who are currently pregnant, increase their folate intake.

II. Serum Folate Levels Associated with Embryotoxicity

The implication of many of these studies is that folate deficiency is capable of producing NTD. However, serum and red cell folate levels have seldom shown a deficient state in women with NTD infants (reviewed in Wald 1993).

Mills et al. (1992) found no differences in serum folate or B_{12} levels between controls and mothers of children with NTD. Most of the samples were collected during the first trimester of pregnancy, but only about 4% were collected at less than 6 weeks of pregnancy.

Yates et al. (1987) examined serum and red cell folate levels in women who had previously had two or more NTD pregnancies. They found no differences between cases and controls in serum B_{12} or folate. However, they did observe a significant decrease in erythrocyte folate levels among cases, and there was a significant trend for lower erythrocyte folate levels to be related to an increase in the number of NTD pregnancies. Wild et al. (1994) observed no differences in serum or erythrocyte folate levels between women who had two NTD pregnancies and control women. They noted, however, that these women had been counseled and offered periconceptional folate supplementation for their next pregnancy. They may have changed aspects of their lifestyle in response to this counseling, thus accounting for the nonsignificant difference in red cell folate levels.

Molloy et al. (1985) observed no differences in the distributions of serum folate or vitamin B_{12} levels between randomly selected control women and 32 women who had children with NTD. Kirke et al. (1993) did observe significantly decreased levels of plasma vitamin B_{12} and plasma and red cell folate levels in women with NTD pregnancies.

The exact mechanism for the protective effect of folates on NTD is unknown, but it does not appear to be due simply to a dietary deficiency of the vitamin.

III. Animal Studies

Since folate derivatives are involved in a number of biochemical reactions required by rapidly growing cells, it is not surprising that embryonic cells might be adversely affected by a lack of folate. One of the first indications that folate deficiency might be embryotoxic was reported in animals. Treatment of rats with a folic acid-deficient diet or a diet containing a crude folate antagonist resulted in little effect on implantation, but increased the number of dead pups at the end of gestation (Nelson and Evans 1949). Later studies by this group also demonstrated malformations among rat embryos exposed to folate-deficient diets (Nelson et al. 1952, 1955).

Many human NTD are thought to be the result of gene–environment interactions (Copp 1994). Several mouse mutant strains have been developed that have varying incidences of NTD. The potential protective effect of folic acid has been examined in several of these strains. Folic acid, 5-formyltetrahydrofolate, Pregnavite Forte F vitamins, homocysteine, and methionine were not able to decrease the incidence of NTD in the *curly tail* strain of mice (Seller 1994). Folate also did not affect the incidence or severity of defects in the *Axd* strain of mice (Essian and Wannberg 1993). These negative results suggest that folate may have an effect on different genetic loci than those that have been examined to date.

IV. Role of Other Compounds in Embryotoxicity

Folate deficiency is not sufficient to produce NTD in mice (HEID et al. 1992). Mice were fed a folate-deficient diet with folate added back to various levels for 35 days prior to mating and continuing throughout gestation. Nearly all of the mice eating diets containing 566 nmol folic acid/kg chow or less produced few liveborn fetuses, with most of the implants being resorbed. At all higher folate levels, development was normal, with significant decreases in fetal weight in some groups of animals. Culture of rat embryos on serum from folate-deficient rats produced abnormal embryonic development (MILLER et al. 1989). These embryos had abnormal axial rotation, poor development of the mid- and forebrain regions, and retarded development of the optic vesicles and branchial arches. However, the neural tubes in these embryos were closed.

Several recent studies have implicated homocysteine as a risk factor for NTD. In folate deficiency, serum homocysteine (STABLER et al. 1988) and cystathionine (STABLER et al. 1993) may be increased. Dietary folate deficiency in rats increased plasma homocysteine levels, which were then decreased following resupplementation with folate (MILLER et al. 1994).

In a preliminary study, STEEGERS-THEUNISSEN et al. (1991) reported no differences in serum B_{12} and serum or red cell folate levels between control women and those with a child with a NTD. These assays were done after the birth of the NTD child. However, the results of a methionine-loading experiment demonstrated that a subgroup (five out of 16) of NTD women had elevated serum homocysteine levels, suggesting a role for altered homocysteine metabolism in the production of NTD. Elevated homocysteine levels could be due to decreased activity of either cystathionine β-synthase or methionine synthase (Fig. 2). This study was later expanded, and they observed that fibroblast cystathionine β-synthase activity was within normal limits in the women with elevated homocysteine concentrations, suggesting that the increase was due to a decrease in the remethylation of homocysteine by methionine synthase (STEEGERS-THEUNISSEN et al. 1994).

Independently, MILLS et al. (1995) demonstrated increased homocysteine levels in women during a pregnancy with a child with a NTD. This group had earlier found that folate and B_{12} were independent risk factors for NTD and had hypothesized that this was due to a defect in methionine synthase activity, since this is the only metabolic reaction dependent on both folate and B_{12} (KIRKE et al. 1993). Regression analyses showed a significant positive correlation between vitamin B_{12} and folate levels. For any plasma folate level, increasing the level of B_{12} decreased the risk of having a child with a NTD; at any plasma B_{12} level, increasing plasma folate decreased the risk of a NTD. Because decreased methionine synthase activity could result in increased plasma homocysteine concentrations, the recent results of MILLS et al. (1995) could also be explained on the basis of a defect in methionine synthase activity (SCOTT et al. 1994).

Schorah et al. (1993) have also suggested that a defect in methionine synthase activity may play a role in producing NTD. They observed a decrease in incorporation of the methyl group from ^{14}C-labeled 5-methyltetrahydrofolate in nucleic acids in trophoblast cells from women who had two NTD pregnancies. Their results could also be explained by impaired cellular absorption of 5-methyltetrahydrofolate. Lucock et al. (1994) observed a relationship between dietary folate intake and disposition of 5-methyltetrahydrofolate in control women, but not in women who had two previous NTD pregnancies. They suggested that a defect might exist in 5,10-methylenetetrahydrofolate reductase.

Alternatively, nitrous oxide inactivates the B_{12} group of methionine synthase (Banks et al. 1968), inhibiting enzyme activity and producing alterations in embryonic folate distribution and decreases in total embryonic folate levels after long-term exposure (Hansen and Billings 1985a). The anesthetic is embryotoxic, producing resorptions, fetal death, decreased fetal weight, and visceral and skeletal malformations and altering body laterality (Fink et al. 1967; Ramazzotto et al. 1979; Lane et al. 1980; Mazze et al. 1984; Fujinaga et al. 1990), but it does not produce NTD. The mechanism whereby nitrous oxide produces its embryotoxic effects is unknown, as is the role of decreased methionine synthase activity in producing NTD.

An alternative explanation for increased homocysteine concentrations in NTD pregnancies involves the ability of homocysteine to independently produce NTD on its own. High concentrations of homocysteine (1.8 m*M*) were embryotoxic to gestation day 10 rat embryos cultured in vitro, but NTD were not observed (VanAerts et al. 1993). This stage of development may be too late to observe NTD, so the experiments were repeated using earlier-stage embryos (gestation day 9.5; VanAerts et al. 1994). Under these experimental conditions, 2.0 m*M* homocysteine was not embryotoxic, but 4 m*M* was embryolethal. Additionally, 2.0 m*M* homocysteine was actually beneficial to rat embryos grown on human serum; control embryos did not grow well in human serum, and the addition of homocysteine improved growth and development. The authors also observed that L-methionine at a concentration of 100 μ*M* was nearly as effective as 2 m*M* L-homocysteine in improving embryonic growth and development, and 1 m*M* 5-formyltetrahydrofolate was ineffective (VanAerts et al. 1994). They suggested that homocysteine does not appear to be responsible for producing NTD, but methionine may be involved.

Several other studies have suggested that methionine is required for normal neural tube closure. In rat embryos grown in a whole embryo culture system on cow serum, methionine was required for proper closure of the neural tube and normal development (Coelho et al. 1989; Coelho and Klein 1990). Addition of methionine to canine serum improved the growth of rat embryos (Flynn et al. 1987). It reduced the incidence of flexion abnormalities, but did not decrease the number of embryos with open anterior neural tubes.

Methionine supplementation was also able to decrease the incidence of NTD in mouse embryos carrying the *Axd* (axial defects) mutation, but folinic acid (5-formyltetrahydrofolate) was ineffective (ESSIAN 1992; ESSIAN and WANNBERG 1993). Even though the incidence of NTD was altered by methionine supplementation, the incidence of curly tails (presumably *Axd* heterozygotes) was not decreased by methionine in backcrosses to BALB/c mice, but was decreased in outcrosses to CF-1 mice. These results suggest that genetic background plays a role in the protective effect.

D. Compounds Which Adversely Affect Folate Levels

Several compounds have been reported to produce megaloblastic anemia, but there is little data suggesting developmental toxicant effects. Since pregnancy increases the demand for folates (CHANARIN 1986), women who may have marginally adequate folate intake may be especially susceptible to adverse effects of these compounds.

I. Triamterene

Triamterene is an antidiuretic compound that at high doses can produce megaloblastosis (JOOSTEN and PELEMANS 1991; CORCINO et al. 1970). A more recent study found that chronic treatment with lower doses was not associated with folate deficiency and megaloblastic anemia (MASON et al. 1991). The drug not only inhibits dihydrofolate reductase (MAASS et al. 1967), but also inhibits intestinal folate transport (ZIMMERMAN et al. 1986).

II. Trimethoprim

From 1969 to 1980, trimethoprim was present in combination with sulphamethoxazole; since 1980 it has been available alone (REEVES 1982). The sulphonamide component blocks conversion of *p*-aminobenzoic acid to dihydrofolate, an enzymatic step that is not present in humans. The conversion of dihydrofolate to tetrahydrofolate is carried out by dihydrofolate reductase, and trimethoprim has a much greater affinity for the bacterial enzyme than for the mammalian enzyme (KING 1985). Patients with normal folate stores who are given the drug generally do not develop any adverse hematological effects (FRISCH 1973). However, patients with suboptimal folate stores are more likely to develop megaloblastic changes (SPECTOR et al. 1973; HILL and KERR 1973; CHAN et al. 1980). The drug combination has been reported to be embryotoxic in rats, but not in rabbits (UDALL 1969; HELM et al. 1976). Trimethoprim alone produced resorptions, small fetuses, cleft palate, and skeletal abnormalities in rats only at very high doses; no adverse effects were observed in rabbits (STEPHAN-GULDNER 1993). NTD were not produced.

III. Sulfasalazine

Folate deficiency commonly occurs in patients with inflammatory bowel disease and may be worsened by sulfasalazine, a drug commonly used to treat this condition (Clinical Nutrition Cases 1988). This drug inhibits intestinal absorption of folate (Halsted et al. 1981; Franklin and Rosenberg 1973) and also inhibits several hepatic folate-metabolizing enzymes such as dihydrofolate reductase, 5,10-methylenetetrahydrofolate reductase, and serine hydroxymethylase (Selhub et al. 1978). The drug has been associated with embryotoxicity in humans with case reports of malformations in fetuses born to women on chronic sulfasalazine treatment (Craxi and Pagliarello 1980; Newman and Correy 1983; Hoo et al. 1988). However, there is no clear evidence that this compound is a developmental toxicant (Mogadam et al. 1980).

IV. 2-Methoxyethanol

2-Methoxyethanol is an organic solvent that has been reported to be embryotoxic in a number of species and to produce paw malformations (syndactyly, oligodactyly, and stunted digit number 1) in mice (Horton et al. 1985). The compound apparently is metabolized to methoxyacetic acid, which is the proximate teratogen (Brown et al. 1984). Although the compound has not been reported to produce a folate deficiency, several of the compounds involved in the one-carbon transfer pathway have been shown to attenuate 2-methoxyethanol-induced digit malformations. These compounds include formate, glycine (Welsch et al. 1987), and L- or D-serine (Clarke et al. 1991). The mechanism for this protective effect is believed to occur via the ability of these compounds to restore normal DNA synthesis (Stedman and Welsch 1989). Such attenuation may not occur as a direct effect of the compounds on folate levels, but may occur by altering flux through the one-carbon pool.

E. Developmental Toxicants Which May Act Via Folate Perturbations

A number of compounds are developmental toxicants in humans and/or animal models and may act via alterations in folate metabolism. Several of these will be reviewed below.

I. Aminopterin and Methotrexate

The folate antagonist aminopterin was first observed to produce embryolethality and malformations in rats and mice in 1950 (Thiersch and Philips 1950). Two years later came the first report of its use as an abortifacient; however, malformations observed in some of the fetuses were attributed to the drug (Thiersch 1952). A number of reports followed which indicated that the drug was not always successful as an abortifacient and produced severe abnormalities in many of those cases in which it was unsuccessful (reviewed in Warkany 1978). Methotrexate is a methyl derivative of aminopterin and was found to act similarly to aminopterin; it produced embryolethality and malformations in rats and rabbits (JORDAN et al. 1977), mice (SKALKO and GOLD 1974) and humans (MILUNSKY et al. 1968).

Recent work has suggested that the defects in animals induced by aminopterin and methotrexate are due to perturbations in one-carbon metabolism and folate homeostasis. DESESSO and GOERINGER (1991) observed a significant decrease in malformations induced in rabbits by methotrexate when 5-formyltetrahydrofolate was administered as long as 24 h after the antifolate. The functional analogue of tetrahydrofolate, 1-(*p*-tosyl)-3,4,4-trimethylimidazolidine (TTI), was also effective in rabbits in decreasing methotrexate-induced (DESESSO and GOERINGER 1992) or aminopterin-induced (GOERINGER et al. 1994) embryotoxicity.

II. Phenytoin

Over 40 years ago, megaloblastic anemia was first reported to occur in epileptic patients on anticonvulsant therapy. This anemia was later found to be corrected by folate administration and was therefore assumed to be due to folate deficiency. This effect was found not only with polytherapy but also with phenytoin (PHT) monotherapy (KLIPSTEIN 1964). A number of other studies demonstrated subnormal serum and/or erythrocyte folate concentrations in epileptic patients on anticonvulsant drugs (reviewed in RIVEY et al. 1984).

The mechanism whereby PHT might cause folate deficiency is not understood. Several mechanisms have been suggested, including impairment of folate absorption either by a PHT-induced change in intestinal pH (BENN et al. 1971) or by inhibition of intestinal conjugases (ROSENBERG et al. 1968; HOFFBRAND and NECHELES 1968), impairment of folate transport into tissues (KRUMDIECK et al. 1978), hepatic enzyme induction and a consequent increased utility of folate as a cofactor (MAXWELL et al. 1972; LABADARIOS et al. 1978), and increased folate catabolism (KELLY et al. 1979). While there is evidence which supports each of these hypotheses, there is also evidence which does not support most of them (reviewed in RIVEY et al. 1984). The mechanism is complex, and multiple mechanisms may be involved. There is general agreement that, whatever the mechanism, PHT does produce folate deficiency (SAUBERLICH 1995).

Folate deficiency does not appear to be the mechanism whereby PHT is teratogenic in animal models (reviewed in HANSEN 1991). Folic acid administered to pregnant mice had no effect on the incidences of PHT-induced resorptions and malformations (MERCIER-PAROT and TUCHMANN-DUPLESSIS 1974; MARSH and FRASER 1973). Results from studies using 5-formyltetrahydrofolate (folinic acid) have been somewhat more confusing. There was a nonlinear response when increasing doses of folinic acid were administered with a teratogenic dose of PHT (SCHARDEIN et al. 1973). Low doses (1 or 10 mg/kg) had no effect on the incidence of PHT-induced cleft palates in mice; high doses (40 or 100 mg/kg) actually increased the frequency somewhat, and the intermediate doses (20 or 30 mg/kg) appreciably lowered the incidence. The authors concluded that 5-formyltetrahydrofolate had little protective effect and at higher doses may have interfered with PHT metabolism or excretion. In another study, SULLIVAN and MCELHATTON (1975) observed no effect of 5-formyltetrahydrofolate when PHT was administered by gavage; they observed a potentiation of PHT-induced clefting when the anticonvulsant was administered in the diet. In another study, dietary administration of folic acid decreased PHT-induced embryotoxicity in rats, as determined by fetal weight and length as well as by the number of ossification centers in sternebrae (ZHU and ZHOU 1989).

The effects of PHT on embryonic folate metabolism have also been investigated. A teratogenic dose of PHT administered by i.p. injection decreased embryonic folate levels on gestational day 11 (NETZLOFF et al. 1979), whereas chronic dietary administration of a teratogenic dose of PHT had no effect on embryonic folate levels on gestational day 10, 12, or 14 (HANSEN and BILLINGS 1985b). Although maternal hepatic 5,10-methylenetetrahydrofolate reductase activity was decreased in PHT-treated animals, there was no decrease in this enzyme activity in PHT-treated embryos.

Overall, there does not appear to be sufficient data to support a role for folate deficiency in PHT-induced embryotoxicity in rodents.

III. Valproic acid

The anticonvulsant drug valproic acid (VPA) is known to produce NTD, especially spina bifida in humans (ROBERT and GUIBAUD 1982; BJERKEDAL et al. 1982; LINDHOUT and SCHMIDT 1986). These defects are estimated to occur in 1%–2% of treated pregnancies (LAMMER et al. 1987).

The mechanism for this embryotoxic effect is unknown, but alterations in folate metabolism have been suspected. SMITH and CARL (1982) have shown decreases in maternal plasma folate levels in rats treated chronically with VPA. HENDEL et al. (1984) suggested a possible inhibition of intestinal folate absorption by VPA. Sera from human epileptic patients on VPA monotherapy were embryotoxic to rat embryos in a whole-embryo culture system (CHATOT et al. 1984). Addition of vitamins, including folic acid, did not decrease the embryotoxic effects of these sera, suggesting that folate deficiency was not responsible for this VPA-induced embryotoxicity.

In vivo administration of folinic acid by injection (TROTZ et al. 1987) or by osmotic mini-pump (WEGNER and NAU 1991) was found to decrease the incidence of exencephaly among offspring from VPA-treated NMRI mice. However, these results were not reproduced in CD-1 mice (HANSEN et al. 1992 and in press), nor in NMRI mice (ELMAZAR et al. 1992). Additionally, 5-formyltetrahydrofolate did not decrease the incidence of VPA-induced open neural tubes among rat embryos treated in vitro (HANSEN and GRAFTON 1991); a number of other folate derivatives were also ineffective in vitro (HANSEN 1993).

In addition to folate derivatives, other compounds involved in the one-carbon pool have been examined for their ability to decrease the incidence of VPA-induced NTD. In vitro, L-methionine, L- and D-serine, and sodium formate were unable to alter the incidence of VPA-induced open neural tubes (HANSEN et al. 1995). Treatment with L-methionine also had no effect on VPA-induced embryotoxicity in vitro (NOSEL and KLEIN 1992). However, if embryos or serum from in vivo methionine-treated rats were cultured in vitro with VPA, there was significant improvement in embryonic development (NOSEL and KLEIN 1992). Improvement was greater when embryos from methionine-treated rats were used. The mechanism for this protective effect is not understood, but may relate to early methylation reactions that may be required for proper neural tube closure. COELHO and KLEIN (1990) observed decreased methylation of amino acids in neural tube proteins from methionine-deficient embryos.

Feeding a folate-deficient diet to NMRI mice prior to VPA administration increased embryotoxicity when compared to the folate sufficient diet plus VPA group (WEGNER et al. 1987). Additionally, the folate antagonist trimethoprim increased the incidences of VPA-induced exencephaly in NMRI mice (ELMAZAR and NAU 1993). The data are less clear with respect to the combination of methotrexate and VPA (ELMAZAR and NAU 1992). The combination of methotrexate and 400 mg VPA/kg demonstrated an increased incidence of exencephaly, but there were no differences with increasing methotrexate dosages and 300 mg VPA/kg. The changes observed at 400 mg/kg were not dose dependent, and in fact the combination of 5.0 mg methotrexate /kg and 400 mg VPA/kg (the highest concentrations used) produced no more exencephaly than 400 mg VPA/kg alone.

WEGNER and NAU (1992) observed no difference in the total folate level of control embryos and those treated with a teratogenic dose of VPA. However, they did find differences in the distribution of folates with decreases in formylated folates (5-formyltetrahydrofolate and 10-formyltetrahydrofolate) and an increase in tetrahydrofolate. They speculated that VPA was able to inhibit glutamate formyltransferase, which is responsible for interconversion of tetrahydrofolate and 5-formyltetrahydrofolate. They observed a decrease in this enzyme activity in vitro. The samples taken for folate analysis consisted of embryonic and decidual tissue; depending on what was collected, the decidua may have consisted, in part, of maternal tissues which would complicate the analysis. They also did not determine whether sufficient VPA reached the embryo to inhibit embryonic glutamate formyltransferase in vivo.

In summary, although the data are not conclusive, folate appears not to play a role in VPA-induced embryotoxicity.

IV. Alcohol

Ethanol is clearly a developmental toxicant in humans (JONES et al. 1973; JONES and SMITH 1975; CLARREN and SMITH 1978) and in a variety of animal species (reviewed in Schardein 1993). Much of the experimental work has been done in the mouse, and this species appears to be a good choice as an animal model (Sulik et al. 1981). Although the mechanism for ethanol-induced embryotoxicity is unknown, ethanol can directly alter normal rodent development in vitro (Brown et al. 1979; Priscott 1982), with open neural tubes observed in rat embryos treated in vitro (Wynter et al. 1983; Hunter et al. 1994).

Chronic alcohol abuse is associated with folate deficiency (Herbert et al. 1963). There are several possible explanations for such a deficiency, including poor dietary intake (Eichner et al. 1972), intestinal malabsorption (Halsted et al. 1973; Romero et al. 1981), increased urinary excretion (McMartin 1984; Eisenga et al. 1989), or altered hepatic metabolism of folate (reviewed in Halsted and Heise 1987).

Ethanol-induced embryotoxicity in rats treated in vivo is exacerbated by folate deficiency (Lin 1988, 1991a); additionally, methanol-induced developmental toxicity was increased in mice on low-folate diets (Sakanashi et al. 1994). This effect appears not to be due to decreased placental transport of folates (Lin 1991b), but may be due to altered embryonic metabolism of folates by ethanol (Lin et al. 1992). In the latter study, although total folates were not different in ethanol-treated fetal liver, they were decreased in fetal brain. There were also differences in folate distribution. In fetal liver, formyltetrahydrofolates and 5-formiminotetrahydrofolate were decreased, while tetrahydrofolate was increased and 5-methyltetrahydrofolate was unaffected. In fetal brain, tetrahydrofolate and formyltetrahydrofolates were increased and 5-methyltetrahydrofolate was decreased. The overall significance of these changes is unclear. However, 10-formyltetrahydrofolate is involved in purine synthesis, so decreases in this derivative in fetal liver could have significant consequences for DNA synthesis.

Interestingly, in light of the possible role of methionine synthase in producing neural tube defects, chronic ethanol consumption has been shown to decrease methionine synthase activity in rats (Koblin and Everman 1991), but acute administration had no effect on this enzyme activity in mice (Koblin and Tomerson 1989). Acute administration of ethanol to pregnant mice resulted in hypomethylation of fetal DNA and suggested that ethanol decreased DNA methyltransferase activity (Garro et al. 1991).

Alcohol-induced embryotoxicity is probably induced by a number of different mechanisms. The role played by folate deficiency is unknown and deserves greater research attention.

V. Pyrimethamine

Compounds other than aminopterin and methotrexate may inhibit dihydrofolate reductase activity and produce a folate deficiency. This category includes pyrimethamine, which is used as an antimalarial drug. This compound was first reported to produce malformations in rats (Thiersch 1954), and the mechanism for the effect may be due to its production of macrocytosis (Petter and Bourbon 1975), which led to hematomas and hemorrhages in areas that were later malformed (Tangapregassom et al. 1985; Horvath et al. 1989). The drug has also been reported to produce malformations in humans (Harpey et al. 1983).

Treatment of rats with various doses of pyrimethamine decreased embryonic folate levels without affecting maternal plasma folate concentrations (Raynaud and Horvath 1994). However, nonembryotoxic doses of pyrimethamine decreased embryonic folate concentrations to the same extent as an embryotoxic dose. It is questionable what role, if any, the decreased embryonic folate levels played in pyrimethamine-induced embryotoxicity.

When folic acid was administered concomitantly with pyrimethamine, there were increased incidences of visceral (dilation of lateral ventricles) and skeletal (fused ribs) defects compared to the pyrimethamine-only group (Chung et al. 1993). Embryotoxicity produced by a higher dosage of pyrimethamine was prevented by i.p. injection of 5-formyltetrahydrofolate (Chung et al. 1993). Oral administration of folic acid with pyrimethamine increased pyrimethamine-induced embryotoxicity, whereas intraperitoneal injection of the same dose of folic acid decreased pyrimethamine-induced embryotoxicity (Kudo et al. 1993). The authors attributed this effect to differences in plasma 5-methyltetrahydrofolate levels produced by the different routes of administration of folic acid, but it is not clear whether this fully explains the results.

It is not clear what role perturbations in folate may play in pyrimethamine-induced embryotoxicity.

F. Conclusions

It is clear that folate therapy can decrease the incidence of NTD in humans. The exact mechanism for this protective effect is unknown. Folates interact with a variety of compounds in one-carbon metabolism, including methionine and homocysteine. These two compounds may also participate in the protection afforded by folate. Methionine in particular lies at a junction between various metabolic pathways and may play a significant role in normal neural tube closure.

A variety of embryotoxic compounds interact with components of these pathways. For most of these compounds, the evidence is not convincing that their embryotoxic mechanism is mediated solely by folate. Further research is needed to clarify the role of folate and methionine in normal neural tube closure and the mechanisms whereby perturbations in the metabolism of these compounds may produce abnormal embryonic development.

References

Bailey LB (1992) Evaluation of a new recommended dietary allowance for folate. J Am Diet Assoc 92: 463–471

Bailey LB (1995) Folate requirements and dietary recommendations. In: Bailey LB (ed) Folate in health and disease. Dekker, New York, pp 123–151

Banks RGS, Henderson RJ, Pratt JM (1968) Reactions of gases in solution. III. Some reactions of nitrous oxide with transition-metal complexes. J Chem Soc A: 2886–2889

Benn A, Swan CHJ, Cooke WT, Blair JA, Matty AJ, Smith ME (1971) Effect of intraluminal pH on the absorption of pteroylmonoglutamic acid. Br Med J 1: 148–150

Bird OD, McGlohon VM, Vaitkus JW (1969) A microbiological assay system for naturally occurring folates. Can J Microbiol 15: 465–472

Bjerkedal T, Czeizel A, Goujard J, Kallen B, Mastroiacova P, Nevin N, Oakley G Jr, Robert E (1982) Valproic acid and spina bifida. Lancet ii: 1096

Block G, Abrams B (1993) Vitamin and mineral status of women of childbearing potential. Ann NY Acad Sci 678: 244–254

Bower C, Stanley FJ (1989) Dietary folate as a risk factor for neural tube defects: evidence from a case-control study in Western Australia. Med J Aust 150: 613–619

Bower C, Stanley FJ, Nicol DJ (1993) Maternal folate status and the risk for neural tube defects. The role of dietary folate. Ann NY Acad Sci 678: 146–155

Brown NA, Goulding EH, Fabro S (1979) Ethanol embryotoxicity: direct effects on mammalian embryos in vitro. Science 206: 573–575

Brown NA, Holt D, Webb M (1984) The teratogenicity of methoxyacetic acid in the rat. Toxicol Lett 22: 93–100

Chan MK, Beale D, Moorhead JF (1980) Acute megaloblastosis due to cotrimoxazole. Br J Clin Pract 34: 187–188

Chanarin I (1986) Folate deficiency. In: Blakley RL, Whitehead VM (eds) Folates and pterins, vol 3. Nutritional, pharmacological, and physiological aspects. Wiley, New York, pp 75–146

Chatot CL, Klein NW, Clapper ML, Resor SR, Singer WD, Russman BS, Holmes GL, Mattson RH, Cramer JA (1984) Human serum teratogenicity studied by rat embryo culture: epilepsy, anticonvulsant drugs, and nutrition. Epilepsia 25: 205–216

Chung M-K, Han S-S, Roh J-K (1993) Synergistic embryotoxicity of combination of pyrimethamine and folic acid in rats. Reprod Toxicol 7: 463–468

Clarke DO, Mebus CA, Miller FJ, Welsch F (1991) Protection against 2-methoxyethanol-induced teratogenesis by serine enantiomers: studies of potential alteration of 2-methoxyethanol pharmacokinetics. Toxicol Appl Pharmacol 110: 514–526

Clarren SK, Smith DW (1978) The fetal alcohol syndrome. N Engl J Med 298: 1063–1067

Clinical Nutrition Cases (1988) Sulfasalazine inhibits folate absorption. Nutr Rev 46: 320–323

Coelho CND, Klein NW (1990) Methionine and neural tube closure in cultured rat embryos: morphological and biochemical analyses. Teratology 42: 437–451

Coelho CN, Weber JA, Klein NW, Daniels WG, Hoagland TA (1989) Whole rat embryos require methionine for neural tube closure when cultured on cow serum. J Nutr 119: 1716–1725

Copp AJ (1994) Genetic models of mammalian neural tube defects. Ciba Found Symp 181: 118–134

Corcino J, Waxman S, Herbert V (1970) Mechanism of triamterene-induced megaloblastosis. Ann Intern Med 73: 419–424

Cossins EA (1984) Folates in biological materials. In: Blakley RL, Benkovic SJ (eds) Folates and pterins, vol 1. Chemistry and biochemistry of folates. Wiley, New York, pp 1–59

Craxi A, Pagliarello F (1980) Possible embryotoxicity of sulfasalazine. Arch Intern Med 140: 1674
Czeizel AE (1995) Folic acid in the prevention of neural tube defects. J Pediatr Gastroenterol Nutr 20: 4–16
Czeizel AE, Dudas I (1992) Prevention of the first occurrence of neural-tube defects by periconceptional vitamin supplementation. Lancet 327: 1832–1835
DeSesso JM, Goeringer GC (1991) Amelioration by leucovorin of methotrexate developmental toxicity in rabbits. Teratology 43: 201–215
DeSesso JM, Goeringer GC (1992) Methotrexate-induced developmental toxicity in rabbits is ameliorated by 1-(p-tosyl)-3,4,4-trimethylimidazolidine, a functional analog for tetrahydrofolate-mediated one-carbon transfer. Teratology 45: 271–283
Duch DS, Bowers SW, Nichol CA (1983) Analysis of folate cofactor levels in tissues using high-performance liquid chromatography. Anal Biochem 130: 385–392
Eichner ER, Buchanan B, Smith JW, Hillman RS (1972) Variations in the hematologic and medical status of alcoholics. Am J Med Sci 263: 35–42
Eisenga BH, Collins TD, McMartin KE (1989) Differential effects of acute ethanol on urinary excretion of folate derivatives in the rat. J Pharmacol Exp Ther 248: 916–922
Elmazar MMA, Nau H (1992) Methotrexate increases valproic acid-induced developmental toxicity, in particular neural tube defects in mice. Teratogen Carcinog Mutagen 12:203–210
Elmazar MMA, Nau H (1993) Trimethoprim potentiates valproic acid-induced neural tube defects (NTDs) in mice. Reprod Toxicol 7: 249–254
Elmazar MMA, Thiel T, Nau H (1992) Effect of supplementation with folinic acid, vitamin B_6, and vitamin B_{12} on valproic acid-induced teratogenesis in mice. Fundam Appl Toxicol 18: 389–394
Essian FB (1992) Maternal methionine supplementation promotes the remediation of axial defects in Axd mouse neural tube mutants. Teratology 45: 205–212
Essian FB, Wannberg SL (1993) Methionine but not folinic acid or vitamin B-12 alters the frequency of neural tube defects in Axd mutant mice. J Nutr 123: 27–34
Fink BR, Shepard TH, Blandau RJ (1967) Teratogenic activity of nitrous oxide. Nature 214: 146–148
Finkelstein JD (1990) Methionine metabolism in mammals. J Nutr Biochem 1: 228–237
Flynn TJ, Friedman L, Black TN, Klein NW (1987) Methionine and iron as growth factors for rat embryos cultured in canine serum. J Exp Zool 244: 319–324
Food and Nutrition Board (1980) Recommended dietary allowances. 9th edn. National Academy of Sciences, Washington
Food and Nutrition Board (1989) Recommended dietary allowances. 10th edn. National Academy of Sciences, Washington
Franklin JL, Rosenberg IH (1973) Impaired folic acid absorption in inflammatory bowel disease: effects of salicylazosulfapyridine (Azulfidine). Gastroenterology 64: 517–525
Frisch JM (1973) Clinical experience with adverse reactions to trimethoprim-sulfamethoxazole. J Infect Dis 128: 607–612
Fujinaga M, Baden JM, Shepard TH, Mazze RI (1990) Nitrous oxide alters body laterality in rats. Teratology 41: 131–135
Garro AJ, McBeth DL, Lima V, Lieber CS (1991) Ethanol consumption inhibits fetal DNA methylation in mice: implications for the fetal alcohol syndrome. Alcohol Clin Exp Res 15: 395–398
Goeringer GC, Scialli AR, DeSesso JM (1994) Developmental toxicity in rabbits induced by aminopterin is decreased by 1- (p-tosyl)-3,4,4-trimethyl imidazolidine (TTI), a functional analog for tetrahydrofolate-mediated one-carbon transfer. Teratology 49: 368 (abstr)
Halsted CH, Heise C (1987) Ethanol and vitamin metabolism. Pharmacol Ther 34: 453–464

Halsted CH, Robles EA, Mezey E (1973) Intestinal malabsorption in folate-deficient alcoholics. Gastroenterology 64: 526–532
Halsted CH, Gandhi G, Tamura T (1981) Sulfasalazine inhibits the absorption of folates in ulcerative colitis. N Engl J Med 305: 1513–1517
Hansen DK (1991) The embryotoxicity of phenytoin: an update on possible mechanisms. Proc Soc Exp Biol Med 197: 361–368
Hansen DK (1993) In vitro effects of folate derivatives on valproate-induced neural tube defects in mouse and rat embryos. Toxicol In Vitro 7: 735–742
Hansen DK, Billings RE (1985a) Effects of nitrous oxide on maternal and embryonic folate metabolism in rats. Dev Pharmacol Ther 8:43–54.
Hansen DK, Billings RE (1985b) Phenytoin teratogenicity and effects on embryonic and maternal folate metabolism. Teratology 31: 363–371 [erratum published in 1986 in Teratology 34: 487]
Hansen DK, Grafton TF (1991) Lack of attenuation of valproic acid-induced effects by folinic acid in rat embryos in vitro. Teratology 43: 575–582
Hansen DK, Dial SL, Grafton TF (1992) Effect of folinic acid on valproate-induced neural tube defects in mice treated in vivo or in vitro. Teratology 45: 493 (abstr)
Hansen DK, Dial SL, Grafton TF (1995) Lack of attenuation of valproic acid-induced embryotoxicity by compounds involved in one-carbon transfer reactions. Toxicol In Vitro 9: 615–621
Hansen DK, Grafton TF, Dial SL, Gehring TA, Siitonen PH Effect of supplemental folic acid on valproate-induced embryotoxicity and tissue zinc levels in vivo. Teratology (in press)
Harpey J-P, Darbois Y, Lefebvre G (1983) Teratogenicity of pyrimethamine. Lancet ii: 399
Heid MK, Bill ND, Hinrichs SH, Clifford AJ (1992) Folate deficiency alone does not produce neural tube defects in mice. J Nutr 122: 888-894
Helm VF, Kretzschmar R, Leuschner F, Neumann W (1976) Untersuchungen über den Einfluβ der Kombination Sulfamoxao/Trimethoprim (CN3123) auf Fertilität und Embryonalentwicklung an Ratten und Kaninchen. Arzneimittelforschung 26: 643–651
Hendel J, Dam M, Gram L, Winkel P, Jorgensen I (1984) The effects of carbamazepine and valproate on folate metabolism in man. Acta Neurol Scand 69: 226–231
Herbert V, Zalusky R, Davidson CS (1963) Correlation of folate deficiency with alcoholism and associated macrocytosis, anemia, and liver disease. Arch Intern Med 58: 977–988
Hibbard BM, Hibbard ED, Jeffcoate TNA (1965) Folic acid and reproduction. Acta Obstet Gynecol Scand 44: 375–400
Hibbard ED, Smithells RW (1965) Folic acid metabolism and human embryopathy. Lancet i: 1254
Hill AV, Kerr DN (1973) Toxicity of co-trimoxazole in nutritional haematinic deficiency. Postgrad Med J 49: 596–598
Hoffbrand AV, Necheles TF (1968) Mechanism of folate deficiency in patients receiving phenytoin. Lancet ii: 528–530
Hoo JJ, Hadro TA, vonBehren P (1988) Possible teratogenicity of sulfasalazine. N Engl J Med 318:1128
Horne DW, Patterson D (1988) *Lactobacillus casei* microbiological assay of folic acid derivatives in 96-well microtiter plates. Clin Chem 34: 2357–2359
Horton VL, Sleet RB, John-Greene JA, Welsch F (1985) Developmental phase-specific and dose-related teratogenic effects of ethylene glycol monomethyl ether in CD-1 mice. Toxicol Appl Pharmacol 80: 108–118
Horvath C, Tangapregassom AM, Tangapregassom MJ, Trecul M, Boucher-Ehrensperger M, Compagnon A, Petter C (1989) Pathogenesis of limb and facial malformations induced by pyrimethamine in the rat. Acta Morph Hung 36: 53–61
Hunter III ES, Tugman JA, Sulik KK, Sadler TW (1994) Effects of short-term exposure to ethanol on mouse embryos in vitro. Toxicol In Vitro 3: 413–421

Jones KL, Smith DW, Ulleland CJ (1973) Pattern of malformation in offspring of chronic alcoholic mothers. Lancet i: 1267–1271

Jones KL, Smith DW (1975) The fetal alcohol syndrome. Teratology 12: 1–10

Joosten E, Pelemans W (1991) Megaloblastic anaemia in an elderly patient treated with triamtrene. Neth J Med 38: 209–211

Jordan RL, Wilson JG, Schumacher HJ (1977) Embryotoxicity of the folate antagonist methotrexate in rats and rabbits. Teratology 15: 73–80

Kelly D, Weir D, Reed B, Scott J (1979) Effect of anticonvulsant drugs on the rate of folate catabolism in mice. J Clin Invest 64: 1089–1096

King K (1985) Trimethoprim. Med J Aust 142: 350–352

Kirke PN, Molloy AM, Daly LE, Burke H, Weir DG, Scott JM (1993) Maternal plasma folate and vitamin B_{12} are independent risk factors for neural tube defects. Q Rev Med 86: 703–708

Klipstein FA (1964) Subnormal serum folate and macrocytosis associated with anticonvulsant drug therapy. Blood 23: 68–86

Koblin DD, Tomerson BW (1989) Methionine synthase activities in mice following acute exposures to ethanol and nitrous oxide. Biochem Pharmacol 38: 1353–1358

Koblin DD, Everman BW (1991) Vitamin B_{12} and folate status in rats after chronic administration of ethanol and acute exposure to nitrous oxide. Alcohol Clin Exp Res 15: 543–548

Krumdieck CL, Fukushima K, Fukushima T, Shiota T, Butterworth CE Jr (1978) A long-term study of the excretion of folate and pterins in a human subject after ingestion of ^{14}C-folic acid, with observations on the effect of diphenylhydantoin administration. Am J Clin Nutr 31: 88–93

Kudo G, Tsunematsu K, Shimoda M, Kokue E (1993) Effects of folic acid on pyrimethamine teratogenesis in rats. Adv Exp Med Biol 338: 469–472

Labadarios D, Dickerson JWT, Parke DV, Lucas EG, Obuwa GH (1978) The effects of chronic drug administration on hepatic enzyme induction and folate metabolism. Br J Clin Pharmacol 5: 167–173

Lammer EJ, Sever LE, Oakley GP Jr (1987) Teratogen update: valproic acid. Teratology 35: 465–493

Lane GA, Nahrwold ML, Tait AR, Taylor-Busch M, Cohen PJ, Beaudoin AR (1980) Anesthetics as teratogens: nitrous oxide is fetotoxic, xenon is not. Science 210: 899–901

Laurence KM, James N, Miller MH, Tennant GB, Campbell H (1981) Double-blind randomised controlled trial of folate treatment before conception to prevent recurrence of neural tube defects. Br Med J 282: 1509–1511

Lin GW-J (1988) Folate deficiency and acute ethanol treatment on pregnancy outcome in the rat. Nutr Res 8: 1151–1160

Lin GW-J (1991a) Effect of dietary folic acid levels and gestational ethanol consumption on tissue folate contents and rat fetal development. Nutr Res 11: 223–230

Lin GW-J(1991b) Maternal-fetal folate transfer: effect of ethanol and dietary folate deficiency. Alcohol 8: 169–172

Lin GW-J, McMartin KE, Collins TD (1992) Effect of ethanol consumption during pregnancy on folate coenzyme distribution in fetal, maternal, and placental tissues. J Nutr Biochem 3: 182–187

Lindenbaum J, Allen RH (1995) Clinical spectrum and diagnosis of folate deficiency. In: Bailey LB (ed) Folate in health and disease. Dekker, New York, pp 43–73

Lindhout D, Schmidt D (1986) In-utero exposure to valproate and neural tube defects. Lancet i: 1392–1393

Lucock MD, Wild J, Schorah CJ, Levene MI, Hartley R (1994) The methylfolate axis in neural tube defects: in vitro characterisation and clinical investigation. Biochem Med Metab Biol 52: 101–114

Maass AR, Wiebelhaus VD, Sosnowski G, Jenkins B, Gessner G (1967) Effect of triamterene on folic reductase activity and reproduction in the rat. Toxicol Appl Pharmacol 10: 413–423

MacKenzie RE (1984) Biogenesis and interconversion of substituted tetrahydrofolates. In: Blakley RL, Benkovic SJ (eds) Folates and pterins, vol 1. Chemistry and biochemistry of folates. Wiley, New York, pp 255–306

Marsh L, Fraser FC (1973) Studies on Dilantin-induced cleft palate in mice. Teratology 7: 23A (abstr)

Mason JB, Zimmerman J, Otradovec CL, Selhub J, Rosenberg IH (1991) Chronic diuretic therapy with moderate doses of triamterene is not associated with folate deficiency. J Lab Clin Med 117: 365–369

Maxwell JD, Hunter J, Stewart DA, Ardeman S, Williams R (1972) Folate deficiency after anticonvulsant drugs: an effect of hepatic enzyme induction? Br Med J 1: 297–299

Mazze RI, Wilson AI, Rice SA, Baden JM (1984) Reproduction and fetal development in rats exposed to nitrous oxide. Teratology 30: 259–265

McMartin KE (1984) Increased urinary folate excretion and decreased plasma folate levels in the rat after acute ethanol treatment. Alcohol Clin Exp Res 8: 172–178

McMartin KE, Virayotha V, Tephly TR (1981) High-pressure liquid chromatography separation and determination of rat liver folates. Arch Biochem Biophys 209: 127–136

Mercier-Parot L, Tuchmann-Duplessis H (1974) The dysmorphogenic potential of phenytoin: experimental observations. Drugs 8: 340–353

Miller JW, Nadeau MR, Smith J, Smith D, Selhub J (1994) Folate-deficiency-induced homocysteinaemia in rats: disruption of S-adenosylmethionine's co-ordinate regulation of homocysteine metabolism. Biochem J 298: 415–419

Miller PN, Pratten MK, Beck F (1989) Growth of 9.5-day rat embryos in folic-acid-deficient serum. Teratology 39: 375–385

Mills JL, Rhoads GG, Simpson JL, Cunningham GC, Conley MR, Lassman MR, Walden ME, Depp OR, Hoffman HJ (1989) The absence of a relation between the periconceptional use of vitamins and neural-tube defects. N Engl J Med 321: 430–435

Mills JL, Tuomilehto J, Yu KL, Colman N, Blaner WS, Koskela P, Rundle WE, Forman M, Toivanen L, Rhoads GG (1992) Maternal vitamin levels during pregnancies producing infants with neural tube defects. J Pediatr 120: 863–871

Mills JL, McPartlin JM, Kirke PN, Lee YJ, Conley MR, Weir DG, Scott JM (1995) Homocysteine metabolism in pregnancies complicated by neural-tube defects. Lancet 345: 149–151

Milunsky A, Graef JW, Gaynor MF (1968) Methotrexate-induced congenital malformations. J Pediatr 72: 790–795

Milunsky A, Jick H, Jick SS, Bruell CL, Maclaughlin DS, Rothman KJ, Willett W (1989) Multivitamin/folic acid supplementation in early pregnancy reduces the prevalence of neural tube defects. J Am Med Assoc 262: 2847–2852

Mogadam M, Dobbins WO, Krelitz BI (1980) The safety of corticosteroids and sulfasalazine in pregnancy associated with inflammatory bowel disease. Gastroenterology 78: 1224

Molloy AM, Kirke P, Hillary I, Weir DG, Scott JM (1985) Maternal serum folate and vitamin B_{12} concentrations in pregnancies associated with neural tube defects. Arch Dis Child 60: 660–665

Morbidity and Mortality Weekly Report (1992) Recommendations for the use of folic acid to reduce the number of cases of spina bifida and other neural tube defects. Morbidity and Mortality Weekly Report 41: 1–7

MRC Vitamin Study Research Group (1991) Prevention of neural tube defects: results of the Medical Research Council Vitamin Study. Lancet 338: 131–137

Mulinare J, Cordero JF, Erickson JD, Berry RJ (1988) Periconceptional use of multivitamins and the occurrence of neural tube defects. J Am Med Assoc 260: 3141–3145

Natsuhori M, Shimoda M, Kokue E-I, Hayama T, Takahashi Y (1991) Tetrahydrofolic acid as the principal congener of plasma folates in pigs. Am J Physiol 261: R82–86

Nelson MM, Evans HM (1949) Pteroylglutamic acid and reproduction in the rat. J Nutr 38: 11–24

Nelson MM, Asling CW, Evans HM (1952) Production of multiple congenital abnormalities in young by maternal pteroylglutamic acid deficiency during gestation. J Nutr 48: 61–79

Nelson MM, Wright HV, Asling SW, Evans HM (1955) Multiple congenital abnormalities resulting from transitory deficiency of pteroylglutamic acid during gestation in the rat. J Nutr 56: 349–363

Netzloff ML, Streiff RR, Frias JL, Rennert OM (1979) Folate antagonism following teratogenic exposure to diphenylhydantoin. Teratology 19: 45–50

Newman EM, Tsai JF (1986) Microbiological analysis of 5-formyltetrahydrofolic acid and other folates using an automatic 96-well plate reader. Anal Biochem 154: 509–515

Newman NM, Correy JF (1983) Possible teratogenicity of sulphasalazine. Med J Aust 1: 528–529

Nosel PG, Klein NW (1992) Methionine decreases the embryotoxicity of sodium valproate in the rat: in vivo and in vitro observations. Teratology 46: 499–507

O'Rahilly R, Muller F (1994) Neurulation in the normal human embryo. Ciba Found Symp 181: 70–82

Petter C, Bourbon J (1975) Foetal red cell macrocytosis induced by pyrimethamine: its teratogenic role. Experientia 31: 369–370

Priest DG, Doig MT (1986) Tissue folate polyglutamate chain length determination by electrophoresis as thymidylate synthase-fluorodeoxyuridylate ternary complexes. Methods Enzymol 122: 313–319

Priest DG, Bunni MA, Mulllin RJ, Duch DS, Galivan J, Rhee MS (1992) A comparison of HPLC and ternary complex-based assays of tissue reduced folates. Anal Lett 25: 219–230

Priscott PK (1982) The effects of ethanol on rat embryos developing in vitro. Biochem Pharmacol 31: 3641–3643

Ramazzotto LJ, Carlin RD, Warchalowski GA (1979) Effects of nitrous oxide during organogenesis in the rat. J Dent Res 58: 1940–1943

Raynaud F, Horvath C (1994) Folate deficiency and congenital malformations induced by pyrimethamine in the rat. Reprod Nutr Dev 34: 461–471

Reeves D (1982) Sulphonamides and trimethoprim. Lancet ii: 370–373

Rivey MP, Schottelius DD, Berg MJ (1984) Phenytoin-folic acid: a review. Drug Intell Clin Pharmacol 18: 292–301

Robert E, Guibaud P (1982) Maternal valproic acid and congenital neural tube defects. Lancet ii: 937

Romero JJ, Tamura T, Halsted CH (1981) Intestinal absorption of [^{3}H] folic acid in the chronic alcoholic monkey. Gastroenterology 80: 99–102

Rosenberg IH, Selhub J (1986) Intestinal absorption of folates. In: Blakley RL, Whitehead VM (eds) Folates and pterins, vol 3. Nutritional, pharmacological, and physiological aspects. Wiley, New York, pp 147–176

Rosenberg IH, Streiff RR, Godwin HA, Castle WB (1968) Impairment of intestinal deconjugation of dietary folate – a possible explanation of megaloblastic anemia associated with phenytoin therapy. Lancet ii: 530–532

Sakanashi TM, Rogers JM, Keen CL (1994) Influence of folic acid intake on the developmental toxicity of methanol in the CD-1 mouse. Teratology 49: 368 (abstr)

Sauberlich HE (1995) Folate status of U.S. population groups. In: Bailey LB (ed) Folate in health and disease. Dekker, New York, pp 171–194

Schardein JL (1993) Chemically-induced birth defects. Schardein JL (ed) 2nd edn. Dekker, New York, pp 598–641
Schardein JL, Dresner AJ, Hentz DL, Petrere JA, Fitzgerald JE, Kurtz SM (1973) The modifying effect of folinic acid on DPH-induced teratogenicity in mice. Toxicol Appl Pharmacol 24: 150–158
Schorah CJ, Babibzadeh N, Wild J, Smithells RW, Seller MJ (1993) Possible abnormalities of folate and vitamin B_{12} metabolism associated with neural tube defects. Ann NY Acad Sci 678: 81–91
Scott J, Kirke P, Molloy A, Daly L, Weir D (1994) The role of folate in the prevention of neural-tube defects. Proc Nutr Soc 53: 631–636
Selhub J (1989) Determination of tissue folate composition by affinity chromatography followed by high-pressure ion pair liquid chromatography. Anal Biochem 182: 84–93
Selhub J, Dhar GJ, Rosenberg IH (1978) Inhibition of folate enzymes by sulfasalazine. J Clin Invest 61: 221–224
Seller MJ (1994) Vitamins, folic acid and the cause and prevention of neural tube defects. Ciba Found Symp 181: 161–173
Shane B, Stokstad ELR (1984) Folates in the synthesis and catabolism of histidine. In: Blakley RL, Benkovic SJ (eds) Folates and pterins, vol 1. Chemistry and biochemistry of folates. Wiley, New York, pp 433–455
Skalko RG, Gold MP (1974) Teratogenicity of methotrexate in mice. Teratology 9: 159–164
Smith DB, Carl GF (1982) Interactions between folates and carbamazepine or valproate in the rat. Neurology 32: 965–969
Smithells RW, Sheppard S, Schorah CJ (1976) Vitamin deficiencies and neural tube defects. Arch Dis Child 51: 944–950
Smithells RW, Sheppard S, Schorah CJ, Seller MJ, Nevin NC, Harris R, Read AP, Fielding DW (1980) Possible prevention of neural-tube defects by periconceptional vitamin supplementation. Lancet i: 339–340
Smithells RW, Nevin NC, Seller MJ, Sheppard S, Harris R, Read AP, Fielding DW, Walker S, Schorah CJ, Wild J (1983) Further experience of vitamin supplementation for prevention of neural tube defect recurrences. Lancet i: 1027–1031
Spector I, Green R, Bowes D, Cohen H, Miller S, Metz J (1973) Trimethoprim-sulphamethoxazole therapy and folate nutrition. S Afr Med J 47: 1230–1232
Stabler SP, Marcell PD, Podell ER, Allen RH, Savage DG, Lindenbaum J (1988) Elevation of total homocysteine in the serum of patients with cobalamin or folate deficiency detected by capillary gas chromatography-mass spectrometry. J Clin Invest 81: 466–474
Stabler SP, Lindenbaum J, Savage DG, Allen RH (1993) Elevation of serum cystathionine levels in patients with cobalamin and folate deficiency. Blood 81: 3404-3413
Stedman DB, Welsch F (1989) Inhibition of DNA synthesis in mouse whole embryo culture by 2-methoxyacetic acid and attenuation of the effects by simple physiological compounds. Toxicol Lett 45: 111–117
Steegers-Theunissen RPM, Boers GHJ, Trijbels FJM, Eskes TKAB (1991) Neural-tube defects and derangement of homocysteine metabolism. N Engl J Med 324: 199
Steegers-Theunissen RPM, Boers GHJ, Trijbels FJM, Finkelstein JD, Blom HJ, Thomas CMG, Borm GF, Wouters MGAJ, Eskes TKAB (1994) Maternal hyperhomocysteinemia: a risk factor for neural-tube defects? Metabolism 43: 1475–1480
Stephan-Guldner M (1993) Preclinical toxicology and safety pharmacology of brodimoprim in comparison to trimethoprim and analogs. J Chemother 5: 400–410
Sulik KK, Johnston MC, Webb MA (1981) Fetal alcohol syndrome: embryogenesis in a mouse model. Science 214: 936–938

Sullivan FM, McElhatton PR (1975) Teratogenic activity of the antiepileptic drugs phenobarbital, phenytoin, and primidone in mice. Toxicol Appl Pharmacol 34: 271–282

Tangapregassom A-M, Tangapregassom M-J, Horvath C, Trecul M, Boucher-Ehrensperger M, Petter C (1985) Vascular anomalies and pyrimethamine-induced malformations in the rat. Teratogen Carcinog Mutagen 5: 55–62

Thiersch JB (1952) Therapeutic abortions with a folic acid antagonist, 4-aminopteroylglutamic acid (4-amino P.G.A.) administered by the oral route. Am J Obstet Gynecol 63: 1298–1304

Thiersch JB (1954) Effect of certain 2,4-diaminopyrimidine antagonists of folic acid on pregnancy and rat fetus. Proc Soc Exp Biol Med 87: 571–577

Thiersch JB, Philips FS (1950) Effect of 4-aminopteroylglutamic acid (aminopterin) on early pregnancy. Proc Soc Exp Biol Med 74: 204–208

Trotz M, Wegner C, Nau H (1987) Valproic acid-induced neural tube defects: reduction by folinic acid in the mouse. Life Sci 41: 103–110

Udall V (1969) Toxicology of sulphonamide-trimethoprim combinations. Postgrad Med J 45 [Suppl]: 42–45

VanAerts LAGJM, Klaasboer HH, Postma NS, Pertijs JCLM, Peereboom JHJC, Eskes TKAB, Noordhoek J (1993) Stereospecific in vitro embryotoxicity of L-homocysteine in pre- and post- implantation rodent embryos. Toxicol In Vitro 7: 743–749

VanAerts LAGJM, Blom HJ, DeAbreu RA, Trijbels FJM, Eskes TKAB, Peereboom-Stegeman JHJC, Noordhoek J (1994) Prevention of neural tube defects by and toxicity of L-homocysteine in cultured post-implantation rat embryos. Teratology 50: 348–360

Van der Weide J, Homan HC, Cozijnsen-van Rheenen E, Vivie-Kipp Y, Poortman J, Kraaijenhagen RJ (1992) Nonisotopic binding assay for measuring vitamin B_{12} and folate in serum. Clin Chem 38: 766–768

Wagner C (1995) Biochemical role of folate in cellular metabolism. In: Bailey LB (ed) Folate in health and disease. Dekker, New York, pp 23–42

Wald N (1993) Folic acid and the prevention of neural tube defects. Ann NY Acad Sci 678: 112–129

Warkany J (1978) Aminopterin and methotrexate: folic acid deficiency. Teratology 17: 353–357

Wegner C, Nau H (1991) Diurnal variation of folate concentrations in mouse embryo and plasma: the protective effect of folinic acid on valproic acid-induced teratogenicity is time dependent. Reprod Toxicol 5: 465–471

Wegner C, Nau H (1992) Alteration of embryonic folate metabolism by valproic acid during organogenesis: implications for mechanism of teratogenesis. Neurology 42 [Suppl 5]: 17–24

Wegner C, Trotz M, Nau H (1986) Direct determination of folate monoglutamates in plasma by high-performance liquid chromatography using an automatic pre-column-switching system as sample clean-up procedure. J Chromatogr 378: 55–65

Wegner C, Trotz M, Nau H (1987) Folate supplementation and deficiency in experimental valproate-induced teratogenesis. In: Nau H, Scott WJ (eds) Pharmacokinetics in teratogenesis: experimental aspects in vivo and in vitro, vol 2. CRC, Boca Raton, pp 4–10

Welsch F, Sleet RB, Greene JA (1987) Attenuation of 2-methoxyethanol and methoxyacetic acid-induced digit malformations in mice by simple physiological compounds: implications for the role of further metabolism of methoxyacetic acid in developmental toxicity. J Biochem Toxicol 2: 225–240

Werler MM, Shapiro S, Mitchell AA (1993) Periconceptional folic acid exposure and risk of occurrent neural tube defects. J Am Med Assoc 269: 1257–1261

Wild J, Seller MJ, Schorah CJ, Smithells RW (1994) Investigation of folate intake and metabolism in women who have had two pregnancies complicated by neural tube defects. Br J Obstet Gynaecol 101: 197–202

Wynter JM, Walsh DA, Webster WS, McEwen SE, Lipson AH (1983) Teratogenesis after acute alcohol exposure in cultured rat embryos. Teratogen Carcinog Mutagen 3: 421–428

Yates JRW, Ferguson-Smith MA, Shenkin A, Guzman-Rodriguez R, White M, Clark BJ (1987) Is disordered folate metabolism the basis for the genetic predisposition to neural tube defects? Clin Genet 31: 279–287

Zhu M, Zhou S (1989) Reduction of the teratogenic effects of phenytoin by folic acid and a mixture of folic acid, vitamins, and amino acids: a preliminary trial. Epilepsia 30: 246–251

Zimmerman J, Selhub J, Rosenberg IH (1986) Competitive inhibition of folic acid absorption in rat jejenum by triamterene. J Lab Clin Med 108: 272–276

CHAPTER 16

Prostaglandin Metabolism

M.P. GOTO and A.S. GOLDMAN

A. Introduction

Prostaglandins are 20-carbon polyunsaturated fatty acids containing a cyclopentane ring and two side chains, and they are distinguished by their very potent biological activities. The basic skeleton, prostanoic acid, and a typical naturally occurring prostaglandin, prostaglandin E_2 (PGE_2), are depicted in Fig. 1. Each prostaglandin is designated by a letter of the alphabet, A–J, according to the composition of its cyclopentane ring, and by a numerical subscript indicating the number of double bonds in the alkyl side chains. In prostaglandins of the F series, subscripts α or β indicate the orientation of the hydroxyl group at C-9 in the ring.

Prostaglandins are thought to be continuously synthesized from arachidonic acid in the membrane, and they are released to the cell exterior as soon as they are synthesized, without being stored in the cell. At the same time, they are also continuously inactivated by dehydrogenases in the extracellular fluid. Thus the turnover of prostaglandins is extremely short, and the biological effects of prostaglandins are thought to be confined to the immediate vicinity of their site of production.

Prostaglandins are members of a group of unsaturated lipids collectively called eicosanoids. Eicosanoids are metabolites of arachidonic acid or similar polyunsaturated fatty acid precursors, and include prostaglandins, thromboxanes, leukotrienes, lipoxins, and various other hydroxy- and hydroperoxy-fatty acids. Prostaglandins were the first of the eicosanoids to be discovered and have so far been reported to be the most ubiquitous, though there has been an increasing number of reports on the importance of the other eicosanoids.

Prostaglandins are known to be closely involved in many aspects of embryonic development. For example, the oocyte–cumulus complex of early bovine embryos produces prostaglandins $F_{2\alpha}$ and E_2 immediately after fertilization. Groups of oocytes with low cleavage rates produce far less prostaglandins than groups with high cleavage rates, while an addition of prostaglandin E_2 to the medium increases the rate of cleavage (GUREVICH et al. 1993). Epidermal growth factor-induced DNA synthesis in Syrian hamster embryonic cells is modulated by prostaglandin E_2 (COWLEN and ELING 1992). In mice, prostaglandin E_2 is detected immunohistochemically in the un-

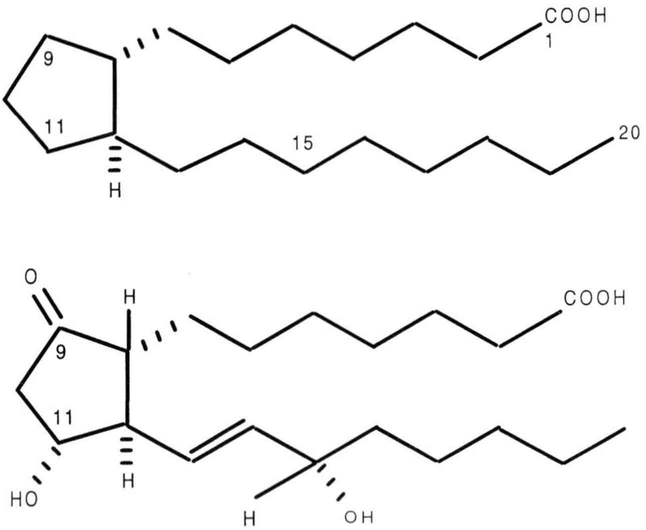

Fig. 1. Structure of prostanoic acid (*top*) and prostaglandin E_2 (PGE2, *bottom*)

fertilized egg and in the embryo from the one-cell stage up to the blastocyst stage (Niimura and Ishida 1987), and day-4 ovine embryos are able to convert arachidonic acid to prostaglandins (Sayre and Lewis 1993). The rate of eruption of ovine embryos from the zona pellucida is significantly increased by prostaglandin E_2, and the increase is inhibited by coadministration of indomethacin, an inhibitor of prostaglandin synthesis (Sayre and Lewis 1993). Prostaglandins are also involved in osteogenesis (Raisz et al. 1993; Scutt et al. 1994) and myogenesis (Zalin 1987). Some prostaglandins are known to increase cyclic adenosine monophosphate (cAMP) in various mammalian cells (Samuelsson et al. 1978; Cowlen and Eling 1992), including embryonic palatal mesenchymal cells (Green and Garbarino 1984). As palatal levels of prostaglandins E_2 and $F_{2\alpha}$ and activity of adenyl cyclase correlate temporally with the transient elevation of palatal cAMP levels during palatal reorientation and epithelial differentiation, it was suggested that prostaglandins modulate palatal medial edge differentiation and shelf elevation by inducing cAMP in the palate (Green et al. 1992; Jones et al. 1986).

Prostaglandins also have protective properties against teratogenic agents. Misoprostol, an analogue of prostaglandin E_1, has been found to be a very effective protector against ionizing radiations in Syrian hamster embryos (LaNasa et al. 1994). When linoleic and γ-linolenic acids, dietary precursors of prostaglandins, are administered to pregnant rats, they protect the embryos from ethanol-induced anomalies of early implantation, placentation, and extraembryonic membrane formation (Varma and Persaud 1982).

If prostaglandins are involved in many aspects of normal development as suggested above, a deficiency or excess of prostaglandins may be expected to

Table 1. The effect of prostaglandins on embryonic development. (From PERSAUD 1974, 1975)

Species	Prostaglandin E_2	Prostaglandin $F_{2\alpha}$
Humans	Abortion	Abortion
Rabbits	Antifertility	Fetal death Abortion
Rats	Antifertility Fetal death Resorption Growth retardation Extensive edema Hemorrhagic lesions	Antifertility Fetal resorption Short tail Cleft palate
Mice	Fetal death Resorption Growth retardation Skeletal defects Microcephalus Polycystic kidney Absence of urinary bladder	Antifertility Fetal death Resorption Reduction abnormalities of the tail and extremities Cleft palate

produce abnormal development, and there have been several reports to support this possibility. The effects of maternal administration of prostaglandins on mammalian embryos and fetuses, according to a review and a report both by PERSAUD (1974, 1975), are listed in Table 1. The table is simplified; the response of each species to administration of prostaglandins varies depending on the dose, route, and timing of administration. The main effect of prostaglandins on a variety of mammalian species is the termination of pregnancy (PERSAUD 1974). Inhibition of implantation, fetal death, fetal resorption, and abortion have been reported in many species. Teratogenesis produced by prostaglandins has also been reported in animal experiments. PERSAUD (1975) reported that maternal injection of a large dose (25 μg) of prostaglandin E_2 is teratogenic and growth inhibitory to mouse fetuses. The anomalies found include meromelia, hydrocephalus, microphthalmia, spina bifida, polycystic kidneys, absence of urinary bladder, and omphalocele. Rioprostil (a synthetic prostaglandin E_1 analogue) was not fetotoxic or teratogenic in rats, but produced increased resorptions, low fetal weight, and increased malformations (HARTNAGEL et al. 1989). There is evidence to show that an excess of prostaglandins may have direct teratogenic consequences on embryos. Prostaglandin E_2 introduced in the culture medium at a concentration of 100 ng/ml produces anomalies, mostly open neural tubes, in 68% of mouse embryos cultured; however, at 10 ng/ml prostaglandin E_2 does not produce anomalies (GOTO et al. 1992). RANDALL and ANTON (1984) have reported that the teratogenic effects of maternal administration of ethanol are reduced by aspirin and suggest that ethanol may increase the production of prostaglandins in the tissue and damage the embryos through an excess of prostaglandins. These results indicate that an excess of prostaglandin E_2 is teratogenic.

With respect to a deficiency of prostaglandins, cyclooxygenase, the enzyme which converts arachidonic acid to prostaglandins, is inhibited by aspirin, indomethacin, and other nonsteroidal anti-inflammatory drugs. Aspirin is a well-known teratogen (Klein et al. 1981a). Montenegro and Palomino (1990) studied the effects of five different nonsteroidal anti–inflammatory drugs on embryonic palatal fusion in vivo and in vitro (palatal shelf organ culture) in experiments using mice. All of the five drugs except indomethacin produced cleft palate when injected, while in paired palatal shelf culture all of the drugs including indomethacin prevented fusion of the opposing shelves. The difference in the effect of indomethacin in vivo and in culture is probably the result of poor placental transfer of indomethacin, as reported by Klein et al. (1981b). Indomethacin also produces defects in neural tube fusion and anomalies of facial arches in mouse embryos in culture (Kay et al. 1988). These reports that inhibitors of cyclooxygenase are teratogenic indicate that a deficiency of prostaglandins may have teratogenic consequences.

B. Signal Transduction

Work from our laboratory and elsewhere suggests that some teratogens produce congenital anomalies through an inhibition of the synthesis of prostaglandins and a resulting local deficiency of prostaglandins. The apparent involvement of an inhibition of the signal transduction pathway involving phosphatidylinositol turnover and protein kinase C activation (Greene et al. 1975, 1989; Winegrad 1987) in complications of diabetes led us to investigate whether an inhibition of this signal transduction pathway and an inhibition of the arachidonic acid cascade are involved in the mechanism of the diabetic embryopathy.

The results from these investigations, described further below (see Sect. D.II), led us to the hypothesis that normal development requires a functioning *signal transduction pathway* involving phosphatidylinositol turnover and protein kinase C linked to *the arachidonic acid cascade* (Fig. 2) and that an inhibition of this pathway at any step may be teratogenic. We will explain the basis for this hypothesis by describing first the signal transduction pathway and the arachidonic acid cascade and then discussing specific teratogens which appear to produce anomalies through an inhibition of this pathway.

Signal transduction refers to the process in which an extracellular signal leads to intracellular responses. During the past decade, the importance of the signal transduction pathway through protein kinase C in embryonic development has come to be recognized (Otte et al. 1990; Winkel et al. 1990). In this pathway, when a ligand, such as a growth factor, a neurotransmitter, or a hormone, is bound to an appropriate cell surface receptor, the ligand–receptor complex activates G protein, which in turn activates membrane-bound phospholipase C. Phospholipase C cleaves membrane phosphatidylinositol into two messengers, diacylglycerol and inositol 1,4,5-trisphosphate. Di-

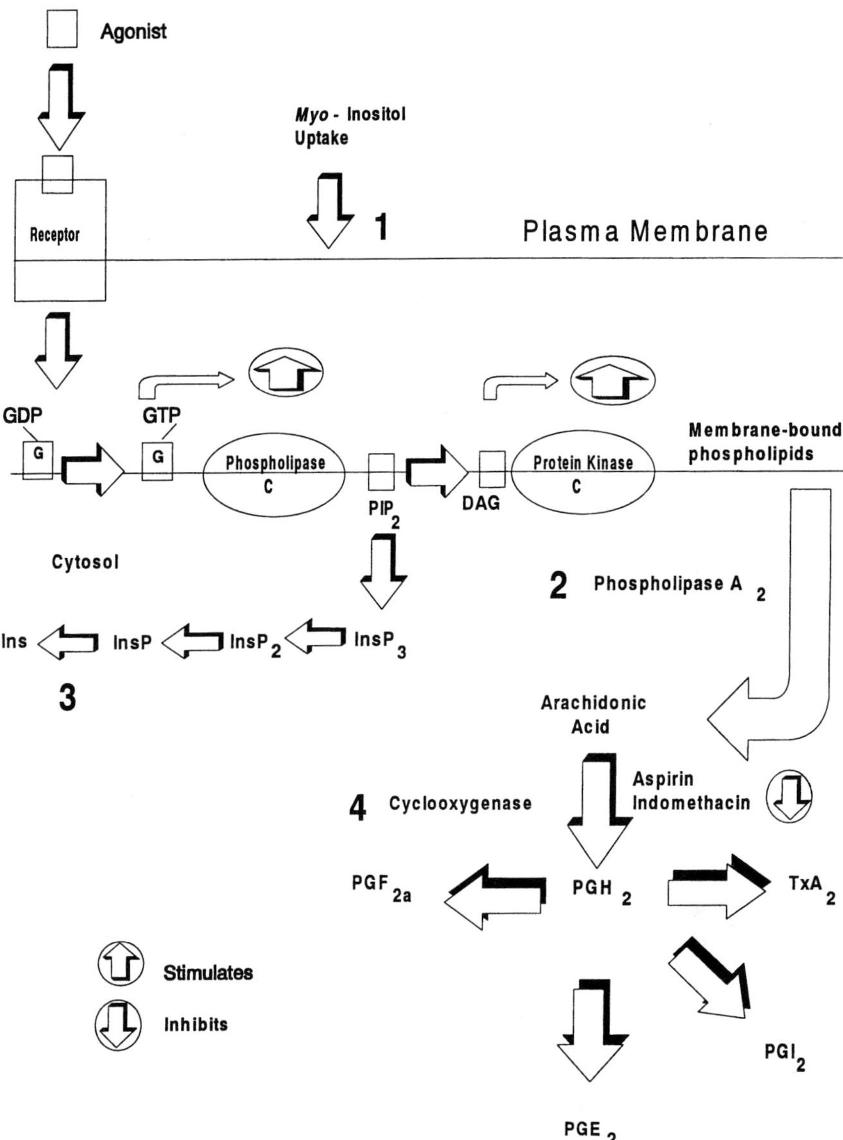

Fig. 2. Signal transduction pathway involving phosphatidylinositol turnover and the arachidonic acid cascade. *GDP*, guanosine 5′ diphosphate; *GTP*, guanosine 5′ triphosphate; *G*, G protein; *DAG*, *sn*-1,2-diacylglycerol; *PIP_2*, phosphatidylinositol 4,5-bisphosphate; *$InsP_3$*, inositol 1,4,5–triphosphate; *$InsP_2$*, inositol 1,4-bisphosphate; *InsP*, inositol phosphate; *Ins*, myo-inositol; PGH_2, prostaglandin H_2; *PGE_2*, prostaglandin E_2; *$PGF_{2\alpha}$*, prostaglandin $F_{2\alpha}$; *PGI_2*, prostaglandin I_2; *TxA_2*, thromboxane A_2. The *numbers* indicate putative sites of inhibition of the pathway. *1*, *myo*-inositol uptake (diabetes); *2*, arachidonic acid release by phospholipase A_2 (glucocorticoids, cyclosporin A); *3*, hydrolysis of inositol monophosphate (lithium); *4*, conversion of arachidonic acid to prostaglandins by cyclooxygenase (aspirin)

acylglycerol is a potent activator of protein kinase C, while inositol 1,4,5-trisphosphate stimulates the release of calcium from its intracellular stores in the endoplasmic reticulum. Protein kinase C requires calcium for its activity, and this release of endogenous calcium by inositol 1,4,5-trisphosphate also activates protein kinase C. Protein kinase C phosphorylates specific serine and threonine residues in target proteins, and activation of protein kinase C leads to phosphorylation of other kinases and finally to the required cellular response.

It has been shown that in many cells and tissues an activation of protein kinase C leads to an increased synthesis of prostaglandins, and an inhibition of protein kinase C to a decreased synthesis of prostaglandins (HUANG et al. 1993; ZAKAR et al. 1994). Our data using protein kinase inhibitors H-7 [1-(5-isoquinolinesulfonyl)-2-methylpiperazine-dihydro-chloride] and HA1004 [*N*-(2-guanidinoethyl)-5-isoquinoline-sulfonamide hydrochloride] (Seikagaku America, Rockville, MD) show that, when protein kinase C is inhibited in mouse embryos in culture, the embryos grow abnormally (GOLDMAN and GOTO 1994). H-7, which is reported to inhibit both protein kinase C and protein kinase A (NISHIKAWA et al. 1986), produces anomalies such as neural tube defects and facial arch anomalies in a dose-related manner at concentrations of 0.1–10 μM in Swiss Webster mouse embryos cultured from day 8 for 24 h. HA1004, which is reported to be more specifically inhibitory to protein kinase A (NISHIKAWA et al. 1986), is not teratogenic at equivalent concentrations, indicating that the anomalies produced by H-7 are the result of an inhibition of protein kinase C. The teratogenic effect of H-7 is attenuated by the addition of 10 ng prostaglandin E_2/ml in the culture medium, suggesting the involvement of an inhibition of production of prostaglandins. These experiments provide evidence that a protein kinase C leading to the production of prostaglandins is involved in normal development of the neural tube and craniofacial region and that an inhibition of this pathway at this site leads to malformations of these structures.

Staurosporine, another protein kinase C inhibitor, has also been reported to be teratogenic in rat embryos in culture (FUJINAGA et al. 1994). The anomalies produced by staurosporine differ from those in our experiments in that they have a high percentage of situs inversus and fewer open neural tubes. The difference may be due to the differences in the species and the method, but it also may be due to the fact that, as the authors point out, staurosporine also inhibits several other protein kinases besides protein kinase C.

C. Arachidonic Acid Cascade

Most prostaglandins are derived from arachidonic acid. Linoleic acid, an essential fatty acid found in plants, is the metabolic precursor of arachidonic acid and is considered to be the primary dietary source of tissue arachidonic acid. Arachidonic acid is primarily found esterified in the *sn*2-position of phospholipids in mammalian cell membranes, both on the cell surface and in intracellular membrane structures. In plasma, arachidonic acid is tightly bound to protein and is mostly inactive. The intracellular level of unesterified arachidonic acid is very low, as it is very quickly incorporated into phospholipids. Thus the main source of arachidonic acid for the production of prostaglandins is considered to be membrane phospholipids.

The arachidonic acid cascade starts when arachidonic acid is released by the enzyme phospholipase A_2 from the *sn*2-position of the membrane phospholipids. This is considered to be a rate-limiting step in the synthesis of prostaglandins. Free arachidonic acid thus released from membrane phospholipids is then metabolized by cyclooxygenase to prostaglandin G_2 and then to prostaglandin H_2. Prostaglandin H_2 is quickly converted to the 2-series of prostaglandins, such as prostaglandin E_2 and prostaglandin $F_{2\alpha}$, by reductase and isomerase, to prostacyclin (prostaglandin I_2) by prostacyclin synthase, and to thromboxanes by thromboxane synthase. Arachidonic acid is also metabolized by lipoxygenase to the 4-series of leukotrienes and to lipoxins.

Prostaglandins of the 1-series and leukotrienes of the 3-series are synthesized from dihomo-γ-linolenic acid, which is also a metabolite of linoleic acid and a precursor of arachidonic acid, through pathways identical to those involving arachidonic acid. Prostaglandins of the 3-series and leukotrienes of the 5-series are synthesized from eicosapentaenoic acid.

Mammalian cells contain at least three different kinds of phospholipase A_2, including group I, which is found in the pancreas, group II, which is distributed throughout the body, and a recently found cytosolic phospholipase A_2. Group I and II phospholipases A_2 have a molecular mass of about 14 kDa and are called secretory phospholipase A_2. The cytosolic phospholipase A_2, which is reported to have a mass of about 60–110 kDa, has a higher affinity for phospholipids containing arachidonic acid and thus is speculated to be the main regulatory enzyme for the arachidonic acid cascade.

The potential importance of the arachidonic acid cascade in development is underscored by reports that some growth factors stimulate gene expression and synthesis of phospholipase A_2 and cyclooxygenases. The existence of a hitherto unknown cyclooxygenase, now called cyclooxygenase-2 (COX-2), which is inducible very rapidly, has been reported recently. Depending on the type of cells used, epidermal growth factor (Chepenik et al. 1994), platelet-derived growth factor (Ryseck et al. 1992), interleukin-1α (Szczepanski et al. 1994), interleukin-1β (Newman et al. 1994), cAMP (Ryseck et al. 1992), phorbol ester (Kujubu and Herschman 1992; Ryseck et al. 1992), endotoxin (Silver et al. 1995), and serum (DeWitt and Meade 1993) have all been reported to induce COX-2. However, studies show that the classical, "constitutive" cyclooxygenase (COX-1) is also inducible, though the time course and magnitude of induction are often different from that of COX-2. Serum stimulation of quiescent 3T3 cells increases COX-1 mRNA three- to fourfold in 3 h, and COX-2 mRNA 70-fold in 1 h (DeWitt and Meade 1993). In human embryo lung fibroblasts and calf pulmonary artery endothelial cells, transforming growth factor-β has been shown to increase levels of COX-1 mRNA in both cell types and phospholipase A_2 mRNA in the latter, and interleukin-1β has been shown to increase COX-1 mRNA in the former cell type and phospholipase A_2 mRNA in both cell types. Neither of the two cytokines affects expression of the COX-2 gene in these cells (Jackson et al. 1993). Chepenik and his coworkers (1994) report that treatment of mouse embryo palate mesenchyme cells with epidermal growth factor stimulates synthesis and accumulation of cytosolic phospholipase A_2 and both COX-1 and COX-2 and also increases their transcripts.

Thus the rate-limiting enzymes of prostaglandin synthesis, phospholipase A_2 and the two cyclooxygenases, all appear to be regulated by many major growth factors as well as other biologically important factors. These observations suggest that prostaglandins are closely involved in cellular responses to those growth factors and are therefore important in the process of normal development.

D. Individual Teratogens

I. Glucocorticoid- and Diphenylhydantoin-Induced Embryopathy

Glucocorticoids are potent teratogens in laboratory animals including mice, rats, rabbits, hamsters, guinea pigs, dogs, and primates (Schardein 1985; Hendrickx and Binkerd 1990). The primary defects induced are cleft lip/palate, although anomalies of the central nervous system, growth retardation, and inhibition of lung maturation have also been reported. Glucocorticoids inhibit the production of prostaglandins in tissue, and this inhibitory effect used to be attributed to a stimulation of the production of a class of proteins, lipocortins (now also called annexins), by glucocorticoids. Lipocortins inhibit the activity of phospholipase A_2, and its inhibitory action was assumed to be direct (Blackwell et al. 1980; Hirata et al. 1980). However, later studies revealed that lipocortin inhibits phospholipase A_2 only at low substrate conditions in vitro and that it does not bind the enzyme (Davidson et al. 1990). It is now thought that lipocortin inhibits phospholipase A_2 in vitro by binding the substrate and reducing its availability to the enzyme (Russo-Marie 1990). Most studies on lipocortin have been performed in in vitro systems, and there have been conflicting reports on whether or not lipocortin inhibits phospholipase A_2 in vivo (Russo-Marie 1990). It now appears that glucocorticoids inhibit prostaglandin production by their inhibition of gene expression and synthesis and/or activity of cyclooxygenase (Bailey et al. 1988; Riley et al. 1992; Newman et al. 1994; Fig. 2, site 4) and phospholipase A_2 (Nakano et al. 1990; Fig. 2, site 2).

Whether glucocorticoids inhibit phospholipase A_2 directly or through lipocortin, arachidonic acid release will be inhibited, resulting in a deficiency of prostaglandins. Tzortzatou et al. (1981) investigated whether coadministration of arachidonic acid prevents glucocorticoid-induced cleft palate in rodents. They found that, when Sprague-Dawley rats received injections of 3.75 mg dexamethasone/kg on days 12–15 of pregnancy, 82.9% of the fetuses had cleft palate at day 19. Coinjection of 200 mg arachidonic acid/kg decreased the incidence of cleft palate to 40%, and the protective effects of arachidonic acid were prevented by indomethacin, an inhibitor primarily of cyclooxygenase (Fig. 2, site 4) rather than lipoxygenase. Similar results were also obtained in mice (PIDDINGTON et al. 1983). Based on these observations of the prevention of glucocorticoid-induced cleft palate by arachidonic acid and a demonstration of glucocorticoid-inhibition of arachidonic acid release and prostaglandin biosynthesis directly in embryonic palates (GUPTA et al. 1985), it was proposed that glucocorticoids exert their teratogenic effects by an inhibition of the production of prostaglandins. The production of thromboxanes would also be reduced by a diminished availability of arachidonic acid, but since the embryo contains little or no thromboxanes and cannot produce them at this stage (ALAM et al. 1982), the effects of glucocorticoids are most likely mediated through changes in the metabolism of prostaglandins.

Diphenylhydantoin, a widely used anticonvulsant, produces craniofacial anomalies in animals (FINNELL 1981; KAY et al. 1988) and humans (fetal hydantoin syndrome) (HANSON and SMITH 1975; HANSON 1986; GOLDMAN et al. 1987). It has been estimated that the fully fledged fetal hydantoin syndrome is present in approximately 10% of the offspring of female patients receiving diphenylhydantoin (HANSEN and BILLINGS 1985), but an additional 30% has been identified as affected by radiographic investigations (KELLY 1984; VANLANG et al. 1984).

Diphenylhydantoin and glucocorticoids may share a common pathway in the mechanism of their teratogenic actions. For example, agents sharing a common teratogenic pathway do not produce additive effects when administered together, but those utilizing different pathways do. Thus when diphenylhydantoin is given together with cortisone, it does not change the frequency of isolated cleft palate induced by cortisone (FRITZ 1976). Administration of a single dose of either cortisone or diphenylhydantoin to pregnant mice on days 11–14 of gestation produces a definite parallel correlation between fetal weight reduction and cleft palate incidence in the two groups, suggesting that both agents may have an identical mechanism of action (MCDEVITT et al. 1981). Furthermore, susceptibility to glucocorticoid- and diphenylhydantoin-induced cleft palate in mice is influenced by the same genes in the H-2 histocompatibility complex on chromosome 17 and in the H-3 histocompatibility complex on chromosome 2 (BONNER and SLAVKIN 1975; GUPTA and GOLDMAN 1982; GOLDMAN et al. 1983a).

In this chapter, only the possible relationship of diphenylhydantoin to prostaglandin deficiency is examined. There are many other hypotheses of the mechanism of diphenylhydantoin-induced anomalies. Several concern the production of toxic oxidative intermediates, which interact irreversibly with embryonic tissue. These include formation of an arene oxide metabolite by cytochrome P-450-mediated oxidation (MARTZ et al. 1977; FINNELL et al.

1992) and bioactivation by prostaglandin synthase (Wells and Vo 1989). Others include reduced glutathione levels in the liver (Wong and Wells 1988), changes in folate metabolism (Netzloff et al. 1979), decreased free radical scavenging enzymes involved in detoxification of diphenylhydantoin (Yerby et al. 1992), and vascular disruption through fetal hypoxia (Danielsson et al. 1992). For a more detailed review of the effects of anticonvulsants in the embryo and alternative mechanistic hypotheses, please refer to Chap. 27.

Based on the observations of the similarities in the mode of action between glucocorticoids and diphenylhydantoin, it was proposed that the embryonic dysmorphogenesis triggered by diphenylhydantoin may be mediated by the glucocorticoid receptor (Goldman et al. 1983b). Evidence of involvement of the glucocorticoid receptor in diphenylhydantoin-induced anomalies is provided by the findings with the use of cortexolone (11-desoxy-17α-hydroxycorticosterone), a glucocorticoid receptor blocker which binds to the glucocorticoid receptor and blocks glucocorticoid action, but is otherwise inert (Kaiser et al. 1972). Cortexolone prevents glucocorticoid-induced and diphenylhydantoin–induced inhibition of both the medial edge epithelial breakdown of mouse palatal shelves in organ culture (Goldman et al. 1983b) and of neural tube fusion and facial arch formation in mouse whole embryo culture (Kay et al. 1990). Moreover, diphenylhydantoin is an alternative ligand for the glucocorticoid receptor in inhibition of the production of prostaglandins, as demonstrated by competitive binding (Katsumata et al. 1982). Thus it is possible that diphenylhydantoin, a drug of different chemical structure from that of glucocorticoids, is capable of binding to the glucocorticoid receptor and effecting or preventing a hormone-like response. Another drug, trifluorperazine, also competitively binds to the glucocorticoid receptor and blocks the induction of tyrosine aminotransferase in rat hepatoma cells (VanBohemen and Rousseau 1981). The finding that levels of glucocorticoid receptor are elevated in the lymphocytes of children with the fetal hydantoin syndrome (Goldman et al. 1987) suggests that the susceptibility of these children to the teratogenic effects of diphenylhydantoin may be mediated through an increased level of the glucocorticoid receptor. The idea of a common teratogenic pathway between glucocorticoids and diphenylhydantoin is further supported by the finding that facial arch anomalies and neural tube defects produced by diphenylhydantoin in mice in culture can be prevented by arachidonic acid added to the culture medium (Kay et al. 1990), suggesting an inhibition of arachidonic acid cascade in the mechanism of the teratogenicity of diphenylhydantoin.

II. Diabetic Embryopathy

Although several hypotheses have been proposed on the etiology of diabetic embryopathy, none can satisfactorily explain the mechanism by which maternal diabetes induces teratogenicity in the offspring. The proposed hypotheses include deficiencies of arachidonic acid and prostaglandins (Gold-

MAN et al. 1985; BAKER et al. 1990; GOTO et al. 1992), changes in the levels of trace elements (URIU-HARE et al. 1985), an increased production of free oxygen radicals (ERIKSSON and BORG 1993), and an altered production of extracellular matrix components such as laminin B and fibronectin (CAGLIERO et al. 1993).

In order to determine whether an inhibition of the arachidonic acid cascade may be involved in the mechanism of diabetic embryopathy, as it appears to be in glucocorticoid- and diphenylhydantoin-induced anomalies, GOLDMAN et al. (1985) studied the effect of arachidonic acid on the outcome of pregnancies of streptozotocin-treated diabetic rats. When the rats received subcutaneous injections of arachidonic acid during the period of organ differentiation (days 6–12), the incidence of congenital anomalies, such as neural tube defects, micrognathia, and cleft palate, was significantly decreased in a dose-related manner. Furthermore, supplementation of the culture medium with arachidonic acid prevented the teratogenic effects of hyperglycemia on cultured mouse embryos. Based on these observations, the authors proposed that diabetic embryopathy may be mediated by a functional deficiency of arachidonic acid. These findings are supported by the reports by PINTER and coworkers (1986) that supplementation of the medium with arachidonic acid prevents hyperglycemia-induced embryopathy and yolk sac damage in rat embryos in culture.

Prostaglandins also prevent hyperglycemia-induced anomalies in mouse embryos in culture (GOLDMAN et al. 1989; BAKER et al. 1990; GOTO et al. 1992). When D-glucose is present at a concentration of 9.5 mg/ml in the culture medium, 83% of the embryos develop open neural tubes. Prostaglandin E_2 in concentrations from 1 pg/ml to 1 ng/ml decreases the incidence of open neural tube in a dose–related fashion. Furthermore, prostaglandin E_2 also prevents anomalies produced by diabetic rat serum as well as hyperglycemia in mouse embryos in culture (GOTO et al. 1992). Prostaglandin E_2 has been used in this and other experiments with embryo culture for three reasons: (1) it is one of the major naturally occurring prostaglandins in tissues, (2) it is relatively stable in vitro and can be expected to be active during the course of the culture over a 24-h period, and (3) it was more efficient in preventing diabetic embryopathy in mouse embryo culture than $PGF_{2\alpha}$ or prostacyclin (BAKER et al. 1990).

Thus deficiencies of arachidonic acid and prostaglandins may be involved in the mechanism of diabetic embryopathy. The next question is how maternal diabetes causes an inhibition of the arachidonic acid cascade in the embryo.

Myo-inositol is needed for resynthesis of phosphatidylinositol in the membrane (BERRIDGE et al. 1989). As phosphatidylinositol turnover is an essential component of signal transduction through protein kinase C, as described above (see Sect. B), a depletion of *myo*-inositol leads to reduced phosphatidylinositol turnover, resulting in an inhibition of protein kinase C. Protein kinase C has been reported to be inhibited in diabetes, and an inhibition of protein kinase C is implicated in the mechanism of diabetic neuropathy (Greene et al. 1989). Since it is known that an inhibition of protein kinase C leads to a decrease in the synthesis of prostaglandins (Burch 1987; Huang et al. 1993), and since prostaglandins are the product of the arachidonic acid cascade, a diabetes-induced depletion of *myo*-inositol leading to an inhibition of protein kinase C is a plausible link between diabetes and inhibition of the arachidonic acid cascade.

Diabetes and hyperglycemia have both been associated with a deficiency of *myo*-inositol in tissue. Glucose reduces *myo*-inositol uptake by embryos in culture in a dose-dependent fashion, and the tissue content of *myo*-inositol in embryos malformed by hyperglycemic culture is reduced (Weigensberg et al. 1990). The decrease of *myo*-inositol was initially considered to be a result of increased conversion of glucose to sorbitol by aldose reductase. However, the malformations and reductions of tissue *myo*-inositol content are not corrected by aldose reductase inhibitors (Hod et al. 1986), though the accumulation of sorbitol in the tissue is decreased. It is now considered that the decrease in *myo*-inositol is mainly a result of a competitive uptake with glucose (Fig. 2, site 1). The anomalies in the cultured embryos are prevented by supplementation of *myo*-inositol in the medium (Hod et al. 1986; Baker et al. 1990; Hashimoto et al. 1990). Oral supplementation with *myo*-inositol also prevents neural tube defects in the fetuses of streptozotocin-treated diabetic rats (Baker et al 1986, 1990). The preventive effect of *myo*-inositol is blocked by indomethacin, an inhibitor primarily of cyclooxygenase rather than of lipoxygenase, both in culture and in vivo. This observation indicates that the deficiency is related to the arachidonic acid cascade in the mechanism of hyperglycemia-induced teratogenesis and that the deficiency of *myo*-inositol is proximal to the arachidonic acid cascade in the pathway (Baker et al. 1986, 1990).

The hypothesis of a depletion of *myo*-inositol as a cause of diabetic embryopathy is also supported by our studies of the teratogenic effects of lithium, which is used for treatment of bipolar affective disorders. Lithium has been reported to be teratogenic in humans (Warkany 1988; Moore 1994) and rodents in vivo (Jurand 1988; Moore 1994) and in vitro (Hansen et al. 1990). Lithium inhibits hydrolysis of inositol monophosphate to *myo*-inositol (Berridge et al. 1989), and the teratogenic effects of lithium have been suggested to result from a depletion of *myo*-inositol (Fig. 2, site 3; Garcia-Palmer et al. 1988). It has been shown that *myo*-inositol prevents lithium-induced alterations in pattern formation in *Xenopus* embryos (Busa and Gimlich 1989). We have shown that lithium teratogenicity may be prevented by *myo*-inositol (Goldman and Goto 1993). Our unpublished preliminary data show that prostaglandin E_2 may also prevent the teratogenic effects of lithium.

The level of glucose needed to produce anomalies in cultured embryos is much higher than in diabetic serum, suggesting that a high concentration of glucose may not be the sole cause of diabetic embryopathy. This is consistent with the findings that three components of diabetic serum, β-hydroxybutyrate, somatomedin inhibitor, and D-glucose, act synergistically or additively to produce malformations (Sadler et al. 1989). In our studies, prostaglandin E_2 is also effective in preventing anomalies produced by diabetic rat serum in mouse embryo culture in which the D-glucose level is maintained at 4 mg/ml, a level which is similar to that of diabetic serum and which is not teratogenic by itself (Goto et al. 1992). Apparently, whatever the teratogenic synergism of another factor or factors besides D-glucose in diabetic serum, the synergism of all these factors in diabetic serum also appears to be reversed by prostaglandins. If diabetic embryopathy is indeed of multifactorial origin, it is reasonable to consider the existence of a common pathway where the effects of all the factors involved converge, and our data on the reversal of the teratogenic effects of diabetic serum by prostaglandin E_2 suggest that the arachidonic acid cascade may be such a pathway.

III. Cyclosporin A

Many reports indicate that cyclosporin A inhibits the arachidonic acid cascade and decreases prostaglandin production (Kurtz et al. 1987; Voss et al. 1988). Some adverse effects of cyclosporin A are ascribed to a secondary deficiency of prostaglandins (Bennett et al. 1988). This inhibitory effect of cyclosporin A on prostaglandin synthesis is thought to be through an inhibition of phospholipase A_2 (Fig. 2, site 2; Fan and Lewis 1985). Currently available literature shows that cyclosporin A is fetotoxic but not teratogenic in mice (Mason et al. 1985; Fein et al. 1989). Only a small portion of cyclosporin A in maternal blood appears to cross the placenta in mice (Backman et al. 1988). In humans, however, cyclosporin A crosses the placenta easily (Venkatarmanan et al. 1988). So far, however, only a small number of cases with exposure to cyclosporin A during pregnancy have been reported (Pickrell et al. 1988; Cockburn et al. 1989; Pujals et al. 1989).

We hypothesized that cyclosporin A may produce anomalies in mouse embryos in culture, where the drug can affect the embryos directly, through its inhibitory action on the arachidonic acid cascade. In order to test this hypothesis, we studied the effects of cyclosporin A on Swiss Webster mouse embryos cultured from day 8 for 24 h (Uhing et al. 1993). At concentrations of 0.1–10 μg/ml in the culture medium, cyclosporin A produced anomalies in 28.6%–78% of the embryos in a dose-dependent manner, in contrast to 6.8% in the controls. The anomalies included defects of neural tube fusion, facial arch anomalies, misshaped or asymmetric head folds, and a failure to rotate. When arachidonic acid was added to the culture medium at a concentration of 20 μg/ml, the incidence of anomalies was reduced from 56% to 20% by 1 μg cyclosporin A/ml from 78% to 33% by 10 μg/ml. When 10 ng prostaglandin E_2/ml was added to the medium with 10 μg cyclosporin A/ml, the incidence of anomalies decreased from 78% to 31%.

Thus our studies show that cyclosporin A, which is reported to inhibit phospholipase A_2 directly and thereby decrease prostaglandin synthesis (Fan and Lewis 1985), is teratogenic in mouse embryos in culture at concentrations overlapping the therapeutic plasma concentrations in humans (0.05–0.25 μg/ml). The anomalies are preventable by supplementation with either arachidonic acid or prostaglandin E_2, supporting the idea that an inhibition of the arachidonic acid cascade is involved in the mechanism of cyclosporin A-induced embryopathy.

Chronic treatment of children with cyclosporin A produce abnormal features similar to those found in children chronically treated with diphenylhydantoin, such as coarsening of the facial features, gingival hyperplasia, and hypertrichosis (Reznik et al. 1987), thus suggesting a similar mode of action of these two drugs.

E. Conclusion

Prostaglandins are closely involved in various aspects of embryonic development. The evidence available is consistent with our hypotheses that normal embryonic development requires a functioning signal transduction pathway leading to the arachidonic acid cascade and prostaglandin synthesis and that an inhibition of this pathway at any step may lead to teratogenesis. Recently, further supportive evidence has been provided by the identification of the gene responsible for the Aarskog-Scott syndrome, which is characterized by facial, skeletal, and urogenital anomalies (Pasteris et al. 1994). The protein product of the gene shows strong homology to the Rho/Rac guanine nucleotide exchange factors, which are involved in the signal transduction pathway and growth regulation. The fact that the syndrome, which involves multiple organs, arises from a mutation of this gene suggests that an anomaly in the signal transduction pathway is responsible for the altered pattern of embryonic development of this syndrome. Taken together, these findings present a picture of a common pathway of normal growth and differentiation leading from signal transduction via protein kinase C to prostaglandin synthesis, which when interrupted by different teratogenic agents results in anomalous development.

When one considers the possibility of prevention of drug-induced anomalies, one exciting aspect of prostaglandin metabolism in development is that embryonic levels of prostaglandins may be manipulated by dietary means. Maternal intake of fatty acids affects the lipid content of the embryo (Kubow and Koski 1995) and prostaglandin production by the newborns (Faloran et al. 1994). Complications of diabetes have been reported to be corrected by dietary supplementations with linoleic acid and its immediate metabolite, γ-linolenic acid, in both animal models and in humans (Kinsell et al. 1959; Houtsmuller et al. 1981; Jamal 1990; Thomlinson et al. 1990). With more information on the ideal balance of lipids and prostaglandins in the tissue, it may become possible to lower the risk of anomalies in the offspring through manipulation of the maternal diet. Supplementation with *myo*-inositol is another possible method of dietary prevention of anomalies. Oral administration of *myo*-inositol has been reported to prevent both diabetic embryopathy (BAKER et al. 1986; GOLDMAN and GOTO 1990) and nerve conduction anomalies of diabetes in animal models (GREENE et al. 1975). Dietary supplementation is a non-invasive and easy method of treatment, and although sufficient perliminary investigation is needed to guarantee its safety, we believe that it may be a very promising potential method of preventing anomalies.

References

Alam I, Capitanio AM, Smith JB, Chepenik IP, Greene RM (1982) Radiological identification of prostaglandins produced by serum-stimulated mouse embryo palate mesenchyme cells. Biochim Biophys Acta 712: 408–413

Backman L, Brandt I, Dallner G, Ringden O (1988) Tissue distribution of [^{3}H] cyclosporin A in mice. Transplant Proc 20: 684–691

Bailey JM, Makheja AN, Pash J, Verma M (1988) Corticosteroids suppress cyclooxygenase messenger RNA levels and prostanoid synthesis in cultured vascular cells. Biochem Biophys Res Commun 157: 1159–1163

Baker L, Piddington R, Goldman AS, Dahlem S, Egler J (1986) Myo-inositol and arachidonic acid are linked in the mechanism of diabetic embryopathy. Pediatr Res 35 Suppl 1: 12A

Baker L, Piddington R, Goldman A, Egler J, Moehring J (1990) *Myo*–inositol and prostaglandins reverse the glucose inhibition of neural tube fusion in cultured mouse embryos. Diabetologia 33: 593–596

Bennett WM, Elzinger L, Kelley V (1988) Pathophysiology of cyclosporin nephrotoxicity: role of eicosanoids. Transplant Proc 20 Suppl 3: 628–633

Berridge MJ, Downes CP, Hanley MR (1989) Neural and developmental actions of lithium: a unifying hypothesis. Cell 59: 411–419

Biddle FG, Fraser FC (1977) Cortisone-induced cleft palate in the mouse. A search for the genetic control of the embryonic response trait. Genetics 85: 289–302

Blackwell GJ, Carnuccio R, DiRosa M, Flower RJ, Parente L, Persico P (1980) Macrocortin: a polypeptide causing the antiphospholipase effect of glucocorticoids. Nature 287: 147–149

Bonner JJ, Slavkin HC (1975) Cleft palate susceptibility linked to histocompatibility-2 (*H-2*) in the mouse. Immunogenetics 2: 213–218

Burch RM (1987) Protein kinase C mediates endotoxin and zymosan–induced prostaglandin synthesis. Eur J Pharmacol 142: 431

Busa WB, Gimlich RL (1989) Lithium-induced teratogenesis in frog embryos prevented by a polyphosphoinositide cycle intermediate or a diacylglycerol analog. Dev Biol 132: 315–324

Cagliero E, Forsberg H, Sala R, Lorenzi M, Eriksson UF (1993) Maternal diabetes induces increased expression of extracellular matrix components in rat embryos. Diabetes 42: 975–979

Chepenik KP, Diaz A, Jaminez SA (1994) Epidermal growth factor coordinately regulates the expression of prostaglandin G/H synthase and cytosolic phospholipase A_2 genes in embryonic mouse cells. J Biol Chem 269(34): 21786–21792

Cockburn I, Krupp P, Monka C (1989) Present experience of sandimmun in pregnancy. Transplant Proc 21: 3730–3732

Cowlen MS, Eling TE (1992) Modulation of c-jun and jun-B messenger RNA and inhibition of DNA synthesis by prostaglandin E_2 in Syrian hamster embryo cells. Cancer Res 52(24): 6912–6916

Danielsson MK, Danielsson BR, Marchner H, Lundin M, Rundqvist E, Reiland S (1992) Histopathological and hemodynamic studies supporting hypoxia and vascular disruption as explanation to phenytoin teratogenicity. Teratology 46: 485–497

Davidson FF, Lister MD, Dennis EA (1990) Binding and inhibition studies on lipocortins using phosphatidylcholine vesicles and phospholipase A_2 from snake venom, pancreas, and a macrophage-like cell line. J Biol Chem 285: 5602–5609

DeWitt DL, Meade EA (1993) Serum and glucocorticoid regulation of gene transcription and expression of the prostaglandin H synthase-1 and prostaglandin H synthase-2 isozymes. Arch Biochem Biophys 306: 94–102

Eriksson UJ, Borg LAH (1993) Diabetes and embryonic malformations: role of substrate-induced free-oxygen radical production for dysmorphogenesis in cultured rat embryos. Diabetes 42: 411–419

Faloran LR, Goto M, Myers TF, Anderson CL, Zeller WP (1994) Perinatal diet enriched with omega-3 polyunsaturated fatty acids (w-3PUFA) is beneficial in suckling rat endotoxic shock. Pediatr Res 35: 311A

Fan TPD, Lewis GP (1985) Effect of cyclosporin A and inhibition of arachidonic acid metabolism on blood flow and cyclooxygenase products in rat skin allografts. Br J Pharmacol 81: 361–371

Fein A, Vechoropoulos M, Nebel L (1989) Cyclosporine-induced embryotoxicity in mice. Biol Neonate 56: 165–173

Finnell RH (1981) Phenytoin-induced teratogenesis: a mouse model. Science 211: 483–484

Finnell RH, Buehler BA, Kerr BM, Ager BL, Levy RH (1992) Clinical and experimental studies linking oxidative metabolism to phenytoin–induced teratogenesis. Neurology 42 Suppl 5: 25–31

Fritz H (1976) The effect of cortisone on the teratogenic action of acetylsalicylic acid and diphenylhydantoin in the mouse. Experimentia 32: 721–722

Fujinaga M, Park HW, Shepard TH, Mirkes PE, Baden JM (1994) Staurosporine does not prevent adrenergic-induced situs inversus, but causes a unique syndrome of defects in rat embryos grown in culture. Teratology 50: 261–274

Garcia-Palmer F, Weigensberg M, Freinkel N (1988) Lithium ($Li+$) and the early post-implantation embryo: embryotoxicity and changes in *myo*-inositol metabolism. Clin Res 36: 482A

Goldman AS, Baker MK, Gasser DL (1983a) Susceptibility to phenytoin-induced cleft palate in mice is influenced by genes linked to *H-2* and *H-3*. Immunogenetics 18: 17–22

Goldman AS, Baker MK, Piddington R, Herold R (1983b) Inhibition of programmed cell death in mouse embryonic palate in vitro by cortisol and phenytoin: receptor involvement and requirement of protein synthesis. Proc Soc Exp Biol Med 174: 239–243

Goldman AS, Baker L, Piddington R, Marx B, Herold R, Egler J (1985) Hyperglycemia-induced teratogenesis is mediated by a functional deficiency of arachidonic acid. Proc Natl Acad Sci USA 82: 8227–8231

Goldman AS, Goto MP (1993) Lithium embryopathy can be prevented by *myo*-inositol in mouse embryo culture. Teratology 47: 391

Goto MP, Goldman AS (1992) Prostaglandin E_2 prevents anomalies induced by hyperglycemia or diabetic serum in mouse embryos. Diabetes 41: 1644–1650

Goto MP, Goldman AS (1993)

Green RM, Lloyd MR, Pisano MM (1992) Cyclic AMP-dependent protein kinase in human embryonic palate mesenchymal cells. In Vitro Cell Dev Biol 28A(11–12): 755–762

Greene DA, De Jesus PV Jr, Winegrad AI (1975) Effects of insulin and dietary myoinositol on impaired peripheral motor nerve conduction velocity in acute streptozotocin diabetes. J Clin Invest 55: 1326–1336

Greene DA, Lattimer-Greene S, Sima AA (1989) Pathogenesis of diabetic neuropathy: role of altered phosphoinositide metabolism. Crit Rev Neurobiol 5: 143–219

Greene RM, Garbarino MP (1984) Role of cyclic AMP, prostaglandins, and catecholamines during normal palate development. Curr Topics Dev Biol 19: 65–79

Gupta C, Goldman AS (1982) H-2 histocompatibility region: influence on the murine glucocorticoid receptor and its response. Science 216: 994–996

Gupta C, Katsumata M, Goldman AS (1985) H-2 histocompatibility region influences the inhibition of arachidonic acid cascade by dexamethasone and phenytoin in mouse embryonic palates. J Craniofac Genet Dev Biol 5: 277–285

Gurevich M, Harel-Markowitz E, Marcus S, Shore LS, Shemesh M (1993) Prostaglandin production by the oocyte cumulus complex around the time of fertilization and the effect of prostaglandin E on the development of the early bovine embryo. Reprod Fertil Dev 5(3): 281–283

Hansen DK, Billings RE (1985) Phenytoin teratogenicity and effects on embryonic and maternal folate metabolism. Teratology 31: 363–371

Hansen DK, Walker RC, Grafton (1990) Effect of lithium carbonate on mouse and rat embryos in vitro. Teratology 41: 155–160

Hanson JW (1986) Teratogen update: fetal hydantoin syndrome. Teratology 33: 349–353

Hanson JW, Smith DW (1975) Fetal hydantoin syndrome. J Pediatr 87: 285–290

Hartnagel RE, Porter MC, Clemens GR, Abrutyn D, Kille JW, Kowalski RL, Norbury KC, Schlueter G, Herbold B (1989) A review of the toxicology profile of rioprostil. Scand J Gastroenterol Suppl 164: 42–45

Hashimoto M, Akazawa S, Akazawa M, Akashi M, Yamamoto H, Maeda Y, Yamaguchi Y, Yamasaki H, Tahara D, Nakanishi T, Nagataki S (1990) Effects of hyperglycemia on sorbitol and *myo*-inositol contents of cultured embryos: treatment with aldose reductase inhibitor and *myo*-inositol supplementation. Diabetologia 33: 597–602

Hendrickx AG, Binkerd PE (1990) Nonhuman primates and teratological research. J Med Primatol 19: 81–108

Hirata F, Schiffmann E, Venkatasubramanian K, Salomon D, Axelrod J (1980) A phospholipase A_2 inhibitory protein in rabbit neutrophils induced by glucocorticoids. Proc Natl Acad Sci USA 77: 2533–2536

Hod M, Star S, Passonneau JV, Unterman TG, Freinkel N (1986) Effect of hyperglycemia on sorbitol and *myo*-inositol content of cultured rat conceptus: failure of aldose reductase inhibitors to modify *myo*-inositol depletion and dysmorphogenesis. Biochem Biophys Res Commun 140: 974–980

Houtsmuller AJ, van Hal-Ferwerda J, Zahn KJ, Henkes JE (1981) Favorable influences of linoleic acid on the progression of diabetic micro- and macroangiopathy in adult onset diabetes mellitus. Prog Lipid Res 20: 377–386

Huang NN, Wang DJ, Heppel LA (1993) Stimulation of aged human lung fibroblasts by extracellular ATP via suppression of arachidonate metabolism. J Biol Chem 268(15): 10789–10795

Jackson BA, Goldstein RH, Roy R, Cozzani M (1993) Effects of transforming growth factor beta and interleukin-1 on expression of cyclooxygenase 1 and 2 and phospholipase A_2 mRNA in lung fibroblasts and endothelial cells in culture. Biochem Biophys Res Commun 197(3): 146–174

Jamal GA (1990) Prevention and treatment of diabetic distal polyneuropathy by the supplementation of gamma-linolenic acid. In: Horrobin D (ed) Omega-6 essential fatty acids. Liss, New York, pp 487–504

Jones JL, Shanfeld J, Davidovitch Z, Greene RM (1986) Immunohistochemical localization of prostaglandins E and F2 in the developing murine palate. J Craniofac Genet Dev Biol 6: 63–71

Jurand A (1988) Teratogenic activity of lithium carbonate: an experimental update. Teratology 38: 101–111

Kaiser N, Milholland RJ, Turnell RW, Rosen F (1972) Cortexolone: binding to glucocorticoid receptors in rat thymocytes and mechanism of its antiglucocorticoid action. Biochem Biophys Res Commun 49: 516–521

Katsumata M, Baker MK, Sussdorf CE, Goldman AS (1982) Diphenylhydantoin. An alternate ligand of a glucocorticoid receptor affecting prostaglandin generation in A/J mice. Science 218: 1313–1315

Kay ED, Goldman AS, Daniel JC (1988) Arachidonic acid reversal of phenytoin-induced neural tube and craniofacial defects in vitro. J Craniofac Genet Dev Biol 8: 179–86

Kay ED, Goldman AS, Daniel JC (1990) Common biochemical pathway of dysmorphogenesis in murine embryos: use of the glucocorticoid pathway by phenytoin. Teratogenesis Carcinog Mutagen 10: 31–39

Kelly TE (1984) Teratogenicity of anticonvulsant drugs. III. Radiographic hand analysis of children exposed in utero to diphenylhydantoin. Am J Med Genet 49: 445–450

Kinsell LW, Michaels GD, Walker G, Wheeler P, Splitter S, Flynn P (1959) Dietary linoleic acid and linoleate: effects in diabetic and nondiabetic subjects with and without vascular disease. Diabetes 8: 179

Klein KL, Scott WJ, Wilson JG (1981a) Aspirin-induced teratogenesis: a unique pattern of cell death and subsequent polydactyly in the rat. J Exp Zool 216: 107–112

Klein KL, Scott WJ, Clark KE, Wildon JG (1981b) Indomethacin-placental transfer, cytotoxicity, and teratology in the rat. Am J Obstet Gynecol 141: 448–453

Kubow S, Koski KG (1995) Maternal dietary glucose-lipid interactions modulate embryological development in vivo and in embryo culture. Biol Reprod 52(1): 145–155

Kujubu DA, Herschman HR (1992) Dexamethasone inhibits mitogen induction of the TIS10 prostaglandin synthase/cyclooxygenase gene. J Biol Chem 267: 7991–7994

Kurtz A, Pfeilschifter J, Kuhn K, Koch KM (1987) Cyclosporin A inhibits PGE_2 release from vascular smooth muscle cells. Biochem Biophys Res Commun 147: 542–549

LaNasa P, Miller RC, Hanson WR, Hall EJ (1994) Misoprostol-induced radioprotection of oncogenic transformation. Int J Radiat Oncol Biol Phys 29: 273–275

Martz F, Failinger C, Blake D (1977) Phenytoin teratogenesis: correlation between embryopathic effect and covalent binding of putative arene oxide metabolite in gestational tissue. J Pharmacol Exp Ther 203: 231–239

Mason RJ, Thomson AW, Whiting PH, Gray ES, Brown PAJ, Catto GRD, Simpson JG (1985) Cyclosporine-induced fetotoxicity in the rat. Transplantation 39: 9–12

McDevitt JM, Gautieri RF, Mann DE Jr (1981) Comparative teratogenicity of cortisone and phenytoin in mice. J Pharm Sci 70: 631–634

Montenegro MA, Palomino H (1990) Induction of cleft palate in mice by inhibitors of prostaglandin synthesis. J Craniofac Genet Dev Biol 10: 83–94

Moore JA (1994) An assessment of lithium using the IEHR evaluative process for assessing human developmental and reproductive toxicity of agents. Reprod Toxicol 9(2): 175–210

Nakano T, Ohara O, Teraoka H, Arita H (1990) Glucocorticoids suppress group II phospholipase A_2 production by blocking mRNA synthesis and post-transcriptional expression. J Biol Chem 265: 12745–12748

Netzloff ML, Streiff RR, Frias JL, Rennert OM (1979) Folate antagonism following teratogenic exposure to diphenylhydantoin. Teratology 19: 45–50

Newman SP, Flower RJ, Croxtall JD (1994) Dexamethasone suppression of IL-1 beta-induced cyclooxygenase 2 expression is not mediated by lipocortin-1 in A549 cells. Biochem Biophys Res Commun 202: 931–939

Niimura S, Ishida K (1987) Immunohistochemical demonstration of prostaglandin E-2 in preimplantation mouse embryos. J Reprod Fertie 80: 505–508

Nishikawa W, Uemura Y, Hidaka H, Shirakawa S (1986) 1-(5-isoquinoline-sulfonyl)-2-methylpiperazine (H-7), a potent inhibitor of protein kinases, inhibits the differentiation of HL-60 cells induced by phorbol diester. Life Sci 39: 1101

Otsuki H, Yamada K, Yuguchi T, Taneda M, Hayakawa T (1994) Prostaglandin E_1 induced c-Fos and Myc proteins and protects rat hippocampal cells against hypoxic injury. J Cereb Blood Flow Metab 14: 150–155

Otte AP, Kramer IM, Mannesse M, Lambrechts C, Durston AJ (1990) Characterization of protein kinase C in early Xenopus embryogenesis. Development 110: 461–470

Pasteris NG, Cadle A, Logie LG, Porteous MEM, Schwartz CE, Stevenson RE (1994) Isolation and characterization of the faciogenital dysplasia (Aarskog-Scott syndrome) gene: a putative Rho/Rac guanine nucleotide exchange factor. Cell 79: 669–675

Persaud TVN (1974) Embryonic and fetal development. In: Ramwell P (ed) The prostaglandins, vol 2. Plenum, New York

Persaud TVN (1975) The effects of prostaglandin E_2 on pregnancy and embryonic development in mice. Toxicology 5: 97–101

Pickrell MD, Sawers R, Michael J (1988) Pregnancy after renal transplantation: severe intrauterine growth retardation during treatment with cyclosporin A. Br Med J 296: 825

Piddington R, Herold R, Goldman AS (1983) Further evidence for a role of arachidonic acid in glucocorticoid teratogenic action in the palate. Proc Soc Exp Biol Med 174: 336–342

Pinter E, Reece EA, Leranth CZ, Garcia Segura M, Hobbins JC, Mahoney MJ, Naftolin F (1986) Arachidonic acid prevents hyperglycemia-associated yolk sac damage and embryopathy. Am J Obstet Gynecol 155: 691–702

Pujals JM, Figueras G, Puig JM, Lloveras J, Aubia J, Masramon J (1989) Osseous malformation in baby born to woman on cyclosporin. Lancet 1: 667

Raisz LG, Fall PM, Gabbitas BY, McCarthy TL (1993) Effects of prostaglandin E_2 on bone formation in cultured fetal rat calvariae: role of insulin-like growth factor-I. Endocrinology 133(4): 1504–1510

Randall CL, Anton RF (1984) Aspirin reduces alcohol-induced prenatal mortality and malformations in mice. Alcohol Clin Exp Res 8: 513–515

Reznik VM, Jones KL, Durham BL, Mendoza SA (1987) Changes in facial appearance during cyclosporin treatment. Lancet 1: 145S-150S

Riley SC, Greer IA, Schembri LA, Challis JRG (1992) Dexamethasone inhibits basal and stimulated prostaglandin E_2 output from human placental cells by inhibition of prostaglandin H synthase. Gynecol Obstet Invest 33: 85–89

Russo-Marie F (1990) Lipocortins as antiphospholipase A_2 and anti-inflammatory proteins. In: Mukherjee AB (ed) Biochemistry, molecular biology, and physiology of phospholipase A_2 and its regulatory factors. Plenum, New York, pp 197–210

Ryseck R-P, Raynoschek C, Macdonald-Bravo H, Dorfman K, Matei M-G, Bravo R (1992) Identification of an immediate early gene, *pghs*-B, whose protein product has prostaglandin synthase/cyclooxygenase activity. Cell Growth Diff 3: 443–450

Sadler TW, Hunter ES, Wynn RE, Phillips LS (1989) Evidence for multifactorial origin of diabetes-induced embryopathies. Diabetes 38: 70–74

Samuelsson G, Goldyne M, Granstrom E, Hamberg M, Hammarstrom S, Malmsten C (1978) Prostaglandins and thromboxames. Annu Rev Biochem 47: 997–1029

Sayre BL, Lewis GS (1993) Arachidonic acid metabolism during early development of ovine embryos: a possible relationship to shedding of the zone pellucida. Prostaglandins 46(6): 557–569

Schardein JL (1985) Chemically induced birth defects. Dekker, New York, pp 306–310 (Drug and chemical toxicology, vol 2)

Scutt A, Duvos C, Lauber J, Mayer R (1994) Time-dependent effects of parathyroid hormone and prostaglandin E_2 on DNA synthesis by periosteal cells from embryonic chick calvaria. Calcif Tissue Int 55(5): 208–215

Silver RM, Edwin SS, Trautman MS, Simmons DL (1995) Bacterial lipopolysaccharide-mediated fetal death. Production of a newly recognized form of inducible cyclooxygenase (COX-2) in murine decidua in response to lipopolysaccharide. J Clin Invest 95: 725–731

Szczepanski A, Moatter T, Carley WW, Gerritsen ME (1994) Induction of cyclooxygenase II in human synovial microvessel endothelial cells by interleukin-1. Arthritis Rheum 37: 495–503

Thomlinson DR, Robinson JP, Compton AM (1990) The effects of gamma-linolenic acid treatment on motor nerve conduction velocity and axonal transport of substance P in diabetic rats. In: Horrobin D (ed) Omega-6 essential fatty acids. Liss, New York, pp 457–463

Tzortzatou GG, Goldman AS, Boutwell WC (1981) Evidence for a role of arachidonic acid in glucocorticoid-induced cleft palate in rats. Proc Soc Exp Biol Med 166: 321–324

Uhing MR, Goldman AS, Goto MP (1993) Cyclosporin A-induced embryopathy in embryo culture is mediated through inhibition of the arachidonic acid pathway. Proc Soc Exp Biol Med 202: 307–314

Uriu-Hare JY, Stern JS, Reaven G, Keen CL (1985) The effect of maternal diabetes on trace element status and fetal development in the rat. Diabetes 34: 1031–1040

VanBohemen G, Rousseau GG (1981) Calmodulin antagonists competitively inhibit dexamethasone binding to the glucocorticoid receptor. FEBS Lett 143: 21–24

VanLang QN, Tassineri MS, Keith LB, Holmes LB (1984) Effects of in vitro exposure to anticovulsants on craniofacial development and growth. J Craniofac Genet Dev Biol 4: 115–133

Varma PK, Persaud TVN (1982) Protection against ethanol-induced embryonic damage by administering gamma-linolenic and linoleic acids. Prostaglandins Leukot Med 8: 641–645

Venkatarmanan R, Koneru B, Wang CP, Burkart GJ, Caritis SN, Starzl TE (1988) Cyclosporine and its metabolites in mother and baby. Transplantation 46: 468–469

Voss BL, Hamilton KK, Samara EN, McKee PA (1988) Cyclosporin suppression of endothelial prostacyclin generation. Transplantation 45: 793–796

Warkany J (1988) Teratogen update: lithium. Teratology 38: 593–596

Wells PG, Vo HPN (1989) Effects of the tumor promoter 12-0–tetradecanoylphorbol-13-acetate on phenytoin-induced embryopathy in mice. Toxicol Appl Pharmacol 97: 398–405

Weigensberg MJ, Garcia-Palmer F-J, Freinkel N (1990) Uptake of *myo*-inossitol by early-somite rat conceptus. Diabetes 39: 575–582

Winegrad AI (1987) Does a common mechanism induce the diverse complications of diabetes? Diabetes 36: 396–406

Winkel GK, Ferguson JE, Takeuchi M, Nuccitelli R (1990) Activation of protein kinase C triggers permature compaction in the four-cell stage mouse embryo. Dev Biol 138: 1–15

Wong M, Wells PG (1988) Effects of N-acetylcysteine on fetal development and on phenytoin teratogenicity in mice. Teratogenesis Carcinog Mutagen 8: 65–71

Yerby MS, Leavitt A, Erickson MD, McCormick KB, Loewenson RB, Sells CJ, Benedetti TJ (1992) Antiepileptics and the development of congenital anomalies. Neurology 42 Suppl 5: 132–140

Zakar T, Teixeira FJ, Hirst JJ, Guo F (1994) Regulation of prostaglandin endoperoxide H synthase by glucocorticoids and activators of protein kinase C in the human amnion. J Reprod Fertil 100: 43–50

Zalin RJ (1987) The role of hormones and prostanoids in the in vivo proliferation and differentiation of human myoblasts. Exp Cell Res 172: 265

CHAPTER 17

Reactive Intermediates

P.G. Wells, P.M. Kim, C.J. Nicol, T. Parman, and L.M. Winn

A. Introduction

The teratogenicity of a number of xenobiotics is thought to depend upon their metabolism, termed "bioactivation", to highly reactive and potentially toxic intermediary metabolites (Fig. 1). Thus, while the parent xenobiotic, or "proteratogen", is relatively inert and non-toxic, reactive intermediates, depending upon their chemical nature, can irreversibly ("covalently") bind to and/or oxidise embryonic molecular targets such as DNA, protein and lipid. These irreversible molecular lesions are thought to initiate embryopathic responses such as in utero death, growth retardation and teratogenesis. This mechanism is distinct from the reversible binding of other xenobiotics or their stable metabolites to a specific protein receptor evoking an embryopathic response (Fig. 1). Receptor-mediated toxicity generally requires exposure to an excessive concentration of the xenobiotic, the effect increasing with rising concentration and dissipating as the concentration decreases. In contrast, the toxicity of a reactive intermediate can occur in biochemically predisposed individuals at therapeutic drug doses or supposedly safe concentrations of environmental chemicals. Such toxicological predisposition occurs because the amount of reactive intermediate formed during the metabolism of most xenobiotics generally constitutes only a small fraction, perhaps 1%–10% of the overall xenobiotic disposition. Accordingly, relatively small changes in overall xenobiotic metabolism, particularly in increasing the formation or decreasing the detoxification of the reactive intermediate (Fig. 2), can have a major effect on tissue exposure and toxicity. The toxicological potential of reactive intermediates, unlike receptor-mediated toxicity, is further compounded by the irreversible nature of the molecular damage, which results in the accumulation over time of molecular lesions. This irreversibility also underlies another difference from most receptor-mediated toxicities, in that the expression of reactive intermediate-mediated toxicity may occur hours, days or, in the case of transplacental carcinogenesis, years after exposure, let alone after the time of peak tissue concentration of the parent xenobiotic. Predisposition is further determined by an array of cytoprotective pathways that detoxify reactive oxygen species (ROS) and related toxic products initiated by the reaction of xenobiotic free radical reactive intermediates with molecular oxygen and cellular lipids (Fig. 2). Individual genetic or environmentally induced dysfunction

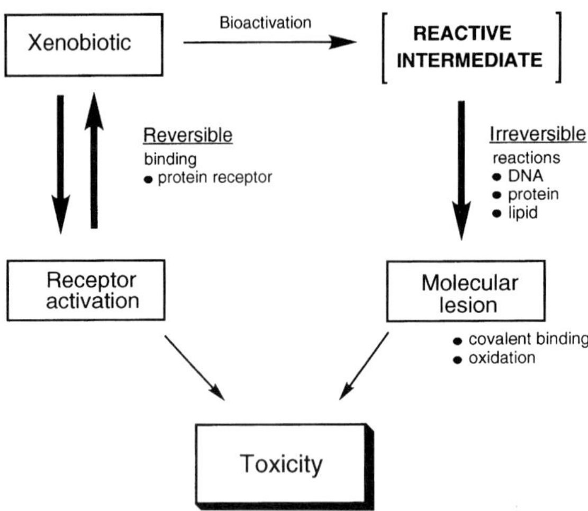

Fig. 1. Chemical teratogenesis initiated by the irreversible reaction of a reactive intermediate with molecular targets, as distinct from the reversible interaction of a xenobiotic with a receptor. Some teratogens may have both effects

in cytoprotective pathways or in pathways involved in the repair of molecular targets can profoundly increase toxicological susceptibility.

This chapter focuses upon reactive intermediate-mediated mechanisms of chemical teratogenesis. Receptor-mediated mechanisms of chemical teratogenesis have been reviewed elsewhere (Juchau 1981; Juchau et al. 1992; Hansen 1991; Nau 1994). A limited number of teratogens are used to exemplify toxicological principles which, despite accumulating evidence, remain largely speculative, particularly in human chemical teratogenesis. While most evidence to date supporting a teratological role for reactive intermediates has been derived from animal models and in vitro systems, human data are discussed where available.

B. Elimination

Many xenobiotics and/or their metabolites are primarily conjugated and eliminated via so-called phase II pathways, which may substantially reduce both the amount reaching the embryo from the mother and the amount bioactivated in the embryo to a toxic reactive intermediate. Phase II pathways include sulfation and glucuronidation, the latter of which is catalysed by a family of enzymes known as the uridine diphosphate (UDP)-glucuronosyltransferases (UGT) (Burchell et al. 1991; Burchell and Coughtrie 1989; Brierly and Burchell 1993). Gunn rats with a hereditary deficiency in UGT exhibit decreased glucuronidation of benzo[a]pyrene metabolites, with conversely enhanced bioactivation and embryotoxicity (Wells et al. 1989c; Hu

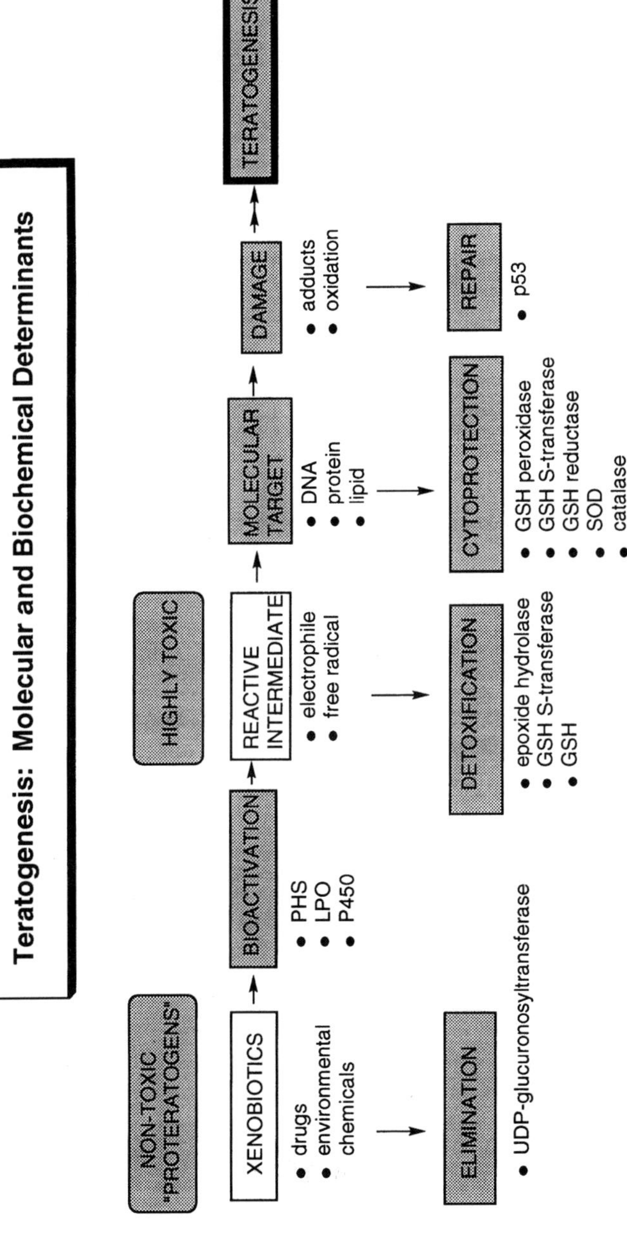

Fig. 2. Balance of pathways determining the embryotoxicity of xenobiotics that are bioactivated to a reactive intermediate. *PHS*, prostaglandin H synthase; *LPO*, lipoxygenase; *P450*, cytochromes P450; *UDP*, uridine diphosphate; *GSH*, glutathione; *SOD*, superoxide dismutase; *G6PD*, glucose-6-phosphate dehydrogenase. (From Winn and Wells 1995b)

Table 1. Enzymes potentially catalysing the bioactivation of teratogenic xenobiotics to electrophilic reactive intermediates (WELLS and WINN 1996)[a]

Enzyme	Substrate[b]	Citation[c]
1. P450[d]		
Not identified[e]	Phenytoin	MARTZ et al. 1977; PANTAROTTO et al. 1982; ROY and SNODGRASS 1990; FINNELL et al. 1994
	Cyclophosphamide	HALES 1981b; FANTEL et al. 1979; SANYAL et al. 1979
	Thalidomide	GORDON et al. 1983; BRAUN et al. 1986
	2-Acetylaminofluorene	FAUSTMANN-WATTS et al. 1983, 1986; JUCHAU et al. 1985a,b;
	Diethylstilbestrol	BALLING et al. 1985
	Thiabendazole	YONEYAMA and ICHIKAWA 1986
	Acetaminophen	HARRIS et al. 1989
	Rifampin	GREENAWAY and FANTEL 1983
	N-Methyl-*N*-(7--propoxynaphthalene-2--ethylhydroxylamine	TERLOUW et al. 1993
	Naphthalene	IYER et al. 1991
CYP1A1	Benzo[a]pyrene	FILLER and LEW 1981; SHUM et al. 1979
CYP1B1	Benzanthracene[f]	POTTENGER et al. 1991; SAVAS et al. 1994; OTTO et al. 1991,1992
CYP2C3	Phenytoin	DOECKE et al. 1991
CYP2C9	Phenytoin	VERONESE et al. 1991
	Dextromethorphan	RAUCY and CARPENTER 1993
CYP2D6	Trichloroethylene Oxide	FORT et al. 1993
CYP2E1	Ethanol	CARPENTER and RAUCY 1995
CYP3A3[g]	Aflatoxin B_1	RAUCY and CARPENTER 1993
	Benzo[a]pyrene	
CYP3A4[g]	Benzo[a]pyrene-7,8-diol	GUENGERICH and SHIMADA 1991
CYP3A5[g]	Aflatoxin B_1	RAUCY and CARPENTER 1993
CYP3A7[g]	Warfarin	YANG et al. 1994
	Benzyloxyresorufin	
2. PHS	Phenytoin	KUBOW and WELLS 1986, 1989
	Benzo[a]pyrene	MARNETT et al. 1975
	Benzo[a]pyrene-7,8-diol	MARNETT et al. 1975; BYCZKOWSKI and KULKARNI 1989;
	2-Naphthylamine	BOYD and ELING 1987
	2-Acetylaminofluorene	BOYD and ELING 1984; BOYD et al. 1983

Table 1. (*Contd.*)

Enzyme	Substrate[b]	Citation[c]
3. Lipoxygenases	Phenytoin	KUBOW and WELLS 1988; YU and WELLS 1995
	Aflatoxin B_1	DATTA and KULKARNI, 1994
	Benzo[a]pyrene-7,8-diol	BYCZKOWSKI and KULKARNI 1989

P450, cytochromes P450; CYP, cytochrome P450; PHS, prostaglandin H synthase; LPO, lipoxygenase.
[a]Electrophilic reactive intermediates are quantified by measuring covalent binding of the xenobiotic to target cellular macromolecules (usually DNA and/or protein). In some cases, covalent binding may result from free radical reactive intermediates, as discussed in the text.
[b]Where available, teratogenic substrates are illustrated, but not all substrates are known teratogens (e.g. dextromethorphan).
[c]Citations here and throughout this review are only representative and should not be considered to be comprehensive.
[d]In general, activities of most P450 (except CYP1B1) in rodent embryos during organogenesis are low to negligible compared with adult hepatic activities, while human fetal activities are substantially higher, constituting 20%–40% or more of adult activities. The term P450 refers nonspecifically to activities for which the nature of the isoenzyme(s) was (were) not determined.
[e]In these studies, the P450 isoenzymes were not identified.
[f]Substrates include benzanthracene, but the teratological relevance is unknown.
[g]Fetal 3A isoenzymes have been identified in humans, but are reported not to be found in rodents and rabbits unless transplacentally induced (WRIGHTEN and STEVENS 1992; YANG et al. 1994).

and WELLS 1992, 1994). Similarly, in rat skin fibroblast culture, UGT-deficient cells are more susceptible to genotoxicity, evidenced by micronucleus formation, initiated by phenytoin, its major para-hydroxylated metabolite and benzo[a]pyrene (VIENNEAU et al. 1995; WELLS and KIM 1996; KIM and WELLS 1996c). Hereditary deficiencies in other phase II pathways, and similar deficits due to environmental causes, may also prove to modulate xenobiotic teratogenicity.

C. Bioactivation

I. Cytochromes P450

1. Embryological Considerations

Cytochromes P450 (P450) are a superfamily of enzymes (NEBERT 1983) that potentially can bioactivate a number of proteratogens (Table 1) to embryotoxic reactive intermediates (MANSON and KANG 1994; JUCHAU 1981; JUCHAU et al. 1992). This is likely to be more teratologically important in humans than in lower species. Depending upon the P450 isoenzyme, human embryonic activities in liver during organogenesis (weeks 3–9) may constitute as much as

40% or more of adult hepatic activities (KITADA and KAMATAKI 1994; RAUCY and CARPENTER 1993; MANSON 1986), while rodent embryonic activities of most P450 isoenzymes during organogenesis (days 8–15) are low or negligible, particularly in uninduced animals (JUCHAU et al. 1992). This substantial species difference necessitates caution when attempting to extrapolate an absence of P450-dependent teratogenicity for a given xenobiotic in rodent models to predictions in humans. It remains to be determined what minimal level of P450 bioactivating activity is necessary to be teratologically relevant, and this threshold may vary both for the P450 isoenzyme and the proteratogen. A recently identified fetal P450 isoenzyme, P450 1B1 (CYP1B1) (POTTENGER and JEFCOATE 1990; SAVAS et al. 1994; SUTTER et al. 1994), is preferentially expressed in the fetus rather than postnatally and may prove to be an important bioactivating isoenzyme. Similarly, it appears that the fetal P450 3A7 enzyme is exclusively expressed in human fetal tissues and is not detectable in adult tissues (RAUCY and CARPENTER 1993). P450-dependent xenobiotic bioactivation may occur via four mechanisms: monooxygenase activity, peroxidase activity, peroxygenase activity or free radical production (ORTIZ DE MONTELLANO 1986). The general peroxidase reaction is discussed later (see Sect. C.II).

2. Mixed-Function Monooxygenase Activity

The mixed-function monooxygenase pathway, also known as the mixed-function oxidase (MFO) pathway, is the most widely appreciated mechanism for the oxidation of proteratogens to electrophilic reactive intermediates that covalently bind to embryonic cellular macromolecules (Fig. 3). This reaction requires molecular oxygen and reduced nicotinamide adenine dinucleotide (phosphate) [NAD(P)H] as an electron donor, with the reduction of the P450–O_2 complex by NAD(P)H-dependent P450 reductase constituting the rate-limiting step (Fig. 4) (WHITE and COON 1980).

3. Peroxygenase Activity

The peroxygenase activity of P450, in contrast to the MFO monooxygenase pathway, is independent of molecular oxygen, NAD(P)H and P450 reductase (Fig. 4). In the presence of synthetic organic hydroperoxides, as well as endogenous hydrogen peroxide (H_2O_2) and lipid hydroperoxides, many xenobiotics (Table 2) can be oxidised via the P450-dependent peroxygenase pathway to the same reactive intermediates and metabolites produced via the MFO pathway, including electrophilic epoxides and arene oxides (Fig. 3). Specifically, P450-peroxygenase activity catalyses the insertion of an oxygen from the hydroperoxide into the substrate, which is distinct from peroxidase activity linking hydroperoxide reduction to one-electron oxidation of electron donating substrates (ORTIZ DE MONTELLANO 1986). While the contribution of the peroxygenase pathway to the embryonic bioactivation of proteratogens has not been investigated, it is more efficient than the MFO pathway and can

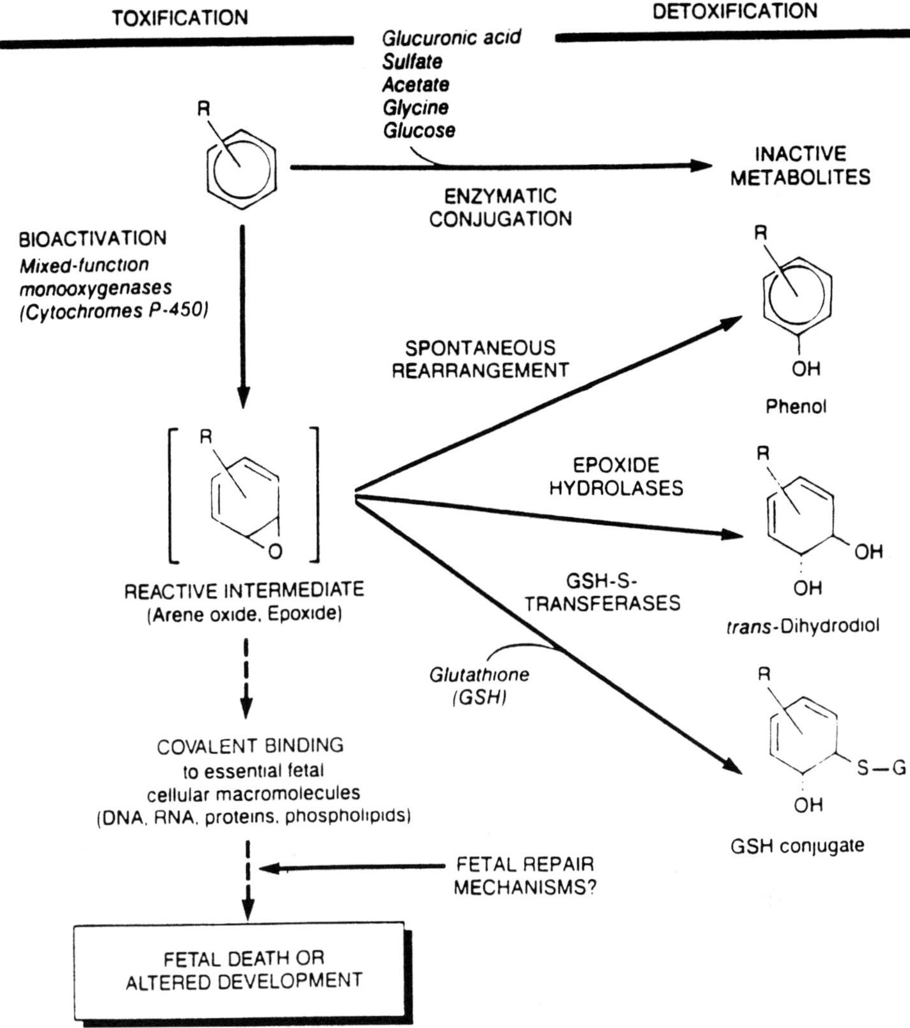

Fig. 3. Postulated bioactivation of xenobiotics to a teratogenic electrophilic reactive intermediate catalysed by cytochromes P450 (From WELLS 1989)

constitute a considerably more active oxidising pathway for some substrates (ANARI et al. 1995). Furthermore, in in vitro studies, physiological concentrations of endogenous lipid hydroperoxides have been shown to enhance the respective hepatic microsomal P450- and CYP1A1-catalysed oxidation of retinoic acid (MUINDI and YOUNG 1993) and diethylstilbestrol (ROY et al. 1992).

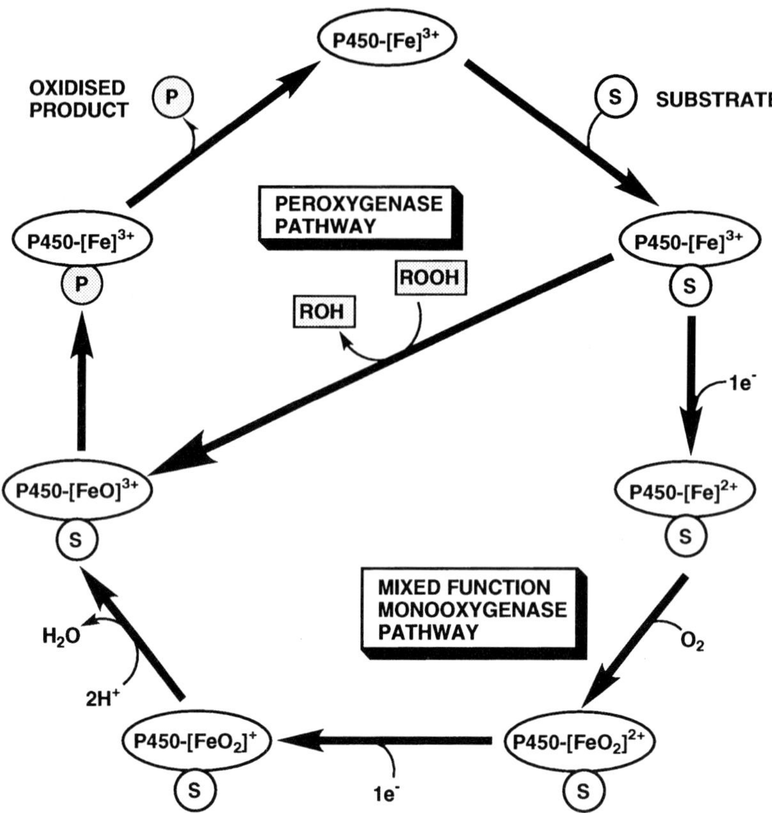

Fig. 4. Mixed-function monooxygenase and peroxygenase pathways of cytochromes P450-dependent xenobiotic oxidation. *ROOH* is an endogenous oxygen donating hydroperoxide that is reduced to its alcohol ROH. *P*, product; *S*, substrate. (Modified from GUENGERICH and MACDONALD 1990)

4. Free Radical Production

Using free radical spin-trapping agents or fluorescent probes, several studies have implicated unspecified P450, CYP2C11 and CYP3A4 in the bioactivation of a number of xenobiotics, including ethanol and toluene, resulting in free radical production and the formation of ROS (KUBOW et al. 1983; TURNER et al. 1991; BONDY and NADERI 1994; SUGIYAMA et al. 1994). The chemical mechanisms and biological importance of these results remain to be established.

II. Peroxidases

1. Prostaglandin H Synthase and Other Peroxidases

In contrast to many P450s, the content and activity of a number of enzyme systems with peroxidase activity, such as prostaglandin H synthase (PHS), and

Table 2. Representative hydroperoxide substrates and xenobiotic co-substrates for cytochromes P450-catalysed peroxygenase activity. (See Fig. 4)

Hydroperoxide substrate (ROOH)	P450 system	Xenobiotic reducing co-substrate	Representative citation
Synthetic hydroperoxides			
CuOOH	Pig LM	Aminopyrine	Kadlubar et al. 1973
t-Butyl OOH	Rat LM[a]	Benzene	Burke and Mayer 1975
p-Menthane OOH	Rabbit LM[b]	Benzo[a]pyrene	Rahimtula and O'Brien 1977
Peracetic acid	Rat Hepatocytes	Benzphetamine	Capdevila et al. 1980
Diacetyl peroxide	Hamster L/KM	Biphenyl	Koop and Hollenberg 1980
t-Butyl perbenzoic acid		Bromobenzene	Kedderis et al. 1983
		Catechol estrogens	Roy and Liehr 1992
		DES	Roy et al. 1992
		Dimethylaniline	Anari et al. 1995
		Ethanol	
		Ethoxyresorufin	
		Ethylmorphine	
		Hydroquinone	
		Metamphetamine	
		Naphthalene	
		1-Naphthol	
		N-Methylcarbazole	
		Phenol	
		p-Nitroanisole	
		Propoxyphene	
		1-Propanol	
Endogenous hydroperoxides PGH_2	Pig AM[c]	Catechol estrogens	Hrycay and O'Brien 1972
Linoleic acid OOH	Human blood PM[d]	DES	Hrycay et al. 1976
13-Hydroperoxy [S-(E,Z)]-9,11 octadecadienoic acid	Thromboxane synthase	Ethanol	Rahimtula and O'Brien 1977
		1-Propanol	Ullrich and Graf 1984
H_2O_2	Rat LM	Retinoic acid	Haurand and Ullrich 1985
Progesterone 17α-OOH	Human CLM[e]	TMPD	Roy et al. 1992
Pregnenolone 17α-OOH			Muindi and Young 1993
Allopregnanolone 17α-OOH			
Cholesterol 7β-OOH			
Cholest-4-ene-3-one 6β-OOH			
Cholesterol 7α-OOH			
Cholesterol 25-OOH			
Cholesterol 26-OOH			
Cholesterol 20α-OOH			

Table 2. (*Contd.*)

Hydroperoxide substrate (ROOH)	P450 system	Xenobiotic seducing co-substrate	Representative citation
Dietary hydroperoxides			
Citronellol OOH	Pig LM	Aminopyrine	KADLUBAR et al. 1973
Geraniol OOH		Benzphetamine	
Myrcene OOH		Dimethylaniline	
		Ethylmorphine	
		Metamphetamine	
		Propoxyphene	

ROOH, general hydroperoxide; CuOOH, cumene hydroperoxide; LM, liver microsomes; L/KM, liver and kidney microsomes; DES, diethylstilbestrol; TMPD, *N*, *N*,*N*′*N*′-tetramethyl-*p*-phenylene diamine; PGH_2, prostaglandin H_2 or 9,11-endoperoxy-15-hydroxy arachidonic acid; H_2O_2, hydrogen peroxide; PM, platelet microsomes; AM, aortic microsomes; CLM, cell line microsomes.
[a]Rat microsomal pretreatments: saline, corn oil, phenobarbital, β-naphthoflavone, isosafrole, ethanol, pregnenolone-16α-carbonitrile, 5,6-benzoflavone or 3-methylcholanthrene.
[b]Rabbit microsomal pretreatments: phenobarbital.
[c]Contains prostacyclin synthase.
[d]Contains thromboxane synthase.
[e]Human cell line microsomes investigated: native microsomes and microsomes from cells transfected with human P450 genes (containing human CYP1A1/1A2/2B6/2D6/2E1 or 3A4).

enzymes with associated peroxidase activity, such as lipoxygenases (LPO), are high in rodent and human embryos during organogenesis (MITCHELL et al. 1985; DATTA et al. 1993; DATTA and KULKARNI 1994; WELLS et al. 1995). These embryonic peroxidases are thought to bioactivate a number of teratogens to reactive electrophilic (Table 1) and/or free radical (Table 3) intermediates that directly or indirectly covalently bind to, and/or oxidise, cellular macromolecules and initiate teratogenesis (Fig. 5). In mice, the teratogenicity of phenytoin and a structurally related anticonvulsant drug and metabolite, trimethadione and dimethadione, are decreased in vivo (WELLS et al. 1989a,b) and in embryo culture (MIRANDA et al. 1994) by the respective PHS and dual PHS/LPO inhibitors acetylsalicylic acid (ASA, aspirin) and eicosatetraynoic acid (ETYA). Similarly in rabbits, teratogenicity of the sedative/hypnotic drug thalidomide, which shares some structural features with phenytoin, is decreased by ASA pretreatment (ARLEN and WELLS 1989, 1996a). Conversely, dietary manipulations that increase the arachidonic acid content of lipid membranes also enhances the susceptibility of such mice to phenytoin teratogenicity (HIGH and KUBOW 1994). In the latter case, arachidonic acid release is thought to be the initiation step in the bioactivation of xenobiotics by both PHS and LPO (Fig. 5). In in vitro models, PHS and other peroxidases such as thyroid peroxidase, horseradish peroxidase, myeloperoxidase and soybean lipoxygenase have been shown to catalyse the bioactivation and molecular target damage initiated by the anticonvulsant drug phenytoin and a

Table 3. Peroxidase enzymes that may catalyse the bioactivation of xenobiotics to teratogenic free radical intermediates and/or the subsequent formation of reactive oxygen species (ROS) (WELLS and WINN 1996)

Enzyme	Substrate	Citation
Prostaglandin H synthase (PHS)	Phenytoin	KUBOW and WELLS 1986, 1989; WELLS et al. 1989a; LIU and WELLS 1995a
	Mephenytoin	LIU and WELLS 1995a
	Nirvanol	LIU and WELLS 1995a
	Trimethadione	WELLS et al. 1989b; LIU and WELLS 1995a
	Dimethadione	WELLS et al. 1989b; LIU and WELLS 1995a
	Thalidomide	ARLEN and WELLS 1990; LIU and WELLS 1995a
	Benzo[a]pyrene	MARNETT et al. 1977, 1978; WINN and WELLS 1994
	13-*cis*-Retinoic acid (isotretinoin)	SAMOKYSZYN et al. 1984 KUBOW 1992
	Diethylstilbestrol	DEGEN et al. 1982
	Cyclophosphamide	KANEKAL and KEHRER 1993
	Acetaminophen	POTTER and HINSON 1987; KELLER and HINSON 1991
	2-Naphthylamine	BOYD and ELING 1987
	2-Acetylaminofluorene	BOYD and ELING 1984
Lipoxygenases (LPO)	Phenytoin	KUBOW and WELLS 1988; YU and WELLS 1995
	Cyclophosphamide	KANEKAL and KEHRER 1993
Thyroid peroxidase	Phenytoin	KUBOW and WELLS 1989
Myeloperoxidase	Phenytoin	UETRECHT and ZAHID 1988
Horseradish peroxidase	Phenytoin	KUBOW and WELLS 1989; LIU and WELLS 1995b
	2-Naphthylamine	BOYD and ELING 1987
	2-Acetylaminofluorene	BOYD and ELING 1984

number of structurally related drugs, as well as some teratogenic environmental chemicals such as benzo[a]pyrene (B[a]P; Tables 1, 3). Peroxidase-dependent xenobiotic bioactivation may occur via three mechanisms: peroxidase-mediated bioactivation, peroxyl radical-mediated bioactivation or via a co-substrate-derived oxidant.

2. Mechanisms of Bioactivation

a) Peroxidase-Mediated Bioactivation

During the reduction of hydroperoxides (ROOH; Fig. 6A), peroxidases undergo two successive one-electron oxidations (loss of one electron), and the enzyme is returned to the active ground state by consecutive one-electron reductions, whereby the enzyme abstracts an electron from two molecules of a reducing co-substrate (AH; Fig. 6A), resulting in the formation of two co-substrate free radical products. In an identical fashion, xenobiotics can serve

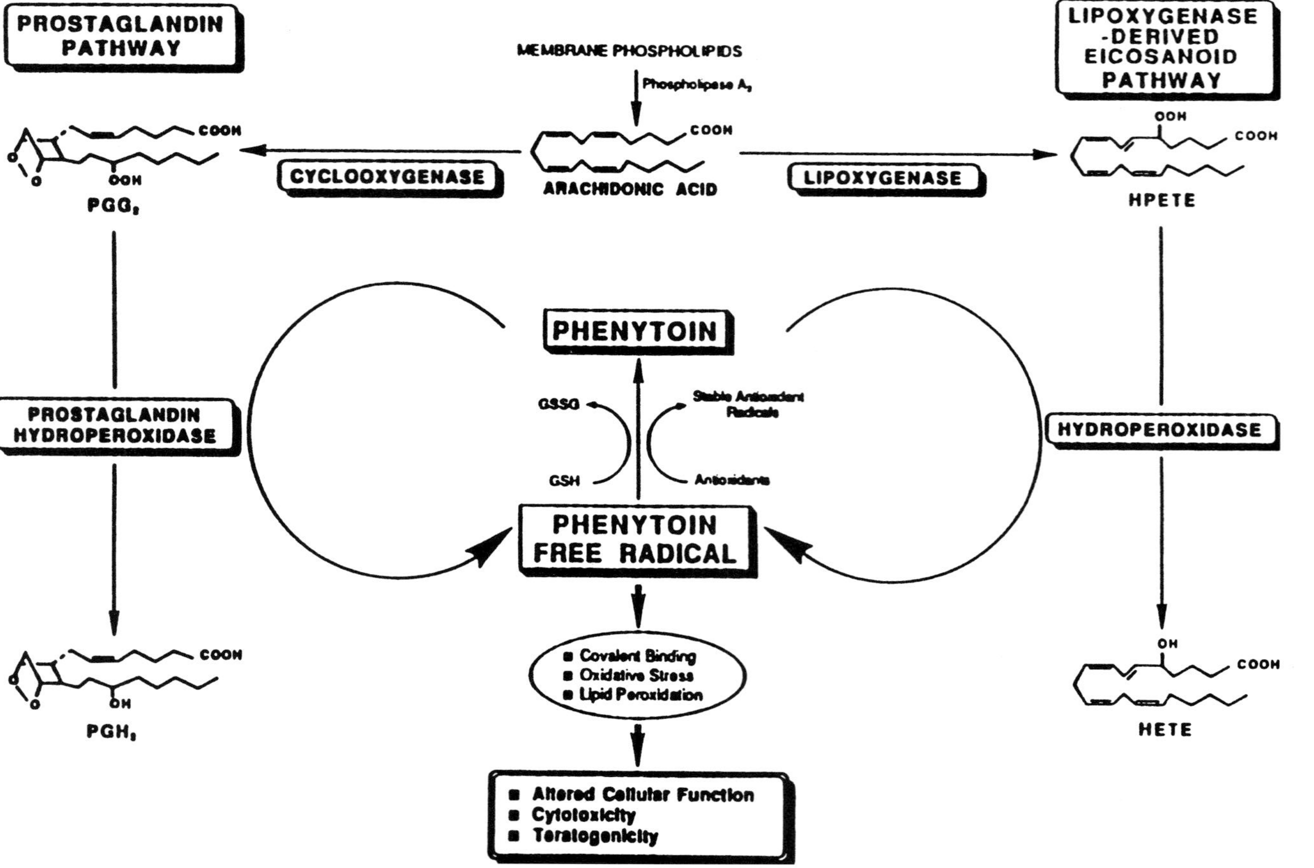

Fig. 5. Postulated bioactivation of phenytoin to a teratogenic free radical intermediate by embryonic enzymes with peroxidase activity. Cyclooxygenase and hydroperoxidase are the components of prostaglandin H synthase (PHS). *PGG_2*, prostaglandin G_2; *PGH_2*, prostaglandin H_2; *GSSG*, glutathione (*GSH*) disulphide; *HPETE*, hydroperoxyeicosatetraenoic acid; *HETE*, hydroxyeicosatetraenoic acid. (From YU and WELLS 1995)

A. Peroxidase-mediated bioactivation

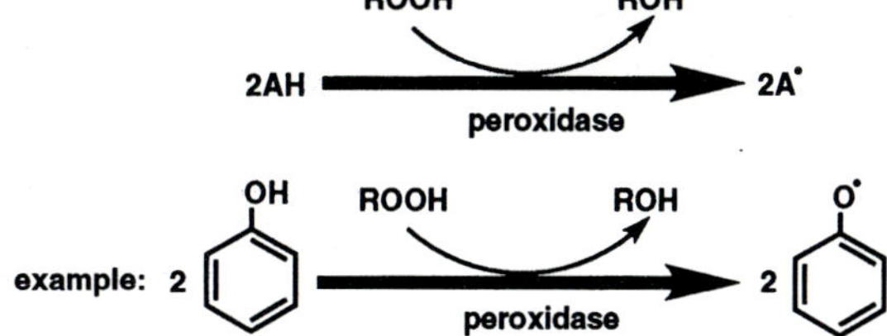

B. Peroxyl radical-mediated bioactivation

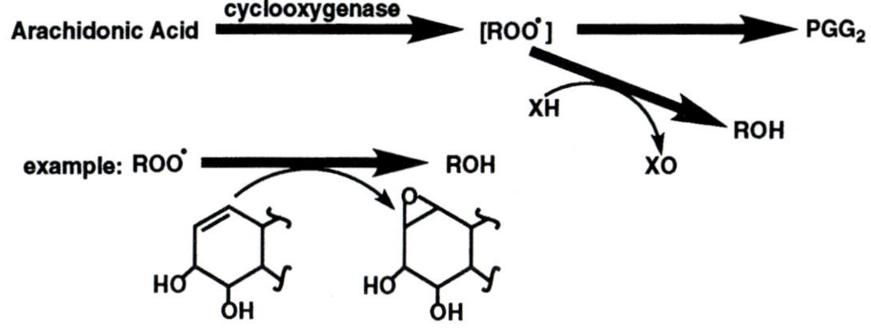

C. Cosubstrate-derived oxidant

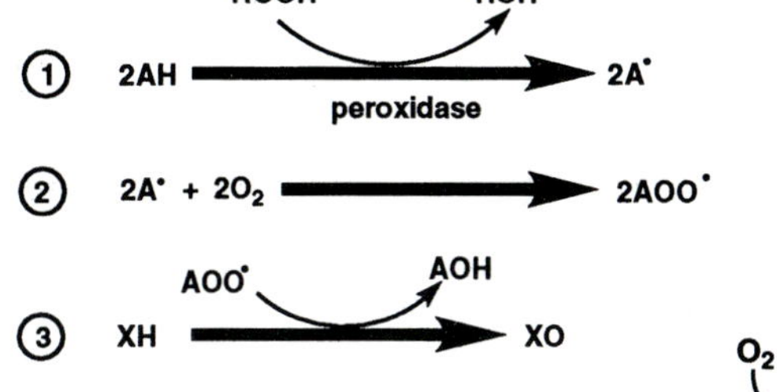

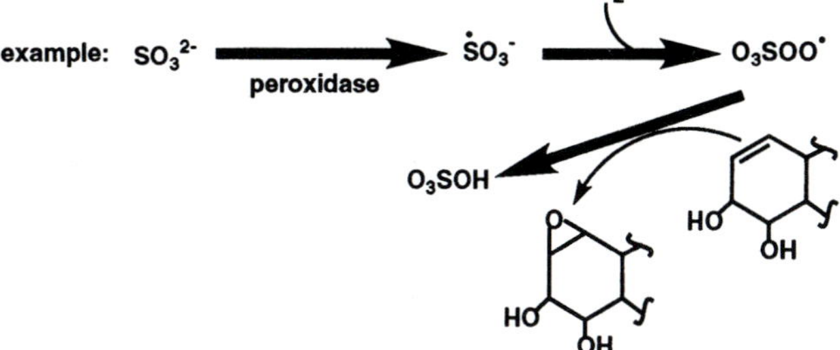

Fig. 6A–C. Mechanisms of peroxidase-dependent bioactivation of xenobiotics. **A** Peroxidase-mediated bioactivation. **B** Peroxyl radical-mediated bioactivation. *PGG*$_2$, prostaglandin G_2. **C** Co-substrate-derived oxidant. (Modified from SMITH et al. 1991)

as co-substrates for peroxidases such as PHS during the reduction of the hydroperoxide prostaglandin (PG) G_2 to the PGH_2 alcohol, with the xenobiotic being bioactivated to potentially toxic, reactive free radical intermediates (SMITH et al. 1991). This mechanism has been postulated to be involved in the bioactivation of phenytoin and related proteratogens by embryonic peroxidases (Fig. 5).

b) Peroxyl Radical-Mediated Bioactivation

During the oxygenation of endogenous fatty acids such as arachidonic acid by the cyclooxygenase component of PHS (Fig. 6B), peroxyl radicals ($ROO^{\bullet}$) are formed that can epoxidise xenobiotics, thereby forming an electrophilic reactive intermediate. This mechanism can contribute to the bioactivation of B[a]P (REED 1988) and hence may play a role in the embryonic bioactivation of B[a]P and similar proteratogens to a teratogenic electrophilic reactive intermediate.

c) Co-substrate-Derived Oxidant

As was described above for xenobiotics (Fig. 6A), during the peroxidase-catalysed reduction of endogenous hydroperoxides (ROOH) to alcohols (ROH), some xenobiotics and endogenous substrates such as SO_3^{2-} can be oxidised to a free radical intermediate such as ${}^{\bullet}SO_3^{2-}$ (Fig. 6C) (SMITH et al. 1991). In this case, however, the free radical reacts with molecular oxygen to produce a peroxyl radical such as $O_3SOO^{\bullet}$ that can epoxidise xenobiotics such as B[a]P, which could be relevant to chemical teratogenesis.

D. Reactive Intermediates

I. Electrophiles

During the bioactivation of several xenobiotics by P450 and/or PHS, an electrophilic reactive intermediate is formed (see Table 4 for definitions). These intermediates are either positively charged carbon atoms (carbocations), as in the bioactivation of 4-(methylnitrosamino)-1-(3-pyridyl)-1-butanone (NNK), or epoxides, as in the bioactivation of B[a]P. The stability of a carbocation determines the type of reactions involved. If the carbocation that may be formed is a tertiary (3°) cation, a substitution nucleophilic unimolecular (SN_1) reaction will occur; the order of stability for carbocations is given in Fig. 7 (FESSENDEN and FESSENDEN 1986a). The order of stability of carbocations for bimolecular (SN_2) reactions is opposite to that for SN_1 reactions. Steric hinderance of a molecule or a nucleophile, the solvent in which the reaction is taking place and the type of leaving group on the molecule are all determinants of the type of reaction that will occur (MARCH 1992).

One of the major reactions of a carbon–carbon double bond is with oxidizing agents to form an epoxide:

$$\overset{O}{\underset{H_2C-CH_2}{/\ \backslash}}$$

These intermediates normally are highly reactive because of the polarity of the C–O bonds and the ring strain present (FESSENDEN and FESSENDEN 1986b). An epoxide formed at the double bond of a benzene ring is called an arene oxide:

[arene oxide structure: benzene ring fused to epoxide O]

Generally, epoxides undergo SN_2 reactions, with nucleophiles attacking the less substituted carbon of the epoxide or arene oxide in basic solutions. In

Table 4. Chemical definitions pertaining to electrophilic reactive intermediates and their reactions with molecular targets

Type of the reaction	Definition
Nucleophile	A reagent that donates an electron pair is called a nucleophile, and the reaction is called nucleophilic.
Electrophile	A reagent that accepts an electron pair is called an electrophile, and the reaction is electrophilic.
Substitution nucleophilic unimolecular reaction (SN_1)	This reaction consists of two steps: 1. Formation of an electrophile $R\text{-}X \rightleftharpoons [R^+] + X$ 2. Addition of the electrophile to a nucleophile $[R^+] + Y \rightleftharpoons R\text{-}Y$
Substitution nucleophilic bimolecular reaction (SN_2)	This is a reaction involving a back side attack. The nucleophile approaches the substrate from a position 180° away from the leaving group. The reaction is a one-step process with no intermediate, wherein the C–Y bond is formed as the C–X bond is broken. $\ddot{Y} + -C-X \longrightarrow Y\text{--}C\text{--}X \longrightarrow Y-C + \ddot{X}$
Borderline reactions	There are reactions that are neither SN_1 nor SN_2, but are in between. This borderline behaviour is found where the rates of formation and destruction of the ion pair are of the same order of magnitude. $RX \overset{k_1}{\rightleftharpoons} R^+X^- \overset{k_2}{\rightleftharpoons} \text{Product}$

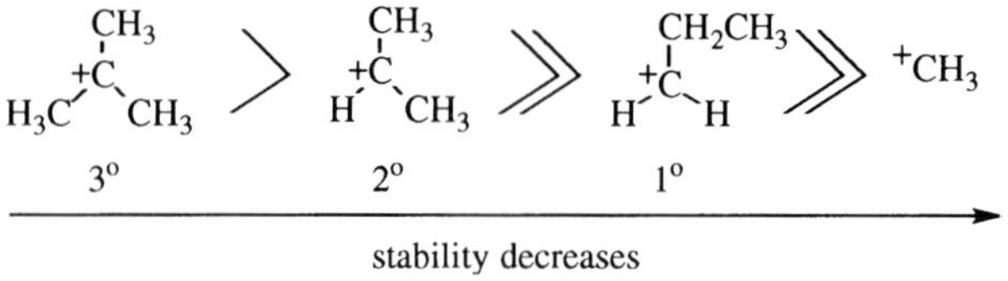

$3° \; (CH_3)_3C^+ > 2° \; (CH_3)_2CH^+ \gg 1° \; CH_3CH_2CH_2^+ \gg {}^+CH_3$

stability decreases

Fig. 7. Order of stability of carbocations

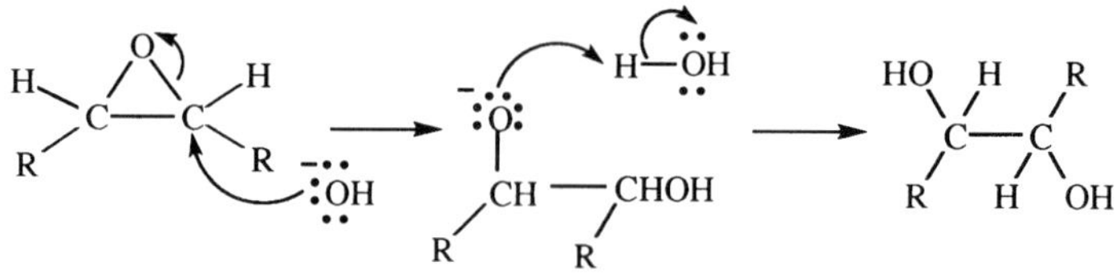

Fig. 8. Reaction of an epoxide with a hydroxide ion to give *trans* dihydrodiols

acidic solutions, the nucleophile attacks the most substituted carbon of the epoxide or arene oxide (FESSENDEN and FESSENDEN 1986b). The reaction of an epoxide with hydroxide ion gives *trans* diols, as shown in Fig. 8.

Epoxides also can be formed via the reaction of peroxyl radicals with the double bond of alkene or arene molecules, wherein the peroxyl radical adds an oxygen atom across the double bond. This is the mechanism by which B[a]P-7,8-diol is bioactivated by PHS (Fig. 9A; MARNETT and ELING 1983). Cytochrome P450 isoenzymes also bioactivate the 7,8-diol of B[a]P to a diol-epoxide. However, when this chemical is bioactivated by PHS, a peroxyl radical of unsaturated fatty acid adds an oxygen atom to form an anti-diol-epoxide, while bioactivation by P450 forms a syn-diol-epoxide (Fig. 9B; MARNETT 1990).

II. Free Radicals

Free radicals are molecules, in this context potential teratogens, with one or more unpaired electrons in their outer orbitals. A free radical is formed by either homolytic cleavage of a covalent bond or by abstraction (loss) of an electron from a molecule. Alternatively, gain of an electron by a non-radical molecule will form a radical:

1. $A - A \longrightarrow 2A^{\cdot}$ (radical)
2. $A \xrightarrow{-e^-} A^{\cdot +}$ (radical cation)
3. $A \xrightarrow{e^-} A^{\cdot -}$ (radical anion)

In general, electrons within atoms and molecules occupy a region of space known as orbitals. Each orbital can hold a maximum of two electrons. A single electron in an orbital is said to be "unpaired". For instance, molecular

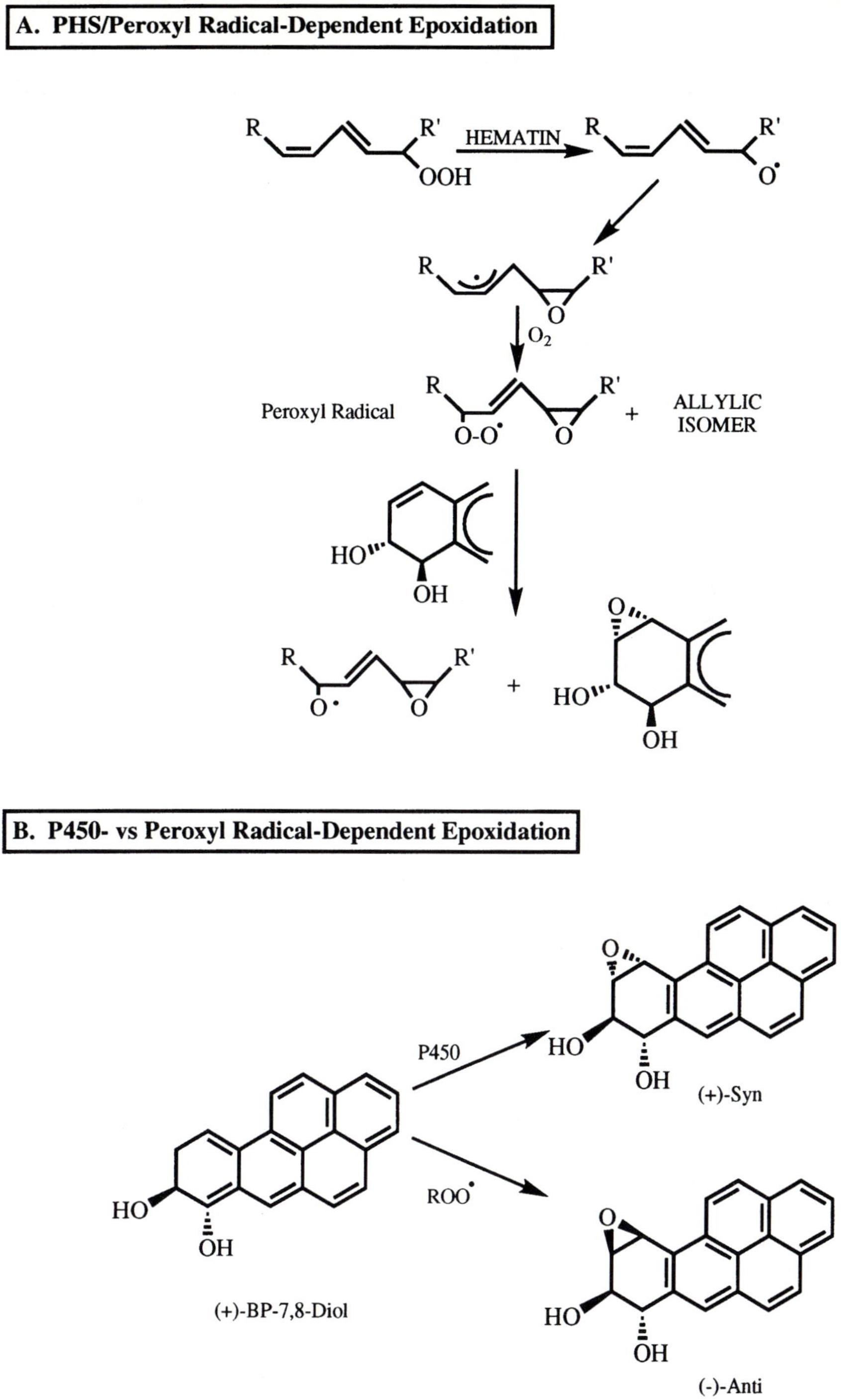

A. PHS/Peroxyl Radical-Dependent Epoxidation

R R' HEMATIN R R'
OOH O•
R R'
O
O_2
Peroxyl Radical R R' + ALLYLIC ISOMER
O-O• O
HO OH
R R' + HO OH
O• O O

B. P450- vs Peroxyl Radical-Dependent Epoxidation

P450 → HO OH (+)-Syn
HO OH (+)-BP-7,8-Diol
ROO• → HO OH (-)-Anti

Fig. 9A,B. Bioactivation of benzo[a]pyrene (B[a]P)-7,8-diol. **A** Bioactivation of B[a]P-7,8-diol by prostaglandin H synthase (PHS). (From MARNETT and ELING 1983). **B** Comparison of the stereochemistry of epoxidation of B[a]P-7,8-diol by cytochromes P450 and peroxyl radicals. (From MARNETT 1990)

oxygen at its ground state is a radical; if the two unpaired electrons in ground-state oxygen move to the same orbital, singlet oxygen is formed, which is not a radical, but nevertheless highly reactive. Addition of an electron to ground-state oxygen results in the formation of superoxide anion radical ($O_2^{\cdot-}$) (HALLIWELL and GUTTERIDGE 1989b). A molecule with an unpaired electron (i.e. paramagnetic) is aligned in a magnetic field and can thus be detected by electron paramagnetic resonance (EPR) spectroscopy, also known as electron spin resonance (ESR) spectroscopy, which is described below. There are different types of radicals depending on the atom (centre) upon which the unpaired electron is residing. For example, there are nitrogen-centred, oxygen-centred, carbon-centred and even metal-centred radicals. For simplicity, only radicals centred upon nitrogen, oxygen and carbon will be discussed here.

It is important to know what factors affect the stability of a radical of interest. For example, unstable radicals are extremely reactive and can be scavenged immediately by other intracellular molecules, such as oxygen or proteins. Therefore, an unstable radical may be too reactive to reach potential molecular targets in other regions of the cell or may not be able to leave the cell to reach other cells in distant tissues. In general, the stability of a radical depends on the position of the unpaired electron on the molecule. A carbon-centred radical is more stable if the unpaired electron is residing on the tertiary (3°) carbon of a molecule (Fig. 10, top).

Stability of a carbon-centred radical increases if there are groups such as phenyl rings adjacent to the carbon-bearing unpaired electron (Fig. 10, bottom). This is due to the ability of the unpaired electron to delocalise over the π orbitals of phenyl rings.

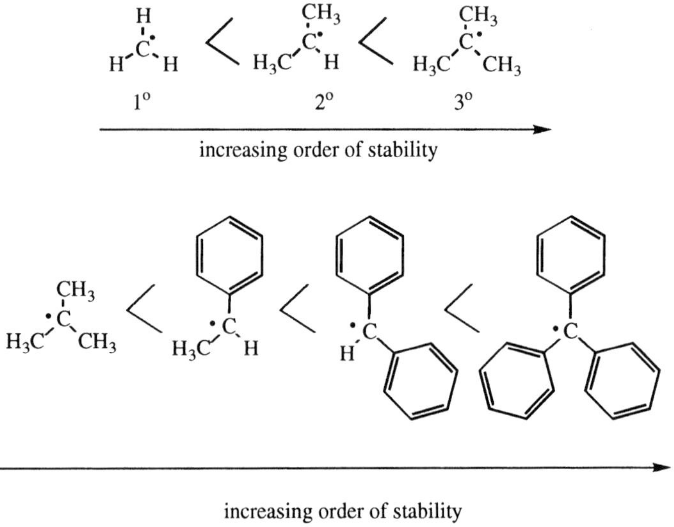

H
H C H
H₃C C H
CH₃
H₃C C CH₃
CH₃
1° < 2° < 3°

increasing order of stability

H₃C C CH₃
CH₃
H₃C C H
H C

increasing order of stability

Fig. 10. Order of stability of carbon-centered radicals

Table 5. Reactions of free radicals

Reaction	Equation
Hydrogen abstraction	$\dot{A} + R\text{-}H \longrightarrow \dot{R} + A\text{-}H$ Radical
Electron transfer	$\dot{A}^{-} + Y \longrightarrow Y^{\bullet -} + A$ Radical anion
Addition to multiple bond	$\dot{A} + R\text{-}CH{=}CH\text{-}R \longrightarrow$ R–CH(A)–ĊH–R Alkyl radical
Addition to oxygen	$\dot{A} + O_2 \longrightarrow A\text{-}O\text{-}\dot{O}$ Peroxyl radical

Typical reactions involving free radicals are given in Table 5 (MARCH 1992). Generally, a free radical reaction has three steps: (1) initiation, (2) propagation and (3) termination. In the initiation step, a radical is formed which, in a propagation reaction, will react with a non-radical molecule to produce a second radical, while the first radical becomes neutral. In the termination step, radicals couple to form neutral molecules:

1. Initiation $R-R \longrightarrow 2R^{\bullet}$
2. Propagation $A-A + R^{\bullet} \longrightarrow R - A + A^{\bullet}$

$$3.\ \text{Termination}\begin{cases} 2R^{\bullet} \longrightarrow R-R \\ A^{\bullet} + R^{\bullet} \longrightarrow R - A \\ A^{\bullet} + A^{\bullet} \longrightarrow A - A \end{cases}$$

Free radicals in biology are potential mediators of a broad spectrum of chemical toxicities. Free radical intermediates are formed via bioactivation of many xenobiotics by P450 isoenzymes and by peroxidases. B[a]P is a good example of a carcinogen/teratogen that is bioactivated by P450 isoenzymes to a carbon-centred radical cation that can covalently bind to DNA. Carbon tetrachloride, an agent that causes liver necrosis, undergoes oxidative bioactivation by CYP2E1 (GUENGRICH and SHIMADA 1991) to a trichloromethyl radical ($\bullet CCl_3$) (ALBANO et al. 1982). Paraquat is a herbicide and a teratogen (JUCHAU et al. 1986) that is reductively bioactivated by P450 to a paraquat radical. Ethanol, which is both hepatotoxic and teratogenic, is oxidatively bioactivated by CYP2E1 to an α-hydroxyethyl radical (ALBANO et al. 1994). Peroxidases, such as PHS, and enzymes with associated peroxidase activity, such as LPO, also can bioactivate teratogenic xenobiotics to a reactive free radical intermediate, including B[a]P (Marnett et al. 1975), phenytoin (Kubow and Wells 1989), 2-naphthylamine (Boyd and Eling 1987) and 2-aminofluorene (Boyd and Eling 1984; Table 3).

Once formed in a cell, a xenobiotic radical can reduce molecular oxygen, forming a superoxide anion radical. In addition, the radical can add across the double bond of molecular oxygen, forming a peroxyl radical. In the latter case, the radical is said to be "scavenged" by molecular oxygen. These reactions subsequently form other toxic ROS (Fig. 11) including alkoxyl radicals, hydrogen peroxide and hydroxyl radicals (see Sect. H). Xenobiotics such as B[a]P-3,6-quinone can also undergo reductive bioactivation, catalysed by P450 reductase, followed by reoxidation by molecular oxygen, which also produces ROS. Such xenobiotics are known as "redox cyclers" and, in accordance with their redox potential, can accept electrons from a number of biological reducing agents (Fig. 12; Kappus 1986; Juchau et al. 1986). Radical intermediates as well as ROS can cause lipid peroxidation, DNA oxidation, and protein degradation, fragmentation, decarboxylation, carbonyl formation, cross-linking and disulphide bond formation in embryonic tissues. These lesions alter or destroy macromolecular function and may contribute to the teratogenicity of a xenobiotic (see Sect. H).

E. Detoxification

I. Glutathione

Glutathione (GSH) is a central component in the detoxification of both electrophilic and free radical reactive intermediates of xenobiotics (Larsson et al. 1983). With electrophiles, the thiol electrons of GSH are attracted to the positively charged atoms of the reactive intermediate, forming a covalent (irreversible) bond (Fig. 13). Depending upon the degree to which the electrophile is "soft" or "hard" (see Sect. H), this reaction generally requires or is accelerated by the enzyme GSH S-transferase (GST). For most reactive intermediates, this glutathione–xenobiotic conjugate is non-toxic and is excreted unchanged or after further metabolism, ultimately to an *N*-acetylcysteine–xenobiotic conjugate (mercapturic acid). If the conjugation of the reactive intermediate with GSH exceeds its rate of synthesis, GSH is depleted and the electrophile is free to react with cellular macromolecules such as protein and DNA (see Sect. H). Should GSH be depleted due to concomitant chemical exposure or other genetic or environmental conditions, the teratogenicity of a xenobiotic bioactivated to an electrophilic reactive intermediate is enhanced. This interaction is illustrated by the enhancement of phenytoin embryopathy in mice in vivo or in embryo culture by pretreatment with the GSH depletors diethylmaleate (DEM) (Harbison 1978; Wong et al. 1989) and acetaminophen (Lum and Wells 1986) or the inhibitor of GSH synthesis, buthionine sulfoximine (BSO) (Wong et al. 1989; Miranda et al. 1994). On the other hand, repletion with GSH or a precursor such as *N*-acetylcysteine may not

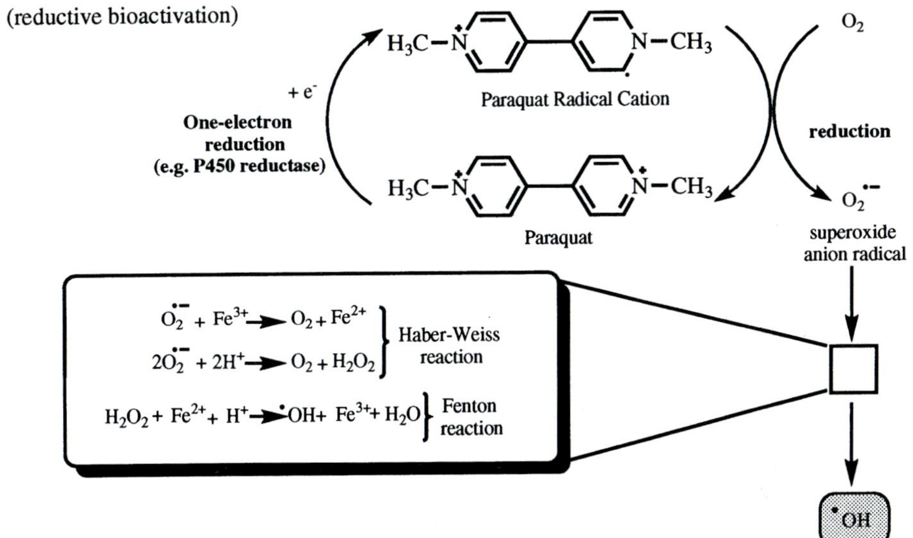

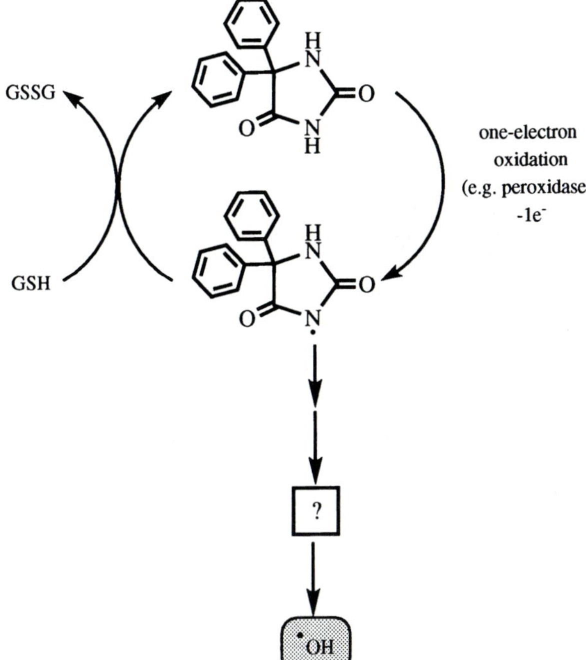

Fig. 11A,B. Potential mechanisms for the formation of hydroxyl radicals (•OH) by free radical intermediates of paraquat (**A**) and phenytoin (**B**). *GSH*, glutathione; *GSSG*, glutathione disulphide

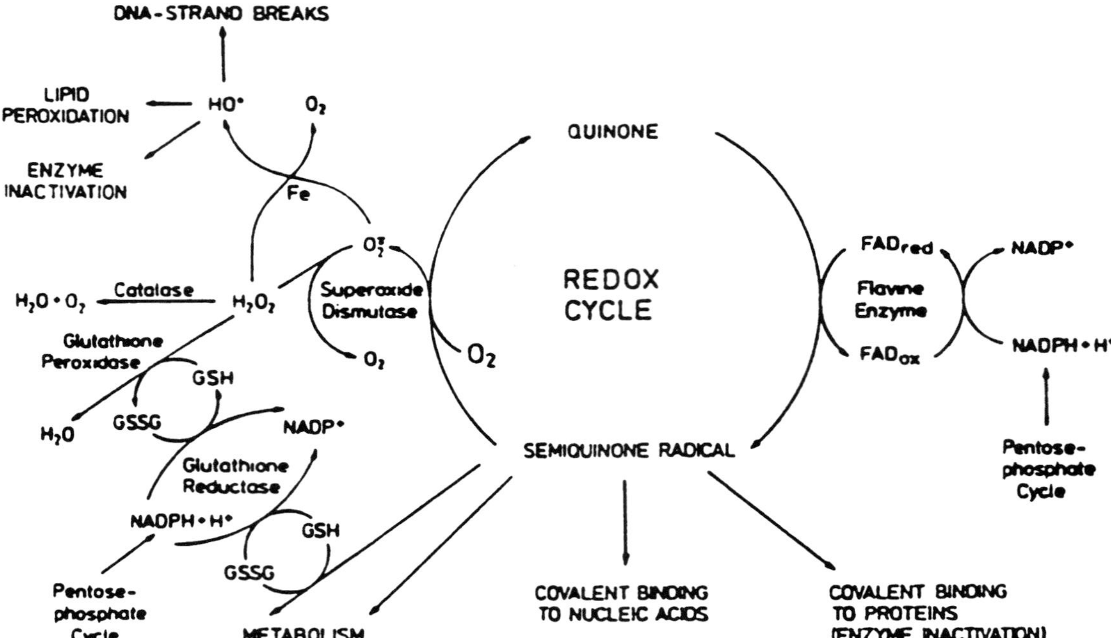

Fig. 12. Redox cycling of quinonoide compounds, oxy radical formation and cytoprotective inactivation. *GSH*, glutathione; *GSSG*, glutathione disulphide; *NADP*, nicotinamide adenine dinucleotide phosphate; *NADPH*, reduced NADP; *FAD*, flavine adenine dinucleotide (From KAPPUS 1986)

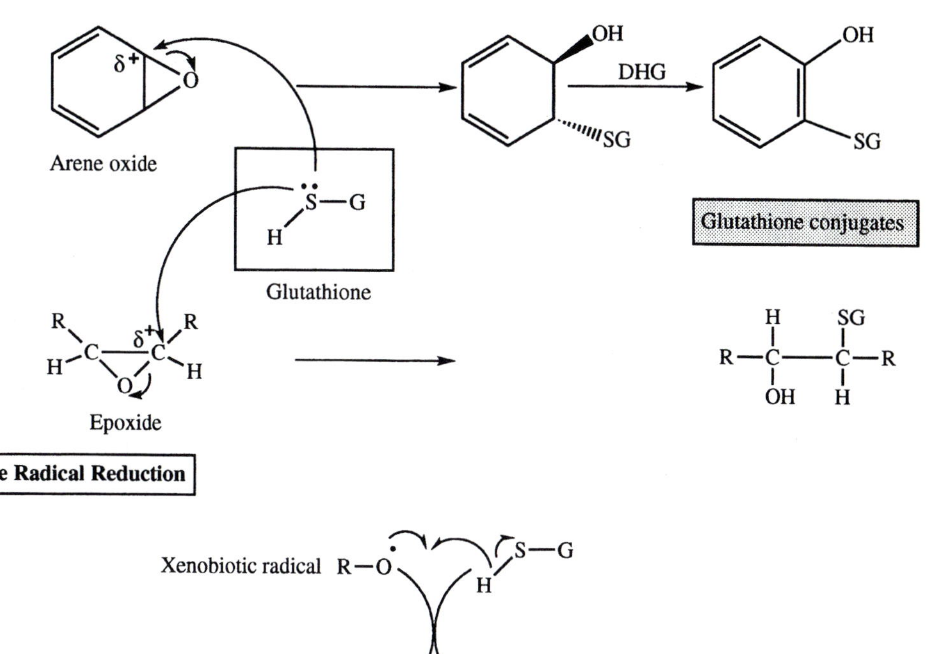

Fig. 13. Detoxification of electrophilic (*top*) and free radical (*bottom*) reactive intermediates by glutathione (GSH). *DHG*, dehydrogenase; *GSSG*, GSH disulphide

provide the expected protection against electrophilic reactive intermediates (Wong and Wells 1988), which may be due to inadequate uptake and/or synthesis of GSH by embryonic target tissues. In some cases, this may be overcome by the use of GSH esters, which are readily absorbed into all cell types, where esterases release free GSH in supraphysiological concentrations (Anderson et al. 1985; Anderson and Meister 1989). For example, pretreatment with the methyl or ethyl esters of GSH, but not *N*-acetylcysteine, can reduce the embryopathy of the teratogenic anticancer drug cyclophosphamide, which is bioactivated on an electrophilic reactive intermediate (Hales 1981a).

In the case of free radical reactive intermediates, GSH can reduce the free radical intermediate back to the nontoxic parent xenobiotic, with a radical (GS•) being created in the process (Fig. 13). Two thiyl radicals combine non-enzymatically to form GSH disulphide or oxidised GSH (GSSG). In this redox reaction, GSH is a co-factor rather than a substrate and is not removed from the cell like the GSH conjugate of electrophilic reactive intermediates. GSH in this case is maintained by the enzyme GSH reductase, which reduces GSSG back to GSH (see Sect. G). In oxidative stress initiated by free radical reactive intermediates, GSH also may serve as a co-factor for the cytoprotective enzymes GSH reductase and GSH peroxidase (see Sect. G), as well as maintaining protein thiols in a reduced state. Depletion and repletion of GSH using various modulators accordingly may respectively enhance and inhibit both free radical-mediated and electrophile-mediated xenobiotic embryopathies, and such effects by themselves, while implicating a reactive intermediate, do not discriminate between these two mechanisms.

However, the oxidation of GSH and other thiols (e.g. cysteine) to their corresponding disulphides (GSSG and cystine) and mixed disulphides occurs only via an oxidative, as distinct from an electrophilic, reactive intermediate, and such disulphides are readily measured by high-performance liquid chromatography (HPLC). Xenobiotics such as paraquat and *tert*-butylhydroperoxide, which are known to initiate free radical-mediated oxidative stress, produce a characteristic disulphide profile, and such characteristic disulphide profiles in rodents and rabbits have been observed in vivo (ARLEN and WELLS 1990, 1996b) and/or in embryo culture (WINN and WELLS 1994, 1996b) with phenytoin and the sedative-hypnotic drug thalidomide, suggesting the teratological involvement of free radical-mediated oxidative stress.

II. Glutathione S-transferase

GST catalyses the conjugation of electrophiles with GSH (Fig. 13; LARSSON et al. 1983). While related xenobiotic toxicities such as cancer have been shown to be enhanced in people with hereditary deficiencies in particular GST isoenzymes (SATO 1988), there have been no direct studies of the teratological relevance of this enzyme family. One rodent study involving selenium deprivation suggested that the induction of a GST isoenzyme with Se-independent GSH peroxidase activity could protect the embryo against phenytoin embryopathy (see Sect. G).

III. Epoxide Hydrolase

Epoxide hydrolase (previously named epoxide hydrase or hydratase) catalyses the insertion of a hydroxide ion into the electrophilic epoxide or arene oxide intermediate of xenobiotics, forming a non-toxic *trans*-dihydrodiol metabolite (Fig. 3). In children exposed in utero to phenytoin, those who postnatally developed the fetal hydantoin syndrome (FHS), a constellation of teratological anomalies, had a lower activity of epoxide hydrolase, suggesting that a genetic inability to detoxify the phenytoin arene oxide intermediate resulted in enhanced embryonic damage (BUEHLER et al. 1990). Furthermore, epoxide hydrolase activity in heteropaternal dizygotic twins exposed in utero to phenytoin, who were discordant for the clinical features of FHS, was found to be much lower in the affected twin compared to the unaffected twin (BUEHLER 1984). This is consistent with results from rodent models in which pretreatment of dams with the epoxide hydrolase inhibitors trichloropropene oxide (TCPO) or cyclohexene oxide enhanced phenytoin covalent binding and teratogenicity (MARTZ et al. 1977; HARBISON 1978). Nevertheless, for these and other studies, it is worth remembering that such chemical probes may have multiple effects, and results from the use of only one probe should be interpreted with caution. For example, TCPO also can deplete GSH and may have other as yet unappreciated, but relevant effects.

F. Oxidative Stress

I. Embryological Considerations

By either direct or indirect reactions with molecular oxygen or via the oxidation of membrane lipids, free radical reactive intermediates can initiate the formation of a number of highly toxic ROS and lipid hydroperoxides (LOOH) (Figs. 12, 14; see Sect. H). These ROS and LOOH, which can oxidise many embryonic molecular targets (see Sect. H), may contribute to a lesser or greater degree, depending upon the xenobiotic, to the reactive intermediate-mediated component of teratogenesis (Fig. 14). In this chapter, detoxifying reactions for these toxic reactive biological species are referred to as cytoprotective pathways (see below), as distinct from the direct detoxification of xenobiotic reactive intermediates (Fig. 2), referred to here as detoxification.

II. Measurements of Oxidative Stress

Oxidative stress can be measured directly or indirectly using a variety of techniques that characterise free radical production, molecular target oxidation and compensatory gene/protein responses to oxidative stress (Table 6). Several of these approaches are discussed below.

1. Salicylate Hydroxylation

The hydroxylation of a number of aromatic probes, including salicylate, has been used as a measure of both in vitro and in vivo formation of hydroxyl radical (•OH) (Table 6; HALLIWELL and GROOTVELD 1988). In the case of salicylate, which has been used both in vivo and in vitro (GROOTVELD and HALLIWELL 1986; INGELMAN-SUNDBERG et al. 1991; KIM and WELLS 1993, 1996a; COUDRAY et al. 1995), •OH hydroxylates salicylate (2-hydroxybenzoic acid) at both the C3 and C5 positions, forming both the 2,3- and 2,5-isomers of dihydroxybenzoic acid (DHBA), whereas only the C5 position is hydroxylated

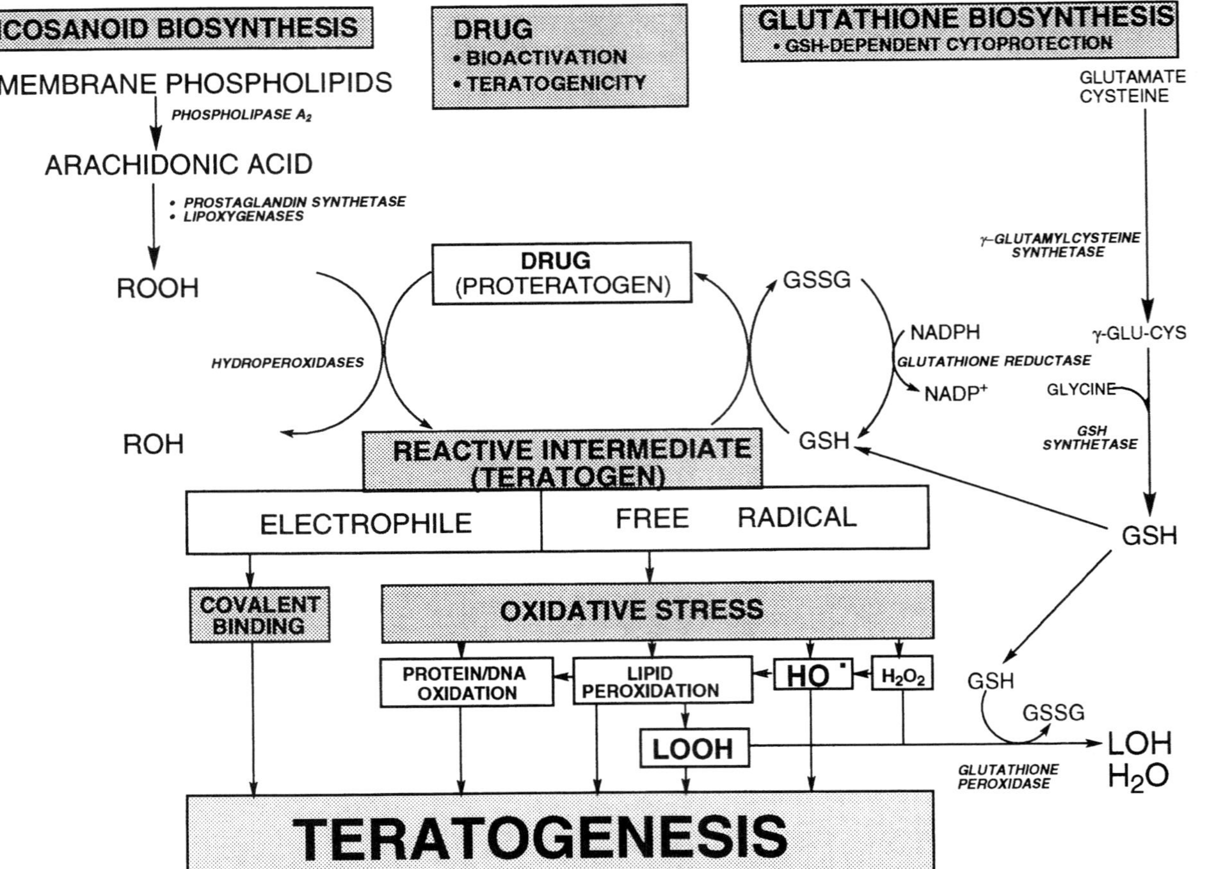

Fig. 14. Postulated roles of xenobiotic covalent binding and free radical-initiated oxidative stress in chemical teratogenesis. *GSH*, glu-tathione; *GSSG*, *GSH* disulphide; *NADP*, nicotinamide adenine dinucleotide phosphate; *NADPH*, reduced NADP (From MIRANDA et al. 1994)

Table 6. Measurement of oxidative stress and its teratologic relevance

	End product(s) detected	Oxidative agent(s)	Reference(s)
Radical detection			
1. Aromatic hydroxylation			
Benzoic acid	Ortho-, meta-, and para-hydroxybenzoates	Gamma radiation	HALLIWELL and GROOTVELD 1988
Salicylic acid	2,3- and 2,5-DHBA	Phenytoin	KIM and WELLS 1993, 1996a
		Paraquat	KIM and WELLS 1995a
Terephthalic acid	Monohydroxy terephthalate	Gamma and UV radiation, Cu/H_2O_2	BARRETO et al. 1995
2. Electron paramagnetic (spin) resonance			
PBN	PBN radical adduct	Phenytoin	KUBOW and WELLS 1989; PARMAN et al. 1996
	PBN radical adduct	Mephenytoin	PARMAN et al. 1996
		Nirvanol	PARMAN et al. 1996
		Trimethadione	PARMAN et al. 1996
		Thalidomide	PARMAN et al. 1996
	PBN radical adduct	Ethanol	REINKE and MOORE 1994
	Antioxidative properties[a]	Cocaine	ZIMMERMAN et al. 1994
	Antioxidative properties[a]	Phenytoin	LIU and WELLS 1994b; WELLS et al. 1989a
POBN	POBN radical adduct	Ethanol	REINKE and MOORE 1994
DMPO	DMPO radical adduct	Ethanol	REINKE and MOORE 1994
3. Fluorescence			
DCFH-DA	DCFH	Ascorbate/$FeSO_4$	LEBEL and BONDY 1991
		Doxorubicin	UBEZIO and CIVOLI 1994
Oxidative damage			
GSH oxidation	GSSG/GS-Protein	Phenytoin	WELLS and WILLIAMS 1994; MIRANDA et al. 1994
		Thalidomide	ARLEN and WELLS 1990
		Paraquat	ARLEN and WELLS 1990
		Tertiary-butyl-hydroperoxide	GARDINER and REED 1994; ARLEN and WELLS 1990

Table 6. (*Contd.*)

	End product(s) detected	Oxidative agent(s)	Reference(s)
Lipid peroxidation	TBARS	Phenytoin	LIU and WELLS 1994b, 1995b
		Ethanol	NORDMANN et al. 1990
DNA oxidation	8-OH-2′-dG	Benzo[a]pyrene	WINN and WELLS 1994; KIM and WELLS 1995b; KIM and WELLS 1996b
		Phenytoin	WINN and WELLS 1995a; LIU and WELLS 1995a
		HPPH	LIU and WELLS 1995a
		Dimethadione	LIU and WELLS 1995a
		d-and *l*-Nirvanol	LIU and WELLS 1995a
		Thalidomide	LIU and WELLS 1995a
		Diethylstilbestrol	ROY et al. 1991
Protein oxidation	Carbonyl groups	Benzo[a]pyrene	WINN and WELLS 1994, 1996b DEBONI et al. 1996; KIM and WELLS 1996b
		Phenytoin	WELLS et al. 1995; WELLS and WINN 1996; LIU and WELLS 1994b, 1995b
Antioxidative gene/protein expression			
Superoxide dismutase	Increased enzyme	Xanthine/xanthine oxidase	LAVAL 1988
		H_2O_2	LU et al. 1993
		$O_2^{\cdot-}$	GREENBERG and DEMPLE 1989[b]
Catalase	Increased enzyme	H_2O_2	LU et al. 1993
		Ozone	WHITESIDE and HASSAN 1987[b]
G6PD	Increased enzyme	$O_2^{\cdot-}$	GREENBERG and DEMPLE 1989[b] ROWLEY et al. 1991

Table 6. (*Contd.*)

	End product(s) detected	Oxidative agent(s)	Reference(s)
GSH peroxidase	Increased enzyme	H_2O_2	Lu et al. 1993
Heat shock proteins	Increased protein	H_2O_2	Spitz et al. 1987
		Reperfusion	Currie 1987
c-Fos and c-myc	Increased protein	Xanthine/xanthine oxidase	Crawford et al. 1988
		H_2O_2	Puri et al. 1995

Cu, copper; DCFH, 2′,7′-dichlorofluorescein; DCFH-DA, 2′,7′-dichlorofluorescin; DHBA, dihydroxybenzoic acid; $O_2^{\cdot-}$, superoxide anion; H_2O_2, hydrogen peroxide; PBN, α-phenyl-*N*-*t*-butylnitrone; POBN, α-[4-pyridyl 1-oxide]-*N*-*t*-butylnitrone; DMPO, 5,5-dimethylpyroline-*N*-oxide; GSH, glutathione; GSSG, glutathione disulfide; TBARS, thiobarbituric acid reactive substances; 8-OH-2′-dG, 8-hydroxy-2′-deoxyguanosine; HPPH, 5-(*p*-hydroxyphenyl)-5-phenylhydantoin, the major phenytoin metabolite; G6PD, glucose-6-phosphate dehydrogenase.
[a]Study used PBN as a free radical scavenging antioxidant, rather than for direct detection of free radical by electron paramagnetic resonance.
[b]Study in *Escherichia coli.*

via in vivo enzymatic reactions (Fig. 15; Ingelman-Sundberg et al. 1991). Thus measurement of 2,3-DHBA can be used to estimate •OH formation. The salicylate hydroxylation assay has a number of advantages, including the following: (a) its simplicity; (b) its sensitivity in the femtomolar range when using HPLC with electrochemical detection; (c) it is free from the confounding contribution of enzymatic hydroxylation; (d) salicylate and its metabolites are not endogenous; and (e) salicylate, particularly when given as acetylsalicylic acid (ASA, aspirin), in therapeutic doses is safe for human studies.

Two potential problems arise: (1) measurement of 2,5-DHBA rather than the 2,3-isomer (Halliwell et al. 1991) and (2) in in vivo studies, potential confounding effects of interindividual variability in glucuronidation of 2,3-DHBA, catalysed by the uridine diphosphate (UDP)-glucuronosyltransferases (UGT). In vivo studies based upon 2,5- rather than 2,3-DHBA formation (Powell and Hall 1990; Onodera and Ashraf 1991; Udassin et al. 1991) may be significantly confounded by P450-catalysed hydroxylation; formation of 2,3-DHBA, but not 2,5-DHBA, has been shown to accurately reflect in vivo •OH production initiated by paraquat and phenytoin (Kim and Wells 1993, 1995a, 1996a). Interindividual variability in 2,3-DHBA glucuronidation is unlikely to be a substantial confounding factor, since 2,3-DHBA is glucuronidated in vivo only to a minor extent (about 10%) in mice (Kim and Wells 1995a, 1996a), which like humans are substantially more avid than rats in glucuronidating similar drugs such as acetaminophen (Kalabis and Wells 1990; de Morais and Wells 1989; de Morais et al. 1992). The concentration of DHBA isomers was not decreased by in vitro incubation with GSH (Udassin et al. 1991), suggesting that GSH conjugation is not likely to be a confounding factor either. Although sulphation and other conjugating path-

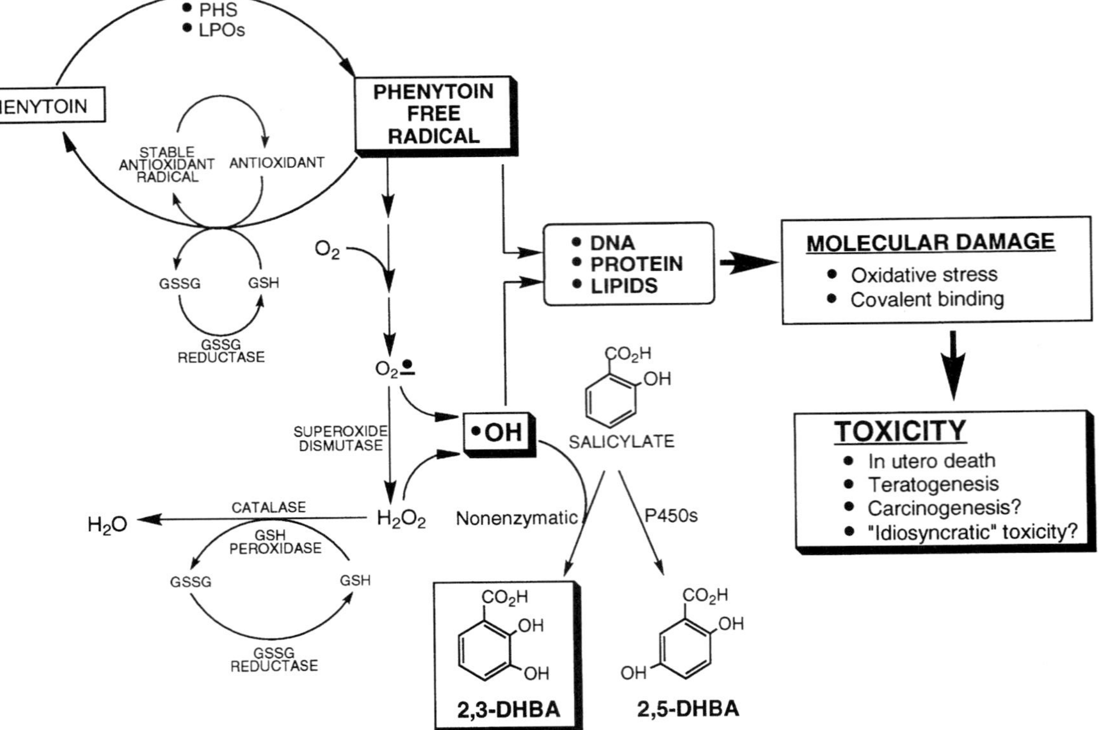

Fig. 15. Postulated basis for the use of salicylate hydroxylation to measure phenytoin-initiated formation of hydroxyl radicals (•OH). Phenytoin may be bioactivated by peroxidases such as prostaglandin H synthase (*PHS*) and via lipoxygenases (*LPO*) to a free radical intermediate that initiates the formation of reactive oxygen species (ROS) such as superoxide anion ($O_2^-\cdot$), hydrogen peroxide (H_2O_2) and •OH. Both the phenytoin free radical and ROS could initiate molecular damage to DNA, protein and lipids, potentially leading to teratogenesis. Phenytoin-initiated •OH can also non-enzymatically hydroxylate salicylate, forming both the 2,3- and 2,5-dihydroxybenzoic acids (*DHBA*), while P450 cytochromes (*P450*) form only the 2,5-isomer. Thus measurement of plasma concentrations of the 2,3-isomer can approximate •OH formation in vivo. *GSH*, glutathione; *GSSG*, oxidised glutathione. (From KIM and WELLS 1996a)

ways for 2,3-DHBA have yet to be determined, potentially confounding contributions were not evident in dose-dependent 2,3-DHBA formation initiated by phenytoin or paraquat (KIM and WELLS 1995a, 1996a).

In addition to characterising xenobiotic-initiated •OH formation, however, investigators should be aware that the use of salicylate and its ASA precursor may nevertheless have additional biological effects in their studies. For example, while ASA administered after phenytoin to mice results in an accurate phenytoin dose-dependent increase in 2,3-DHBA concentration reflecting •OH formation, if ASA is given before phenytoin, 2,3-DHBA formation, while still evident, is substantially reduced (KIM and WELLS 1996a). This effect is to be expected, since ASA is a potent inhibitor of PHS, which is thought to catalyse the bioactivation of phenytoin to a free radical intermediate that initiates •OH formation (Fig. 15). ASA also inhibits phenytoin-initiated DNA oxidation (LIU and WELLS 1995a) and teratogenesis (WELLS et al. 1989a).

In summary, determinations of salicylate hydroxylation and similar approaches, when conducted appropriately, may prove useful not only in investigating the role of oxidative stress in chemically initiated and "spontaneous" in utero death and teratogenesis, but also in characterising interindividual differences in human teratologic susceptibility.

2. Electron Paramagnetic (Spin) Resonance Spectrometry

The advantages of electron paramagnetic resonance (EPR), also known as electron spin resonance (ESR) spectrometry, include direct and definitive evidence for a free radical intermediate and its chemical nature. ESR may be used to identify both in vitro and in vivo formation of free radical intermediates in animals, although the use (and safety) of free radical spin-trapping agents in pregnant women has not been addressed. ESR requires access to an ESR spectrometer and a spectroscopist skilled in the interpretation of spectral data.

A free radical is a paramagnetic molecule with an unpaired electron in its outer orbital having a spin of $-1/2$ or $+1/2$ (HALLIWELL and GUTTERIDGE 1989b). In a magnetic field, electrons line up parallel or antiparallel to the magnetic field. This causes formation of two energy levels. By applying electromagnetic radiation, electrons will be moved to the higher energy level. During this process, energy is absorbed in the microwave region. This energy can be calculated using the following equation:

$$\Delta \mathrm{E} = g\beta H$$

where ΔE is the energy gap between the two energy levels, and g is a constant value called the "splitting factor"; for a free electron and all biologically important radicals, the value of g is 2.00232. H is the applied magnetic field, and β is the Bohr magneton constant. The ESR spectrometer is set to display not the absorbance, but rather the rate of change of absorbance as the first derivative spectrum. Hyperfine splitting is the number of lines in the ESR spectrum of a radical. A radical can be identified by looking at the g value, hyperfine structure and line shape. The number of lines for a radical can be determined by using the following equation:

$$\text{number of lines} = 2nI + 1$$

where n is the number of interacting nuclei and I is the nuclear spin quantum number of an atom. For example, the I value for ^{1}H, ^{13}C, ^{19}F and ^{31}P is 1/2, and the I value for ^{14}N is 1. Atoms such as ^{16}O, ^{12}C and ^{32}S have an I value of zero (HALLIWELL and GUTTERIDGE 1989b). ESR spectroscopy is a sensitive method, detecting radicals at concentrations as low as 10^{-10} M. However, if the radical is not sufficiently stable, it may not remain long enough to be detected. Typical free radicals in biological systems are usually both unstable and formed in low concentrations, which makes them difficult to detect. Accordingly, spin trapping was developed to detect short-lived reactive free radicals. The spin-trapping reaction takes advantage of the stability of the nitroxyl free radical function:

•O
|
R—N—R

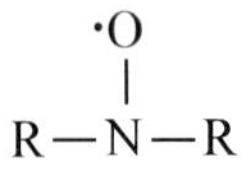

This function is stable because the unpaired electron is resonating between the nitrogen and the oxygen as shown below:

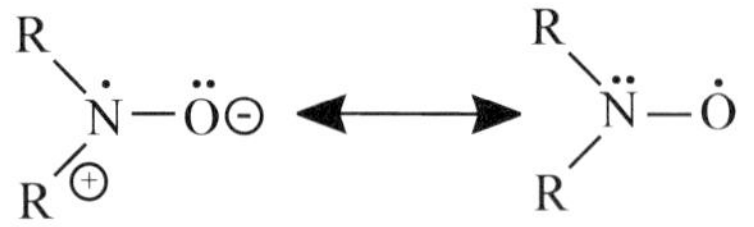

$R_2\dot{N}^{+}{-}\ddot{O}^{\ominus} \longleftrightarrow R_2\ddot{N}{-}\dot{O}$

Nitrones are spin-trapping agents which contain a nitroxyl free radical function as part of the double bond in a molecule. Highly substituted nitrones such as α-phenyl-*N*-*tert*-butylnitrone (PBN) and 5,5-dimethyl pyrroline-*N*-oxide (DMPO) are commonly used as the spin traps of choice in biological systems. The structure of these two trapping agents is given below:

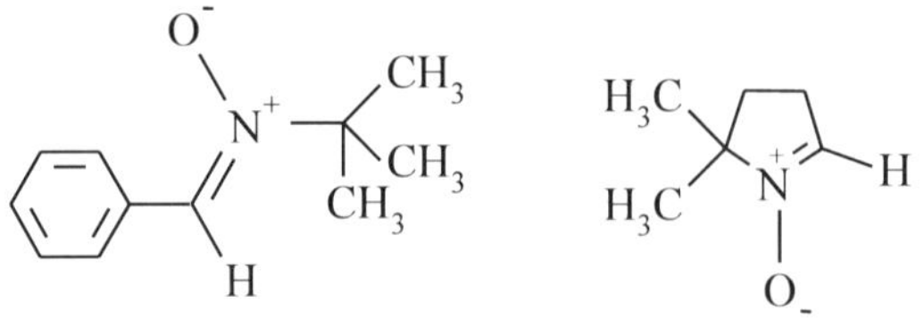

PBN

DMPO

Examples of radicals trapped by these agents are given in Table 7. The number of ESR lines for a carbon-centred radical trapped by PBN is six. For a nitrogen-centred radical trapped by PBN, the number of lines is 18, and for a hydroxyl radical trapped by DMPO, the number of lines is four.

Table 7. Electron spin resonance (ESR) spectra for representative free radicals trapped by spin trapping agents

Radical type (source)	Radical	Trapping agent	ESR spectrum	Reference
Carbon-centred (e.g., ethanol)	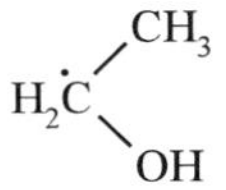 $H_2\dot{C}(CH_3)OH$ α-Hydroxyethyl radical	PBN	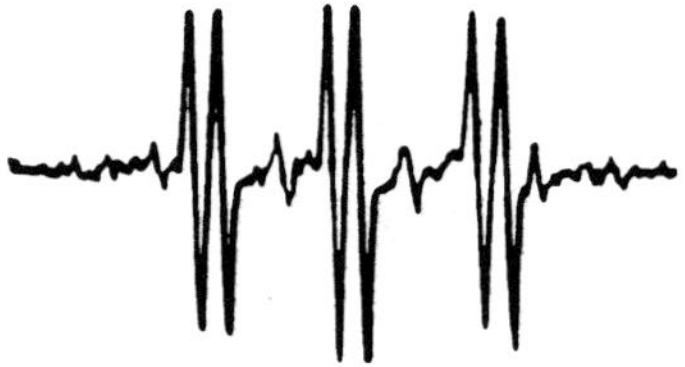	REINKE and MOORE 1994
Nitrogen-centred (e.g., 3-methylindole)	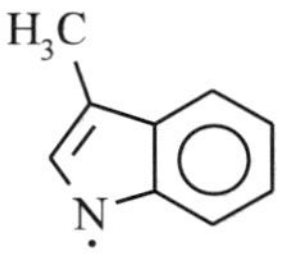 H_3C, N• 3-Methylindole radical	PBN		CHEN et al. 1994
Oxygen-centred (e.g., ROS initiators[a])	·OH Hydroxyl radical	DMPO	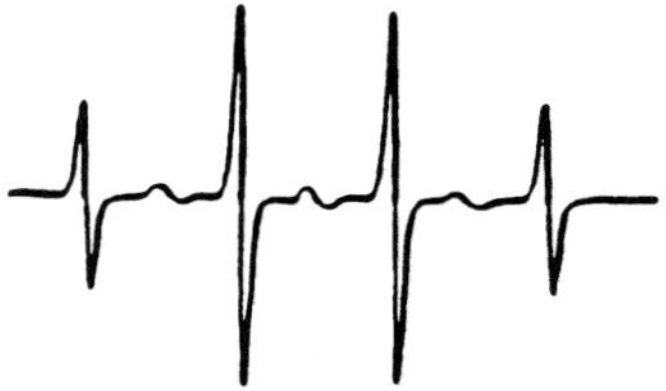	HASELOFF et al. 1990

PBN, α-phenyl-*N*-*tert*-butylnitrone; DMPO, 5,5-dimethylpyrroline-*N*-oxide; ROS, reactive oxygen species.
[a]ROS initiators include ionising radiation and many xenobiotics (e.g., phenytoin, benzo[a]pyrene, paraquat, ethanol).

3. Fluorescence Detection of Free Radicals and Oxidative Damage

Fluorescent detection methods are useful since they allow for detection of both free radical production and oxidative damage via both qualitative (e.g. microscopic visualisation) and quantitative (e.g. flow cytometry) methods (Table 6). Additionally, unlike ESR, fluorescence detection has the added advantage of permitting a direct measure of free radical formation in living cells (intra- and extracellular), allowing for mechanistic studies with regard to xenobiotic-initiated toxicity, such as cell death (BURGHARDT et al. 1994). Fluorescence detection has been used for the measurement of various parameters including: (a) production of reactive oxygen species such as hydrogen peroxide and lipid hydroperoxides with the fluorescent probe 2,7-dichlorofluorescein diacetate (CATHCART et al. 1983; BASS et al. 1983; SCOTT et al. 1988); (b) lipid damage via malondialdehyde production or decreased membrane fluidity (JAIN 1988; TANGORRA et al. 1991); and (c) protein damage via increased tryptophan release or carbonyl formation (DAVIES et al. 1987; LEBEL and BONDY 1991; MURPHY and SIES 1990). These and other fluorescence techniques, such as detection of GSH/protein thiol depletion/alteration, and their applications are reviewed by BURGHARDT et al. (1994).

4. Oxidative Damage

Oxidative damage to potential macromolecular targets such as DNA, protein and lipid (see Sect. H) provide sensitive indices of oxidative stress, as well as providing insights into potential molecular mechanisms of teratological initiation (Table 6).

5. Protein/Gene Expression

Enhanced expression of cytoprotective enzymes such as superoxide dismutase, catalase, glucose-6-phosphate dehydrogenase and GSH peroxidase (see Sect. G), stress proteins such as heat shock proteins, and oncogene products such as c-fos and c-myc proteins can provide sensitive indices of enhanced oxidative stress (Table 6), and, in the case of heat shock proteins and oncogene expression, also may provide insights into teratological mechanisms.

G. Cytoprotection

A number of cytoprotective pathways for the detoxification of ROS (Figs. 12, 14) have been shown to be embryoprotective, as exemplified below for phenytoin.

I. Glutathione

Glutathione is essential in a number of cytoprotective pathways that protect cellular macromolecules from oxidative damage caused by ROS and lipid hydroperoxides (Figs. 12, 14). In addition to maintaining membrane thiols in their reduced form (RSH), GSH serves as a co-factor for GSH peroxidase and GSH reductase (see below). Depletion of GSH using diethyl maleate or acetaminophen (HARBISON 1978; LUM and WELLS 1986; WONG et al. 1989), or inhibition of GSH synthesis using buthionine sulphoximine (WONG et al. 1989; MIRANDA et al. 1994), enhances the embryopathic effects of phenytoin in vivo and/or in embryo culture. The further teratological relevance of GSH and the differences between the cytoprotective and detoxifying roles of GSH are discussed in Sect. E.

II. Antioxidants

The cell has a variety of antioxidants distributed in the cytosol (e.g. vitamin C) and within lipid membranes (e.g. vitamin E) to provide comprehensive protection via the direct reductive detoxification of hydroxyl and lipid peroxyl radicals, forming relatively stable and non-toxic vitamin radicals in the process. In pregnant CD-1 mice, pretreatment with either caffeic acid, a water-soluble vitamin in vegetables similar to vitamin C, or vitamin E significantly reduced phenytoin teratogenicity (WELLS et al. 1989a; SANYAL and WELLS 1993). There are no human data demonstrating an embryoprotective effect of antioxidative vitamins in chemical teratogenesis.

III. Glutathione Peroxidase

GSH peroxidases catalyse the reduction of toxic hydrogen peroxide (H_2O_2) and lipid hydroperoxides (LOOH) to their corresponding stable alcohols, water and LOH (Figs. 14, 15). There are Se-dependent and Se-independent GSH peroxidase activities (HALLIWELL and GUTTERIDGE 1989c), the former of which can be reduced in pregnant rodents by dietary deprivation of Se (OZOLINS et al. 1996). Inhibition of Se-dependent GSH peroxidase activity in pregnant CD-1 mice by dietary deprivation for 15 days enhanced phenytoin embryopathy, and this teratological enhancement was eliminated by concomitant selenite administration, which restores GSH peroxidase activity (OZOLINS et al. 1996). These results suggest an important embryoprotective role for Se-dependent GSH peroxidase and, by association, a role for H_2O_2 and for LOOH in the mechanism of phenytoin teratogenicity.

More prolonged dietary deprivation of Se, for at least 40 days, while further reducing Se-dependent GSH peroxidase activity, also induces Se-independent GSH peroxidase activity, provided by a GST isoenzyme, that can reductively detoxify LOOH (OZOLINS et al. 1996). Pregnant CD-1 mice treated at this time appear more resistant to phenytoin embryopathy, suggesting that an antioxidative GST may be embryoprotective and that LOOH may contribute to the mechanism of phenytoin teratogenicity.

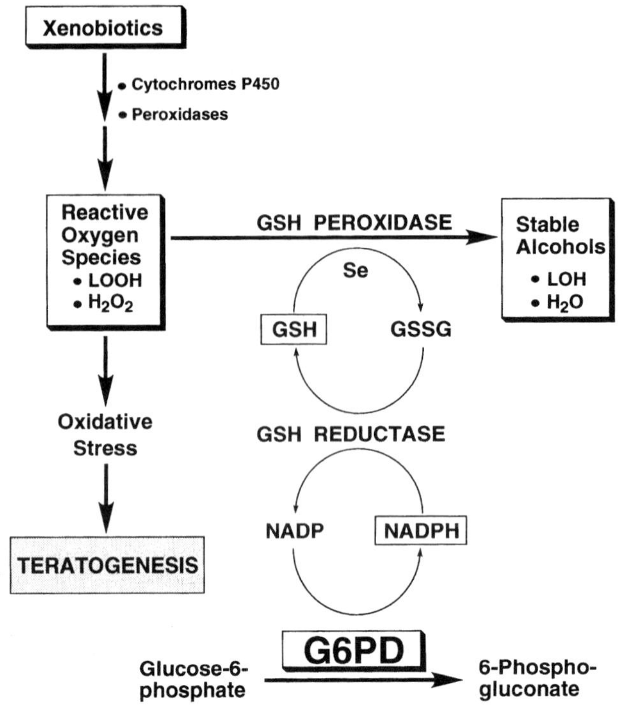

Fig. 16. Postulated teratological relevance of glutathione (*GSH*) reductase and glucose-6-phosphate dehydrogenase (*G6PD*). *GSSG*, GSH disulphide; *NADP*, nicotinamide adenine dinucleotide phosphate; *NADPH*, reduced NADP. (From NICOL and WELLS 1996)

IV. Glutathione Reductase

GSH reductase is essential for the reduction of GSSG to GSH, thereby maintaining intracellular GSH concentrations in the face of both physiological and xenobiotic-initiated oxidative stress (Fig. 16). The teratological importance of this enzyme is suggested by murine studies in which pregnant CD-1 mice pretreated with a non-teratogenic dose of 1,3-bis(2-chloroethyl)-1-nitrosourea (BCNU), a potent inhibitor of GSH reductase, were more susceptible to phenytoin teratogenicity (WONG and WELLS 1989). It is not known whether human genetic deficiencies in GSH reductase activity, or enzyme inhibition by concomitant drug therapy or environmental chemical exposure, may play a role in human teratological predisposition.

V. Glucose-6-phosphate Dehydrogenase

In the reduction of GSSG, glucose-6-phosphate dehydrogenase (G6PD) is required to produce the co-factor NADPH necessary for GSH reductase activity (Fig. 16). Genetic deficiencies in G6PD are X-linked and common, affecting almost 10% of the general population (about 400 million people worldwide) and up to 25% of racially predisposed populations, particularly from the Mediterranean region, Africa and the Orient (PANICH 1986; MEHTA 1994). It is generally believed that G6PD deficiencies are of toxicological importance only in red blood cells, which, lacking nuclei, are unable to synthesise more G6PD in the face of xenobiotic-initiated oxidative stress, therefore succumbing to free radical-mediated hemolysis. The susceptibility of red blood cells is further augmented in utero and neonatally by lower levels of vitamin E, GSH reductase and catalase (GROSS 1976; MATTHAY and MENTZER 1976). However, embryonic synthesis of many enzymes is low to negligible compared to adult activities, and hence embryos may be at risk from genetic G6PD deficiencies. While there is no human evidence, pregnant mice with heterozygous or homozygous G6PD deficiencies have a substantially higher incidence of phenytoin-initiated fetal resorptions (in utero death) and postpartum lethality, independent of gender, compared to congenic G6PD-normal controls (NICOL and WELLS 1996). The human teratological relevance of such deficiencies, if important, may have escaped notice in part because affected embryos died in utero before the pregnancy was recognised.

VI. Superoxide Dismutase and Catalase

In the reduction of molecular oxygen, which occurs during both physiological and xenobiotic-initiated oxidative stress, superoxide dismutase (SOD) catalyses the dismutation of toxic superoxide anions ($O_2^{\bullet -}$) to H_2O_2, and catalase catalyses the subsequent reductive detoxification of H_2O_2 to water (Figs. 12, 14, 17). Detoxification of these toxic ROS is essential for cellular viability, and addition of either of these antioxidative enzymes to a rodent embryo culture system substantially increases embryonic antioxidative activity (ERIKSSON and BORG 1991, 1993; WINN and WELLS 1996a) and completely inhibits phenytoin-initiated DNA oxidation and phenytoin- and B[a]P-initiated embryopathy (WINN and WELLS 1995a, 1996a), as well as reducing hyperglycemic embryopathies (ERIKSSON and BORG 1991, 1993). These results suggest that SOD and catalase, while low in embryos (EL-HAGE and SINGH 1990), nevertheless may have considerable teratological relevance. Conversely, these and related results discussed below (see Sect. H) also implicate ROS in the teratological mechanism of phenytoin and a number of related chemicals, including the environmental carcinogen/teratogen B[a]P.

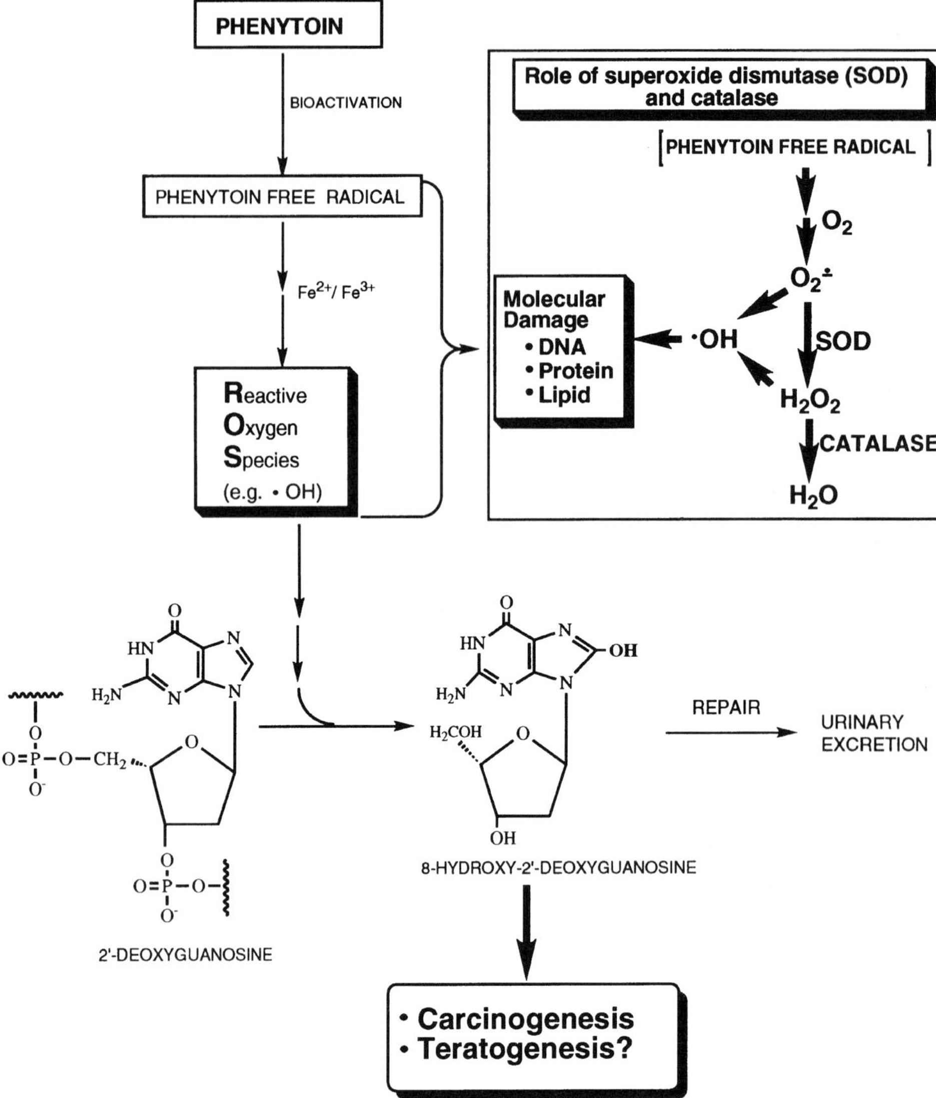

Fig. 17. Role of superoxide dismutase (*SOD*) and catalase in the detoxification of reactive oxygen species. (From WINN and WELLS 1995a)

H. Molecular Target Damage

I. Covalent Binding

The positively charged atom of an electrophilic reactive intermediate readily accepts electron pairs from electron-rich nucleophilic sites on small molecules such as GSH and on cellular macromolecules such as proteins and DNA. This reaction results in a covalent bond that is generally irreversible, except by enzymatic cleavage, and the xenobiotic covalently bound to such molecules is referred to as an adduct. The formation of a covalent bond often is referred to as arylation for aryl xenobiotics that bind via an aromatic ring, and alkylation for alkyl xenobiotics (Fig. 18).

1. DNA

DNA damage has been the focus of several recent studies due to its central role in information transfer between generations of somatic cells (Richter et al. 1988; Adelman et al. 1988).

a) Electrophilic Reactive Intermediates

Many carcinogens, mutagens and teratogens are bioactivated to electrophilic intermediates that covalently bind to nucleotides (Table 8), and DNA- adducts of several xenobiotics have been characterised (e.g. 4-(methylnitrosamino)-1-(3-pyridyl)-1-butanone, NNK; polycyclic aromatic hydrocarbons, PAH; anthracene arylhalides; ethylene oxide; Hemminki et al. 1994). The teratogenic anticonvulsant drug phenytoin also has been shown to covalently bind to embryonic DNA (Liu and Wells 1994a). The alkylation and arylation of nucleotides is site selective in aqueous solutions; arylating carcinogens react selectively with exocyclic amino groups on DNA bases, in contrast to simple SN_2 alkylating agents (see Table 4 for definitions), which react primarily with the pyridine-type ring nitrogen sites within DNA bases. The aqueous reactivity of the arylating agents also contrasts with that of alkylating agents, which react in water by a mechanism closer to an SN_1 reaction (e.g. *N*-nitroso compounds such as *N*-ethyl-*N*-nitrosourea; Moschel 1994). The interaction of alkylating and arylating agents with DNA bases depends on two chemical factors: (1) the ionic character of the reagent (i.e. SN_1 or SN_2 character) and (2) whether the charge on the ionic intermediate is localised (hard electrophile) or delocalised (soft electrophile) (Dipple and Moschel 1990). In these studies, guanosine was the base of choice, although reactions with the other bases will follow the same rules. Simple alkylating agents such as *N*-nitrosodimethylamine (NDMA) react with guanosine largely through a SN_2 mechanism and principally alkylate the N7 position of guanosine. In general, it has been shown that nitrosamines predominantly alkylate the N7 position of guanosine. In the case of NDMA, the relative yield of methylated N7 position of guanosine in rats treated with radiolabelled NDMA was 1.000, while the relative yield of O^6-methylation was 0.010 (Shuker and Bartsch 1994). Xenobiotics such as polycyclic aromatic hydrocarbons (PAH; e.g. B[a]P), which are ionic and delocalise the charge on the reaction centre quite readily around the aromatic ring, react almost exclusively at the N2 position (an exocyclic amino group) of guanosine. Xenobiotics such as alkylnitrosoureas form a hard (localised) ion and therefore react via an SN_1 mechanism; however, alkylnitrosoureas can also react via an SN_2 mechanism, alkylating both the N7 and O6

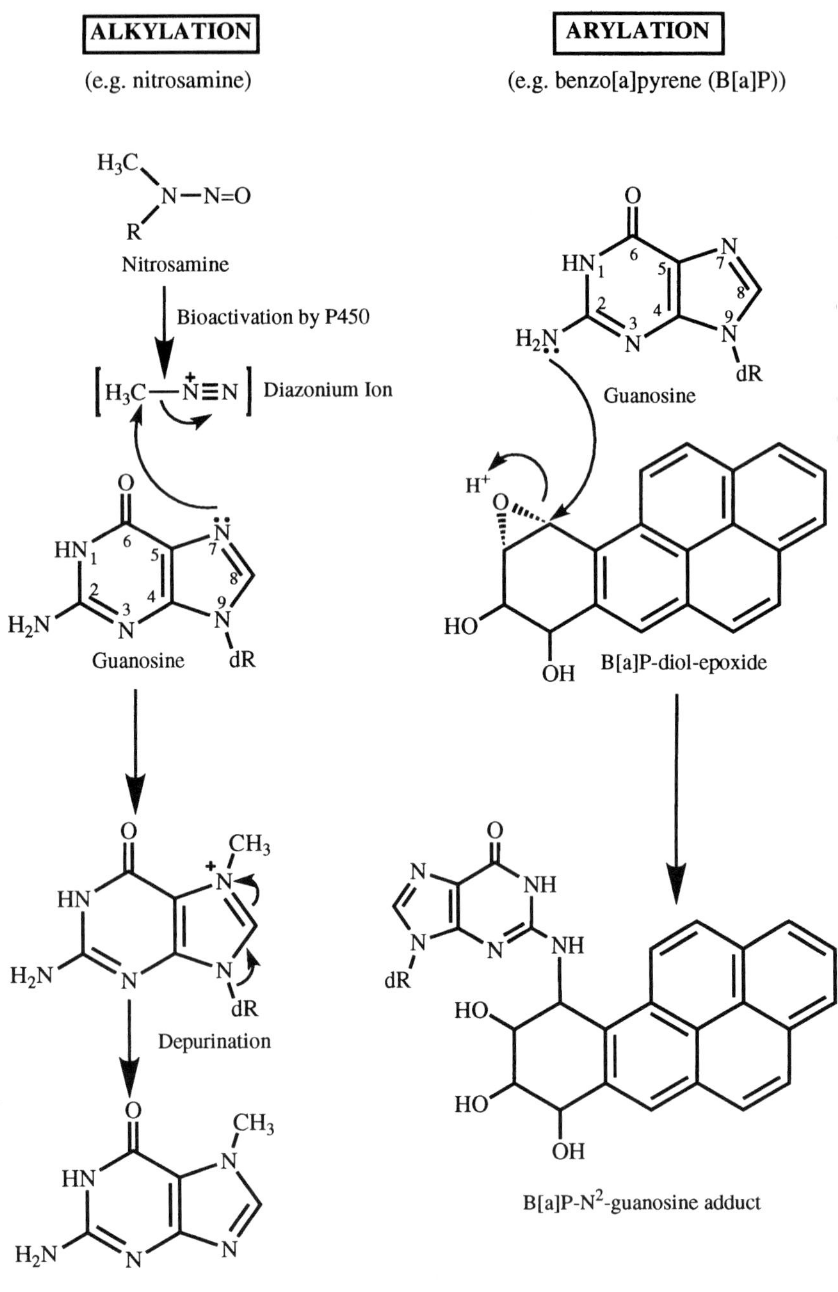

Fig. 18. Alkylation (*left*; e.g. nitrosamine) and arylation (*right*; e.g. benzo[a]pyrene, B[a]P) of guanosine by electrophilic reactive intermediates of nitrosamines and B[a]P. R, alkyl or aryl substituents; dR, deoxyribose sugar

Table 8. Molecular damage of potential relevance to chemical teratogenesis (WELLS and WINN 1996)

Molecular damage	Target	Substrate	Citation
Covalent binding	Protein	Phenytoin	MARTZ et al. 1977
		Benzo[a]pyrene	SHUM et al. 1979
		Thiabendazole	YONEYAMA et al. 1985; YONEYAMA and ICHIKAWA 1986
		2-Acetylaminofluorene	ROY and KULKARNI 1991
		Benzo[a]pyrene-7,8-diol	MARNETT et al. 1975
	DNA	Phenytoin	LIU and WELLS 1995a; WINN and WELLS 1995a
		Benzo[a]pyrene	SHUM et al. 1979; WANG and LU 1990; LU et al. 1993
		2-Acetylaminofluorene	ROY and KULKARNI 1991
		2-Naphthylamine	BOYD and ELING 1987
		Benzo[a]pyrene-7,8-diol	MARNETT et al. 1975
		Aflatoxin B_1	HSIEH and HSIEH 1993
		Cyclophosphamide	BENSON et al. 1988;
		N-Methyl-*N*-nitrosourea	PLATZEK et al. 1988; FRANK et al. 1993
		N-Acetoxy-2-acetylaminofluorene	MIRKES et al. 1991
		Ethylmethanesulfonate	PLATZEK et al. 1994a
		Acetoxymethyl-methylnitrosamine	PLATZEK et al. 1993
		6-Mercaptopurine riboside	PLATZEK et al. 1994b
Oxidation	Protein/glutathione	Phenytoin	LIU and WELLS 1994b, 1995a; WELLS and WILLIAMS 1994; WINN and WELLS 1994; WELLS et al. 1995
		Thalidomide	ARLEN and WELLS 1990;
		Diamide	HIRANRUENGCHOK and HARRIS 1993
	DNA	Phenytoin	LIU and WELLS 1995a
		Mephenytoin	LIU and WELLS 1995a
		Nirvanol	LIU and WELLS 1995a
		Trimethadione	LIU and WELLS 1995a
		Thalidomide	LIU and WELLS 1995a
		Benzo[a]pyrene	WINN and WELLS 1994
		Cyclophosphamide (strand breaks)	PILLANS et al. 1989
		Phosphoramide mustard (strand breaks)	LITTLE and MIRKES 1987
	Lipid	Phenytoin	LIU and WELLS 1994b, 1995b
		Cocaine	ZIMMERMAN et al. 1994
		Cyclophosphamide	LEAR et al. 1992

positions of guanosine. Styrene oxide and benzylating agents are examples of borderline agents which fall in the centre of the spectrum and can alkylate guanosine at the N2, N7 and O6 positions (DIPPLE and MOSCHEL 1990).

b) Reaction of Free Radicals with DNA and Its Nucleotides

Many free radicals, like other reactive intermediates, will react with DNA to form a radical adduct. Several xenobiotics are bioactivated to alkyl free radical intermediates which can add across the double bonds of DNA bases. For example, guanine is methylated by *tert*-butylhydroperoxide (TBH) in the presence of ferrous ion to give exclusively C8-methylation (MAEDA et al. 1974). Adenine also is methylated by TBH to give the corresponding mono- and dimethyl derivatives where methyl groups were substituted in positions C2, C8 or both (MAEDA et al. 1974). Hydrazine is bioactivated by P450 to a methyl radical that adds across the double bonds of DNA nucleotides specifically at the C8 position of guanosine (AUGUSTO 1993). DNA can also be arylated by aromatic radical cations formed from polycyclic aromatic hydrocarbons such as B[a]P. One-electron oxidation of B[a]P by horseradish peroxidase resulted in formation of a B[a]P radical cation (B[a]P$\bullet^{+}$), and DNA arylation occurred at both the N7 and C8 position of purines due to both electrophilic and radical properties of the aromatic radical cation (CAVALIERI et al. 1988). The depurinated adducts of B[a]P bound via the C6 of B[a]P to the N7 or C8 positions of guanosine were found in the urine and faeces of treated animals (ROGAN et al. 1990). More recently, adducts of B[a]P covalently bound via the C6 of B[a]P to C8 of guanosine also have been identified in the skin of mice treated with B[a]P. While a radical reactive intermediate of B[a]P binds to guanosine specifically at the C8 position, since B[a]P is also a cationic intermediate, it also can bind to the N7 position of guanosine. In the case of hydrazine, the methyl radical intermediate is very reactive and has been postulated to react with more proximate cytosolic natural scavengers, such as GSH, rather than diffusing into the nucleus and reacting with DNA. Therefore, toxicologically relevant oxidation of hydrazine to a free radical intermediate may occur in the nucleus, catalysed by transition metal ions such as iron chelated to DNA. DNA and iron (III) form a strong soluble chelate that is capable of catalysing the oxidation of mono-substituted hydrazine deriv-atives (AUGUSTO 1993).

c) Detection of DNA Adducts

Numerous structurally diverse carcinogens and teratogens bind covalently to DNA, forming adducts. In the absence of DNA repair, such lesions may result in damage or cause mutations at important genomic sites, ultimately leading to cancer, in utero death or teratogenesis. Therefore, the characterisation of such adducts provides mechanistic information about both the initiating potential of xenobiotics and potential determinants of individual teratological susceptibility.

α) Exhaustive Washing. In animal models and in in vitro systems, radiolabelled forms of teratogens, if available, can be administered, followed by removal of target tissues, extraction of DNA and exhaustive washing with solvents to remove any non-covalently bound xenobiotic. The DNA is then placed in a liquid scintillation spectrometer to quantify the associated xenobiotic, which is presumed to be covalently bound. However, radiolabelled xenobiotics often are unavailable and in any event cannot be used in human studies.

β) ^{32}P-Postlabelling. For studies with pregnant women, an assay must be sensitive enough to detect low levels of adducts in only microgram quantities of DNA; it must also be able to characterise adducts of any potentially teratogenic xenobiotic and must not require additional administration of a radiolabelled form of the teratogen in question. The ^{32}P-postlabelling technique (GUPTA 1993), which can detect as little as one adduct in 10^{10} nucleotides, often satisfies these conditions. In the first step of this technique, called adduct enrichment, after digestion of alkylated/arylated DNA, selective enzymatic dephosphorylation of only the normal nucleotide 3′-monophosphates forms nucleosides (nucleotides with no phosphate substituents) that are separated from the alkylated/arylated nucleotides, which are resistant to dephosphorylation. In some cases, adducts may alternatively be separated from normal nucleotides by differential solvent extraction. Subsequently, alkylated/arylated nucleotides are selectively 5′-phosphorylated by a mutant T4 polynucleotide kinase, which cannot phosphorylate nucleosides (lacking a 3′-phosphate). 3′,5′-Diphosphate adducts are separated by thin-layer chromatography (TLC) and detected by autoradiography. If the levels of adducts are sufficiently high, HPLC and a tandom mass spectrometer can be used to characterise the adducts. Adducts are quantified by either liquid scintillation spectrometry or HPLC equipped with a radioactivity detector (GORELICK 1993).

2. Protein and Lipids

a) Binding of Electrophiles to Proteins

Electrophilic reactive intermediates of many teratogens such as phenytoin and B[a]P also covalently bind to proteins (Table 8). Alkylation and arylation of proteins by electrophilic reactive intermediates occur via the same mechanism as for DNA. A nucleophilic site such as sulfhydryl groups or amino groups in proteins attack the positively charged carbon of the reactive intermediate and form an irreversible covalent bond. Covalent binding to proteins cause changes to secondary, tertiary and quaternary structure, resulting in functional changes or complete loss of function, which in turn can alter cellular function or initiate cellular death (NELSON and PEARSON 1990).

b) Binding of Free Radicals to Proteins

Free radicals covalently bind to proteins by adding across the double bond of amino acids. An example of this is B[a]P, which can be bioactivated to a toxic reactive intermediate by cytochromes P450 (GUENGERICH 1987) and peroxidases (MARNETT et al. 1978; KIM and WELLS 1995b) to a radical cation intermediate. If the reactive intermediate is not detoxified, it can arylate protein (BOROUJERDI et al. 1981; DIPPLE et al. 1984; FRENKEL 1989).

c) Detection of Protein Adducts

α) Radiolabelled Substrate. Several approaches are available to measure protein covalent binding. In the most common method, the radiolabelled teratogen is administered to animals or added to an in vitro preparation that has a functional bioactivating system. Organs or tissues are removed, and the protein is precipitated by addition of an ice-cold organic solvent, such as ethanol or methanol, and exhaustively washed with heated solvent to remove excess of radioactivity. The amount of radioactivity incorporated into the protein, which is measured by liquid scintillation spectrometry, is assumed to approximate the amount of xenobiotic covalently bound to protein (GUENGERICH 1994).

β) Antibodies. Another potential method involves the development of antibodies that recognise protein–xenobiotic covalent adducts (HINSON and ROBERTS 1992). While this method is more technically demanding, it permits the measurement of xenobiotic covalent binding to protein without the use of radioisotope, which is particularly useful in human studies and/or when a radiolabelled form of the xenobiotic is unavailable. Antibody-dependent characterisation of adducts also can be used to identify localised molecular lesions in specific embryonic tissues. Using this highly sensitive approach, antibodies have been used to detect phenytoin–protein adducts in people with idiosyncratic adverse drug reactions to phenytoin, in which case at least one of the target proteins was determined to be prostacyclin synthase (LEEDER et al. 1995).

II. Oxidation

1. DNA

Even in cells of normal untreated animals, there is a considerable intensity and frequency of oxidative damage (CLAYSON et al. 1994). AMES and GOLD (1991) estimated that the genome from adult rat liver cells contains about one million lesions per cell and that one hundred thousand new lesions are added daily. However, further oxidative DNA damage may become threatening for the cell under conditions of oxidative stress. If xenobiotic-initiated ROS are not detoxified by cellular cytoprotective pathways, they can cause irreversible modifications to DNA (SHIGENAGA et al. 1990). Oxidative modifications to nucleic acid polymers have been shown to disrupt transcription, translation and DNA replication and to cause mutations and ultimately cell death (Spitz et al. 1987; Ames 1989; Simic et al. 1989; Spector et al. 1989). The damage caused by such molecular modifications has been proposed to contribute to aging, cancer and other age-related degenerative diseases (Cathcart et al. 1984; Adelman et al. 1988; Ames 1989) as well as teratogenesis (Liu and Wells 1995a; Winn and Wells 1995a).

Oxidative DNA damage may occur either directly from the interaction of ROS with various groupings in the DNA helix or indirectly from the activation of endonucleases (Halliwell and Aruoma 1991). There are several types of DNA damage including the following: (a) strand breaks (single or double); (b) sister chromatid exchange; (c) DNA–DNA and DNA–protein cross-links; and (d) base modifications (Pacifici and Davies 1991). DNA bases can undergo ring saturation, ring opening, ring contraction and hydroxylation, all of which can cause local distortions in the double helix. The phosphodiester backbone of DNA also may be damaged by ROS, resulting in strand breaks (Simic et al. 1989; Teebor et al. 1988). Lipid peroxidation also can ultimately damage DNA via oxidative attack by lipid radicals and/or lipid peroxyl radicals on DNA. This produces DNA radicals, which can ultimately lead to the formation of DNA adducts, DNA strand breaks and DNA–protein cross-linking (Vaca et al. 1988).

Generally, the hydroxyl radical, generated chemically or by ionizing radiation, can add across the double bonds of a DNA base, forming a hydroxylated product, or abstract a hydrogen atom from either the DNA base or the deoxyribose sugar, forming DNA-centred free radicals of different types (Pryor 1988). For example, pyrimidine nucleotides can be oxidised to pyrimidine radicals by hydroxyl radicals, which are generated from the reaction of hydrogen peroxide with different transition metal ions at different pH values. In these studies, hydroxyl radicals attack predominantly at the C5 and/ or C6 carbon atoms in the pyrimidine ring, forming mono- and dihydroxylated products. For example, the reaction of hydroxyl radical with thymidine, as shown in Fig. 19, forms first a radical intermediate and subsequently a dihydroxylated product (thymine glycol), which is a measure of oxidative damage to DNA (Catterall et al. 1993).

Free radicals can also cause DNA strand scission, as shown for radicals generated from hydrazine derivatives such as 2-phenylethylhydrazine (Augusto et al. 1984).

Another mechanism by which radicals can damage DNA is via oxidation of the phosphate backbone, resulting in the loss of phosphate from the C2′-sugar-derived radical. It has been postulated that at pH 7 greater fragmentation occurs due to enhancement in the rate of hydroxyl radical formation from complexed metal ions which may be bonded to the nucleic acid itself under the conditions of the experiment. From ESR studies, it has been suggested that the first site of damage in pyrimidines is the addition of hydroxyl radical to the C5–C6 double bond. Subsequent protonation of the hydroxyl

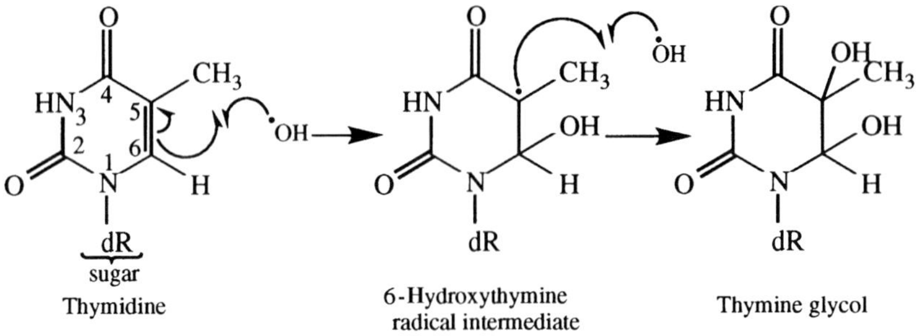

Fig. 19. Reaction of a hydroxyl radical with thymidine

radical adduct forms the radical cation. The radical centre is rapidly transferred to the C2′ in the ribose ring, leading to loss of the phosphate group.

The oxidised guanine analogue 8-hydroxy-2′-deoxyguanosine (8-OH-2′-dG) is thought to be formed in DNA at the C8 position by the hydroxylation of deoxyguanosine residues by •OH, which is formed by various oxygen radical-producing agents (KASAI et al. 1986). The production of 8-OH-2′-dG is thought to represent one of approximately 20 oxidative DNA modifications resulting from oxygen radical-initiated DNA damage (SHIGENAGA and AMES 1991; Table 9). After enzymatic digestion, the 8-OH-2′-dG adduct in DNA can be readily measured among the mononucleosides using HPLC with electrochemical detection (FLOYD et al. 1986). Accordingly, 8-OH-2′-dG formation can be used as a biological marker of oxidative DNA damage, as well as providing insight into potential molecular mechanisms of toxicological initiation. In vivo studies have shown a correlation between the production of 8-OH-2′-dG and tumor promotion (KASAI et al. 1986) and carcinogenesis (FLOYD 1990). In mice in vivo and in mouse embryo culture, both phenytoin and B[a]P initiate substantial formation of 8-OH-2′-dG, implicating ROS-initiated DNA oxidation in their molecular mechanisms of teratological initiation (LIU and WELLS 1995a; WINN and WELLS 1994, 1995a, 1996b).

2. Lipids

The process of lipid peroxidation is initiated by the oxidative attack of a free radical on lipids, whereby an electron is abstracted from the double bond of an unsaturated fatty acid to form a lipid radical. This leads to a chain reaction that culminates in the formation of lipid breakdown products such as alcohols and potentially toxic aldehydes (HALLIWELL and GUTTERIDGE 1989a), alkanes and alkenes (MULLER and SIES 1984; VACA et al. 1988). Malondialdehyde (MDA), along with many other carbonyl compounds, is formed as the result of lipid peroxidation. Dialdehydes such as MDA can attack amino groups on protein molecules and cause cross-linking between two proteins or intramolecular cross-links within one protein molecule (HALLIWELL and GUTTERIDGE

Table 9. Oxidative base modifications resulting from reactive oxygen species (ROS)

DNA base	Oxidized product	Reference
Thymidine	*cis*-5,6-Dihydro-5,6-dihydroxythymine *trans*-5,6-Dihydro-5,6-dihydroxythymine 5-Hydroxy-5,6-dihydrothymine 6-Hydroxy-5,6-dihydrothymine 5,6-Dihydrothymine 5-Hydroxymethyluracil Thymidine glycol	INFANTE et al. 1973; NISHIMOTO et al. 1983
Cytosine	5-Hydroxycytosine 5,6-Dihydroxycytosine Uracil glycol 6-Hydroxycytosine Cytosine glycol	DIZDAROGLU and SIMIC 1985
Guanosine	N^6-(2′Deoxy-α-D-erythro-pentopyranosyl)-2,6-diamo-5-formamidopyrimid-4-one N^6-(2′-Deoxy-β-D-erythro-pentopyranosyl)-2,6-diamo-5-formamidopyrimid-4-one 5′,8-Cyclo-2′,5′-dideoxyguanosine 8′-Hydroxy-2′-deoxyguanosine 9-(2′-Deoxy-α-D-erythro-pentopyranosyl) guanosine 9-(2′-Deoxy–β-D-erythro-pentofuranosyl) guanosine 9-(2′-Deoxy–α-D-erythro-pentofuranosyl) guanosine 9-(2′-Deoxy-α-L–threo-pentofuranosyl) guanosine 9-(2′-Deoxy-β-D-erythro-pento-1,5-dialdo-1,4-furanosyl)-guanosine	BERGER and CADET 1985
Adenosine	4,6-Diamino-5-formamidopyrimidine 8-Hydroxyadenosine	DIZDAROGLU 1985

1989a). Alkanes such as ethane and pentane, as well as diene conjugates, are formed during lipid peroxidation, and their formation has been proposed to be a sensitive index for lipid peroxidation (MULLER and SIES 1984). Furthermore, alkanes can undergo aliphatic hydroxylation by P450, producing alcohols that can be oxidised to toxic aldehydes. For example, ethanol is bioactivated by P450 to a reactive free radical intermediate (α-hydroxyethyl radical) that initiates lipid peroxidation (TIMBRELL 1991). As described previously, lipid peroxidation can also ultimately damage DNA via direct attack by the lipid radical on DNA.

3. Protein

Free radicals such as hydroxyl radicals generated from the Fenton reaction (Fig. 11A), or xenobiotic free radicals, such as the α-hydroxyethyl radical generated enzymatically from ethanol, react with proteins such as albumin, histones surrounding DNA, and enzymes. As the result of this interaction, protein radicals are formed via: (a) hydrogen (one-electron) abstraction from the sulphhydryl groups to form a sulphur-centred radical; (b) oxidative decarboxylation to form a carbon-centred radical; and (c) hydrogen abstraction from the side chain of amino acids to form a carbon-centred radical. Formation of a sulphur-centred radical will result in the formation of mixed-disulphide bonds, which can activate or inactivate enzymes or alter the function of structural or transport proteins. Formation of any kind of protein radical will cause protein cross-linking and changes in amino acid composition, as well as changes in secondary, tertiary and quaternary structure of proteins (DAVIES et al. 1991). Purified proteins that are exposed to ROS-generating systems undergo a variety of changes, which are summarised in Table 10. Oxidative modification to the structure of a protein results in increased proteolytic susceptibility and ultimately loss of function (DAVIES 1987; DAVIES et al. 1987; DAVIES and DELSIGNORE 1987). Oxidation of protein has been associated with several physiological and pathological processes, including aging, arthritis, pulmonary diseases (LEVINE et al. 1990) and chemical teratogenesis (LIU and WELLS 1994b, 1995b; WINN and WELLS 1994).

In general, hydroxyl radicals can abstract a hydrogen from the side chain of an amino acid such as L-glutamic acid to give a secondary carbon-centred radical, •CHR′R″. However, in polypeptides, hydrogen abstraction occurs on the α-carbon of the amino acid backbone to give a different carbon-centred radical (-NH-•CH-CO-). Abstraction of hydrogen from poly-L-glutamic acid occurs at both the inactivated C–H bond (C2) and the activated α-carbon (C1), as shown below. The α-carbon is activated because of the presence of an adjacent nitrogen and the fact that radical formed can delocalise onto both the nitrogen and carbonyl group adjacent to it.

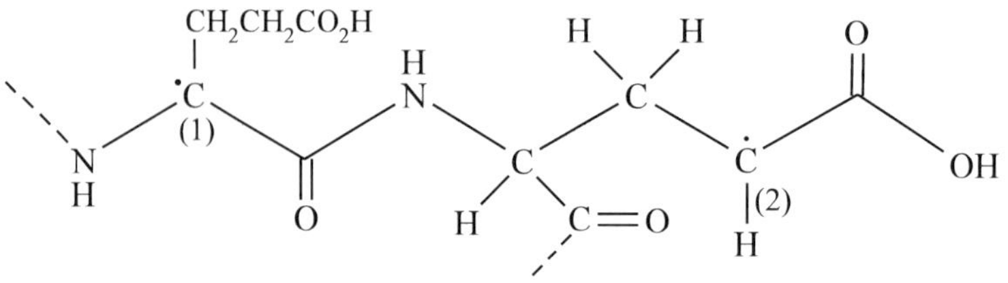

Poly-L-glutamic acid

For amino acids such as methionine, *S*-methylcysteine and phenylalanine, oxidative decarboxylation has been proposed as the pathway through which an α-aminoalkyl radical (R-•CH-NH) is formed. The proposed mechanism of oxidative decarboxylation is thought to involve an initial attack by hydroxyl radicals largely at the sulphur atom of methionine, resulting in formation of a hydroxylated product, also known as a hydroxyl-adduct. The reaction of this adduct is proposed to be pH dependent; at pH values greater than 2, hydroxyl

Table 10. Changes to purified proteins after exposure to reactive oxygen species (ROS)

Observation	Cause
Decrease in native fluorescence	Due to destruction of tryptophan and tyrosine
Shifts in isoelectric points	Due to charge modifications to amino acid R groups
Increases in molecular weight	Due to formation of covalent and/or hydrophobic intermolecular associations
Decreases in molecular weight	Due to the scission of the peptide backbone to generate peptide fragments

From Pacifici and Davies (1991).

anion is lost from this adduct and a sulphur radical cation is formed, which subsequently binds the carboxylate function to form a cyclic sulphur-centred radical cation. This initiates decarboxylation to give the α-aminoalkyl radical (Davies et al. 1983).

One marker of oxidative damage to protein involves the formation of carbonyl groups on amino acid residues. A possible mechanism of carbonyl formation mediated by metal ion-catalysed oxidation of proteins starts with Fe (II) binding to a lysyl residue to form an Fe(II)–protein complex. Hydrogen peroxide reacts with this complex to form hydroxyl radical, hydroxyl anion and an Fe(III)–protein complex. The hydroxyl radical then abstracts a hydrogen atom from the amino group of lysine to form a carbon-centred radical. The amino group will donate its unpaired electrons to Fe(III), reducing the complex to a Fe(II)–protein complex. The amino derivative formed will then react with a molecule of water to form a carbonyl group on the lysine (aldehyde derivative) and release ammonia. At this point the metal ion dissociates from the protein (Stadtman 1990). Several methods have been developed to detect protein carbonyl groups. These methods involve: (a) reaction of 2,4-dinitrophenylhydrazine with the carbonyl group on the protein and detection of 2,4-dinitrophenylhydrazine derivatives spectrophotometrically; (b) reduction of the carbonyl groups to tritiated alcohols by tritiated sodium borohydride ($[^3H]NaBH_4$) and measurement of the radioactivity of protein; (c) reaction of carbonyl groups with fluoresceinamine to produce stable secondary amines (Schiff bases) and spectrophotometric measurement of these stable amines after reduction of the bases with sodium cyanoborohydride ($NaCNBH_3$) (Stadtman 1990); and (d) derivatisation of oxidised proteins with 2,4-dinitrophenylhydrazine and subsequent detection by immunochemical methods (Keller et al. 1993). Protein oxidation and degradation are initiated in embryo culture and in vivo by both phenytoin and B[a]P (Liu and Wells 1994b, 1995b; Winn and Wells 1994, 1995a, 1996b), which may contribute to teratological initiation.

I. Repair

I. Protein

The cell has developed a system of proteolytic enzymes responsible for the recognition and selective degradation of oxidatively denatured proteins (PACIFICI and DAVIS 1991). Oxidised proteins exhibit increases in denaturation leading to protein aggregates and protein cross-links, which give rise to insoluble aggregates, and ultimately to lipofuscin and other inclusion bodies. PACIFICI et al. (1989) have shown that most of the proteolytic activity against oxidatively modified protein substrates was expressed by an unusually large proteolytic complex that they called macroxyproteinase (MOP), which can selectively degrade oxidatively denatured proteins, preventing their accumulation and potential cytotoxicity. Oxidatively modified proteins may be spared from proteolytic degradation by direct repair mechanisms, such as direct enzymatic re-reduction of sulphhydryl and heme groups, which may restore proteins to their native conformation.

II. DNA

DNA repair is under genetic control, and the number of genes regulating this process in humans is estimated to exceed 100. This property of self-repair is unique to DNA and emphasises the importance of DNA error recognition and correction processes for survival. There are multiple DNA repair pathways, including direct repair (alkyltransferases and photolyases), excision repair (glycosylases, endonucleases and excinucleases) and mismatch repair, each of which specialises in certain kinds of DNA damage (ALLAN and GARNER 1994; Table 11).

The predominant oxidative molecular alterations requiring repair are base alterations, single-strand breaks and double-strand breaks (VON SONNTAG 1987). The primary repair mechanism for xenobiotic-initiated DNA oxidative damage is excision repair. Excision repair relies on the fact that genetic information is stored on both strands of the DNA helix, allowing for removal of a damaged base or nucleotide and then replacing it with a normal base by using the complementary strand as a template. *Base excision* repair targets single, damaged bases, where the damaged base is removed by DNA glycosylase enzymes that cleave the base–sugar bond to leave an apurinic/apyrimidinic (AP) site. A DNA AP endonuclease enzyme can then recognise the AP site and nicks the strand at that point. The damaged part of the strand is removed and new DNA synthesis fills the gap, which is joined by a DNA ligase enzyme (HALLIWELL and ARUOMA 1991). In *nucleotide excision* repair, an enzyme system with excision nuclease activity hydrolyses two phosphodiester bonds, one on either side of the damaged site, to generate an oligonucleotide carrying the damage (SANCAR 1994). The excised oligonucleotide is then released from the duplex, and the resulting gap is filled in and ligated to com-

Table 11. DNA repair proteins and their substrates

Enzyme and substrate	Reference
Alkyltransferase	
O^6-methylguanine	Souliotis and Kyrtopolous 1989; Souliotis et al. 1990; Krytopolous et al. 1990
O^4-methylthymimine	Wilkinson et al. 1989; Sassanfar et al. 1991
Alkyl phosphotriesters	
Photolyase	
Thymine dimers	Thomas et al. 1991
DNA glycosylases	
Uracil	Varshney et al. 1991; Verri et al. 1992
3- and 7-Methyladenine	Habraken et al. 1991; Singer et al. 1992
Formamidopyrimidines	Graves et al. 1992
Hypoxanthine	Dianov and Lindahl 1991
8-Hydroxyguanine	Tchou et al. 1991
Redoxyendonucleases	
Thymine glycols	Huq et al. 1992
AP endonucleases	
Abasic sites	Parrish et al. 1992; Hori et al. 1992
Uvr ABC excinuclease	
Thymine dimers	Mazur and Grossman 1991
Bulky DNA adducts	Thomas et al. 1991; Bertrand-Burggraff et al. 1991; Oleykowski et al. 1993; Nouso et al. 1993; Tang et al. 1992
Low molecular weight adduct	Voight et al. 1989; Visse et al. 1992; Roberts et al. 1989

From Allan and Garner (1994).

plete the repair (Sancar 1994). In humans, three diseases are associated with defects in nucleotide excision repair: xeroderma pigmentosum, Cockayne's syndrome and trichothiodystrophy.

The role of DNA repair in chemical teratogenesis has been investigated using transgenic mice with a heterozygous deficiency in the p53 tumor suppressor gene. This gene is important because it can directly or indirectly facilitate DNA repair. In vivo studies have shown that mice deficient in p53 are more susceptible to both B[a]P- and phenytoin-initiated teratogenesis (Nicol et al. 1995; Laposa and Wells 1995). Therefore, xenobiotic-initiated DNA damage in the developing embryo may constitute an important molecular mechanism in chemical teratogenesis.

Embryos in vivo appear to have a significant capacity to repair damaged DNA (Wells and Winn 1996). In fact, it appears that mouse embryos exposed to phenytoin in vivo remove oxidised guanosine bases at a rate comparable to that in maternal liver (Liu and Wells 1995a). Conversely, in mouse embryo culture studies, embryos incubated with phenytoin for 24 h showed no evidence of repair mechanisms (Winn and Wells 1995a), suggesting that embryonic repair mechanisms may be compromised in vitro.

Acknowledgement. This work was supported by grants to P.G. Wells from the Medical Research Council of Canada and the Hospital for Sick Children Foundation of Toronto.

References

Adelman R, Saul RL, Ames BN (1988) Oxidative damage to DNA: relation to species metabolic rate and life span. Proc Natl Acad Sci USA 85: 2706–2708

Albano E, Lott KAK, Slater TF, Stier A, Sympsons MCR, Tomasi A (1982) Spin-trapping studies on the free-radical products formed by metabolic activation of carbon tetrachloride in rat liver microsomal fractions isolated hepatocytes and in vivo in the rat. Biochem J 204: 593–603

Albano E, Tomasi A, Ingelman-Sunberg M (1994) Spin-trapping of alcohol-derived radicals in microsomes and reconstituted systems by electron spin resonance. Methods Enzymol 233: 117–127

Allan JM, Garner CR (1994) The use of purified DNA repair proteins to detect DNA damage. Mutat Res 313: 165–174

Ames BN (1989) Endogenous oxidative DNA damage, ageing, and cancer. Free Radic Res Commun 7: 121–128

Ames BN, Gold LS (1991) Endogenous mutagens and the causes of aging and cancer. Mutat Res 250: 3–16

Anari MR, Khan S, Liu ZC, O'Brien PJ (1995) Cytochrome P450 peroxidase/peroxygenase mediated xenobiotic metabolic activation and cytotoxicity in isolated hepatocytes. Chem Res Toxicol 8: 997–1004

Anderson ME, Meister A (1989) Glutathione monoesters. Anal Biochem 183: 16–20

Anderson ME, Powrie F, Puri RN, Meister A (1985) Glutathione monoethyl ester: preparation, uptake by tissues, and conversion to glutathione. Arch Biochem Biophys 239: 538–548

Arlen RR, Wells PG (1989) Effect of the cyclooxygenase inhibitor acetylsalicylic acid on thalidomide teratogenicity in New Zealand white rabbits. FASEB J 3: A1025

Arlen RR, Wells PG (1990) Extracellular thiol-disulfide status reflecting chemical toxicity mediated via oxidative stress: Validation with paraquat and t-butyl-hydroperoxide, and implications for phenytoin and thalidomide teratogenicity. FASEB J 4: A608

Arlen RR, Wells PG (1996a) In vivo evidence for prostaglandin synthetase-catalysed bioactivation of thalidomide to a teratogenic reactive intermediate in the New Zealand white rabbit. J Pharmacol Exp Ther 277: 1649–1658

Arlen RA, Wells PG (1996b) Extracellular thiol-disulfide status reflecting chemical toxicity mediated via oxidative stress: Validation with paraquat and t-butyl-hydroperoxide, and implications for phenytoin and thalidomide teratogenicity (submitted)

Augusto O (1993) Alkylation and cleavage of DNA by carbon-centered radical metabolites. Free Radic Biol Med 15: 329–336

Augusto O, Faljoni-Alario A, Leite LCC, Nobrega FG (1984) DNA strand scission by the carbon radical derived from 2-phenylethylhydrazine metabolism. Carcinogenesis 5: 781–784

Balling R, Haaf M, Maydl R, Metzler M, Beier HM (1985) Oxidative and conjugative metabolism of diethylstilbestrol by rabbit preimplantation embryos. Dev Biol 109: 370–378

Barreto JC, Smith GS, Strobel NHP, McQuillin PA, Miller TA (1995) Terephthalic acid: a dosimeter for the detection of hydroxyl radicals in vitro. Life Sci 56: 89–96

Bass DA, Parce JW, Dechatelet LR, Szejda P, Seeds MC, Thomas M (1983) Flow cytometric studies of oxidative product formation by neutrophils: a graded response to membrane stimulation. J Immunol 130: 1910–1917

Benson AJ, Martin CN, Garner RC (1988) N-(2-Hydroxyethyl)-N-[2-(7-guaninyl)-ethyl], the putative major DNA adduct of cyclophosphamide in vitro and in vivo in the rat. Biochem Pharmacol 37: 2979–2985

Berger M, Cadet J (1985) Isolation and characterization of the radiation-induced degradation products of 2′-deoxyguanosine in oxygen free aqueous solutions. Z Naturforsch 40b: 1519–1531

Bertrand-Burggraff B, Keyse SM, Lindahl T, Wood RD (1991) Identification of the different intermediates in the interaction of (A)BC excinuclease with its substrates by DNase I footprinting on two uniquely modified oligonucleotides. J Mol Biol 219: 27–36

Bondy SC, Naderi S (1994) Contribution of hepatic cytochrome P450 systems to the generation of reactive oxygen species. Biochem Pharmacol 48: 155–159

Boroujerdi M, Kung H, Wilson A, Anderson MW (1981) Metabolism and DNA binding of benzo[a]pyrene in vivo in the rat. Cancer Res 41: 951–957

Boyd JA, Eling TE (1984) Evidence for a one-electron mechanism of 2-aminofluorene oxidation by prostaglandin H synthase and horseradish peroxidase. J Biol Chem 259: 13885–13896

Boyd JA, Eling TE (1987) Prostaglandin H synthase-catalyzed metabolism and DNA binding of 2-naphthylamine. Cancer Res 47: 4007–4014

Boyd JA, Harvan DJ, Eling TE (1983) The oxidation of 2-aminofluorene by prostaglandin endoperoxide synthetase. J Biol Chem 258: 8246–8254

Braun AG, Harding FA, Weinreb SL (1986) Teratogen metabolism: thalidomide activation is mediated by cytochrome P-450. Toxicol Appl Pharmacol 82: 175–179

Brierly CH, Burchell B (1993) Human UDP-glucuronosyl transferases: chemical defence, jaundice and gene therapy. Bioessays 15: 749–754

Buehler BA (1984) Epoxide hydrolase activity and fetal hydantoin syndrome. Proc Greenwood Genet Ctr 3: 109–110

Buehler BA, Delimont D, VanWaes M, Finnell RH (1990) Prenatal prediction of risk of the fetal hydantoin syndrome. N Engl J Med 322: 1567–1572

Burchell B, Coughtrie MWH (1989) UDP-glucuronosyltransferases. Pharmacol Ther 43: 261–289

Burchell B, Nebert DW, Nelson DR, Bock KW, Iyanagi T, Jansen PLM, Lancet D, Mulder GJ, Roy Chowdhury J, Siest G, Tephley TR, Mackenzie PI (1991) The UDP glucuronosyl-transferase gene superfamily: suggested nomenclature based on evolutionary divergence. DNA Cell Biol 10: 487–494

Burghardt RC, Barhoumi R, Doolittle DJ, Philips TD (1994) Application of fluorescence imaging for in vitro toxicology testing. In: Hayes WA (ed) Principles and methods of toxicology 3rd edn. Raven, New York, pp 1231–1258

Burke MD, Mayer RT (1975) Inherent specificities of purified cytochromes P-450 and P-448 toward biphenyl hydroxylation and ethoxyresorufin deethylation. Drug Metab Dispos 3: 245–253

Byczkowski JZ, Kulkarni AP (1989) Lipoxygenase-catalyzed epoxidation of benzo[a]pyrene-7,8-dihydrodiol. Biochem Biophys Res Commun 159: 1199–1205

Capdevila J, Estabrook RW, Prough RA (1980) Differences in the mechanism of NADPH- and cumene hydroperoxide-supported reactions of cytochrome P-450. Arch Biochem Biophys 200: 186–195

Carpenter SP, Raucy JL (1995) CYP2E1: a putative role in chemical-mediated teratogenesis. Proc Int Soc Study Xenobiot 8: 238

Cathcart R, Schwiers E, Ames BN (1983) Detection of picomole levels of hydroperoxides using a fluorescent dichlorofluorescein assay. Anal Biochem 134: 111–116

Cathcart R, Schwiers E, Saul RL, Ames BN (1984) Thymine glycol and thymidine glycol in human and rat urine: a possible assay for oxidative DNA damage. Proc Natl Acad Sci USA 81: 5633–5637

Catterall H, Davies MJ, Gilbert BC, Polack N (1993) EPR spin-trapping studies of the reaction of the hydroxyl radical with pyrimidine nucleobases. Nucleosides and nucleotides, polynucleotides and RNA. Direct evidence for sites of initial attack for strand breakage. J Chem Soc Perkin Trans II: 2039–2047

Cavalieri EL, Devanesan PD, Rogan EG (1988) Radical cations in horseradish peroxidase and prostaglandin H synthase mediated metabolism and binding of benzo[a]pyrene to deoxyribonucleic acid. Biochem Pharmacol 37: 2183–2187

Clayson DB, Mehta R, Iverson F (1994) Oxidative DNA damage – the effects of certain genotoxic and operationally non-genotoxic carcinogens. Mutat Res 317: 25–42

Coudray C, Talla M, Martin S, Fatome M, Favier A (1995) High-performance liquid chromatography-electrochemical determination of salicylate hydroxylation products as an in vivo marker of oxidative stress. Anal Biochem 227: 101–111

Crawford D, Zbinden I, Amstad P, Cerutti P (1988) Oxidant stress induces the protooncogenes c-fos and c-myc in mouse epidermal cells. Oncogene 3: 27–32

Currie WR (1987) Effects of ischemia and reperfusion temperature on the synthesis of stress-induced (heat shock) proteins in isolated and perfused rat hearts. J Mol Cell Cardiol 19: 795–808

Datta K, Kulkarni AP (1994) Oxidative metabolism of aflatoxin B_1 by lipoxygenase purified from human term placenta and intrauterine conceptal tissues. Teratology 50: 311–317

Datta K, Roy SK, Joseph, P, Mitra AK, Srinivasan NS, Kulkarni AP (1993) Xenobiotic metabolism in 10-weeks old human conceptus. Toxicologist 13: 60.

Davies KJA (1987) Protein damage and degradation by oxygen radicals: General aspects. J Biol Chem 262: 9895–9901

Davies KJA, Delsignore ME (1987) Protein damage and degradation by oxygen radicals: modification of secondary and tertiary structure. J Biol Chem 262: 9908–9913

Davies KJA, Delsignore ME, Lin SW (1987) Protein damage and degradation by oxygen radicals: modification of amino acids. J Biol Chem 262: 9902–9907

Davies MJ, Gilbert BC, Haywood RM (1991) Radical-induced damage to proteins: ESR spin-trapping studies. Free Radic Res Commun 15: 111–127

Davies MJ, Gilbert BC, Norman ROC (1983) Electron spin resonance studies, part 64. The hydroxyl radical induced decarboxylation of methionine and some related compounds. J Chem Soc Perkin Trans II: 731–738

DeBoni U, Kim PM, Wells PG (1996) Induction by 2,3,7,8-tetrachlorodibenzo-p-dioxin (TCDD) of prostaglandin H synthase and peroxidase-dependent benzo[a]pyrene (B[a]P)-initiated protein oxidation in the mechanism of B[a]P genotoxicity. Fundam Appl Toxicol 30 [Suppl 1, Part 2]: 246

Degen GH, Eling TE, McLachlan JA (1982) Oxidative metabolism of diethylstilbestrol by prostaglandin synthetase. Cancer Res 42: 919–923

deMorais SMF, Wells PG (1989) Enhanced acetaminophen toxicity in rats with bilirubin glucuronyl transferase deficiency. Hepatology 10: 163–167

deMorais SMF, Uetrecht JP, Wells PG (1992) Decreased glucuronidation and increased bioactivation of acetaminophen in Gilbert's syndrome. Gastroenterology 102: 577–586

Dianov G, Lindahl T (1991) Preferential recognition of I-T base-pairs in the initiation of excision repair by hypoxanthine-DNA glycosylase. Nucleic Acids Res 19: 3829–3834

Dipple A, Moschel RC (1990) Chemistry of DNA alkylation and arylation. In: Mendelsohn M, Albertini RJ (eds) Mutation and the environment, part A. Basic mechanisms. Wiley-Liss, New York, pp 71–80

Dipple A, Moschel RC, Bigger CAH (1984) Polycyclic aromatic carcinogens. In: Searle CE (ed) Chemical carcinogens 2nd edn. American Chemical Society, Washington DC, pp 41–163

Dizdaroglu M (1985) Formation of an 8-hydroxyguanine moiety in deoxyribonucleic acid on gamma-irradiation in aqueous solution. Biochemistry 24: 4476–4481

Dizdaroglu M, Simic M (1985) Radiation-induced crosslinking of pyrimidine oligonucleotides. Radiat Phys Chem 26: 309–316

Doecke CJ, Veronese ME, Pond SM, Miners JO, Birkett DJ, Sansom LN, and McManus ME (1991) Relationship between phenytoin and tolbutamide hydroxylations in human liver microsomes. Br J Clin Pharmacol 1: 125–130

El-Hage S, Singh SM (1990) Temporal expression of genes encoding free radical-metabolizing enzymes is associated with higher mRNA levels during in utero development in mice. Dev Genet 11: 149–159

Eriksson UJ, Borg LAH (1991) Protection by free oxygen radical scavenging enzymes against glucose-induced embryonic malformations in vitro. Diabetologia 34: 325–331

Eriksson UJ, Borg LAH (1993) Diabetes and embryonic malformations: role of substrate-induced free-oxygen radical production for dysmorphogenesis in cultured rat embryos. Diabetes 42: 411–419

Fantel AG, Greenaway JC, Juchau MR, Shepard TH (1979) Teratogenic bioactivation of cyclophosphamide in vitro. Life Sci 25: 67–72

Faustman-Watts EM, Greenaway JC, Namkung MJ, Fantel AG, Juchau MR (1983) Teratogenicity in vitro of 2-acetylaminofluorene: role of biotransformation. Teratology 27: 19–28

Faustman-Watts EM, Giachelli CM, Juchau MR (1986) Carbon monoxide inhibits monooxygenation by the conceptus and embryotoxic effects of proteratogens in vitro. Toxicol Appl Pharmacol 83: 590–595

Fessenden RJ, Fessenden JS (1986a) Alkyl halides: substitution and elimination reaction. In: Organic chemistry, 3rd edn. Brooks/Cole, Monterey, pp 175–198

Fessenden RJ, Fessenden JS (1986b) Organic chemistry, 3rd edn. Brooks/Cole, Monterey, pp 294–298

Filler R, Lew KJ (1981) Developmental onset of mixed-function oxidase activity in preimplantation mouse embryos. Proc Natl Acad Sci USA 78: 6991–6995

Finnell RH, Kerr BM, van Waes M, Steward RL, Levy RH (1994) Protection from phenytoin-induced congenital malformations by coadministration of the antiepileptic drug stiripentol in a mouse model. Epilepsia 35: 141–148

Floyd RA (1990) The role of 8-hydroxyguanine in carcinogenesis. Carcinogenesis 11: 1447–1450

Floyd RA, Watson JJ, Wong PK, Altmiller DH, Richard RC (1986) Hydroxyl free radical adduct of deoxyguanosine: sensitive detection and mechanism of formation. Free Radic Res Commun 1: 163–172

Fort DJ, Stover EL, Rayburn JR, Huli M, Bantle JA (1993) Evaluation of the developmental toxicity of trichloroethylene and detoxification metabolites using xenopus. Teratogen Carcinogen Mutagen 13: 35–45

Frank AA, Collier JM, Forsyth CS, Zeng W, Stoner GD (1993) Ellagic acid embryoprotection in vitro: distribution and effects on DNA adduct formation. Teratology 47: 275–280

Frenkel K (1989) Oxidation of DNA bases by tumor promoter-activated processes. Environ Health Perspect 81: 45–54

Gardiner CS, Reed DJ (1994) Status of glutathione during oxidant-induced oxidative stress in the preimplantation mouse embryo. Biol Reprod 51: 1307–1314

Gordon GB, Spielberg SP, Blake DA, Balasubramanian V (1983) Thalidomide teratogenesis: evidence for a toxic arene oxide metabolite. Proc Natl Acad Sci USA 78: 2545–2548

Gorelick NJ (1993) Application of HPLC in the ^{32}P-postlabelling assay. Mutat Res 288: 5–18

Graves R, Laval J, Pegg AE (1992) Sequence specific repair by *Escherichia coli* Fpg protein. Carcinogenesis 13: 1455–1459

Greenaway JC, Fantel AG (1983) Enhancement of rifampin teratogenicity in cultured rat embryos. Toxicol Appl Pharmacol 69: 81–88

Greenberg JT, Demple B (1989) A global response induced in *Escherichia coli* by redox cycling agents overlap with that induced by peroxide stress. J Bacteriol 171: 3933–3939

Grootveld M, Halliwell B (1986) Aromatic hydroxylation as a potential measure of hydroxyl radical formation in vivo. Identification of hydroxylated derivatives of salicylate in human body fluids. Biochem J 237: 499–504

Gross S (1976) Hemolytic anemia in premature infants: relationship to vitamin E, selenium, glutathione peroxidase and erythrocyte lipids. Semin Hematol 3: 187–199

Guengerich FP (1987) Cytochrome P450 enzymes and drug metabolism. Prog Drug Metab 10: 1–54

Guengerich FP (1994) Analysis and characterization of enzymes. In: Hayes AW (ed) Principles and methods of toxicology 3rd edn. Raven, New York, pp 1279–1280

Guengerich FP, MacDonald TL (1990) Mechanisms of cytochrome P-450 catalysis. FASEB J 4: 2453–2459

Guengerich FP, Shimada T (1991) Oxidation of toxic and carcinogenic chemicals by human cytochrome P-450 enzymes. Chem Res Toxicol 4: 391–407

Gupta RC, (1993) ^{32}P-Postlabelling analysis of bulky aromatic adducts. In: Phillips DH, Castegnero M, Bartsch H (eds) Postlabelling methods for detection of DNA adducts. IARC Scientific Publications, Lyon, pp 11–23

Habraken Y, Carter CA, Kirk M, Ludlum DB (1991) Release of 7-alkyl-guanines from N-(2-chlorethyl)-N′-cyclohexyl-N-nitrosourea modified DNA by 3-methyl-adenine DNA glycosylase II. Cancer Res 51: 199–203

Hales BF (1981a) Modification of the teratogenicity and mutagenicity of cyclophosphamide with thiol compounds. Teratology 23: 373–381

Hales BF (1981b) Modification of the mutagenicity and teratogenicity of cyclophosphamide in rats with inducers of the cytochromes P-450. Teratology 24: 1–11

Halliwell B, Aruoma O (1991) DNA damage by oxygen derived species: its mechanism and measurement in mammalian systems. FEBS Lett 281: 9–19

Halliwell B, Grootveld M (1988) Methods for the measurement of hydroxyl radicals in biochemical systems: deoxyribose degradation and aromatic hydroxylation. Methods Biochem Anal 33: 59–90

Halliwell B, Gutteridge JMC (eds) (1989a) Lipid peroxidation: a radical chain reaction. In: Halliwell B, Gutteridge JMC (eds) Free radicals in biology and medicine, 2nd edn. Claredon, Oxford, pp 188–276

Halliwell B, Gutteridge JMC (eds) (1989b) The chemistry of oxygen-derived species. In: Halliwell B, Gutteridge JMC (eds) Free radicals in biology and medicine, 2nd edn. Claredon, Oxford, pp 47–53

Halliwell B, Gutteridge JMC (1989c) Protection against oxidants in biological systems: the superoxide theory of oxygen toxicity. In: Halliwell B, Gutteridge JMC (eds) Free radicals in biology and medicine. Halliwell, Oxford University Press, London, pp 86–187

Halliwell B, Kaur H, Ingelman-Sundberg M (1991) Hydroxylation of salicylate as an assay for hydroxyl radicals: a cautionary note. Free Radic Biol Med 10: 439–441

Hansen DK (1991) The embryotoxicity of phenytoin: an update on possible mechanisms. Proc Soc Exp Biol Med 197: 361–368

Harbison RD (1978) Chemical-biological reactions common to teratogenesis and mutagenesis. Environ Health Perspect 24: 87–100

Harris C, Stark KL, Luchtel DL, Juchau MR (1989) Abnormal neurulation induced by 7-hydroxy-2-acetylaminofluorene and acetaminophen: evidence for catechol metabolites as proximate dysmorphogens. Toxicol Appl Pharmacol 101: 432–436

Haurand M, Ullrich V (1985) Isolation and characterization of thromboxane synthase from human platelets as a cytochrome P-450 enzyme. J Biol Chem 260: 15059–15067

Hemminki K, Dipple A, Shuker DEG, Kadlubar FF, Segerback D, Bartsch H (eds) (1994) DNA adducts: Identification and biological significance. IARC Scientific, Lyon, p 125

High KA, Kubow S (1994) n-3 fatty acids inhibit defects and fatty acid changes caused by phenytoin in early gestation in mice. Lipids 29: 771–778

Hinson JA, Roberts DW (1992) Role of covalent and noncovalent interactions in cell toxicity: effects on proteins. Annu Rev Pharmacol Toxicol 32: 471–510

Hiranruengchok R, Harris C (1993) Glutathione oxidation and embryotoxicity elicited by diamide in the developing rat conceptus in vitro. Toxicol Appl Pharmacol 120: 62–71

Hori N, Doi T, Karaki Y, Kikuchi M, Ikehara M, Ohtsuka E (1992) Participation of glutamic acid 23 of T4 endonuclease V in the beta-elimination reaction of an abasic site in a synthetic duplex DNA. Nucleic Acids Res 29: 4761–4764

Hrycay EG, O'Brien PJ (1972) Cytochrome P-450 as a microsomal peroxidase in steroid hydroperoxide reduction. Arch Biochem Biophys 153: 480–494

Hrycay EG, Gustafsson J-A, Ingelman-Sundberg M, Ernster L (1976) The involvement of cytochrome P-450 in hepatic microsomal steroid hydroxylation reactions supported by sodium periodate, sodium chlorite, and organic hydroperoxides. Eur J Biochem 61: 43–52

Hsieh L-L, Hsieh T-T (1993) Detection of Aflatoxin B_1-DNA adducts in human placenta and cord blood. Cancer Res 53: 1278–1280

Hu Z, Wells PG (1992) In vitro and in vivo biotransformation and covalent binding of benzo[a]pyrene in Gunn and RHA rats with genetic deficiency in bilirubin uridine diphosphate-glucuronosyltransferase. J Pharmacol Exp Ther 263: 334–342

Hu Z, Wells PG (1994) Modulation of benzo[a]pyrene bioactivation by glucuronidation in lymphocytes and hepatic microsomes from rats with a hereditary deficiency in bilirubin UDP-glucuronosyltransferases. Toxicol Appl Pharmacol 127: 306–313

Huq I, Haukanes BI, Helland DE (1992) Purification to homogeneity and characterisation of a redoxyendonuclease from calf thymus. Eur J Biochem 206: 833–839

Infante GA, Jirathana P, Fendler JH, Fendler EJ (1973) Radiolysis of pyrimidines in aqueous solutions. 1. Product formation in the interaction of e^-_{aq} •H, •OH, and Cl_2^- with thymine. J Chem Soc Faraday Trans 169: 1586–1596

Ingelman-Sundberg M, Kaur H, Terelius Y, Persson J, Halliwell B (1991) Hydroxylation of salicylate by microsomal fractions and cytochrome P-450. Lack of production of 2,3-dihydroxybenzoate unless hydroxyl radical formation is permitted. Biochem J 276: 753–757

Iyer P, Martin JE, Irvin TR (1991) Role of biotransformation in the in vitro preimplantation embryotoxicity of naphthalene. Toxicology 66: 257–270

Jain SK (1988) Evidence for membrane lipid peroxidation during the in vivo aging of human erythrocytes. Biochim Biophys Acta 937: 205–210

Juchau MR (1981) Enzymatic bioactivation and inactivation of chemical teratogens and transplacental carcinogens/mutagens. In: Juchau MR (ed) The biochemical basis of chemical teratogenesis. Elsevier, North Holland, New York, pp 63–94

Juchau MR, Bark DH, Shewey LM, Greenaway JC (1985a) Generation of reactive dysmorphogenic intermediates by rat embryos in culture: effects of cytochrome P-450 inducers. Toxicol Appl Pharmacol 81: 533–544

Juchau MR, Giachelli CM, Fantel AG, Greenaway JC, Shepard TH, Faustman-Watts EM (1985b) Effects of 3-methylcholanthrene and phenobarbital on the capacity of embryos to bioactivate teratogens during organogenesis. Toxicol Appl Pharmacol 80: 137–147

Juchau MR, Fantel AG, Harris C, Beyer BK (1986) The potential role of redox cycling as a mechanism for chemical teratogenesis. Environ Health Perspect 70: 131–136

Juchau MR, Lee QP, Fantel AG (1992) Xenobiotics biotransformation/bioactivation in organogenesis-stage conceptal tissues: implications for embryotoxicity and teratogenesis. Drug Metab Rev 24: 195–238

Kadlubar FF, Morton KC, Ziegler DM (1973) Microsomal-catalyzed hydroperoxide-dependent C-oxidation of amines. Biochem Biophys Res Commun 54: 1255–1261

Kalabis GM, Wells PG (1990) Biphasic modulation of acetaminophen bioactivation and hepatotoxicity by pretreatment with the interferon inducer polyinosinic-polycytidylic acid. J Pharmacol Exp Ther 255: 1408–1419

Kanekal S, Kehrer JP (1993) Mechanism of cyclophosphamide by lipoxygenases. Drug Metab Dispos 22: 74–78
Kappus H (1986) Overview of enzyme systems involved in bio-reduction of drugs and in redox cycling. Biochem Pharmacol 35: 1–6
Kasai H, Crain PF, Kuchino Y, Nishimura S, Ootsuyama A, Tanooka H (1986) Formation of 8-hydroxyguanine moiety in cellular DNA by agents producing oxygen radicals and evidence for its repair. Carcinogenesis 7: 1849–1851
Kedderis GL, Dwyer LA, Rickert DE, Hollenberg PF (1983) Source of the oxygen atom in the product of cytochrome P-450-catalyzed N-demethylation reactions. Mol Pharmacol 23: 758–760
Keller RJ, Hinson JA (1991) Mechanism of acetaminophen-stimulated NADPH oxidation catalyzed by the peroxidase-H_2O_2 system. Drug Metab Dipos 19: 184–187
Keller RJ, Halmes CN, Hinson JA, Pumford NR (1993) Immunochemical detection of oxidized proteins. Chem Res Toxicol 6: 430–433
Kim PM, Wells PG (1993) Salicylate hydroxylation as an in vivo probe for hydroxyl radical formation in CD-1 mice: relevance to phenytoin teratogenicity. Toxicologist 13: 253.
Kim PM, Wells PG (1995a) Paraquat-initiated hydroxyl radical (HO•) formation: characterisation in vivo by salicylate hydroxylation and glucuronidation. Toxicologist 15: 276
Kim PM, Wells PG (1995b) Evidence for cytochrome P4501A1- and prostaglandin H synthase-dependent DNA oxidation in the mechanism of benzo[a]pyrene-initiated micronucleus formation in cultured Wistar rat skin fibroblasts. Proc Am Assoc Cancer Res 36: 600
Kim PM, Wells PG (1996a) Phenytoin-initiated hydroxyl radical formation: characterisation by enhanced salicylate hydroxylation. Mol Pharmacol 49: 172–181
Kim PM, Wells PG (1996b) Peroxidase-dependent bioactivation and DNA oxidation in benzo[a]pyrene-initiated micronucleus formation. (submitted)
Kim PM, Wells PG (1996c) Genoprotection by UDP-glucuronosyltransferases in peroxidase-dependent, reactive oxygen species-mediated micronucleus initiation by the carcinogens 4-(methylnitrosamino)-1-(3-pyridyl)-1-butanone and benzo[a]pyrene. Cancer Res 56: 1526–1532
Kitada M, Kamataki T (1994) Cytochrome P450 in human fetal liver: significance and fetal-specific expression. Drug Metab Rev 26: 2305–323
Koop DR, Hollenberg PF (1980) Kinetics of the hydroperoxide-dependent dealkylation reactions catalyzed by rabbit liver microsomal cytochrome P-450. J Biol Chem 255: 9685–9692
Kubow S (1992) Inhibition of isotretinoin teratogenicity by acetylsalicylic acid pretreatment in mice. Teratology 45: 55–63
Kubow S, Wells PG (1986) In vitro evidence for prostaglandin synthetase-catalysed bioactivation of phenytoin to a free radical intermediate. Pharmacologist 28: 195
Kubow S, Wells PG (1988) In vitro evidence for lipoxygenase-catalysed bioactivation of phenytoin. Pharmacologist 30: A74
Kubow S, Wells PG (1989) In vitro bioactivation of phenytoin to a reactive free radical intermediate by prostaglandin synthetase, horseradish peroxidase and thyroid peroxidase. Mol Pharmacol 35: 504–511
Kubow S, Dubose CM Jr, Janzen EG, Carlson JR, Bray TM (1983) The spin-trapping of enzymatically and chemically catalysed free radicals from indolic compounds. Biochem Biophys Res Commun 114: 168–174
Kyrtopoulos SA, Ampatzi P, Davaris P, Haritopoulos N, Golematis B (1990) Studies in gastric carcinogenesis. IV. O^6-Methylguanine and its repair in normal and atrophic biopsy specimens of human gastric mucosa: correlation of O^6-alkyltransferase activities in gastric mucosa and circulating lymphocytes. Carcinogenesis 11: 431–436
Laposa RR, Wells PG (1995) Preliminary evaluation of phenytoin teratogenicity in transgenic mice deficient in the p53 tumor suppressor gene. Toxicologist 15: 161

Larsson A, Orrenius S, Holmgren A, Mannervik B (1983) Functions of glutathione: biochemical, physiological, toxicological, and clinical aspects. Raven, New York

Laval F (1988) Pretreatment with oxygen species increases the resistance to hydrogen peroxide in Chinese hamster fibroblasts. J Cell Physiol 201: 73–79

Lear L, Nation RL, Stupans I (1992) Effects of cyclophosphamide and adriamycin on rat hepatic microsomal glucuronidation and lipid peroxidation. Biochem Pharmacol 44: 747–753

Lebel CP, Bondy SC (1991) Persistent protein damage despite reduced oxygen radical formation in the aging rat brain. Int J Dev Neurosci 9: 139–146

Leeder JS, Gaedigk A, Lu X (1995) Identification of prostacyclin synthase (CYP 8) and not human CYPs3A as the target of "anti-CYP3A" antibodies in patients with anticonvulsant hypersensitive reactions. ISSX Proc 8: 233

Levine RL, Garland D, Oliver CN, Amici A, Climent I, Lenz A, Ahn B, Shaltiel S, Stadtman ER (1990) Determination of carbonyl content in oxidatively modified proteins. Methods Enzymol 186: 464–478

Little SA, Mirkes PE (1987) DNA cross-linking and single-strand breaks induced by teratogenic concentrations of 4-hydroperoxycyclophosphamide and phosphoramide mustard in postimplantation rat embryos. Cancer Res 47: 5421–5426

Liu L, Wells PG (1994a) In vitro and in vivo cytochromes P450- and peroxidase-catalysed formation of phenytoin-DNA adducts in murine maternal hepatic and embryonic tissues. Proceedings of the 10th international symposium on microsomes and drug oxidations, Toronto, Canada July 1994, p 555

Liu L, Wells PG (1994b) In vivo phenytoin-initiated oxidative damage to proteins and lipids in murine maternal hepatic and embryonic tissue organelles: potential molecular targets mediating chemical teratogenesis. Toxicol Appl Pharmacol 125: 247–255

Liu L, Wells PG (1995a) DNA oxidation as a potential molecular mechanism mediating drug-induced birth defects: phenytoin and structurally related teratogens initiate the formation of 8-hydroxy-2′-deoxyguanosine in vitro and in vivo in murine maternal hepatic and embryonic tissues. Free Radic Biol Med 19: 639–648

Liu L, Wells PG (1995b) Potential molecular targets mediating chemical teratogenesis: in vitro peroxidase-catalysed phenytoin bioactivation and oxidative damage to proteins and lipids in murine maternal and embryonic tissues. Toxicol Appl Pharmacol 134: 71–80

Lu D, Maulik N, Moraru II, Kreutzer DL, Das DK, (1993) Molecular adaptation of vascular endothelial cells to oxidative stress. Am J Phys 264: 715–722

Lu L-JW, Anderson LM, Jones AB, Moskal TJ, Salazar JJ, Hokanson JA, Rice JM (1993) Persistence, gestation stage-dependent formation and interrelationship of benzo[a]pyrene-induced DNA adducts in mothers' placentae and fetuses of erythroceubs patas monkeys. Carcinogenesis 14: 1805–1813

Lum JT, Wells PG (1986) Pharmacological studies on the potentiation of phenytoin teratogenicity by acetaminophen. Teratology 33: 53–72

Maeda M, Nushi K, Kawazoe Y (1974) Studies on chemical alterations of nucleic acids and components VII. C-Alkylation of purine bases through free radical process catalyzed by ferrous ion. Tetrahedron 30: 2677–2682

Manson J (1986) Teratogens. In: Klaassen CD, Amdur MD, Doull J (eds) Toxicology, the basic science of poisons. MacMillan, New York, pp 195–220

Manson JM, Kang YJ (1994) Test methods for assessing female reproductive and developmental toxicology. In: Hayes AW (ed) Principles and methods of toxicology 3rd edn. Raven, New York, pp 989–1037

March J (1992) Aliphatic nucleophilic substitution. In: March J (ed) Advanced organic chemistry mechanism and structure, 4th (edn.) Wiley, Toronto, pp 293–369

Marnett LJ (1990) Prostaglandin synthase mediated metabolism of carcinogens and a potential role for peroxyl radical as reactive intermediates. Environ Health Perspect 88: 5–12

Marnett LJ, Eling TE (1983) Hodgson E, Bend JR, Philpot RM (eds) Cooxidation during prostaglandin biosynthesis: a metabolic activation of xenobiotics. In: Reviews in biochemical toxicology, vol 5. Elsevier/North Holland, Biomedical, New York, pp 135–172

Marnett LJ, Wlodawer P, Samuelsson B (1975) Co-oxygenation of organic substrates by prostaglandin synthetase of sheep vesicular gland. J Biol Chem 250: 8510–8517

Marnett LJ, Reed GA, Johnson JT (1977) Prostaglandin synthase-dependent benzo[a]pyrene oxidation: products of the oxidation and inhibition of their formation by antioxidants. Biochem Biophys Res Commun 79: 569–576

Marnett LJ, Reed GA, Dennison DJ (1978) Prostaglandin synthase dependent activation of 7,8-dihydro-7,8-dihydroxybenzo[a]pyrene to mutagenic derivatives. Biochem Biophys Res Commun 82: 210–216

Martz F, Failinger C, Blake DA (1977) Phenytoin teratogenicity: correlation between embryopathic effect and covalent binding of putative arene oxide metabolite in gestational mouse. J Pharmacol Exp Ther 203: 231–239

Matthay KK, Mentzer WC (1976) Erythrocyte enzymopathies in the newborn. Clin Haematol 10: 31–55

Mazur SJ and Grossman L (1991) Dimerization of Escherichia coli Uvr A and its binding to undamaged and ultraviolet light damaged DNA. Biochemistry 30: 4432–4443

Mehta AB (1994) Glucose-6-phosphate dehydrogenase deficiency. Postgrad Med J 70: 871–877

Mitchell MD, Brennecke SP, Saed SP, Strickland DM (1985) Arachidonic acid metabolism in the fetus and neonate. In: Cohen MM (ed) Biological protection with prostaglandins. CRC Press, Boca Raton pp 27–44

Miranda AF, Wiley MJ, Wells PG (1994) Evidence for embryonic peroxidase-catalysed bioactivation and glutathione-dependent cytoprotection in phenytoin teratogenicity: modulation by eicosatetraynoic acid and buthionine sulfoximine in murine embryo culture. Toxicol Appl Pharmacol 124: 230–241

Mirkes PE, Little SA, Beland FA, Huitfeldt HS, Poirier MC (1991) Quantitation and immunohistochemical localization of DNA adducts in rat embryos and associated yolk sac membranes exposed in vitro to N-acetoxy-2-acetylaminofluorene (N-Ac-AAF). Teratogen Carcinogen Mutagen 11: 93–102

Moschel RC (1994) Reaction of aralkyl halides with nucleic acid components and DNA. In: Hemminki K, Dipple A, Shuker DEG, Kadlubar FF, Segerback D, Bartsch H (eds) DNA adducts: identification and biological significance. IARC Sci Publ 125: 25–36

Muindi JF, Young CW (1993) Lipid hydroperoxides greatly increase the rate of oxidative catabolism of all-trans-retinoic acid by human cell culture microsomes genetically enriched in specified cytochrome P-450 isoforms. Cancer Res 53: 1226–1229

Muller A, Sies H (1984) Assay of ethane and pentane from isolated organs and cells. Methods Enzymol 105: 311–319

Murphy ME, Sies H (1990) Visible-range low-level chemiluminescence in biological systems. In: Packer L, Glazer AN (eds) Methods in enzymology vol 186, part B. Academic, New York, pp 595–610

Nau H (1994) Retinoid teratogenesis: toxicokinetics and structure-specificity. Arch Toxicol [Suppl] 16: 118–127

Nebert DW (1983) Genetic differences in drug metabolism: proposed relationship to human birth defects. In: Johnson EM, Kochhar DM (eds) Teratogenesis and reproductive toxicology, Handbook of Experimental Pharmacology, vol 65. Springer, Berlin Heidelberg New York, pp 49–62

Nelson SD, Pearson PG (1990) Covalent and noncovalent interactions in acute lethal cell injury caused by chemicals. Annu Rev Pharmacol Toxicol 30: 169–195

Nicol CJ, Wells PG (1996) Glucose-6-phosphate dehydrogenase (G6PD): an embryoprotective role in endogenous oxidative stress and phenytoin teratogenesis. Fundam Appl Toxicol 30 [Suppl 1, Part 2]: 197

Nicol CJ, Harrison ML, Laposa RR, Gimelshtein IL, Wells PG (1995) A teratologic suppressor role for p53 in benzo[a]pyrene-treated transgenic p53-deficient mice. Nature Genet 10: 181–187

Nishimoto S, Ide H, Wada T, Kagiya T (1983) Radiation-induced hydroxylation of thymine promoted by electron-affinic compounds. Int J Radiat Biol 44: 585–600

Nordmann R, Ribiere C, Rouach H (1990) Ethanol-induced lipid peroxidation and oxidative stress in extrahepatic tissues. Alcohol Alcohol 25: 231–237

Nouso K, Bohr VA, Schut HAJ, Sniderwine EG (1993) Quantitation of 2-amino-3-methylimidazo[4,5-f]-quinoline and 2-amino-3,8-dimethylimidazo[4,5-f]quinoxaline DNA adducts in specific sequences using alkali or Uvr ABC exonuclease. Mol Carcinogen 7: 126–134

Oleykowski CA, Mayernik JA, Lim SE, Groopman JD, Grossman L, Wogan GN, Yeung AT (1993) Repair of aflatoxin B1 DNA adducts by the UvrABC endonuclease of Escherichia coli. J Biol Chem 268: 7990–8002

Onodera T, Ashraf M (1991) Detection of hydroxyl radicals in post-ischemic reperfused heart using salicylate as a trapping agent. J Mol Cell Cardiol 23: 365–370

Ortiz de Montellano PR (1986) Oxygen activation and transfer. In: Ortiz de Montellano PR (ed) Cytochrome P-450: structure, mechanism and biochemistry. Plenum, New York, pp 217–271

Otto S, Marcus C, Pidgeon C, Jefcoate C (1991) A novel adrenocorticotrophin-inducible cytochrome P450 from rat adrenal microsomes catalyzes polycyclic aromatic hydrocarbon metabolism. Endocrinology 129: 970–982

Otto S, Bhattacharyya KK, Jefcoate CR (1992) Polycyclic aromatic hydrocarbon metabolism in rat adrenal, ovary and testis microsomes is catalyzed by the same novel cytochrome P450 (P450RAP). Endocrinology 131: 3067–3076

Ozolins TRS, Siksay DLA, Wells PG (1996) Modulation of embryonic glutathione peroxidase activity and phenytoin teratogenicity by dietary deprivation of selenium in CD-1 mice. J Pharmacol Exp Ther 277: 945–953

Pacifici RE, Davies KJA (1991) Protein, lipid and DNA repair systems in oxidative stress: the free-radical theory of aging revisited. Gerontology 37: 166–180

Pacifici RE, Salo DC, Davies KJA (1989) Macroxyproteinase (MOP): a 670-kDa proteinase complex that degrades oxidatively denatured proteins in red blood cells. Free Radic Biol Med 7: 521–536

Panich V (1986) G6PD variants in southern Asian populations. In: Yoshida A, Beutler E (eds) Glucose-6-phosphate dehydrogenase. Academic, New York, pp 195–241

Pantarotto C, Arbix M, Sezzano P, Abbruzzi R (1982) Studies on 5,5-diphenyl hydantoin irreversible binding to rat liver microsomal proteins. Biochem Pharmacol 31: 1501–1507

Parman T, Chen G, Bray TM, Wells PG (1996) Bioactivation of phenytoin, thalidomide and related teratogens to a free radical intermediate using prostaglandin H synthase (PHS) or hepatic microsomes: characterisation by electron spin resonance (ESR) spectrometry. Fundam Appl Toxicol 30 [Suppl 1, Part 2]: 246

Parrish DD, Lambert WC, Lambert MW (1992) Xeroderma pigmentosum endonuclease complexes show reduced activity on and affinity for psoralen crosslinked nucleosomal DNA. Mutat Res 273: 157–170

Pillans PI, Ponzi SF, Parker MI (1989) Cyclophosphamide induced DNA strand breaks in mouse embryo cephalic tissue in vivo. Carcinogenesis 10: 83–85

Platzek T, Bochert G, Pauli B, Meister R, Neubert D (1988) Embryotoxicity induced by alkylating agents. 5. Dose-response relationships of teratogenic effects of methylnitrosourea in mice. Arch Toxicol 62: 411–423

Platzek T, Bochert G, Meister R, Neubert D (1993) Embryotoxicity induced by alkylating agents. 7. Low dose prenatal-toxic risk estimation based on NOAEL risk factor approach, dose-response. Relationships, and DNA adducts using methylnitrosourea as a model compound. Teratogen Carcinogen Mutagen 13: 101–125

Platzek T, Bochert G, Rahm T (1994a) Embryotoxicity induced by alkylating agents. 8. DNA adduct formation induced by ethylmethanesulfonate in mouse embryos. Teratogen Carcinogen Mutagen 14: 65–73

Platzek T, Schwabe R, Rahm U, Bochert G (1994b) DNA modification induced by 6-mercaptopurine riboside in murine embryos. Chem Biol Interact 93: 59–71

Pottenger LH, Jefcoate CR (1990) Characterization of a novel cytochrome P450 from the transformable cell line, C3H/10T1/2. Carcinogenesis 11: 321–327

Pottenger LH, Christou M, Jefcoate CR (1991) Purification and immunological characterization of a novel cytochrome P450 from C3H/10T1/2 Cells. Arch Biochem Biophys 286: 488–497

Potter DW, Hinson JA (1987) Mechanisms of acetaminophen oxidation to N-acetyl-p-benzoquinone imine by horseradish peroxidase and cytochrome P450. J Biol Chem 262: 966–973

Powell SR, Hall D (1990) Use of salicylate as a probe for HO• formation in isolated ischemic rat hearts. Free Radic Biol Med 9: 133–141

Pryor W (1988) Why is hydroxyl radical the only radical that commonly adds to DNA? Hypothesis: it has a rare combination of high electrophilicity, high thermochemical reactivity and a mode of production that can occur near DNA. Free Radic Biol Med 4: 219–223

Puri PL, Avantaggiati ML, Burgio VL, Chirillo P, Collepardo D, Natoli G, Balsano C, Levrero M (1995) Reactive oxygen intermediates (ROIs) are involved in the intracellular transduction of angiotensin II signal in C2C12 cells. Ann NY Acad Sci 752: 394–405

Rahimtula AG, O'Brien PJ (1977) The role of cytochrome P-450 in the hydroperoxide-catalyzed oxidation of alcohols by rat-liver microsomes. Eur J Biochem 77: 201–208

Raucy JL, Carpenter SJ (1993) The expression of xenobiotic-metabolizing cytochromes P450 in fetal tissues. J Pharmacol Toxicol Methods 29: 121–128

Reed GA (1988) Oxidation of environmental carcinogens by prostaglandin H synthase. Environ Carc Rev C6: 223–259

Reinke LA, Moore DR (1994) Metabolism of ethanol to 1-hydroxyethyl radicals in rat liver microsomes: comparative studies with three spin trapping agents. Free Radic Res 21: 213–222

Richter C, Park JW, Ames BN (1988) Normal oxidative damage to mitochondrial and nuclear DNA is extensive. Proc Natl Acad Sci USA 85: 6465–6467

Roberts JD, Van Houton B, Qu Y, Farrell NP (1989) Interaction of novel (platinum) complexes with DNA. Nucleic Acids Res 17: 9719–9734

Rogan EG, RamaKrishna NVS, Higginbotham S, Cavalieri EL, Jeong H, Jankowiak R, Small GJ (1990) Identification and quantitiation of 7-(benzo[*a*]pyren-6-yl) guanine in the urine and feces of rats treated with benzo[*a*]pyrene. Chem Res Toxicol 3: 441–445

Rowley DL, Pease AJ, Wolf RE Jr (1991) Genetic and physical analysis of the growth rate-dependent regulation of *Escherichia coli ZWF* expression. J Bacteriol 173: 4660–4667

Roy D, Liehr JG (1992) Target organ-specific inactivation of drug metabolizing enzymes in kidney of hamsters treated with estradiol. Mol Cell Biol 110: 31–39

Roy D, Snodgrass WR (1990) Covalent binding of phenytoin and modulation of diphenylhydantoin metabolism by thiols in A/J mouse liver microsomes. J Pharmacol Exp Ther 252: 895–900

Roy D, Floyd RA, Liehr JG (1991) Elevated 8-hydroxyguanosine levels in DNA of diethylstilbestrol-treated Syrian hamsters: covalent DNA damage by free radicals generated by redox cycling of diethylstilbestrol Cancer Res 51: 3882–3885

Roy D, Bernhardt A, Strobel HW, Liehr JG (1992) Catalysis of the oxidation of steroid and stilbene estrogens to estrogen quinone metabolites by the ß-naphthoflavone-inducible cytochrome P-450 IA family. Arch Biochem Biophys 296: 450–456

Roy SK, Kulkarni AP (1991) Lipoxygenase: a new pathway for 2-aminofluorene bioactivation. Cancer Lett 60: 33–39

Samokyazyn VM, Sloane BF, Honn KV, Marnett LJ (1984) Cooxidation of 13-cis-retinoic acid by prostaglandin H synthase. Biochem Biophys Res Commun 124: 430–436

Sancar A (1994) Mechanisms of DNA excision repair. Science 266: 1954–1956

Sanyal MK, Kitchin KT, Dixon RL (1979) Anomalous development of rat embryos cultured in vitro with cyclophosphamide and microsomes. Pharmacologist 21: 231

Sanyal S, Wells PG (1993) Reduction in phenytoin teratogenicity by pretreatment with the antioxidant d-α-tocopherol acetate (vitamin E) in CD-1 mice. Toxicologist 13: 252

Sassanfar M, Dosanjih MK, Essigmann JM, Samson L (1991) Relative efficiencies of the bacterial, yeast, and human DNA methyltransferases of the repair of O^6-methylthymine: suggestive evidence for O^6-methylthymine repair by eukaryotic methyltransferase. J Biol Chem 266: 2767–2771

Sato J (1988) Glutathione S-transferases and hepatocarcinogenesis. Jpn J Cancer Res 79: 556–572

Savas U, Bhattacharyya KK, Christou M, Alexander DL, Jefcoate CR (1994) Mouse cytochrome P-450EF, representative of a new lB subfamily of cytochrome P-450s. J Biol Chem 269: 14905–14911

Scott JA, Homcy CJ, Khaw BA, Rabito CA (1988) Quantitation of intracellular oxidation in a renal epithelial cell line. Free Radic Biol Med 4: 79–83

Shigenaga MK and Ames BN (1991) Assays for 8-hydroxy-2′-deoxyguanosine: a biomarker of in vivo oxidative DNA damage. Free Radic Biol Med 10: 211–216

Shigenaga MK, Park JW, Cundy KC, Gimeno CJ, Ames BN (1990) In vitro oxidative DNA damage: measurement of 8-hydroxy-2′ -deoxyguanosine in DNA and urine by high-performance liquid chromatography with electrochemical detection. Methods Enzymol 186: 521–530

Shuker DEG, Bartsch H (1994) DNA adducts of nitrosoamines. In: Hemminki K, Dipple A, Shuker DEG, Kadlubar FF (eds) DNA adducts: identification and biological significance. IARC Sci Publ 125: 73–89

Shum S, Jensen NM, Nebert DW (1979) The murine aH locus: in utero toxicity and teratogenesis associated with genetic differences in benzo[a]pyrene metabolism. Teratology 20: 365–376

Simic MG, Bergtold DS, Karam LR (1989) Generation of oxy radicals in biosystems. Mutat Res 214: 3–12

Singer B, Antoccia A, Basy AK, Dosanjih MK, Fraenkel CH, Gallagher PE, Kusmierek JT, Qui ZH, Rydberg B (1992) Both purified human 1,N-6-ethano-adenine binding protein and 3-methyladenine-DNA glycosylase act on 1,N-6-ethano-adenine and 3-methyladenine. Proc Natl Acad Sci USA 89: 9386–9390

Smith BJ, Curtis JF, Eling TE (1991) Bioactivation of xenobiotics by prostaglandin H synthase. Chem Biol Interact 79: 245–264

Souliotis VL, Kyrtopoulos SA (1989) A novel, sensitive assay for O^6-methyl- and O^6-ethylguanine in DNA, based on repair by the enzyme O^6-alkylguanine-DNA-alkyltransferase in competition with an oligonucleotide containing O^6-methylguanine. Cancer Res 49: 6997–7001

Souliotis VL, Kaila S, Boussiotis VA, Pangalis GA, Kyrtopoulos SA (1990) Accumulation of O^6-methylguanine in human blood leukocyte DNA during exposure to prokarbazine and its relationship with dose and repair. Cancer Res 50: 2759–2764

Spector A, Kleiman NJ, Huang RC, Wang RR (1989) Repair of H_2O_2-induced DNA damage in bovine lens epithelial cell cultures. Exp Eye Res 49: 685–698

Spitz DR, Dewey WC, Li GC (1987) Hydrogenperoxide or heat shock induces resistance to hydrogen peroxide in Chinese hamster fibroblasts. J Cell Physiol 131: 364–373

Stadtman ER (1990) Metal ion-catalyzed oxidation of proteins: biochemical mechanism and biological consequences. Free Radic Biol Med 9: 315–325

Sugiyama K, Correia MA, Thummel KE, Nagata K, Darbyshire JF, Osawa Y, Gillette JR (1994) pH - dependent one- and two-electron oxidation of 3,5-dicarbethoxy-2,6-dimethyl-4-ethyl-1,4-dihydropyridine catalyzed by horseradish peroxidase. Chem Res Toxicol 7: 633–642

Sutter TR, Tang YM, Hayes CL, Wo Y, Jabs WE, Li X, Yin H, Cody CW, Greenlee WF (1994) Complete cDNA sequence of a human dioxin-inducible mRNA identifies a new gene subfamily of cytochrome P450 that maps to chromosome 2. J Biol Chem 269: 13092–13099

Tang M, Pierce JR, Doisy RP, Nazimiec ME, Macleod MC (1992) Differences and similarities in the repair of two benzo[a]prene diol epoxide isomers induced DNA adducts by Uvr A, Uvr B and Uvr C gene products. Biochemistry 31: 8429–8436

Tangorra A, Curatola G, Bertoli E (1991) Evaluation of antiepileptic drug effect on membrane fluidity. Exp Mol Pathol 55: 180–189

Tchou J, Kasai H, Shibutani S, Chung MH, Laval J, Grollman AP, Nisimura S (1991) 8-Oxoguanine (8-hydroxyguanine) DNA glycosylase and its substrate specificity. Proc Natl Acad Sci USA 88: 4690–4694

Teebor GW, Boorstein RJ, Cadet J (1988) The repairability of oxidative free radical mediated damage to DNA: a review. Int J Radiat Biol 54: 131–150

Terlouw GDC, Namkung MJ, Juchau MR, Bechter RF (1993) In vitro embryotoxicity of N-methyl-N-(7-propoxynaphthalene-2-ethyl) hydroxylamine (QAB): evidence for N-dehydroxylated metabolite as a proximate dysmorphogen. Teratology 48: 431–439

Thomas DC, Husain I, Chaney SG, Panigrahi GB, Walker IG (1991) Sequence effect on incision by (A)BC excinuclease of 4NQO adducts and UV photoproducts. Nucleic Acids Res 19: 365–370

Timbrell JA (1991) Toxic response to foreign compounds. In: Timbrell JA (ed) Principles of biologichemical toxicology, 2nd edn. Taylor and Francis, London, pp 217–218

Turner MJ III, Fields CE, Everman DB (1991) Evidence for superoxide formation during hepatic metabolism of tamoxifen. Biochem Pharmacol 41: 1701–1705

Udassin R, Ariel I, Haskel Y, Kitrossky N, Chevion M (1991) Salicylate as an in vivo free radical trap: studies on ischemic insult to the rat intestine. Free Radic Biol Med 10: 1–6

Ullrich V, Graf H (1984) Prostacyclin and thromboxane synthase as P-450 enzymes. TIPS 5: 352–355

Ubezio P, Civoli F (1994) Flow cytometric detection of hydrogen peroxide production by doxorubicin in cancer cells. Free Radic Biol Med 16: 509–516

Uetrecht J, Zahid N (1988) N-chlorination of phenytoin by myeloperoxidases to a reactive metabolite. Chem Res Toxicol 1: 148:151

Vaca CE, Wilhelm J, Harms-Ringdahl M (1988) Interaction of lipid peroxidation products with DNA. A review. Mutat Res 195: 137–149

Varshney U, Van De Sande JH (1991) Specificities and kinetics of uracil excision from uracil containing DNA oligomers by *Escherichia coli* uracil DNA glycosylase. Biochemistry 30: 4055–4061

Veronese ME, Mackenzie PI, Doecke CJ, McManus ME, Miners JO, Birkett DJ (1991) Tolbutamide and phenytoin hydroxylations by cDNA-expressed human liver cytochrome P4502C9. Biochem Biophys Res Commun 3: 1112–1118

Verri A, Mazzarello P, Spadari S, Focher F (1992) Uracil DNA glycosylases preferentially excise mispaired uracil. Biochem J 287: 1007–1010

Vienneau DS, DeBoni U, Wells PG (1995) Potential genoprotective role for UDP-glucuronosyltransferases in chemical carcinogenesis: initiation of micronuclei by benzo[a]pyrene and benzo[e]pyrene in UDP-glucuronosyltransferase-deficient cultured rat skin fibroblasts. Cancer Res 55: 1045–1051

Visse R, de Ruijter M, Moolenaar GF, van de Putte P (1992) Analysis of UvrABC endonuclease reaction intermediates on cisplatin-damaged DNA using mobility shift gel electrophoresis. J Biol Chem 267: 6736–6742
Voight JM, Van Houten B, Sancar A, Topal MD (1989) Repair of O^6-methylguanine by ABC excinuclease of *Escherichia coli* in vitro. J Biol Chem 264: 5172–5176
von Sonntag C (1987) The chemical basis of radiation biology. Taylor and Francis, London
Wang MY, Lu LJW (1990) Differential effect of gestation stage on benzo[a]pyrene-induced micronucleus formation and/or covalent DNA modifications in mice. Cancer Res 50: 2146–2151
Wells PG (1989) Chemical teratogenesis. In: Kalant H, Roschlau WHE (eds) Principles of medical pharmacology, 5th edn. Decker, Toronto, pp 644–657
Wells PG, Kim PM (1996) UDP-glucuronosyltransferase (UGT)- mediated genoprotection in phenytoin- and 5-(p-hydroxyphenyl)-5-phenylhydantoin (HPPH)-initiated DNA oxidation and micronucleus formation. Fundam Appl Toxicol 30 [Suppl 1, Part 2]: 234
Wells PG, Williams LM (1994) Effect of phenytoin on embryonic soluble thiols and disulfides in murine embryo culture. Toxicologist 14: 164
Wells PG, Winn LM (1996) Biochemical toxicology of chemical teratogenesis. Crit Rev Biochem Mol Biol 31: 1–40
Wells PG, Zubovits JT, Wong ST, Molinari LM, Ali S (1989a) Modulation of phenytoin teratogenicity and embryonic covalent binding by acetylsalicylic acid, caffeic acid and alpha-phenyl-N-t-butylnitrone: implications for bioactivation by prostaglandin synthetase. Toxicol Appl Pharmacol 97: 192–202
Wells PG, Nagai MK, Spano Greco G (1989b) Inhibition of trimethadione and dimethadione teratogenicity by the cyclooxygenase inhibitor acetylsalicylic acid: a unifying hypothesis for the teratologic effects of hydantoin anticonvulsants and structurally related compounds. Toxicol Appl Pharmacol 97: 406–414
Wells PG, Obilo FC, de Morais SMF (1989c) Benzo(a)pyrene embryopathy in rats genetically deficient in bilirubin UDP-glucuronyl transferase. FASEB J 3: A1025
Wells PG, Leeder JS, Winn LM (1995) Phenytoin-initiated protein oxidation in murine embryo culture: a potential molecular mechanism mediating phenytoin teratogenicity. Toxicologist 15: 276
White RE, Coon MJ (1980) Oxygen activation by cytochrome P-450. Annu Rev Biochem 49: 315–356
Whiteside C, Hassan H (1987) Induction and inactivation of catalase and superoxide dismutase of *Escherichia coli* by ozone. Arch Biochem Biophys 257: 464–471
Wilkinson MC, Potter PM, Cawkwell L, Georgiadis P, Patel D, Swann PF, Margison GP (1989) Purification of the Escherichia coli ogt gene product to homogeneity and its rate of action of O^6-methylguanine, O^6-ethylguanine and O^4-methylthymine in dodecadeoxyribonucleotides. Nucleic Acids Res 17: 8475–8484
Winn LM, Wells PG (1994) Benzo[a]pyrene-initiated DNA and protein oxidation in murine embryo culture: a potential molecular mechanism mediating benzo[a]pyrene teratogenicity. Proceedings of the 27th annual symposium of the Society of Toxicology of Canada
Winn LM, Wells PG (1995a) Phenytoin-initiated DNA oxidation in murine embryo culture, and embryo protection by the antioxidative enzymes superoxide dismutase and catalase: evidence for reactive oxygen species mediated DNA oxidation in the molecular mechanism of phenytoin teratogenicity. Mol Pharmacol 48: 112–120
Winn LM, Wells PG (1995b) Free radical-mediated mechanisms of anticonvulsant teratogenicity. Eur J Neurology 2 [Suppl 4]: 5–29
Winn LM, Wells PG (1996a) Enhancement of embryonic activity of superoxide dismutase (SOD) and inhibition of benzo[a]pyrene embryopathy in embryo culture by exogenous addition of SOD. Fundam Appl Toxicol 30 [Suppl 1, Part 2]: 244
Winn LM, Wells PG (1996b) Evidence for embryonic prostaglandin H synthase-catalysed bioactivation and reactive oxygen species-mediated oxidation of macro-

molecules in phenytoin and benzo[a]pyrene teratogenesis. Free Radic Biol Med (in press)

Wong M, Wells PG (1988) Effects of N-acetylcysteine on fetal development and phenytoin teratogenicity in mice. Teratogen Carcinogen Mutagen 8: 65–79

Wong M, Wells PG (1989) Modulation of embryonic glutathione reductase and phenytoin teratogenicity by 1,3-bis(2-chloroethyl)-1-nitrosourea. J Pharmacol Exp Ther 250: 336–342

Wong M, Helston LMJ, Wells PG (1989) Enhancement of murine phenytoin teratogenicity by the gamma-glutamylcysteine synthetase inhibitor buthionine sulfoximine and the glutathione depletor diethylmaleate. Teratology 40: 127–141

Wrighton SA, Stevens JC (1992) The human hepatic cytochromes P450 involved in drug metabolism. Crit Rev Toxicol 22: 1–21

Yang H-YL, Lee QP, Rettie AE, Juchau MR (1994) Functional cytochrome P4503A isoforms in human embryonic tissues: expression during organogenesis. Mol Pharmacol 46: 922–928

Yoneyama M, Ichikawa H (1986) Irreversible in vivo and in vitro binding of thiabendazole to tissue protein of pregnant mice. Food Chem Toxicol 24: 1283–1286

Yoneyama M, Ogata A, Hiraga K (1985) Irreversible in vivo binding of thiabendazole to macromolecules in pregnant mice and its relation to teratogenicity. Food Chem Toxicol 23: 733–736

Yu WK, Wells PG (1995) Evidence for lipoxygenase-catalyzed bioactivation of phenytoin to a teratogenic reactive intermediate: in vitro studies using linoleic acid-dependent soybean lipoxygenase, and in vivo studies using pregnant CD-1 mice. Toxicol Appl Pharmacol 131: 1–12

Zimmerman EF, Potturi RB, Resnick E, Fisher JE (1994) Role of oxygen free radicals in cocaine-induced vascular disruption in mice. Teratology 49: 192–201

CHAPTER 18

Hypoxia and Altered Redox Status in Embryotoxicity

C. Harris

A. Introduction

The developing mammalian conceptus undergoes a number of important biochemical and physiological changes as it expresses its program for growth and differentiation. Many of these transitions occur in concert with, or as a result of, changes in the physical environment of the conceptus. The early organogenesis-stage embryo has been shown to exist in a relatively hypoxic environment in which the predominant metabolic pathways for glucose utilization in energy production are anaerobic (Shepard et al. 1970; Tanimura and Shepard 1970; Freinkel et al. 1984). The significance of these metabolic pathways are discussed in detail in Chap. 8 of this volume. As development continues, an active heartbeat and expansive vasculature carry O_2 to previously diffusion-limited cell populations as these cells begin to convert to predominantly aerobic metabolic pathways. In vitro and in vivo studies have shown that the embryo has little tolerance to changes in its O_2 environment, with embryotoxicity and dysmorphogenesis occurring under conditions of too much as well as too little O_2 (Miki et al. 1988; New and Coppola 1970a,b; Jenkinson et al. 1986; Fantel et al. 1989). Survival of the conceptus during these extremes will likely depend on the ability of the conceptus to protect itself from oxidative damage (as in the case of hyperoxia), to compensate for inadequate energy production (as in the case of hypoxia), to reestablish the correct redox environment for cell function, and to repair damage. Conditions of too much or too little O_2 pose different problems for the developing organism and ostensibly work their deleterious actions via different mechanisms. Regardless of causation or response in these extreme conditions, it is clear that the cellular redox status is affected in either case. How the embryo establishes and controls its redox status, and how it responds to stress conditions are topics which have only recently been investigated, but an understanding of which may help to explain a number of seemingly unrelated mechanisms of embryotoxicity.

The cellular redox state or "status" is defined by the net oxidizing or reducing potential of a number of endogenous biochemical oxidation and reduction couples. The most important of these involves the most powerful of cellular oxidants, O_2, and its conversion to the very poor reductant H_2O. At the opposite end of the spectrum lies the weak oxidant H^+ and its conversion to the strong reductant H_2. The latter reaction is carried out to the greatest extent by bacteria. Intermediate to these extremes are a number of other metabolic cellular components that undergo oxidation–reduction reactions and contribute to the total reducing or oxidizing environment. These other systems include several compounds known to be of importance in the cellular response to drugs, toxic chemicals, and any extreme fluctuations in ambient O_2, temperature, and the physical environment. Important contributors to the maintenance of the redox status include pyridine nucleotides (NAD[P]$^+$ and NAD[P]H), reduced (GSH) and oxidized glutathione (GSSG), ascorbic acid, cytochromes a, c, b, b_2, f, pyruvate/lactate, and oxaloacetate/malate (METZLER 1977; LOACH 1976). Excess O_2 or inhibition of O_2 utilization can result in a net change in the cellular environment to that of a predominantly oxidizing condition. This can have profound deleterious consequences for the cell, because it may affect many of the critical biochemical and physiological processes that are designed to operate under reducing conditions. In addition, direct oxidation of cellular macromolecules may occur, leading to altered genetic and biochemical functions (PHILLIPS et al. 1984; EMERIT et al. 1982; DAVIES 1986). Conditions of relative hypoxia, which are defined as anything less than a normal O_2 tension and are dependent on both the absolute concentration of O_2 and duration of the condition, produce different sets of problems, such as altered energy production and metabolism pathways, and will result in a different subset of cellular responses with their accompanying changes in the redox status. Hypoxia was one of the first teratogenic stimuli to be studied systematically in animal models and was found to produce a broad spectrum of developmental anomalies encompassing the entire developmental spectrum (DARESTE 1877; GRABOWSKI 1970). More recent studies indicate that chemical induction of general or localized hypoxia can occur via different mechanisms following chemical exposure. The factors will be discussed first in terms of the spectrum of developmental effects elicited, followed by recent discoveries demonstrating the relationship between hypoxic damage and responses of the intracellular redox status.

B. Hypoxia

The effects of hypoxia on embryonic development have been investigated for many years, and the work in this area represents some of the first systematic studies ever conducted to elucidate mechanisms of abnormal development. An excellent review of these early studies was provided by GRABOWSKI (1970). Experimentally, the hypoxic state can be achieved by producing a partial vacuum, thus lowering barometric pressure, by replacing the usual O_2 with nitrogen at the normal barometric pressure, or by several other means, including disruption of blood flow and the chemical inhibition of O_2 uptake and utilization. Most of the early studies on the effects of hypoxia on developing organisms utilized the chick as an animal model and made direct comparisons between the two aforementioned methods that were used to produce hypoxia. These comparisons showed that few differences existed in the final outcome (CURLEY and INGALLS 1957). Results of chick and numerous subsequent mammalian studies have now shown that hypoxia produces a very broad spectrum of development effects. Virtually all organs and structures of the developing conceptus can be affected by hypoxia to some degree, including the nervous system, cardiovascular system, major soft organs (liver and kidney), reproductive organs, skeleton, cartilage, and others (GRABOWSKI and PAAR 1958; RÜBSAAMEN 1952). The spectrum and severity of effects can be controlled by the duration, extent, and timing of the hypoxic event. In general, it has been reported that sensitivity of embryos to hypoxia follows a cephalic to caudal gradient within the embryo and that overall sensitivity increases with gestational age (RÜBSAAMEN 1952; BÜCHNER 1955). Because the underlying metabolic processes in the early embryo appear to be anaerobic, it is logical to assume that hypoxia would be less detrimental in these early stages of embryogenesis. Deleterious consequences of hypoxia become progressively more critical as the need for, and availability of, O_2 increases and the activity of anaerobic pathways decline. In the gestational day (GD)-3 chick embryo, complete lethality is observed at relative O_2 concentrations of less than 2%, while teratogenic effects are manifest as a near normal distribution ranging from 2%–16% (GRABOWSKI and PAAR 1958; GRABOWSKI 1961). The tissues of the developing conceptus show a clear difference in terms of sensitivity to the reduction in O_2 levels. Neural tissues and head mesenchyme appear to be most susceptible, while the mesonephros, heart, skin, and selected endodermal derivatives are most resistant (GRABOWSKI 1961, 1970; RÜBSAAMEN 1952; GREENAWAY 1986). It is generally hypothesized that the cells and tissues with the highest metabolic rates and, therefore, the highest demand for O_2 and nutrients are the most susceptible to adverse effects of hypoxia. This theory, however, fails to account for a number of other important variables, including nutrient supply and the ability to maintain an adequate redox status as would be expected for the heart, a resistant tissue, which certainly has a high metabolic demand. The mechanisms by which hypoxia is elicited may also play an important role, because the excessive and nonproductive consumption of intracellular oxygen, perhaps with an increase of reactive oxygen species (ROS), may have more far-reaching effects than a simple lowering of extracellular O_2.

Proposed mechanisms of hypoxia-induced malformations suggest that a "critical period" exists for embryonic cell populations during which rapidly proliferating and respiring cells are most at risk for consequences related to reductions of available O_2. The populations of cells at risk change during development and result in a wide spectrum of affected cells and tissues. Reductions in energy production and resultant "direct" hypoxic effects should affect only those rapidly respiring cells with a disproportionately high O_2 demand.

One of the major effects of hypoxia on cells involves the alterations in mitochondrial function and disruption of the protomotive force (p) as has been reviewed for aerobic cells by D.P. JONES et al. (1990). The force is obtained from the normal mitochondrial functions of substrate oxidation via Krebs cycle dehydrogenases and processing of fuel substrates through the electron transport chain. The energy derived from this activity is utilized to make adenosine triphosphate (ATP) from adenosine diphosphate (ADP) and inorganic phosphorus. The ATP thus formed is utilized to drive a number of critical cellular processes, including the supply and regulation of metabolic precursors and ions, maintenance of membrane integrity, regulation of specific transport functions, synthesis of critical micro- and macromolecules (including many which are essential for the maintenance of intracellular redox status), removal of waste products, and maintenance of osmotic balances through control of H^+ and K^+ exchange. It has been demonstrated that some cells that are less susceptible to hypoxia have the capacity to inhibit the flux of ions and, therefore, protect against large changes in pH and the loss of normal osmotic balance (AW et al. 1987a,b). Chemical agents that inhibit specific targets in the mitochondrial electron transport chain (such as cyanide) have different overall effects than those seen with hypoxia alone. When administered under normoxic conditions, cyanide produces mitochondrial swelling, loading of phosphate, and loss of mitochondrial membrane potential much faster than occurs under conditions of hypoxia alone (D.P. JONES et al. 1990). The protective changes appear to occur in response to oxygen depletion per se and not the selective inhibition of ATP synthesis. Although the stabilizing mechanisms that occur during hypoxia may protect cells from direct damage, the reduced cellular function and indirect effects may predispose cells to effects related to other factors such as chemical exposure and extreme environmental conditions. Several chemical oxidants which are known to inhibit Na^+, K^+ ATPases, and Ca^+ ATPases can have enhanced toxic and killing effects due to the changes elicited by hypoxia (D.P. JONES 1985; D.P. JONES et al. 1990). In addition, the ATP-dependent process of glutathione synthesis has been shown to be reduced under conditions of hypoxia, and the inability to respond to chemical insult by replenishing GSH oxidized or lost in detoxication processes can be an important consideration (SHAN et al. 1989).

I. Hypoxia as a Cause of Birth Defects

By the late 1940s, teratology experiments were being carried out in earnest in a number of laboratories worldwide. Maintenance of pregnant animals in anoxic environments that simulated conditions of high altitude was one of the earlier observations known to result in congenital malformations. It was recognized that anoxia produced a spectrum of effects, ranging from death to various degrees of malformation, related to the developmental timing of the anoxic episode. INGALLS et al. (1950) showed that anencephaly (cranioschisis) was produced in mice by a single anoxic exposure around the eighth or ninth day of gestation; cleft palate was produced on GD 14 and 15 and "open eyes" on GD 15–17. Later experiments in the same laboratory reported a number of additional malformations, including irregularities and fusions of the ribs and vertebrae, cryptorchidism, and microagnathia (INGALLS et al. 1952). Both the duration of anoxia and the relative height of simulated altitude (25 000–30 000 feet for 1/4–5 h) were found to affect the severity of malformations in a cephalocaudal sequence (INGALLS and CURLEY 1957), results which were later confirmed by MURAKAMI and KAMEYAMA (1963). The cause of malformations was proposed to be due to the direct lowering of maternal arterial O_2, resulting in reduction of embryonic or fetal O_2 tensions which were critical for the rapidly growing and differentiating conceptus. Of interest, however, was the anecdotal suggestion that a secondary consequence of relative hypoxia may be the lowering of conceptal temperature, which could easily result from a compromise of the energy production pathways. A stay of several hours in the anoxia chamber of Ingalls was reported to result in a lowering of maternal body temperature by 10 °F (INGALLS et al. 1952). The implications of maternal thermoregulation have been discussed in terms of the general impact of these changes in their effect on acute toxicological studies (WATKINSON and GORDON 1993).

II. Vascular Clamping – Experimentation

Some possible mechanisms of hypoxia-related embryonic death and malformations could be inferred from in vivo experiments such as those described above, but little evidence was found to prove that hypoxia per se was directly responsible for the alterations in normal development. One of the first attempts to provide the type of experimental rigor necessary to understand the etiology of malformations was provided in a rat model developed by BRENT and FRANKLIN (1960). Laprotomized GD-9 rats provided visual access to the intact uterus and associated vasculature. One horn of the uterus was clamped at the cervical and ovarian ends and across the uterine mesentery for time periods from ½ to 3h to provide complete isolation of the maternal circulation (Fig. 1; Table 1). The opposite horn of the uterus exposed in the same way was used as a positive control. After the prescribed period of anoxia, the incision was closed and the pregnancy was continued almost to term (day 21), at which time the animal was killed and fetuses removed for assessment. Mortality and malformations increased with increasing duration of the anoxic episode and also changed with advancing gestational age, as seen in in vivo experiments involving whole animal exposure to altitude (INGALLS et al. 1952). Again, a cephalocaudal gradient of effects was seen. Based on these initial experimental manipulations, additional hypotheses were introduced to explain the spectrum of teratogenic effects seen (mortality/resorption, anophthalmia, micropthalmia, renal aplasia, renal agenesis, uterine agenesis, pancake adrenal, anencephaly, omphalocele, absent external ear, hydrocephalus, and missing fourth aortic arch). These included suggestions of an inability to produce anabolic and metabolic products necessary for embryogenesis and an excess of catabolic products unable to communicate with maternal circulation (BRENT

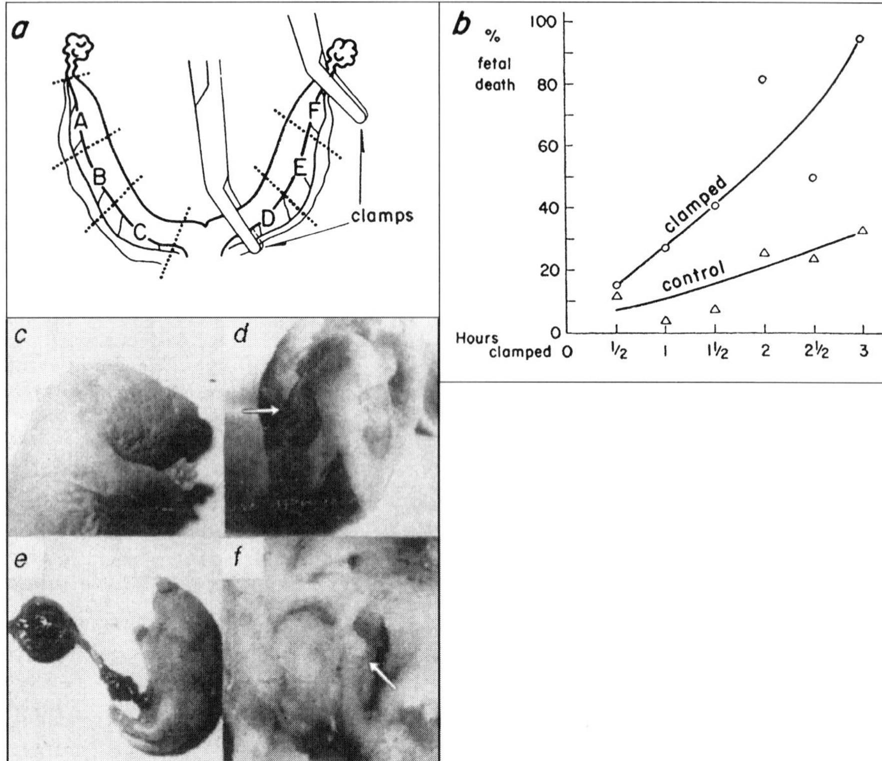

Fig. 1a–f. Uterine clamping. **a** Method of clamping the uterus. Radioactivity (in counts/min per mg) of clamped and unclamped horns: *A*, 37.774; *B*, 52.881; *C*, 53.829; *D*, 0.100; *E*, 0.000; *F*, 0.278. **b** Percentage mortality among experimental and control fetuses after various periods of clamping. The number of pregnant rats in each case was five. **c** Anopthalmia and absent pinna in animal from litter of a rat clamped for 3 h. **d** Anopthalmia on the *left* (*arrow*) with normal eye on the *right* in an animal from the 2-h experimental group. Skin over the orbit has been removed. **e** Anencephaly and omphalocele in a fetus from the 2-h experimental group. **f** Dissection of a fetus from the litter of a rat clamped for 1½ h, showing aplastic left kidney (*arrow*) and corresponding changes in the left adrenal gland. (From BRENT and FRANKLIN 1960, with permission)

and FRANKLIN 1960). Now, over 20 years later the proximal causative agent responsible for the malformation has still not been conclusively identified. Direct effects of hypoxia, changes in pH, reductions of energy production, excessive carbon dioxide concentrations, or other physiological disturbances alone or in concert might elicit the observed malformations. Changes in fetal pathology resulting from uterine vascular clamping (30–90 min) was evaluated in GD-14 pregnant rats and determined to involve a progression of effects that included edema of the mesenchymal tissues in the limbplates (occurring 3 h

Table 1. Fetal weights of clamped and control animals (Brent and Franklin 1960)

Time (h)	Fetuses (*n*)		Weight (g) ± S.D.[a]
	Day 9	Day 21	
Controls			
½	25	22	4.54 ± 0.55
1	21	20	4.91 ± 0.36
1½	24	22	4.90 ± 0.61
2	27	20	4.91 ± 0.32
2½	25	19	4.90 ± 0.77
3	18	13	4.50 ± 0.86
Clamped animals			
½	33	28	4.42 ± 0.58
1	29	21	4.57 ± 0.46
1½	39	23	4.50 ± 0.63
2	35	6	3.79 ± 0.99
2½	34	17	4.20 ± 0.64
3	21	1	–

The control and clamped groups are the same as in Fig. 1b. Clamping occurred on day 9 for the times specified. Fetuses were evaluated on day 21 (which is usually term in the rat).
[a]On day 21, surviving fetuses only.

after unclamping), which was accompanied by a marked dilation of the fetal vessels and finally necrosis. Observed limb malformations following hypoxia during this period of development were consistent with necrotic changes that also occurred in surrounding tissues, including the developing bone, and are most likely responsible for the limb defects. Large blisters and hematomas were also observed in the region of the snout and were suggested to be partly responsible for the inability of the palate to close (Leist and Grauwiler 1974). More definitive studies by Webster et al. (1987) addressed concerns of other studies that also showed an increase in fetal hemorrhage in the unclamped (control) uterine horn and that general trauma to the uterus could be introduced by handling for 5 min, clamping of the uterine wall or uterine fat (not containing major blood vessels), or stretching the uterine blood vessels on GD 14–16. These manipulations also resulted in fetal hemorrhage. The authors speculate that any stress that results in a generalized or local sympathetic nervous system discharge, such as shock, will produce compromised perfusion and hypoxia in broader areas of the reproductive tract and not just the isolated uterine horn. It has been suggested that stresses other than hypoxia induced by direct vascular clamping can also cause uterine blood vessels to contract to the point of causing developmental disruptions of the distal extremities (Webster et al. 1987). Uterine trauma may be the key event in producing the broadest spectrum of deleterious effects. Millicovsky and DeSesso (1980) reported that in vivo uterine clamping in the rabbit resulted in severe cardiovascular derangements in the embryo within 3–4 min, but that these effects were not seen if the umbilical cords were also clamped. Such results indicate that factors in addition to acute bouts of hypoxia are necessary to elicit embryonic damage and may be mediated by uterine ischemia. In this experimental model, the laprotomy may serve as a priming event to contribute to subsequent hypoxia-related manifestations. The deleterious effects caused by these manipulations of the uterus are very similar to the types of alterations seen in whole embryo culture following bouts of hypoxia.

III. Edema Syndrome

The most common effects of hypoxia in terms of elicited malformations may, however, be related to indirect effects of the hypoxic condition. The collective indirect effects of hypoxia have been summarized in description of the edema syndrome as outlined in detail by Grabowski (1970).

In addition to the obvious direct effects of hypoxia on cells of the developing conceptus that could interfere with functional, differentiative, or proliferative capacities, several effects of hypoxia appear to have a much more indirect origin. Acute hypoxia has been shown to result in direct effects, with the common result being cell death, which occurs in selective regions of susceptibility in the conceptus. Moderate hypoxia for longer periods of time produces considerably less cell death and an embryo of normal appearance, but despite the lack of immediate effects, these embryos will still develop malformations subsequently. A common malformation and alteration can be characterized by a dramatic increase in internal and external hemorrhage, accompanied by the formation of diffuse hematomas in the head mesenchyme and areas of the neural epithelium. Hemorrhage is believed to be responsible for most of the embryolethal effects of hypoxia. Other manifestations of the syndrome include the production of clear, fluid-filled blisters in subcutaneous areas, which also may be closely associated with some of the dysmorphogenic effects (Grabowski 1964, 1970). Disruption of ion regulation and osmotic control at the cellular level results in the large-scale accumulation of fluids in the conceptus. Conspicuous are the enlargement of the heart and enlargement of embryonic blood vessels. Because extraembryonic blood vessels in the chick expand very little, most of the changes associated with hypoxia occur primarily in the embryo proper. The extent of hypervolemia is illustrated by the 40%–60% increase in plasma volume measured in the GD-3 and GD-5 chick embryo, which has been suggested to result in rupture of the vessel, causing the hemorrhage (Grabowski 1964; Grabowski and Schroeder 1968). As a result of these changes, blood pressure increases (more than fourfold) and discrete changes occur in the flexure radius of the embryo. Decreases in plasma glucose, sodium, and chloride are accompanied by increases in lactic acid, free amino acids, and potassium (Grabowski 1966). Several chemical agents which have been shown to elicit some or all of the characteristics of the edema syndrome in chicks and mammalian embryos are discussed below. It is possible that consequences of excessive redox cycling (nonrespiratory O_2 consumption) or the chemical inhibition of ion regulation is responsible for the identical ultimate effects. In either of these latter two cases, it is the ability to maintain normal cellular redox status, including the GSH and pyridine nucleotide redox cycles, that is critical for determining whether the regulatory pathways operate normally or result in the misregulation of osmotic balance (D.P. JONES et al. 1990; TRIBBLE and JONES 1990).

IV. Chemicals and Response in Hypoxia

Several examples of chemicals producing the characteristic manifestations of the edema syndrome can be found in the literature, although it has not been shown whether localized or generalized hypoxia has been induced and is, in fact, responsible for the teratogenic or dysmorphogenic effects. Recent experiments conducted in the rat whole embryo culture system have shown that the widely used pesticide lindane (γ-hexachlorocyclohexane) produces a pronounced dilation of cephalic vasculature, accompanied by large, clear blisters around the prosencephalon and marked edema as part of its spectrum of abnormalities (Fig. 2; MCNUTT and HARRIS 1994). This result is of particular interest with regard to a discussion of hypoxia because of the known pathways of lindane metabolism. Controlled evaluations of lindane metabolism under conditions of changing ambient gases showed that a reduction in O_2 results in the shift of lindane metabolism in a direction favoring the increased production of potentially damaging reactive intermediates (YAMAMOTO et al. 1983). Based on these results, we postulated that the GD-10 rat embryo may be more sensitive to lindane toxicity due to hypoxia resulting from the relatively long diffusion distance from the maternal O_2 supply that exists prior to the onset of an active vitelline circulation. The tissues of the embryo proper, being in a more hypoxic environment, would, therefore, be expected to generate higher concentrations of reactive intermediates and elicit greater damage. Indirect measurements of reactive metabolite production involving the cellular redox status have been demonstrated by a selective depletion of GSH and cysteine in the embryo, but not by the visceral yolk sac, as is the usual case for chemical exposures in vitro (MCNUTT and HARRIS 1995). Other examples of teratogenic/dysmorphogenic chemicals eliciting edema syndrome-like effects include valproic acid, aminocarb, cocaine, and phenytoin, some examples of which are discussed in more detail below. As with the edema syndrome itself, chemical exposures may result in the production of a broad spectrum of lesions depending on the developmental stage, dose, and duration of exposure.

In recent years, the pioneering techniques of BRENT and FRANKLIN (1960) and other in vitro and in vivo approaches have been employed and modified in a number of different contexts to probe more specific aspects of the effects of hypoxia, especially those involving exposure to known embryotoxic or teratogenic chemical agents.

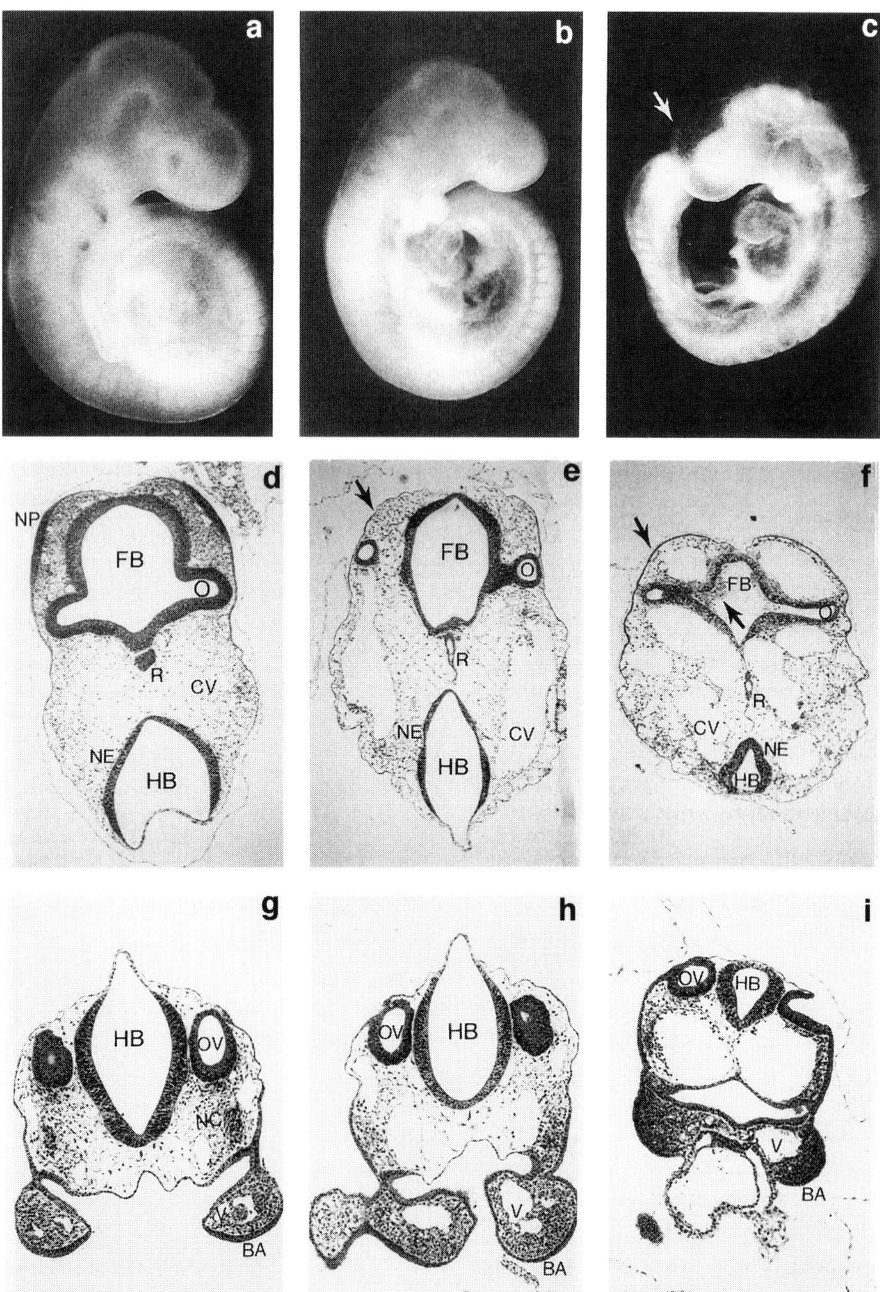
a
b
c
d
NP
FB
O
R
CV
NE
HB
e
FB
O
R
NE
HB
CV
f
FB
R
CV
NE
HB
g
HB
OV
NC
V
BA
h
OV
HB
V
BA
i
OV
HB
V
BA

1. Smoking and Nicotine

Human epidemiological data suggested that mothers who smoked were predisposed to delivering prematurely and that the offspring were prone to respiratory distress (HERON 1962). Injection of nicotine into pregnant guinea pigs also caused a blanching and constriction of the uterine vasculature, indicating a possible role of reduced oxygenation or nutrient delivery as a result of insufficient perfusion (HOWREN 1965). Chronic exposure of maternal rats to daily doses of nicotine (3.0 mg/kg, i.p.) or bouts of hypoxic stress throughout the gestational period each produced smaller offspring at birth. In addition, nicotine-exposed offspring were born later and resulted in fewer live surviving pups (BECKER and MARTIN 1971). Recent studies by MITCHELL and HAMMER (1985) reported that exposure to nicotine prior to implantation resulted in the significant reduction of embryonic cell number and rates of cell proliferation. These changes were accompanied by significant and sustained reductions in oviduct blood flow. Current evidence does not, however, support the role of nicotine as a major teratogen producing persistent morphological abnormalities via chemically induced hypoxia, but does suggest that it may constitute an important risk factor in mediating other hypoxia-induced developmental lesions. In this regard, the relationship between smoking, repeated nicotine-induced ischemia-reperfusion injury of the heart, and an increased incidence of sudden infant death syndrome (SIDS) may require a reevaluation of the effects of maternal tobacco use during pregnancy (TOLSON et al. 1995; SLOTKIN et al. 1995).

Fig. 2a–i. Effects of lindane on rat conceptuses in whole embryo culture. **a–c** Photomicrographs of representative gestational day (GD)-11 whole rat conceptuses cultured under in vitro conditions for 26 h in **a** control, **b** 100 μ*M* lindane, or **c** 200 μ*M* lindane added directly to the culture medium. Note the dose-related abnormal axial rotation and cephalic edema (*arrow*). The extraembryonic membranes have been removed for photography. × 20. **d–i** Light photomicrographs of transverse histological sections of **d–f** the cephalic and **g–i** the branchial arch regions showing selective morphologic defects in GD-11 embryos exposed to lindane for 26 h in vitro. **d,g** Control. **e,h** 100 μ*M* lindane. **f,i** 200 μ*M* lindane. Embryos were embedded in glycol methacrylate and stained with hematoxylin and eosin. **d–f** Rathke's pouch was used as a landmark to orient sections. Controls illustrate normally developing structures. *FB,* forebrain; *HB,* hindbrain; *O,* optic vesicles; *NP,* nasal placodes; *OV,* otic vesicles; *CV,* anterior cardinal vein; *NE,* neuroepithelium. Lindane-exposed embryos display a compressed forebrain and hindbrain region, distended anterior cardinal veins, absence of nasal placodes (*arrows*), and less densely packed mesoderm. Note the blebbing of cellular material into the ventricular lumen in 200 μ*M* lindane-treated embryos (*arrow* in **f**) × 60. **g–i** Sections through the first brachial arch illustrate the abnormal brachial arch (*BA*) development, fewer migrating neural crest cells (*NC*), and distended vessels (*V*) in the arch of lindane-exposed embryos (**h,i**) compared to control (**g**). × 95. (From MCNUTT and HARRIS 1994, with permission)

2. Cocaine

Dramatic increases in the use of cocaine in women of childbearing age has led to a number of case studies reporting an increase in miscarriages, placental abruption, and prematurity. In order to evaluate the possible mechanistic implications of these observations, the effects of cocaine on uterine blood flow and fetal oxygenation were studied in pregnant ewes (WOODS et al. 1987; MOORE et al. 1986). Cocaine administered intravenously as bolus doses produced dose-dependent increases in maternal blood pressure and concomitant decreases in uterine blood flow. The single-dose exposure resulted in significant decreases in uterine blow flow for durations in excess of 15 min. The decreased uterine blood flow resulted in fetal hypoxemia, hypertension, and tachycardia which were more severe than what was seen following direct administration of the drug to the fetus. It was suggested that either the direct effects of cocaine, release of fetal catecholamines, or a combination of the two may be responsible for the vasoconstrictive effects. Maternal catecholamine determinations showed that maternal plasma norepinephrine levels rose by 210% following drug administration (MOORE et al. 1986). Subsequent reports have now suggested that additional consequences of cocaine exposure may include low birth weight and gross anatomical malformations in the utero-exposed offspring, including ileal atresia, hypospadias, and limb malformations. Limb malformations were asymmetrical and consisted of a missing third and fourth digit of the left hand (Fig. 3; CHASNOFF et al. 1988). This malformation was very unusual but of considerable interest, because similar defects were seen in uterine stress and vascular clamping experiments conducted in the rat (WEBSTER et al. 1987). A teratogenic dose of cocaine (50 mg/kg) administered to fetuses on GD 14–16 resulted in severe hemorrhage and edema in their extremities (footplates, tail, genital tubercle, and upperlip/nose) when evaluated after 48h. Later examination on the GD 21 showed the only observable malformation to be reduction defects of the limbs and tail. As with previous discussions of the effects of hypoxia in the late organogenic and postorganogenic conceptus, teratogenic effects of cocaine were initiated

Fig. 3. Examples of the effects of cocaine as a cause of congenital malformations. **1** Left hand of an infant born to a mother described as a heavy cocaine user (CHASNOFF et al. 1988). Note the absence of the third and fourth digits. (Courtesy of Dr. I. Chasnoff, Department of Pediatrics, Northwestern Memorial Hospital, Chicago, Illinois 60611). **2** Rat fetus from a dam given a single dose of cocaine (60 mg/kg) 48 h earlier. Note the severe hemorrhage affecting the footplates and tail and the large fluid-filled blister on the nose and upper lip. **3** Palmer surface of the forelimb of a rat fetus from a dam given cocaine (60 mg/kg) 48 h earlier on day 16. Note the large blood-filled blister, affecting the second, third, and fourth digits. **4** Genital tubercle of an 18-day-old rat fetus from a dam given cocaine (60 mg/kg) 48 h earlier. There is a large hemorrhage in the tubercle. **5** Left and right forelimbs of a 5-week-old rat from a dam given cocaine (60 mg/kg) on day 16 of gestation. Note the reduction of second and third digits of the right limb. **6** Right and left hindlimbs of a 5-week-old rat from a dam given cocaine (60 mg/kg) on day 16 of gestation. Note the reduction of the second to fourth digits of the left limb. (From WEBSTER and BROWN-WOODMAN 1990, with permission)

through a progressive sequence of hemorrhage, vasodialation, edema, and necrosis (WEBSTER et al. 1990). The asymmetric nature of cocaine-induced terata prompted speculation that cellular mechanisms may be related to the effects of other chemicals which have been shown to elicit asymmetric defects in vitro. Using the rat whole embryo culture system, FANTEL et al. (1990)

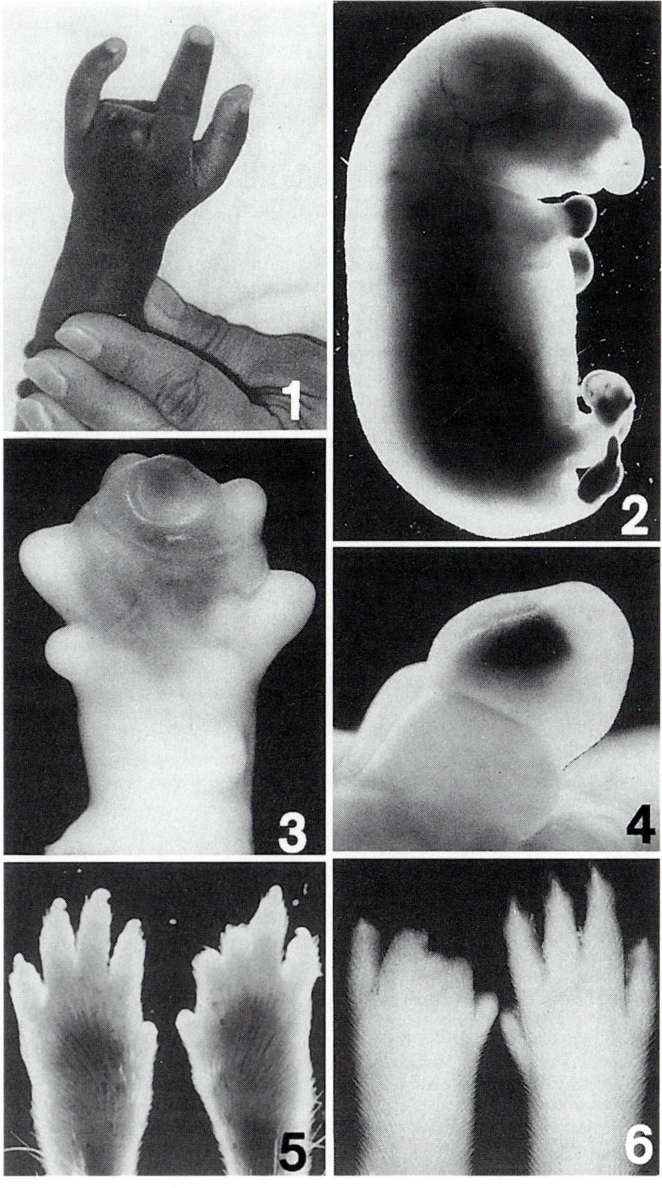

investigated the direct effects of cocaine exposure in midgestation rat conceptuses. Under these conditions, moderate hypoxia (10%–12% O_2), in conjunction with cocaine exposure, resulted in significant reduction of growth and development, including reductions in the diameter of vitelline arteries. In addition, moderate hypoxia and cocaine resulted in the production of axially asymmetric defects similar to those reported previously with severe hypoxia or reactive chemicals such as niridazole (Greenaway et al. 1985). The mechanism of cocaine action was to involve the selective inhibition of the mitochondrial terminal electron transport system (Fantel et al. 1990). Reduced O_2 in the conceptus is also believed to result in the increased uptake of glucose to meet resultant increased energy demands, which, in this case, are potentiated by the concomitant inhibition of mitochondrial function. It is also possible to speculate that increasing sensitivity of hypoxia with advancing gestational age may be related to the changing pattern of energy production pathways, as anaerobic energy pathways have been found to decrease with advancing gestational age.

3. Niridazole and Related Nitroheterocyclic Agents

In addition to cocaine, a number of other chemical agents have been shown to be affected by changes in O_2 concentration. One of the more interesting and most thoroughly studied is the in vitro dysmorphogen niridazole. This nitroheterocyclic, antihistosomal agent has been shown to elicit an unusual asymmetric malformation of the embryonic forebrain region (Greenaway et al. 1986) when the conceptus is exposed in the presence of lower than usual O_2 concentrations (5% O_2, 5% CO_2, 90% N_2). The ability to elicit this defect was directly related to the compound's relatively high redox potential and its proposed ability to deplete intracellular O_2 through redox cycling (Fantel et al. 1989; Barber and Fantel 1993). Selective localized hypoxia produced by excessive utilization of O_2 via redox cycling has, therefore, been shown to be directly related to the production of this unusual defect, one which can also be elicited by simply lowering ambient O_2 tensions. Demonstration of the role of hypoxia, created as a consequence of xenobiotic metabolism or due to reductions in available O_2, still does not explain the basis for asymmetry. If related to redox cycling and the ability to deplete O_2 by this means alone, it would not be predicted that a greater relative hypoxia would have the same effect, nor that excess O_2 would result in an asymmetrical necrotic damage on the opposite (left) side. The logical suggestion of a functional difference was investigated by Fantel et al. (1991) by removing the hearts and by separation of the right and left sides of the GD-11 embryo by hemi-microdissection along the midsagittal axis and assaying individual fractions for NADH oxidase activity. This approach allowed comparison of mitochondrial maturity based on relative O_2 utilization e.g., the relative activity of the mitochondrial electron transport system (Fantel et al. 1991). Results showed that a precocious development of mitochondrial function occurs on the embryo's left side, causing relative deficiencies in the ability of the cells on the right side to meet energy demands, resulting in a loss of cell integrity and leading to necrosis. Conversely, the necrosis produced on the left side under conditions of hyperoxia is believed to result from superoxide leakage from the functionally mature mitochondria.

4. Phenytoin, Vasodilators, and Vasoconstrictors

Phenytoin and nifedipine, as well as other vasodialating agents including nitrendipine, felodipine, and hydralazine, have been shown to produce nearly identical phalangeal defects in rabbits (Danielsson et al. 1990). As in other in vitro and in vivo studies of hypoxia, these distal malformations were preceded by edema, hemorrhage, and necrosis in the cartilage of the developing digits. Whether hypoxia was induced by vascular disruption, as with nifedipine, or by vascular disruption in conjunction with excessive O_2 consumption and the production of relative hypoxia due to the biotransformation and possible ROS generation following phenytoin exposure was not established. Early descriptions of phenytoin metabolism to an epoxide would not lead one to expect sufficient O_2 consumption to support this hypothesis (Martz et al. 1977). Recent investigations, however, have shown that a more likely and more extensive metabolism of phenytoin to free radicals occurs via peroxidative pathways (Kubow and Wells 1989). These forms of reactive metabolite generation could easily lead to redox cycling, ROS production, and significantly higher rates of O_2 consumption. Liu and Wells (1993, 1994) have now demonstrated that a significant oxidation of macromolecules occurs during phenytoin embryotoxicity and teratogenicity to support this notion. Millicovsky and Johnston (1981) have reported that the deleterious developmental effects produced in mice by phenytoin can be greatly diminished by maintaining the pregnant animals under conditions of hyperbaric O_2. It can be speculated that the reduced ambient O_2 results in the reduction of phenytoin-induced oxidative damage to critical macromolecules. The means by which phenytoin may directly disrupt vascular function has yet to be clearly established.

Intraperitoneal injection of epinephrine or vasopressin produce bradycardia in fetal hearts during the last 6 days of gestation in the rat and are associated with decreased fetal blood pressure. Direct injection of the vasoconstrictors into fetal pericardial sacs produced the opposite effects of increased fetal blood pressure, bradycardia, elevated K^+, and decreased Na^+, characteristic of experimentally induced hypoxia (Chernoff and Grabowski 1971).

C. Hypoxia and Redox Status

A number of investigations conducted in vivo and in vitro have demonstrated the production of embryolethality, terata, dysmorphogenesis, and overt embryotoxicity under conditions of hypoxia (FANTEL et al. 1991; INGALLS et al. 1952; MURAKAMI and KAMEYAMA 1963) and hyperoxia (FERM 1964; MIKI et al. 1988; MORRISS and NEW 1979), although the mechanisms of these outcomes have not yet been elucidated. Light and electron microscopic studies show distinct histological and ultrastructural differences between embryos exposed to too little or too much O_2. Under hypoxic conditions, the most prominent alteration was the swelling and distortion of mitochondria and the absence of similar changes in other organelles (MIKI et al. 1988). Too much O_2 under the same whole embryo culture conditions, produced an increase in phagolysosomes in the cytoplasm in the absence of morphological alterations of the nucleus or mitochondria. Little information is currently available to describe the biochemical and physiological disturbances that are elicited under these conditions or their specific mechanisms of action.

A major challenge encountered by cells of the developing conceptus involves the ability of the organism to maintain and regulate adequate amounts of intracellular reducing equivalents, particularly in the form of GSH and the pyridine nucleotides NADPH and NAD^+. The former is normally the most abundant reducing equivalent responsible for maintaining the overall reducing environment, while the latter are involved in numerous homeostatic pathways related to O_2 utilization, mitochondrial function, energy production, biosynthesis, and regulation of GSH status. Drastic changes in the redox condition can result from fluctuation in the O_2 environment. When O_2 is in excess, metabolic generation of free radicals or reactive chemical intermediates can result in the collective condition of "oxidative stress" (SIES 1985). This occurs when the production of intracellular oxygen free radicals exceeds the ability of the cells' antioxidant enzyme capacities to remove the oxidants and restore homeostasis. The condition can occur as a result of hyperoxia or due to the decrease in antioxidant defences or repair mechanisms whereby normal levels of oxidants are no longer removed and thus accumulate to elicit cellular toxicity. Glutathione depletion and/or oxidation has been shown to create this type of stress, leading to a prooxidizing environment and an increase in oxidative damage (KETTERER et al. 1983; KOSOWER et al. 1969; SIES 1985). The $O_2^{\dot{-}}$ derived radicals known to be responsible for cellular damage are generated through a univalent reduction of O_2 to produce the superoxide anion radical $O_2^{\dot{-}}$ or a divalent reduction to produce H_2O_2. If adequate activities and cofactors for the antioxidant enzymes superoxide dismutase (SOD), catalase, and glutathione peroxidase (GSH-Px) are available, $O_2^{\dot{-}}$ and H_2O_2 are detoxified to H_2O and O_2. The latter reaction (GSH-Px) depends on the availability of GSH, which becomes oxidized to GSSG in the catalytic process. When GSH is lacking, excess O_2 radical production ensues (SIES 1985; T.W. JONES et al. 1986; D.P. JONES 1985).

The reductive capacity of cells can provide a buffer against oxidative challenges or, in some cases, facilitate the enhanced toxicity of some chemicals through the initiation of a process known as "redox cycling," especially under conditions of relative hypoxia (KAPPUS 1986; JUCHAU et al. 1986). Chemicals that are capable of undergoing single electron reductions to form free radical intermediates are considered to be likely redox cycling agents. Depending on their redox potentials, these chemicals can accept a single electron from endogenous reducing agents (NADPH, NADH, GSH, ascorbate, reduced flavoproteins, and ferredoxin) to form the free radical. If the strong oxidant (O_2) is present, the reduced chemical (free radical) is reoxidized to the parent compound and the O_2 is converted to superoxide anion radical ($O_2^{\dot{-}}$). This reaction can result in the rapid utilization of cellular O_2, which in turn potentiates the already detrimental cellular hypoxic condition. In the absence of O_2, or under hypoxic conditions, the free radical can interact directly with GSH to restore the parent compound at the expense of generating glutathione disulfide (GSSG). The GSSG, thus formed, is converted back to two molecules of GSH by the action of the NADPH-dependent enzyme GSSG -reductase (GSSG-Rd). Under conditions of relative hypoxia and in the presence of adequate GSH and NADPH, the reaction cycle is "futile" due to the regeneration capacities of the respective redox cycles. The combined effects of hypoxia result in the cell becoming a more reducing environment, thus increasing the likelihood of additional redox cycling (D.P. JONES et al. 1990). Chemicals known to undergo redox cycling include the nitrogen heterocycles (niridazole), adriamycin, mitomyocin C, and others for which the teratogenicity of some can be potentiated by further lowering intracellular O_2 tension (GREENAWAY et al. 1985; BARBER and FANTEL 1993). Some of the important interrelationships between redox cycling, GSH, and pyridine nucleotide redox couples are shown schematically in Fig. 4.

I. Glutathione and Related Low-Molecular-Weight Thiols

One of the most important intracellular redox cycles in the developing conceptus involves the tripeptide intracellular protectant glutathione (GSH). In vitro studies have shown that GSH appears to be the major intracellular thiol in both the preimplantation and postimplantation rodent conceptus and, as has been shown for other mature cells and tissues, is responsible for maintaining the cellular reducing potential. In the preimplantation embryo, GARDINER and REED (1994) have shown that GSH levels are relatively high in the unfertilized oocyte, but that they decrease by approximately ten fold up to the blastocyst stage. Other dramatic fluctuations of GSH content and redox status occur throughout early embryogenesis. Many of the changes may be related to specific developmental events which dictate, or are dictated by, changes in intracellular redox status. The increased formation of GSSG and protein–GSH mixed disulfides indicate that controlled oxidative stress may be responsible for some of the regulatory events in embryogenesis.

In the postimplantation conceptus undergoing organogenesis (GD 10–11), concentrations of GSH have been calculated to be in excess of 30 nmol/mg protein in the visceral yolk sac and 20 nmol/mg protein in the embryo proper, which are equal to concentrations published for GSH content in the liver when

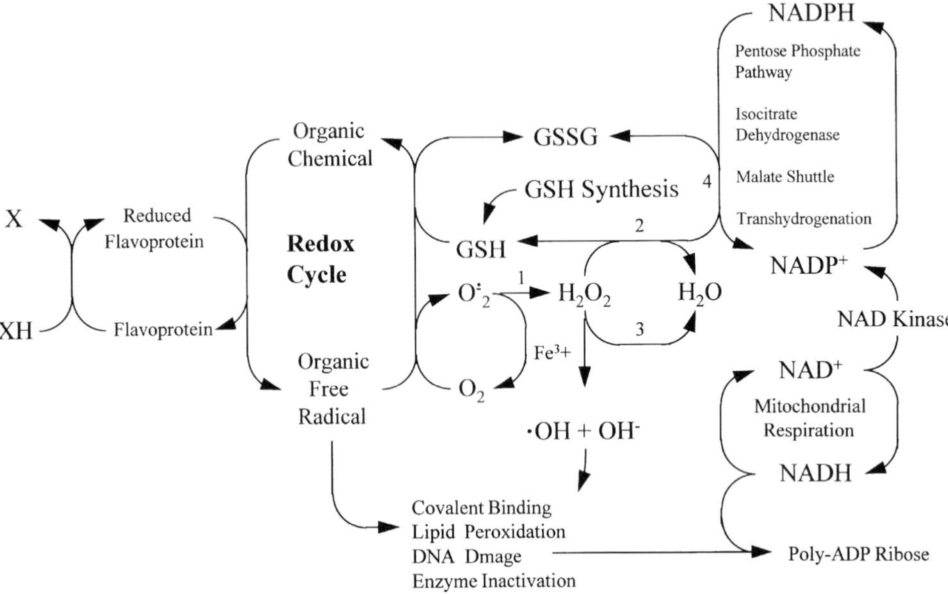

Fig. 4. Interrelationship between redox cycling, glutathione, and pyridine nucleotide redox status. Key enzymes in the maintenance of cellular homeostasis through their antioxidant functions include: *1*, superoxide dismutase; *2*, glutathione peroxidase; *3*, catalase; *4*, glutathione disulfide reductase. *ADP*, adenosine diphosphate; *NAD*, nucleotide adenine diphosphate; *NADP*, NAD phosphate; *NADPH*, NADP, reduced form. Sources of additional NADPH are listed, as well as the possibility of forming additional $NADP^+$ through the activity of NAD kinases

also expressed in units per mg protein (Harris et al. 1986). Intracellular GSH is found in the organogenesis-stage rat conceptus predominantly in the reduced form, although the ratio of reduced to oxidized glutathione (2:1) appears to be considerably lower than in most mature organs such as liver and kidney (GSH to GSSG, > 100:1). Under normal conditions, very little of the total GSH pool is found as protein or soluble thiol mixed disulfides, but under conditions of oxidative stress such as those produced by the thiol oxidant diamide, the percentage of total GSH found as protein-GSH mixed disulfide can increase to greater than 20% of total GSH equivalents (Hiranruengchok and Harris 1995a). Glutathione depleted due to net oxidation or as a result of adduct formation can be replenished via reduction by the NADPH-dependent GSSG reductase pathway, in the former case, or via de novo GSH synthesis in either case. In some instances, the ability to actively synthesize GSH appears to be of greater protective importance than the absolute levels of GSH present. Glutathione concentrations and their role in embryotoxicity have now been studied throughout the gestational spectrum, from the unfertilized ovum to the neonate. Depletion of GSH has been shown to be deleterious at every stage of development, either through losses due to production of reactive intermediates and/or oxidative stress mechanisms (GARDINER and REED 1994; HIRANRUENGCHOK and HARRIS 1995a; HARRIS et al. 1987; WONG et al. 1989).

Not only do temporal variations in GSH occur throughout the developmental spectrum, but spatial changes are evident as well. The visceral yolk sac has the highest concentrations of GSH measured to date, which may not be surprising considering that this tissue barrier contains significantly higher activities for most biotransformation enzymes and is the first line of defense for the conceptus against environmental changes and xenobiotic insult (HARRIS et al. 1986). Addition of any number of xenobiotics and drugs to the culture medium results in increased utilization of GSH, resulting in the formation of covalent adducts or activation of the glutathione peroxidase and related chemical pathways that result in the net oxidation of GSH to form the disulfide (GSSG). Overproduction of GSSG has also been shown to result in protein thiol oxidation or S-thiolation of critical enzymes and proteins, thus contributing to the oxidative load (HIRANRUENGCHOK and HARRIS 1995a). These GSH or cysteine–protein mixed disulfides can result in alterations in important processes such as intermediary metabolism, membrane transport, and ligand binding functions (HIRANRUENGCHOK and HARRIS 1995b; AMMON and MARK 1985). Restoration of the reducing GSH conditions in cells is accomplished by reducing GSSG back to GSH in an NADPH-dependent pathway involving the GSSG reductase (GSSG-Rd). In addition, when GSH is depleted by any means not resulting in the net oxidation of GSH, repletion of intracellular stores is accomplished by de novo biosynthesis of new GSH. Initial characterizations of this process in the organogenesis-stage conceptus in vitro have shown that synthesis is also regulated in a temporally and spatially independent manner (HARRIS 1993). Equal depletion of GSH below 30% of controls in both embryo and visceral yolk sac by α,β-unsaturated carbonyl diethyl maleate (DEM) was followed by a rapid repletion of GSH in the visceral yolk sac over the first 3h. Only after this period did the embryo begin to replenish lost GSH, but appeared to do so at a slower rate than in the visceral yolk sac. We and others have now clearly demonstrated that the direct embryotoxicity in vivo and in vitro of a number of teratogenic or dysmorphogenic agents is affected by the differential GSH status in the developing conceptus. These agents include 7-OH-acetylaminofluorene, 2-nitrosofluorene, phenytoin, cytochalasin D, valproic acid, 5-fluorouracil, acetaminophen, acrolein, arsenic, lindane, and diamide, to name a few (HARRIS et al. 1987, 1988; MCNUTT and HARRIS 1995; WONG et al. 1989; HIRANRUENGCHOK and HARRIS 1993; SLOTT and HALES 1987).

Based on the foregoing discussion, the role of GSH protection of the conceptus during oxidative stress appears straightforward. The importance of maintaining GSH redox status during conditions of hypoxia, however, is more difficult to envision, except when the special case of redox cycling is considered. Figure 4 illustrates some of the relationships between GSH and pyridine nucleotide redox couples, as well as the pathways for redox cycling.

Under conditions of hypoxia, organic free radicals, rather than ROS, are preferentially generated, for which GSH and the antioxidant NADPH-dependent GSSG-Rd are of prime importance in determining cell survival (D.P. Jones et al. 1990; Kappus 1986). The necessity for maintaining GSH status and its regeneration from GSSG is underscored by the observations that replenishment of GSH via de novo synthesis is compromised during hypoxia due to reductions in ATP (Shan et al. 1989). It must also be considered that the normal supply of amino acid precursors, especially cysteine and glutamate, may also be compromised during hypoxia.

II. Pyridine Nucleotide Status

With regards to metabolic processes and redox status, the role and flux of pyridine nucleotides is an important consideration. It has been well documented that the developing rodent conceptus undergoes a dramatic shift in energy production pathways from the early somite stage to the close of organogenesis (Shepard et al. 1970; Tanimura and Shepard 1970). Early embryos are essentially anaerobic, relying on glycolysis as the major source of energy. While these conceptuses have been shown to be less sensitive to the direct effects of hypoxia, the ability to maintain normal cellular homeostasis through control of intracellular redox status may be seriously compromised.

The pyridine nucleotides NAD(H) and NADP(H) constitute an important biological redox couple involved in glycolysis, the Krebs cycle, mitochondrial respiration, and the complete oxidative metabolism of glucose to CO_2 and H_2O. The rate of reoxidation of NADH (to NAD^+) in the mitochondrial electron transport chain usually determines the overall rate of oxidation in the Krebs cycle, although O_2 may also be limiting under some conditions. Hypoxia or chemicals that inhibit the terminal cytochrome oxidases of the electron transport chain result in the net reduction of NAD^+ to NADH, because they prevent the enzymatic reoxidation of the electron carrier. The net reduction of NAD^+ likely contributes significantly to the overall reductive cellular environment produced during hypoxia. The NADPH redox couple is important for a number of biosynthesis reactions in mature and developing tissues. Utilization of NADPH in such processes as β-oxidation of fatty acids, GSSG reduction, cytochrome P450 monooxygenation, and many other reactions result in the net oxidation of NADPH to $NADP^+$, from which the reduced form can be regenerated via the activity of the pentose phosphate pathway, malate shuttle, isocitrate dehydrogenase, and others (Eggelston and Krebs 1974; Thurman and Kauffman 1980). It is the availability of NADPH that determines the extent to which GSH status can be maintained by providing reducing equivalents for GSSG-Rd. Relatively little is known about the synthesis and turnover of pyridine nucleotides during development. Small changes in their redox status could conceivably have a profound impact on the normal course of development due to the disturbance of energy metabolism, biosynthesis, and redox-related regulatory mechanisms.

Figure 5 shows the relative concentrations of the major oxidized and reduced pyridine nucleotides in the visceral yolk sac of the GD-10 to -11 conceptus grown in whole embryo culture (THORSRUD and HARRIS 1995). We have now determined that there are significant changes in the content and

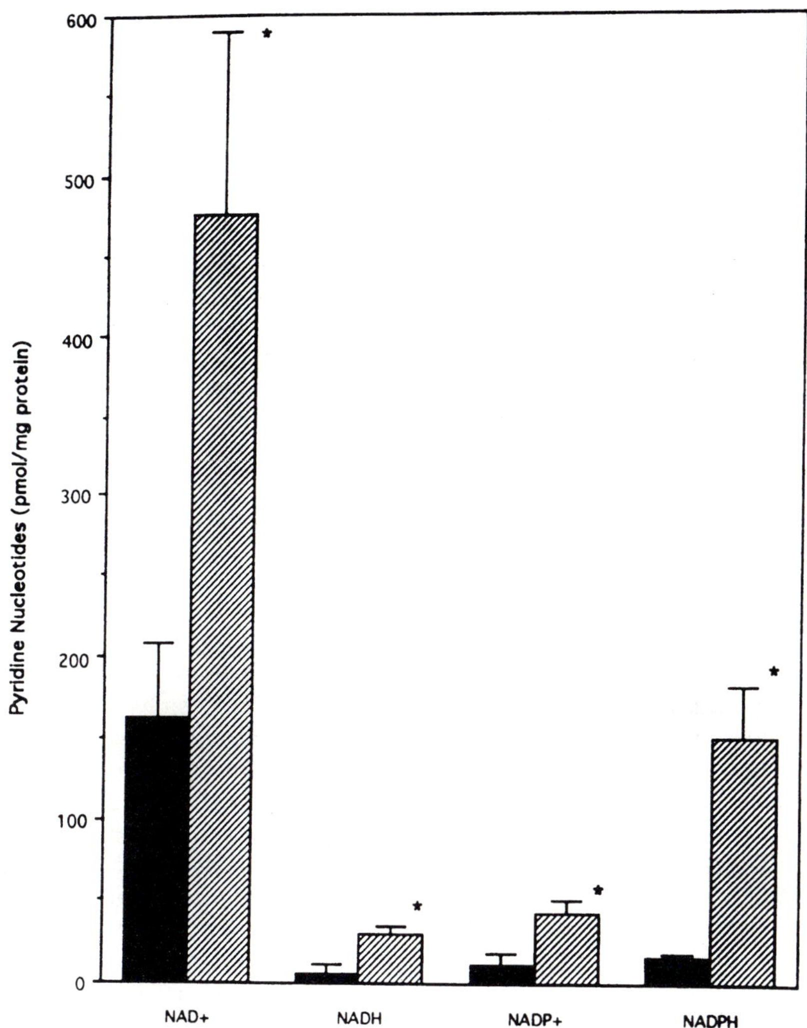

Fig. 5. Comparison of the reduced and oxidized pyridine nucleotide species in the visceral yolk sac of gestational day (GD) 10 (*black bars*) and 11 (*shaded bars*) rat conceptuses under control conditions, as evaluated by high-performance liquid chromatography (HPLC). *NAD*, nucleotide adenine diphosphate; *NADP*, NAD phosphate; *NADPH*, NADP, reduced form. *Asterisks* indicate a difference in the various nucleotide concentrations from GD 10 to GD 11 ($p < 0.05$). (From THORSRUD and HARRIS 1995, with permission)

distribution of the pyridine nucleotides which coincide with the reported change from a glycolytic energy dependence on GD 10 to a predominant Krebs cycle dependence on GD 11. On GD 10, NAD^+ is the major species, found in the visceral yolk sac at concentrations of 164 pmol NAD^+/mg protein and in the embryo proper at much higher concentrations of 609 pmol NAD^+/mg protein. By GD 11, visceral yolk sac concentrations were found to increase and embryonic levels decrease to become nearly equal at around 476 pmol NAD^+/mg protein. On either day, NADH constitutes only 2%–8% of the total NAD(H) equivalents. During this same period, a significantly different pattern of change occurs with the NADP(H) redox cycle. Concentrations of NADPH are relatively low in the GD-10 embryo and visceral yolk sac at 39 and 15 pmol NADPH/mg protein, respectively, with the oxidized form of this pyridine nucleotide constituting 18% and 42% of the total NADP(H) in embryo and visceral yolk sac, respectively. By GD 11, total NADP(H) equivalents increase significantly and dramatically in both embryo and visceral yolk sac by one order of magnitude, while the percentages found in the oxidized form in either tissue were roughly equal at 22% (C. Harris and B.A. Thorsrud, unpublished data). The significantly smaller reserves of NADPH in the embryo and visceral yolk sac of the GD-10 conceptus may have profound implications regarding the survival and protection of the conceptus following insult. Wholesale oxidation of NADPH through the activity of GSSG reductase, for example, could rob other NADPH-dependent biosynthesis pathways and account for a significant portion of the embryotoxicity seen at this early stage of development.

Alterations of these steady state pyridine nucleotide levels in response to conditions of hypoxia and cyanide exposure were characterized in real time in the visceral yolk sac using microfiberoptic probes as has been described by us previously (Fig. 6; Thorsrud and Harris 1993) and also through end-point high-performance liquid chromatography (HPLC) determinations, as described for the conditions and concentrations reported above. Net increases in reduced pyridine nucleotide species following exposures were accompanied by relatively small changes in the overall ratios of total pyridine nucleotides. It was also of interest to note that hypoxia produced a significant increase in all four pyridine nucleotide species (oxidized and reduced) in the embryo on GD 10 but not on GD 11. In contrast, cyanide exposure resulted in net overall reductions in pyridine nucleotides in GD-10 embryos, but an increase in concentrations significantly greater than controls following exposures to the same concentrations of cyanide on GD 11. The physiological basis and developmental importance of these observations have not yet been determined. Spatial and temporal differences in pyridine nucleotide concentration and redox status in the embryo proper and in the visceral yolk sac are not entirely unexpected based on constantly changing metabolic and functional differences. It is also highly likely that concentrations are selectively distributed in the various tissues of the embryo proper and that these differences are related to cell-specific sensitivity and resistance to hypoxia and chemical insults.

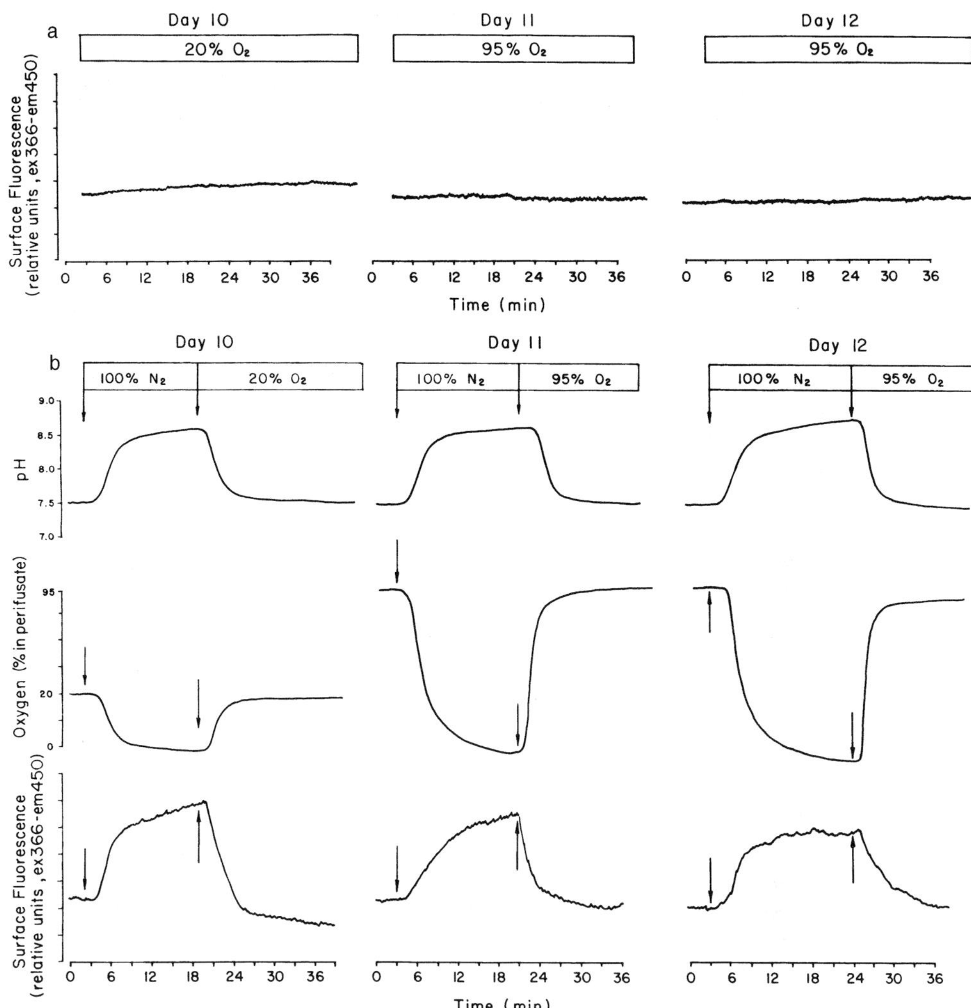

Fig. 6a,b. Simultaneous, real-time determinations of effluent pH, O_2, and microfiberoptic surface fluorescence in viable, intact, gestational day-10 to -12 rat conceptuses maintained in vitro in a modified perfusion apparatus. Surface fluorescence changes for total reduced pyridine nucleotides (NAD[P]H) were monitored at an excitation of 366 nm and emission at 450 nm. **a** No net changes under normoxic conditions. **b** Hypoxia was induced by replacing the normoxic O_2 with 100% nitrogen, which results in a significant increase in reduced pyridine nucleotide species on all 3 gestational days. Surface fluorescence returned to normal following restoration of normoxic conditions. (From THORSRUD and HARRIS 1993, with permission)

Preliminary studies of the effects of hypoxia on pyridine nucleotide status in the organogenesis stage rat conceptus in vitro have been reported (THORSRUD and HARRIS 1993). Real-time microfiberoptic fluorescence monitoring of reduced pyridine nucleotide redox status (NADPH and NADH) show similar responses in the GD-10 and GD-11 visceral yolk sac in response to hypoxia, indicative of aerobically mature cells which respond to hypoxia by inhibiting the normal oxidation of NADH, resulting in the increase of this, the fluorescent, species, as has been reported for mature tissues and organs such as the liver (Thurman and Lemasters 1988). Microfiberoptic measurements of real-time fluorescence changes associated with pyridine nucleotide status are not yet possible in the extraembryonic membrane-enclosed embryo proper. Absolute concentrations of pyridine nucleotides were, however, determined in the tissues of embryos and corresponding visceral yolk sacs on GD 10 and GD 11 from rat conceptuses cultured in vitro. When normalized as molar concentrations per mg protein, the GD-10 embryo contains nearly four times as much NAD(H) and twice as much NADP(H) as the corresponding visceral yolk sac. As expected for actively respiring tissues, NADPH to $NADP^+$ and NAD^+ to NADH ratios are high, indicating that a redox steady state exists that appears to be not unlike mature respiring tissues. By GD 11, visceral yolk sac nucleotide concentrations increase significantly to a point at which NAD(H) in the visceral yolk sac equals that found in the embryo. Steady state NADP(H) concentrations still remain less than 50% of those found in the embryo. The significance and mechanisms of these differential pyridine nucleotide status changes and the possible responses to disturbance via hypoxia or chemical exposure have not yet been determined in the developing mammalian conceptus.

Decreases in NAD^+ are not always highly correlated with similar increments in NADH. Perhaps the reduced form is rapidly utilized by other cellular pathways. For example, it has been suggested that NADH may be the preferred electron donor for embryonic, but not adult, cytochrome P450 monooxygenases. This cofactor is also believed to play an important role in DNA repair mediated by poly-ADP ribosylation which consumes large quantities of NAD[H] (Stubberfield and Cohen 1988). In spite of these possibilities, however, our data obtained from the microfiberoptic sensors in the visceral yolk sac do not show an extremely rapid decline in surface fluorescence, indicating that the reduced species (NADH) are not rapidly oxidized or removed via other pathways.

One additional point of interest as a result of these studies relates to the dramatic differences in pyridine nucleotide levels and redox ratios seen on successive days of gestation and in comparison between embryos and the visceral yolk sac on any single gestational day. The differences can be further accentuated by conditions of hypoxia. In the GD-10 embryo, intracellular concentrations of NADPH, $NADP^+$, and NAD^+ were found to increase significantly above controls by 70%, 400%, and 35% respectively, in response to hypoxia. In conjunction with these increases, NADH was found to be reduced by 70%. These changes were very much unlike the concomitant alterations in the visceral yolk sac from the same exposure which showed a net increase in the reduced species (NADPH and NADH), which relate to the increased surface fluorescence measured with the fluorescence probe. On GD

11, when the embryo is believed to have converted to aerobic metabolism, net pyridine nucleotide concentrations in the visceral yolk sac had increased to equal those found in the corresponding embryos. In response to hypoxia, the increases in surface fluorescence were not as dramatic as seen on GD10, and the greatest changes in absolute pyridine nucleotide levels measured by HPLC appeared to be characterized more by decreases in the oxidized species rather than increases in the reduced forms. The molecular and biochemical bases for these changes have not yet been clearly elucidated.

In addition to the determination of pyridine nucleotide status, we have also sought to further characterize intracellular redox status by determining the relative abundance and phosphorylation state of the adenine nucleotides, ATP, ADP, and AMP. The relative concentrations of the three forms and overall abundance of total adenine nucleotides were nearly equal in GD-10 and GD-11 embryos (6.9 and 5.1, respectively) when expressed as nmol/mg protein. In the GD-10 visceral yolk sac, however, total adenine nucleotide concentrations were significantly lower than those in the corresponding embryo (6.9 versus 2.1). By GD 11, these differences disappeared, with concentrations in embryo and visceral yolk sac being almost equal. Conditions of hypoxia produced a 27% increase in ATP in the GD-10 embryo, while causing a 46% reduction in this same nucleotide in the GD-11 embryo. Slight overall reductions in ATP were seen relative to controls in the visceral yolk sac on both days as a result of hypoxia. We interpret these results to indicate that similar, relatively hypoxia-insensitive, energy production pathways exist in the visceral yolk sac on GD 10 and 11. The GD-10 embryo proper, however, showed that it was able to respond to conditions of hypoxia by increasing the amount of ATP, most likely through an O_2-insensitive pathway such as glycolysis, which has been shown to predominate at this stage of development (Tanimura and Shepard 1970). Significant reductions in ATP due to hypoxia in the GD-11 embryo reflect the observed shift to Krebs cycle dependence at this developmental stage, which results in decreased ATP production in the absence of sufficient O_2.

III. Control of pH and Hypoxia

A great deal of research has been directed toward describing the role of pH in controlling and regulating the events associated with fertilization and early cleavage embryogenesis. It is well known that conditions of hyperoxia and oxidative stress will interfere with fertilization, cell division, and differentiation (Gardiner and Reed 1994; Goto et al. 1993; Umaoka et al. 1992). In the postimplantation rat conceptus, it has been determined that conceptal pH decreases with advancing gestational age from an alkaline to a more acidic milieu (Scott et al. 1986). How these changes in redox status affect intracellular pH in various tissues of the organogenesis-stage conceptus and how they are affected by hypoxia is not yet known. Utilizing the GD-12 rat conceptus in whole embryo culture and an extremely small microfiberoptic pH sensor (TAN et al. 1994), it was determined that the pH of exocoelomic fluid did not change significantly during acute changes in extraembryonic pH (6.6 to 8.6) or during alteration of O_2 tensions in the bathing perifusate from 0% O_2 (100% N_2) to 95% O_2 (5% N_2). The pH values were measured only in exocoelomic fluid, and it is not yet known what, if any, changes occur in the cells of the embryo or visceral yolk sac. Under conditions of chemically induced oxidative stress, where both GSH and pyridine nucleotide redox ratios are known to be disturbed, rapid and dramatic alterations in conceptal pH are observed. Treatment of conceptuses with the thiol oxidant diamide (100μ*M*) results in a rapid (30s) decrease of conceptal pH of 0.3 units. The magnitude of change has been shown in other systems to be capable of complete activation of some important enzymes involved in intermediate metabolism.

References

Ammon HPT, Mark H (1985) Thiols and pancreatic B-cell function: a review. Cell Biochem Funct 3: 157–171
Aw TY, Anderson BS, Jones DP (1987a) Suppression of mitochondrial respiratory function following short-term anoxia. Am J Physiol 252: C362–C368
Aw TY, Anderson BS, Jones DP (1987b) Mitochondrial transmembrane ion distribution during anoxia. Am J Physiol 252: C356–C361
Barber CV, Fantel AG (1993) The role of oxygenation in embryotoxic mechanisms of three bioreducible agents. Teratology 47: 209–223
Becker RF, Martin JC (1971) Vital effects of chronic nicotine absorption and chronic hypoxic stress during pregnancy and the nursing period. Am J Obstet Gynec 110: 522–533
Brent RL, Franklin JB (1960) Uterine vascular clamping: new procedure for the study of congenital malformations. Science 132: 88–91
Büchner F (1955) Differenzierungsstörungen im mittleren und hinteren Korperdrittel des Hühnchens nach experimentellen Sauerstoffmangel in der Frühentwicklung. Beitr Pathol Anat 115: 617–643
Chasnoff IJ, Chisum GM, Kaplan WE (1988) Maternal cocaine use and genitourinary tract malformations. Teratology 37: 201–204
Chernoff N, Grabowski CT (1971) Responses of the rat foetus to maternal injections of adrenaline and vasopressin. Br J Pharmacol 43: 270–278
Curley FJ, Ingalls TH (1957) Hypoxia at normal atmospheric pressure as a cause of congenital malformations in mice. Proc Soc Exp Biol Med 94: 87–88
Danielson BRG, Danielson M, Reiland S, Rundqvist E, Dencker L, Regard CG (1990) Histological and in vitro studies supporting decreased uteroplacental blood flow as explanation for digital defects after administration of vasodilators. Teratology 41: 185–193
Dareste MC (1877) La production artificielle des monstruosités. Reinwald, Paris
Davies KJA (1986) Intracellular proteolytic systems may function as secondary antioxidant defenses: a hypothesis. J Free Biol Med 2: 155–173
Eggleston LV, Krebs HA (1974) Regulation of the pentose phosphate cycle. Biochem J 138: 425–435
Emerit I, Feingold J, Keck M, Levy A, Michelson AM (1982) Activated oxygen species at the origin of chromosome breakage and sister-chromosome exchanges. Mutat Res 93: 165–172
Fantel AG, Juchau MR, Burroughs CJ, Person RE (1989) Studies of embryotoxic mechanisms of niridazole: evidence that oxygen depletion plays a role in dysmorphogenicity. Teratology 39: 243–251

Fantel AG, Person RE, Burroughs-Gleim CJ, Mackler B (1990) Direct embryotoxicity of cocaine in rats: effects on mitochondrial activity, cardiac function, and growth and development in vitro. Teratology 42: 35–43

Fantel AG, Person RE, Burroughs-Gleim C, Shepard TH, Juchau MR, Mackler B (1991) Asymmetric development of mitochondrial activity in rat embryos as a determinant of the defect patterns induced by exposure of hypoxia, hyperoxia, and redox cyclers in vitro. Teratology 44: 355-362

Ferm V (1964) Teratogenic effects of hyperbaric oxygen. Proc Soc Exp Biol Med 116: 975–976

Freinkel N, Lewis NJ, Akazawa S, Roth SI, and Gorman L (1984) The honeybee syndrome – implications of the teratogenicity of mannose in rat-embryo culture. New Engl J Med 310: 223–230

Gardiner CS, Reed DJ (1994) Status of glutathione during oxidant-induced oxidative stress in the preimplantation mouse embryo. Biol Reprod 51: 1307–1314

Goto Y, Noda Y, Mori T, Nakano M (1993) Increased generation of reactive oxygen species in embryos cultured in vitro. Free Radic Biol Med 15: 69–75

Grabowski CT (1961) A quantitative study of the lethal and teratogenic effects of hypoxia on the three-day chick embryo. Am J Anat 109: 25–36

Grabowski CT (1964) The etiology of hypoxia-induced malformations in the chick embryo. J Exp Zool 157: 307–326

Grabowski CT (1966) Physiological changes in the bloodstream of chick embryos exposed to teratogenic doses of hypoxia. Dev Biol 13: 199–213

Grabowski CT (1970) Embryonic oxygen deficiency – a physiological approach to analysis of teratological mechanisms. Adv Teratol 4: 125–169

Grabowski CT, Paar JA (1958) The teratogenic effects of graded doses of hypoxia on the chick embryo. Am J Anat 103: 313–348

Grabowski CT, Schoeder RE (1968) A time-lapse photographic study of chick embryos exposed to teratogenic doses of hypoxia. J Embryol Exp Morphol 19: 347–362

Greenway JC, Mirkes PE, Walker EA, Juchau MR, Shepard TH, Fantell AG (1985) The effect of oxygen concentration on the teratogenicity of salicylate, niridazole, cyclophosphamide, and phosphoramide mustard in rat embryos in vitro. Teratology 23: 287–295

Greenway JC, Fantel A, Juchau MR (1986) On the capacity of nitroheterocyclic compounds to elicit an unusual axial asymmetry in cultured rat embryos. Toxicol Appl Pharmacol 82: 307–315

Harris C (1993) Glutathione biosynthesis in the postimplantation rat conceptus in vitro. Toxicol Appl Pharmacol 120: 247–256

Harris C, Fantel AG, Juchau MR (1986) Differential glutathione depletion by L-buthionine-S,R-sulfoximine in rat embryo versus visceral yolk sac in vivo and in vitro. Biochem Pharmacol 35: 4437–4441

Harris C, Namkung MJ, Juchau MR (1987) Regulation of intracellular glutathione in rat embryos and visceral yolk sacs and its effects on 2-nitrosofluorene-induced malformations in the whole embryo culture system. Toxicol Appl Pharmacol 88: 141–152

Harris C, Stark KL, Juchau MR (1988) Glutathione status and the incidence of neural tube defects elicited by direct acting teratogens in vitro. Teratology 37: 577–590

Heron HJ (1962) The effect of smoking during pregnancy: a review with a preview. New Zeal Med J 61: 545–548

Hiranruengchok R, Harris C (1993) Glutathione oxidation and embryotoxicity elicited by diamide in the developing rat conceptus in vitro. Toxicol Appl Pharmacol 120: 62–71

Hiranruengchok R, Harris C (1995a) Diamide-induced alterations of intracellular thiol status and the regulation of glucose metabolism in the developing rat conceptus in vitro. Teratology 52: 205–214

Hiranruengchok R, Harris C (1995b) Formation of protein-glutathione mixed disulfides in the developing rat conceptus following diamide treatment in vitro. Teratology 52: 196–204

Howren HH (1965) A review of the literature concerning smoking during pregnancy. Virginia Med Month 92: 274–279

Ingalls TH, Curley FJ (1957) Principles governing the genesis of congenital malformations induced in mice by hypoxia. New Engl J Med 257: 1121–1127

Ingalls TH, Curley FJ, Prindle RA (1950) Anoxia as a cause of fetal death and congenital defect in mouse. Am J Dis Child 80: 34–45

Ingalls TH, Curley FJ, Prindle RA (1952) Experimental production of congenital anomalies: timing and degree of anoxia as factors causing fetal deaths and congenital anomalies in the mouse. New Engl J Med 247: 758–767

Jenkinson PC, Anderson D, Gangolli SD (1986) Malformations induced in cultured rat embryos by enzymically generated active oxygen species. Teratogen Carcinogen Mutagen 6: 547–554

Jones DP (1985) The role of oxygen concentration in oxidative stress: hypoxic and hyperoxic models. In: Sies H (ed) Oxidative stress. Academic, New York, p 151

Jones DP, Aw TY, Shan X, Tribble DL (1990) Characteristics of hypoxic cells that enhance their susceptibility to chemical injury. Plenum, New York, pp 1–9

Jones TW, Thor H, Orrenius S (1986) Cellular defense mechanisms against toxic substances. Arch Toxicol Suppl 9: 259–271

Juchau MR, Fantel AG, Harris C, Beyer BK (1986) The potential role of redox cycling as a mechanism for chemical teratogenesis. Environ Health Perspect 70: 131–136

Kappus H (1986) Overview of enzyme systems involved in bio-reduction of drugs in redox-cycling. Biochem Pharmacol 35: 1–7

Ketterer B, Coles B, Meyer DL (1983) The role of glutathione in detoxication. Environ Health Perspect 49: 59–69

Kosower NS, Song K-R, Kosower EM (1969) Glutathione. IV. Intracellular oxidation and cellular injury. Biochem Biophys Acta 192: 23–28

Kubow S, Wells PG (1989) In vitro bioactivation of phenytoin to a reactive free radical intermediate by prostaglandin synthetase, horseradish peroxidase and thyroid peroxidase. Mol Pharmacol 35: 504–511

Leist KH, Grauwiler J (1974) Fetal pathology in rats following uterine vessel clamping on day 14 of gestation. Teratology 10: 55–68

Liu L, Wells PG (1993) Formation of 8-hydroxy-2′-deoxyguanosine: in vivo evidence for phenytoin-initiated oxidative DNA damage in murine maternal hepatic and embryonic tissues, and in vitro horseradish peroxidase-catalysed bioactivation and 2′-deoxyguanosine oxidation by phenytoin and structurally related teratogens. Proc Int Soc Stud Xenobiot 4: 113

Liu L, Wells PG (1994) In vivo phenytoin-initiated oxidative damage to proteins and lipids in murine maternal hepatic and embryonic tissue organelles: potential molecular targets of chemical teratogenesis. Toxicol Appl Pharmacol 125: 247–255

Loach PA (1976) Oxidation-reduction potentials, absorbance bands and molar absorbance of compounds used in biochemical studies. In: Fasman GD (ed) Handbook of biochemistry and molecular biology. CRC press, Cleveland pp 122–130

Martz F, Failinger C III, Blake DA (1977) Correlation between embryopathic effect and covalent binding of putative arene oxide metabolite in gestational tissue. J Pharmacol Exp Ther 203: 231–239

McNutt TL, Harris C (1994) Lindane embryotoxicity and differential alteration of cysteine and glutathione levels in rat embryos and visceral yolk sacs. Reprod Toxicol 8: 351–362

Metzler DE (1977) Electrode potentials and free energy changes for oxidation-reduction reactions. In: Metzler DE (ed) Biochemistry: the chemical reactions of living cells. Academic, New York, pp 172–174

Miki A, Fujimoto E, Ohsaki T, Mizoguti H (1988) Effects of oxygen concentration on embryonic development in rats: a light and electron microscopic study using whole-embryo culture techniques. Anat Embryol 178: 337–343

Millicovsky G, DeSesso JM (1980) Differential embryonic cardiovascular responses to acute maternal uterine ischemia: an in vivo microscopic study of rabbit embryos with either intact or clamped umbilical cords. Teratology 22: 335–343

Millicovsky G, Johnston MC (1981) Maternal hypoxia greatly reduces the incidence of phenytoin-induced cleft lip and palate in A/J mice. Science 212: 671–672

Mitchell JA, Hammer RE (1985) Effects of nicotine on oviductal blood flow and embryo development in the rat. J Reprod Fertil 74: 71–76

Moore TR, Sorg J, Miller L, Key TC, Resnik R (1986) Hemodynamic effects of intravenous cocaine on the pregnant ewe and fetus. Am J Obstet Gynecol 155: 883–888

Morriss GM, New DAT (1979) Effect of oxygen concentration on morphogenesis of cranial neural folds and neural crest in cultured rat embryos. J Embryol Exp Morphol 54: 17–35

Murakami U, Kameyama Y (1963) Vertebral malformation in the mouse foetus caused by maternal hypoxia during early stages of pregnancy. J Embryol Exp Morphol 11: 107–118

New DAT, Coppola PT (1970a) Effects of different oxygen concentrations on the development of rat embryos in culture. J Reprod Fertil 21: 109–118

New DAT, Coppola PT (1970b) Development of explanted rat fetuses in hyperbaric oxygen. Teratology 3: 153–162

Phillips BJ, James TEB, Anderson D (1984) Genetic damage in CHO cells exposed to enzymatically generated active oxygen species. Mutat Res 126: 265–271

Rübsaamen H (1952) Uber die teratogenetische Wirkung des Sauerstoffmangels in der Frühentwicklung. Ein Beitrag zur Kausalgenese der Missbildungen bei Mench und Tier. Beitr Pathol Anat 112: 336–379

Scott WJ, Duggan CA, Schreiner CM, Collins MD, Nau H (1986) Intracellular pH of rodent embryos and its association with teratogenic response. In: Welsch F (ed) Approaches to elucidate mechanisms in teratogenesis. Hemisphere, New York, pp 99–107

Shan X, Aw TW, Shapira R, Jones DP (1989) O_2 dependence of glutathione synthesis in hepatocytes. Toxicol Appl Pharmacol 101: 261

Shepard TH, Tanimura T, Robkin MA (1970) Energy metabolism in early mammalian embryos. Dev Biol Suppl 4: 42–58

Sies H (1985) Oxidative stress. Academic, New York

Slotkin TA, Lappi EC, McCook EC, Lorber BA, Seidler FJ (1995) Loss of neonatal hypoxia tolerance after prenatal nicotine exposure: implications for sudden infant death syndrome. Brain Res Bull 38: 69–75

Slott VL, Hales BF (1987) Enhancement of the embryotoxicity of acrolein, but not phosphoramide mustard, by glutathione depletion in rat embryos in vitro. Biochem Pharmacol 36: 2019–2025

Stubberfield CR, Cohen GM (1988) NAD^+ depletion and cytotoxicity in isolated hepatocytes. Biochem Pharmacol 37: 3967–3974

Tan W, Shi Z-Y, Thorsrud BA, Harris C, Kopelman R (1994) Near field fiberoptic sensors and biological applications. Scan Probe Microscop II SPIE 2068: 59–68

Tanimura T, Shepard TH (1970) Glucose metabolism by rat embryos in vitro. Proc Soc Exp Biol Med 135: 51–54

Thorsrud BA, Harris C (1993) Real time micro-fiberoptic monitoring of endogenous fluorescence in the rat conceptus during hypoxia. Teratology 48: 343–353

Thorsrud BA, Harris C (1995) Real time micro-fiberoptic redox fluorometry: modulation of the pyridine nucleotide status of the organogenesis-stage rat visceral yolk sac with cyanide and alloxan. Toxicol Appl Pharmacol 135: 237–245

Thurman RG, Kauffman FG (1980) Factors regulating drug metabolism in intact hepatocytes. Pharmacol Rev 31: 229–251

Thurman RG, Lemasters JL (1988) New micro-optical methods to study metabolism in periportal and pericentral regions of the liver lobule. Drug Metabol Rev 19: 263–281

Tolson CM, Seidler FJ, McCook EC, Slotkin TA (1995) Does concurrent or prior nicotine exposure interact with neonatal hypoxia to produce cardiac cell damage. Teratology 52: 289–305

Tribble DL, Jones DP (1990) Oxygen dependence of oxidative stress, rate of NADPH supply for maintaining the GSH pool during hypoxia. Biochem Pharmacol 39: 729–736

Umaoka Y, Noda Y, Narimoto K, Mori T (1992) Effects of oxygen toxicity on early development of mouse embryos. Mol Reprod Dev 31: 28–33

Watkinson WP, Gordon CJ (1993) Caveats regarding the use of the laboratory rat as a model for acute toxicological studies: modulation of toxic response via physiological and behavioral mechanisms. Toxicology 81: 15–31

Webster WS, Lipson AH, Brown-Woodman PDC (1987) Uterine trauma and limb defects. Teratology 35: 253–260

Webster WS, Brown-Woodman PDC (1990) Cocaine as a cause of congenital malformations of vascular origin: experimental evidence in the rat. Teratology 41: 689–697

Wong M, Helston LMJ, Wells PJ (1989) Enhancement of murine phenytoin teratogenicity by the γ-glutamylcysteine synthetase inhibitor L-buthionine-(S,R)-sulfoximine and by the glutathione depletor diethyl maleate. Teratology 40: 127–141

Woods JR, Plessinger MA, Clark KE (1987) Effect of cocaine on uterine blood flow and fetal oxygenation. JAMA 257: 957–961

Yamamoto T, Egashira T, Yamanaka Y, Yoshida T, Kuroiwa Y (1983) Initial metabolism of gamma-hexachlorocyclohexane (γ-HCH) by rat liver microsomes. J Pharm Dyn 6: 721–728

CHAPTER 19

Altered Embryonic pH

S.M. BELL, C.M. SCHREINER, and W.J. SCOTT, JR.

A. Introduction

In this chapter we present information which has led us to believe that alteration of intracellular pH (pH_i) in the mammalian embryo is a potential mechanism by which exogenous chemical and physical insults lead to congenital malformations. Historically, there has been little interest in pH as an important developmental parameter, and those measurements which were made in lower forms must be viewed with great caution, as reviewed by BRACHET (1968).

More recently, concern regarding a negative influence of lower pH from acid rain and other industrial pollution sources on the development of vertebrate and invertebrate embryos has surfaced (e.g., FREDA and DUNSON 1984; PAGANO et al. 1985). Our own investigations have focused on the intracellular pH of mammalian embryos, especially the mouse, with special attention given to the measurement of pH in target tissues known to be affected by teratogen exposures.

B. Historical Perspective of Agents Hypothesized to Act by Altering Embryonic Intracellular pH

I. Acetazolamide and CO_2

The concept that alteration of pH inside embryonic cells may be a mechanism of teratogenesis was first put forward by WEAVER and SCOTT (1984b). They had shown that the unusual malformation syndrome induced by acetazolamide, postaxial, right-sided, forelimb ectrodactyly (Fig. 1), could be mimicked by maternal exposure to CO_2 (WEAVER and SCOTT 1984a). This evidence supported the view that acetazolamide acts to induce teratogenesis by inhibiting the enzyme carbonic anhydrase within the embryo. More importantly, it permitted the focus of mechanistic research to center on CO_2, which is raised when carbonic anhydrase is inhibited, and one inevitable consequence of increased CO_2 tension is a reduction of pH_i (for a review, see ROOS and BORON 1981). A second line of evidence supporting altered pH_i as a mechanism of teratogenesis involves the coadministration of amiloride along

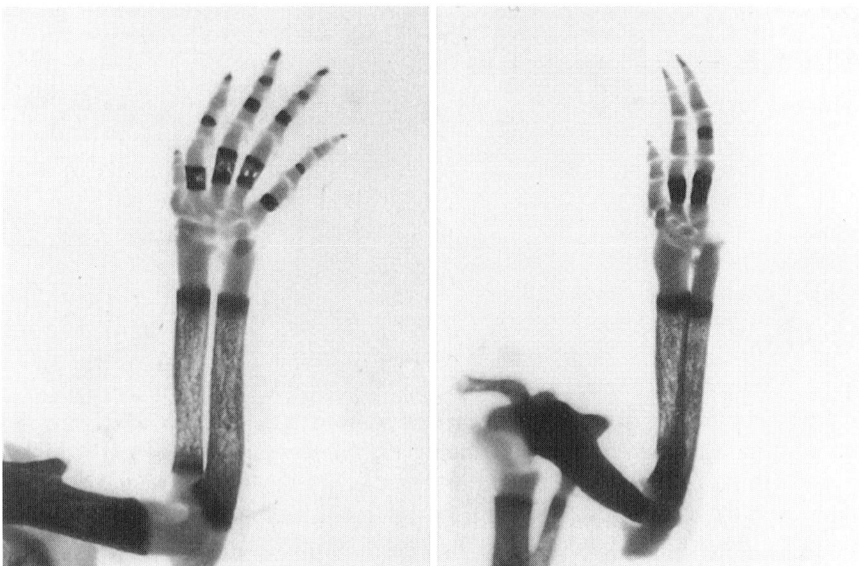

Fig. 1. Normal right forelimb (*left*). Right forelimb exhibiting postaxial ectrodactyly 4,5 (*right*)

with acetazolamide. The result of this dual treatment is an increased frequency and severity of postaxial forelimb ectrodactyly (Ellison and Maren 1972; Scott et al. 1990). Amiloride is a K^+-sparing diuretic which inhibits a membrane transporter that exchanges extracellular Na^+ for intracellular protons (see p. 559 for more detail). By inhibiting this antiporter, amiloride interferes with regulation of pH_i, and we postulate that this pathway of pH_i regulation is especially critical when $[H^+]$ is raised intracellularly by xenobiotics. Subsequently, pH_i has been measured in mouse embryos (Scott et al. 1990) and in mouse embryo forelimb buds (Schreiner et al. 1993, 1995) after treatment with acetazolamide or acetazolamide plus amiloride. In both cases, whole embryo and forelimb bud, a slight decrease of pH_i was measurable after acetazolamide treatment and a larger decrease was evident after the combination treatment. These results are compatible with the hypothesis of altered embryonic pH_i as a mechanism of teratogenesis, but doubts about its veracity must remain until the pH_i changes can be causally connected with subsequent cellular changes leading directly to the malformation.

II. Anticonvulsants

1. Valproic Acid

Using the information collected in studies of acetazolamide teratogenesis, we developed three criteria as indicators that a particular agent might induce congenital malformations by altering embryonic pH_i: (1) the agent must induce postaxial, right forelimb ectrodactyly in the C57BL/6 strain of mouse; (2) SWV mice must be relatively resistant to this effect; and (3) coadministration of amiloride in C57BL/6 mice must exacerbate the frequency and severity of forelimb ectrodactyly. Table 1 lists the teratogens which meet the first criteria and indicate those for which it has been determined that a difference in sensitivity exists between the C57BL/6 and SWV mouse strains. Data relative to the third criterion is presented in Table 2. Since a number of these agents are teratogenic to varying extents both in other mouse strains and other species, we hypothesize that alteration of embryonic pH_i may be a common mechanism by which some compounds induce congenital malformations. Although our focus has been on drug-dosing scenarios which predominantly induce limb malformations, administration of these agents at other developmental time points induces a variety of other embryonic malformations presumably also caused by alterations in pH_i.

The anticonvulsant valproic acid (VPA) meets all three criteria. It was first shown to induce postaxial, right forelimb ectrodactyly in NMRI mice (Nau and Scott 1987). Subsequently, Collins et al. (1991) reported on VPA-induced ectrodactyly, and in a preliminary report (Scott et al. 1989) amiloride was shown to exacerbate VPA-induced right forelimb ectrodactyly. We have also shown that a comparable dosage of VPA is unable to induce ectrodactyly in SWV mice (W.J. Scott et al., unpublished observations).

These findings have led us to measure pH_i in embryos exposed to VPA. A preliminary report (Schreiner et al. 1994) indicates that pH_i of embryos and

Table 1. Teratogens known to induce postaxial right-sided forelimb ectrodactyly in the C57Bl/6 mouse strain

Teratogen	Resistant to ectrodactyly in the SWV mouse strain
Weak acids	
CO_2	X
Dimethadione	X
Methoxy acetic acid	Not determined
Phenytoin	Not determined
Retinoic acid	X
Valproic acid	X
Non weak acids	
Acetazolamide	X
Cadmium	X
Ethanol	X
Hyperthermia	X

Table 2. Potentiation of teratogenesis by intracellular pH inhibitors

Teratogen	pH_i inhibitor	Implantation sites (*n*)	Resorbed (%)	Ectrodactyly[b] (%)	Bilateral ectrodactyly (%)
Acetazolamide, 500 mg/kg (s.c)[a]	–	168	13	50	10
Acetazolamide, 500 mg/kg (s.c)	Amiloride, 4 mg/kg (s.c)[a]	85	28	92	57
	Amiloride, 4 mg/kg (s.c)[a]	74	10	0	0
Acetazolamide, 500 mg/kg (s.c)	DIDS, 50 mg/kg (i.p.)	45	73	92	67
	DIDS, 50 mg/kg (i.p.)	61	11	2	0
Valproic acid, 300 mg/kg (i.p.)	–	73	25	9	0
Valproic acid, 300 mg/kg (i.p.)	Amiloride, 5 mg/kg	52	23	75	43
Retinoic acid, 25 mg/kg (i.p.)	–	68	63	20	8
Retinoic acid, 25 mg/kg (i.p.)	Amiloride, 4 mg/kg	74	49	45	10

DIDS, 4,4′-disothiocyanato-2,2′-stilbene disulfonic acid.
[a]As reported in SCOTT et al. (1990).
[b]Includes individuals with unilateral and bilateral ectrodactyly.

limb buds is reduced 1 h after treatment with a teratogenic dose of VPA and has returned to control values by 4 h after treatment. Once again, these results are consistent with the hypothesis that alteration of embryonic pH_i is a mechanism of teratogenesis. However, more rigorous testing of the concept is needed, including a credible mechanism by which VPA would reduce pH_i. Can diffusion of a weak acid such as VPA carry enough protons into the cell to reduce pH_i? Nonteratogenicity of the 2-en analogue of VPA (NAU and SCOTT 1987) suggests otherwise. Or does VPA affect some other cellular machinery such as Na^+ channels or the H^+ lactate transporter and alter pH_i through such an indirect mechanism?

2. Trimethadione

Trimethadione (TMO) was often prescribed for the treatment of petit mal epilepsy. It has proven to be a potent human teratogen, as summarized in the ANONYMOUS (1979). This review evaluated 53 cases, of which 13 ended in abortion and 33 of the 40 live infants had at least one major congenital malformation. TMO is rapidly demethylated to dimethadione (DMO), which is not further metabolized. DMO is a weak acid with a pK_a of 6.13, and the control of petit mal seizures has been postulated to result from the induction of an intracellular acidosis (BUTLER et al. 1966). Both TMO and DMO are teratogenic in a variety of laboratory animals (for a review, see SCOTT et al. 1988), but for the purpose of this review COLLINS et al. (1991) showed that DMO can induce right forelimb ectrodactyly when given to C57BL/6 mice. In unpublished studies, we have shown that coadministration of amiloride can increase the frequency of DMO-induced ectrodactyly, but that DMO plus amiloride given to SWV mice is ineffective in inducing limb malformations. DMO does reduce pH_i, as shown for example by LUCAS et al. (1988), but no studies have yet been done in mammalian embryos after maternal administration of a teratogenic dose of DMO.

III. Cadmium

Cadmium is an environmental pollutant known to induce postaxial forelimb ectrodactyly in rats (BARR 1973), mice (LAYTON and LAYTON 1979; MESSERLE and WEBSTER 1982), and hamsters (FERM and CARPENTER 1967; FERM 1971). The sensitivity of C57BL/6 mice and relative resistance of SWV mice was demonstrated by KUCZUK and SCOTT (1984). Subsequently, these latter authors proposed that cadmium-induced ectrodactyly was due to a lowering of pH in the embryo (FEUSTON and SCOTT 1985). However, the mechanism by which cadmium might acidify the embryo was quite tenuous. Since then it has become quite clear that cadmium is an inhibitor of proton channels (for a review, see DECOURSEY and CHERNY 1994). These channels carry protons from the cell interior to the outside and are inhibited by 1m*M* cadmium or less. Thus a plausible mechanism by which cadmium could alter embryonic pH_i is now available, and future studies need to search for the presence and function of these channels during embryogenesis.

IV. Ethanol

The fetal alcohol syndrome (FAS), seen in the offspring of alcoholic women, has led to numerous animal studies attempting to recreate the clinical condition. Many hypotheses have been put forward to explain how ethanol might induce abnormal embryogenesis, but none can be considered as verified. Alteration of embryonic pH has not been postulated as a potential mechanism of alcohol embryopathy, but should be considered for a variety of reasons. Clinically, alcoholics are frequently acidotic through the metabolic consequences of ethanol breakdown. Moreover, ethanol is postulated to perturb plasma membrane physiology where pH_i regulatory proteins reside, providing a second means of lowering intracellular $[H^+]$. Ethanol does induce postaxial, forelimb ectrodactyly in C57BL/6 mice, but not in SWV mice (ZIMMERMAN et al. 1990). These diverse pieces of evidence suggest that the alteration of embryonic pH warrants further consideration as one potential mechanism of ethanol-induced embryotoxicity.

V. Hyperthermia

Increased body temperature is a clear danger to the developing mammalian embryo (EDWARDS 1986). Whether sufficient increase or duration of elevated temperature can account for malformations clinically remains debatable. As with ethanol, numerous mechanisms by which hyperthermia might alter embryogenesis have been proposed, but based on the following evidence altered embryonic pH_i should also be considered. Elevated temperature has been shown to reduce pH_i in many cell types (e.g., KIANG et al. 1990), and we presume that this insult could lower pH_i in the mammalian embryo. C57BL/6 mice respond to hyperthermic exposure on day 9 or 10 of gestation by having offspring with postaxial, forelimb ectrodactyly. SWV mice were nearly totally resistant to the same exposures (BENNETT et al. 1989). Thus strong circumstantial evidence is available suggesting that reduction of embryonic pH_i may be involved in hyperthermia-induced teratogenicity.

C. pH of Embryo Tissues and Fluids

Over the years a number of approaches have been used to measure pH_i in a variety of tissues, body fluids, and cell lines. Because of its location and size, the mammalian embryo severely limits the approaches available to measure embryonic pH_i. Other than our own efforts, we are aware of only a few other studies in which pH_i of organogenesis-stage vertebrate embryos has been measured. GILLESPIE and GREENWELL (1988) measured pH_i and cellular regulating mechanisms in somatic cells of the early chick embryo using the pH-sensitive dye 2,7-bis (carboxyethyl)-5(6)-carboxyfluorescence acetoxymethyl ester (BCECF-AM). Following intercalation into the cell, BCECF-AM is cleaved to BCECF by cellular esterases and trapped within the cell. By relating the ratio of the emission scans obtained at 532 nm following excitation at 440 and 506 nm to those of cells bathed in solutions of known pH, an unknown pH value can be obtained. Over the pH range of 6.4–7.4, a linear relationship exists between cellular pH and the fluorescent emission at 532 nm following excitation at the pH-sensitive wavelength of 506 nm. Emission values are normalized by excitation at the isobestic wavelength of 440 nm (RINK et al. 1982). In freshly dissected somites from 2-day-old chicks, GILLESPIE and GREENWELL (1988) found a pH_i of 7.18 $\pm$ 0.02 and a Na^+-dependent HCO_3^- transport present to recover from an acid load. If these somites were kept in 10% fetal calf serum for 2–5 h, pH_i rose to 7.36 $\pm$ 0.02 and recovery from an acid load was attributed to Na^+/H^+ exchange. More recently, this same laboratory measured pH_i in chick embryo neural crest cells in vitro (DICKENS et al. 1990). They found that pH_i was 7.30 $\pm$ 0.02 in closely packed cells, but rose to 7.48 $\pm$ 0.02 in cells which had migrated away from the neural tube explant. They further demonstrated Na^+-HCO_3^--dependent regulatory control of pH_i in all cells, whereas 66% also have a Na^+/H^+ exchange mechanism. They speculated that Na^+/H^+ exchange might be activated as neural crest cells begin to migrate, leading to the higher pH_i in cells which have left the explant.

Two recent manuscripts have reported pH_i in explants of dorsal ectoderm from gastrulating *Xenopus* embryos (SATER et al. 1994; GUTKNECHT et al. 1995). Both studies utilized BCECF to measure pH_i and found values ranging from 7.6 to 7.88 in the dorsal ectoderm of stage-10 embryos. GUTKNECHT et al. (1995) found that dorsal ectoderm pH_i decreased with advancing embryonic age from 7.88 at stage 10, to 7.59 at stage 11, to 7.26 at stage 13. They provided evidence that this decrease of pH_i could be attributed to a changing set point of pH_i-regulatory proteins, including a Na^+/H^+ exchanger (NHE). Furthermore, these authors found that prevention of this age-related cytoplasmic acidification led to suppression of posterior development, although neural induction and anterior development were unaffected. In contrast, SATER et al. (1994) found an alkalinization of dorsal ectoderm in response to the signal for neural induction. The basis of these conflicting results is unclear, but may have resulted from the different methodologies used. Both studies indicated that pH_i is very high in early embryonic cells and may have morphogenetic influences.

The method we chose to measure pH_i in mammalian embryos utilizes the transplacental distribution of a tracer dose of the ^{14}C-radiolabeled weak acid 5,5-dimethyloxazolidine-2,4-dione (^{14}C-DMO). This compound distributes throughout the body only in response to the concentration of water and pH in the tissues (WADDELL and BUTLER 1959). The nonionized form of the compound freely crosses cell membranes and attains the same concentration in all body water; the ionized form of the compound, which has very limited permeability across cell membranes, accumulates in compartments with high pH.

Initial measurements of pH_i were done using whole-embryo homogenates from rats (COLLINS et al. 1989). It was found that the average pH_i of rat embryos decreased steadily from 7.47 ± 0.03 on day 11.5 to 7.11 ± 0.03 on day 14. This pattern was evident in the embryos of two other mammalian species, mouse and monkey (COLLINS et al. 1996), although a slightly different model was used to calculate pH_i in these latter two species (Table 3). These results indicate to us that the average pH_i of mammalian embryos declines during organogenesis, a finding in general agreement with the recent studies described above performed in *Xenopus*. This decline can be correlated closely with declining growth rate, but whether this association is causal remains to be proven. pH_i of comparably staged mouse and rat embryos is nearly identical, whereas the pH_i of monkey embryos is 0.1–0.3 pH units lower. We speculate that the lower pH_i of the primate embryo compared to that of rodent embryos is associated with the slower growth rate of primate embryos during organogenesis.

During these studies, the pH of embryo blood and extraembryonic fluids was also measured. In rats, embryonic blood pH was 7.54, 7.53, and 7.52 on days 11.5, 12, and 13, respectively, and then declined to 7.43 ± 0.02 on day 14. These values were all close to those of maternal blood, which ranged from 7.47

Table 3. Intracellular pH of mouse, rat, and monkey embryos at comparable developmental stages

Species[a]	Day of gestation	No. of animals	Stage[b]	Maternal blood pH	Embryonic pH_i	DMO ratio[g]
Mouse	9	4		7.27 ± 0.10[c]	7.65 ± 0.02[f]	1.37 ± 0.12
Rat	11	6		7.33 ± 0.10[c]	7.66 ± 0.05[f]	1.42 ± 0.10
Monkey	24–29	5	10–12	7.33 ± 0.04[c]	7.35 ± 0.18[f]	1.36 ± 0.32
Mouse	11	4		7.24 ± 0.06[d]	7.22 ± 0.06[c]	1.27 ± 0.07
Rat	13	13		7.44 ± 0.07[d]	7.27 ± 0.10[c]	0.97 ± 0.07
Monkey	30–31	4	14–16	7.33 ± 0.04[d]	7.15 ± 0.15[c]	1.02 ± 0.07
Mouse	12	5		7.30 ± 0.10[e]	7.12 ± 0.10[f]	1.15 ± 0.09
Rat	14	8		7.44 ± 0.07[e]	7.01 ± 0.09[f]	0.87 ± 0.09
Monkey	35–36	3	18–20	7.35 ± 0.04[e]	6.79 ± 0.21[f]	0.79 ± 0.09

Taken from Collins et al. (1996). pH values and dimethadione (DMO) ratios are means ± SD.

[a] Mouse values taken from Scott et al. (1987); the rat values represent a larger number of samples which include the data from Scott et al. (1987).

[b] Comparable developmental stages in rodent embryos were derived from Butler and Juurlink (1987). Stages 10–12 were gestational days 8.5–9.5 in the mouse, and gestational days 10–11 in the rat. Stages 14–16 were gestational days 10.5–11.5 in the mouse, and 12–13 in the rat. Stages 18–20 were gestational days 12.5–13.5 in the mouse, and 14–15 in the rat.

[c] Not significantly different at the $\alpha = 0.05$ level using the general linear model analysis of variance (ANOVA).

[d] Using Duncan's multiple range test ($\alpha = 0.05$), mouse pH_e was significantly different from monkey pH_e, which was significantly different from rat pH_e.

[e] Using Duncan's multiple range test ($\alpha = 0.05$), the rat pH_e was significantly different from either the monkey or mouse pH_e (which are not significantly different from each other).

[f] Using Duncan's multiple range test, it was determined that both mouse and rat pH_i (which are not significantly different from each other) are significantly different from monkey pH_i at $p \leq 0.05$.

[g] DMO concentration ratio in extraembryonic fluid over maternal plasma.

to 7.53. In C57BL/6 mice, embryo blood pH ranged from 7.35 to 7.45 on days 10 and 11 of gestation. In contrast to the rat, the pH of mouse embryo blood was 0.1–0.2 pH units higher than maternal blood, indicating a potential gradient for differential distribution of weak electrolytes between mother and embryo. We were unable to collect blood from monkey embryos. In comparison to rodents, the pH of chick embryo blood measured by microelectrode ranged from 7.66 to 8.0 on days 4 and 6 of development, except for a value of 7.42 in the jugular vein of a 6-day-old chick (MEUER et al. 1989).

Interstitial (extracellular) pH (pH_e) of chick embryos has been measured by microelectrode, and an interesting craniocaudal gradient was observed in day-2 (GILLESPIE and MCHANWELL 1987) and day-4, but not day-6 embryos (MEUER et al. 1989). In the former study, pH was very high in the cranial interstitial regions (approximately 8) and dropped off markedly in the unsegmented caudal region, the segmental plate (approximately 7.6), and it was suggested that high pH facilitated cell migration.

Most recently, our own efforts have been directed toward pH_i measurement within small, discrete regions of the mammalian embryo. Based on our earlier experience we have selected radiolabeled DMO measurement by computer-assisted image analysis of autoradiographic sections. We have found that DMO distribution is heterogeneous in day-10 or -11 mouse embryos, suggesting fluctuating regional pH_i (SCHREINER et al. 1993, 1995). For example, in gestation day-10 embryos, pH_i is higher in the proximal core mesoderm of the limb than in the peripheral mesoderm just underlying the ectoderm. As embryonic age increases, this pH_i pattern shifts. By day 11 of gestation, the area of high pH_i has shifted to the peripheral mesoderm, so that the lower pH_i is characteristic of the proximal core mesoderm. Presently, we are unable to make any correlations between this fluctuating pH_i profile and the special sensitivities of the distal, postaxial limb bud to malformations induced by pH_i-altering teratogens.

D. Pharmacokinetics

The exchange of drugs and chemicals between mother and embryo is a subject of critical importance in deciding whether a particular exposure might have teratologic consequences. Many physiological and physicochemical factors have a role in determining the rate and extent of placental transfer, and these have been reviewed frequently (e.g., NAU and LIDDIARD 1978; MIRKIN 1973; WADDELL and MARLOWE 1981; LEVY and HAYTON 1973; KRAUER et al. 1980). One factor involved in the *extent* of transplacental distribution of weak electrolytes (weak acids and bases) is the degree of ionization, which is directly related to the pH of fluid in which the electrolyte is dissolved. Generally, the ionized form of a weak electrolyte cannot pass through the plasma membrane because of its high polarity, whereas the nonionized, nonpolar form can readily diffuse through this barrier. Thus the nonionized form is considered to be in equilibrium on either side of the cell membrane, but the amount of ionized electrolyte is dependent on the pH of the fluid in each compartment and the p*K* (the pH at which half of the electrolyte is ionized and half is nonionized) of the agent under study.

Perhaps the first discussion of "ion trapping" as a factor in teratogenic response evolved during the studies of KEBERLE et al. (1965) and FABRO (1973), who examined the distribution of thalidomide following maternal administration. Although thalidomide administration at early stages of rabbit development was not demonstrated to induce limb malformations, both groups showed that thalidomide, in the form of metabolites, most of which are weak acids, accumulated in the rabbit blastocyst when compared to maternal plasma or uterine fluid. The suggested mechanism of this accumulation is: (a) diffusion of nonpolar thalidomide into the blastocyst; (b) subsequent spontaneous hydrolysis to polar metabolites in the blastocyst fluid and/or embryonic cells due to high pH; (c) trapping of these metabolites due to inability to cross membranes associated with their high polarity. These findings remain a potentially important part of the unsolved mystery of thalidomide-induced teratogenesis. A shortcoming was the lack of accurate measurements of the pH within the compartments under study.

The idea that pH is an important factor in determining the extent of placental transfer was taken a step further in the studies by NAU and SCOTT (1986). In this work, pH of maternal blood was measured directly and pH of the embryo was calculated according to the transplacental distribution of DMO, a probe which distributes between compartments based solely on pH (WADDELL and BUTLER 1959). NAU and SCOTT (1986) demonstrated that, on day 9 of mouse gestation, pH of the embryo is about 0.4 pH units higher than maternal blood. These authors went on to show: (a) that several weak acids, including VPA and methoxyacetic acid accumulate in the mouse embryo; (b) that the accumulation is dependent on the p*K* of the acid, i.e., the lower the p*K*, the higher the accumulation; (c) that after equilibrium is reached, the disparity between compartments remains steady over time; (d) the accumulation is dependent on the pH gradient, as shown by an absence of accumulation later in pregnancy when the pH of mother and embryo is about the same; and (e) the accumulation in the embryo is dependent on an ionizable hydrogen. When the acid function of VPA was transformed to a nearly neutral amide, the resulting drug, valpromide, did not accumulate in the embryo.

Many of these principles were confirmed by BROWN (1987), who investigated the distribution of carboxylic acids in whole-embryo culture using rat embryos explanted on day 12 (plug date, day 1). He showed that xenobiotics such as DMO, VPA, and methoxyacetic acid distributed according to pH gradient between exocoelomic fluid and the culture medium. Endogenous acids such as propionate and butyrate exceeded the pH-driven accumulation by incorporation into embryonic macromolecules.

Thus pH can be a driving force in embryonic accumulation of drugs and environmental chemicals. It has been suggested that the preponderance of weak acids as human teratogens (NAU and SCOTT 1986) is at least partially dependent on the accumulation of acidic substances in the alkaline milieu of the early mammalian embryo. In contrast, the transplacental distribution of weak bases is not strictly governed by the pH gradient (ROBERTS et al. 1989).

The pH gradient between mother and embryo can be an integral factor in a number of phenomena related to teratologic response. This gradient changes temporally during organogenesis as the pH of embryos becomes lower with age (SCOTT et al. 1990; BROWN 1987; NAU and SCOTT 1986), whereas maternal pH remains relatively constant. Thus the higher potency of many teratogens during early organogenesis may in part be related to higher exposure. Species difference in teratologic response may also be related to the size of the maternal/embryonic pH gradient. Although embryonic pH at comparable developmental stages is nearly identical in mouse and rat embryos, pH of maternal blood is significantly higher in the rat (SCOTT et al. 1987). Recently it has been found that primate embryos during early organogenesis have pH values nearly equal to maternal plasma (COLLINS et al. 1996), so that the potential for pH gradients to be a determinant of species difference in teratologic response is further enhanced (Table 3).

E. Cellular Regulation of Intracellular pH

In the early part of this century, it was thought that small ions such as H^+ were distributed across the plasma membrane according to a Donnan equilibrium. The interested reader is referred to ROOS and BORON (1981) for a historical review leading to the present view that pH_i is maintained well above equilibrium by plasma membrane-regulatory processes. The sum of these regulatory processes must account for H^+ passively entering the cell, the acidifying effects of metabolism, and the fluxes of ionized weak acids and bases. Although the eukaryotic cell has a variety of intracellular buffers, these offer only partial, short-term solutions to acid loading. In the long run, excess acid must be extruded from the cell, and we next discuss the more common means by which this is accomplished.

As depicted in Fig. 2, a variety of ion translocating systems have been shown to maintain cellular pH homeostasis, including the NHE, both Na^+-dependent and -independent Cl^-/HCO_3^- exchangers, H^+ channels, Na^+ channels, and adenosine triphosphate (ATP)-dependent H^+ pumps. Of these, the NHE and Cl^-/HCO_3^- exchange systems are thought to be the primary regulators of pH_i, although the relative contribution of the other aforementioned systems is cell type specific.

I. Na^+/H^+ Exchange

Members of the NHE gene family electroneutrally translocate extracellular Na^+ into the cell in exchange for an intracellular H^+. Recent molecular

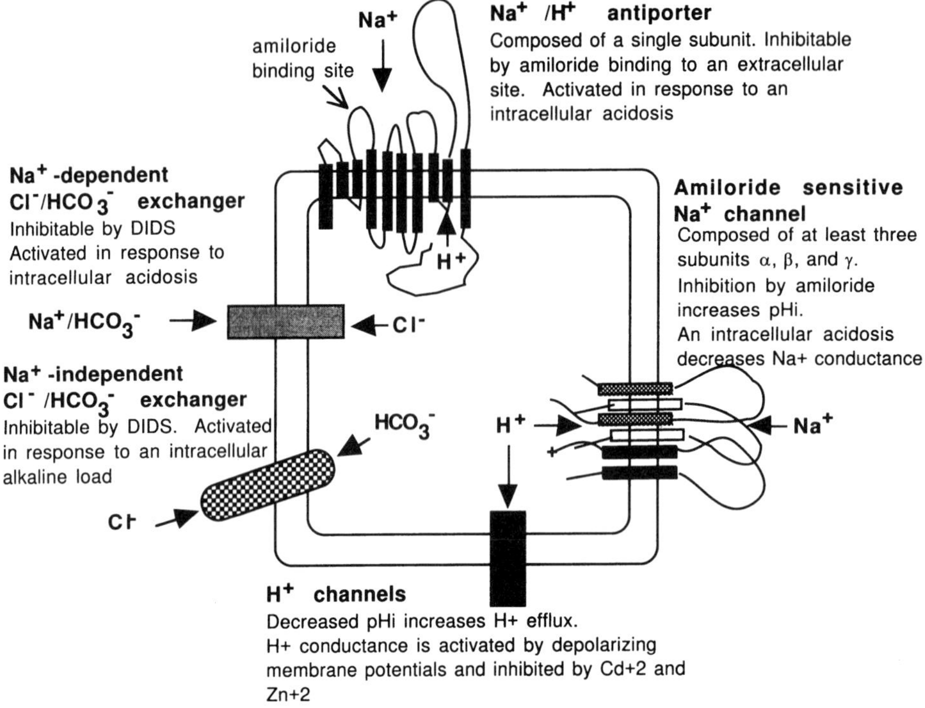

Fig. 2. Cellular ion translocating systems

biology approaches have identified four distinct members of this gene family in a variety of species, including human, rat, rabbit, porcine, hamster, mouse, and guinea pig (TSE et al. 1993). The characterized cDNA have been denoted NHE-1, -2, -3, and -4. All predict plasma membrane-localized proteins comprised of a similar domain architecture ranging in size from 717 to 832 amino acids. The greatest degree of homology is observed between the N-terminal hydrophobic domains of the isoforms, which are thought to contain an N-terminal signal peptide, a single conserved putative N-linked glycosylation site, and ten to 12 membrane-spanning domains involved in ion transport. The carboxy terminal hydrophilic domain exhibits the greatest isoform diversity and is believed to differentially regulate ion transport by isoform-specific phosphorylation events by such kinases as protein kinase C, chorioallantoic membrane (CAM) kinase II, and/or the cyclic adenosine monophosphate (cAMP)-dependent kinase.

The activity of all four exchangers is regulated by a cytoplasmically localized allosteric H^+ modifier site, the "set point" of which is cell type and exchanger specific (WAKABAYASHI et al. 1992). Truncation experiments have revealed that this site is located in a cytoplasmic loop of the amino terminus of NHE-1 (WAKABAYASHI et al. 1992). Recent experiments in which the cytoplasmic domain of human NHE-1 has been mutagenized suggest that the affinity of the allosteric site for H^+ is regulated by phosphorylation events, but also by calmodulin-binding sites present in the middle of the carboxy-terminal cytoplasmic domain. The calmodulin-binding sites serve to inhibit NHE activity through an autoinhibitory mechanism which is released following Ca^{+2} binding leading to NHE activation (WAKABAYASHI et al. 1994).

Northern blot analysis and RNAse protection assays of adult rat and rabbit tissues revealed that NHE-2 expression is present in a variety of tissues, including uterus, kidney, stomach, colon, jejunum, ileum, testes, heart, brain, lung, and skeletal muscle (TSE et al. 1993; COLLINS et al. 1993; WANG et al. 1993). NHE-3 expression is detectable only in kidney, stomach, small intestine, and large intestine (ORLOWSKI et al. 1992). NHE-4 also exhibits a very restricted expression pattern detectable only in uterus, brain, kidney, stomach, small intestine, and large intestine (ORLOWSKI et al. 1992). In contrast to isoforms 2, 3, and 4, NHE-1 expression has been detected in every tissue examined, albeit at varying levels (ORLOWSKI et al. 1992). The ubiquitous nature of NHE-1 expression has resulted in NHE-1 being dubbed the "housekeeping exchanger," whose primary functions are to maintain pH_i and cell volume regulation. The other exchanger isoforms are thought to be involved in more specialized roles, such as the renal transepithelial reabsorption of Na^+ and the luminal secretion of H^+ necessary for HCO_3^- reabsorption.

Examination of NHE activity has relied on the use of known inhibitory compounds including amiloride, cimetidine, clonidine, and harmaline. Relatively few methods have been used to study NHE activity. The common methods involve cellular acid loading with NH_4^+ followed by determination of the difference between H^+-activated $^{22}Na^+$ uptake in the presence and absence of the K^+-sparing diuretic amiloride or one of its 5-amino substituted analogues. Alternatively, cellular pH_i recovery has been evaluated in the presence and absence of inhibitors using the pH-sensitive fluorescent indicator BCECF. These approaches have been utilized in a variety of cell lines and types; however, it is only recently that associations have been made between specific NHE isoforms, expression in cell types, and inhibition by amiloride and its analogues. Transfection studies in which the different NHE contained in expression plasmids have been introduced into exchanger-deficient cell lines have revealed that NHE-1 is the isoform most sensitive to inhibition by amiloride and its derivatives (ORLOWSKI 1993). Although the $K_{i(0.5)}$ for NHE-1 is approximately the same as for NHE-2 (1.5×10^{-6} *M*), NHE-1 is more sensitive to inhibition by the more potent NHE inhibitors 5-(*N*-ethyl-*N*-isopropyl amiloride (EIPA; K_i, 1.5×10^{-8} *M*) and dimethyl amiloride (DMA; K_i, 2.3×10^{-8} *M*), whereas the determined K_i values for NHE-2 are 7.9×10^{-8} and 2.5×10^{-7} *M*, respectively (YU et al. 1993).

Studies in our own laboratory utilizing a cloned mouse NHE-1 cDNA suggest that NHE-1, but not NHE-2, is expressed in the day-10 (plug date, day 0), developing mouse embryo and forelimb bud a gestational time point at which a variety of pH_i-altering teratogens are capable of inducing forelimb malformations. Consistent with the expression data of NHE-1 in the day-10 forelimb bud, the recovery of acid-loaded mouse limb bud cells was inhibited in the presence of amiloride, suggesting that functional NHE protein is expressed in these cells (Duggan and Scott 1989). NHE activity may also be critical to such early developmental processes as blastocoele expansion, since this process was inhibitable by EIPA (Manejwala et al. 1989).

Several recent lines of evidence suggest that genetic variability may exist at the NHE-1 locus in mouse as well as in human. In mouse, three different NHE-1 alleles have been identified by Morahan and Rakar (1993). Interestingly, additional work by Morahan et al. (1994) has linked the development of type 1 diabetes in the mouse with the NHE-1 locus. We are currently investigating which of the alleles identified by Morahan and Rakar (1993) are present in the teratogenically sensitive mouse strains such as C57BL/6J versus the less sensitive strains such as CD-1 and SWV and whether or not the identified allelic differences are found within the coding regions of the NHE-1 gene. Consistent with the possibility of an allelic difference having functional consequences, in vitro examination of NHE activity in limb bud cell cultures derived from C57Bl/6 versus SWV mice suggest that SWV mice possess a more active NHE or other pH_i regulator than C57B1/6 mice (Duggan and Scott 1989). Differences in the human population at this locus may also play a role in human susceptibility to teratogenesis, since differences in NHE activity have been documented in cells isolated from individuals with and without diabetic neuropathy and essential hypertension (Ng et al. 1994; Williams and Howard 1994).

II. Cl^-/HCO_3^- Exchange

The Cl^-/HCO_3^- exchangers comprise two additional mechanisms of pH_i regulation, both of which are inhibitable by 4,4′-diisothiocyanato-2,2′-stilbene disulfonic acid (DIDS). The Na^+-independent Cl^-/HCO_3^- exchanger effluxes HCO_3^- in response to an intracellular alkaline load. Three distinct members of this anion exchanger gene family have recently been characterized and cloned, denoted AE1, AE2, and AE3 (Tanner 1993; Kopito 1990). AE1, alias band 3, is the exchanger found predominantly in red blood cells. AE3 has been found predominantly in neural tissues, and AE2 is found ubiquitously (Kopito 1990). Although not likely involved in the cellular response to teratogens which induce an intracellular acidosis, it is interesting to note that the mouse AE2 gene has been mapped to chromosome 5 in the vicinity of two mouse limb mutants, hemimelic extra toes (Hx) and hammertoe (Hm) (White et al. 1994).

Of greater interest is the Na^+-dependent Cl^-/HCO_3^- exchange system, which exchanges extracellular Na^+ and HCO_3^- ions for an intracellular Cl^-, resulting in the neutralization of an intracellular acidosis. This DIDS-inhibitable exchange system has been identified in a variety of cell types, including smooth muscle (KAHN et al. 1991), fibroblasts (CASSEL et al. 1988), glomerular mesangial cells (BOYARSKY et al. 1988), hepatic epithelial cells (GRUBMAN et al. 1994), and neurons (THOMAS 1982; SCHWIENING and BORON 1994). Like the NHE, this exchange system is activated in response to an intracellular acidosis; its point of inactivation, however, seems to be cell type specific, ranging from becoming inactive in the normal pH_i range (BOYARSKY et al. 1988) to not being inactivated until intracellular pH values are obtained as high as pH 7.4 (BORON et al. 1978; THOMAS 1982; KAHN et al. 1991). Whether or not the Na^+-dependent Cl^-/HCO_3^- exchange system is also comprised of a multigene family is currently unknown and awaits their identification and characterization at the molecular level.

III. Na^+ Channels

Na^+ channels are not characteristically thought of as a regulator of pH_i. These channels, comprised of a multigene family, fall into two primary groupings, the voltage-dependent channels involved in neurosensory processes and voltage-independent channels characteristic of tight epithelial layers. Recent reports suggest that a relationship exists between activity of the epithelial (voltage-independent) Na^+ channels and pH_i. The studies by CHURARD and DURAND (1992) using a *Xenopus* epithelial cell line indicate that a reciprocal relationship exists between Na^+ transport and pH_i; inhibition of Na^+ transport by amiloride concentrations which would not affect NHE activity induced cellular acidifications, and Na^+ transport at the apical membrane declined with decreasing pH_i. The work of HARVEY et al. (1991) and LYALL et al. (1993) suggest that apical epithelial Na^+ channels, in addition to conducting Na^+ ions, also serve as H^+-conductive pathways which are sensitive to inhibition by amiloride and its analogues. Thus, Na^+ channel activity may be altered by the action of pH_i-altering teratogens in two ways. First, teratogen-induced reductions in embryonic pH_i may be inhibiting the normal activity of these channels, resulting in the disruption of the recently characterized fields of ionic gradients thought to be involved in several developmental processes (METCALF et al. 1994; SHI and BORGENS 1995). In support of this hypothesis, the amiloride-sensitive Na^+ channels have been shown to generate the transneural tube potential present in the developing axolotl embryo at the time of neural tube closure. Inhibition of Na^+ channel activity by iontophoresing either amiloride or benzamil into the lumen of the closing neural tube inhibits activity of the epithelial Na^+ channels, disrupts the transneural tube potential, and phenotypically produces embryos with severe abnormalities of the central nervous system and cranium (SHI and BORGENS 1994). Alternatively, inhibiting the activity of Na^+ channels by known inhibitors may also serve to potentiate teratogen-induced acidotic conditions by inhibiting the leak of H^+ through these channels, thus prolonging the acidotic milieu. Consistent with this hypothesis is the observed ability of amiloride and benzamil to potentiate the effects of pH_i-altering teratogens such as acetazolamide. In addition, triamterene, another agent known to inhibit the activity of epithelial Na^+ channels, was previously demonstrated to also potentiate the teratogenic action acetazolamide in the rat (ELLISON and MAREN 1972).

Physiological approaches have identified two types of epithelial Na^+ channels, denoted H and L, referring to a high (H) or low (L) amiloride-binding affinity, respectively (for a review, see OH and BENOS 1992). H-type channels are inhibited by amiloride with a K_i less than 0.5 μM, are found in both epithelial and nonepithelial tissues such as the kidney, bladder, sweat ducts, thyroid, colon enterocytes, and human B lymphoid cells, and are thought to be composed of six nonidentical protein subunits. The rank order of potency of amiloride analogue binding to H-type channels is phenamil, benzamil > amiloride > > EIPA, in contrast to the interaction of these agents with the L-type channels, in which the order is phenamil, benzamil ≥ amiloride = EIPA. Relative to amiloride, the L-type channel has a K_i greater than 1 μM. L-type channels have also been characterized in a variety of cell types, including kidney proximal tubules, brain endothelial cells, cochlear hair cells, lung aveolar type II cells, and trophectoderm cells of the blastocyst.

The interrelationship observed between Na^+ channel activity and pH_i may also be an indicator of teratogenic sensitivity in the human population. cDNA for three subunits of an H-type Na^+ channel have been characterized in the rat and human (MCDONALD et al. 1994; CANESSA et al. 1993, 1994). Interestingly, the presence of a premature stop codon in the human β-gene has already been linked to one human syndrome, Liddles' syndrome, in which increased Na^+ absorption in the kidney distal nephron results in severe hypertension (SHIMKETS et al. 1994).

IV. H^+ Channels

Although direct molecular verification as to the existence of a voltage-activated H^+ channel is still currently lacking, a wealth of information has been accumulated over the last 10 years regarding the regulation of H^+ currents from a variety of cell types, including oocytes, neurons, myocytes, macrophages, granulocytes, and epithelial cell types. As reviewed by DECOURSEY and CHERNY (1994) and LUKACS et al. (1993), the identification and characterization of these putative channels have relied on patch clamp- and voltage clamp-based experiments, although a direct clamp on a single channel has not been obtainable. Properties unique to this current include the following: (a) there is a high selectivity of H^+ over other cations (≥10^6 to 1), (b) intracellular acidification increases H^+ efflux, (c) H^+ conductance is reduced by acidotic extracellular pH, and (d) depolarizing membrane potentials activate conductance. Therefore, it has been proposed that this putative H^+ channel is another cellular acid extrusion mechanism which may function following metabolic acid bursts. Currently no true organic inhibitors of this putative channel have been identified; however, both Cd^{+2} and Zn^{+2} prohibit conductance at concentrations of 1 m*M* or less.

The only data linking these putative channels to the embryo are the detection of Na^+-independent, amiloride-insensitive H^+ currents activated by depolarizing immature urodele oocytes (Barish and Baud 1984). Baltz et al. (1990) also found that the recovery of acid-loaded two-cell mouse embryos in HCO_3^--free medium was insensitive to amiloride, EIPA, or the presence of extracellular Na^+, also suggestive of acidic pH_i activating a proton channel.

Although possessing all of the properties of interest as a regulator of pH_i, evaluating the putative role of this channel in the embryonic response to pH_i -altering teratogens will require its identification at the molecular level.

F. Potentiation of Teratogenesis by Inhibitors of Intracellular pH Recovery

To further explore the hypothesis that a variety of teratogens mediate their actions by altering embryonic pH_i levels, several studies have been conducted in which pH_i-altering teratogens have been administered in conjunction with compounds known to have an inhibitory action on one of the primary pH_i-regulatory systems. The agents which have been used include amiloride and DIDS (see Table 2). Amiloride is a weak base whose protonated form is known to inhibit both NHE activity and that of amiloride-sensitive Na^+ channels. DIDS is a potent inhibitor of both Na^+-dependent and -independent Cl^-/HCO_3^- exchangers.

Previous work by Ellison and Maren (1972) demonstrated that a high dose of amiloride (4 mg/kg), administered in conjunction with acetazolamide, potentiated the teratogenic action of acetazolamide in the rat. Studies in our own laboratory have revealed that the potentiating action of amiloride also extends to acetazolamide-induced teratogenesis of the C57BL/6 mouse. Measurement of pH_i levels in whole-embryo homogenates as well as forelimb buds following treatment with either acetazolamide alone or in combination with amiloride revealed that amiloride coadministration further reduced pH_i levels and prolonged the acetazolamide-induced acidosis (Scott et al. 1990; Schreiner et al. 1995). An increase in the frequency and severity of postaxial ectrodactyly following coadministration of amiloride is not a phenomenon unique to acetazolamide teratogenesis, since an increase is also observed upon coadministration with both VPA and retinoic acid (Scott et al. 1989; Table 2). In the case of acetazolamide, whether the increased acidotic condition induced by amiloride is attributable to inhibition of H^+ ion currents effluxing from Na^+ channels or due to inhibition of NHE activity is currently unknown. Two approaches are currently being taken to distinguish between these possibilities. The first approach is to evaluate the teratogenicity of a variety of amiloride analogues when coadministered with acetazolamide. The analogues used were chosen based on the fact that a variety of amiloride analogues have been formulated which possess greater affinities for one of these two systems (Kleyman and Cragoe 1988). Structure–activity relationships have revealed that derivatives bearing hydrophobic substitutions on the terminal nitrogen atom of the guanidine moiety of amiloride have a greater affinity for epithelial Na^+ channels. The analogues most frequently used in the study of epithelial Na^+ channels are benzamil and phenamil. In contrast, the attachment of hydrophobic groups on the 5-amino nitrogen atom of amiloride increases its affinity for NHE. Derivatives in this class which have been extensively used in the study of NHE activity include 5-(*N*-methyl-*N*-isobutyl) amiloride (MIBA), EIPA, 5-(*N*,*N*-dimethyl) amiloride (DMA), and 5-(*N*,*N*-hexamethylene) amiloride (HMA). Excluding HMA, all of the analogues have been capable of potentiating the action of acetazolamide (Bell and Scott 1994). Interestingly, the acidotic environment produced by acetazolamide must be present for these agents to act, since their administration alone fails to induce the malformation.

The second approach being taken is to directly evaluate the role of the proteins which maintain pH homeostasis by generating transgenic mouse lines deficient in the different systems and evaluating their subsequent teratogenic sensitivity.

Although DIDS is a known inhibitor of both Na^+-dependent and -independent Cl^-/HCO_3^- exchange systems, the system of interest is likely only the Na^+-dependent one, since this system alleviates an acid load. Interestingly, as in the case of amiloride, coadministration of DIDS with acetazolamide also resulted in an increase in the frequency and severity of forelimb malformations. These results suggest that circumstances which will prolong a teratogen-induced acidotic condition will increase the severity of the effect.

G. Cellular Activities Associated with pH

If alteration of pH_i is a mechanism by which exogenous agents induce malformations, then we would like to know which cellular activities are disrupted leading to abnormal development. We have focused our efforts on cell proliferation for two reasons. First, the malformation under scrutiny, postaxial, forelimb ectrodactyly, is a tissue reduction outcome with no indication that cell death is involved (Holmes and Trelstad 1979). Thus a logical route for the absence of skeletal elements is a localized reduction of cell proliferation. Second, a large, sometimes controversial body of literature exists indicating an association between pH_i, activation of Na^+/H^+ exchange, and cell proliferation. Grinstein et al. (1989) critically reviewed the supporting and contradictory evidence for a causal association between activation of Na^+/H^+ exchange and subsequent cytoplasmic alkalinization leading to cell proliferation. Their strongest conclusion was that exchanger activity can be permissive to the proliferative response, but they admit that the importance of antiport activity to the processes leading to cellular proliferation remains controversial.

We have examined proliferative rate in the limb bud of mouse embryos exposed to acetazolamide or acetazolamide plus amiloride using immunohistochemistry of bromodeoxyuridine (BrdU) to document the labeling index (SCHREINER and SCOTT 1988; SCHREINER et al. 1995). A decrease of the labeling index in the postaxial limb bud mesoderm of embryos exposed to acetazolamide plus amiloride has been documented in this work. However, this decrease was small (about 15% lower than controls) and was only seen at a single time point (15 h after treatment). We are skeptical that the acidotic changes due to this treatment regimen lead to malformation by interfering directly with cell proliferation.

If, in this single case, decrease of proliferative rate is not part of the abnormal pathogenesis, there are many other cellular functions sensitive to pH alterations which could plausibly play an important developmental role. Some of these include gap-junctional conductance, receptor-mediated endocytosis, apoptosis, enzyme activity (especially as related to glycolysis), protein synthesis, membrane permeability and conductance, cell migration, cellular differentiation, and secretion and uptake of intercellular messengers, i.e., insulin, serotonin, and intracellular messengers, i.e., cAMP. Just as with cellular proliferation, the exact role of pH_i in the regulation of these cellular processes has not been defined. Generally, a permissive role has been assigned to pH regulation of cellular events and this concept has been expanded by BUSA and NUCCITELLI (1984), who suggest a central role of protons in the regulation of cellular processes, since "all life is based on aqueous chemistry and because water spontaneously ionizes."

H. Conclusion

Ten years ago we hypothesized that reduction of pH_i might be a mechanism by which exogenous agents induce congenital malformations. During this time we have done many experiments which generally support the hypothesis, but unequivocal proof remains elusive. We hope to take a large step forward using targeted gene knockout strategy to compromise the ability of the embryo to respond to acidic insults. Our initial efforts are focused on disruption of the NHE-1 locus, since this exchanger is present in the embryo and thought to be a primary cellular regulator of pH_i. Due to the many compensatory mechanisms of pH_i regulation potentially present in the developing embryo, disruption of one locus may not be sufficient to predispose the embryo to teratogenic insults. This line of investigation will begin to describe the ionic regulatory machinery available to the mammalian embryo, but whether it verifies our hypothesis remains to be seen.

References

Anonymous (1979) Fetal trimethadione syndrome: report of an additional family and further delineation of this syndrome. In: Oski FA, Stockman JA III (eds) Yearbook of pediatrics. Yearbook Medical, Chicago, pp 417–419

Baltz JM, Biggers JD, Lechene C (1990) Apparent absence of Na^+/H^+ antiport activity in the two-cell mouse embryo. Dev Biol 138: 421–429

Barish M, Baud C (1984) A voltage-gated hydrogen ion current in the oocyte membrane of the axolotl, Ambystoma. J Physiol (Lond) 352: 243–263

Barr M (1973) The teratogenicity of cadmium chloride in two stocks of Wistar rats. Teratology 7: 237–242

Bell SM, Scott WJ (1994) Inhibitors of Na/H exchanger activity potentiate the teratogenic action of acetazolamide. Teratology 49: 368

Bennett G, Scott W, Collins M (1989) Hyperthermia-induced postaxial forelimb ectrodactyly. Teratology 39: 441–442

Boron W, Russell J, Brodwick M, Keifer D, Roos A (1978) Influence of c-AMP on intracellular pH regulation and chloride fluxes in barnacle muscle fibers. Nature 276: 511–513

Boyarsky G, Ganz M, Sterzel R, Boron W (1988) pH regulation in single glomerular mesangial cells. II. Na^+-dependent and -independent Cl^-/HCO_3^- exchangers. Am J Physiol 255: C857–C869

Brachet J (1968) Chemical embryology. Hafner, New York

Brown N (1987) Teratogenicity of carboxylic acids: distribution studies in whole embryo culture. In: Nau H, Scott W (eds) Pharmacokinetics in teratogenesis. CRC, Boca Raton, pp 153–163

Busa W, Nuccitelli R (1984) Metabolic regulation via intracellular pH. Am J Physiol 246: R409–R438

Butler T, Kuroiwa Y, Waddell W, Poole D (1966) Effects of 5,5-dimethyl-2,4-oxazolidinedione (DMO) on acid-base and electrolyte equilibria. J Pharmacol Exp Ther 152: 62–66

Canessa CM, Horisberger J-D, Rossier BC (1993) Epithelial sodium channel related to proteins involved in neurodegeneration. Nature 361: 467–469

Canessa CM, Schild L, Buell G, Thoren B, Gautschi I, Horisberger J-D, Rossier BC (1994) Amiloride-sensitive epithelial Na^+ channel is made of three homologous subunits. Nature 367: 463–466

Cassel D, Scharf O, Rotman M, Cragoe E, Maayan K (1988) Characterization of Na^+-linked and Na^+-independent Cl^-/HCO_3^- exchange systems in Chinese hamster lung fibroblasts. J Biol Chem 263: 6122–6127

Churard F, Durand J (1992) Coupling between the intracellular pH and the active transport of sodium in an epithelial cell line from Xenopus laevis. Comp Biochem Physiol 102A: 7–14

Collins J, Honda T, Knobel S, Bulus N, Conary J, Dubois R, Ghishan F (1993) Molecular cloning, sequencing, tissue distribution, and functional expression of a Na^+/H^+ exchanger (NHE-2). Proc Natl Acad Sci USA 90: 3938–3942

Collins M, Duggan C, Schreiner C, Scott W (1989) Decreasing pH of rat embryos and fluids estimated by transplacental distribution of DMO. Am J Physiol 257: R542–R549

Collins M, Walling K, Resnick E, Scott W (1991) Effect of administration time on malformations induced by three anticonvulsant agents in C57BL/6 mice with emphasis on forelimb ectrodactyly. Teratology 44: 617–627

Collins M, Scott W, Hendrickx A, Peterson P, Nau H (1996) Estimated intracellular pH of monkey embryos at various stages of organogenesis by dimethadione distribution. Reprod Fertil Dev (in press)

DeCoursey TE, Cherny VV (1994) Voltage-activated hydrogen ion currents. J Membr Biol 141: 203–223

Dickens C, Gillespie J, Greenwell J (1990) Measurement of intracellular calcium and pH in avian neural crest cells. J Physiol (Lond) 428: 531–544

Duggan CA, Scott W (1989) Na^+/H^+ antiporter activity in cultured limb bud cells from acetazolamide sensitive (C57) and resistant (SWV) mouse embryos. Teratology 39: 450

Edwards M (1986) Hyperthermia as a teratogen: a review of experimental studies and their clinical significance. Teratog Carcinog Mutagen 6: 563–582

Ellison AC, Maren TH (1972) The effect of potassium metabolism on acetazolamide-induced teratogenesis. Johns Hopkins Med J 130: 105–115

Fabro S (1973) Passage of drugs and other chemicals into the uterine fluids and preimplantation blastocyst. In: Boreus L (ed) Fetal pharmacology. Raven, New York, pp 443–461

Ferm V (1971) Developmental malformations induced by cadmium. Biol Neonate 19: 101–107

Ferm VH, Carpenter SJ (1967) Teratogenic effect of cadmium and its inhibition by zinc. Nature 216: 1123

Feuston M, Scott W (1985) Cadmium-induced forelimb ectrodactyly: a proposed mechanism of teratogenesis. Teratology 32: 407–419

Freda J, Dunson W (1984) Sodium balance of amphibian larvae exposed to low environmental pH. Physiol Zool 57: 435–443

Gillespie J, McHanwell S (1987) Measurement of intra-embryonic pH during the early stages of development in the chick embryo. Cell Tissue Res 247: 445–451

Gillespie J, Greenwell J (1988) Changes in intracellular pH and pH regulating mechanisms in somitic cells of the early chick embryo: a study using fluorescent pH-sensitive dye. J Physiol (Lond) 405: 385–395

Grinstein S, Rotin D, Mason M (1989) Na^+/H^+ exchange and growth factor-induced cytosolic pH changes. Role in cellular proliferation. Biochim Biophys Acta 988: 73–97

Grubman SA, Perrone RD, Lee DW, Murray SL, Rogers LC, Wolkoff LI, Mulberg AE, Cherington V, Jefferson DM (1994) Regulation of intracellular pH by immortalized human intrahepatic biliary epithelial cell lines. Am J Physiol 266: G1060–G1070

Gutknecht DR, Koster CH, Tertoolen LGJ, de Laat SW, Durston AJ (1995) Intracellular acidification of gastrula ectoderm is important for posterior axial development in Xenopus. Development 121: 1911–1925

Harvey B, Lacoste I, Ehrenfield J (1991) Common channels for water and protons at apical and basolateral cell membranes of frog skin and urinary bladder epithelia: effect of oxytocin, heavy metals and inhibitors of H^+ adenosine triphosphatase. J Gen Physiol 97: 749–776

Holmes L, Trelstad R (1979) The early limb deformity caused by acetazolamide. Teratology 20: 289–296

Kahn AM, Cragoe JEJ, Allen JC, Seidel C, Shelat H (1991) Effects of pH_i on Na^+-H^+, Na^+-dependent, and Na^+-independent Cl^--HCO_3^- exchangers in vascular smooth muscle. Am J Physiol 261: C837–C844

Keberle H, Faigle J, Fritz H, Knusel F, Loustalat P, Schmid K (1965) Theories on the mechanism of action of thalidomide. In: Robson J, Sullivan F, Smith R (eds) Embryopathic activity of drugs. Churchill, London, pp 210–226

Kiang J, McKinney L, Gallin E (1990) Heat induces intracellular acidification in human A-431 cells: Role of Na^+-H^+ exchange and metabolism. Am J Physiol 259: C727–C737

Kleyman TR, Cragoe EJ (1988) Amiloride and its analogs as tools in the study of ion transport. J Membr Biol 105: 1–21

Kopito R (1990) Molecular biology of the anion exchanger gene family. Int Rev Cytol 123: 177–199

Krauer B, Krauer F, Hytten F (1980) Drug disposition and pharmacokinetics in the maternal-placental-fetal unit. Pharmacol Ther 10: 301–328

Kuczuk MH, Scott WJ (1984) Potentiation of acetazolamide-induced ectrodactyly in SWV and C57BL/6J mice by cadmium sulfate. Teratology 29: 427–435

Layton WM, Layton MW (1979) Cadmium induced limb defects in mice: strain associated differences in sensitivity. Teratology 19: 229–236

Levy G, Hayton W (1973) Pharmacokinetic aspects of placental drug transfer. In: Boreus L (ed) Fetal pharmacology. Raven, New York, pp 29–39

Lucas C, Gillies R, Olson J, Giuliano K, Martinez R, Sneider J (1988) Intracellular acidification inhibits the proliferative response in BALB/c-3T3 cells. J Cell Physiol 136: 161–167

Lukacs GL, Kapus A, Nanda A, Romanek R, Grinstein S (1993) Proton conductance of the plasma membrane: properties, regulation, and functional role. Am J Physiol 265: C3–C14

Lyall V, Belcher TS, Biber TUL (1993) Na^+ channel blockers inhibit voltage-dependent intracellular pH changes in principal cells of frog (Rana pipiens) skin. Comp Biochem Physiol 105A: 503–511

Manejwala FM, Cragoe EJ, Schultz RM (1989) Blastocoel expansion in the preimplantation mouse embryo: role of extracellular sodium and chloride and possible apical routes of their entry. Dev Biol 133: 210–220

McDonald FJ, Snyder PM, McCray PB, Welsh MJ (1994) Cloning, expression, and tissue distribution of a human amiloride-sensitive Na^+ channel. Am J Physiol 266: L728–L734

Messerle K, Webster W (1982) The classification and development of cadmium-induced limb defects in mice. Teratology 25: 61–70

Metcalf MEM, Shi R, Borgens RB (1994) Endogenous ionic currents and voltages in amphibian embryos. J Exp Zool 268: 307–322

Meuer H, Sieger U, Baumann R (1989) Measurement of pH in blood vessels and interstitium of 4 and 6 days-old chick embryos. J Dev Physiol 11: 354–359

Mirkin BL (1973) Drug distribution in pregnancy. In: Boreus L (ed) Fetal pharmacology. Raven, New York, pp 1–26

Morahan G, Rakar S (1993) Localization of the mouse Na^+/H^+ exchanger gene on distal Chromosome 4. Genomics 115: 231–232

Morahan G, McClive P, Huang D, Little P, Baxter A (1994) Genetic and physiological association of diabetes susceptibility with raised Na^+/H^+ exchange activity. Proc Natl Acad Sci USA 91: 5898–5902

Nau H, Liddiard C (1978) Placental transfer of drugs during early human pregnancy. In: Neubert D, Merker H-J, Nau H, Langman J (eds) Role of pharmacokinetics in prenatal and perinatal toxicology. Thieme, Stuttgart, pp 465–482

Nau H, Scott W (1986) Weak acids may act as teratogens by accumulating in the basic milieu of the early mammalian embryo. Nature 323: 276–278

Nau H, Scott WJ (1987) Teratogenicity of valproic acid and related substances in the mouse: drug accumulation and pH_i in the embryo during organogenesis and structure-activity considerations. Arch Toxicol Suppl 11: 128–139

Ng LL, Davies JE, Siczkowski M, Sweeney FP, Quinn PA, Krolewski B (1994) Abnormal Na^+/H^+ antiporter phenotype and turnover of immortalized lymphoblasts from type 1 diabetic patients with nephropathy. J Clin Invest 93: 2750–2757

Oh Y, Benos D (1992) Amiloride-sensitive sodium channels. In: Cragoe E, Kleyman T, Simchowitz L (eds) Amiloride and its analogs. VCH, Weinheim, pp 41–56

Orlowski J (1993) Heterologous expression and functional properties of amiloride high affinity (NHE-1) and low affinity (NHE-3) isoforms of the rat Na/H exchanger. J Biol Chem 268: 16369–16377

Orlowski J, Kandasamy RA, Shull GE (1992) Molecular cloning of putative members of the Na/H exchanger gene family. J Biol Chem 267: 9331–9339

Pagano G, Cipollaro M, Corsale G, Esposito A, Ragucci E, Giordano G (1985) pH-induced changes in mitotic and developmental patterns in sea urchin embryogenesis. Teratog Carcinog Mutagen 5: 101–112

Rink TJ, Tsein RY, Pozzan T (1982) Cytoplasmic pH and free Mg^{2+} in lymphocytes. J Cell Biol 95: 189–196

Roberts LG, Luck W, Holder CL, Scott WJ, Nau H, Slikker W Jr (1989) Embryo-maternal distribution of basic compounds in the CD-1 mouse: doxylamine and nicotine. Toxicol Appl Pharmacol 97: 134–140

Roos A, Boron WF (1981) Intracellular pH. Physiol Rev 61: 297–421

Sater AK, Alderton JM, Steinhardt RA (1994) An increase in intracellular pH during neural induction in Xenopus. Development 120: 433–442

Schreiner C, Scott W (1988) Effects of acetazolamide on proliferation in the developing mouse limb in vitro. Teratology 37: 490

Schreiner CM, Scott WJ, Collins MD, Colvin J, McCandless D (1993) Estimation of intracellular pH by computer assisted imaging in the developing mouse forelimb bud exposed to acetazolamide. In: Fallon JF, Goetinck PF, Kelley RO, Stocum DL (eds) Limb development and regeneration. Wiley-Liss, New York, pp 403–408

Schreiner CM, Scott WJ, Colvin J, McCandless D (1994) Changes of intracellular pH in the mouse embryo forelimb bud following administration of valproic acid. Teratology 49: 363

Schreiner CM, Collins MD, Scott WJ, Vorhees CV, Colvin J, McCandless D (1995) Estimating intracellular pH in developing rodent embryos using a computer imaging technique: changes in embryonic pH and proliferation rates following maternal treatment with acetazolamide. Teratology 52: 160–168

Schwiening CJ, Boron WF (1994) Regulation of intracellular pH in pyramidal neurones from the rat hippocampus by Na^+-dependent Cl^-/HCO_3^- exchange. J Physiol (Lond) 475: 59–67

Scott W, Duggan C, Schreiner C, Collins M, Nau H (1987) Intracellular pH of rodent embryos and its association with teratogenic response. In: Welsch F (ed) Approaches to elucidate mechanisms in teratogenesis. Hemisphere, Washington, pp 99–108

Scott W, Fradkin R, Wilson J (1988) Trimethadione teratogenicity in rats and rhesus monkeys. In: Neubert D, Merker H, Hendrickx A (eds) Non-human primates. Developmental biology and toxicology. Ueberreuter Wissenschaft, Vienna, pp 431–441

Scott W, Kobrin D, McDowell S, Collins M (1989) Potentiation of teratogenesis due to weak acids by coadministration of amiloride. Teratology 39: 480

Scott WJ, Duggan CA, Schreiner CM, Collins MD (1990) Reduction of embryonic intracellular pH: a potential mechanism of acetazolamide-induced limb malformations. Toxicol Appl Pharmacol 103: 238–254

Shi R, Borgens RB (1994) Embryonic neuroepithelial sodium transport, the resulting physiological potential, and cranial development. Dev Biol 165: 105–116

Shi R, Borgens RB (1995) Three-dimensional gradients of voltage during development of the nervous system as invisible coordinates for the establishment of embryonic pattern. Dev Dyn 202: 101–114

Shimkets RA, Warnock DG, Bositis CM, Nelson-Williams C, Hansson JH, Schambelan M, Gell JJR, Ulick SRVM, Findling JW, Canessa CM, Rossier BC, Lifton RP (1994) Liddle's syndrome: heritable human hypertension caused by mutations in the β subunit of the epithelial sodium channel. Cell 79: 407–414

Tanner M (1993) Molecular and cellular biology of the erythrocyte anion exchanger (AE1). Semin Hematol 30: 34–57

Thomas R (1982) Snail neuron intracellular pH regulation. In: Nucitelli R, Deamer D (eds) Intracellular pH: its measurement, regulation and utilization in cellular functions. Liss, New York, pp 189–204

Tse C-M, Levine SA, Yun CHC, Montrose MH, Little PJ, Pouyssegur J, Donowitz M (1993) Cloning and expression of a rabbit cDNA encoding a serum-activated ethylisopropylamiloride-resistant epithelial Na^+/H^+ exchanger isoform (NHE-2). J Biol Chem 268: 11917–11924

Waddell W, Butler T (1959) Calculation of intracellular pH from the distribution of 5,5 dimethyl-2,4 oxazolidine (DMO). Application to skeletal muscle of the dog. J Clin Invest 38: 720–729

Waddell W, Marlowe C (1981) Transfer of drugs across the placenta. Pharmacol Ther 14: 375–390

Wakabayashi S, Fafournoux P, Sardet C, Pouysségur J (1992) The Na^+/H^+ antiporter cytoplasmic domain mediates growth factor signals and controls "H^+-sensing". Proc Natl Acad Sci USA 89: 2424–2428

Wakabayashi S, Bertrand B, Ikeda T, Pouyssegur J, Shigekawa M (1994) Mutation of calmodulin-binding site renders the Na^+/H^+ exchanger (NHE1) highly H^+-sensitive and Ca^{2+} regulation-defective. J Biol Chem 269: 13710–13715

Wang Z, Orlowski J, Shull G (1993) Primary structure and functional expression of a novel gastrointestinal isoform of the rat Na/H exchanger. J Biol Chem 268: 11925–11928

Weaver TE, Scott JWJ (1984a) Acetazolamide teratogenesis: interaction of maternal metabolic and respiratory acidosis in the induction of ectrodactyly in C57BL/6J mice. Teratology 30: 195–202

Weaver TE, Scott WJ (1984b) Acetazolamide teratogenesis: association of maternal respiratory acidosis and ectrodactyly in C57BL/6J mice. Teratology 30: 187–193

White RA, Geissler EN, Adkison LR, Dowler LL, Alper SL, Lux SE (1994) Chromosomal location of the murine anion exchanger genes encoding AE2 and AE3. Mamm Genome 5: 827–829

Williams B, Howard RL (1994) Glucose-induced changes in Na^+/H^+ antiport activity and gene expression in cultured vascular smooth muscle cells. J Clin Invest 93: 2623–2631

Yu F, Shull G, Orlowski J (1993) Functional properties of the rat Na/H exchanger NHE-2 isoform expressed in Na/H exchanger-deficient Chinese hamster ovary cells. J Biol Chem 268: 25536–25541

Zimmerman E, Scott W, Collins M (1990) Ethanol-induced limb defects in mice: effect of strain and RO-4513. Teratology 41: 453–462

CHAPTER 20
Maternal Physiological Disruption

E.W. Carney

A. Introduction

The concept of homeostasis during pregnancy is almost an oxymoron due to the continuously and profoundly dynamic nature of the maternal/conceptus relationship. Maternal physiological adaptations to pregnancy observed in humans and other mammals include a progressive rise of up to 50% in plasma volume, a 30%–50% increase in cardiac output, decreased systemic vascular resistance, increased pulmonary ventilation, increases of up to 50% in glomerular filtration rate, and changes in mineral metabolism, just to name a few (Parisi and Creasy 1992). The development of the placenta is another dramatic example of a pregnancy-associated change, as the placenta is essentially a temporary organ formed solely to fill a need during gestation. Finally, the conceptus appears to recapitulate in a relatively short number of days or weeks what evolution took millenia to accomplish in its development from a fertilized ovum to a complex individual composed of multiple organ systems.

These changes exhibited by the maternal system, the placenta, and developing conceptus do not occur independently, but are exquisitely orchestrated to achieve successful reproduction. Essential to this orchestration is an extensive, multidirectional communication network composed of endocrine and paracrine signaling molecules which allow all three units to adapt cooperatively. Although the cellular and molecular details of this network are just beginning to be understood, its mere existence implies that chemically induced perturbations to any one of these components could easily upset the balance of the entire system. Given this, it should not be surprising that maternal physiological disruptions, whether due to chemical exposure or disease, can often have adverse developmental consequences (DeSesso 1987; Khera 1987; Chernoff et al. 1989). In fact, the most common outcome of standard developmental toxicity hazard identification tests is one in which developmental toxicity occurs only in the presence of maternal toxicity (Khera 1985).

Despite the fact that approximately 75% of chemicals exhibit the latter toxicity profile (Khera 1985), very little mechanistic information concerning the relationship between maternal and developmental toxicity is available. Historically, teratogenicity screening tests were initially designed to identify compounds which act selectively on the embryo. Even today, standard hazard identification tests (Environmental Protection Agency 1984; European

Table 1. Comparison of maternal and fetal end points in standard developmental toxicity hazard identification studies

Maternal end points	Fetal end points
Body weight/body weight gain	Body weight
Feed consumption	Litter size
Clinical signs of toxicity	Resorption rate
Liver/kidney weights	Sex
Gross pathology	External variations, malformations
	Visceral variations, malformations
	Skeletal delays in ossification (more than 200 fetal bones)
	Skeletal malformations

ECONOMIC COMMUNITY 1988) remain heavily focused on developmental end points (SCHWETZ and MOORMAN 1987), as evidenced by the highly detailed and comprehensive nature of the fetal skeletal evaluation. Over 200 fetal bones are typically assessed not only for structural morphology, but for subtle delays in ossification as well. In contrast, maternal parameters of toxicity tend to be relatively nonspecific in these studies (Table 1).

Investigators have gradually come to acknowledge a potential role for maternal toxicity in contributing to developmental toxicity. Initial efforts to study this issue were aimed at identifying a common syndrome of developmental effects in response to diverse forms of maternal toxicity (KAVLOCK et al. 1985; KHERA 1985, 1987; BEYER and CHERNOFF 1986; CHERNOFF et al. 1990). Although evidence for such a syndrome was generally limited, these studies did heighten awareness of maternal/developmental interactions in the context of developmental toxicity testing. Gradually the concept of maternal toxicity has begun to evolve from this fairly generic one to one which takes into consideration the highly specific nature of chemical toxicants, the diverse repertoire of maternal responses to toxic insult during pregnancy, and the stage-specific homeostatic capabilities of the placenta and developing embryo/fetus. It is clear that further progress in this area will come from mechanistic studies which seek to establish direct cause and effect relationships between specific types of maternal physiological perturbations and development of the embryo/fetus.

Before delving into the details of specific types of maternal toxicity, it is useful to consider whether there is any practical value in knowing whether a chemical acts directly on embryos or, instead, acts indirectly via maternal toxicity. Ostensibly, this type of mechanistic information has little bearing on traditional procedures used to calculate exposure limits for regulated chemicals, as these can be mathematically derived from the lowest no-observable-effect level, regardless of whether it is based on maternal or developmental toxicity. However, animal tests often employ dosages many times higher than potential human exposures and induce maternal perturbations which often are not observed at these lower exposure levels. If one is faced with the question of whether low-dose human exposures not associated with maternal toxicity can still be developmentally toxic, it is very important to know whether maternal physiological disruption, in and of itself, is a cause of developmental toxicity. As such, a better understanding of the maternal/developmental relationship would be of practical benefit to physicians managing drug therapies during pregnancy, to clinical teratologists counseling exposed patients, to policy makers developing scientifically valid regulations, and to legal professionals evaluating claims that a particular chemical exposure was responsible for an adverse developmental outcome. The importance of this issue is also exemplified by a major workshop convened by the United States Environmental Protection Agency to address this issue (KIMMEL et al. 1987).

This chapter will summarize the current state of knowledge concerning maternal physiological disruption as a potential cause of abnormal embryo development. The approach will be to describe how specific challenges to maternal physiology alter the maternal biochemical milieu, to evaluate the ability of various conceptus defenses (e.g., placenta, yolk sac) to maintain embryonic homeostasis in the face of these challenges, and to discuss the ramifications on the embryo should homeostasis be overwhelmed. The highly integrated and ever changing nature of the maternal/developmental relationship has made research in this area particularly challenging. Therefore, research strategies and methods which have found success or which hold promise for the future also will be highlighted.

B. Specific Maternal Physiological Disruptions

I. Acid–Base Imbalance

Maintenance of extracellular fluid pH within fairly narrow limits (approximate pH, 6.8–8.0) is essential to maternal survival. Fortunately, the maternal system has a number of regulatory mechanisms, among them the bicarbonate/carbonic acid/CO_2, plasma protein, phosphate, and hemoglobin buffers, as well as the lungs and kidneys, which are effective in dealing with a variety of acid–base disturbances. However, the capacity of these systems to maintain pH is finite, and extracellular fluid pH can become excessively basic (alkalemia) or acidic (acidemia) following exposure to acids or bases, in response to toxic injury of the lungs or kidneys, or in conjunction with several disease states (SESTOFT and BARTELS 1983; GUYTON 1986).

Perhaps the most likely acid–base disorder to be encountered in the context of toxicology studies is metabolic acidosis, which is caused by a net gain of acid (e.g., dosing with an acid) or net loss of base (e.g., failure of bicarbonate reabsorption in the kidney). Metabolic acidosis can also occur in diabetes mellitus, due to production of β-hydroxybutyric and acetoacetic acids, or through loss of bicarbonate-containing pancreatic juice as a sequel to diarrhea. The compensatory strategy in metabolic acidosis is neutralization of excess acid with endogenous base, resulting in a fall in blood base concentration. The exquisite sensitivity of respiratory control centers also leads to increased pulmonary ventilation, and blood PCO_2 quickly falls to restore the normal ratio of CO_2 to HCO_3^- (BLECHNER 1993). Therefore, it is common to have a metabolic acidosis which is characterized by very small changes in blood pH, but with significantly depressed PCO_2 and HCO_3^- concentrations. The PCO_2 and HCO_3^- changes may persist for several hours or more, as restoration of normal levels is limited by the rate of bicarbonate reabsorption in the kidney (SESTOFT and BARTELS 1983).

In contrast, agents which inhibit pulmonary ventilation can cause a respiratory acidosis which is characterized by increased PCO_2 and which is corrected by increased renal HCO_3^- reabsorption. Inhibition of carbonic anhydrase, the enzyme which interconverts CO_2, carbonic acid, and HCO_3^-, results in a mixed respiratory/metabolic acidosis characterized by increased PCO_2, slightly increased HCO_3^-, and decreased pH (WEAVER and SCOTT 1984). Thus the primary homeostatic strategy in each of these disorders is maintenance of normal extracellular fluid pH, but this is achieved through very different compensatory changes in PCO_2 and HCO_3^- levels. Metabolic and respiratory alkaloses also have their own characteristic profiles (SESTOFT and BARTELS 1983; BLECHNER 1993).

Acid–base imbalance as a specific cause of developmental toxicity has been investigated using ethylene glycol (EG) and sodium salicylate as model acid–base disrupters (KHERA 1991). Large oral bolus doses of EG induce a shift in EG metabolism, leading to the accumulation of a weak acid metabolite, glycolic acid. High concentrations (mmol range) of this metabolite in maternal blood cause a corresponding depletion of plasma HCO_3^-, along with significant declines in plasma pH and PCO_2 (JACOBSEN et al. 1984; HEWLETT et al. 1989). When $NaHCO_3$ was coadministered with EG, maternal acid–base changes were ameliorated, as were fetal body weight reductions, many skeletal defects, and placental histological lesions. Sodium salicylate produced a slightly different acid–base disturbance characterized by decreased maternal plasma PCO_2, HCO_3^-, and phosphate, but with no appreciable change in pH. Again, fetal and placental effects were attenuated by $NaHCO_3$ coadministration and exacerbated with simultaneous NH_4Cl treatment (KHERA 1991).

These in vivo data have recently been complemented by two whole-embryo culture studies which examined the role of culture medium pH in embryotoxicity. In a study of formate–pH interactions, ANDREWS et al. (1993) cultured gestation day-9.5 (presomite) rat conceptuses in medium-titrated with HCl to pH 8.13, 7.75, 7.00, 6.50, or 6.00. Embryolethality and growth inhibition occurred at pH 6.5 or less, but development was normal at the higher pH values. CARNEY et al. (1996), using day-10.5 (early somite) rat embryos, observed that embryos grew normally at pH 6.91–7.41, while embryo and yolk sac protein content, along with head length, were slightly inhibited at pH 6.74. Both the in vivo and in vitro data suggest that extracellular pH per se is not embryotoxic over the range of pH 6.9–8.1.

How is it that in vivo neutralization of acidosis ameliorated developmental toxicity, yet in vitro "acidosis" had very little effect? In attempting to explain this phenomenon, one must consider that PCO_2 and HCO_3^- levels in whole-embryo culture are maintained at "normal" levels. Hence, simply reducing culture medium H^+ concentration does not adequately mimic metabolic acidosis in vivo. Instead, it may be the changes in PCO_2 which occur during metabolic acidosis which might be teratogenic. Such a role for CO_2 is supported by the observation that exposure to an hypercarbonic atmosphere causes terata in rats (HARING 1960) and rabbits (GROTE 1965). In addition, in sensitive strains of mice, maternal hypercapnia is associated with a highly specific malformation (right forelimb ectrodactyly), which also is observed in response to acetozolamide and other carbonic anhydrase inhibitors (WEAVER and SCOTT 1984). Conversely, the low PCO_2 levels characteristic of metabolic acidosis appear to correspond with axial skeletal defects (Table 2); however, the effects of hypocarbonic atmospheres on embryos have not been directly tested. Finally, there may be a biological basis for the teratogenicity of altered PCO_2 based on the ability of CO_2 to modulate intracellular pH and the importance of intracellular pH in cellular differentiation, function, and morphogenesis (BUSA and NUCCITELLI 1984; CARNEY and BAVISTER 1987). Interestingly, modulation of intracellular pH has been proposed as a mechanism of action for several weak acid teratogens, such as valproic acid and 2-ethylhexanoic acid (SCOTT et al. 1987; see Chap. 11).

Table 2. Relationships between maternal acid–base disruption and developmental toxicity

Agent	Type of acid–base imbalance	Developmental effects[a] (species)
Ethylene glycol[b,c]	Metabolic acidosis (decreased PCO_2, HCO_3^-, pH)	Axial skeleton defects (rodents)
Sodium salicylate[c]	Metabolic acidosis (decreased PCO_2, HCO_3^-, no pH change)	Axial skeleton defects, resorptions, hydrocephaly (rat)
Acetozolamide[d]	Metabolic/respiratory acidosis (increased PCO_2, HCO_3^-, decreased pH)	Right-sided ectrodactyly (mice) Axial skeleton defects (rabbit)
Diabetes mellitus[e]	Ketoacidosis	Craniofacial, axial skeleton defects (human)
Hypercapnia[f]	Increased PCO_2	Forelimb ectrodactyly (mice) Cardiac malformations (rats)

[a]List is not all-inclusive.
[b]Reviewed in CARNEY (1994).
[c]From KHERA (1991).
[d]From LAYTON and HALLESY (1965); LAYTON (1971); GREEN et al. (1973); NAKATSUKA et al. (1992).
[e]Reviewed in KHERA (1987).
[f]From HARING (1960).

In addition to the possibility of direct developmental toxicity due to maternal acid–base imbalance, maternal acid–base changes also have the potential to upset maternal–embryo pH gradients, which in turn can influence the distribution of a toxicant to the embryo (Scott et al. 1987). This hypothesis is consistent with the whole-embryo culture study by Andrews et al. (1993), in which acidic pH induced a leftward shift in the dose–response curve of sodium formate. For a weak acid such as formate, a decrease of one pH unit brings about a ten fold increase in the percentage of nonionized formic acid. Compounding this increased proportion of uncharged formic acid would be a greater pH gradient driving it across the yolk sac and into the more alkaline intracellular environment of the embryo (Scott et al. 1987). A study using the drugs bupivacaine and meperidine also indicated that acidosis may affect placental transfer, as more drug was delivered to rabbit fetuses when the perfusate was adjusted to pH 7.0 as compared with pH 7.5 (Gaylard et al. 1990).

Although maternal acid–base disruption is likely to play an important role in developmental toxicity, its mechanism of action is far from clear. One suggested area of investigation would be to manipulate individual components of the acid–base regulatory system, particularly PCO_2 and bicarbonate levels, to determine effects on development both in vivo and in vitro. Further, very little is known about the mechanisms for regulation of pH in various embryonic and extraembryonic tissues. Classical cell biology approaches such as treatment with the Na^+/H^+ antiporter inhibitor amiloride would be useful in this regard. Mapping the expression patterns of various antiporter and other intracellular pH-regulatory genes during embryogenesis may also provide clues to explain differences in regional sensitivity to acid–base disruption.

II. Osmotic Disruption

Changes in the osmotic pressure of maternal plasma and extracellular fluids can be brought about in a number of ways. Rapid administration of osmotically active material may directly lead to hyperosmolality (Glasser et al. 1973), while the same condition can be brought about indirectly due to dehydration, increased salt intake, loss of body water through vomiting, or inhibition of water reabsorption in the renal collecting duct (e.g., diabetes insipidus). Conversely, compounds such as morphine, nicotine, and certain tranquilizers and anesthetics increase water reabsorption in the kidney and thus have the potential to induce hypo-osmolality (Guyton 1986).

Mechanisms for maternal regulation of osmolality are intimately tied to those involved in the maintenance of total blood volume. The main components of this regulatory system are the thirst centers in the brain, regulating water intake, and the posterior pituitary hormone arginine vasopressin (also called antidiuretic hormone), which controls reabsorption of water by the renal collecting ducts. At the cellular level, the concentrations of sodium and chloride ions are far and away the major determinants of total osmotic pressure. Proteins in plasma and interstitial fluid are also an important source of osmotic pressure, usually referred to as oncotic or colloid osmotic pressure (GUYTON 1986). This is particularly important in the movement of solutes across capillary walls.

Despite the major cardiovascular and renal changes during normal pregnancy, maternal plasma osmolality remains relatively constant (290–300 mOsmol/kg H_2O) throughout gestation (PARISI and CREASY 1992; WOODS 1986). Studies on osmoregulation during fetal life have shown that the osmotic pressure of maternal and fetal plasma are normally very similar, while the amniotic fluid is relatively hypotonic. Furthermore, injection of hypotonic or hypertonic solutions into the mother brings about equivalent changes in maternal and fetal plasma osmolality in numerous species, including rabbits (DANCIS et al. 1957), rats (ADOLPH and HOY 1963), sheep (WOODS 1986), nonhuman primates (BRUNS et al. 1964), and humans (BATTAGLIA et al. 1960). Thus, changes in maternal plasma osmolality appear to be readily transmitted to the fetus. However, once the fetal kidneys become functional, osmotic challenges can be dealt with by concentration or dilution of the fetal urine and excretion of this urine into the amniotic cavity.

Much less is known about osmoregulation during embryonic life, despite the fact that embryonic edema, blistering, and hemorrhage ("edema syndrome") commonly occur in response to many teratogenic agents (GRABOWSKI 1977a). Indirect clues regarding the effects of maternal hyperosmolality come from previously mentioned studies with EG. KHERA (1991) found that a single high dose (3333 mg/kg) of EG given to gestation day-11 rats caused maternal plasma osmolality to rise from the control value of 285 mOsmol/kg H_2O to as high as 359 mOsmol/kg H_2O within 1 h after administration. Because hyperosmolality can be cytotoxic or even mutagenic (BRUSICK 1986), KHERA hypothesized that maternal hyperosmolality might play a role in EG-induced teratogenesis.

Additional perspective on this hypothesis was recently obtained in a rat whole-embryo culture study of EG (CARNEY et al. 1996). In this study, gestation day-10.5 CD rat embryos contained within their visceral yolk sacs were cultured with up to 50 mmol EG/l for 46 h and then evaluated for growth and morphological development. Embryo development was essentially normal in the EG-containing media, despite the fact that the medium was extremely hypertonic (418 mOsmol/kg H_2O). Furthermore, there was not the slightest evidence of shrinkage in the conceptuses, as indicated by a lack of effect on visceral yolk sac diameter. This indirect evidence suggests that the midgestation conceptus is able to cope with extreme hyperosmotic challenges.

In particular, the visceral yolk sac seems to be a crucial barrier for maintaining osmolality of the conceptus' internal environment. Consistent with this notion are studies in which chemically induced visceral yolk sac damage led to alterations in the osmolality of the exocoelomic fluid surrounding the embryo. For example, trypan blue treatment of embryos in vivo and in vitro caused lysosomal damage to the visceral yolk sac endoderm, a decrease in exocoelomic fluid osmolality, and a corresponding increase in the incidence of embryonic edema (Rogers et al. 1985). The opposite effect was observed following treatment with leupeptin, a specific inhibitor of yolk sac lysosomal proteinase activity. Leupeptin treatment of whole rat embryos in vitro caused an increase in exocoelomic fluid osmolality and malformations characterized by decreased neural tube volume (Daston et al. 1991a).

Taken in aggregate, these studies suggest that embryos (enclosed within yolk sac) and fetuses are fairly resistant to maternal hyperosmolality. However, more work is needed to determine the effects of hyper- or hypo-osmolality at different stages of development, in different tissues, in different species, and induced by different types of osmotic disruptors (e.g., osmotic versus oncotic pressure disruptors). Ironically, a substantial body of in vitro data already published could have helped address this question. However, culture medium osmolality tends not to be reported by many investigators.

III. Maternal Cardiovascular Disturbances

1. Introduction

The cardiovascular system of pregnancy plays an essential role in development, as it is responsible for delivery of oxygen, other gases, and nutrients to the developing conceptus. On the maternal side, the heart and uterine vessels are the major determinants of blood flow to the conceptus. Blood flow initially is directed to the conceptus by way of the uterine decidua and visceral yolk sac and gradually shifts (on gestation days 11–13 in the rat) to favor the chorio-allantoic placenta as a route of entry (Buelke-Sam et al. 1982). Although yolk sac and placental tissues exhibit increased angiogenesis and other forms of remodeling in response to certain maternal cardiovascular conditions, their compensatory capabilities are limited (Jackson et al. 1988; Strick et al. 1991). Given these limitations and the paramount importance of the yolk sac and placenta in development, it is not surprising that many maternal cardiovascular disturbances can adversely influence development. The types of cardiovascular conditions that have been studied for effects on development include maternal heart rate, uterine vasoconstriction, and various anemias.

2. Maternal Cardiac Function

An association between maternal bradycardia and cleft lip/palate has been noted in mice given the anticonvulsant drug phenytoin, in that the A/J mouse strain is susceptible to both types of effects, while neither of these effects is seen in the C57/BL6 strain (Watkinson and Millicovsky 1983). Exposing A/J mice to a hyperoxic atmosphere post-treatment decreased the incidence of cleft lip/palate. Thus it was suggested that the bradycardia was sufficient to cause an embryonic hypoxia, which, in turn, was responsible for the cleft lip/palate seen in these mice (Millicovsky and Johnston 1981). It should be noted that exposure to hyperoxic or hypoxic atmospheres is also accompanied by changes in blood PCO_2, lactic acid, and phosphate as well as other fluid disturbances and that these may also play a causal role (GRABOWSKI 1977b; BLECHNER 1993).

Cocaine is another agent for which an association between maternal cardiotoxicity and developmental effects, such as placental abruption and miscarriage (CHASNOFF et al. 1985), has been suggested. Interestingly, myocardium from pregnant or progesterone-treated nonpregnant rats has been shown to be one to four orders of magnitude more sensitive to the cardiotoxic effects of cocaine than myocardium of nonpregnant rats (SHARMA et al. 1992), thus exemplifying an important interaction between a maternal physiological adaptation to pregnancy and systemic toxicity.

3. Uterine Vasoconstriction

The effects of uterine vasoconstriction have been studied using the experimental technique of uterine artery clamping (BRENT and FRANKLIN 1960) or following administration of vasoactive drugs. Cocaine, in addition to its cardiotoxic effects, also causes uterine vasoconstriction (CHASNOFF et al. 1985). A maternal etiology for cocaine's developmental toxicity is suggested by the 25%–50% decrease in uterine blood flow and severe fetal hypoxia which followed injection of cocaine into pregnant ewes and the lack of change in fetal oxygenation when cocaine was injected directly into the fetus (WOODS et al. 1987). Similarly, injection of epinephrine or vasopressin into pregnant rats induced uterine vasoconstriction and produced signs of fetal hypoxia (CHERNOFF and GRABOWSKI 1971).

The placenta also exhibits a number of important adaptive responses related to its roles in gas exchange, nutrient transfer, and endocrine function. For example, thinning of the human placental villous membrane occurs in placentae at high altitudes (low oxygen), presumably in an effort to increase diffusion capacity (JACKSON et al. 1988). Similarly, vascularity of the chick chorioallantoic membrane is inversely related to oxygen tension (STRICK et al. 1991). These placental changes can be detected and even quantitated morphometrically (JACKSON et al. 1988) and may be a useful diagnostic indicator of maternal hypoxia-mediated developmental toxicity.

4. Maternal Anemias

Another potential mechanism for decreased oxygen delivery to the fetus is a decrease in maternal hemoglobin concentration (anemia). A wide variety of anemias can result from chemical exposure, either through decreased red cell production or increased loss of red cells (e.g., hemorrhage, hemolysis). Evidence of teratogenicity caused by severe maternal anemia was provided by a study of diflunisal, an anti-inflammatory analgesic. A standard developmental hazard identification study in rabbits revealed increased incidences of axial skeletal malformations only at doses which also caused a severe (more than 50% decreased hemoglobin and erythrocyte count) maternal hemolytic anemia (CLARK et al. 1984). To determine whether the maternal anemia was responsible for the fetal malformations, a subsequent experiment was performed in which a single dose of diflunisal was given on gestation day 5. This treatment caused an anemia persisting until gestation day 15, yet the drug was cleared from the maternal circulation prior to gestation day 9. As gestation day 9 is the critical day for induction of axial skeletal malformations by hypoxia treatment, the authors concluded that maternal anemia was the probable teratogen. Consistent with this interpretation is the known teratogenicity of severe hypovolemia following acute hemorrhage (WILSON 1953).

An issue of practical importance is the degree of anemia required to adversely affect developmental outcome. Slight to moderate anemias have been reported in pregnant rats and/or rabbits exposed to propylene dichloride (HANLEY et al. 1990) or EG monobutyl ether (TYL et al. 1984). No treatment-related teratogenicity was reported in either of these studies, although delayed ossification in isolated skeletal districts was seen in most of these cases. In sheep, maternal hematocrit had to decline more than 50% before fetal oxygen consumption was affected (PAULONE et al. 1987). These data indicate that a fairly large margin of safety exists for induction of teratogenic effects due to anemia, although it seems likely that slight to moderate anemias can be responsible for more subtle effects on growth and maturation.

An important consideration in extrapolating animal data for human risk assessment are the large differences between species in susceptibility to various anemias. For example, rat red blood cells are extremely sensitive to hemolysis induced by butoxyethanol, while human red blood cells are highly resistant to this effect (GHANAYEM and SULLIVAN 1993). Comparison of human and animal blood responses may be a simple, yet powerful tool for extrapolating between species.

IV. Body Temperature

1. Introduction

Maintenance of body temperature is one function which is exclusively the responsibility of the mother throughout in utero life (IMAI-MATSUMURA et al. 1990). Because of the fundamental influence of temperature on chemical reaction rates, one might expect deviations from normal body temperature to have broad consequences, particularly to a developing organism.

2. Hyperthermia

Elevations in maternal core body temperature of approximately 3 °–5 °C have been shown in a number of animal species to induce neural tube defects, axial skeletal malformations, and other developmental problems. Severe hyperthermia may also be developmentally toxic in humans, although this issue has been difficult to resolve due to the coexistance of underlying disease conditions responsible for the hyperthermia (WARKANY 1986). As the issue of hyperthermia during pregnancy has been extensively addressed by others, readers are referred to several excellent reviews on this subject (GERMAN 1984; MIRKES 1985; WARKANY 1986; KIMMEL et al. 1993; see also Chap. 15).

3. Hypothermia

Although body temperature measurement is perhaps the most fundamental monitor of human health as practiced in clinical medicine, it is rarely measured in animal toxicity studies. However, several studies in nonpregnant rats and mice have shown that numerous diverse agents administered by various routes of exposure at acutely toxic levels consistently induce what appears to be a centrally regulated, moderate hypothermia (STAUCH et al. 1969; WATKINSON and GORDON 1993). This hypothermic response is characterized by a decrease in core body temperature of 1 °–3 °C, which is brought about by decreases in heart rate and blood pressure and increases in peripheral blood flow (i.e., increased heat dissipation). That these changes are centrally regulated is suggested by the observation that exposed animals placed in an ambient temperature gradient will choose a temperature 1 °–3 °C lower than that of unexposed animals. Furthermore, a significant survival advantage was conferred on animals which were allowed to elaborate this hypothermic response versus those in which the response was blocked by various experimental means (WATKINSON and GORDON 1993).

The hypothermic response in pregnant animals has only been studied to a limited extent. Early work on the teratogenicity of hypothermia during pregnancy showed that exposure to extremely low temperatures led to increased incidences of resorptions and morphological abnormalities in mice and hamsters (SMITH 1957; MUNRO and BARNETT 1969). In chick embryos, hypothermia of a more physiological nature (35.9 °C) increased the incidence of embryolethality and cardiac malformations (de la CRUZ et al. 1966). In vitro, 24-h exposure of day-9 (neurulating) mouse embryos to 32 °C or 35 °C resulted in decreased growth and protein content, as well as increased incidences of incomplete neural tube closure, poor brain expansion, and rotational abnormalities (SMOAK and SADLER 1991). Gestation day-10 (early limb bud stage) embryos did not exhibit dysmorphogenesis in response to hypothermia, but did have fewer somites and lower total protein contents. These results suggest a basis for the teratogenicity of even mild hypothermia.

Other studies, however, have failed to show adverse effects of maternal hypothermia. RANDALL et al. (1988) exposed gestation day-10 mice to a teratogenic and hypothermia-inducing dose of ethanol and then, in an effort to keep the mice normothermic, maintained the mice for 6 h post-treatment in an incubator set at 32 °C. A second group of mice were given ethanol and kept in a 22 °C incubator, while control mice received isocaloric sucrose and were housed at 22 °C or 32 °C. Although the 32 °C environment attenuated the decline in core body temperature (maximum decline of 2 °C in the 32 °C/ethanol groups versus 4 °C in the 22 °C/ethanol group), there was no reduction in malformation rate or fetal weight effects.

To complicate the issue further, another group of studies have demonstrated a protective effect of hypothermia. The incidence of malformations and other forms of developmental toxicity caused by radiation exposure (MAROIS 1966) or uterine vascular clamping (GEORGE et al. 1967) is significantly reduced by maternal hypothermia. In mouse whole-embryo cultures carried out at 37 °C, exposure to hypoglycemic conditions for 4–6 h caused a high rate of dysmorphogenesis. However, the incidence of malformed embryos was significantly decreased when these same exposures were done at 32 °C or 35 °C (SMOAK and SADLER 1991). The authors of this study postulated that hypothermia is protective against hypoglycemia-induced dysmorphogenesis because of a decrease in embryonic glycolytic activity at lower temperatures. The lowered metabolic activity would serve to balance the decreased substrate availability and, therefore, the hypoglycemic state would not be as detrimental. This hypothesis, although attractive, has not been examined further.

Obviously, there is no clear answer supporting or refuting a causal role for hypothermia in teratogenicity. However, some of the approaches used by WATKINSON and GORDON (1993) could be applied to pregnant animals in order to shed more light on this issue. It may be particularly useful to study associations between maternal hypothermia and end points other than teratogenicity, such as growth retardation and/or delayed ossification of the fetal skeleton, as these are likely to be very sensitive to maternal hypothermia. Of potential importance to human risk assessment is the fact that the hypothermic response to toxic insult observed in small laboratory mammals is not seen in humans and other large mammals due to their greater mass and increased thermal inertia (WATKINSON and GORDON 1993).

V. Stress

The stress response, first described by SELYE (1950), is characterized by elevated catecholamine and glucocorticoid levels, as well as a variety of other important systemic adaptations, and may be a ubiquitous factor in a wide variety of toxic exposures. Many have hypothesized that maternal stress could be responsible for some of the developmental toxicity which so often accompanies maternal toxicity. Investigators addressing this issue have displayed a considerable degree of imagination through their use of alarms, buzzers, flashing lights, forced exercise, jet engine noise, foot shock, feed restriction or restraint for inducing stress. Decreased litter size birth weight, embryo mortality and/or abortion were seen in many of these studies (GEBER 1966; GEBER and ANDERSON 1967; EUKER and RIEGLE 1973; KHOLKUTE and UDUPA 1978; MATSUZAWA et al. 1981), although other procedures which could be considered stressful failed to induce adverse effects (NAWROT et al. 1980; TYL et al. 1994). Among the most consistent response to maternal stress administered in the last trimester of pregnancy in rodents is a demasculinization of sexual differentiation (WARD and WEISZ 1984; VOM SAAL et al.

1990). Effects of maternal stress on anogenital distance, estrous cyclicity, mating behavior, and other traits bear striking resemblance to those caused by certain estrogenic or anti-androgenic compounds (KELCE et al. 1994). Stress associated with chemical exposure has also been shown to cause increased incidences of developmental variations, such as supernumerary ribs and dilated renal pelvis, as well as terata, such as cleft palate (BARLOW et al. 1975; KAVLOCK et al. 1985; BEYER and CHERNOFF 1986; CHERNOFF et al. 1987, 1990).

Whether the above protocols adequately model the stress which occurs in chemically induced maternal toxicity is a critical question. Physical confounders such as abdominal pressure during restraint or hyperthermia during forced exercise are difficult to avoid with these protocols and could produce artifactual effects in the offspring (SCIALLI 1988). In addition, it must be appreciated that there is no such thing as a universal stress response. Different stressors activate widely different neuronal pathways (HARBUZ and LIGHTMAN 1989) and elaborate diverse combinations of endogenous opioids, hypothalamic releasing factors, and cytokines (SZEKELY 1990). Thus the maternal endocrine profile induced by an anxiogenic stimulus such as foot shock may be considerably different from that occurring in response to chemical toxicity. Finally, levels of endogenous corticosterone which are seen during maternal toxicity are considered by some to be too low to directly alter development (DASTON 1994; SLOTKIN et al. 1994), although others have argued that they are sufficient to affect embryo development (BARLOW et al. 1980; HANSEN et al. 1988; ELDEIB and REDDY 1990).

Studies in rodents showing that stress exacerbates the teratogenicity of agents such as vitamin A (HARTEL and HARTEL 1960), salicylate (GOLDMAN and YAKOVAC 1963) and sodium arsenite (RASCO and HOOD 1994) suggest that glucocorticoids may act as potentiators, rather than direct developmental toxicants. Consistent with this idea, SLOTKIN et al. (1994) observed that maternal exposure to the synthetic glucocorticoid dexamethasone stimulated fetal adenylate cyclase activity and, most importantly, potentiated the adenylate cyclase response to the β-adrenergic agonist isoproteranol. Because cyclic adenosine monophosphate (cAMP) is a widely recognized intracellular regulator of cell differentiation, these workers proposed that glucocorticoids may lower the teratogenic threshold for agents which act via the cAMP system.

It is also important to recognize that the hypothalamic–pituitary axis, of which glucocorticoids are a part, is tightly linked with the immune system and its network of cytokines. In particular, it is known that interleukin-1, interleukin-6, and tumor necrosis factor (TNF)-α liberated during acute phase responses can activate the hypothalamic–pituitary axis (ESKAY et al. 1990). Cytokines such as these are increasingly being found to play active roles in preimplantation development (ZOLTI et al. 1991), implantation (CROSS et al. 1994), placental growth and endocrine function (BRIGSTOCK et al. 1989; WEGMANN 1990) and organogenesis (ROBERTSON et al. 1992; BEN-RAFAEL and ORVIETO 1992). In addition, high levels of TNF-α are teratogenic in mice

(TAUBENECK et al. 1994a). Therefore, these components of the stress response may be important in maternal stress-induced developmental toxicity.

Recent studies in rats and mice suggest that zinc may be a common effector molecule involved in developmental responses associated with activation of the hypothalamic–pituitary axis and/or immune system (see also Chap. 12). Increased sequestration of zinc in maternal liver and decreased distribution of zinc to the conceptus has been described for a number of diverse xenobiotic (6-mercaptopurine, valproic acid, urethane), naturally occurring (α-hederin), or endogenous (TNF-α) agents following administration of acutely toxic doses (AMEMIYA et al. 1986, 1989; KEEN et al. 1989; TAUBENECK et al. 1994a,b). A causal link between decreased zinc distribution to the conceptus and developmental toxicity is indicated by the amelioration of such toxicity following zinc supplementation either through the diet or in serum used for whole-embryo culture (DASTON et al. 1991b, 1994).

VI. Other Physiological Disruptions

The examples cited in this chapter represent just a small fraction of the many different maternal perturbations which might be encountered as a result of chemical toxicity. In addition to the maternal physiological disruptions discussed here, other types of maternal toxicity have been studied to varying degrees. An extensive body of literature on the role of maternal nutritional imbalances in teratogenesis has been accumulated over the last 20 years. In addition to dietary deficiency or excesses (see Chap. 12), several studies relating to standardized developmental tests also have been done. These include a study showing the developmental effects of altering rabbit intestinal flora (CLARK et al. 1986) and another demonstrating that diuretic-induced maternal hypokalemia was responsible for wavy ribs and scapular and humeral malformations in rats (ROBERTSON et al. 1981). Disruption of thyroid, pancreatic, pituitary, or other endocrine organ function is another area which has received significant attention (SHEPARD 1977). In contrast, there is a paucity of information on the relationship between altered kidney function and development (KAVLOCK et al. 1993), while the impact of liver toxicity apparently has not been examined at all. This is somewhat ironic, given that these two organs are among the most common targets for toxicity.

C. Strategies and Methods for Future Research

Although we must always appreciate the active communication which occurs between mother, placenta, and embryo, research methods designed to assess the role of altered maternal physiology on development often require that these components be studied in isolation. An example of success in this approach was provided in a study of maternal causes of poor reproductive outcomes in women. Rat postimplantation embryos were cultured in sera from women with histories of spontaneous abortion. Maternal nutritional imbalance was implicated as a possible cause of these abortions, as supplementation of the serum with certain vitamins or amino acids ameliorated many of the embryonic alterations which were observed in embryos cultured in unsupplemented sera (CHATOT et al. 1984). Others have also used embryo culture to systematically evaluate the effects of various maternal biochemical characteristics associated with diabetes mellitus (SADLER et al. 1987). Critical to the success of these approaches is their "reconstitution" of the in vitro system with physiologically relevant maternal factors.

These in vitro strategies need to be complemented by in vivo approaches in future studies. Physiological monitoring is one methodology which is just beginning to surface as a potentially powerful tool in many areas of toxicology. Particularly alluring are telemetric techniques, which allow for continuous monitoring of specific variables using implantable monitoring devices, thus allowing free movement of the animal and less potential for confounding factors (KIMMEL et al. 1993). Pharmacological agents could also be employed to restore normal physiology in a specific manner. For example, the opioid antagonist naloxone has been used to study the mechanisms by which stress inhibits various reproductive functions in rats (HULSE et al. 1982; HULSE and COLEMAN 1983). One feature of the pharmacological approach is that the test agent remains present; thus one may be able to distinguish between direct chemical effects and those due to altered maternal physiology. Finally, we simply need more extensive surveillance of maternal physiology during standard developmental toxicity tests. This practice would undoubtedly reveal more associations between specific types of maternal toxicity and developmental outcome, which could be examined more definitively in subsequent mechanistic studies.

D. Maternal Toxicity and Risk Assessment

The issue of maternal toxicity and its role in developmental toxicity risk assessment has been the subject of much debate. Although a complete review of this topic is outside the purview of this chapter, certain key issues are relevant. One such issue concerns the concept of a maximum tolerated dose for developmental toxicity studies. Currently, relatively nonspecific indicators of maternal health, such as a statistically significant effect on maternal body weight, are required as evidence that a maximum tolerated dose was achieved. In the future, hazard identification studies might provide much more useful information if specific end points of maternal physiology which are known to affect development are incorporated and accepted as satisfactory criteria for achievement of a maximum tolerated dose.

The second issue focuses on the interpretation of data in which maternal and developmental toxicity coexist. One simple approach to this issue has been to calculate numerical indices comparing maternal/developmental effect levels.

Measures such as the relative teratogenic index (RTI; Fabro et al. 1982) or adult/developmental ratio (A/D ratio; Johnson et al. 1987) may be useful for early priority setting in a tiered testing approach, but have been criticized for use in risk assessment due to problems of predictability across species and across the dose–response curve (Setzer and Rogers 1991; Daston et al. 1991c). Given the questions concerning the utility of such indices, there currently remains a considerable burden of proof lying with the investigator if developmental effects are suspected to be secondary to altered maternal physiology. This burden is justifiable in that maternal toxicity is not always associated with developmental toxicity (Khera 1985). Thus a cause and effect relationship between the two is not automatic. In supporting conclusions that developmental effects are secondary to maternal toxicity, the specific type and the severity of maternal physiological disruption first need to be defined. In so doing, it is important to go beyond examination of group trends and examine the relationship between individual maternal responses and correlate them with developmental outcomes of her offspring. Second, relevant mechanistic data, whether previously published or generated for the case at hand, should be available to establish a relationship between the specific maternal alterations and developmental toxicity. Ideally, data indicating that the test material or its metabolites are not directly toxic to embryos would be provided through the use of whole-embryo culture or related techniques. Last, one must always make robust use of the underlying biology of the mother–embryo relationship in assessing the biological plausibility of the conclusions. Distilling all of the available information, the key question to ask is whether the developmental effects could have occurred in the absence of maternal toxicity.

Finally, the issue of extrapolation between animals and humans is one in which a better understanding of the maternal/developmental relationship is likely to have considerable impact. This is particularly true because, unlike human *embryos*, human maternal physiology is very accessible to monitoring in a clinical setting. Thus comparable data for specific maternal end points can be readily obtained across species. Many physiological functions, such as acid–base homeostasis, employ highly conserved regulatory mechanisms and interspecies differences would appear to be minimal. Conversely, distinct species differences have been shown for parameters such as susceptibility to hemolysis (Ghanayem and Sullivan 1993). In any event, the ability to obtain comparative data on maternal physiology offers great potential for increasing the certainty with which animal data are used to predict human risk.

E. Conclusions

From the previous discussion it is apparent that maternal toxicity is not a single entity, but a large group of maternal alterations which differ according to the chemical agent administered, dose, stage of gestation, and other factors. This chapter has reviewed some specific maternal physiological perturbations, such as acid–base imbalance, thermoregulatory disorders, osmotic disturbances, and various anemias, for which some information is available concerning effects on development. Much more of this information will be required if the lingering controversy over the maternal/developmental toxicity issue is ever to be resolved. Progress in this area can be achieved through a systematic analysis of specific maternal perturbations and corrective responses and a corresponding assessment of the embryo and/or placenta's ability to withstand such challenges. By understanding the dynamic interplay between mother, placenta, and embryo under normal and abnormal conditions, a better understanding of the relationship between maternal health and development in humans can be obtained.

Acknowledgement. The author is grateful to Ms. Jaime Gilles for secretarial assistance and to Dr. W. Breslin, Dr. J. Bus, Dr. J. Mattsson, and Mr. J. Pitt for critical review of this manuscript.

References

Adolph EF, Hoy PA (1963) Regulation of electrolyte composition of fetal rat plasma. Am J Physiol 204: 392–400

Amemiya K, Keen CL, Hurley LS (1986) 6-Mercaptopurine-induced alterations in mineral metabolism and teratogenesis in the rat. Teratology 34: 321–334

Amemiya K, Hurley LS, Keen CL (1989) Effect of 6-mercaptopurine on ^{65}Zn distribution in the pregnant rat. Teratology 39: 387–393

Andrews JE, Ebron-McCoy M, Kavlock RJ, Rogers JM (1993) Lowering pH increases embryonic sensitivity to formate in whole embryo culture. Toxicol In Vitro 7: 757–762

Barlow SM, McElhatton PR, Sullivan FM (1975) The relationship between maternal restraint and food deprivation, plasma corticosterone, and induction of cleft palate in the offspring of mice. Teratology 12: 97–104

Barlow SM, Knight AF, Sullivan FM (1980) Diazepam-induced cleft palate in the mouse: the role of endogenous maternal corticosterone. Teratology 21: 149–155

Battaglia F, Prystowsky H, Smisson C, Hellegers AE, Burns P (1960) Fetal blood studies. XIII. The effect of the administration of fluids intravenously to mothers upon the concentrations of water and electrolytes in plasma of human fetuses. Pediatrics 25: 2–10

Ben-Rafael Z, Orvieto R (1992) Cytokines – involvement in reproduction. Fertil Steril 58: 1093–1099

Beyer PE, Chernoff N (1986) The induction of supernumerary ribs in rodents: role of the maternal stress. Teratogen Carcinogen Mutagen 6: 419–429

Blechner JN (1993) Maternal-fetal acid-base physiology. Clin Obstet Gynecol 36: 3–12

Brent RL, Franklin JB (1960) Uterine vascular clamping: new procedure for the study of congenital malformation. Science 132: 89–91

Brigstock DR, Heap RB, Brown KD (1989) Polypeptide growth factors in uterine tissues and secretions. J Reprod Fertil 85: 747–758

Bruns PD, Hellegers AE, Seeds AE Jr, Behrman RE, Battaglia FC (1964) Effects of osmotic gradients across the primate upon fetal and placental water contents. Pediatrics 34: 407–411

Brusick D (1986) Genotoxic effects in cultured mammalian cells produced by low pH treatment conditions and increased ion concentration. Environ Mutagen 8: 879–886

Buelke-Sam J, Golson JF, Nelson CJ (1982) Blood flow during pregnancy in the rat: dynamics of and litter variability in uterine flow. Teratology 26: 279–288

Busa WB, Nuccitelli R (1984) Metabolic regulation via intracellular pH. Am J Physiol 246 (Regul Integr Comp Physiol 15): R409–R438
Carney EW (1994) An integrated perspective on the developmental toxicity of ethylene glycol. Reprod Toxicol 8: 99–113
Carney EW, Bavister BD (1987) Regulation of hamster embryo development in vitro by carbon dioxide. Biol Reprod 36: 1155–1163
Carney EW, Liberacki AB, Bartels MJ, Breslin WJ (1996) Identification of proximate toxicant for ethylene glycol-induced developmental toxicity using rat whole embryo culture. Teratology 53: 38–46
Chasnoff IJ, Burns WJ, Schnoll SH, Burns KA (1985) Cocaine use in pregnancy. N Engl J Med 313: 666–669
Chatot CL, Klein NW, Clapper ML, Resor SR, Singer WD, Russman BS, Homes GL, Mattson RH, Cramer JA (1984) Human serum teratogenicity studied by rat embryo culture: epilepsy, anticonvulsant drugs, and nutrition. Epilepsia 25: 205–216
Chernoff H, Grabowski CT (1971) Responses of the rat foetus to maternal injections of adrenaline and vasopressin. Br J Pharmacol 43: 270–278
Chernoff N, Kavlock RJ, Beyer PE, Miller D (1987) The potential relationship of maternal toxicity, general stress, and fetal outcome. Teratogen Carcinogen Mutagen 7: 241–253
Chernoff N, Rogers JM, Kavlock RJ (1989) Review paper: an overview of maternal toxicity and prenatal development: considerations for developmental toxicity hazard assessments. Toxicology 59: 111–125
Chernoff NW, Setzer R, Miller DB, Rosen MB, Rogers JM (1990) Effects of chemically induced maternal toxicity on prenatal development in the rat. Teratology 42: 651–658
Clark RL, Robertson RT, Minsker DH, Cohen SM, Tocco DJ, Allen HL, James ML, Bokelmam DL (1984) Diflunisal-induced maternal anemia as a cause of teratogenicity in rabbits. Teratology 30: 319–332
Clark RL, Robertson RT, Chennekatu PP, Bland JA, Nolan TE, Oppenheimer L, Bokelman DL (1986) Association between adverse maternal and embryo-fetal effects in norfloxacin-treated and food-deprived rabbits. Fund Appl Toxicol 7: 272–286
Cross JC, Werb Z, Fisher SJ (1994) Implantation and the placenta: key pieces of the development puzzle. Science 266: 1508–1518
Dancis J, Worth M Jr, Schneidall PB (1957) Effect of electrolyte disturbances in the pregnant rabbit on the fetus. Am J Physiol 188: 535–537
Daston GP (1994) Relationships between maternal and developmental toxicity. In: Kimmel CA, Buelke-Sam J (eds) Developmental toxicology, 2nd edn. Raven, New York, pp 189–212
Daston GP, Baines D, Yonker JE, Lehman-McKeeman L (1991a) Effects of lysosomal proteinase inhibition on the development of the rat embryo in vitro. Teratology 43: 253–261
Daston GP, Overmann GJ, Taubeneck MW, Lehman-McKeeman LD, Rogers JM, Keen CL (1991b) The role of metallothionein induction and altered zinc status in maternally mediated developmental toxicity: comparison of the effects of urethane and stryene in rats. Toxicol Appl Pharmacol 110: 450–463.
Daston GP, Rogers JM, Versteeg DJ, Sabourin TD, Baines D, Marsh SS (1991c) Interspecies comparisons of A/D ratios: A/D ratios are not constant across species. Fund Appl Toxicol 17: 696–722
Daston GP, Overmann GJ, Baines D, Taubeneck MW, Lehman-McKeeman LD, Rogers JM, Keen CL (1994) Altered Zn status by α-hederin in the pregnant rat and its relationship to adverse developmental outcome. Reprod Toxicol 8: 15–24
de la Cruz MV, Campillo-Sainz C, Monoz-Armas S (1966) Congenital heart defects in chick embryos subjected to temperature variations. Circ Res 18: 257–262
DeSesso JM (1987) Maternal factors in developmental toxicity. Teratogen Carcinogen Mutagen 7: 225–240

Eldeib MR, Reddy CS (1990) Role of maternal plasma corticosterone elevation in the teratogenicity of secalonic acid D in mice. Teratology 41: 137–146

Environmental Protection Agency (1984) Pesticide assessment guidelines, subdivision F, hazard evaluation: human and domestic animals. NTIS report PB86-108958, pp 126–130

Eskay RL, Grino M, Chen HT (1990) Interleukins, signal transduction, and the immune system-mediated stress response. In: Porter JC, Jezova D (eds) Circulating regulatory factors and neuroendocrine function. Plenum, New York, pp 331–343

Euker JS, Riegle GD (1973) Effects of stress on pregnancy in the rat. J Reprod Fertil 34: 343–346

European Economic Community (1988) Methods for the determination of toxicity. Official journal of the European Communities, vol 31, no L 133, 30 May

Fabro S, Schull G, Brown NA (1982) The relative teratogenic index and teratogenic potency. Proposed components of developmental toxicity risk assessment. Teratogen Carcinogen Mutagen 2: 61–76

Gaylard DG, Carson RJ, Reynolds F (1990) Effect of umbilical perfusate pH and controlled maternal hypotension on placental drug transfer in the rabbit. Anesth Analg 71: 42–8

Geber WF (1966) Developmental effects of chronic maternal audiovisual stress on the rat fetus. J Embryol Exp Morphol 16: 1–16

Geber WF, Anderson TA (1967) Abnormal fetal growth in the albino rat and rabbit induced by maternal stress. Biol Neonat 11: 209–215

George DG, Franklin JB, Brent RL (1967) Altered embryonic effects of uterine vascular clamping in the pregnant rat by uterine temperature control. Proc Soc Exp Biol Med 124: 257–260

German J (1984) Embryonic stress hypothesis of teratogenesis. Am J Med 76: 293–301

Ghanayem BI, Sullivan CA (1993) Assessment of the haemolytic activity of 2-butoxyethanol and its major metabolite, butoxyacetic acid, in various mammals including humans. Hum Exp Toxicol 12: 305–311

Glasser L, Sternglanz PD, Combie J, Robinson A (1973) Serum osmolality and its applicability to drug overdose. Am J Coll Phys 60: 695–699

Goldman AS, Yakovac WC (1963) The enhancement of salicylate teratogenicity by maternal immobilization in the rat. J Pharmacol Exp Ther 142: 351–357

Grabowski CT (1977a) Altered electrolyte and fluid balance. In: Wilson JG, Fraser FC (eds) Handbook of teratology, vol II. Plenum, New York, pp 153–170

Grabowski CT (1977b) Atmospheric gases. Variations in concentration and some common pollutants. In: Wilson JG, Fraser FC (eds) Handbook of teratology, vol I. Plenum, New York, pp 405–420

Green MA, Azar CA, Maren TH (1973) Strain differences in susceptibility to the teratogenic effect of acetazolamine in mice. Teratology 8: 143–146

Grote W (1965) Disorders of embryonic development induced by increased CO_2 and O_2 partial pressure and reduced atmospheric pressure. Z Morphol Anthropol 56: 165–194

Guyton AC (1986) Textbook of medical physiology. Saunders, Philadelphia

Hanley TR, Kirk HD, Berdasco NM, Johnson KA (1990) Evaluation of the developmental toxicity of propylene dichloride in rats and rabbits. Teratology 41: 562

Hansen DK, Holson RR, Sullivan PA, Grafton TF (1988) Alterations in maternal plasma corticosterone levels following treatment with phenytoin. Toxicol Appl Pharmacol 96: 24–32

Harbuz MS, Lightman SL (1989) Responses of hypothalamic and pituitary mRNA to physical and psychological stress in the rat. J Endocrinology 122: 7505–711

Hartel A, Hartel G (1960) Experimental study of teratogenic effect of emotional stress in rats. Science 132: 1483–1484

Haring OM (1960) Cardiac malformations in rats induced by exposure of the mother to carbon dioxide during pregnancy. Circ Res 8: 1218–1227

Hewlett TP, Jacobsen D, Collins TD, McMartin KE (1989) Ethylene glycol and glycolate kinetics in rats and dogs. Vet Hum Toxicol 31: 116–120

Hulse GK, Coleman GJ (1983) The role of endogenous opioids in the blockade of reproductive function in the rat following exposure to acute stress. Pharmacol Biochem Behav 19: 795–799

Hulse G, Coleman G, Nicholas J, Greenwood K (1982) Reversal of the anti-ovulatory action of stress in rats by prior administration of naloxone hydrochloride. J Reprod Fertil 66: 451–456

Imai-Matsumura K, Morimoto A, Murakami N, Nakayama T (1990) Maternal thermal stimulation changes metabolic activity in fetal hypothalamus. Brain Res 513: 295–298

Jacobsen D, Ovrebo S, Ostborg J, Serjersted OM (1984) Glycolate causes the acidosis in ethylene glycol poisoning and is effectively removed by hemodialysis. Acta Med Scand 216: 409–416

Jackson MR, Mayhew TM, Haas JD (1988) On the factors which contribute to thinning of the villous membrane in human placentae at high altitude. I. thinning and regional variation in thickness of trophoblast. Placenta 9: 1–8

Johnson EM, Christian MS, Dansky L, Gabel BEG (1987) Use of the adult developmental relationship in prescreening for developmental hazards. Teratogen Carcinogen Mutagen 7: 273–285

Kavlock RJ, Chernoff N, Rogers EH (1985) The effect of acute maternal toxicity on fetal development in the mouse. Teratogen Carcinogen Mutagen 5: 3–13

Kavlock RJ, Logsdon T, Gray J (1993) Fetal development in the rat following disruption of maternal renal function during pregnancy. Teratology 48: 247–258

Keen CL, Peters JM, Hurley LS (1989) The effect of valproic acid on ^{65}Zn distribution in the pregnant rat. J Nutr 119: 607–611

Kelce WR, Monosson E, Gamcsidk MP, Laws SC, Gray E Jr (1994) Environmental hormone disruptors: evidence that vinclozolin developmental toxicity is mediated by antiandrogenic metabolites. Toxicol Appl Pharmacol 126: 276–285

Khera KS (1985) Maternal toxicity: a possible etiological factor in embryo-fetal deaths and fetal malformations of rodent-rabbit species. Teratology 31: 129–153

Khera KS (1987) Maternal toxicity of drugs and metabolic disorders – a possible etiologic factor in the intrauterine death and congenital malformation: a critique on human data. CRC Crit Rev Toxicol 17: 345–375

Khera KS (1991) Chemically induced alterations in maternal homeostasis and histology of conceptus: their etiologic significance in rat fetal anomalies. Teratology 44: 259–297

Kholkute SD, Udupa KN (1978) Effect of immobilization stress on implantation and pregnancy in the rat. Ind J Exp Biol 16: 799–800

Kimmel CA, Cuff JM, Kimmel GL, Heredia DJ, Tudor N, Silverman PM, Chen J (1993) Skeletal development following heat exposure in the rat. Teratology 47: 229–242

Kimmel GL, Kimmel CA, Francis EZ (1987) Evaluation of maternal and developmental toxicity. Proceedings of the consensus workshop on the evaluation of maternal and developmental toxicity held in Rockville, Maryland, 12–14 May 1986. Teratogen Carcinogen Mutagen 7(3): 203

Layton WM (1971) Teratogenic action of acetazolamide in golden hamsters. Teratology 4: 95–102

Layton WM, Hallesy D (1965) Deformity of forelimb in rats; association with high doses of acetazolamide. Science 149: 306–308

Marois MM (1966) Action teratogene des rayons X sur l'embryon de rat. II. Effet radioprotecteur de l'hypothermie. Bull Acad Natl Med 150: 186–193

Matsuzawa T, Nakata M, Foto I, Tsushima M (1981) Dietary deprivation induces fetal loss and abortion in rabbits. Toxicology 22: 255–259

Millicovsky G, Johnston MC (1981) Maternal hyperoxia greatly reduces the incidence of phenytoin-induced cleft lip and palate in A/J mice. Science 212: 671–672

Mirkes PE (1985) Effects of acute exposures to elevated temperatures on rat embryo growth and development in vitro. Teratology 32: 259–266

Munro KM, Barnett SA (1969) Variation of the lumbar vertebrae of mice at two environmental temperatures. J Embryol Exp Morphol 21: 97–103

Nakatsuka T, Komatsu T, Fujii T (1992) Axial skeletal malformations induced by acetazolamide in rabbits. Teratology 45: 629–636

Nawrot PS, Cook RO, Staples RE (1980) Embryotoxicity of various noise stimuli in the mouse. Teratology 22: 279–289

Paulone ME, Edelstone DI, Shedd A (1987) Effects of maternal anemia on uteroplacental and fetal oxidative metabolism in sheep. Am J Obstet Gynecol 156: 230–236

Parisi VM, Creasy RK (1992) Maternal biologic adaptations to pregnancy. In: Reece EA, Hobbins JC, Mahoney MJ, Petrie RH (eds) Medicine of the fetus and mother. Lippincott, Philadelphia, pp 831–848

Randall CL, Anton RF, Becker HC (1988) Role of alcohol-induced hypothermia in mediating the teratogenic effects of alcohol in C57BL/6J mice. Alcohol Clin Exp Res 12: 412–416

Rasco JF, Hood RD (1994) Effects of maternal restraint stress and sodium arsenate in mice. Reprod Toxicol 8: 49–54

Robertson RT, Minsker DH, Bokelman DL, Durand G, Conquet P (1981) Potassium loss as a causative factor for skeletal malformations in rats produced by indacrinone: a new investigational loop diuretic. Toxicol Appl Pharmacol 60: 142–150

Robertson SA, Brannstrom M, Seamark RF (1992) Cytokines in rodent reproduction and the cytokine-endocrine interaction. Curr Opinion Immunol 4: 585–590

Rogers JM, Daston GP, Ebron MT, Carver B, Stefanadis JG, Grabowski CT (1985) Studies on the mechanism of trypan blue teratogenicity in the rat developing in vivo and in vitro. Teratology 31: 389–399

Sadler TW, Hunter ES III, Balkan W, Wynn RE (1987) The role of maternal serum factors in diabetes-induced embryopathies as studied in whole-embryo culture. In: Welsch F (ed) Approaches to elucidate mechanisms in teratogenesis. Hemisphere, Washington, pp 109–122

Schwetz BA, Moorman MP (1987) Assessment of adult toxicity in developmental versus prechronic toxicology studies. Teratogen Carcinogen Mutagen 7: 211–223

Scialli AR (1988) Is stress a developmental toxin? Reprod Toxicol 1: 163–171

Scott WJ Jr, Duggan CA, Schreiner CM, Collins MD, Nau H (1987) Intracellular pH of rodent embryos and its association with teratogenic response. In: Welsch F (ed) Approaches to elucidate mechanisms in teratogenesis. Hemisphere, Washington, pp 99–108

Selye H (1950) The physiology and pathology of exposure to stress. Acta, Montreal

Sestoft L, Bartels PD (1983) Biochemistry and differential diagnosis of metabolic acidoses. Clin Endocrin Metab 12: 287–302

Setzer RW, Rogers JM (1991) Assessing developmental hazard: the reliability of the A/D ratio. Teratology 44: 653–665

Sharma A, Plessinger MA, Sherer DM, Liang C-S, Miller RK, Woods JR Jr (1992) Pregnancy enhances cardiotoxicity of cocaine: role of progesterone. Toxicol Appl Pharmacol 113: 30–35

Shepard TH (1977) Maternal metabolic and endocrine imbalances. In: Wilson JG, Fraser FC (eds) Handbook of teratology, vol I. Plenum, New York, pp 387–404

Slotkin TA, Lau C, McCook EC, Lappi SE, Seidler FJ (1994) Glucocorticoids enhance intracellular signaling via adenylate cyclase at three distinct loci in the fetus: a mechanism for heterologous teratogenic sensitization? Toxicol Appl Pharmacol 127: 64–75

Smith AU (1957) The effect on foetal development of freezing pregnant hamsters. J Embryol Exp Morphol 5: 311–323

Smoak IW, Sadler TW (1991) Hypothermia: teratogenic and protective effects on the development of mouse embryos in vitro. Teratology 43: 635–641

Stauch BS, Felig P, Baxter JD, Schimpff SC (1969) Hypothermia in hypoglycemia. JAMA 210: 354–346

Strick DM, Waycaster RL, Montani J, Gay WJ, Adair TH (1991) Morphometric measurements of chorioallantoic membrane vascularity: effects of hypoxia and hyperoxia. Am J Physiol 260: H1385–H1389

Szekely JI (1990) Opioid peptides and stress. Crit Rev Neurobiol 6: 1–12

Taubeneck MW, Daston GP, Rogers JM, Gershwin ME, Ansari A, Keen CL (1994a) Tumor necrosis factor-α-conditioned serum results in abnormal development of cultured embryos. Teratology 49: 382

Taubeneck MW, Daston GP, Rogers JM, Keen CL (1994b) Altered maternal zinc metabolism following exposure to diverse developmental toxicants. Reprod Toxicol 8: 25–40

Tyl RW, Millicovsky G, Dodd DE, Pritts IM, France KA, Fisher LC (1984) Environ Health Perspect 57: 47–68

Tyl RW, Ballantyne B, Fisher LC, Fait DL, Savine TA, Pritts IM, Dodd DE (1994) Evaluation of exposure to water aerosol or air by nose-only or whole-body inhalation procedures for CD-1 mice in developmental toxicity studies. Fund Appl Toxicol 23: 251–260

Vom Saal FS, Quadagno DM, Even MD, Keisler LW, Keisler DH, Khan S (1990) Paradoxical effects of maternal stress on fetal steroids and postnatal reproductive traits in female mice from different intrauterine positions. Biol Reprod 43: 751–761

Ward IL, Wiesz J (1984) Differential effects of maternal stress on circulating levels of corticosterone, progesterone, and testosterone in male and female rat fetuses and their mothers. Endocrinology 114: 1635–1644

Warkany J (1986) Teratogen update: hyperthermia. Teratology 33: 365–371

Watkinson WP, Millicovsky G (1993) Effect of phenytoin on maternal heart rate in A/J mice: possible role in teratogenesis. Teratology 28: 1–8

Watkinson WP, Gordon CJ (1993) Caveats regarding the use of the laboratory rat as a model for acute toxicological studies: modulation of the toxic response via physiological and behavioral mechanisms. Toxicology 81: 15–31

Weaver TE, Scott WJ Jr (1984) Acetazolamide teratogenesis: interaction of maternal metabolic and respiratory acidosis in the induction of ectrodactyly in C57BL/6J mice. Teratology 30: 195–202

Wegmann TG (1990) The cytokine basis for cross-talk between the maternal immune and reproductive systems. Curr Opinion Immunol 2: 765–769

Wilson JG (1953) Influence of severe hemorrhagic anemia during pregnancy on development of the offspring in the rat. Proc Soc Exp Biol Med 84: 66–69

Woods JR, Plessinger MA, Clark KE (1987) Effect of cocaine on uterine blood flow and fetal oxygenation. JAMA 257: 957–961

Woods LL (1986) Fetal renal contribution to amniotic fluid osmolality during maternal hypertonicity. Am J Physiol 250: 235–239

Zolti M, Ben-Rafael Z, Meirom R, Shemesh M, Bider D, Mashiach S, Apte R (1991) Cytokine involvement in oocytes and early embryos. Fertil Steril 56: 265–272

Subject Index

Springer and the environment

At Springer we firmly believe that an international science publisher has a special obligation to the environment, and our corporate policies consistently reflect this conviction.

We also expect our business partners – paper mills, printers, packaging manufacturers, etc. – to commit themselves to using materials and production processes that do not harm the environment. The paper in this book is made from low- or no-chlorine pulp and is acid free, in conformance with international standards for paper permanency.

Printing: Saladruck, Berlin
Binding: Buchbinderei Lüderitz & Bauer, Berlin